Lecture Notes in Artificial Intelligence 4481

Edited by J. G. Carbonell and J. Siekmann

Subseries of Lecture Notes in Computer Science

T0223162

JingTao Yao   Pawan Lingras
Wei-Zhi Wu   Marcin Szczuka
Nick J. Cercone   Dominik Ślęzak (Eds.)

Rough Sets and Knowledge Technology

Second International Conference, RSKT 2007
Toronto, Canada, May 14-16, 2007
Proceedings

Springer

Series Editors

Jaime G. Carbonell, Carnegie Mellon University, Pittsburgh, PA, USA
Jörg Siekmann, University of Saarland, Saarbrücken, Germany

Volume Editors

JingTao Yao
University of Regina, Regina, Saskatchewan, Canada
E-mail: jtyao@cs.uregina.ca

Pawan Lingras
Saint Mary's University, Halifax, Nova Scotia, Canada
E-mail: pawan@cs.smu.ca

Wei-Zhi Wu
Zhejiang Ocean University, Zhejiang, P.R. China
E-mail: wuwz@zjou.edu.cn

Marcin Szczuka
Warsaw University, Warsaw, Poland
E-mail: szczuka@mimuw.edu.pl

Nick J. Cercone
York University, Toronto, Ontario, Canada
E-mail: ncercone@yorku.ca

Dominik Ślęzak
Infobright Inc., Toronto, Ontario, Canada
E-mail: slezak@infobright.com

Library of Congress Control Number: 2007925941

CR Subject Classification (1998): I.2, H.2.4, H.3, F.4.1, F.1, I.5, H.4

LNCS Sublibrary: SL 7 – Artificial Intelligence

ISSN   0302-9743
ISBN-10   3-540-72457-5 Springer Berlin Heidelberg New York
ISBN-13   978-3-540-72457-5 Springer Berlin Heidelberg New York

Springer is a part of Springer Science+Business Media

springer.com

© Springer-Verlag Berlin Heidelberg 2007
Printed in Germany

Typesetting: Camera-ready by author, data conversion by Scientific Publishing Services, Chennai, India
Printed on acid-free paper   SPIN: 12061785   06/3180   5 4 3 2 1 0

Preface

This volume contains the papers selected for presentation at the International Conference on Rough Sets and Knowledge Technology (RSKT 2007), a part of the Joint Rough Set Symposium (JRS 2007) organized by Infobright Inc. and York University. JRS 2007 was held for the first time during May 14–16, 2007 in MaRS Discovery District, Toronto, Canada. It consisted of two conferences: RSKT and the 11th International Conference on Rough Sets, Fuzzy Sets, Data Mining, and Granular Computing (RSFDGrC 2007).

The two conferences that constituted JRS 2007 investigated rough sets as an emerging methodology established more than 25 years ago by Zdzisław Pawlak. Rough set theory has become an integral part of diverse hybrid research streams. In keeping with this trend, JRS 2007 encompassed rough and fuzzy sets, knowledge technology and discovery, soft and granular computing, data processing and mining, while maintaining an emphasis on foundations and applications.

The RSKT series was launched in 2006 in Chongqing, China. The RSKT conferences place emphasis on exploring synergies between rough sets and knowledge discovery, knowledge management, data mining, granular and soft computing as well as emerging application areas such as biometrics and ubiquitous computing, both at the level of theoretical foundations and real-life applications.

In RSKT 2007, a special effort was made to include research spanning a broad range of applications. This was achieved by including in the conference program, special sessions on multiple criteria decision analysis, biometrics, Kansei engineering, autonomy-oriented computing, soft computing in bioinformatics, as well as tutorials and sessions related to other application areas.

Overall, we received 319 submissions to the Joint Rough Set Symposium. Every paper was examined by at least two reviewers. The submission and review processes were performed jointly for both conferences that together constituted JRS 2007, i.e., RSFDGrC 2007 and RSKT 2007.

Out of the papers initially selected, some were approved subject to revision and then additionally evaluated. Finally, 139 papers were accepted for JRS 2007. This gives an acceptance ratio slightly over 43% for the joint conferences.

Accepted papers were distributed between the two conferences on the basis of their relevance to the conference themes.

The JRS 2007 conference papers are split into two volumes (LNAI 4481 for RSKT and LNAI 4482 for RSFDGrC). The regular, invited, and special session papers selected for presentation at RSKT 2007 are included within 12 chapters and grouped under specific conference topics.

This volume contains 70 papers, including 3 invited papers presented in Chap. 1. The remaining 67 papers are presented in 11 chapters related to multiple criteria decision analysis, logical and rough set foundations, biometrics, Kansei engineering, soft computing in bioinformatics, autonomy-oriented computing, ubiquitous computing and networking, rough set algorithms, genetic algorithms, and rough set applications.

We wish to thank all of the authors who contributed to this volume. We are very grateful to the chairs, advisory board members, program committee members, and other reviewers not listed in the conference committee, for their help in the acceptance process.

We are grateful to our Honorary Chairs, Setsuo Ohsuga and Lotfi Zadeh, for their support and visionary leadership. We also acknowledge the scientists who kindly agreed to give the keynote, plenary, and tutorial lectures: Andrzej Bargiela, Mihir K. Chakraborty, Bernhard Ganter, Sushmita Mitra, Sadaaki Miyamoto, James F. Peters, Andrzej Skowron, Domenico Talia, Xindong Wu, Yiyu Yao, Chengqi Zhang, and Wojciech Ziarko. We also wish to express our deep appreciation to all special session organizers.

We greatly appreciate the co-operation, support, and sponsorship of various companies, institutions and organizations, including: Infobright Inc., MaRS Discovery District, Springer, York University, International Rough Set Society, International Fuzzy Systems Association, Rough Sets and Soft Computation Society of the Chinese Association for Artificial Intelligence, and National Research Council of Canada.

We wish to thank several people whose hard work made the organization of JRS 2007 possible. In particular, we acknowledge the generous help received from: Tokuyo Mizuhara, Clara Masaro, Christopher Henry, Julio V. Valdes, April Dunford, Sandy Hsu, Lora Zuech, Bonnie Barbayanis, and Allen Gelberg.

Last but not least, we are thankful to Alfred Hofmann of Springer for support and co-operation during preparation of this volume.

May 2007

JingTao Yao
Pawan Lingras
Wei-Zhi Wu
Marcin Szczuka
Nick Cercone
Dominik Ślęzak

RSKT 2007 Conference Committee

JRS Honorary Chairs	Setsuo Ohsuga, Lotfi A. Zadeh
JRS Conference Chairs	Dominik Ślęzak, Guoyin Wang
JRS Program Chairs	Nick Cercone, Witold Pedrycz
RSKT 2007 Chairs	JingTao Yao, Pawan Lingras, Wei-Zhi Wu, Marcin Szczuka
JRS Organizing Chairs	Jimmy Huang, Miriam G. Tuerk
JRS Publicity Chairs	Aboul E. Hassanien, Shoji Hirano, Daniel Howard, Igor Jurisica, Tai-hoon Kim, Duoqian Miao, Bhanu Prasad, Mark S. Windrim

RSKT 2007 Steering Committee

Yiyu Yao (Chair)	Jiming Liu	Zbigniew Suraj
Malcolm Beynon	Stan Matwin	Shusaku Tsumoto
Nick Cercone	Ernestina Menasalvas	Julio V. Valdes
Salvatore Greco	Duoqian Miao	Guoyin Wang
Jerzy Grzymała-Busse	Ewa Orłowska	S.K. Michael Wong
Etienne Kerre	Sankar K. Pal	Bo Zhang
Yuefeng Li	Andrzej Skowron	Ning Zhong

RSKT 2007 Program Committee

Philippe Balbiani	Jimmy Huang	Artur Przelaskowski
Mohua Banerjee	Taghi M. Khoshgoftaar	Anna M. Radzikowska
Jan Bazan	Dai-Jin Kim	Vijay Raghavan
Theresa Beaubouef	Bożena Kostek	Zbigniew Raś
Santanu Chaudhury	Geuk Lee	Kenneth Revett
Davide Ciucci	Yee Leung	Henryk Rybiński
Jianhua Dai	Deyu Li	Gerald Schaefer
Martine DeCock	Fanzhang Li	Lin Shang
Lipika Dey	Jiuzhen Liang	Piotr Synak
Jiali Feng	Jiye Liang	Noboru Takagi
Bernhard Ganter	Benedetto Matarazzo	Hideo Tanaka
Xinbo Gao	Rene Mayorga	Xizhao Wang
Vladimir Gorodetsky	Sushmita Mitra	Anita Wasilewska
Daryl Hepting	Tatsuo Nishino	Szymon Wilk
Shoji Hirano	Keyun Qin	Arkadiusz Wojna
Xiaohua Hu	Yuhui Qiu	Zhaohui Wu

Table of Contents

Biometrics

Kansei Engineering

Autonomy-Oriented Computing

Soft Computing in Bioinformatics

Ubiquitous Computing and Networking

Rough Set Algorithms

Knowledge Representation and Reasoning

Genetic Algorithms

Rough Set Applications

Decision-Theoretic Rough Set Models

Yiyu Yao

Department of Computer Science, University of Regina,
Regina, Saskatchewan, Canada S4S 0A2
yyao@cs.uregina.ca

Abstract. Decision-theoretic rough set models are a probabilistic extension of the algebraic rough set model. The required parameters for defining probabilistic lower and upper approximations are calculated based on more familiar notions of costs (risks) through the well-known Bayesian decision procedure. We review and revisit the decision-theoretic models and present new results. It is shown that we need to consider additional issues in probabilistic rough set models.

Keywords: Bayesian decision theory, decision-theoretic rough sets, probabilistic rough sets, variable precision rough sets.

1 Introduction

Ever since the introduction of rough set theory by Pawlak in 1982 [10,11,13], many proposals have been made to incorporate probabilistic approaches into the theory [14,29,30,31]. They include, for example, rough set based probabilistic classification [24], 0.5 probabilistic rough set model [15], decision-theoretic rough set models [32,33], variable precision rough set models [6,35], rough membership functions [12], parameterized rough set models [14,17], and Bayesian rough set models [2,18,19]. The results of these studies increase our understanding of the rough set theory and its domain of applications.

The decision-theoretic rough set models [29,32,33] and the variable precision rough set models [6,35,36] were proposed in the early 1990's. The two models are formulated differently in order to generalize the 0.5 probabilistic rough set model [15]. In fact, they produce the same rough set approximations [29,30]. Their main differences lie in their respective treatment of the required parameters used in defining the lower and upper probabilistic approximations. The decision-theoretic models systematically calculate the parameters based on a loss function through the Bayesian decision procedure. The physical meaning of the loss function can be interpreted based on more practical notions of costs and risks. In contrast, the variable precision models regard the parameters as primitive notions and a user must supply those parameters. A lack of a systematic method for parameter estimation has led researchers to use many ad hoc methods based on trial and error.

The results and ideas of the decision-theoretic model, based on the well established and semantically sound Bayesian decision procedure, have been successfully applied to many fields, such as data analysis and data mining [3,7,16,21,22,23,25,34]

J.T. Yao et al. (Eds.): RSKT 2007, LNAI 4481, pp. 1–12, 2007.

information retrieval [8,20], feature selection [27], web-based support systems [26], and intelligent agents [9]. Some authors have generalized the decision-theoretic model to multiple regions [1].

The main objective of this paper is to revisit the decision-theoretic rough set model and to present new results. In order to appreciate the generality and flexibility of the model, we show explicitly the conditions on the loss function under which other models can be derived, including the standard rough set model, 0.5 probabilistic rough set model, and both symmetric and asymmetric variable precision rough set models. Furthermore, the decision-theoretic model is extended from a two-class classification problem into a many-class classification problem. This enables us to observe that some of the straightforward generalizations of notions and measures of the algebraic rough set model may not necessarily be meaningful in the probabilistic models.

2   Algebraic Rough Set Approximations

Let U be a finite and nonempty set and E an equivalence relation on U. The pair $apr = (U, E)$ is called an approximation space [10,11]. The equivalence relation E induces a partition of U, denoted by U/E. The equivalence class containing x is given by $[x] = \{y \mid xEy\}$.

The equivalence classes of E are the basic building blocks to construct algebraic rough set approximations. For a subset $A \subseteq U$, its lower and upper approximations are defined by [10,11]:

$$\underline{apr}(A) = \{x \in U \mid [x] \subseteq A\};$$
$$\overline{apr}(A) = \{x \in U \mid [x] \cap A \neq \emptyset\}. \tag{1}$$

The lower and upper approximations, $\underline{apr}, \overline{apr} : 2^U \longrightarrow 2^U$, can be interpreted as a pair of unary set-theoretic operators [28]. They are dual operators in the sense that $\underline{apr}(A) = (\overline{apr}(A^c))^c$ and $\overline{apr}(A) = (\underline{apr}(A^c))^c$, where A^c is set complement of A. Other equivalent definitions and additional properties of approximation operators can be found in [10,11,28].

Based on the rough set approximations of A, one can divide the universe U into three disjoint regions, the positive region POS(A), the boundary region BND(A), and the negative region NEG(A):

$$\text{POS}(A) = \underline{apr}(A),$$
$$\text{BND}(A) = \overline{apr}(A) - \underline{apr}(A),$$
$$\text{NEG}(A) = U - \text{POS}(A) \cup \text{BND}(A) = U - \overline{apr}(A) = (\overline{apr}(A))^c. \tag{2}$$

Some of these regions may be empty. One can say with certainty that any element $x \in \text{POS}(A)$ belongs to A, and that any element $x \in \text{NEG}(A)$ does not belong to A. One cannot decide with certainty whether or not an element $x \in \text{BND}(A)$ belongs to A.

We can easily extend the concepts of rough set approximations and regions of a single set to a partition of the universe. Consider first a simple case. A set

$\emptyset \neq A \neq U$ induces a partition $\pi_A = \{A, A^c\}$ of U. The approximations of the partition π_A are defined by:

$$\underline{apr}(\pi_A) = (\underline{apr}(A), \underline{apr}(A^c)) = (\underline{apr}(A), (\overline{apr}(A))^c);$$
$$\overline{apr}(\pi_A) = (\overline{apr}(A), \overline{apr}(A^c)) = (\overline{apr}(A), (\underline{apr}(A))^c). \tag{3}$$

Based on the positive, boundary and negative regions of A and A^c, we have the corresponding three disjoint regions of π_A:

$$\mathrm{POS}(\pi_A) = \mathrm{POS}(A) \cup \mathrm{POS}(A^c),$$
$$\mathrm{BND}(\pi_A) = \mathrm{BND}(A) \cup \mathrm{BND}(A^c) = \mathrm{BND}(A) = \mathrm{BND}(A^c),$$
$$\mathrm{NEG}(\pi_A) = U - \mathrm{POS}(\pi_A) \cup \mathrm{BND}(\pi_A) = \emptyset. \tag{4}$$

In general, let $\pi = \{A_1, A_2, \ldots, A_m\}$ be a partition of the universe U. We have:

$$\underline{apr}(\pi) = (\underline{apr}(A_1), \underline{apr}(A_2), \ldots, \underline{apr}(A_m));$$
$$\overline{apr}(\pi) = (\overline{apr}(A_1), \overline{apr}(A_2), \ldots, \overline{apr}(A_m)). \tag{5}$$

We can extend the notions of three regions to the case of a partition:

$$\mathrm{POS}(\pi) = \bigcup_{1 \leq i \leq m} \mathrm{POS}(A_i),$$
$$\mathrm{BND}(\pi) = \bigcup_{1 \leq i \leq m} \mathrm{BND}(A_i),$$
$$\mathrm{NEG}(\pi) = U - \mathrm{POS}(\pi) \cup \mathrm{BND}(\pi) = \emptyset. \tag{6}$$

It can be verified that $\mathrm{POS}(\pi) \cap \mathrm{BND}(\pi) = \emptyset$ and $\mathrm{POS}(\pi) \cup \mathrm{BND}(\pi) = U$.

The positive and boundary regions can be used to derive two kinds of rules, namely, certain and probabilistic rules, or deterministic and non-deterministic rules [5,11]. More specifically, for $[x] \subseteq \mathrm{POS}(A_i)$ and $[x'] \subseteq \mathrm{BND}(A_i)$, we have the two kinds of rules, respectively, as follows:

$$[x] \xrightarrow{c=1} A_i, \qquad [x'] \xrightarrow{0<c<1} A_i, \tag{7}$$

where the confidence measure c of a rule is defined by:

$$c = \frac{|A_i \cap [x]|}{|[x]|}, \tag{8}$$

and $|\cdot|$ is the cardinality of a set. It can be verified that $\underline{apr}(A_i) \cap \underline{apr}(A_j) = \emptyset$ for $i \neq j$. An element of U belongs to at most one positive region of A_i's. On the other hand, an element may be in more than one boundary region. Thus, an element may satisfy at most one certain rule, or satisfy more than one probabilistic rule.

To quantify the degree of dependency of the partition π on the partition U/E, many authors use only the positive region. For example, Pawlak suggests the following measure [11]:

$$r(\pi|U/E) = \frac{|\mathrm{POS}(\pi)|}{|U|}. \tag{9}$$

This measure only considers the coverage of certain rules. The effects of uncertain rules are not considered. Since $\text{NEG}(\pi) = \emptyset$, $\text{POS}(\pi) \cap \text{BND}(\pi) = \emptyset$, and $\text{POS}(\pi) \cup \text{BND}(\pi) = U$, it may be sufficient to consider only $\text{POS}(\pi)$. As we will show later, when those notions are generalized into a probabilistic version, it is necessary to consider all three regions.

3 Probabilistic Rough Set Approximations

The Bayesian decision procedure deals with making decision with minimum risk based on observed evidence. We present a brief description of the procedure from the book by Duda and Hart [4], and apply the procedure for the construction of probabilistic approximations [32,33].

3.1 The Bayesian Decision Procedure

Let $\Omega = \{w_1, \ldots, w_s\}$ be a finite set of s states, and let $\mathcal{A} = \{a_1, \ldots, a_m\}$ be a finite set of m possible actions. Let $P(w_j|\mathbf{x})$ be the conditional probability of an object x being in state w_j given that the object is described by $\mathbf{x}$. Let $\lambda(a_i|w_j)$ denote the loss, or cost, for taking action a_i when the state is w_j. For an object with description $\mathbf{x}$, suppose action a_i is taken. Since $P(w_j|\mathbf{x})$ is the probability that the true state is w_j given $\mathbf{x}$, the expected loss associated with taking action a_i is given by:

$$R(a_i|\mathbf{x}) = \sum_{j=1}^{s} \lambda(a_i|w_j)P(w_j|\mathbf{x}). \tag{10}$$

The quantity $R(a_i|\mathbf{x})$ is also called the conditional risk.

Given a description $\mathbf{x}$, a decision rule is a function $\tau(\mathbf{x})$ that specifies which action to take. That is, for every $\mathbf{x}$, $\tau(\mathbf{x})$ takes one of the actions, $a_1, \ldots, a_m$. The overall risk $\mathbf{R}$ is the expected loss associated with a given decision rule. Since $R(\tau(\mathbf{x})|\mathbf{x})$ is the conditional risk associated with action $\tau(\mathbf{x})$, the overall risk is defined by:

$$\mathbf{R} = \sum_{\mathbf{x}} R(\tau(\mathbf{x})|\mathbf{x})P(\mathbf{x}), \tag{11}$$

where the summation is over the set of all possible descriptions of objects. If $\tau(\mathbf{x})$ is chosen so that $R(\tau(\mathbf{x})|\mathbf{x})$ is as small as possible for every $\mathbf{x}$, the overall risk $\mathbf{R}$ is minimized. Thus, the Bayesian decision procedure can be formally stated as follows. For every $\mathbf{x}$, compute the conditional risk $R(a_i|\mathbf{x})$ for $i = 1, \ldots, m$ defined by equation (10) and select the action for which the conditional risk is minimum. If more than one action minimizes $R(a_i|\mathbf{x})$, a tie-breaking criterion can be used.

3.2 Probabilistic Rough Set Approximation Operators

In an approximation space $apr = (U, E)$, an equivalence class $[x]$ is considered to be the description of x. The partition U/E is the set of all possible descriptions.

The classification of objects according to approximation operators can be easily fitted into the Bayesian decision-theoretic framework. The set of states is given by $\Omega = \{A, A^c\}$ indicating that an element is in A and not in A, respectively. We use the same symbol to denote both a subset A and the corresponding state. With respect to the three regions, the set of actions is given by $\mathcal{A} = \{a_1, a_2, a_3\}$, where a_1, a_2, and a_3 represent the three actions in classifying an object, deciding POS(A), deciding NEG(A), and deciding BND(A), respectively. Let $\lambda(a_i|A)$ denote the loss incurred for taking action a_i when an object belongs to A, and let $\lambda(a_i|A^c)$ denote the loss incurred for taking the same action when the object does not belong to A.

The probabilities $P(A|[x])$ and $P(A^c|[x])$ are the probabilities that an object in the equivalence class $[x]$ belongs to A and A^c, respectively. The expected loss $R(a_i|[x])$ associated with taking the individual actions can be expressed as:

$$R(a_1|[x]) = \lambda_{11} P(A|[x]) + \lambda_{12} P(A^c|[x]),$$
$$R(a_2|[x]) = \lambda_{21} P(A|[x]) + \lambda_{22} P(A^c|[x]),$$
$$R(a_3|[x]) = \lambda_{31} P(A|[x]) + \lambda_{32} P(A^c|[x]), \tag{12}$$

where $\lambda_{i1} = \lambda(a_i|A)$, $\lambda_{i2} = \lambda(a_i|A^c)$, and $i = 1, 2, 3$. The Bayesian decision procedure leads to the following minimum-risk decision rules:

(P)   If $R(a_1|[x]) \le R(a_2|[x])$ and $R(a_1|[x]) \le R(a_3|[x])$,   decide POS($A$);

(N)   If $R(a_2|[x]) \le R(a_1|[x])$ and $R(a_2|[x]) \le R(a_3|[x])$,   decide NEG($A$);

(B)   If $R(a_3|[x]) \le R(a_1|[x])$ and $R(a_3|[x]) \le R(a_2|[x])$,   decide BND($A$).

Tie-breaking criteria should be added so that each element is classified into only one region. Since $P(A|[x]) + P(A^c|[x]) = 1$, we can simplify the rules to classify any object in $[x]$ based only on the probabilities $P(A|[x])$ and the loss function λ_{ij} ($i = 1, 2, 3$; $j = 1, 2$).

Consider a special kind of loss functions with $\lambda_{11} \le \lambda_{31} < \lambda_{21}$ and $\lambda_{22} \le \lambda_{32} < \lambda_{12}$. That is, the loss of classifying an object x belonging to A into the positive region POS(A) is less than or equal to the loss of classifying x into the boundary region BND(A), and both of these losses are strictly less than the loss of classifying x into the negative region NEG(A). The reverse order of losses is used for classifying an object that does not belong to A. For this type of loss function, the minimum-risk decision rules (P)-(B) can be written as:

(P)   If $P(A|[x]) \ge \gamma$ and $P(A|[x]) \ge \alpha$, decide POS(A);

(N)   If $P(A|[x]) \le \beta$ and $P(A|[x]) \le \gamma$, decide NEG(A);

(B)   If $\beta \le P(A|[x]) \le \alpha$, decide BND($A$);

where

$$\alpha = \frac{\lambda_{12} - \lambda_{32}}{(\lambda_{31} - \lambda_{32}) - (\lambda_{11} - \lambda_{12})},$$
$$\gamma = \frac{\lambda_{12} - \lambda_{22}}{(\lambda_{21} - \lambda_{22}) - (\lambda_{11} - \lambda_{12})},$$

$$\beta = \frac{\lambda_{32} - \lambda_{22}}{(\lambda_{21} - \lambda_{22}) - (\lambda_{31} - \lambda_{32})}. \tag{13}$$

By the assumptions, $\lambda_{11} \leq \lambda_{31} < \lambda_{21}$ and $\lambda_{22} \leq \lambda_{32} < \lambda_{12}$, it follows that $\alpha \in (0, 1]$, $\gamma \in (0, 1)$, and $\beta \in [0, 1)$.

For a loss function with $\lambda_{11} \leq \lambda_{31} < \lambda_{21}$ and $\lambda_{22} \leq \lambda_{32} < \lambda_{12}$, more results about the required parameters α, β and γ are summarized as follows [29]:

1. If a loss function satisfies the condition:

$$(\lambda_{12} - \lambda_{32})(\lambda_{21} - \lambda_{31}) \geq (\lambda_{31} - \lambda_{11})(\lambda_{32} - \lambda_{22}), \tag{14}$$

   then $\alpha \geq \gamma \geq \beta$.
2. If a loss function satisfies the condition:

$$\lambda_{12} - \lambda_{32} \geq \lambda_{31} - \lambda_{11}, \tag{15}$$

   then $\alpha \geq 0.5$.
3. If a loss function satisfies the conditions,

$$\lambda_{12} - \lambda_{32} \geq \lambda_{31} - \lambda_{11},$$
$$(\lambda_{12} - \lambda_{32})(\lambda_{21} - \lambda_{31}) \geq (\lambda_{31} - \lambda_{11})(\lambda_{32} - \lambda_{22}), \tag{16}$$

   then $\alpha \geq 0.5$ and $\alpha \geq \beta$.
4. If a loss function satisfies the condition:

$$(\lambda_{12} - \lambda_{32})(\lambda_{32} - \lambda_{22}) = (\lambda_{31} - \lambda_{11})(\lambda_{21} - \lambda_{31}), \tag{17}$$

   then $\beta = 1 - \alpha$.
5. If a loss function satisfies the two sets of equivalent conditions,

$$\text{(i).} \quad (\lambda_{12} - \lambda_{32})(\lambda_{21} - \lambda_{31}) \geq (\lambda_{31} - \lambda_{11})(\lambda_{32} - \lambda_{22}),$$
$$(\lambda_{12} - \lambda_{32})(\lambda_{32} - \lambda_{22}) = (\lambda_{31} - \lambda_{11})(\lambda_{21} - \lambda_{31}); \tag{18}$$
$$\text{(ii).} \quad \lambda_{12} - \lambda_{32} \geq \lambda_{31} - \lambda_{11},$$
$$(\lambda_{12} - \lambda_{32})(\lambda_{32} - \lambda_{22}) = (\lambda_{31} - \lambda_{11})(\lambda_{21} - \lambda_{31}); \tag{19}$$

   then $\alpha = 1 - \beta \geq 0.5$.

The condition of Case 1 guarantees that the probabilistic lower approximation of a set is a subset of its probabilistic upper approximation. The condition of Case 2 ensures that a lower approximation of A consists of those elements whose majority equivalent elements are in A. The condition of Case 4 results in a pair of dual lower and upper approximation operators. Case 3 is a combination of Cases 1 and 2. Case 5 is the combination of Cases 1 and 4 or the combination of Cases 3 and 4.

When $\alpha > \beta$, we have $\alpha > \gamma > \beta$. After tie-breaking, we obtain the decision rules:

(P1)   If $P(A|[x]) \geq \alpha$, decide POS(A);

(N1)   If $P(A|[x]) \leq \beta$, decide NEG(A);

(B1)   If $\beta < P(A|[x]) < \alpha$, decide BND($A$).

When $\alpha = \beta$, we have $\alpha = \gamma = \beta$. In this case, we use the decision rules:

(P2)   If $P(A|[x]) > \alpha$, decide POS(A);

(N2)   If $P(A|[x]) < \alpha$, decide NEG(A);

(B2)   If $P(A|[x]) = \alpha$, decide BND(A).

For the second set of decision rules, we use a tie-breaking criterion so that the boundary region may be nonempty.

3.3   Derivations of Other Probabilistic Models

Based on the general decision-theoretic rough set model, it is possible to construct specific models by considering various classes of loss functions. In fact, many existing models can be explicitly derived.

The standard rough set model [15,24]. Consider a loss function:

$$\lambda_{12} = \lambda_{21} = 1, \quad \lambda_{11} = \lambda_{22} = \lambda_{31} = \lambda_{32} = 0. \tag{20}$$

There is a unit cost if an object in A^c is classified into the positive region or if an object in A is classified into the negative region; otherwise there is no cost. From equation (13), we have $\alpha = 1 > \beta = 0$, $\alpha = 1 - \beta$, and $\gamma = 0.5$. From decision rules (P1)-(B1), we can compute the approximations as $\underline{apr}_{(1,0)}(A) = $ POS$_{(1,0)}(A)$ and $\overline{apr}_{(1,0)}(A) = $ POS$_{(1,0)}(A) \cup $ BND$_{(1,0)}(A)$. For clarity, we use the subscript $(1,0)$ to indicate the parameters used to define lower and upper approximations. The standard rough set approximations are obtained as [15,24]:

$$\underline{apr}_{(1,0)}(A) = \{x \in U \mid P(A|[x]) = 1\},$$
$$\overline{apr}_{(1,0)}(A) = \{x \in U \mid P(A|[x]) > 0\}, \tag{21}$$

where $P(A|[x]) = |A \cap [x]|/|[x]|$.

The 0.5 probabilistic model [15]. Consider a loss function:

$$\lambda_{12} = \lambda_{21} = 1, \quad \lambda_{31} = \lambda_{32} = 0.5, \quad \lambda_{11} = \lambda_{22} = 0. \tag{22}$$

A unit cost is incurred if an object in A^c is classified into the positive region or an object in A is classified into the negative region; half of a unit cost is incurred if any object is classified into the boundary region. For other cases, there is no cost. By substituting these λ_{ij}'s into equation (13), we obtain $\alpha = \beta = \gamma = 0.5$. By using decision rules (P2)-(B2), we obtain the 0.5 probabilistic approximations [15]:

$$\underline{apr}_{(0.5,0.5)}(A) = \{x \in U \mid P(A|[x]) > 0.5\},$$
$$\overline{apr}_{(0.5,0.5)}(A) = \{x \in U \mid P(A|[x]) \geq 0.5\}. \tag{23}$$

The 0.5 model corresponds to the application of the simple majority rule.

The symmetric variable precision rough set model [35]. The symmetric variable precision rough set model corresponds to Case 5 of the decision-theoretic model discussed in the last subsection. As suggested by many authors [17,35], the value of α should be in the range $(0.5, 1]$. This condition can be satisfied by a loss function with $\lambda_{11} \leq \lambda_{31} < \lambda_{21}$ and $\lambda_{22} \leq \lambda_{32} < \lambda_{12}$ and the condition:

$$\lambda_{12} - \lambda_{32} > \lambda_{31} - \lambda_{11}. \tag{24}$$

The condition of symmetry, i.e., $\beta = 1 - \alpha$, is guaranteed by the additional condition:

$$(\lambda_{12} - \lambda_{32})(\lambda_{32} - \lambda_{22}) = (\lambda_{31} - \lambda_{11})(\lambda_{21} - \lambda_{31}). \tag{25}$$

By decision rules (P1)-(B1), we obtain the probabilistic approximations of the symmetric variable precision rough set model [35]:

$$\underline{apr}_{(\alpha,1-\alpha)}(A) = \{x \in U \mid P(A|[x]) \geq \alpha\},$$
$$\overline{apr}_{(\alpha,1-\alpha)}(A) = \{x \in U \mid P(A|[x]) > 1 - \alpha\}. \tag{26}$$

They are defined by a single parameter $\alpha \in (0.5, 1]$. The symmetry of parameters α and $\beta = 1 - \alpha$ implies the duality of approximations, i.e., $\underline{apr}_{(\alpha,1-\alpha)}(A) = (\overline{apr}_{(\alpha,1-\alpha)}(A^c))^c$ and $\overline{apr}_{(\alpha,1-\alpha)}(A) = (\underline{apr}_{(\alpha,1-\alpha)}(A^c))^c$.

As an example, consider a loss function:

$$\lambda_{12} = \lambda_{21} = 4, \quad \lambda_{31} = \lambda_{32} = 1, \quad \lambda_{11} = \lambda_{22} = 0. \tag{27}$$

It can be verified that the function satisfies the conditions given by equations (24) and (25). From equation (13), we have $\alpha = 0.75$, $\beta = 0.25$ and $\gamma = 0.5$. By decision rules (P1)-(B1), we have:

$$\underline{apr}_{(0.75,0.25)}(A) = \{x \in U \mid P(A|[x]) \geq 0.75\},$$
$$\overline{apr}_{(0.75,0.25)}(A) = \{x \in U \mid P(A|[x]) > 0.25\}. \tag{28}$$

In general, higher costs of mis-classification, namely, λ_{12} and λ_{21}, increase the α value [29].

The asymmetric variable precision rough set model [6]. The asymmetric variable precision rough set model corresponds to Case 1 of the decision-theoretic model. The condition on the parameters of the model is given by $0 \leq \beta < \alpha \leq 1$. In addition to $\lambda_{11} \leq \lambda_{31} < \lambda_{21}$ and $\lambda_{22} \leq \lambda_{32} < \lambda_{12}$, a loss function must satisfy the following conditions:

$$(\lambda_{12} - \lambda_{32})(\lambda_{21} - \lambda_{31}) > (\lambda_{31} - \lambda_{11})(\lambda_{32} - \lambda_{22}). \tag{29}$$

By decision rules (P1)-(B1), we obtain the probabilistic approximations of the asymmetric variable precision rough set model [6]:

$$\underline{apr}_{(\alpha,\beta)}(A) = \{x \in U \mid P(A|[x]) \geq \alpha\},$$
$$\overline{apr}_{(\alpha,\beta)}(A) = \{x \in U \mid P(A|[x]) > \beta\}. \tag{30}$$

They are no longer dual operators.

Consider a loss function:

$$\lambda_{12} = 4, \quad \lambda_{21} = 2, \quad \lambda_{31} = \lambda_{32} = 1, \quad \lambda_{11} = \lambda_{22} = 0. \tag{31}$$

The function satisfies the conditions given in equation (29). From equation (13), we have $\alpha = 0.75$, $\beta = 0.50$ and $\gamma = 2/3$. By decision rules (P1)-(B1), we have:

$$\underline{apr}_{(0.75,0.50)}(A) = \{x \in U \mid P(A|[x]) \geq 0.75\},$$
$$\overline{apr}_{(0.75,0.50)}(A) = \{x \in U \mid P(A|[x]) > 0.50\}. \tag{32}$$

They are not dual operators.

4 Probabilistic Approximations of a Partition

The decision-theoretic rough set model examined in the last section is based on the two-class classification problem, namely, classifying an object into either A or A^c. In this section, we extend the formulation to the case of more than two classes.

Let $\pi = \{A_1, A_2, \ldots, A_m\}$ be a partition of the universe U, representing m classes. For this m-class problem, we can solve it in terms of m two-class problems. For example, for the class A_i, we have $A = A_i$ and $A^c = U - A_i = \bigcup_{i \neq j} A_j$. For simplicity, we assume the same loss function for all classes A_i's. Furthermore, we assume that the loss function satisfies the conditions: (i) $\lambda_{11} \leq \lambda_{31} < \lambda_{21}$, (ii) $\lambda_{22} \leq \lambda_{32} < \lambda_{12}$, and (iii) $(\lambda_{12} - \lambda_{32})(\lambda_{21} - \lambda_{31}) > (\lambda_{31} - \lambda_{11})(\lambda_{32} - \lambda_{22})$. It follows that $\alpha > \gamma > \beta$. By decision rules (P1)-(B1), we have the positive, boundary and negative regions:

$$\text{POS}_{(\alpha,\beta)}(A_i) = \{x \in U \mid P(A|[x]) \geq \alpha\},$$
$$\text{BND}_{(\alpha,\beta)}(A_i) = \{x \in U \mid \beta < P(A|[x]) < \alpha\},$$
$$\text{NEG}_{(\alpha,\beta)}(A_i) = \{x \in U \mid P(A|[x]) \leq \beta\}. \tag{33}$$

The lower and upper approximations are given by:

$$\underline{apr}_{(\alpha,\beta)}(A_i) = \text{POS}_{(\alpha,\beta)}(A_i) = \{x \in U \mid P(A|[x]) \geq \alpha\},$$
$$\overline{apr}_{(\alpha,\beta)}(A_i) = \text{POS}_{(\alpha,\beta)}(A_i) \cup \text{BND}_{(\alpha,\beta)}(A_i) = \{x \in U \mid P(A|[x]) > \beta\}. \tag{34}$$

Similar to the algebraic case, we can define the approximations of a partition $\pi = \{A_1, A_2, ..., A_m\}$ based on those approximations of equivalence classes of π. For a partition π, the three regions can be defined by:

$$\text{POS}_{(\alpha,\beta)}(\pi) = \bigcup_{1 \leq i \leq m} \text{POS}_{(\alpha,\beta)}(A_i),$$
$$\text{BND}_{(\alpha,\beta)}(\pi) = \bigcup_{1 \leq i \leq m} \text{BND}_{(\alpha,\beta)}(A_i),$$
$$\text{NEG}_{(\alpha,\beta)}(\pi) = U - \text{POS}_{(\alpha,\beta)}(\pi) \cup \text{BND}_{(\alpha,\beta)}(\pi). \tag{35}$$

In contrast to the algebraic case, we have the following different properties of the three regions:

1. The three regions are not necessarily pairwise disjoint. Nevertheless, the family $\{\text{POS}_{(\alpha,\beta)}(\pi), \text{BND}_{(\alpha,\beta)}(\pi), \text{NEG}_{(\alpha,\beta)}(\pi)\}$ is a covering of U.
2. The family of positive regions $\{\text{POS}_{(\alpha,\beta)}(A_i) \mid 1 \leq i \leq m\}$ does not necessarily contain pairwise disjoint sets. That is, it may happen that $\text{POS}_{(\alpha,\beta)}(A_i) \cap \text{POS}_{(\alpha,\beta)}(A_j) \neq \emptyset$ for some $i \neq j$.
3. If $\alpha > 0.5$, the family $\{\text{POS}_{(\alpha,\beta)}(A_i) \mid 1 \leq i \leq m\}$ contains pairwise disjoint sets.
4. If $\beta > 0.5$, the three regions are pairwise disjoint. That is, $\{\text{POS}_{(\alpha,\beta)}(\pi), \text{BND}_{(\alpha,\beta)}(\pi), \text{NEG}_{(\alpha,\beta)}(\pi)\}$ is a partition of U. Furthermore, the family boundary regions $\{\text{BND}_{(\alpha,\beta)}(A_i) \mid 1 \leq i \leq m\}$ contains pairwise disjoint sets.
5. If $\beta = 0$, we have $\text{NEG}_{(\alpha,\beta)}(\pi) = \emptyset$.

When generalizing results from the algebraic rough set model, it is necessary to consider the implications of those properties.

The positive and boundary regions give rise to two kinds of rules. For $[x] \subseteq \text{POS}_{(\alpha,\beta)}(A_i)$ and $[x'] \subseteq \text{BND}_{(\alpha,\beta)}(A_i)$, we have:

$$[x] \xrightarrow{c > \alpha} A_i, \quad [x'] \xrightarrow{\beta < c < \alpha} A_i. \tag{36}$$

Unlike the algebraic rough set model, the probabilistic positive region may also produce non-deterministic rules. The negative region $\text{NEG}_{(\alpha,\beta)}(\pi)$ consists of all those objects that cannot be classified by the above rules.

In the application of probabilistic rough set models, some authors proposed straightforward generalizations of the notions of the algebraic rough set model. For example, the dependency of partition π on U/E is still quantified by using only the probabilistic positive region. The measure used in the variable precision model is [35]:

$$r_{(\alpha,\beta)}(\pi|U/E) = \frac{|\text{POS}_{(\alpha,\beta)}(\pi)|}{|U|}. \tag{37}$$

This may not necessarily be meaningful for the following two reasons. First, the probabilistic positive region may also produce non-deterministic rules. Second, the negative region is no longer empty. It is therefore necessary to consider the impact of the negative region. Similar comments can also be made regarding the generalizations of other measures.

5   Conclusion

A revisit to the decision-theoretic rough set model brings new insights into the probabilistic approaches to rough sets. Different probabilistic models, proposed either before or after the decision-theoretic models, can be easily derived from the decision-theoretic model. More importantly, instead of introducing ad hoc parameters, the Bayesian decision procedure systematically computes the required parameters based on a loss function.

The two-class decision-theoretic model is extended into a many-class model. The results show that some of the straightforward generalizations of the algebraic rough set model may not necessarily be meaningful. From our analysis, it becomes clear that we need to examine new and different measures for the probabilistic rough set models. The decision-theoretic rough set model opens an avenue for future research. A promising direction may be to study various measures based on the loss function within the decision-theoretic model.

References

1. Abd El-Monsef, M.M.E. and Kilany, N.M. Decision analysis via granulation based on general binary relation, *International Journal of Mathematics and Mathematical Sciences*, **2007**, Article ID 12714, 2007.
2. Greco, S., Matarazzo, B. and Slowinski, R. Rough membership and bayesian confirmation measures for parameterized rough sets, *LNAI 3641*, 314-324, 2005.
3. Deogun, J.S., Raghavan, V.V., Sarkar, A. and Sever, H. Data mining: trends in research and development, in: T.Y. Lin and Cercone, N. (Eds.), *Rough Sets and Data Mining*, Klower Academic Publishers, Boston, 9-45, 1997.
4. Duda, R.O. and Hart, P.E. *Pattern Classification and Scene Analysis*, Wiley, New York, 1973.
5. Grzymala-Busse, J.W. LERS – a system for learning from examples based on rough sets, in: Slowinski, R. (Ed.), *Intelligent Decision Support: Handbook of Apllications and Advances of the Rough Sets Theory*, Kluwer Academic Publishers, Dordrecht, 3-18, 1992.
6. Katzberg, J.D. and Ziarko, W. Variable precision rough sets with asymmetric bounds, in: Rough Sets, Fuzzy Sets and Knowledge Discovery, Ziarko, W. (Ed), Springer, London, 167-177, 1994.
7. Kitchener, M., Beynon, M. and Harrington, C. Explaining the diffusion of medicaid home care waiver programs using VPRS decision rules, *Health Care Management Science*, **7**, 237244, 2004.
8. Li, Y., Zhang, C. and Swanb, J.R. Rough set based model in information retrieval and filtering, *Proceeding of the 5th International Conference on Information Systems Analysis and Synthesis*, 398-403, 1999.
9. Li, Y., Zhang, C. and Swanb, J.R. An information fltering model on the Web and its application in JobAgent, *Knowledge-Based Systems*, **13**, 285-296, 2000.
10. Pawlak, Z. Rough sets, *International Journal of Computer and Information Sciences*, **11**, 341-356, 1982.
11. Pawlak, Z., *Rough Sets: Theoretical Aspects of Reasoning About Data*, Kluwer Academic Publishers, Boston, 1991.
12. Pawlak, Z. and Skowron, A. Rough membership functions, in: R.R. Yager and M. Fedrizzi and J. Kacprzyk (Eds.), *Advances in the Dempster-Shafer Theory of Evidence*, John Wiley and Sons, New York, 251-271, 1994.
13. Pawlak, Z. and Skowron, A. Rudiments of rough sets, *Information Sciences*, **177**, 3-27, 2007.
14. Pawlak, Z. and Skowron, A. Rough sets: some extensions, *Information Sciences*, **177**, 28-40, 2007.
15. Pawlak, Z., Wong, S.K.M. and Ziarko, W. Rough sets: probabilistic versus deterministic approach, *International Journal of Man-Machine Studies*, **29**, 81-95, 1988.

16. Qiu, G.F., Zhang, W.X., and Wu, W.Z. Characterizations of attributes in generalized approximation representation spaces, *LNAI 3641*, 84-93, 2005.
17. Skowron, A. and Stepaniuk, J. Tolerance approximation spaces, *Fundamenta Informaticae*, **27**, 245-253, 1996.
18. Slezak, D. Rough sets and Bayes factor, *LNAI 3400*, 202-229, 2005.
19. Slezak, D. and Ziarko, W. Attribute reduction in the Bayesian version of variable precision rough set model, *Electronic Notes in Theoretical Computer Science*, **82**, 263-273, 2003.
20. Srinivasan, P., Ruiz, M.E., Kraft, D.H. and Chen, J. Vocabulary mining for information retrieval: rough sets and fuzzy sets, *Information Processing and Management*, **37**, 15-38, 2001.
21. Tsumoto, S. Accuracy and coverage in rough set rule induction, *LNAI 2475*, 373-380, 2002.
22. Tsumoto, S. Statistical independence from the viewpoint of linear algebra, *LNAI 3488*, 56-64, 2005.
23. Wei, L.L. and Zhang, W.X. Probabilistic rough sets characterized by fuzzy sets, *International Journal of Uncertainty Fuzziness and Knowledge-Based Systems*, **12**, 47-60, 2004.
24. Wong, S.K.M. and Ziarko, W. Comparison of the probabilistic approximate classification and the fuzzy set model, *Fuzzy Sets and Systems*, **21**, 357-362, 1987.
25. Wu, W.Z. Upper and lower probabilities of fuzzy events induced by a fuzzy set-valued mapping, *LNAI 3461*, 345-353, 2005.
26. Yao, J.T. and Herbert, J.P. Web-based Support Systems based on Rough Set Analysis, manuscript, 2007.
27. Yao, J.T. and Zhang, M. Feature selection with adjustable criteria, *LNAI 3641*, 204-213, 2005.
28. Yao, Y.Y. Two views of the theory of rough sets in finite universes, *International Journal of Approximation Reasoning*, **15**, 291-317, 1996.
29. Yao, Y.Y. Information granulation and approximation in a decision-theoretical model of rough sets, in: Polkowski, L., Pal, S.K., and Skowron, A. (Eds), *Rough-neuro Computing: Techniques for Computing with Words*, Springer, Berlin, 491-516, 2003.
30. Yao, Y.Y. Probabilistic approaches to rough sets, *Expert Systems*, **20**, 287-297, 2003.
31. Yao, Y.Y. Probabilistic rough set approximations, manuscript, 2006.
32. Yao, Y.Y. and Wong, S.K.M. A decision theoretic framework for approximating concepts, *International Journal of Man-machine Studies*, **37**, 793-809, 1992.
33. Yao, Y.Y., Wong, S.K.M. and Lingras, P. A decision-theoretic rough set model, in: Z.W. Ras, M. Zemankova and M.L. Emrich (Eds.), *Methodologies for Intelligent Systems*, **5**, North-Holland, New York, 17-24, 1990.
34. Zhang, W.X., Wu, W.Z., Liang, J.Y. and Li, D.Y. *Rough Set Theory and Methodology* (in Chinese), Xi'an Jiaotong University Press, Xi'an, China, 2001.
35. Ziarko, W. Variable precision rough set model, *Journal of Computer and System Sciences*, **46**, 39-59, 1993.
36. Ziarko, W. Acquisition of hierarchy-structured probabilistic decision tables and rules from data, *Expert Systems*, **20**, 305-310, 2003.
37. Ziarko, W. Probabilistic rough sets, *LNAI 3641*, 283-293, 2005.

Efficient Attribute Reduction Based on Discernibility Matrix

Zhangyan Xu[1,2], Chengqi Zhang[3,4], Shichao Zhang[1], Wei Song[2], and Bingru Yang[2]

[1] Department of Computer, Guangxi Normal University, 541004, Guilin China
xyzwlx72@yahoo.com.cn, zhangsc@mailbox.gxnu.edu.cn
[2] School of Information Engineering, Univ. of Science and Technology Beijing, China
{sgyzfr,bryang_kD}@yahoo.com.cn
[3] Faculty of Information Technology, University of Technology, Sydney, Australia
chengqi@it.uts.edu.au
[4] Department of Information Systems,City University of Hong Kong,Hong Kong, China
chengqi@it.uts.edu.au

Abstract. To reduce the time complexity of attribute reduction algorithm based on discernibility matrix, a simplified decision table is first introduced, and an algorithm with time complexity $O(|C||U|)$ is designed for calculating the simplified decision table. And then, a new measure of the significance of an attribute is defined for reducing the search space of simplified decision table. A recursive algorithm is proposed for computing the attribute significance that its time complexity is of $O(|U/C|)$. Finally, an efficient attribute reduction algorithm is developed based on the attribute significance. This algorithm is equal to existing algorithms in performance and its time complexity is $O(|C||U|) + O(|C|^2|U/C|)$.

Keywords: Attribute reduction, simplified decision table, discernibility matrix.

1 Introduction

Recently, some efforts on attribute reduction have focused on dealing with inconsistency in decision information systems. Since Pawlak proposed the attribute reduction based on positive region [1], there has been some work developed for improving the efficiency of the attribute reduction based on positive region [2,3]. Latter, Skowron and Hu proposed the attribute reduction based on discernibility matrix [4,5]. There have been also other types of knowledge reduction [6,7].

In this paper, we study the attribute reduction based on discernibility matrix and to design correspondence attribute reduction algorithm. The attribute reduction algorithm based on discernibility matrix given in [8-14] starts with an empty set of attributes and heuristically adds new attributes one by one, in a greedy way, until a super reduction is constructed. In each loop, if the attribute a_k frequently occurs in the discernibility matrix, it will be added. This is equivalent to choosing the attribute that 'discerns' the largest number of pairs of objects with different decisions. Full details of the algorithm can be found in [8-14]. In these algorithms, the elements of discernibility matrix are used as the heuristic information. So it must be first to

J.T. Yao et al. (Eds.): RSKT 2007, LNAI 4481, pp. 13–21, 2007.

calculate the discernibility matrix. The time complexity and space complexity are both $O(|C \|U|^2)$, where C and U are attributes set and objects set of a decision table, respectively. The time complexity of these algorithms proposed by authors in [8-10] is $O(|C|^2|U|^2)$. The time complexity of these algorithms proposed by authors in [11-14] is cut down to $O((|C|+\log|U|)|U|^2)$. Hence if the elements of discernibility matrix are used as the heuristic information to design attribute reduction algorithm, the best time complexity of this kind of algorithm is not lower than $O(|C\|U|^2)$. On the other hand, it needs large space to store the discernibility matrix. When the data set is very large, algorithm is difficult to operate.

To lower the time complexity of attribute reduction algorithm based on discernibility matrix, we design an algorithm based on the significance of attribute and its time complexity is $O(|C\|U|)+O(|C|^2|U/C|)$.

The rest of this paper is as follows. In Section 2, we introduce some basic concepts. We present our algorithm in Section 3 and illustrate the use of the algorithm with an example in Section 4. We summarize this paper in Section 5.

2 Concepts and Definitions

In this section, we introduce the basic concepts and correspondence definitions.

Definition 1. A decision table is defined as $S = (U,C,D,V,f)$, where $U = \{x_1, x_2, \cdots, x_n\}$ is the set of objects, $C = \{c_1, c_2, \cdots, c_r\}$ is the set of condition attributes, D is the set of decision attributes, and $C \cap D = \varnothing$; $V = \bigcup_{a \in C \cup D} V_a$, where V_a is the value range of attribute a. $f: U \times C \cup D \rightarrow V$ is an information function, in which an information value for each attribute of an object, i.e., $\forall a \in C \cup D, x \in U, f(x,a) \in V_a$. Every attribute subset $P \subseteq (C \cup D)$ determines a binary indiscernibility relation $IND(P)$:

$$IND(P) = \{(x,y) \in U \times U \mid \forall a \in P, f(x,a) = f(y,a)\}$$

$IND(P)$ determines a partition of U, which is denoted by $U/IND(P)$ (in short U/P). Any element $[x]_P = \{y \mid \forall a \in P, f(x,a) = f(y,a)\}$ in U/P is called equivalence class.

Definition 2. For a decision table $S = (U,C,D,V,f)$, let $U/D = \{D_1, D_2, \cdots, D_k\}$ be the partition of D to U, and $U/C = \{C_1, C_2, \cdots, C_m\}$ be the partition of C to U, where $C_i (i = 1,2,\cdots,m)$ is basic block, then $POS_C(D) = \bigcup_{D_i \in U/D} C_-(D_i)$ is called positive region of C on D. If $POS_C(D) = U$, then the decision table is called consistent, else it is called inconsistent.

Theorem 1. For a decision table $S = (U,C,D,V,f)$, there is

$$POS_C(D) = \bigcup_{X \in U/C \wedge \forall x, y \in X \Rightarrow f(x,D) = f(y,D)} X$$

Proof: According to the definition of positive region of C for D, it is easy to know the proposition is right.

According to Theorem 1, we have the following definition of simplicity decision table.

Definition 3. For a decision table $S = (U,C,D,V,f)$, let $U/C = \{[x'_1]_C,[x'_2]_C,\cdots,[x'_m]_C\}$ and $U' = \{x'_1,x'_2,\cdots,x'_m\}$. For the definition of positive region, there is $POS_C(D) = [x'_{i_1}]_C \cup [x'_{i_2}]_C \cup \cdots \cup [x'_{i_t}]_C$, where $\{x'_{i_1},x'_{i_2},\cdots,x'_{i_t}\} \subseteq U'$ and $\forall x,y \in [x'_{i_s}]_C$ $(s=1,2,\cdots,t)$, there are $f(x,D) = f(y,D)$; let $U'_{pos} = \{x'_{i_1},x'_{i_2},\cdots,x'_{i_t}\}$ and $U'_{neg} = U'-U'_{pos}$. It is said that the 5-tuple $S' = (U',C,D,V,f)$ is a simplicity decision table.

Definition 4. For a decision table $S = (U,C,D,V,f)$, we define discernibility matrix $M = (m_{ij})$, whose elements are defined as follow:

$$m_{ij} = \begin{cases} \{c_k \mid c_k \in C, f(x_i,c_k) \neq f(x_j,c_k), f(x_i,D) \neq f(x_j,D)\} \\ \varnothing \quad \text{el se} \end{cases}$$

where $k = 1,2,...,r$.

Definition 5. For a decision table $S = (U,C,D,V,f)$, $M = (m_{ij})$ is discernibility matrix, $\forall B \subseteq C$, if B satisfies: (1) $\forall \varnothing \neq m_{ij} \in M$, such that $B \cap m_{ij} \neq \varnothing$; (2) $\forall b \in B$, $B - \{b\}$ is not satisfied (1), then B is called the attribute reduction of C for D based on discernibility matrix.

In next section, we would define a new reasonable formula for measuring the significance of attribute to reduce the search space of simplified decision table.

3 The Significance of Attribute

In these old algorithm based on discernibility matrix, it was first to calculate the discernibility matrix, so the time complexity of the algorithm is not lower than $O(|C||U|^2)$. To cut down the time complexity of the attribute reduction, we designed a new and reasonable formula for measuring the significance of attribute to reduce the search space of the simplicity decision table. In order to propose the significance of attribute, we first introduce the follow proposition.

Definition 6. For a decision table $S = (U,C,D,V,f)$, $S' = (U',C,D,V,f)$ is its simplicity decision table. $\forall B \subseteq C$, we define knowledge of the attribute set B for D as follow:

$$Sig_D(B) = \{ \bigcup_{X \in U'/B \wedge (X \subseteq U'_{pos}) \wedge |X/D|=1} X \} \cup \{ \bigcup_{X \in U'/B \wedge (X \subseteq U'_{neg}) \wedge |X/C|=1} X \}.$$

where we consider $Sig_D(\varnothing) = \varnothing$. We can easily know $Sig_D(C) = U'$.

Theorem 2. For a decision table $S = (U,C,D,V,f)$, $S' = (U',C,D,V,f)$ is its simplicity decision table, $M = (m_{ij})$ is the deiscernibility matrix of the old decision table. $\forall B \subseteq C$, if $Sig_D(B) = Sig_D(C)$, there is $\forall \varnothing \neq m_{ij} \in M$ such that $B \cap m_{ij} \neq \varnothing$.

Theorem 3. For a decision table $S = (U,C,D,V,f)$, $S' = (U',C,D,V,f)$ is its simplicity decision table, and $M = (m_{ij})$ is the discernibility matrix of the old decision table. $\forall B \subseteq C$, if $\forall \emptyset \neq m_{ij} \in M$ such that $B \cap m_{ij} \neq \emptyset$, there has $Sig_D(B) = Sig_D(C)$.

Theorem 4. For a decision table $S = (U,C,D,V,f)$, $S' = (U',C,D,V,f)$ is its simplicity decision table, and $M = (m_{ij})$ is the discernibility matrix of the old decision table. $\forall B \subseteq C$, if $Sig_D(B) \neq Sig_D(C)$, there must exist $\emptyset \neq m_{i_0 j_0} \in M$ such that $B \cap m_{i_0 j_0} = \emptyset$.

Theorem 5. For a decision table $S = (U,C,D,V,f)$, $S' = (U',C,D,V,f)$ is its simplicity decision table. $\forall B \subseteq C$, if $Sig_D(B) = Sig_D(C)$ and $\forall b \in B$ there is $Sig_D(B-\{b\}) \neq Sig_D(C)$, then B is an attribute reduction C for D based on the discernibility matrix.

Definition 7. (The significance of attribute) For a decision table $S = (U,C,D,V,f)$, $S' = (U',C,D,V,f)$ is its simplicity decision table. For $P \subseteq C$, the significance of arbitrary attribute a ($a \in (C-P)$) to attribute set P is defined as follow:

$$I_p(a) = \left| Sig_D(P \cup \{a\}) - Sig_D(P) \right|.$$

Theorem 6[3]. For a decision table $S = (U,C,D,V,f)$, to $\forall P \subseteq C, \forall a \in (C-P)$, there is
$$U/(P \cup \{a\}) = \bigcup_{X \in U/P} (X/\{a\}).$$

Theorem 7. For a decision table $S = (U,C,D,V,f)$, $S' = (U',C,D,V,f)$ is its simplicity decision table. For $P \subseteq C, \forall a \in (C-P)$, there is

$$U'/(P \cup \{a\}) = \bigcup_{X \in U'/P} (X/\{a\}).$$

Theorem 8. For a decision table $S = (U,C,D,V,f)$, $S' = (U',C,D,V,f)$ is its simplicity decision table. For $P \subseteq C, \forall a \in (C-P)$, there is

$$I_p(a) = \left| Sig_D(P \cup \{a\}) - Sig_D(P) \right|$$

$$\bigcup \{ {}_{X \in U'/P \wedge (X \not\subseteq U'_{pos} \wedge X \not\subseteq U'_{neg}) \wedge Y \in X/\{a\} \wedge Y \subseteq U'_{pos} \wedge |Y/D|=1} Y \}$$

$$= | \{ {}_{X \in U'/P \wedge X \subseteq U'_{pos} \wedge |X/D| \neq 1 \wedge Y \in X/\{a\} \wedge |Y/D|=1} Y \}$$

$$\bigcup \{ {}_{X \in U'/P \wedge (X \not\subseteq U'_{pos} \wedge X \not\subseteq U'_{neg}) \wedge Y \in X/\{a\} \wedge Y \subseteq U'_{neg} \wedge |Y/C|=1} Y \}$$

$$\bigcup \{ {}_{X \in U'/P \wedge X \subseteq U'_{neg} \wedge |X/C| \neq 1 \wedge Y \in X/\{a\} \wedge |Y/C|=1} Y \} | .$$

4 Algorithm for Calculating the Significance of Attribute

According to Definition 6, it is first to calculate the simplified decision table before calculating the significance of attribute. So we first propose an efficient algorithm for calculating the simplified decision table. Calculating the simplified decision table is in fact to calculate the $IND(C)$. To our best knowledge, the best algorithm for computing $IND(C)$ is the algorithm of [3] with the time complexity $O(|C||U|\log|U|)$ at present. So we used radix sorting to design a good algorithm for computing $IND(C)$. And its time complexity is cut down to $O(|C||U|)$.

Algorithm 1.   Computing the simplicity decision table

Input: Decision table $S = (U, C, D, V, f)$, $U = \{x_1, x_2, \cdots, x_n\}$, $C = \{c_1, c_2, \cdots, c_r\}$

Output: $U'_{pos}, U'_{neg}, U', M_i, m_i (1 \le i \le s)$.

1. To each $c_i (i = 1, 2, \cdots, r)$, calculate the maximum and minimum of $f(x_j, c_i)$ $(j = 1, 2, \cdots, n)$ and denote M_i and m_i respectively;

2. use state list to store the objects $x_1, x_2, \cdots, x_n$ in turn ; let the head pointer of the list point to x_1 ;

3. for (i=1;i<r+1;i++)

    3.1 the ith "distribution": construct M_i-m_i+1 empty queues, let $front_k$ and end_k ($k=0,1,\ldots, M_i$-m_i) be the head pointer and tail pointer of the kth queue respectively. Distribute the object x of the list U to the $f(x,c_i) - m_i$ th queue according to the elements order of list U.

    3.2 the ith "collection": the head pointer of the list points to the head pointer of the first nonempty queue, modify the tail pointer of each nonempty and let it point to the head object of the next nonempty queue. In this way, recombine $M_i - m_i + 1$ queues to a new list;

4. Let the objects sequence of list from Step 3 be $x'_1, x'_2, \cdots, x'_n$;

    $t=1; B_t = \{x'_1\}$;

    for (j=2;j<n+1;j++)

       if any $c_i \in C(i = 1, 2, \cdots, r)$ there is $f(x'_j, c_i) = f(x'_{j-1}, c_i)$,

       then $B_t = B_t \cup \{x'_j\}$;

       else { $t=t+1$; $B_t = \{x'_j\}$; }

5. $U'_{pos} = \varnothing; U'_{neg} = \varnothing$;

    for (i=1;i<t+1;i++)

       if any $x, y \in B_i$ there is $f(x, D) = f(y, D)$, then we take out the first object of B_i to U'_{pos} ; Else we take out the first object of B_i to U'_{neg} ;

    $U' = U'_{pos} \cup U'_{neg}$;

Complexity analyze of Algorithm 1. the time complexity for the first step of algorithm is $O(|C||U|)$; the time complexity for the second step is $O(|U|)$; the time complexity for Step 3.1 is $O(|U|+M_i - m_i + 1)$, the time complexity for Step 3.2 is $O(M_i - m_i + 1)$, so the time complexity for Step 3 is $O(|C||U|+\sum_{i=1}^{r}(M_i - m_i + 1))$; the time complexity for Step 4 is $O(|C||U|)$; the time complexity for Step 5 is $O(|D||U|)$ (the decision attribution usually is only one). Hence the time complexity of algorithm is $O(|C\|U|+\sum_{i=1}^{r}(M_i-m_i+1))$. In most condition, especially to the large-scale decision table, there is often $\max_{1\le i\le s}(M_i-m_i+1)\le|U|$ (for example, the mushroom in UCI mushroom, it has more 8000 objects and 22 attributions, but these attribution values are single letter, so there is at most 26 kinds difference values in each attribution.) ,hence $(|C\|U|+\sum_{i=1}^{r}(M_i-m_i+1))\le|C\|U|+|C\|U|$, therefore the time complexity of Algorithm 1 is $O(|C\|U|)$. It is easily to know that the space complexity of Algorithm 1 is $O(|U|)$.

According to Theorems 7 and 8, we can design the following efficient algorithm for calculating the significance of attribute.

Algorithm 2. compute $I_p(a)$

Input : $S_p=\{X\mid X\in U'/P\wedge((X\not\subseteq U'_{pos}\wedge X\not\subseteq U'_{neg})$

$\qquad\qquad \vee(X\subseteq U'_{pos}\wedge|X/D|\ne 1)\vee(X\subseteq U'_{neg}\wedge|X/C|\ne 1))\},M_a,\ m_a\ ;$

Output : $S_{P\cup\{a\}}=\{X\mid X\in U'/(P\cup\{a\})\wedge((X\not\subseteq U'_{pos}\wedge X\not\subseteq U'_{neg})$

$\qquad\qquad \vee(X\subseteq U'_{pos}\wedge|X/D|\ne 1)\vee(X\subseteq U'_{neg}\wedge|X/C|\ne 1))\}\ ;\ I_p(a)\ ;$

1.   for (j=1; j<|S_p|+1; j++) $\forall x\in X_j$, let $x.flag=j$; // $X_j\in S_p$

2.   Let $T=X_1\cup X_2\cup\cdots\cup X_{|S_p|}=\{x_1,x_2,\cdots,x_z\}$; where $X_j\in S_p(j=1,2,\cdots,|S_p|)$;

3.   use state list to store the objects $x_1,x_2,\cdots,x_z$ in turn ;

4.   construct M_a-m_a+1 empty queues, let $front_k$ and end_k (k=0,1,..., M_a-m_a) be the head pointer and tail pointer of the kth queue respectively. Distribute the object $x\in T$ of the list to the $f(x,a)-m_a$ th queue according to the elements order of list. //where M_a,m_a are the maximal value and the minimal value of the attribute a in the old decision table.

5. $I_p(a)=0$; $S_{P\cup\{a\}}=\varnothing$;

6.   It deals with the each not empty queue $Col=\{y_1,y_2,\cdots,y_h\}$ as follow acquired from the four step;

    $t=1; B_t=\{y_1\}$;

    for (j=2;j<h+1;j++)

      { if ($y_j.flag == y_{j-1}.flag$)

          $B_t=B_t\cup\{y_j\}$;

        else { $t=t+1; B_t=\{y_j\}$; };

    }

    for (i=1;j<t+1;i++)

        if ($B_i\subseteq U'_{pos}\wedge|B_i/D|=1$) $I_p(a)=I_p(a)\cup|B_i|$;

        else   if ($B_i\subseteq U'_{neg}\wedge|B_i/C|=1$)

              $I_p(a)=I_p(a)\cup|B_i|$;

        else $S_{P\cup\{a\}}=S_{P\cup\{a\}}\cup\{B_i\}$;

Complexity analyses of Algorithm 2. The time complexity of the first step is $O(|T|)$; The time complexity of the fourth step is $O(|T|)$; The time complexity of the first *for* cycle of the sixth step is $O(|Col|)$. The time complexity of the second *for* cycle is also $O(|Col|)$. So the time complexity of the sixth step is $O(|T|)$. Therefore, the time complexity of Algorithm 2 is $O(|T|)$. Because of $O(|T|) \leq O(|U'|)$, the worst time complexity of Algorithm 2 is $O(|U'|)$. The space complexity is $O(|T|)$. For the same reason, the worst space complexity of the algorithm is $O(|U'|)$.

5   Attribute Reduction Algorithm Based on Discernibility Matrix

According to Algorithms 1 and 2, we now can design an efficient attribute reduction algorithm based on discernibility matrix.

Algorithm 3. Attribute reduction algorithm based on discernibility matrix

Input: decision table $S = (U,C,D,V,f)$, $U = \{x_1,x_2,\cdots,x_n\}$, $C = \{c_1,c_2,\cdots,c_s\}$;

Output: attribute reduction R;

1. It uses the algorithm 1 to calculate $U' = \{u_1,u_2,\cdots,u_m\}$, $m_i, M_i (i=1,2,\cdots,s)$, U'_{pos}, U'_{neg};
2. let $R = \varnothing$; $S_R = \{\{u_1,u_2,\cdots,u_m\}\}$; // When algorithm begins, input set S is the only one equivalence, i.e. U';
3. To each attribute $c_i \in C - R$, calculating $I_R(c_i)$; Denote $I_R(c_k) = \max\limits_{c_i \in C-R} I_R(c_i)$; If the attribute like that is not only one, we arbitrary select one;
4. If $S_{R \cup \{c_k\}}$ is an empty set, then stop the algorithm, output the attribute reduction $R \cup \{c_k\}$; // $S_{R \cup \{c_k\}}$ is the corresponding output set to calculate $I_R(C_k)$.
5. If $S_{R \cup \{c_k\}} \neq \varnothing$, then $R = R \cup \{c_k\}$; The algorithm will turn to step 3;

The complexity of Algorithm 3. It can be known from Algorithm 1 that the time and space complexities of the first step are $O(|C||U|)$ and $O(|U|)$ respectively. The worst time and space complexity third step are $O(|U'|) = O(|U/C|)$ from analyzing of Algorithm 2. So the worst time and space complexities of the third step are $O(|U/C||C-R|)$ and $O(|U/C||C-R|)$ respectively. The worst time complexity of the third step to the fifth is $O(|U/C||C|) + O(|U/C||C-1|) + \cdots + O(|U/C|) = O(|U/C||C|^2)$. The worst space complexity of the third step to the fifth is $O(|U/C||C|)$. Therefore the worst time and space complexities of Algorithm 3 are $(O(|C||U|) + O(|C|^2|U/C|)$ and $O(|U|) + O(|C||U/C|)$ respectively.

6   Conclusion

At present, the elements of discernibility matrix are used as the heuristic information by all the existing attribute reduction algorithms based on discernibility matrix. In these algorithms, it is first to calculate the discernibility matrix. Because it must be first to calculate the Skowron's discenibility matrix in this kind of attribute reduction algorithm, the best time complexity of this kind algorithm is not lower than $O(|C||U|^2)$. On the other hand, it needs the large space to store the discernibility matrix. Once the data set is very large, algorithm is difficult to operate. To lower the time complexity of attribute reduction algorithm based on discernibility matrix, firstly, the simplified decision table and the significance of attribute are introduced. Then an efficient attribute reduction based on the significance of attribute is proposed. And And it is proved that our algorithm is equivalent to existing algorithms in performance and the time complexity is $O(|C||U|) + O(|C|^2|U/C|)$.

Acknowledgement

This work is partially supported by Australian Large ARC Grants (DP0449535, DP0559536 and DP0667060), a China NSF Major Research Program (60496327), a China NSF Grant (60463003), an Overseas Outstanding Talent Research Program of the Chinese Academy of Sciences (06S3011S01), and an Overseas-Returning High-level Talent Research Program of China Hunan-Resource Ministry.

References

1. Pawlak, Z.: Rough set theory and its application to data analysis. Cybernetics and Systems, 9(1998) 661-668
2. Guan, J.W., Bell ,D.A.: Rough computational methods for information systems, Artificial intelligence, 105(1998) 77-103
3. Nguyen, S.H., Nguyen, H.S.: Some efficient algorithms for rough set methods. In Proceedings of the Conference of Information Processing and Management of Uncertainty in Knowledge-Based Systems IPMU'96, Granada, Spain, (1996) 1451-1456
4. Skowron, A., Rauszer, C.: The discernibilinity matrices and functions in information systems. In: R. Slowincki (ed), Intelligent decision support-handbook of applications and advances of the rough sets theory, Dordrecht:Kluwer,(1992) 331-362
5. Hu, X.H., Cercone, N.: Learning in relational databases: A rough set approach. International journal of computational intelligence, 11(1995) 323-338
6. Kkryszkiewicz, M.: Comparative study of alternative types of knowledge reduction in inconsistent systems. International Journal of Intelligent Systems. 16(2001) 105-120
7. Li, D.R., Zhang, B.: On knowledge reduction inconsistent decision information systems. International Journal of Uncertainty, Fuzziness and Knowledge-based Systems, 50(2004) 651-672
8. Wang, J., Cui, J., Zhao, C.: Investigation on AQ11, ID3 and the Principle of Discernibility Matrix. Journal of Computer Science and Technology, 16(2001)1-12
9. Wang, J., Wang, J.: Reduction algorithms based on discernibility matrix: the ordered attributes method, Journal of Computer Science and Technology, 16(2001)489-504
10. Zhang, J., Wang, J., Li, D.(ed.): A new heuristic reduction algorithm base on rough sets theory. Lecture Notes in Computer Sciences, 2762(2003): 247- 253.

11. Wang, B., Chen, S.B.: A complete algorithm for attribute reduction. T.-J. Tarn et al. (Eds.): Robotic Welding, Intelligence and Automation, Lecture Notes in Computer Sciences, Vol. 299. Springer -Verlag Berlin Heidelberg New York(2004)
12. Bazan, J., Nguyen, H.S., Nguyen, S.H.(ed.): Rough set algorithms in classification problems. In Rough Set Methods and Applications: New Developments in Knowledge Discovery in Information Systems. Vol.56. Physica-VerlagNew York(2000)
13. Hu, K.Y., Diao, L.L., Lu, Y.C.(ed.): A Heuristic Optimal Reduction Algorithm. Leung, K.S., Chan, L.W., Meng, H.(Eds.): IDEAL 2000, LNCS.Vol.1983. Springer-Verlag Berlin Heidelberg New York(2000)
14. Korzen, M., Jaroszewicz, S.: Finding reducts without building the discernibility matrix. In Proc. of the 5th Int. Conf. on Intelligent Systems Design and Applications (ISDA'05)

Near Sets. Toward Approximation Space-Based Object Recognition

James F. Peters

Department of Electrical and Computer Engineering,
University of Manitoba
Winnipeg, Manitoba R3T 5V6 Canada
jfpeters@ee.umanitoba.ca

Abstract. The problem considered in this paper is how to recognize objects that are qualitatively but not necessarily spatially near each other. The term *qualitatively near* is used here to mean closeness of descriptions or distinctive characteristics of objects. The solution to this problem is inspired by the work of Zdzisław Pawlak during the early 1980s on the classification of objects by means of their attributes. In working toward a solution of the problem of the approximation of sets that are qualitatively near each other, this article considers an extension of the basic model for approximation spaces. The basic approach to object recognition is to consider the degree of overlap between families of perceptual neighbourhoods and a set of objects representing a standard. The proposed approach to object recognition includes a refinement of the generalized model for approximation spaces. This is a natural extension of recent work on nearness of objects. A byproduct of the proposed object recognition method is what we call a near set. The contribution of this article is an approximation space-based approach to object recognition formulated in the context of near sets.

Keywords: Approximation space, feature, near set, object recognition, perceptual neighborhood.

> *An approximation space ... serves as a formal*
> *counterpart of perception ability or observation.*
> – Ewa Orlowska, March, 1982.

1 Introduction

The problem considered in this paper is how to recognize objects that are qualitatively but not necessarily spatially near each other. The term *qualitatively near* is used here to mean closeness of descriptions or distinctive characteristics of objects. The term *object* denotes something perceptible. If we choose *shading* as a feature and let $B_{shading}(x) = \{y \mid shading(x) = shading(y)\}$, then the objects in Fig. 1 can be partitioned, where the objects in $B_{shading}(x)$, *i.e.*, equivalence class containing objects that are *descriptively* indiscernible from x,

J.T. Yao et al. (Eds.): RSKT 2007, LNAI 4481, pp. 22–33, 2007.

are not adjacent to each other (see, *e.g.*, $B_{shading}(x1) = \{x1, x11, x15, x16\}$ in Fig. 1.2 or $B_{shading}(g1) = \{g1, g2, g3\}$ in Fig. 1.1).

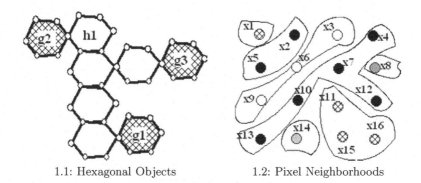

1.1: Hexagonal Objects   1.2: Pixel Neighborhoods

Fig. 1. Non-Adjacent Objects with Matching Descriptions

The solution to this problem is inspired by the work of Zdzisław Pawlak during the early 1980s on the classification of objects [22] and elaborated in [25,26,27]. In working toward a solution of the problem of the approximation of sets that are qualitatively near each other, this article considers an extension of the basic model for approximation spaces. The basic approach is to consider families of perceptual neighborhoods containing objects with matching descriptions that are possibly space-independent. A *perceptual neighborhood* is an equivalence class containing observed sample objects with matching descriptions. The proposed approach to object recognition is a straightforward extension of the rough set approach, where approximation can be considered as formal counterpart of perception [18] in the context of families of perception granules (neighborhoods). The term *perception granule* comes from [48]. A byproduct of the proposed approximation method is what we call a near set.

The approach to classifying objects such as those in Fig. 1.2 contrasts sharply with the approach to defining neighborhoods with an Adjacency relation in [7]. For example, the hexagons with mesh interiors (*g1, g2, g3*) in Fig. 1.1 are descriptively near each other but spatially non-adjacent. The refinement of approximation spaces in [28] is close to what is known as a nearness space [11,48] with the exception of the distinction between attributes and features as well as covering $\mathcal{F}$ (family of neighborhoods) that underly the approach to approximation spaces in this article. In addition, the proposed approach of nearness of objects [28] is not restricted to the neighborhood of a point x and $x \in Cl(A)$ (closure of A) as in [48], since we consider the nearness of objects that are not points. The contribution of this article is an approach to approximating a set based on the union of families of sets of objects with matching descriptions, which provides a foundation for near sets.

This article is organized as follows. The distinction between features and attributes is explored in Sect. 2. An approach to pattern recognition is briefly presented in Sect. 3. A refinement of the generalized approximation space model is given in Sect. 4. Sample near sets extracted from ethogram tables are presented in Sect. 5.

2   Features and Measurements

Underlying the study of near sets is an interest in classifying sample objects by means of probe functions associated with object features. The term *feature* was originally identified with the cast of a face [14]. More recently, the term *feature* is defined as the make, form, fashion or shape (of an object) [19]. This term comes from the Latin term *factura*, i.e., *facture*, which means the action or process of making an object or the result of an action or process (e.g., a work of art, image made with a digital camera). In effect, the term *feature* characterizes some aspect of the makeup of an object. From a philosophical perspective that can be traced back to Kant [15], features highlight an interest in the *appearances* of objects rather than calling attention to the properties or qualities that are somehow inherent in objects. The term *feature* is commonly used in pattern recognition theory [21], statistical learning theory [45], reinforcement learning [34], neural computing [4], science (e.g., ethology [16,33,35]), image processing [13,5], biotechnology, industrial inspection, the internet, radar, sonar, and speech recognition [9]. More recently, the term *feature* has been used in rough set theory [5,33,34,35,28,30].

Historically, semantically, and philosophically, there is a distinction between the terms *feature* and *attribute*. An *attribute* is a quality regarded as characteristic or inherent in an object [19]. In philosophy, an attribute is a property of an object (e.g., spatial extension of a piece of wax). The term *attribute* is commonly used in database theory [44], data mining [47], and philosophy [12]. In rough set theory [25], an attribute is treated as a partial function, which is a relation that associates each element of a set of objects (domain) with at most one element of a value set (codomain) [49].

It was Zdzisław Pawlak who proposed classifying objects by means of their attributes considered in the context of an approximation space [22]. The proposed approach to classifying objects can also be explained in terms of features. Implicit in the original work of Pawlak is a distinction between features (*makeup, appearance*) of objects and knowledge about objects. The knowledge about an object is represented by a measurement associated with each feature of an object. It can observed that a feature is an invariant characteristic of objects belonging to a class [46]. The distinction between features and corresponding measurements associated with features is usually made in the study of pattern recognition (see, e.g., [17,21]). Let A denote a set of features for objects in a set X. For each $a \in A$, we associate a function f_a that maps X to some set V_{f_a} (range of f_a). The value of $f_a(x)$ is a measurement associated with feature a of an object $x \in X$. The function f_a is called a *probe* [21]. By $Inf_B(x)$, where $B \subseteq A$ and $x \in U$ we denote the *signature* of x, i.e., the set $\{(a, f_a(x)) : a \in B\}$. If the set $B = \{a_1, \ldots, a_m\}$, then Inf_B is identified with a vector $(f_{a_1}(x), \ldots, f_{a_m}(x))$ of probe function values for features in B.

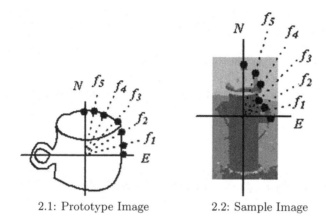

2.1: Prototype Image   2.2: Sample Image

Fig. 2. Image Patterns

3   Approach to Pattern Recognition

The problem considered here is to determine whether there is a correspondence between an object in a prototype image $\mathcal{I}$ (*e.g.*, cup in Fig. 2.1) and an object in a sample image I_1 (*e.g.*, fire hydrant in Fig. 2.2). By way of illustration, consider *contour* as a helpful feature in considering the form of various objects. Let $I_1, \mathcal{I}, f$ denote sample image, prototype image, probe function associated with *contour*, respectively. Then, following the approach suggested in [21], pattern recognition is defined for real-valued probe functions in

$$\mathcal{I} \approx (I_1)T \Leftrightarrow \forall f.|f(\mathcal{I}) - f(I_1)| < \varepsilon, \varepsilon \in [0, 1],$$

where $\mathcal{I}$ is approximately the same as I_1 after some transformation T *iff* the differences between pairs of probe function values is less than some threshold.

4   Approximation Spaces and Object Recognition

This section introduces a view of approximation spaces defined in a slightly modified manner in comparison with the original definition in [38]. Any generalized approximation space (GAS) is a tuple

$$GAS = (U, A, N_r, \nu_B),$$

where U is a universe of objects, A, a set of probe functions, N_r, a neighbourhood family function and ν_B is an overlap function defined by

$$\nu_B : \mathcal{P}(U) \times \mathcal{P}(U) \longrightarrow [0, 1],$$

where $\mathcal{P}(U)$ is the powerset of U. ν_B maps a pair of sets to a number in $[0,1]$ representing the degree of overlap between the sets of objects with features defined by B, and $\mathcal{P}(U)$ is the powerset of U [39]. For each subset $B \subseteq A$ of probe functions, define the binary relation $\sim_B = \{(x, x') \in U \times U : \forall f \in B, f(x) = f(x')\}$. Since each $\sim_B$ is, in fact, the usual Ind_B (indiscernibility) relation, for $B \subset F$ and $x \in U$, let $[x]_B$ denote the equivalence class containing x, i.e.,

$$[x]_B = \{x' \in U : \forall f \in B, f(x') = f(x)\} \subseteq U.$$

If $(x, x') \in \sim_B$ (also written $x \sim_B x'$), then x and x' are said to be *indiscernible* with respect to all feature probe functions in B, or simply, *B-indiscernible*. Then define a family of neighborhoods $N_r(A)$, where

$$N_r(A) = \bigcup_{B \subseteq P_r(A)} [x]_B,$$

where $P_r(A) = \{B \subseteq A \mid |B| = r\}$ for any r such that $1 \leq r \leq |A|$. That is, r denotes the number of features used to construct families of neighborhoods. For the sake of clarity, we sometimes write $[x]_{B_r}$ to specify that the equivalence class represents a neighborhood formed using r features from B. Families of neighborhoods are constructed for each combination of probe functions in B using $\binom{|B|}{r}$, i.e., $|B|$ probe functions taken r at a time. Information about a sample $X \subseteq U$ can be approximated from information contained in B by constructing a $N_r(B)$-lower approximation

$$N_r(B)_* X = \bigcup_{x:[x]_{B_r} \subseteq X} [x]_{B_r},$$

and a $N_r(B)$-upper approximation

$$N_r(B)^* X = \bigcup_{x:[x]_{B_r} \cap X \neq \emptyset} [x]_{B_r}.$$

Then $N_r(B)_* X \subseteq N_r(B)^* X$ and the boundary region $BND_{N_r(B)}(X)$ between upper and lower approximations of a set X is defined to be the complement of $N_r(B)_* X$, i.e.

$$BND_{N_r(B)}(X) = N_r(B)^* X \setminus N_r(B)_* X = \{x \in N_r(B)^* X \mid x \notin N_r(B)_* X\}.$$

Remark 1. **What is a Near Set?** A set X is termed a "near set" relative to a chosen family of neighborhoods $N_r(B)$ *iff* $|BND_{N_r(B)}(X)| \geq 0$. This means every rough set is a near set but not every near set is a rough set. Object recognition and the problem of the nearness of objects have motivated the introduction of near sets (see, *e.g.*, [28,29]).

4.1   Object Recognition

It is now possible to formulate a basis for object recognition, which parallels the traditional formulation of pattern recognition. Assume $N_r(B)_*X$ defines a standard for classifying perceived objects. The notation $B_j(x)$ denotes a member of the family of neighborhoods in $N_r(B)$, where $j \in B$. Put

$$\nu_j(B_j(x), N_r(B)_*X) = \frac{|B_j(x) \cap N_r(B)_*X|}{|N_r(B)_*X|},$$

(called *lower rough coverage*) where ν_j is defined to be 1, if $N_r(B)_*X = \emptyset$. Let $\mathcal{O}, \mathcal{O}_{id}$ denote sample object and standard object, respectively. Then recognition of sample objects that are approximately the same as $\mathcal{O}_{id}$ is defined by comparing overlap function values in

$$\mathcal{O} \approx (\mathcal{O}_{id})T \Leftrightarrow |\nu_j(\mathcal{O}, N_r(B)_*X) - \nu_B(\mathcal{O}_{id}, N_r(B)_*X| < \varepsilon,$$

where $\varepsilon \in [0, 1]$. The sample object $\mathcal{O}$ is approximately the same as $\mathcal{O}_{id}$ after some transformation T *iff* the difference in coverage values is less than some threshold. An image-based model for object recognition is given in [10].

4.2   Percepts and Perception

The set $N_r(B)$ contains a set of percepts. A *percept* is a byproduct of perception, *i.e.*, something that has been observed [19]. For example, a member of $N_r(B)$ represents *what has been perceived about objects belonging to a neighborhood*, *i.e.*, observed objects with matching probe function values. Collectively, $N_r(B)$ represents a *perception*, a product of perceiving. Perception is defined as the extraction and use of information about one's environment [1]. This is basic idea is represented in the *sample objects, perceptual neighborhoods* and *judgemental percepts* columns in Fig. 3. In this article, we are focusing on the perception of acceptable objects.

4.3   Sensing, Classifying, and Peceptual Judgement

Sensing provides a basis for probe function measurements commonly associated with features such as colour, contour, shape, arrangement, entropy, and so on. A probe function can be thought of as a model for a sensor. Classification combines evaluation of a disposition of sensor measurements with judgement (apprehending the significance of a vector of probe measurements for an observed object). The result is a higher level percept, which has been traditionally called a decision. In the context of percepts, the term *judgement* means a conclusion about an object's measurements rather than an abstract idea. This form of judgement is considered *perceptual*. Perceptual judgements provide a basis for the formulation of abstract ideas (models of perception, rules) about a class (type) of object. Let D denote a feature called *decision* with a probe $d_B : X \times B \longrightarrow \{0, 1\}$, where X denotes a set of sample objects; B, a set of probe functions; 0, "reject perceived object" and 1, "accept perceived object". A set of objects d with matching perceptual judgements (*e.g.*, $d_B(x) = 1, x \in X$ for an acceptable object) is a mathematical model representing the abstract notion *acceptable*.

5   Near Sets From Ethograms

This section briefly considers particular near sets derived from an ethogram. An *ethogram* is a set of descriptions of behaviour patterns of a species [2], which is fundamental in ethology [43]. In this work, an ethogram is represented by a decision system that provides a record of observations of episodic behaviour of a swarm. The form of ethogram in Table 1 was introduced in [33,35] and elaborated in [31,34]. An *episode* is a sequence of states that terminates. During a swarm episode, an ethogram table is constructed, which provides the basis for an approximation space such as the one represented in Fig. 3.

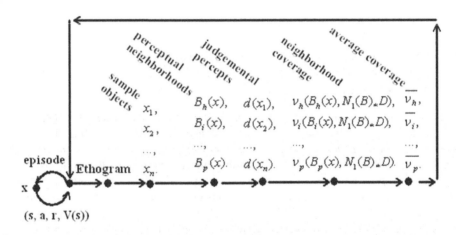

Fig. 3. Approximate Adaptive Learning Cycle

Let $s, a, p(s, a), r$ denote the state, action, action preference and reward associated with a previous action by an observed organism. Define a *behaviour* to be a collection $(s, a, p(s, a), r)$ at any one time t, and let d denote a decision (1 = choose action, 0 = reject action) for acceptance of a behaviour. Let $U_{beh} = \{x_0, x_1, x_2, \ldots\}$ denote a set of behaviours. Decisions to accept or reject an action are made by the actor during the learning process; let d denote a decision (0=reject, 1=accept). Often ethograms also exclude $p(s, a)$ or include a column for "proximate cause" (see [43]). Let $S = \{k, \ell\}$ be the collection of two states, and let $A = \{i, j, k\}$ be the set of possible actions, with $A(k) = \{h, i\}$, $A(\ell) = \{i, j\}$.

The calculations are performed on the feature values shown in the first four columns of Table 1. Put $B = \{s, a, p(s, a), r\}$. Let $U_{beh} = \{x_0, x_1, \ldots, x_9\}$ and let $D = \{x \in U : d(x) = 1\} = \{x_0, x_3, x_4, x_6, x_8\}$ be the decision class. Then

Table 1. Sample Ethogram

x_i	s	a	$p(s,a)$	r	d
x_0	k	h	0.0	0.75	1
x_1	k	i	0.0	0.75	0
x_2	ℓ	i	0.0	0.1	0
x_3	ℓ	j	0.0	0.1	1
x_4	k	h	0.0	0.75	1
x_5	k	i	0.0	0.75	0
x_6	ℓ	i	0.010	0.9	1
x_7	ℓ	j	0.025	0.9	0
x_8	k	h	0.01	0.75	1
x_9	k	i	0.056	0.75	0

Case 1. N_1, **1-Feature Neighborhoods**
Let $D = \{x \in Object \mid d(x) = 1\} = \{x_0, x_3, x_4, x_6, x_8\}$, $B = \{s, a, p(s,a), r\}$,
and observe

$B_{s_k}(x_0) = \{x_0, x_1, x_4, x_5, x_8, x_9\}$, $B_{s_\ell}(x_2) = \{x_2, x_3, x_6, x_7\}$,
$B_{a_h}(x_0) = \{x_0, x_4, x_8\}$, $B_{a_i}(x_1) = \{x_1, x_2, x_5, x_6, x_9\}$, $B_{a_j}(x_3) = \{x_3, x_7\}$,
$B_{p_{0.0}}(x_0) = \{x_0, x_1, x_2, x_3, x_4, x_5\}$,
$B_{p_{0.01}}(x_6) = \{x_6, x_8\}$, $B_{p_{0.025}}(x_7) = \{x_7\}$, $B_{p_{0.056}}(x_9) = \{x_9\}$
$B_{r_{0.1}}(x_2) = \{x_2, x_3\}$, $B_{r_{0.75}}(x_0) = \{x_0, x_1, x_4, x_5, x_8, x_9\}$, $B_{r_{0.9}}(x_6) = \{x_6, x_7\}$,

$(N_1(B))_* D = B_{a_h}(x_0) \cup B_{p_{0.01}}(x_6) = \{x_0, x_4, x_6, x_8\}$,
$(N_1(B))^* D = B_{s_k}(x_0) \cup B_{s_\ell}(x_2) \cup B_{a_h}(x_0) \cup B_{a_i}(x_1) \cup B_{a_j}(x_3) \cup B_{p_{0.0}}(x_0) \cup B_{p_{0.01}}(x_6) \cup$
$B_{p_{0.75}}(x_0) \cup B_{r_{0.1}}(x_2) \cup B_{r_{0.9}}(x_6) = \{x_0, x_1, x_2, x_3, x_4, x_5, x_6, x_7, x_8, x_9\}$,
$BND_{N_1(B)} D = \{x_1, x_2, x_5, x_7, x_9\}$.

Using $(N_1(B))_* D$ together with each block $B_{f_v}(x)$, $f \in B$, $f(x) = v$, we obtain

$\nu_{s_k}(B_{s_k}(x_0), (N_1(B))_* D) = \frac{3}{4}$, $\nu_{s_\ell}(B_s(x_2), (N_1(B))_* D) = \frac{1}{4}$,
$\nu_{a_h}(B_{a_h}(x_0), (N_1(B))_* D) = \frac{3}{4}$, $\nu_{a_i}(B_{a_i}(x_1), (N_1(B))_* D) = \frac{1}{4}$,
$\nu_{a_j}(B_{a_j}(x_3), (N_1(B))_* D) = 0$,
$\nu_{p_{0.0}}(B_{p_{0.0}}(x_0), (N_1(B))_* D) = \frac{1}{2}$, $\nu_{p_{0.01}}(B_{p_{0.01}}(x_6), (N_1(B))_* D) = \frac{1}{2}$,
$\nu_{p_{0.025}}(B_{p_{0.025}}(x_7), (N_1(B))_* D) = 0$, $\nu_{p_{0.056}}(B_{p_{0.056}}(x_9), (N_1(B))_* D) = 0$,
$\nu_{r_{0.1}}(B_{r_{0.1}}(x_2), (N_1(B))_* D) = 0$, $\nu_{r_{0.75}}(B_{r_{0.75}}(x_0), (N_1(B))_* D) = \frac{3}{4}$,
$\nu_{r_{0.9}}(B_{r_{0.9}}(x_6), (N_1(B))_* D) = \frac{1}{4}$.

Recently, we have found that lower coverage obtained in this manner has proved to be useful in solving image pattern recognition problems (see, *e.g.*, [10]). Next, we obtain the average lower coverage for each feature, which indicates that features a and r are more important than s and $p(s,a)$ (it happens that this matches our tuition about the content of an ethogram, where actions and rewards have greater weight).

$$\overline{\nu_s} = \frac{\frac{3}{4} + \frac{1}{4}}{2} = 1, \ \overline{\nu_a} = \frac{\frac{3}{4} + \frac{1}{4} + 0}{3} = 0.3, \ \overline{\nu_p} = \frac{\frac{1}{2} + \frac{1}{2} + 0 + 0}{4} = 0.25 \ \overline{\nu_r} = \frac{0 + \frac{3}{4} + \frac{1}{4}}{3} = 0.3.$$

In sum, notice that all of the features are used to construct families of neighborhoods, but not in the usual way, since features are considered separately to construct feature-based neighborhoods. The lower and upper approximations are obtained by taking into account feature-based families of neighborhoods. The set D is both a near set as well as a rough set. This does not always happen. Average lower coverage has proved to be useful in reinforcement learning (see, e.g., [34,31]). In the remaining cases, only the approximation sets and boundary set are given.

Remark 2. **Significance of the Lower Approximation**
This goes back to Archimedes, who suggested approximating the unknown area of a bounded region in the plane by summing the areas of all of the small rectangles entirely contained inside the bounded region. Each rectangle inside the bounded region is well-understood, since we know that it is inside the bounded region, *i.e.*, there is no part of an inner rectangle that is outside the bounded region (we know an inner rectangle belongs entirely inside the bounded region). Also notice that the bounded region provides a basis for evaluating all rectangles, those inside, overlapping or entirely outside the bounded region. Analogously, each perceptual neighborhood $B_j(x)$ contained in the lower approximation of a set D is well-understood because the objects in $B_j(x)$ are entirely contained inside the set of perceptual judgements D, assuming that $D = \{x \mid d(x) = 1,$ *i.e., accept behaviour associated with* $x\}$. That is, based on knowledge represented by $B_j(x)$, the sample objects in $B_j(x) \subseteq D$ are known to have acceptable behaviours. For this reason, $(N_1(B))_*D$ can be used as a norm or standard in evaluating all of the perceptual neighbourhoods gathered together during an episode. That is, we can measure the extent that the objects in each perceptual neighbourhood overlap with the acceptable objects in $(N_1(B))_*D$.

Case 2. N_2, **2-Feature Neighborhoods**
$(N_2(B))_*D =$
$B_{sa}(x_0) \cup B_{sp}(x_6) \cup B_{sp}(x_8) \cup B_{ap}(x_0) \cup B_{ap}(x_3) \cup B_{ap}(x_6) \cup B_{ap}(x_8) \cup B_{pr}(x_6) \cup B_{pr}(x_8) =$
$\{x_0, x_3, x_4, x_6, x_8\}$,
$(N_2(B))^*D =$
$B_{sa}(x_2) \cup B_{sa}(x_3) \cup B_{sp}(x_0) \cup B_{sp}(x_2) \cup B_{pr}(x_0) \cup B_{pr}(x_2) \cup B_{sr}(x_0) \cup B_{sr}(x_2) \cup B_{sr}(x_6) \cup$
$B_{sr}(x_8) = \{x_0, x_1, x_2, x_3, x_4, x_5, x_6, x_7, x_8, x_9\}$,
$BND_{N_2(B)}D = \{x_1, x_2, x_5, x_7, x_9\}$.

Case 3. N_3, **3-Feature Neighborhoods**
$(N_3(B))_*D =$
$B_{sap}(x_0) \cup B_{sap}(x_3) \cup B_{sap}(x_6) \cup B_{sap}(x_8) \cup B_{apr}(x_0) \cup B_{apr}(x_3) \cup B_{apr}(x_6) \cup B_{apr}(x_8) \cup$
$B_{spr}(x_0) \cup B_{spr}(x_3) \cup B_{spr}(x_6) \cup B_{spr}(x_8) \cup B_{sar}(x_0) \cup B_{sar}(x_3) \cup B_{sar}(x_6) =$
$\{x_0, x_3, x_4, x_6, x_8\}$,
$(N_3(B))^*D = \{x_0, x_1, x_2, x_3, x_4, x_5, x_6, x_8\}$,
$BND_{N_3(B)}D = \{x_1, x_2, x_5\}$.

Case 4. N_4, **4-Feature Neighborhoods**
$(N_4(B))_*D = (N_4(B))^*D =$
$B(x_0) \cup B(x_3) \cup B(x_6) \cup B(x_8) = \{x_0, x_3, x_4, x_6, x_8\}$,
$BND_{N_4(B)}D = \emptyset$.

This case is interesting because D is a near set but not a rough set.

In sum, D is a near set as well as a rough set in cases 1, 2 and 3. D is a near set but not a rough set in case 4 and 4 (quadruple feature families of neighborhoods). The lower approximation in several cases equals D, which means the objects in D are known with certainty for certain but not all feature combinations.

6 Conclusion

It is Zdzisław Pawlak's original 1981 paper on classification of objects by means of attributes that has led to the introduction of near sets and the proposed approach to object recognition. In this approach, the focus is on the comparison between families of perceptual neighborhoods containing observed sample objects with matching descriptions and perception granules representing a standard. The standard we have in mind is the lower approximation of a set of sample objects representing perceptual judgements, *i.e.*, objects judged to be acceptable. This has led to a refinement of the generalized approximation space model to include families of neighborhoods. Object recognition is defined in terms of a measure of the degree of overlap between perceptual neighborhoods and a set of objects constituting a standard. The *feature identification* and *feature extraction* are currently the subject of intense research in connection with solving object recognition, ethology and reinforcement learning problems. It is conjectured that near sets will be useful in solving a number of object recognition problems.

Acknowledgements

The author gratefully acknowledges the suggestions and insights by Andrzej Skowron, Christopher Henry, Dan Lockery and David Gunderson concerning topics in this paper. This research has been supported by the Natural Sciences and Engineering Research Council of Canada (NSERC) grant 185986.

References

1. Audi, R. (Ed.): The Cambridge Dictionary of Philosophy, 2nd. Cambridge University Press, UK (1999).
2. Lehner, P.N.: Handbook of Ethological Methods, 2nd Ed. Cambridge University Press, UK (1979).
3. Bertsekas, D.P., Tsitsiklis, J.N.: *Neuro-Dynamic Programming*, Athena Scientific, Belmont, MA (1996).
4. C.M. Bishop, C.M.: Neural Networks and Pattern Recognition. Oxford University Press, UK (1995).
5. Borkowski, M., Peters, J.F.: Matching 2D image segments with genetic algorithms and approximation spaces. Transactions on Rough Sets V, LNCS 4100 (2006), 63-101.
6. Feature Extraction: http://en.wikipedia.org/wiki/Feature_extraction
7. Galton, A.: The mereotopology of discrete space. In: Freksa, C., Mark, D.M. (Eds.): Spatial Information Theory, LNCS 1661. Springer, Berlin (1999) 251-266.

8. Gibbs, J.W.: Elementary Principles in Statistical Mechanics. Dover, NY (1960).
9. Guyon, I., Gunn, S., Nikravesh, M., Zadeh, L. (Eds.): Feature Extraction. Foundations and Applications. Springer, Heidelberg, Studies in Fuzziness and Soft Computing 207, 2006.
10. Henry, C., Peters, J.F.: Image Pattern Recognition Using Approximation Spaces and Near Sets. Proc. 2007 Joint Rough Set Symposium (JRS07), Toronto, Canada 14-16 May (2007).
11. Herrlich, H.: A concept of nearness. General Topology and Applications 4 (1974) 191-212.
12. Honderich, T. (Ed.): The Oxford Companion to Philosophy, 2nd Ed. Oxford University Press, UK (2005).
13. Jähne, B.: Digital Image Processing, 6th Ed. Springer, Berlin (2005).
14. Johnson, S.: A Dictionary of the English Language, 11th Ed. Clarke and Sons, London (1816).
15. Kant, I.: Critique of Pure Reason, trans. by N.K. Smith. Macmillan, Toronto (1929).
16. Lehner, P.N.: Handbook of Ethological Methods, 2nd Ed. Cambridge University Press, UK (1996).
17. Mendel, J.M., Fu, K.S. (Eds.): Adaptive, Learning and Pattern Recognition Systems. Theory and Applications. Academic Press, London (1970).
18. Orłowska, E.: Semantics of Vague Concepts, *Applications of Rough Sets*, Institute for Computer Science, Polish Academy of Sciences, Report 469, March (1982).
19. The Oxford English Dictionary. Oxford University Press, London (1933).
20. Pal, S. K. Polkowski, L., Skowron, A. (Eds.): Rough-Neural Computing: Techniques for Computing with Words. Cognitive Technologies, Springer, Heidelberg, 2004.
21. Pavel, M.: Fundamentals of Pattern Recognition, 2nd Edition. Marcel Dekker, Inc., NY (1993).
22. Pawlak Z.: *Classification of Objects by Means of Attributes*, Institute for Computer Science, Polish Academy of Sciences, Report 429, March (1981).
23. Pawlak Z.: *Rough Sets*, Institute for Computer Science, Polish Academy of Sciences, Report 431, March (1981).
24. Pawlak Z.: Rough sets, *International J. Comp. Inform. Science*, **11**(1982), 341–356.
25. Pawlak, Z., Skowron, A.: Rudiments of rough sets, Information Sciences, ISSN 0020-0255, vol. 177 (1) (2007) 3-27.
26. Pawlak, Z., Skowron, A.: Rough sets: Some extensions, Information Sciences, ISSN 0020-0255, vol. 177 (1) (2007) 28-40.
27. Pawlak, Z., Skowron, A.: Rough sets and Boolean reasoning, Information Sciences, ISSN 0020-0255, vol. 177 (1) (2007) 41-73 .
28. Peters, J.F.: Near sets. Special theory about nearness of objects, *Fundamenta Informaticae*, vol. 75 (1-4) (2007) 407-433.
29. Peters, J.F., Skowron, A., Stepaniuk, J.: Nearness in approximation spaces. G. Lindemann, H. Schlilngloff et al. (Eds.), Proc. Concurrency, Specification & Programming (CS&P'2006). Informatik-Berichte Nr. 206, Humboldt-Universität zu Berlin (2006) 434-445.
30. Peters, J.F.: Classification of objects by means of features. In: Kacprzyk, J., Skowron, A.: Proc. Special Session on Rough Sets, IEEE Symposium on Foundations of Computational Intelligence (FOCI07) (2007).
31. Peters, J.F., Henry, C.: Approximation spaces in off-policy Monte Carlo learning, *Engineering Applications of Artificial Intelligence* (2007), in press.
32. Peters, J.F.: Approximation space for intelligent system design patterns, *Engineering Applications of Artificial Intelligence*, **17(4)** (2004), 1–8.

33. Peters, J.F.: Rough ethology: Towards a Biologically-Inspired Study of Collective behaviour in Intelligent Systems with Approximation Spaces. *Transactions on Rough Sets*, **III**, LNCS 3400 (2005), 153-174.
34. Peters, J.F., Henry, C.: Reinforcement learning with approximation spaces, *Fundamenta Informaticae* **71** (2-3) (2006) 323-349.
35. Peters, J.F., Henry, C., Ramanna, S.: Rough Ethograms: Study of Intelligent System behaviour. In: M.A. Kłopotek, S. Wierzchoń, K. Trojanowski (Eds.), *New Trends in Intelligent Information Processing and Web Mining (IIS05)*, Gdańsk, Poland, June 13-16 (2005) 117-126.
36. Polkowski, L.: *Rough Sets. Mathematical Foundations*, Springer-Verlag,Heidelberg (2002).
37. Polkowski, L., Skowron, A. (Eds.): Rough Sets in Knowledge Discovery 2, *Studies in Fuzziness and Soft Computing* **19**, Springer-Verlag,Heidelberg (1998).
38. Skowron, A., Stepaniuk, J.: Generalized approximation spaces, in: Lin, T.Y., Wildberger, A.M. (Eds.), *Soft Computing, Simulation Councils*, San Diego (1995) 18–21.
39. A. Skowron, J. Stepaniuk. Tolerance approximation spaces. Fundamenta Informaticae 27(2-3) (1996) 245–253.
40. Skowron, A., Stepaniuk, J.: Information granules and rough-neural computing. In: Pal et al. [20] (2204) 43-84.
41. Skowron, A., Swiniarski, R., Synak, P.: Approximation spaces and information granulation, *Transactions on Rough Sets III* (2005), 175-189.
42. Stepaniuk, J.: Approximation spaces, reducts and representatives, in [37], 109–126.
43. Tinbergen, N.: On aims and methods of ethology, *Zeitschrift für Tierpsychologie* **20** (1963) 410–433.
44. Ullman, J.D.: Principles of Database and Knowledge-Base Systems, vol. 1. Computer Science Press, MD (1988).
45. Vapnik, V.N.: Statistical Learning Theory. Chichester, UK, Wiley-Interscience, 1998.
46. Watanabe, S.: Pattern Recognition: Human and Mechanical. John Wiley & Sons, Chichester, UK (1985).
47. Witten, I.H., Frank, E.: Data Mining, 2nd Ed. Elsevier, Amsterdam (2005).
48. Wolski, M.: Similarity as nearness: Information quanta, approximation spaces and nearness structures. Proc. CS&P 2006. Infomatik-Berichte (2006), 424-433.
49. Partial function: `http://en.wikipedia.org/wiki/Partial_function`

On Covering Rough Sets

Keyun Qin[1], Yan Gao[2], and Zheng Pei[3]

[1] Department of Mathematics, Southwest Jiaotong University,
Chengdu, Sichuan 610031, China
[2] College of Computer Science and Technology, Henan Polytechnic University,
Jiaozuo, Henan, 454000, China
[3] School of Mathematics & Computer Science, Xihua University,
Chengdu, Sichuan, 610039, China
keyunqin@263.net
pqyz@263.net

Abstract. This paper is devoted to the discussion of extended covering rough set models. Based on the notion of neighborhood, five pairs of dual covering approximation operators were defined with their properties being discussed. The relationships among these operators were investigated. The main results are conditions with which these covering approximation operators are identical.

Keywords: Rough set, covering rough set, covering lower approximation, covering upper approximation, representative element.

1 Introduction

Rough set theory(RST), proposed by Pawlak [2], [3], is an extension of set theory for the study of intelligent systems characterized by insufficient and incomplete information. It provides a systematic approach for the study of indiscernibility of objects. Typically, indiscernibility is described using equivalent relations. When objects of a universe are described by a set of attributes, one may define the indiscernibility of objects based on their attribute values. When two objects have the same value over a certain group of attributes, we say they are indiscernible with respect to this group of attributes, or have the same description with respect to the indiscernibility relation. Objects of the same description consist of an equivalence class and all the equivalence classes form a partition of the universe. With this partition, rough set theory approximates any subset of objects of the universe by two sets, called the lower and upper approximations. They can be formally described by a pair of unary set-theoretic operators. It is noticed that equivalence relation or partition, as the indiscernibility relation in Pawlak's original rough set theory, is restrictive for many applications. To address this issue, several interesting and meaningful extensions to equivalent relation have been proposed in the past, such as tolerance relation [4,12], similarity relation [13], and others [14,15,16,17]. This leads to various approximation operators. By adopting the notion of neighborhood systems from topological space, Lin[6,7] proposed a more general framework for the study of approximation operators. Zakowski [20] have used coverings of a

J.T. Yao et al. (Eds.): RSKT 2007, LNAI 4481, pp. 34–41, 2007.

universe for establishing the covering generalized rough set theory and an extensive body of research works have been developed [2,3,11]. In [22], the concept of reducts of coverings is introduced and the conditions for two coverings to generate the same covering lower approximation or the same covering upper approximation were given. In[5], the indiscernibility relation is generalized to any binary reflexive relation and some generalized approximation operators were introduced. The following problems worth paying attention to. Some important properties of Pawlak's lower and upper approximation do not hold for the covering lower and upper approximation, such as Duality, Multiplication and Addition [22]. For covering upper approximation, even Monotonicity does not hold.

This paper is devoted to the discussion of extended covering rough set models. Based on the concept of neighborhood, five pairs of dual covering approximation operators were defined with their properties being discussed. The relationship among these operators were investigated. Some equivalent conditions for covering approximation operators coinciding with each other were given. Furthermore, if each element of the universe is a representative element [2], the Multiplication for Zakowski's lower approximation operator holds.

2   Preliminaries

This section presents a review of some fundamental notions of Pawlak's rough sets and covering rough sets. We refer to [2,9,10] for details.

2.1   Fundamentals of Pawlak's Rough Sets

Let U be a finite set, the universe of discourse, and R an equivalence relation on U, called an indiscernibility relation. The pair (U, R) is called a Pawlak approximation space. R will generate a partition $U/R = \{[x]_R; x \in U\}$ on U, where $[x]_R$ is the equivalence class with respect to R containing x. $\forall X \subseteq U$, the upper approximation $\overline{R}(X)$ and lower approximation $\underline{R}(X)$ of X are defined as [9,10] $\overline{R}(X) = \{x; [x]_R \cap X \neq \emptyset\}$, $\underline{R}(X) = \{x; [x]_R \subseteq X\}$ Alternatively, in terms of equivalence classes of R, the pair of lower and upper approximation can be defined by $\overline{R}(X) = \cup\{[x]_R; [x]_R \cap X \neq \emptyset\}$, $\underline{R}(X) = \cup\{[x]_R; [x]_R \subseteq X\}$. Let $\emptyset$ be the empty set and $\sim X$ the complement of X in U, the following conclusions have been established for Pawlak's rough sets: (1)$\underline{R}(U) = U = \overline{R}(U)$. (2)$\underline{R}(\emptyset) = \emptyset = \overline{R}(\emptyset)$. (3)$\underline{R}(X) \subseteq X \subseteq \overline{R}(X)$. (4)$\underline{R}(X \cap Y) = \underline{R}(X) \cap \underline{R}(X)$, $\overline{R}(X \cup Y) = \overline{R}(X) \cup \overline{R}(Y)$. (5)$\underline{R}(\underline{R}(X)) = \underline{R}(X)$, $\overline{R}(\overline{R}(X)) = \overline{R}(X)$. (6)$\underline{R}(X) = \sim \overline{R}(\sim X)$, $\overline{R}(X) = \sim \underline{R}(\sim X)$. (7)$X \subseteq Y \Rightarrow \underline{R}(X) \subseteq \underline{R}(Y)$, $\overline{R}(X) \subseteq \overline{R}(Y)$. (8)$\underline{R}(\sim \underline{R}(X)) = \sim \underline{R}(X)$, $\overline{R}(\sim \overline{R}(X)) = \sim \overline{R}(X)$. (9)$\overline{R}(\underline{R}(X)) \subseteq X \subseteq \underline{R}(\overline{R}(X))$.

It has been shown that (3), (4) and (8) are the characteristic properties of the lower and upper approximations [8,23,18].

2.2   Concepts and Properties of Covering Rough Sets

Definition 1. *Let U be a universe of discourse, C a family of subsets of U. If no subsets in C is empty, and $\cup C = U$, C is called a covering of U.*

It is clear that a partition of U is a covering of U, so the concept of a covering is an extension of the concept of a partition. In the following discussion, the universe of discourse U is considered to be finite. We will list some definitions and results about covering rough set.

Definition 2. *[2] Let U be a non-empty set, C a covering of U. The pair (U, C) is called a covering approximation space.*

Definition 3. *[2] Let (U, C) be a covering approximation space and $x \in U$. The family of sets $Md(x) = \{K \in C; x \in K \wedge (\forall S \in C)(x \in S \wedge S \subseteq K \Rightarrow K = S)\}$ is called the minimal description of x.*

Definition 4. *[2] Let (U, C) be a covering approximation space and $X \subseteq U$. The family $C_*(X) = \{K \in C; K \subseteq X\}$ is called the covering lower approximation set of X. Set $X_* = \cup C_*(X)$ is called the covering lower approximation of X. Set $X_*^* = X - X_*$ is called the covering boundary of X. The family $Bn(X) = \cup\{Md(x); x \in X_*^*\}$ is called the covering boundary approximation set family of X. The family $C^*(X) = C_*(X) \cup Bn(X)$ is called the covering upper approximation family of X. Set $X^* = \cup C^*(X)$ is called the covering upper approximation of X. If $C^*(X) = C_*(X)$, X is said to be exact, otherwise inexact.*

Theorem 1. *[2] Let (U, C) be a covering approximation space, then $\forall X, Y \subseteq U$ and $\forall x \in U$, (1) $C_*(\emptyset) = C^*(\emptyset) = \emptyset$, $C_*(U) = C^*(U) = C$. (2) $C_*(X) \subseteq C^*(X)$. (3) $C_*(X_*) = C_*(X) = C^*(X_*)$. (4) $X \subseteq Y \Rightarrow C_*(X) \subseteq C_*(Y)$, $X \subseteq Y \Rightarrow X_* \subseteq Y_*$. (5) $C_*(X_*)^* = \emptyset$. (6) $C_*(\{x\}) \neq \emptyset \Leftrightarrow \{x\} \in C$. (7) $C^*(\{x\}) = Md(x)$. (8) $\cap Md(x) = \cap\{K \in C; x \in K\}$.*

Theorem 2. *[2] Let (U, C) be a covering approximation space and $X, Y \subseteq U$, then (1) $U_* = U = U^*$. (2) $\emptyset_* = \emptyset = \emptyset^*$. (3) $X_* \subseteq X \subseteq X^*$. (4) $(X_*)_* = X_*$, $(X^*)^* = X^*$.*

By providing some examples, Zhu [22] shown that the following properties do not hold for the covering lower and upper approximations: (1) $(X \cap Y)_* = X_* \cap Y_*$, $(X \cup Y)^* = X^* \cup Y^*$. (2) $X_* =\sim (\sim X)^*$, $X^* =\sim (\sim X)_*$. (3) $X \subseteq Y \Rightarrow X^* \subseteq Y^*$. (4) $(\sim X_*)_* =\sim X_*$, $(\sim X^*)^* =\sim X^*$.

3   The Extension of the Covering Approximation Operators

Let (U, C) be a covering approximation space. For each $x \in U$, $N(x) = \cap\{K \in C; x \in K\}$ is called the neighborhood of x. By (8) of Theorem 1, $N(x) = \cap Md(x)$. We know that Pawlak's approximation operators can be defined in two different, but equivalent, ways. Similarly, we consider five pairs of dual approximation operators defined by means of neighborhoods as follows: for each $X \subseteq U$, (I1) $\underline{C_1}(X) = X_* = \cup\{K \in C; K \subseteq X\}$, $\overline{C_1}(X) =\sim \underline{C_1}(\sim X) = \cap\{\sim K; K \in C, K \cap X = \emptyset\}$. (I2) $\underline{C_2}(X) = \{x \in U; N(x) \subseteq X\}$, $\overline{C_2}(X) = \{x \in U; N(x) \cap X \neq \emptyset\}$. (I3) $\underline{C_3}(X) = \{x \in U; \exists u(u \in N(x) \wedge N(u) \subseteq X)\}$, $\overline{C_3}(X) =$

$\{x \in U; \forall u(u \in N(x) \rightarrow N(u) \cap X \neq \emptyset)\}$. (I4) $\overline{C_4}(X) = \cup\{N(x); N(x) \cap X \neq \emptyset\}$, $\underline{C_4}(X) = \{x \in U; \forall u(x \in N(u) \rightarrow N(u) \subseteq X)\}$. (I5) $\overline{C_5}(X) = \cup\{N(x); x \in X\}$, $\underline{C_5}(X) = \{x \in U; \forall u(x \in N(u) \rightarrow u \in X)\}$.

Remark 1. The operator $\underline{C_1}$ is just Zakowski's lower approximation[20] and is studied in [2,3,11]. $\overline{C_5}$ is Zhu's upper approximation operator[22].

3.1   On Approximation Operators (I1)

By the definition, $\underline{C_1}(X) = X_*$ for any $X \subseteq U$. Consequently, by Theorem 1, Theorem 2 and the duality of $\overline{C_1}$ and $\underline{C_1}$, we have:

Theorem 3. *Let (U, C) be a covering approximation space and $X, Y \subseteq U$, then*
(1) $\underline{C_1}(U) = U = \overline{C_1}(U)$, $\underline{C_1}(\emptyset) = \emptyset = \overline{C_1}(\emptyset)$. (2) $\underline{C_1}(X) \subseteq X \subseteq \overline{C_1}(X)$.
(3) $X \subseteq Y \Rightarrow \underline{C_1}(X) \subseteq \underline{C_1}(Y) \wedge \overline{C_1}(X) \subseteq \overline{C_1}(Y)$. (4)$\underline{C_1}(\underline{C_1}(X)) = \underline{C_1}(X)$,
$\overline{C_1}(\overline{C_1}(X)) = \overline{C_1}(X)$.

Definition 5. *[2] Let (U, C) be a covering approximation space and $K \in C, x \in K$. x is called a representative element of K if $\forall S \in C(x \in S \Rightarrow K \subseteq S)$.*

By Fact 7 in [2], x is a representative element of K if and only if $Md(x) = \{K\}$, and if and only if $N(x) = K$. We denote by C_0 the set of all representative elements of sets of the covering C, that is $C_0 = \{x \in U; \exists K \in C(x \in K \wedge \forall S \in C(x \in S \Rightarrow K \subseteq S))\}$.

Lemma 1. *Let (U, C) be a covering approximation space and $x \in U$. Then, $x \in C_0$ if and only if $|Md(x)| = 1$.*

Proof. Assume that $x \in C_0$. It follows that $Md(x) = \{N(x)\}$ and $|Md(x)| = 1$. Conversely, assume that $|Md(x)| = 1$. We suppose that $Md(x) = \{K\}$, this means that K is the unique minimal element of $\{S \in C; x \in S\}$ and x is a representative element of K, and consequently $x \in C_0$.

Theorem 4. *Let (U, C) be a covering approximation space. Then, $C_0 = U$ if and only if for any $X, Y \subseteq U$, $\underline{C_1}(X \cap Y) = \underline{C_1}(X) \cap \underline{C_1}(Y)$.*

Proof. ($\Rightarrow$) Suppose that $C_0 = U$. For each $X, Y \subseteq U$ and $x \in \underline{C_1}(X) \cap \underline{C_1}(Y)$, there exist $K_1, K_2 \in C$ such that $x \in K_1, K_1 \subseteq X$ and $x \in K_2, K_2 \subseteq Y$. By $C_0 = U$, there exist $K \in C$ such that x is a representative element of K, it follows that $K \subseteq K_1$, $K \subseteq K_2$ and hence $x \in K \subseteq K_1 \cap K_2 \subseteq X \cap Y$, that is $x \in \underline{C_1}(X \cap Y)$. It follows that $\underline{C_1}(X \cap Y) \supseteq \underline{C_1}(X) \cap \underline{C_1}(Y)$, and hence $\underline{C_1}(X \cap Y) = \underline{C_1}(X) \cap \underline{C_1}(Y)$ by (3) of Theorem 4. Conversely, if $C_0 \neq U$, then there exists $x \in U$ such that $x \notin C_0$. It follows that $|Md(x)| > 1$. Suppose that $K_1, K_2 \in Md(x)$ and $K_1 \neq K_2$, it follows that $x \in K_1 \cap K_2 = \underline{C_1}(K_1) \cap \underline{C_1}(K_2)$. On the other hand, for each $K \in C$ such that $K \subseteq K_1 \cap K_2$, $x \notin K$ and hence $x \notin \cup\{K \in C; K \subseteq K_1 \cap K_2\} = \underline{C_1}(K_1 \cap K_2)$, this contradicts $\underline{C_1}(K_1 \cap K_2) = \underline{C_1}(K_1) \cap \underline{C_1}(K_2)$.

By the duality, we have the following corollary:

Corollary 1. *Let (U, C) be a covering approximation space. The following conditions are equivalent: (1) $C_0 = U$. (2)For each $X, Y \subseteq U$, $\overline{C_1}(X \cup Y) = \overline{C_1}(X) \cup \overline{C_1}(Y)$.*

Theorem 5. *Let (U, C) be a covering approximation space such that $C_0 = U$. Then for each $X \subseteq U$, $\underline{C_1}(X) = \underline{C_2}(X)$, $\overline{C_1}(X) = \overline{C_2}(X)$. This means that approximation operators $(\overline{I1})$ and $(\overline{I2})$ are equivalent, when $C_0 = U$.*

Proof. By the duality, we need only to prove $\underline{C_1}(X) = \underline{C_2}(X)$. If $x \in \underline{C_1}(X)$, then there exists $K \in C$ such that $K \subseteq X$ and $x \in K$, it follows that $N(x) \subseteq K \subseteq X$ and hence $x \in \underline{C_2}(X)$. Conversely, if $x \in \underline{C_2}(X)$, then $N(x) \subseteq X$, suppose that x is a representative element of S, it follows that $S = N(x) \subseteq X$ and hence $x \in \underline{C_1}(X)$ by $x \in S$.

3.2   On Approximation Operators (I2~I5)

By Theorem 4.4 of [5], we have

Theorem 6. *Let (U, C) be a covering approximation space and $X, Y \subseteq U$, then (1) $\underline{C_i}(U) = U = \overline{C_i}(U)$, $\underline{C_i}(\emptyset) = \emptyset = \overline{C_i}(\emptyset)$ for $i = 2, 3, 4, 5$. (2) $\underline{C_i}(X) \subseteq X \subseteq \overline{C_i}(X)$ for $i = 2, 4, 5$. (3) $X \subseteq Y \Rightarrow \underline{C_i}(X) \subseteq \underline{C_i}(Y) \wedge \overline{C_i}(X) \subseteq \overline{C_i}(Y)$ for $i = 2, 3, 4, 5$. (4)$\underline{C_i}(X \cap Y) = \underline{C_i}(X) \cap \underline{C_i}(Y)$, $\overline{C_i}(X \cup Y) = \overline{C_i}(X) \cup \overline{C_i}(Y)$ for $i = 2, 4, 5$.*

Theorem 7. *Let (U, C) be a covering approximation space and $X \subseteq U$, then (1) $\underline{C_i}(\underline{C_i}(X)) = \underline{C_i}(X)$, $\overline{C_i}(\overline{C_i}(X)) = \overline{C_i}(X)$ for $i = 2, 5$. (2) $X \subseteq \underline{C_4}(\overline{C_4}(X))$, $\overline{C_4}(\underline{C_4}(X)) \subseteq X$.*

Proof. We only prove (1) for $i = 2$. By (2) of Theorem 6 , $\underline{C_2}(\underline{C_2}(X)) \subseteq \underline{C_2}(X)$. Conversely, suppose that $x \in \underline{C_2}(X)$. It follows that $N(x) \subseteq X$. Consequently, $N(y) \subseteq N(x) \subseteq X$ for any $y \in N(x)$, that is $y \in \underline{C_2}(X)$ and hence $N(x) \subseteq \underline{C_2}(X)$, $x \in \underline{C_2}(\underline{C_2}(X))$.

The following properties do not hold in general: (1) $X \subseteq \underline{C_2}(\overline{C_2}(X))$. (2) $X \subseteq \underline{C_5}(\overline{C_5}(X))$. (3) $\underline{C_4}(\underline{C_4}(X)) = \underline{C_4}(X)$, $\overline{C_4}(\overline{C_4}(X)) = \overline{C_4}(X)$. (4) $\underline{C_3}(X) \subseteq X \subseteq \overline{C_3}(X)$.

Example 1. Let $U = \{x, y, z\}$, $K_1 = \{x, y\}$, $K_2 = \{y, z\}$, $C = \{K_1, K_2\}$. Clearly, C is a covering of U, $N(x) = \{x, y\}$,$N(y) = \{y\}$, $N(z) = \{y, z\}$. (1) For $X = \{x\}$, $\overline{C_4}(X) = \cup\{N(u); x \in N(u)\} = N(x) = \{x, y\}$, $\overline{C_4}(\overline{C_4}(X)) = \overline{C_4}(\{x, y\}) = N(x) \cup N(y) \cup N(z) = U$. (2) For $X = \{y\}$, $\underline{C_5}(\overline{C_5}(X)) = \overline{C_5}(N(y)) = (\sim N(x)) \cap (\sim N(z)) = \emptyset$. $\underline{C_3}(X) = \{v \in U; \exists u(u \in N(v) \wedge N(u) \subseteq X)\} = U$. (3) For $X = \{z\}$, $\underline{C_2}(\overline{C_2}(X)) = \underline{C_2}(\{z\}) = \emptyset$.

4   Connections of Covering Approximation Operators

Theorem 8. *Let (U,C) be a covering approximation space. For each $X \subseteq U$, (1) $\underline{C_4}(X) \subseteq \underline{C_2}(X)$. (2) $\underline{C_2}(X) \subseteq \underline{C_3}(X)$. (3) $\underline{C_4}(X) \subseteq \underline{C_5}(X)$.*

Proof. (1)Suppose that $x \in \underline{C_4}(X)$. By the definition, for every $u \in U$ such that $x \in N(u)$, $N(u) \subseteq X$ followed. Consequently, $N(x) \subseteq X$ and $x \in \underline{C_2}(X)$. (2)Suppose that $x \in \underline{C_2}(X)$. It follows that $N(x) \subseteq X$. Hence $x \in \underline{C_3}(X)$ by $x \in N(x)$. (3)can be proved similarly.

Corollary 2. *Let (U,C) be a covering approximation space. For each $X \subseteq U$, (1) $\underline{C_4}(X) \subseteq \underline{C_2}(X) \subseteq X \subseteq \overline{C_2}(X) \subseteq \overline{C_4}(X)$. (2) $\underline{C_4}(X) \subseteq \underline{C_5}(X) \subseteq X \subseteq \overline{C_5}(X) \subseteq \overline{C_4}(X)$. (3)$\underline{C_4}(X) \subseteq \underline{C_2}(X) \subseteq \underline{C_3}(X)$, $\overline{C_3}(X) \subseteq \overline{C_2}(X) \subseteq \overline{C_4}(X)$.*

Generally, we cannot substitute $=$ for $\subseteq$ due to the following example.

Example 2. Let $U = \{x,y,z\}$, $K_1 = \{x,y\}$, $K_2 = \{y,z\}$, $C = \{K_1, K_2\}$. Clearly, C is a covering of U, $N(x) = \{x,y\}$, $N(y) = \{y\}$, $N(z) = \{y,z\}$. For $X = \{y\}$, $\underline{C_4}(X) = \cap\{\sim N(u); N(u) \not\subseteq X\} = (\sim N(x)) \cap (\sim N(z)) = \emptyset$, $\underline{C_2}(X) = \{u; N(u) \subseteq X\} = \{y\}$. $\underline{C_3}(X) = \{v \in U; \exists u(u \in N(v) \wedge N(u) \subseteq X)\} = U$. For $X = \{x,z\}$, $\underline{C_4}(X) = \cap\{\sim N(u); N(u) \not\subseteq X\} = (\sim N(x)) \cap (\sim N(y)) \cap (\sim N(z)) = \emptyset$, $\underline{C_5}(X) = \cap\{\sim N(u); u \in (\sim X)\} = \sim N(y) = \{x,z\}$, $\underline{C_2}(X) = \{u; N(u) \subseteq X\} = \emptyset$. $\underline{C_3}(X) = \{v \in U; \exists u(u \in N(v) \wedge N(u) \subseteq X)\} = \emptyset$.

Lemma 2. *Let (U,C) be a covering approximation space. Then, $\{N(x); x \in U\}$ forms a partition of U if and only if for each $x,y \in U$, $x \in N(y) \Rightarrow y \in N(x)$.*

Proof. Suppose that $\{N(x); x \in U\}$ forms a partition of U. For each $x,y \in U$, if $x \in N(y)$, then $N(x) \subseteq N(y)$, and $N(x) \cap N(y) = N(x) \neq \emptyset$, it follows that $N(x) = N(y)$ and $y \in N(y) = N(x)$. Conversely, suppose that $x \in N(y) \Rightarrow y \in N(x)$ for each $x,y \in U$. If $N(x) \cap N(y) \neq \emptyset$, then there exist $z \in U$ such that $z \in N(x)$ and $z \in N(y)$, it follows that $x \in N(z)$ and $y \in N(z)$, consequently, $N(x) = N(z)$, $N(y) = N(z)$, and $N(x) = N(y)$. That is to say, $\{N(x); x \in U\}$ forms a partition of U.

Theorem 9. *Let (U,C) be a covering approximation space. Then, $\{N(x); x \in U\}$ forms a partition of U if and only if for each $X \subseteq U$, $\underline{C_4}(X) = \underline{C_2}(X)$.*

Proof. Suppose that $\{N(x); x \in U\}$ forms a partition of U, $X \subseteq U$ and $x \in U$. If $x \in \underline{C_2}(X)$, then $N(x) \subseteq X$. For each $y \in U$ such that $x \in N(y)$, $y \in N(x)$ followed and hence $N(y) = N(x) \subseteq X$. By the definition, $x \in \underline{C_4}(X)$. This means $\underline{C_4}(X) \supseteq \underline{C_2}(X)$ and hence $\underline{C_4}(X) = \underline{C_2}(X)$ by (1) of Theorem 8. Conversely, suppose that $\underline{C_4}(X) = \underline{C_2}(X)$ for each $X \subseteq U$. For each $x,y \in U$ such that $x \in N(y)$, by $x \in \overline{C_2}(N(x)) = \underline{C_4}(N(x)) = \{v \in U; \forall u(v \in N(u) \to N(u) \subseteq N(x))\}$, it follows that $N(y) \subseteq N(x)$ and hence $y \in N(x)$. By Lemma 2, $\{N(x); x \in U\}$ forms a partition of U.

Theorem 10. *Let (U,C) be a covering approximation space. Then, $\{N(x); x \in U\}$ forms a partition of U if and only if for each $X \subseteq U$, $\overline{C_4}(X) = \overline{C_5}(X)$.*

Proof. Suppose that $\{N(x); x \in U\}$ forms a partition of U, $X \subseteq U$ and $x \in U$. If $x \in \overline{C_4}(X) = \cup\{N(y); N(y) \cap X \neq \emptyset\}$, then there exists $y \in U$ such that $x \in N(y)$ and $N(y) \cap X \neq \emptyset$. Let $z \in N(y) \cap X$, it follows that $z \in X$ and $z \in N(y)$. Consequently, $y \in N(z)$, $N(y) \subseteq N(z)$ and $x \in N(z)$, that is $x \in \cup\{N(u); u \in X\} = \overline{C_5}(X)$, and $\overline{C_4}(X) \subseteq \overline{C_5}(X)$. By (2) of Theorem 8, $\overline{C_4}(X) = \overline{C_5}(X)$. Conversely, suppose that for each $X \subseteq U$, $\overline{C_4}(X) = \overline{C_5}(X)$. For each $x, y \in U$ such that $x \in N(y)$, by $y \in N(y)$, it follows that $y \in \cup\{N(z); N(z) \cap \{x\} \neq \emptyset\} = \overline{C_4}(\{x\})$, and hence $y \in \overline{C_5}(\{x\}) = N(x)$. This means $\{N(x); x \in U\}$ forms a partition of U.

Theorem 11. *Let (U, C) be a covering approximation space, then (1) $\{N(x); x \in U\}$ forms a partition of U if and only if for each $X \subseteq U$, $\overline{C_4}(X) = \overline{C_3}(X)$. (2) $\{N(x); x \in U\}$ forms a partition of U if and only if for each $X \subseteq U$, $\overline{C_2}(X) = \overline{C_3}(X)$. (3) $\{N(x); x \in U\}$ forms a partition of U if and only if for each $X \subseteq U$, $\overline{C_5}(X) = \overline{C_3}(X)$. (4) $\{N(x); x \in U\}$ forms a partition of U if and only if for each $X \subseteq U$, $\overline{C_2}(X) = \overline{C_5}(X)$.*

By Theorem 9, 10 and 11, if any two pairs of operators are identical, then they are all identical.

5   Conclusions

In this paper, five pairs of dual covering approximation operators were defined and their properties have been discussed. Some equivalent conditions about these operators were given. For a covering approximation space (U, C), define a binary relation R on U as follows: for each $x, y \in U$, $(x, y) \in R$ if and only if $\forall K \in C(x \in K \rightarrow y \in K)$. It is trivial to verify that the successor of an element with respect to R coincides with its neighborhood, that is $R_s(x) = \{y \in U; (x, y) \in R\} = N(x)$. With this definition, binary relation based covering rough set can be constructed. We will discuss this problem in our future work.

Acknowledgements

This work has been supported by the National Natural Science Foundation of China (Grant No. 60474022) and the Young Foundation of Sichuan Province (Grant no. 06ZQ026-037).

References

1. Allam, A., Bakeir, M., Abo-Tabl, E.: New approach for basic roset concepts. In: LNCS, Vol. 3641, 2005, 64-73.
2. Bonikowski, Z., Bryniarski, E., Wybraniec, U.: Extensions and intentions in the rough set theory. Information Sciences. 107 (1998) 149-167.
3. Bryniarski, E.: A calculus of rough sets of the first order. Bull. Pol. Acad. Sci. 16 (1989) 71-77.

4. Cattaneo, G.: Abstract approximation spaces for rough theories. In: L.Polkowski, A.Skowron(Eds.), Rough Sets in Knowledge Discovery 1: Methodology and Applications, Physica-Verlag, Heidelberg, 1998, pp. 59-98.
5. Gomolinska, A.: A comparative study of some generalized rough approximations. Fundamenta Informaticae. 51(2002) 103-119.
6. Lin, T.: Neighborhood systems and relational database. Proceedings of CSC'88, 1988.
7. Lin, T.: Neighborhood systems and approximation in database and knowledge base systems. Proceedings of the Fourth International Symposium on Methodologies of Intelligent Systems, Poster Session, CSC'88, 1989.
8. Lin, T., Liu, Q.: Rough approximate operators: axiomatic rough set theory. In: W. P. Ziarko(Ed.), Rough Sets, Fuzzy Sets and Knowledge Discovery, Springer-Verlag, London, 1994, pp. 256-260.
9. Pawlak, Z.: Rough sets. International Journal of Computer and Information Science. 11 (1982) 341-356.
10. Pawlak, Z.: Rough sets: Theoretical Aspects of Reasoning About Data. Kluwer Academic Publishers, Boston, 1991.
11. Pomykala, J. A.: Approximation operations in approximation space. Bull. Pol. Acad. Sci. 9-10 (1987) 653-662.
12. Skowron, A., Stepaniuk, J.: Tolerance approximation spaces. Fundamenta Informaticae. 27 (1996) 245-253.
13. Slowinski, R., Vanderpooten, D.: A generalized definition of rough approximations based on similarity. IEEE Trans. Data Knowledge Eng. 2 (2000) 331-336.
14. Wasilewska, A.: Topological rough algebras: In: Lin, T. Y., Cercone, N.(Eds.), Rough Sets and Data Mining, Kluwer Academic Publishers, Boston, 1997. pp. 411-425.
15. Keyun, Q., Zheng, P.: On the topological properties of fuzzy rough sets. Fuzzy Sets and Systems. 151(3) (2005) 601-613.
16. Yao, Y. Y.: Relational interpretations of neighborhood operators and rough set approximation operators. Information Sciences. 101 (1998) 239-259.
17. Yao, Y. Y.: Constructive and algebraic methods of theory of rough sets. Information Sciences. 109 (1998) 21-47.
18. Yao, Y. Y.: Two views of the theory of rough sets in finite universes. International Journal of Approximate Reasoning. 15 (1996) 291-317.
19. Yao, Y. Y.: On generalizing rough set theory. Lecture Notes in AI. 2639(2003) 44-51.
20. Zakowski, W.: Approximations in the space $(U, \prod)$. Demonstratio Mathematica. 16 (1983) 761-769.
21. Zhu, F., Wang, F.: Some results on covering generalized rough sets. Pattern Recog. Artificial Intell. 15 (2002) 6-13.
22. Zhu, F., Wang, F.: Reduction and axiomization of covering generalized rough sets. Information Sciences. 152 (2003) 217-230.
23. Zhu, F., He, H.: The axiomization of the rough set. Chinese Journal of Computers. 23 (2000) 330-333.
24. Zhu, W.: Topological approaches to covering rough sets. Information Science. in press.
25. Zhu, W., Wang, F.: Relations among three types of covering rough sets. In:IEEE GrC 2006, Atlanda, USA. pp: 43-48.

On Transitive Uncertainty Mappings

Baoqing Jiang[1], Keyun Qin[2], and Zheng Pei[3]

[1] Institute of Data and Knowledge Engineering, Henan University,
Kaifeng, Henan 475001, China
jbq@henu.edu.cn
[2] Department of Mathematics, Southwest Jiaotong University,
Chengdu, Sichuan, 610031, China
keyunqin@263.net
[3] School of Mathematics & Computer Science, Xihua University,
Chengdu, Sichuan, 610039, China
pqyz@263.net

Abstract. This paper is devoted to the discussion of transitive uncertainty mapping in general approximation space. It is proved that the best low-approximation mapping exist if the uncertainty mapping is transitive. Furthermore, the best low-approximation mapping is defined and its properties are discussed.

Keywords: Rough set, covering rough set, covering lower approximation, covering upper approximation, representative element.

1 Introduction

Rough set theory(RST), proposed by Pawlak [2], [3], is an extension of set theory for the study of intelligent systems characterized by insufficient and incomplete information. It provides a systematic approach for the study of indiscernibility of objects. Typically, indiscernibility is described using equivalent relations. When objects of a universe are described by a set of attributes, one may define the indiscernibility of objects based on their attribute values. When two objects have the same value over a certain group of attributes, we say they are indiscernible with respect to this group of attributes, or have the same description with respect to the indiscernibility relation. Objects of the same description consist of an equivalence class and all the equivalence classes form a partition of the universe. With this partition, rough set theory approximates any subset of objects of the universe by two sets, called the lower and upper approximations. They can be formally described by a pair of unary set-theoretic operators. It is noticed that equivalence relation or partition, as the indiscernibility relation in Pawlak's original rough set theory, is restrictive for many applications. To address this issue, several interesting and meaningful extensions to equivalent relation have been proposed in the past, such as tolerance relation [4,12], similarity relation [13], and others [14,15,16,17]. This leads to various approximation operators. By adopting the notion of neighborhood systems from topological space, Lin[6,7] proposed a more general framework for the study of approximation operators.

J.T. Yao et al. (Eds.): RSKT 2007, LNAI 4481, pp. 42–49, 2007.

Zakowski [20] have used coverings of a universe for establishing the covering generalized rough set theory and an extensive body of research works have been developed [2,3,11]. A. Gomolinska [5] provided a new approach for the study of rough approximations where the starting point is a generalized approximation space. The rough approximation operator was regarded as set-valued mapping, called approximation mapping. Two pairs of basic approximation mappings were defined typically and generalized approximation mappings were constructed by the compositions of these basic approximation mappings. Some axioms for approximation mappings were proposed. Based on these axioms, the best low-approximation mapping was studied.

This paper is devoted to the discussion of transitive uncertainty mapping. The motivation is to construct the best, in accordance with Gomolinska's axioms, approximation operators in general approximation space. It is proved that the best low-approximation mapping exist if the uncertainty mapping is transitive. Furthermore, the best low-approximation mapping is defined and its properties are discussed.

2   Preliminaries

This section presents a review of some fundamental notions of Pawlak's rough sets. We refer to [2,9,10] for details.

Let U be a finite set, the universe of discourse, and R an equivalence relation on U, called an indiscernibility relation. The pair (U, R) is called a Pawlak approximation space. R will generate a partition $U/R = \{[x]_R; x \in U\}$ on U, where $[x]_R$ is the equivalence class with respect to R containing x. For each $X \subseteq U$, the upper approximation $\overline{R}(X)$ and lower approximation $\underline{R}(X)$ of X are defined as [9,10]

$$\overline{R}(X) = \{x; [x]_R \cap X \neq \emptyset\}, \tag{1}$$

$$\underline{R}(X) = \{x; [x]_R \subseteq X\}. \tag{2}$$

Alternatively, in terms of equivalence classes of R, the pair of lower and upper approximation can be defined by

$$\overline{R}(X) = \cup\{[x]_R; [x]_R \cap X \neq \emptyset\}, \tag{3}$$

$$\underline{R}(X) = \cup\{[x]_R; [x]_R \subseteq X\}. \tag{4}$$

Let $\emptyset$ be the empty set and $\sim X$ the complement of X in U, the following conclusions have been established for Pawlak's rough sets:

(1)$\underline{R}(U) = U = \overline{R}(U)$.
(2)$\underline{R}(\emptyset) = \emptyset = \overline{R}(\emptyset)$.
(3)$\underline{R}(X) \subseteq X \subseteq \overline{R}(X)$.
(4)$\underline{R}(X \cap Y) = \underline{R}(X) \cap \underline{R}(X)$, $\overline{R}(X \cup Y) = \overline{R}(X) \cup \overline{R}(Y)$.
(5)$\underline{R}(\underline{R}(X)) = \underline{R}(X)$, $\overline{R}(\overline{R}(X)) = \overline{R}(X)$.

$(6)\underline{R}(X) =\sim \overline{R}(\sim X), \overline{R}(X) =\sim \underline{R}(\sim X).$
$(7)X \subseteq Y \Rightarrow \underline{R}(X) \subseteq \underline{R}(Y), \overline{R}(X) \subseteq \overline{R}(Y).$
$(8)\underline{R}(\sim \underline{R}(X)) =\sim \underline{R}(X), \overline{R}(\sim \overline{R}(X)) =\sim \overline{R}(X).$
$(9)\overline{R}(\underline{R}(X)) \subseteq X \subseteq \underline{R}(\overline{R}(X)).$

It has been shown that (3), (4) and (8) are the characteristic properties of the lower and upper approximations [8,23,18].

3   A General Notion of Rough Approximation Mapping

A general approximation space is a triple $A = (U, I, k)$, where U is a non-empty set called the universe, $I : U \rightarrow P(U)$ is an uncertainty mapping, and $k : P(U) \times P(U) \rightarrow [0, 1]$ is a rough inclusion function.

In general approximation space $A = (U, I, k)$, $w \in I(u)$ is understood as w is in some sense similar to u and it is reasonable to assume that $u \in I(u)$ for every $u \in U$. Then $\{I(u); u \in U\}$ forms a covering of the universe U. The role of the uncertainty mapping may be played by a binary relation on U.

We consider mappings $f : P(U) \rightarrow P(U)$. We can define a partial ordering relation, $\leq$, on the set of all such mappings as follows: $f \leq g$ if and only if $\forall \subseteq U(f(x) \subseteq g(x))$, for every $f, g : P(U) \rightarrow P(U)$. By id we denote the identity mapping on $P(U)$. $g \circ f : P(U) \rightarrow P(U)$ defined by $g \circ f(x) = g(f(x))$ for every $x \subseteq U$, is the composition of f and g. We call g dual to f, written $g = f^d$, if $g(x) =\sim f(\sim x)$. The mapping f is monotone if and only if for every $x, y \subseteq U$, $x \subseteq y$ implies $f(x) \subseteq f(y)$.

3.1   Axioms for Rough Approximation Mappings

Theoretically speaking, every rough approximation operator is a mapping from $P(U)$ to $P(U)$, we call it approximation mapping. [5] proposed some fundamental properties that any reasonable rough approximation mapping $f : P(U) \rightarrow P(U)$ should possibly possess. They are the following axioms:

($a1$) Every low-mapping f is decreasing, i.e., $f \leq id$.

($a2$) Every upp-mapping f is increasing, i.e., $id \leq f$.

($a3$) If f is a low-mapping, then $(*)\forall x \subseteq U \forall u \in f(x)(I(u) \subseteq x)$.

($a4$) If f is a upp-mapping, then $(**)\forall x \subseteq U \forall u \in f(x)(I(u) \cap x \neq \emptyset)$.

($a5$) For each $x \subseteq U$, $f(x)$ is definable in A, i.e., there exists $y \subseteq U$ such that $f(x) = \cup\{I(u); u \in y\}$.

($a6$) For each $x \subseteq U$ definable in A, $f(x) = x$.

The motivation behind these axioms was analyzed in[5]. Also, it is noticed that finding appropriate candidates for low- and upp-mappings satisfying these axioms is not an easy matter in general case.

3.2   The Structure of Rough Approximation Mappings

Let $A = (U, I, k)$ be a general approximation space. The approximation mappings $f_0, f_1 : P(U) \rightarrow P(U)$ were defined as[5]: for every $x \subseteq U$,

$$f_0(x) = \bigcup\{I(u); u \in x\}, \tag{5}$$

$$f_1(x) = \{u; I(u) \cap x \neq \emptyset\}. \tag{6}$$

Observe that f_0^d and f_1^d satisfy:

$$f_0^d(x) = \{u; \forall w(u \in I(w) \Rightarrow w \in x)\}, \tag{7}$$

$$f_1^d(x) = \{u; I(u) \subseteq x\}. \tag{8}$$

If $\{I(u); u \in U\}$ is a partition of U, then $f_0 = f_1$, $f_0^d = f_1^d$ and they are the classical rough approximation operators.

Based on f_0, f_1 and their dual mappings, several approximation mappings were defined[5] by means of operations of composition and duality as follows: for every $x \subseteq U$,

$f_2 \doteq f_0 \circ f_1^d$: i.e., $f_2(x) = \bigcup\{I(u); I(u) \subseteq x\}$,
$f_3 \doteq f_0 \circ f_1$: i.e., $f_3(x) = \bigcup\{I(u); I(u) \cap x \neq \emptyset\}$,
$f_4 \doteq f_0^d \circ f_1 = f_2^d$: i.e., $f_4(x) = \{u; \forall w(u \in I(w) \Rightarrow I(w) \cap x \neq \emptyset)\}$,
$f_5 \doteq f_0^d \circ f_1^d = f_3^d$: i.e., $f_5(x) = \{u; \forall w(u \in I(w) \Rightarrow I(w) \subseteq x)\}$,
$f_6 \doteq f_1^d \circ f_1^d$: i.e., $f_6(x) = \{u; \forall w(w \in I(u) \Rightarrow I(w) \subseteq x)\}$,
$f_7 \doteq f_0 \circ f_6 = f_0 \circ f_1^d \circ f_1^d = f_2 \circ f_1^d$: i.e., $f_7(x) = \bigcup\{I(u); \forall w(w \in I(u) \Rightarrow I(w) \subseteq x)\}$,
$f_8 \doteq f_1^d \circ f_1$: i.e., $f_8(x) = \{u; \forall w(w \in I(u) \Rightarrow I(w) \cap x \neq \emptyset)\}$,
$f_9 \doteq f_0 \circ f_8 = f_0 \circ f_1^d \circ f_1 = f_2 \circ f_1$: i.e., $f_9(x) = \bigcup\{I(u); \forall w(w \in I(u) \Rightarrow I(w) \cap x \neq \emptyset)\}$.

Theorem 1. *[5] Consider any $f : P(U) \to P(U)$.*

(1) $f(x)$ is definable for any $x \subseteq U$ iff there is a mapping $g : P(U) \to P(U)$ such that $f = f_0 \circ g$.

(2) The condition $(*)$ is satisfied iff $f \leq f_1^d$.

(3) The condition $(**)$ is satisfied iff $f \leq f_1$.

Theorem 2. *[5] For any sets $x, y \subseteq U$, we have that:*

(1) $f_i(\emptyset) = \emptyset$ and $f_i(U) = U$ for $i = 0, 1, \cdots, 9$. $f_i^d(\emptyset) = \emptyset$ and $f_i^d(U) = U$ for $i = 0, 1$.

(2) f_i and f_j^d are monotone for $i = 0, 1, \cdots, 9$ and $j = 0, 1$.

(3) $f_i(x \cup y) = f_i(x) \cup f_i(y)$ for $i = 0, 1, 3$.

(4) $f_i(x \cap y) = f_i(x) \cap f_i(y)$ and $f_j^d(x \cap y) = f_j^d(x) \cap f_j^d(y)$ for $i = 5, 6$ and $j = 0, 1$.

Theorem 3. *[5] For any sets $x, y \subseteq U$, we have that:*

(1) $f_5 \leq f_1^d \leq f_2 \leq id \leq f_4 \leq f_1 \leq f_3$.

(2) $f_5 \leq f_0^d \leq id \leq f_0 \leq f_3$.

(3) $f_6 \leq f_7 \leq f_1^d$.

(4) $f_8 \leq f_9 \leq f_1$.

(5) $f_i \circ f_i = f_i$ for $i = 2, 4$.

In view of the previous results and in accordance with the axioms, any low- or upp-mapping should have the form $f_0 \circ g$, where $g : P(U) \to P(U)$ satisfies $f_0 \circ g \circ f_0 = f_0$ and, moreover, $f_0 \circ g \leq f_1^d$ in the lower case, while $id \leq$

$f_0 \circ g \leq f_1$ in the upper case[5]. Clearly, $\leq$ −maximal among the low-mappings and $\leq$ −minimal among the upp-mappings would be the best approximation operators. The greatest element among the low-mappings just described is the mapping $h : P(U) \rightarrow P(U)$ where for any $x \subseteq U$,

$$h(x) = \cup\{(f_0 \circ g)(x); g : P(U) \rightarrow P(U) \wedge f_0 \circ g \circ f_0 = f_0 \wedge f_0 \circ g \leq f_1^d\}. \quad (9)$$

It is noticed that an analogous construction, using $\cap$, does not provide us with the least element of the family of upp-mappings[5].

4   The Best Approximation Operators

In this section, we discuss the condition with which the approximation mapping h exist. We noticed that (9) make sense provided $S \neq \emptyset$ where
$$S = \{g; g : P(U) \rightarrow P(U), f_0 \circ g \circ f_0 = f_0, f_0 \circ g \leq f_1^d\}.$$

Theorem 4. *[5] Consider any* $f : P(U) \rightarrow P(U)$. *$f$ satisfies (a5) and (a6) if and only if there is a mapping* $g : P(U) \rightarrow P(U)$ *such that* $f = f_0 \circ g$ *and* $f_0 \circ g \circ f_0 = f_0$.

Proof. ($\Rightarrow$) Assume f satisfies (a5) and (a6). By Theorem 1, there is a mapping $g : P(U) \rightarrow P(U)$ such that $f = f_0 \circ g$. Consider any $x \subseteq U$, by definability of $f_0(x)$, we have $f_0 \circ g \circ f_0(x) = f(f_0(x)) = f_0(x)$. Hence $f_0 \circ g \circ f_0 = f_0$.

($\Leftarrow$) Assume $f = f_0 \circ g$ and $f_0 \circ g \circ f_0 = f_0$ for some $g : P(U) \rightarrow P(U)$. By Theorem 1, f satisfies (a5). If $x \subseteq U$ is definable, then there is $y \subseteq U$ such that $x = \cup\{I(u); u \in y\} = f_0(y)$. Consequently $f(x) = f_0 \circ g(x) = f_0 \circ g(f_0(y)) = f_0(y) = x$. Hence f satisfies (a6).

Theorem 5. *If $S \neq \emptyset$, then $h = f_0 \circ G$ is the greatest element among the low-mappings which satisfies (a1), (a3), (a5) and (a6), where $G : P(U) \rightarrow P(U)$ satisfies: for every $x \subseteq U$,*
$$G(x) = \cup\{g(x); g \in S\}.$$

The proof of this theorem is trivial.

Theorem 6. *If*

$$\forall u \in U \forall v \in U(u \in I(v) \Rightarrow I(u) \subseteq I(v)) \quad (10)$$

is satisfied, then
    (1) $f_0 \circ f_1^d = f_1^d$,
    (2) $f_1^d \circ f_0 = f_0$.

Proof. Assume (10). Consider any $x \subseteq U$ and $u \in U$.
    (1) If $u \in f_0 \circ f_1^d(x) = \cup\{I(v); v \in f_1^d(x)\}$, then there exist $v \in f_1^d(x)$ such that $u \in I(v)$. Hence $I(u) \subseteq I(v) \subseteq x$. By definition, $u \in f_1^d(x)$ and $f_0 \circ f_1^d \leq f_1^d$. It follows that $f_0 \circ f_1^d = f_1^d$ by $f_0 \geq id$.
    (2) If $u \in f_0(x)$, then there exist $v \in x$ such that $u \in I(v)$. Hence $I(u) \subseteq I(v) \subseteq f_0(x)$ and $u \in f_1^d \circ f_0(x)$. In other words, $f_1^d \circ f_0 \geq f_0$. It follows that $f_1^d \circ f_0 = f_0$.

Theorem 7. $S \neq \emptyset$ *if and only if* (10) *is satisfied.*

Proof. Suppose that $S \neq \emptyset$. It follows that there exists $g : P(U) \to P(U)$ such that $f_0 \circ g \circ f_0 = f_0$ and $f_0 \circ g \leq f_1^d$. By

$$f_0 = f_0 \circ g \circ f_0 \leq f_1^d \circ f_0 \leq f_0,$$

$f_1^d \circ f_0 = f_0$ followed. For every $u, v \in U$ with $u \in I(v)$, by

$$I(v) = f_0(\{v\}) = f_1^d \circ f_0(\{v\}) = f_1^d(I(v)) = \{w; I(w) \subseteq I(v)\},$$

it follows that $I(u) \subseteq I(v)$.

Conversely, assume (10). By Theorem 6,

$$f_0 \circ f_1^d \circ f_0 = (f_0 \circ f_1^d) \circ f_0 = f_1^d \circ f_0 = f_0,$$

$$f_0 \circ f_1^d \leq f_1^d.$$

Hence $f_1^d \in S$ and $S \neq \emptyset$.

By Theorem 6 and Theorem 7, if (10) is satisfied, f_1^d is $\leq$ −maximal among the low-mappings which satisfy $(a1)$, $(a3)$, $(a5)$ and $(a6)$. Hence f_1^d is the best low-approximation mapping.

5   The Transitive Uncertainty Mapping

In view of the previous results, the condition (10) plays a central role in general approximation spaces. It is just the transitivity of uncertainty mapping. In this section, we will concentrate on properties specific for this kind of uncertainty mapping.

Theorem 8. *Assume* (10). *For any sets* $x, y \subseteq U$, *we have that:*
(1) $f_i \circ f_i = f_i$ *for* $i = 0, 1$.
(2) $f_2 = f_6 = f_7 = f_1^d$.
(3) $f_4 = f_1$.
(4) $f_8 = f_9$.

Proof. Consider any $x \subseteq U$ and $u \in U$.

(1) If $u \in f_1 \circ f_1(x) = \{v; I(v) \cap f_1(x) \neq \emptyset\}$, then $I(u) \cap f_1(x) \neq \emptyset$. It follows that there exists $v \in I(u)$ such that $I(v) \cap x \neq \emptyset$. By $I(v) \subseteq I(u)$, $I(u) \cap x \neq \emptyset$ followed. By the definition, $u \in f_1(x)$. In other words, $f_1 \circ f_1 \leq f_1$. Consequently, $f_1 \circ f_1 = f_1$ by $f_1 \geq id$.

If $u \in f_0 \circ f_0(x) = \cup\{I(v); v \in f_0(x)\}$, then there exists $v \in U$ such that $v \in f_0(x)$ and $u \in I(v)$. By the definition, there is $w \in x$ such that $v \in I(w)$. Consequently, $u \in I(v) \subseteq I(w)$ and $u \in f_0(x)$. In other words, $f_0 \circ f_0 \leq f_0$ and $f_0 \circ f_0 = f_0$ followed by $f_0 \geq id$.

(2) By Theorem 6, $f_7 = f_0 \circ f_1^d \circ f_1^d = (f_0 \circ f_1^d) \circ f_1^d = f_1^d \circ f_1^d = f_6$, $f_6 = f_1^d \circ f_1^d = (f_1 \circ f_1)^d = f_1^d$, $f_2 = f_0 \circ f_1^d = f_1^d$.

(3) and (4) can be proved similarly.

By (1) of Theorem 8, $f_0 \circ id \circ f_0 = f_0 \circ f_0 = f_0$ and $f_0 = f_0 \circ id$, it follows that f_0 satisfies axiom ($a6$). We summarize the approximation mappings and satisfiability of the axioms in Table 1. By + (resp., −) we denote that a condition is (is not) satisfied, while ⊥ denotes that the result does not count. From the Table we know that f_1^d is the best low-approximation mapping since it satisfies all axioms and f_0 is in our opinion the best candidate for a upp-mapping since it satisfies three axioms ($a1$), ($a5$) and ($a6$).

Table 1. Approximation mappings and satisfiability of the axioms if (10) holds

f	Form	status	$a1$	$a2$	$a3$	$a4$	$a5$	$a6$
f_0		upp	⊥	+	⊥	−	+	+
$f_1 = f_4$		upp	⊥	+	⊥	+	−	−
f_0^d		low	+	⊥	−	⊥	−	−
$f_1^d = f_2 = f_6 = f_7$		low	+	⊥	+	⊥	+	+
f_3	$f_0 \circ f_1$	upp	⊥	+	⊥	−	+	−
f_5	$f_0^d \circ f_1^d$	low	+	⊥	+	⊥	−	−
$f_8 = f_9$	$f_1^d \circ f_1$	upp	⊥	−	⊥	+	−	−

Acknowledgements

This work has been supported by the National Natural Science Foundation of China (Grant No. 60474022) and Henan Innovation Project for University Prominent Research Talents (Grant No. 2007KYCX018) and the Young Foundation of Sichuan Province (Grant no. 06ZQ026-037).

References

1. Allam, A., Bakeir, M., Abo-Tabl, E.: New approach for basic roset concepts. In: LNCS, Vol. 3641, 2005, 64-73.
2. Bonikowski, Z., Bryniarski, E., Wybraniec, U.: Extensions and intentions in the rough set theory. Information Sciences. 107 (1998) 149-167.
3. Bryniarski, E.: A calculus of rough sets of the first order. Bull. Pol. Acad. Sci. 16 (1989) 71-77.
4. Cattaneo, G.: Abstract approximation spaces for rough theories. In: L.Polkowski, A.Skowron(Eds.), Rough Sets in Knowledge Discovery 1: Methodology and Applications, Physica-Verlag, Heidelberg, 1998, pp. 59-98.
5. Gomolinska, A.: A comparative study of some generalized rough approximations. Fundamenta Informaticae. 51(2002) 103-119.
6. Lin, T.: Neighborhood systems and relational database. Proceedings of CSC'88, 1988.
7. Lin, T.: Neighborhood systems and approximation in database and knowledge base systems. Proceedings of the Fourth International Symposium on Methodologies of Intelligent Systems, Poster Session, CSC'88, 1989.

8. Lin, T., Liu, Q.: Rough approximate operators: axiomatic rough set theory. In: W. P. Ziarko(Ed.), Rough Sets, Fuzzy Sets and Knowledge Discovery, Springer-Verlag, London, 1994, pp. 256-260.
9. Pawlak, Z.: Rough sets. International Journal of Computer and Information Science. 11 (1982) 341-356.
10. Pawlak, Z.: Rough sets: Theoretical Aspects of Reasoning About Data. Kluwer Academic Publishers, Boston, 1991.
11. Pomykala, J. A.: Approximation operations in approximation space. Bull. Pol. Acad. Sci. 9-10 (1987) 653-662.
12. Skowron, A., Stepaniuk, J.: Tolerance approximation spaces. Fundamenta Informaticae. 27 (1996) 245-253.
13. Slowinski, R., Vanderpooten, D.: A generalized definition of rough approximations based on similarity. IEEE Trans. Data Knowledge Eng. 2 (2000) 331-336.
14. Wasilewska, A.: Topological rough algebras: In: Lin, T. Y., Cercone, N.(Eds.), Rough Sets and Data Mining, Kluwer Academic Publishers, Boston, 1997. pp. 411-425.
15. Keyun, Q., Zheng, P.: On the topological properties of fuzzy rough sets. Fuzzy Sets and Systems. 151(3) (2005) 601-613.
16. Yao, Y. Y.: Relational interpretations of neighborhood operators and rough set approximation operators. Information Sciences. 101 (1998) 239-259.
17. Yao, Y. Y.: Constructive and algebraic methods of theory of rough sets. Information Sciences. 109 (1998) 21-47.
18. Yao, Y. Y.: Two views of the theory of rough sets in finite universes. International Journal of Approximate Reasoning. 15 (1996) 291-317.
19. Yao, Y. Y.: On generalizing rough set theory. Lecture Notes in AI. 2639(2003) 44-51.
20. Zakowski, W.: Approximations in the space $(U, \prod)$. Demonstratio Mathematica. 16 (1983) 761-769.
21. Zhu, F., Wang, F.: Some results on covering generalized rough sets. Pattern Recog. Artificial Intell. 15 (2002) 6-13.
22. Zhu, F., Wang, F.: Reduction and axiomization of covering generalized rough sets. Information Sciences. 152 (2003) 217-230.
23. Zhu, F., He, H.: The axiomization of the rough set. Chinese Journal of Computers. 23 (2000) 330-333.
24. Zhu, W.: Topological approaches to covering rough sets. Information Science. in press.
25. Zhu, W., Wang, F.: Relations among three types of covering rough sets. In: IEEE GrC 2006, Atlanda, USA. pp: 43-48.

A Complete Method to Incomplete Information Systems

Ming-Wen Shao[1,2]

[1] School of Information Technology, Jiangxi University of Finance & Economics,
Nanchang, Jiangxi 330013, P. R. China
shaomingwen1837@163.com
[2] Jiangxi Key Laboratory of Data and Knowledge Engineering, Jiangxi University of
Finance & Economics, Nanchang, Jiangxi 330013, P. R. China

Abstract. In the paper, we present a novel method for handling incomplete information systems. By the proposed method we can transform an incomplete information system into a complete set-value information system without loss any information, and we discuss the relationship between the reducts of incomplete information system and the reducts of it's complements. For incomplete decision tables, we introduce two complete methods according to different criterions of certain factor of decision rules, i.e., maximal sum complement and maximal conjunction complement of certain factor of decision rules.

Keywords: Rough sets, incomplete information systems, knowledge reduction.

1 Introduction

Rough sets theory, introduced by Pawlak [1], has been conceived as a tool to conceptualize and analyze various types of data, in particular, has important applications to artificial intelligence and cognitive sciences, as a tool for dealing with vagueness and uncertainty of facts, and in classification [2,3,4,5].

A concept related to rough set is information system (attribute-value system). According to whether or not there are missing data (null values), information system can be classified into two categories: complete and incomplete. A incomplete information system contains null value for at last one attribute, a null value may be some value in the domain of the corresponding attribute, however, it is unknown. Here we consider the case in which a null value means an applicable value. Some important results have recently been obtained for incomplete information system by knowledge acquisition methodologies [3,5,6,7,8,9].

There are several ways in which null value may be handled in [10,11,12]. The simplest method to hand null value is to remove objects with unknown values or replace null values with most common values [10] in the original system. More complex approaches which provide strategies to deal with null values in terms of statistics are studied [13], in which it is suggested to predict the null values on the basis of values of other attributes of an object and relevant class information.

J.T. Yao et al. (Eds.): RSKT 2007, LNAI 4481, pp. 50–59, 2007.

The problem of rules extraction from incomplete information system was discussed in the context of the Rough Sets [3,6,7,10,14]. Modeling uncertainty caused by the appearance of unknown values by means of fuzzy sets was described in [3]; A methodology of rules generation from original incomplete systems was discussed in [6,7]; A learning algorithm is proposed [14], which can simultaneously derive rules from incomplete system and estimate the missing values in the learning process.

In the paper, we describe a new method for handling incomplete information system. By the proposed method we can transform an incomplete information system into a complete set-value information system. For decision tables, we show two complete methods according to different criterions of certain factor of decision rules, i.e., maximal sum complement of certain factor and maximal conjunction complement of certain factor. The relationship between the reducts of incomplete information system and the reducts of it's complements are also discussed in details.

2   Incomplete Information Systems

An information system (IS) is an ordered triplet $\mathcal{I} = (U, AT, f)$, where U is a finite nonempty set of objects and AT is a finite nonempty set of attributes, $f_a : U \rightarrow V_a$ for any $a \in AT$, where V_a is the domain of attribute a.

It may happen that some of attribute values for objects are missing. To indicate such a situation a distinguished value, so-called *null value*, is usually assigned to those attributes. We denote special symbol $*$ to indicate that the value of an attribute is unknown. Here, we assume that an object $x \in U$ possesses only one value for an attribute a ($a \in AT$). Thus, if the value of an attribute a is missing, then the real value must be one of value of V_a. An IS in which values of all attributes for all objects from U are known is called complete, it is called incomplete otherwise.

Example 1. An incomplete IS $\mathcal{I} = (U, AT, f)$ is presented in Table 1.

Table 1. An incomplete IS

U	a_1	a_2	a_3
x_1	1	1	1
x_2	1	*	1
x_3	2	1	1
x_4	1	2	*
x_5	1	*	1
x_6	2	2	2
x_7	1	1	1

From Table 1, we have a set-value IS $\mathcal{I}_F = (U, AT, F)$, see Table 2, where $F = \{F_a : U \rightarrow \mathcal{P}(V_a) | \ a \in AT\}$,

$$F_a(x) = \begin{cases} \{f_a(x)\} & f_a(x) \neq *, \\ V_a & f_a(x) = *. \end{cases}$$

Table 2. The set-value IS $\mathcal{I}_F$

U	a_1	a_2	a_3
x_1	{1}	{1}	{1}
x_2	{1}	{1,2}	{1}
x_3	{2}	{1}	{1}
x_4	{1}	{2}	{1,2}
x_5	{1}	{1,2}	{1}
x_6	{2}	{2}	{2}
x_7	{1}	{1}	{1}

In the following, we only consider the set-value IS that derived from incomplete IS.

Let $\mathcal{I}_F = (U, AT, F)$ be a set-value IS and $A \subseteq AT$. Let

$$R_A^* = \{(x,y) \in U \times U | \ \forall \ a \in A, F_a(x) \cap F_a(y) \neq \emptyset\},$$

we denote $[x]_A^* = \{y \in U | (x,y) \in R_A^*\}$. In general, U/R_A^* do not constitute a partition of U, they may overlap.

Let $A \subseteq AT$, we say A is a reduct of $\mathcal{I}_F$, if $R_A^* = R_{AT}^*$ and $R_{A-\{a\}}^* \neq R_{AT}^*$ ($\forall \ a \in A$).

Definition 1. Let $\mathcal{I}_F = (U, AT, F)$ be a set-value IS, $X \subseteq U$, $A \subseteq AT$, a pair of lower and upper approximations, $\underline{A}(X)$ and $\overline{A}(X)$, is defined by

$$\underline{A}(X) = \{x \in U | \ [x]_A^* \subseteq X\}, \quad \overline{A}(X) = \{x \in U | \ [x]_A^* \cap X \neq \emptyset\}.$$

Theorem 1. Let $\mathcal{I}_F = (U, AT, F)$ be a set-value IS, X, $Y \subseteq U$, $A \subseteq B \subseteq AT$, then

(1) $\underline{A}(\emptyset) = \overline{A}(\emptyset) = \emptyset$, $\overline{A}(U) = \underline{A}(U) = U$;

(2) $\underline{A}(X \cap Y) \subseteq \underline{A}(X) \cap \underline{A}(Y)$, $\overline{A}(X \cup Y) \supseteq \overline{A}(X) \cup \overline{A}(Y)$;

(3) $X \subseteq Y \Rightarrow \underline{A}(X) \subseteq \underline{A}(Y)$, $X \subseteq Y \Rightarrow \overline{A}(X) \subseteq \overline{A}(Y)$;

(4) $\underline{A}(X \cup Y) \supseteq \underline{A}(X) \cup \underline{A}(Y)$, $\overline{A}(X \cap Y) \subseteq \overline{A}(X) \cap \overline{A}(Y)$;

(5) $\underline{A}X \subseteq X \subseteq \overline{A}X$;

(6) $\underline{A}X \subseteq \underline{B}X$, $\overline{A}X \supseteq \overline{B}X$.

Proof. It immediately follows from Definition 1.

Definition 2. Let $\mathcal{I}_F = (U, AT, F)$ be a set-value IS. We denote

$$D(x,y) = \{a \in AT | \ F_a(x) \cap F_a(y) = \emptyset \ (x,y \in U)\},$$

then $D(x,y)$ is called discernibility attribute set of $\mathcal{I}_F$, and $\mathcal{D} = (D(x,y) : x, y \in U)$ is called discernibility matrix of $\mathcal{I}_F$.

Theorem 2. Let $\mathcal{I}_F = (U, AT, F)$ be a set-value IS and $A \subseteq AT$, then $R_A^* = R_{AT}^*$ iff $A \cap D(x,y) \neq \emptyset$ ($\forall D(x,y) \neq \emptyset, x, y \in U$).

Proof. Suppose that $A \cap D(x, y) \neq \emptyset$ ($\forall D(x, y) \neq \emptyset, x, y \in U$), then $\exists a \in A$, such that $a \in D(x, y)$, which implies $F_a(x) \cap F_a(y) = \emptyset$, i.e. $(x, y) \notin R_A^*$. Thus, $R_A^* \subseteq R_{AT}^*$. On the other hand, it is evident that $R_{AT}^* \subseteq R_A^*$. Therefore, $R_A^* = R_{AT}^*$.

Conversely, assume that $R_A^* = R_{AT}^*$, then $[x]_A^* = [x]_{AT}^*$ ($\forall x \in U$). If $y \notin [x]_{AT}^*$, then $y \notin [x]_A^*$. Thus $\exists a \in A$, such that $F_a(x) \cap F_a(y) = \emptyset$, which implies $a \in D(x, y)$. Therefore, $A \cap D(x, y) \neq \emptyset$ ($\forall D(x, y) \neq \emptyset$).

We denote

$$\triangle = \bigwedge_{(x,y)\in\ U \times U} \bigvee D(x, y),$$

then $\triangle$'s prime implications determine reducts uniquely for set-value IS (see [15]).

Example 2. Table 3 is the discernibility matrix of Table 2, where, values of $D(x_i, x_j)$ for any pair (x_i, x_j) of objets from U are placed.

Table 3. The discernibility matrix of $\mathcal{I}_F$

x/y	x_1	x_2	x_3	x_4	x_5	x_6	x_7
x_1			a_1	a_2		$a_1 a_2 a_3$	
x_2			a_1			$a_1 a_3$	
x_3	a_1	a_1		$a_1 a_2$	a_1	$a_2 a_3$	a_1
x_4	a_1		$a_1 a_2$			a_1	a_2
x_5			a_1			$a_1 a_3$	
x_6	$a_1 a_2 a_3$	$a_1 a_3$	$a_2 a_3$	a_1	$a_1 a_3$		$a_1 a_2 a_3$
x_7			a_1	a_2		$a_1 a_2 a_3$	

From the Table 3, we have

$$\triangle = a_1 \wedge a_2 \wedge (a_1 \vee a_2 \vee a_3) \wedge (a_1 \vee a_3) \wedge (a_1 \vee a_2) \wedge (a_2 \vee a_3) = a_1 \wedge a_2.$$

Thus, $\{a_1, a_2\}$ is the unique reduct of set-value IS.

Let $\mathcal{I}_F = (U, AT, F)$ be a set-value IS. We denote

$$f' = \{f_a' : U \rightarrow V_a, \ f_a'(x) \in F_a(x), \ (a \in AT, x \in U)\},$$

then f' is called a selection of F.

It is easy to see that $\mathcal{I}_{f'} = (U, AT, f')$ is a complement of the original incomplete IS. Let F^* denotes the set of all selections of F. Then, $S_F = \{(U, AT, f') : f' \in F^*\}$ is the set of all the complements of the original incomplete IS.

Example 3. Table 4 is a selection of set-value IS $\mathcal{I}_F = (U, AT, F)$ presented in Table 2.

Table 4. A selection $\mathcal{I}_{f'} = (U, AT, f')$

U	a_1	a_2	a_3
x_1	1	1	1
x_2	1	2	1
x_3	2	1	1
x_4	1	2	2
x_5	1	1	1
x_6	2	2	2
x_7	1	1	1

Let $\mathcal{I} = (U, AT, f)$ be an incomplete IS. We denote

$$Y^* = \{(x, a) \mid x \in U, \ a \in AT, \ f_a(x) = *\}, \ B^* = \{a \in AT \mid \exists \ x \in U, f_a(x) = *\}.$$

The number of all the complement of $\mathcal{I}$ is $\prod\limits_{(x,a)\in Y^*} |V_a|$, i.e., $|S_F| = \prod\limits_{(x,a)\in Y^*} |V_a|$,

where $|\cdot|$ denotes the cardinal of a set. We can select a complement from S_F according to different criterions.

Example 4. In *Example 1*, since $Y^* = \{(x_2, a_2), \ (x_4, a_3), \ (x_5, a_2)\}$, then $\prod\limits_{(x,a)\in Y^*} |V_a| = 2 \times 2 \times 2 = 8$.

Theorem 3. *Let $\mathcal{I}_{f'} \in S_F$, $A \subseteq AT$. We denote $R_A^{f'} = \{(x, y) \in U \times U \mid \forall \ a \in A, f'_a(x) = f'_a(y)\}$, then $R_A^* = \bigcup\limits_{f' \in F^*} R_A^{f'}$.*

Proof. For any $f' \in F^*$, we have $R_A^{f'} \subseteq R_A^*$. Thus $\bigcup\limits_{f' \in F^*} R_A^{f'} \subseteq R_A^*$. On the other hand, for any $(x, y) \in R_A^*$, we can easily conclude that $F_a(x) \cap F_a(y) \neq \emptyset$ ($\forall \ a \in A$). Hence $\exists \ f' \in F^*$, such that $f'_a(x) = f'_a(y)$ ($\forall \ a \in AT$), which implies $(x, y) \in R_A^{f'}$. Therefore, $R_A^* \subseteq \bigcup\limits_{f' \in F^*} R_A^{f'}$.

We denote $[x]_A^{f'} = \{y \in U \mid (x, y) \in R_A^{f'}\}$, by Theorem 3 we have $[x]_A^* = \bigcup\limits_{f' \in F^*} [x]_A^{f'}$.

Theorem 4. *Let $\mathcal{I}_{f'} \in S_F$ and $A \subseteq AT$, if $\forall \ f' \in F^*$, $R_A^{f'} = R_{AT}^{f'}$, then $R_A^* = R_{AT}^*$.*

Proof. It is immediately from Theorem 3.

Theorem 5. *Let $\mathcal{I}_F = (U, AT, F)$ be a set-value IS and $B^* \subseteq A \subseteq AT$, then $R_A^* = R_{AT}^*$ iff $R_A^{f'} = R_{AT}^{f'}$, $\forall \ f' \in F^*$.*

Proof. It is evident that $R_{AT}^{f'} \subseteq R_A^{f'}$ ($\forall f' \in F^*$), and we only need to prove that $R_A^{f'} \subseteq R_{AT}^{f'}$. For any $(x,y) \in R_A^{f'}$, by Theorem 3 we have $(x,y) \in R_A^*$. Since $R_A^* = R_{AT}^*$, then $\exists R_{AT}^{f^1} \in R_{AT}^*$ such that $(x,y) \in R_{AT}^{f^1}$, i.e. $f_a^1(x) = f_a^1(y)$ ($\forall a \in AT$). It is easy to see that $\forall f^1, f^2 \in F^*$ there always are

$$f_a^1(x) = f_a^2(x), \ (\forall a \in AT - B^*, \ \forall x \in U).$$

Therefore,

$$f_a'(x) = f_a'(y) \ (\forall a \in AT - B^*). \tag{1}$$

On the other hand, since $(x,y) \in R_A^{f'}$, then

$$f_a'(x) = f_a'(y) \ (\forall a \in A). \tag{2}$$

By Eq. (1) (2), we have $f_a'(x) = f_a'(y)$ ($\forall a \in AT$), i.e., $(x,y) \in R_{AT}^{f'}$. Therefore,

$$R_A^{f'} \subseteq R_{AT}^{f'}.$$

Conversely, it follows immediately from Theorem 4.

Theorem 6. *Let $\mathcal{I}_F = (U, AT, F)$ be a set-value IS, A is a reduct of $\mathcal{I}_F$. We denote $C = A \cup B^*$, then $R_C^{f'} = R_{AT}^{f'}$, $\forall f' \in F^*$.*

Proof. Since $R_A^* = R_{AT}^*$ and $A \subseteq C$, then $R_C^* = R_{AT}^*$; on the other hand, since $B^* \subseteq C$, by Theorem 5 we have that $R_C^{f'} = R_{AT}^{f'}$, ($\forall f' \in F^*$).

Theorem 7. *Let $\mathcal{I}_{f'} = (U, AT, f') \in S_F$, A is a reduct of $\mathcal{I}_{f'}$ and $A \cap B^* = \emptyset$. We denote $C = A \cup B^*$, then $R_C^{f'} = R_{AT}^{f'}$, $\forall f' \in F^*$.*

Proof. It is similar to the proof of Theorem 5.

3   Incomplete Decision Tables

A decision table (DT) is an IS $\mathcal{I} = (U, AT \cup \{d\}, f)$, where d ($d \notin AT$ and $* \notin V_d$) is a distinguished attribute called the decision, and the element of AT are called conditions. A DT is called complete, if it is a complete IS; it is incomplete otherwise.

Example 5. An incomplete DT is presented in Table 5, similar to incomplete IS, from Table 5, we have a set-value DT $\mathcal{I}_F = (U, AT \cup \{d\}, F)$, see Table 6.

In the following, we only consider the set-value DT derived from incomplete DT. Let $\mathcal{I}_F = (U, AT \cup \{d\}, F)$ be a set-value DT, we denote

$$f' = \{f_a' : U \to V_a, \ f_a'(x) \in F_a(x) \text{ and } f_d'(x) = F_d(x), \ (a \in AT, x \in U)\},$$

then f' is called a selection of F.

Table 5. An incomplete DT $\mathcal{I} = (U, AT \cup \{d\}, f)$

U	a_1	a_2	a_3	d
x_1	1	1	1	1
x_2	2	*	2	2
x_3	1	2	1	2
x_4	1	1	*	1
x_5	2	2	2	1
x_6	1	1	1	2
x_7	1	*	1	1
x_8	2	1	2	1

Table 6. A set-value DT $\mathcal{I}_F = (U, AT \cup \{d\}, F)$

U	a_1	a_2	a_3	d
x_1	{1}	{1}	{1}	1
x_2	{2}	{1,2}	{2}	2
x_3	{1}	{2}	{1}	2
x_4	{1}	{1}	{1,2}	1
x_5	{2}	{2}	{2}	1
x_6	{1}	{1}	{1}	2
x_7	{1}	{1,2}	{1}	1
x_8	{2}	{1}	{2}	1

Similar to incomplete IS, we can compute the set of all the complements of the original incomplete DT.

In set-value DT, R^*_{AT} is defined as in set-value IS, we denote

$$R_d = \{(x, y) \in U \times U \mid F_d(x) = F_d(y)\},$$

then $\mathcal{I}_F$ is called consistent, if $R^*_{AT} \subseteq R_d$; it is inconsistent otherwise.

Let $\mathcal{I}_F$ be a consistent DT and $A \subseteq AT$, A is called a reduct of DT, if $R^*_A \subseteq R_d$ and $R^*_B \not\subseteq R_d$ ($\forall B \subset A$).

Let $\mathcal{I}_F = (U, AT \cup \{d\}, F)$ be a consistent set-value DT. We denote

$$D_d(x, y) = \begin{cases} \{a \in AT : F_a(x) \cap F_a(y) = \emptyset\}, & F_d(x) \neq F_d(y), \\ \emptyset, & F_d(x) = F_d(y). \end{cases}$$

then $D_d(x, y)$ is called discernibility attribute set of DT, and $\mathcal{D}_d = (D_d(x, y) : x, y \in U)$ is called discernibility matrix of DT.

We denote

$$\triangle = \bigwedge_{(x,y)\in\ U \times U} \bigvee D_d(x, y),$$

then $\triangle$ determine reducts uniquely for consistent set-value DT.

Knowledge hidden in data contained in decision tables may be discovered and expressed in the form of decision rule $t \to s$, where

$$t = \wedge(c, v), c \in AT, v \in V_c \backslash \{*\} \text{ and } s = \vee(d, w), w \in V_d.$$

We denote

$$||t|| = \{x \in U|\ f_a(x) = v, (a, v) \in t\}, \quad ||s|| = \{x \in U|\ f_d(x) = w, (d, w) \in s\},$$

where $||t||$ be the set of objects of property $\wedge(c, v)$ $(c \in AT, v \in V_c)$, and $||s||$ be the set of objects of property $\vee(d, w)$ $(w \in V_d)$, then

$$||t \vee s|| = ||t|| \vee ||s||, \quad ||t \wedge s|| = ||t|| \wedge ||s||.$$

Let $r : \wedge(c, v) \to \vee(d, w)$ be a decision rule, we denote

$$cer_{\mathcal{I}_F}(s \to t) = \frac{card(||s \wedge t||)}{card(||s||)},$$

then $cer_{\mathcal{I}_F}(s \to t)$ is called the *certainty factor* of rule r.

Let $\mathcal{I}_{f'} = (U, AT \cup \{d\}, f') \in S_F$ and $U/R_d = \{D_1, D_2, \cdots, D_n\}$. A membership distribution function $\mu_A^{f'} : U \to [0, 1]^n$ is defined as follows [16]:

$$\mu_A^{f'}(x) = (D(D_1/[x]_A^{f'}), \ldots, D(D_n/[x]_A^{f'})), \quad x \in U$$

where

$$D(D_i/[x]_A^{f'}) = \frac{|D_i \bigcap [x]_A^{f'}|}{[x]_A^{f'}}.$$

It is evident that $D(D_i/[x]_A^{f'})$ is the *certainty factor* of the rule

$$\bigwedge_{a \in A} (a, f_a'(x)) \to (d, f_d'(D_i)).$$

Let $x \in U$, we denote

$$m_A^{f'}(x) = \max\{D(D_i/[x]_A^{f'}) : i \leq n\} = D(D_j/[x]_A);$$

$$\eta_A^{f'}(x) = \{D_j : m_A^{f'}(x) = D(D_j/[x]_A^{f'})\}.$$

Let $D_j \in \eta_A^{f'}(x)$, then $D(D_i/[x]_A^{f'}) \leq D(D_j/[x]_A^{f'})$ $(\forall\ D_i \in U/R_d)$, i.e., the *certainty factor* of rule $\bigwedge_{a \in A}(a, f_a'(x)) \to (d, f_d'(D_j))$ is maximal in all the rules supported by object x. Rule $\bigwedge_{a \in A}(a, f_a'(x)) \to (d, f_d'(D_j))$ is called maximal confidence rule supported by object x.

We denote

$$M_A^{f'} = \sum_{[x]_A \in\ U/R_A} m_A^{f'}(x), \quad m_A^{f'} = \bigwedge_{[x]_A \in\ U/R_A} m_A^{f'}(x).$$

Let

$$M_A^{f^1} = \max\{M_A^f : f \in F^*\}, \quad m_A^{f^2} = \max\{m_A^f : f \in F^*\},$$

then f^1 is called maximal sum selection of *certainty factor*, and f^2 is called maximal conjunction selection of *certainty factor*.

Let $\mathcal{I}_F = (U, AT \cup \{d\}, F)$ be a set-value DT, we select $f^1, f^2 \in F^*$ such that

$$M_A^{f^1} = \max\{M_A^{f'} : f' \in F^*\}, \quad m_A^{f^2} = \max\{m_A^{f'} : f' \in F^*\},$$

then we have the two complete DT

$$(U, AT \cup \{d\}, f^1), \quad (U, AT \cup \{d\}, f^2).$$

In all the selection of F, the sum of *certainty factor* of rules hidden in $(U, AT \cup \{d\}, f_1)$ is maximal, the conjunction of *certainty factor* of rules hidden in $(U, AT \cup \{d\}, f_2)$ is maximal.

Example 6. In *Example 5*, from Table 6, we select $\mathcal{I}_{f^1} \in S_F$ (see Table 7).

Table 7. DT $\mathcal{I}_{f^1} = (U, AT \cup \{d\}, f^1)$

U	a_1	a_2	a_3	d
x_1	1	1	1	1
x_2	2	1	2	2
x_3	1	2	1	2
x_4	1	1	2	1
x_5	2	2	2	1
x_6	1	1	1	2
x_7	1	1	1	1
x_8	2	1	2	1

It can be easily checked that $M_{AT}^{f'} \leq M_{AT}^{f^1}$, $m_{AT}^{f'} \leq m_{AT}^{f^1}$ ($\forall f' \in F^*$). Therefore, f^1 not only is a maximal sum selection of *certainty factor*, but also a maximal conjunction selection of *certainty factor*.

4 Conclusions

In the paper, a new method is proposed to handling incomplete information systems. By the proposed method we transform an incomplete information system into a complete set-value information system, in which we discussed the problems of set approximation and attribute reduction. For incomplete decision tables, we introduce two complete methods according to different criterions of certain factor of decision rules, i.e., maximal sum complement and maximal conjunction complement of certain factor of decision rules. The relationship between the reducts of incomplete information system and the reducts of it's complements are also discussed in details. This paper may provide a new, different understanding and representations to incomplete information systems.

Acknowledgments

This paper has been supported by the research grant No. [2006]321 and [2006]230 from Technology Program of Jiangxi Education Office.

References

1. Pawlak, Z.: Rough sets. International Journal of Computer and Information Science 11 (1982) 341–356.
2. Pawlak, Z.: Rough Sets Theory and It's Application to Data Analysis [J]. Cybernetics Systems, An International Journal 29 (1998) 661-688.
3. Slowinski, R., Strfanowski, J.: Rough-set reasoning about uncertain date. Fund. Inform 27 (1996) 229–244.
4. Skowron, A.: A synthesis of decision rules: Applications of discernibility matrix.In: Proceedings of the International Conference on Intelligent Information Systems, Augustow, Poland (1993) 30–46.
5. Shusaku, T.: Rule discovery in database with missing values based on rough set model. In: Ning Zhong, Lizhu Zhou (Eds.), Methodologies for Knowledge Discover and Data Ming, Springer, Beijing, China (1999) 274–278.
6. Kryszkiewicz, M.: Rough set approach to incomplete information systems. Information Sciences 112 (1998) 39–49.
7. Kryszkiewicz, M.: Rules in incomplete information systems. Information Sciences 113 (1999) 271–292.
8. Roman, S., Daniel V.: A genaralized definition of rough approximations based on similarity. IEEE Transactions on Knowledge and Data Engineering 12 (2000) 331–336.
9. Akira, N.: A rough logic based on incomplete information and its application. International Journal of Approximate Reasoning 15 (1996) 367–378.
10. Chmielewski, M.R., Grzymala-Busse J.W., Peterson, N.W., Than, S.: The rule induction systems LERS-A version for personal computers. Found.Comput. Decision Sci. 18(3/4) (1993) 181–212.
11. Slowinski, R., Stefanowski, J.: Rough classification in incomplete information systems. Math.Comput. Modelling 12(10/12) (1989) 1347–1357.
12. Slowinski, R., Stefanowski, J.: Handling various types of uncertainty in rough set approach. In: Ziarko, W. (ED.), Rough Sets Fuzzy Sets and Knowledge Discovery (RSKD'93), Springer, Berlin (1994).
13. Kononenko, I., Bratko, I., Roskar, E.: Experiments in automatic learning of medical diagnostic rules. Technical report, Jozef Stefan Institute, Ljubljana, Yugoslavia (1984).
14. Tzung, P.H., Li, H.T., Shyue, L.W.: Learning rules from incomplete training examples by rough sets. Expert Systems with Applications, 22 (2002) 285–293.
15. Skowron, A., Rauszer, C.: The discernibility matrices and functions in information systems. In: Slowinski, R.(Ed.), Intrlligent Decision Support : Handbook of Applications and Advances to Rough Sets Theory, Kluwer Academic Publisher, Dordrecht (1992) 331–362.
16. Zhang, W.X., Mi, J.S., Wu, W.Z.: Knowledge reductions in inconsistent information systems. Chinese Journal of Computer 26(1) (2003) 12-18.

Information Concept Lattice and Its Reductions

Wei Xu[1,2], Ping Xu[2], and Wenxiu Zhang[2]

[1] School of Management, Graduate University of Chinese Academy of Sciences,
Chinese Academy of Sciences, Beijing, 100080, China
xuw-06b1@mails.gucas.ac.cn
[2] School of Sciences, Xi'an Jiaotong University, Shannxi, 710049, China
{xuping,wxzhang}@mail.xjtu.edu.cn

Abstract. In this paper, the combination of formal concept analysis and rough set theory is considered. The notion of information concept lattice is presented and some properties are given. We present the reduction theory of information concept lattice and obtain the reduction method. Information concept lattice is compared with rough set theory and concept lattice.

Keywords: rough set, concept lattice, discernibility matrices, information system.

1 Introduction

The theory of rough sets (RS), proposed by Pawlak [1], is a new mathematical tool to deal with uncertain and vague knowledge. The basic concepts are that of an equivalence relation and its lower and upper approximation on a set of objects called the universe. By the lower and upper approximation of sets, it provides a mathematical approach for knowledge discovery, and has many important applications in various fields [2-4], such as knowledge acquisition, data analysis and so on.

Formal concept analysis (FCA), also called concept lattice (CL), is proposed by Wille [5] in 1982. CL is an ordered hierarchical structure of formal concepts that are defined by a binary relation between a set of objects and a set of attributes. Each formal concept is an (objects, attributes) pair, which consists of two parts: the extension (objects covered by the concept) and intension (attributes describing the concept). As an effective tool for data analysis and knowledge processing, it has been applied to various fields [6-8], such as data mining, information retrieval, software engineering, and so on.

Based on the similarities of two theories, how to compare formal concept analysis with rough set theory has attracted increasing attention from academics in the past decade [9-13]. For example, Saquer and Deogun [10] studied approximations of a set of objects and a set of properties based on the system of formal concepts in the concept lattice. Yao [11] presents a comparative study of rough set theory and formal concept analysis, and gives some concept correspondence relation.

In this paper, the combination of FCA and RS is considered. The notion of information concept lattice is presented and some properties are given. Then the

J.T. Yao et al. (Eds.): RSKT 2007, LNAI 4481, pp. 60–67, 2007.

reduction theory of information concept lattice is presented and the reduction method is obtained. Finally, Information concept lattice is compared with rough set theory and concept lattice.

2 Information Concept Lattice in an IFC

The notion of information systems (IS) provides a convenient basis for the representation of objects in terms of their attributes. Information systems (IS) is a triplet (U, AT, F), where $U = \{x_1,...,x_n\}$ is a non-empty finite set of objects and $AT = \{a_1,...,a_m\}$ is a non-empty finite set of attributes, $F = \{F_a : U \rightarrow V_a, \forall a \in AT\}$ is the relationship set of U and AT, where V_a, a finite set, is called domain of an attribute a. If $F_{a_i}(x) = 0, a_i \in AT$, it represent that the object x doesn't have the attribute a_i. In this paper, IS is normal, i.e. $\forall a \in AT$, $\exists x_1 \neq x_2$, $F_a(x_1) = 0$, $F_a(x_2) \neq 0$; $\forall x \in X$, $\exists a_1 \neq a_2$, $F_{a_1}(x) \neq 0$, $F_{a_2}(x) = 0$. A normal IS is presented in table 1.

Table 1. A normal IS (U, AT, F)

U/AT	a	b	c	d	e
x_1	1	2	0	1	1
x_2	2	1	2	0	0
x_3	1	2	1	0	0
x_4	0	0	0	1	0
x_5	1	2	1	0	0
x_6	2	1	0	2	2

Let (U, AT, F) be a normal IS. $(a,v), a \in AT, v \in V_a - \{0\}$ is called information attribute (IA). The set of some IAs is called information attribute set (IAS). The IAS of an object x is denoted by $\|x\|$: $\|x\| = \{(a, F_a(x)) \mid F_a(x) \neq 0, a \in AT\}$ and the object set of an IA (a,v) is denoted by $\|(a,v)\|$: $\|(a,v)\| = \{x \in U \mid F_a(x) = v\}$. The set of all information attributes is denoted by Y: $Y = \{(a_i, v_{ij}), i = 1,2,...,n; j = 1,2,..., |V_{a_i} - \{0\}|\}$, and (U, AT, F, Y) is called an information formal context (IFC).

Definition 1. Let (U, AT, F, Y) be an IFC. A pair of dual operators are defined by: for $X \subseteq U$ and $\Delta \subseteq Y$,

$$X^* = \left\{ (a_i, v_{ij}) \in Y \;\middle|\; X \subseteq \|(a_i, v_{ij})\| \right\} \tag{1}$$

$$\Delta^* = \left\{ x \in U \;\middle|\; \Delta \subseteq \|x\| \right\} \tag{2}$$

X^* is the IAS of all IA shared by all the objects in X, and Δ^* is the set of all the objects that possess all the information attributes in Δ.

Definition 2. Let (U, AT, F, Y) be an IFC. A pair (X, Δ) is called an information concept (IC) of (U, A, I), if and only if, $X^* = \Delta$ and $X = \Delta^*$, $X \subseteq U$, $B \subseteq A$. X is called the extension and Δ is called the intension of the IC (X, Δ).

Property 1. Let (U, AT, F, Y) be an IFC. The following properties hold: for all X_1, X_2, $X \subseteq U$ and all $\Delta_1, \Delta_2, \Delta \subseteq Y$,

1. $X_1 \subseteq X_2 \Rightarrow X_2^* \subseteq X_1^*$,
2. $\Delta_1 \subseteq \Delta_2 \Rightarrow \Delta_2^* \subseteq \Delta_1^*$,
3. $X \subseteq X^{**}, \Delta \subseteq \Delta^{**}$,
4. $X^* = X^{***}, \Delta^* = \Delta^{***}$,
5. $X \subseteq \Delta^* \Leftrightarrow \Delta \subseteq X^*$,
6. $\left(X_1 \cup X_2\right)^* = X_1^* \cap X_2^*, \left(\Delta_1 \cup \Delta_2\right)^* = \Delta_1^* \cap \Delta_2^*$,
7. $\left(X^{**}, X^*\right)$ and $\left(\Delta^*, \Delta^{**}\right)$ are all information concepts,
8. $(\phi, Y), (U, \phi)$ are information concepts.

Proof. It is obvious to get the result.

Let (X, Δ) be an IC, if $\Delta \neq Y$, the IAS Δ doesn't have $\left(a_i, v_{ij}\right)$ and $\left(a_i, v_{ik}\right)$ $i \in \{1, 2, \ldots, n\}$, $v_{ij} \neq v_{ik}$, $v_{ij}, v_{ik} \in V_{a_i}$ at the same time. It is obvious form the normalization of the ICF (U, AT, F, Y) and IC (ϕ, Y). (X, Δ) is an IC in an ICF, and when $\Delta \neq Y$, Δ is expressed by $\left\{\left(a_i, v_{ij}\right) \mid i \in I \subseteq \{1, 2, \ldots, n\}, a_i \neq a_k \text{ if } i \neq k, i, k \in I\right\}$.

The information concepts of an IFC(U, AT, F, Y) are ordered by:

$$(X_1, \Delta_1) \leq (X_2, \Delta_2) \Leftrightarrow X_1 \subseteq X_2 (\Leftrightarrow \Delta_1 \supseteq \Delta_2) \tag{3}$$

where (X_1, Δ_1) and (X_2, Δ_2) are IC. Furthermore, (X_1, Δ_1) is called a subconcept of (X_2, Δ_2), and (X_2, Δ_2) is called a superconcept of (X_1, Δ_1). The set of all IC forms a complete lattice called the ICL of (U, AT, F, Y) and denoted by $L(U, AT, F, Y)$. The infimum and supremum are given by:

$$(X_1, \Delta_1) \wedge (X_2, \Delta_2) = \left(X_1 \cap X_2, (\Delta_1 \cup \Delta_2)^{**}\right) \tag{4}$$

$$(X_1, \Delta_1) \vee (X_2, \Delta_2) = \left((X_1 \cup X_2)^{**}, \Delta_1 \cap \Delta_2\right) \tag{5}$$

Example 1. There are 9 ICs based on the IFC in table1 denoted by $IC_i(i=1,2,...,9)$ and
$IC_1 = (\phi,Y)$, $IC_2 = (\{x_1\},\{(a,1),(b,2),(d,1),(e,1)\})$,
$IC_3 = (\{x_3,x_5\},\{(a,1),(b,2),(c,1)\})$, $IC_4 = (\{x_2\},\{(a,2),(b,1),(c,2)\})$,
$IC_5 = (\{x_6\},\{(a,2),(b,1),(d,2),(e,2)\})$, $IC_6 = (\{x_1,x_4\},\{(d,1)\})$,
$IC_7 = (\{x_1,x_3,x_5\},\{(a,1),(b,2)\})$, $IC_8 = (\{x_2,x_6\},\{(a,2),(b,1)\})$, $IC_9 = (U,\phi)$ .The   ICL   is
presented in Fig. 1.

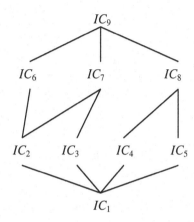

Fig. 1. The ICL of (U, AT, F, Y)

3 An Approach to Reduction in Information Concept Lattice

In this section, an approach to reduction is considered using the idea of discernibility
matrix [14].

Definition 3. Let $L(U,AT_1,F_1,Y_1)$ and $L(U,AT_2,F_2,Y_2)$ be two ICLs. If for any
$(X,\Delta) \in L(U,AT_2,F_2,Y_2)$ there   exists $(X',\Delta') \in L(U,AT_1,F_1,Y_1)$ such
that $X' = X$, $L(U,AT_1,F_1,Y_1)$ is said to be finer than $L(U,AT_2,F_2,Y_2)$, denoted by:

$$L(U,AT_1,F_1,Y_1) \le L(U,AT_2,F_2,Y_2) \tag{6}$$

If $L(U,AT_1,F_1,Y_1) \le L(U,AT_2,F_2,Y_2)$ and $L(U,AT_2,F_2,Y_2) \le L(U,AT_1,F_1,Y_1)$,
these two ICLs are said to be isomorphic to each other, denoted by:

$$L(U,AT_1,F_1,Y_1) \cong L(U,AT_2,F_2,Y_2) \tag{7}$$

Theorem 1. Let (U, AT, F, Y) be an IFC. For any $B \subseteq AT, B \neq \phi$,
$L(U,AT,F,Y) \le L(U,B,F_B,Y_B)$ holds.

Definition 4. Let (U,AT,F,Y) be an ICF. If there exists an attribute set $B \in AT$ such that $L(U,B,F_B,Y_B) \cong L(U,AT,F,Y)$, then B is called a consistent set of (U,AT,F,Y). And further, if $L(U,B-\{b\},F_{B-\{b\}},Y_{B-\{b\}}) \neq L(U,AT,F,Y)$ for all $b \in B$, then B is called a reduction of (U,AT,F,Y). The intersection of all the reductions of (U,AT,F,Y) is called the core of (U,AT,F,Y).

Definition 5. Let $L(U,AT,F,Y)$ be an ICL. $(X_i,\Delta_i),(X_j,\Delta_j) \in L(U,AT,F,Y)$, $D((X_i,\Delta_i),(X_j,\Delta_j)) = \{a|\ (a,v) \in \Delta_i \triangle \Delta_j\}$ is called the discernibility attributes set between (X_i,Δ_i) and (X_j,Δ_j), $\Lambda = \left(D((X_i,\Delta_i),(X_j,\Delta_j)),(X_i,\Delta_i),(X_j,\Delta_j) \in L(U,AT,F,Y) \right)$ is called the discernibility matrix of the ICL $L(U, AT, F, Y)$.

In the discernibility matrix Λ, only those non-empty elements are useful to reduction. We also denote the set of non-empty elements in the matrix by Λ:

$$\Lambda = \left\{ D\left((X_i,\Delta_i),(X_j,\Delta_j)\right) \middle|\ (X_i,\Delta_i),(X_j,\Delta_j) \in L(U,AT,F,Y),(X_i,\Delta_i) \neq (X_j,\Delta_j) \right\} \quad (8)$$

Theorem 2. Let (U, AT, F, Y) be an ICF. For any $B \subseteq AT$ such that $B \neq \phi$, the following assertions are equivalent:

1. B is a consistent set.
2. For all $(X_i,\Delta_i),(X_j,\Delta_j) \in L(U,AT,F,Y)$, if $(X_i,\Delta_i) \neq (X_j,\Delta_j)$, $\Delta_i \cap Y_B \neq \Delta_j \cap Y_B$.
3. For all $(X_i,\Delta_i),(X_j,\Delta_j) \in L(U,AT,F,Y)$, if $D((X_i,\Delta_i),(X_j,\Delta_j)) \neq \phi$, then
$$B \cap D((X_i,\Delta_i),(X_j,\Delta_j)) \neq \phi.$$

Proof. $1 \Rightarrow 2$. Suppose $\forall (X_i,\Delta_i),(X_j,\Delta_j) \in L(U,AT,F,Y)$ and $(X_i,\Delta_i) \neq (X_j,\Delta_j)$. Since B is a consistent set, we have $L(U,B,F_B,Y_B) \leq L(U,AT,F,Y)$. There exist $\Omega_i,\Omega_j \in Y_B$ such that $(X_i,\Omega_i),(X_j,\Omega_j) \in L(U,B,F_B,Y_B)$ and $(X_i,\Omega_i) \neq (X_j,\Omega_j)$. Hence, $\Omega_i = X_i^{*_{Y_B}} = X_i^* \cap Y_B = \Delta_i \cap Y_B$, $\Omega_j = X_j^{*_{Y_B}} = X_j^* \cap Y_B = \Delta_j \cap Y_B$ and $\Omega_i \neq \Omega_j$. Thus, $\Delta_i \cap Y_B \neq \Delta_j \cap Y_B$.

$2 \Rightarrow 3$. For $(X_i,\Delta_i),(X_j,\Delta_j) \in L(U,AT,F,Y)$, and $\Delta_i \cap Y_B \neq \Delta_j \cap Y_B$. There exists an IA (a,v), $(a,v) \in Y_B \cap \Delta_i \cap \overline{\Delta}_j$ or $(a,v) \in Y_B \cap \Delta_j \cap \overline{\Delta}_i$. Thus, we have:

$$B \cap D((X_i,\Delta_i),(X_j,\Delta_j)) = B \cap \left[\{a|\ (a,v) \in (\Delta_i \cup \Delta_j)\} - \{a|\ (a,v) \in (\Delta_i \cap \Delta_j)\} \right]$$
$$= B \cap \left[\{a|\ (a,v) \in (\Delta_i \cup \Delta_j)\} \cap \{a|\ (a,v) \in (\overline{\Delta}_i \cup \overline{\Delta}_j)\} \right]$$
$$= \left(B \cap \{a|\ (a,v) \in \Delta_i\} \cap \{a|\ (a,v) \in \overline{\Delta}_j\} \right) \cup \left(B \cap \{a|\ (a,v) \in \Delta_j\} \cap \{a|\ (a,v) \in \overline{\Delta}_i\} \right) \neq \phi.$$

$3 \Rightarrow 2$. Suppose $(X_i, \Delta_i), (X_j, \Delta_j) \in L(U, AT, F, Y)$ and $B \cap D((X_i, \Delta_i), (X_j, \Delta_j)) \neq \phi$.

Hence, there exists $a \in B$ such that $a \in D((X_i, \Delta_i), (X_j, \Delta_j))$, i.e. $\exists (a, v)$, $(a, v) \in \Delta_i$ or $(a, v) \in \Delta_j$ and $(a, v) \notin \Delta_i \cap \Delta_j$. Thus, if $(a, v) \in \Delta_i, (a, v) \notin \Delta_j$, $(a, v) \in \Delta_i \cap Y_B$ and $(a, v) \notin \Delta_j \cap Y_B$; if $(a, v) \in \Delta_j$ and $(a, v) \notin \Delta_i$, $(a, v) \in \Delta_j \cap Y_B, (a, v) \notin \Delta_i \cap Y_B$. It follows that $\Delta_i \cap Y_B \neq \Delta_j \cap Y_B$.

$2 \Rightarrow 1$. If for all $(X, \Delta) \in L(U, AT, F, Y)$, $(X, Y_B \cap \Delta) \in L(U, B, F_B, Y_B)$ hold, $L(U, B, F_B, Y_B) \leq L(U, AT, F, Y)$. Thus, B is a consistent set. We need only to prove that for $X^{*_{Y_B}} = \Delta \cap Y_B$ and $(\Delta \cap Y_B)^* = X$ hold. Firstly, we have $X^{*_{Y_B}} = X^* \cap Y_B = \Delta \cap Y_B$. Secondly suppose $(\Delta \cap Y_B)^* \neq X$, since $((\Delta \cap Y_B)^*, (\Delta \cap Y_B)^{**}) \in L(U, AT, F, Y)$, we have $(X, \Delta) \neq ((\Delta \cap Y_B)^*, (\Delta \cap Y_B)^{**})$ which implies that $\Delta \cap Y_B \neq (\Delta \cap Y_B)^{**} \cap Y_B$. On the one hand, $\Delta \cap Y_B \subseteq \Delta \Rightarrow (\Delta \cap Y_B)^* \supseteq \Delta^* = X \Rightarrow (\Delta \cap Y_B)^{**} \subseteq X^* = \Delta \Rightarrow (\Delta \cap Y_B)^{**} \cap Y_B \subseteq \Delta \cap Y_B$. On the other hand, $\Delta \cap Y_B \subseteq (\Delta \cap Y_B)^{**} \Rightarrow \Delta \cap Y_B \cap Y_B \subseteq (\Delta \cap Y_B)^{**} \cap Y_B$. Thus, $\Delta \cap Y_B = (\Delta \cap Y_B)^{**} \cap Y_B$, which is a contradiction. Thus, $(\Delta \cap Y_B)^* = X$.

Theorem 2 shows that to find a reduction in an IFC is to find a minimal subset B of attributes set such that $B \cap D((X_i, \Delta_i), (X_j, \Delta_j)) \neq \phi$.

Example 2. Reductions of ICL are presented based on discernibility matrix. There is a non-empty elements set in the discernibility matrix:

$$\Lambda = \{\{c\}, \{d\}, \{a, b\}, \{d, e\}, \{a, b, c\}, \{a, b, d\}, \{a, b, e\}, \{c, d, e\}, \{a, b, c, d\}, \{a, b, d, e\}, AT\}.$$

$B_1 = \{a, c, d\}$, $B_2 = \{b, c, d\} \subseteq AT$ are considered. B_1 satisfies $B_1 \cap H \neq \phi (\forall H \in \Lambda)$, and for every subset of B_1, there exists one set which intersect the subset is empty. For example, $\{c, d\}$ is a subset of B_1 but $\{c, d\} \cap \{a, b\} = \phi$. It is similar to $B_2 = \{b, c, d\}$. Hence, $\{a, c, d\}$ and $\{b, c, d\}$ are reductions of the IFL.

4   Relationships Between ICL and RS

Let (U, AT, F, Y) be a normal IFC. The IAS is extended by $Y_0 = Y \cup \{(a_i, 0), a_i \in AT, i = 1, 2, \dots n\}$ and an IS is denoted by (U, AT, F, Y_0). Respectively, $\|x\| = \{(a, F_a(x)) \mid a \in AT\}$ and $\|(a, v)\| = \{x \in U \mid F_a(x) = v\}$, which is denoted all the objects that have the IA (a, v).

Let (U, AT, F, Y_0) be an IS. According to the definitions and the properties in section2, $L(U, AT, F, Y_0)$ is a complete lattice which denotes all the information concepts in (U, AT, F, Y_0).

Theorem 3. Let (U, AT, F, Y_0) be an IS. $(I_{AT}(x), \|x\|)$ is an IC.

Proof. By the definition of $I_{AT}(x)$ and $\|x\|$, $(I_{AT}(x))^* = \|x\|, (\|x\|)^* = I_{AT}(x)$.

The IC set denoted by $O_L = \{(I_{AT}(x_i), \|x_i\|), i = 1, 2, \ldots, n\} \subseteq L(U, AT, F, Y_0)$, then we have the theorem below:

Theorem 4. The set of the extensions of all IC in O_L forms a partition of U.

Proof. Firstly, $ex(X_i, \Delta_i) \neq \phi$ in IC set O_L. Secondly, $X_i = X_j$ if $(X_i, \Delta_i) = (X_j, \Delta_j)$ and $X_i \neq X_j$ if $(X_i, \Delta_i) \neq (X_j, \Delta_j)$, and then we get $X_i \cap X_j = \phi$ from $I_{AT}(x_i) \cap I_{AT}(x_j) = \phi, i \neq j$. Finally, it is obvious that $\cup I_{AT}(x_i) = U$. Thus, the set of extensions of all IC in O_L forms a partition of U.

We know that it is consistent between the set of extension of all IC set Y_0 and the partition of U.

5   Relationships Between ICL and CL

Let (U, AT, F, Y) be an IFC. When $V_a = \{0, 1\}, a \in AT$, an object x has an attribute a if $F_a(x) = 1$, else the object x doesn't have the attribute. The relations $I \subseteq U \times AT$ are introduced and an object x which has an attribute a is denoted by xIa. All IC set Y can be denoted by $Y = \{(a_i, 1), i = 1, 2, \ldots, n\}$ in IFC (U, AT, F, Y). As every IA $(a_i, 1)$ is unique to attribute a_i, we rewrite the set Y by $Y = \{a_i, i = 1, 2, \ldots, m\}$ which $Y = AT$.

A pair of dual operators are considered, for $X \subseteq U$, $\Delta \subseteq Y (i.e. \Delta \subseteq AT)$:

$$X^* = \{(a_i, v_{ij}) \in Y \mid X \subseteq \|(a_i, v_{ij})\|\}, \Delta^* = \{x \in U \mid \Delta \subseteq \|x\|\}. \text{ As } (a_i, v_{ij}) \text{ is equal to } a_i,$$

we can get $X^* = \{a \in Y \mid \forall x, xIa\}, \Delta^* = \{x \in U \mid \forall a, xIa\}$. It is obvious that when $V_a = \{0, 1\}, a \in AT$ in ICF, ICF is equal to FC and ICL is equal to CL.

6   Conclusions

This paper presents a new notion for knowledge discovery in information systems. The concept of information concept lattice is introduced to information systems. The reduction theory of information concept lattice is presented and the approach to reduction is given using discernibility matrix. This paper extends the concept of the formal context and presents the consistent structure of these reduction approaches between rough set theory and concept lattice.

References

1. Pawlak, Z.: Rough Set. International Journal of Computer and Information Sciences, 11(5) (1982) 341-356
2. Greco, S., Matarazzo B., Slowinski R.: Rough Sets Methodology for Sorting Problems in Presence of Multiple Attributes and Criteria. European Journal of Operational Research, 138 (2002) 247-259
3. Kryszkiewicz, M.: Rules in incomplete information systems. Information Sciences, 113 (1999) 271-292
4. Zhang,W.X., Leung,Y., Wu,W.Z.: Information Systems and Knowledge Discovery. Science Press, Beijing (2003)
5. Wille, R.: Restructuring Lattice Theory: An Approach Based on Hierarchies of Concepts. In: Ordered Sets, Rival, I. (ed.), Reidel, Dordrecht-Boston (1982) 445-470
6. Ganter, B., Wille, R.: Formal Concept Analysis. Mathematical Foundations, Springer-Verlag, New York (1999)
7. Zhang, W.X., L. Wei, J.J. Qi: Attribute Reduction in Concept Lattice Based on Discernibility Matrix, RSFDGrC2005, Lecture Notes in Artificial Intelligence, 3642 (2005) 157-165
8. Oosthuizen, G.D.: The Application of Concept Lattice to Machine Learning. Technical Report, University of Pretoria, South Africa (1996)
9. Kent, R.E.: Rough Concept Analysis: A Synthesis of Rough Sets and Formal Concept Analysis, Fundamenta Informaticae, 27 (1996) 169-181
10. Saquer, J., Deogun, J.S.: Formal Rough Concept Analysis, New Directions in Rough Sets, Data Mining and Granular-Soft Computer Science. Springer, Berlin, (1999) 91-99
11. Yao, Y.Y.: A Comparative Study of Formal Analysis and Rough Set Theory in Data Analysis. Proceedings of Rough Sets and Current Trends in Computing (2004) 59-68
12. Yao, Y.Y.: Concept Lattices in Rough Set Theory. Proceedings of Annual Meeting of the North American Fuzzy Information Processing Society (NAFIPS2004) (2004) 796-801
13. Yao, Y.Y., Chen, Y.H.: Rough Set Approximations in Formal Concept Analysis, Proceedings of Annual Meeting of the North American Fuzzy Information Processing Society (NAFIPS2004) (2004) 73-78
14. Skowron, A., Rauszer, C.: The Discernibility Matrices and Functions in Information Systems. In: Slowinski, R. (ed.), Intelligent Decision Support: Handbook of Applications and Advances of the Rough Set Theory, Kluwer Academic Publishers, Dordrecht (1992) 331-362

Homomorphisms Between Relation Information Systems

Changzhong Wang[1], Congxin Wu[1], and Degang Chen[2]

[1] Department of Mathematics, Harbin Institute of Technology
Harbin, Heilongjiang, 151000, P.R.China
[2] School of Mathematics & Physics, North China Electric Power University
Beijing, 102206, P.R.China
changzhongwang@126.com

Abstract. Information system is one of the important mathematical models in the field of artificial intelligence. The concept of homomorphism is very useful to study the communication between two information systems. In this paper, some properties of relation information systems under homomorphisms are investigated. The concept of a relation mapping between two universes is proposed in order to construct a binary relation on one universe according to the given binary relation on the other universe. The main properties of the mapping are studied. Furthermore, the notion of homomorphism of information systems based on arbitrary binary relations is proposed, and it is proved that the reductions of the original system and image system are equivalent to each other under the condition of homomorphism.

Keywords: Consistent functions, relation mappings, relation information systems, homomorphism, reduction.

1 Introduction

Rough set theory [7], proposed by Pawlak, is an excellent tool for data analysis with important applications in data mining and knowledge discovery. A concept related to rough set is information system. In fact, most applications based on rough set theory, such as classification, decision support and knowledge discovery problems, can fall into the knowledge representation model, i.e. an information system. In recent years, many topics on information systems have been widely investigated by many scholars [1-6,9-11].

The theory of rough sets deals with the approximation of an arbitrary subset of a universe by two definable or observable subset called lower and upper approximations. However, lower and upper approximations are not primitive notions. They are constructed from other concepts, such as binary relations on a universe, partitions and coverings of a universe, and approximation space. For an information system, it can be seen as a composition of some approximation spaces on the same universe. The communication between two information systems is a very important topic in the field of artificial intelligence. In mathematics, it can be explained as a mapping between two information systems.

J.T. Yao et al. (Eds.): RSKT 2007, LNAI 4481, pp. 68–75, 2007.

The notion of homomorphism on information systems as a kind of tool to study the relationship between two information systems was introduced by Grzymala Busse in [1,2]. A homomorphism can be regarded as a special communication between two information systems. Image system is seen as an explanation system of the original system. A homomorphism on information systems is very useful for aggregating sets of objects, attributes, and descriptors of the original system. The notions of superfluousness and reducts of an information system are central notions in decision making, data analysis, reasoning about data and other subfields of artificial intelligence[2-4,6,8,10,11]. In [2], the authors depicted the conditions which make an information system to be selective in terms of endomorphism of the system. In [4], with algebraic approach, the authors discussed the features of superfuousness and reducts of an information system under some homomorphisms. .

However, the requirement of an indiscernibilty relation or a partition in rough set theory is a condition that limits the application domain of rough set theory. So several important generalizations were proposed to solve this problem. One of these generalizations is to relax an equivalence relation to general binary relation [5]. The work in our paper represents a new contribution to the development of the theory of homomorphism between information systems. We develop a method for defining an arbitrary binary on a universe according to a relation on another universe. In this sense, our method is a mechanism for communicating between two information systems. We define the concept of homomorphism between two information systems based on arbitrary binary relations. Under the condition of the homomorphism, some characters of relation operations in the original system and some structure features of the original system are guaranteed in explanation system.

2   Consistent Function and Its Properties

Let U and V be finite and nonempty universes. The class of all binary relations on U (respectively, on V) will be denoted by $\Re(U)$ (respectively, by $\Re(V)$). Let $R \in \Re(U)$, the successor neighborhood of $x \in U$ with respect to R will be denoted by $R_s(x)$, that is, $R_s(x) = \{y \in U : xRy\}$. In this section, we introduce the concepts of consistent functions and investigate their main properties which will be used in the following sections.

Definition 2.1. Let U and V be finite and nonempty universes, $f : U \to V$ a mapping from U to V, and R a binary relation on U. Let

$$[x]_f = \{y \in U : f(y) = f(x)\}, [x]_R = \{y \in U : R_s(y) = R_s(x)\}.$$

Then both of $\left\{[x]_f : x \in U\right\}$ and $\{[x]_R : x \in U\}$ are partitions on U. If $[x]_f \subseteq R_s(y)$ or $[x]_f \cap R_s(y) = \emptyset$ for any $x, y \in U$, then f is called a type-1 consistent function with respect to R on U. If $[x]_f \subseteq [x]_R$ for any $x \in U$, then f is called a type-2 consistent function with respect to R on U.

From definition 2.1, an injection is trivially both a type-1 and a type-2 consistent function.

Theorem 2.2. Let $f : U \to V, R \in \Re(U)$. If f is a type-1 consistent function with respect to R on U, then $\forall x \in U, f^{-1}(f(R_s(x))) = R_s(x)$.

Proof. Because $f^{-1}(f(R_s(x))) \supseteq R_s(x)$ for any $x \in U$ is always true, we only need to prove that $f^{-1}(f(R_s(x))) \subseteq R_s(x), \forall x \in U$.

For any $x_1 \in f^{-1}(f(R_s(x)))$, we have $f(x_1) \in f(R_s(x))$, which implies $\exists x_2 \in R_s(x)$ such that $f(x_1) = f(x_2)$. Since $[x_2]_f = \{y \in U : f(y) = f(x_2)\}$, thus, $x_2 \in [x_1]_f$, which implies $[x_1]_f \cap R_s(x) \neq \emptyset$. Since f is a type-1 consistent function with respect to R on U, we must have $x_1 \in [x_2]_f \subseteq R_s(x)$. Thus $f^{-1}(f(R_s(x))) \subseteq R_s(x)$. Therefore $f^{-1}(f(R_s(x))) = R_s(x), \forall x \in U$.

Corollary 2.3. Let $f : U \to V, R_1, R_2, \cdots, R_n \in \Re(U)$. If f is a type-1 consistent function with respect to each relation $R_i (i \leq n)$ on U, then $\forall x \in$

$$U, f^{-1}\left(f\left(\bigcap_{i=1}^{n} (R_i)_s(x)\right)\right) = \bigcap_{i=1}^{n} (R_i)_s(x).$$

Proof. It is similar to the proof Theorem 2.2.

Theorem 2.4. Let $f : U \to V, R_1, R_2 \in \Re(U)$. If f is a type-1 consistent function with respect to R_1 and R_2 on U, then $\forall x \in U, f((R_1 \cap R_2)_s(x)) = f((R_1)_s(x)) \cap f((R_2)_s(x))$.

Proof. Since $f((R_1 \cap R_2)_s(x)) \subseteq f((R_1)_s(x)) \cap f((R_2)_s(x))$ for any $x \in U$ is always true, we only need to prove the inverse inclusion for any $x \in U$.

For any $y \in f((R_1)_s(x)) \cap f((R_2)_s(x))$, we have $y \in f((R_1)_s(x))$ and $y \in f((R_2)_s(x))$. Thus $f^{-1}(y) \subseteq f^{-1}(f((R_1)_s(x)))$ and $f^{-1}(y) \subseteq f^{-1}(f((R_2)_s(x)))$. Since f is a type-1 consistent function with respect to R_1 and R_2 on U, by Theorem 2.2, $f^{-1}(y) \subseteq (R_1)_s(x)$ and $f^{-1}(y) \subseteq (R_2)_s(x)$. Hence

$$f^{-1}(y) \subseteq (R_1)_s(x) \cap (R_2)_s(x) = (R_1 \cap R_2)_s(x)$$

This implies $y \in f((R_1 \cap R_2)_s(x))$. Therefore $f((R_1 \cap R_2)_s(x)) \supseteq f((R_1)_s(x)) \cap f((R_2)_s(x))$. It follows that $\forall x \in U, f((R_1 \cap R_2)_s(x)) = f((R_1)_s(x)) \cap f((R_2)_s(x))$.

By Corollary 2.3 and Theorem 2.4, we directly get the following corollary.

Corollary 2.5. Let $f : U \to V, R_1, R_2, \cdots, R_n \in \Re(U)$. If f is a type-1 consistent function with respect to each relation R_i on U, then $\forall x \in U, f\left(\bigcap_{i=1}^{n} (R_i)_s(x)\right) = \bigcap_{i=1}^{n} f((R_i)_s(x))$.

3 Relation Mapping and Its Properties

In this section, we define the notions of relation mappings and study their main properties.

Definition 3.1. Let $f : U \to V, x| \to f(x) \in V, x \in U$. f can induce a mapping from $\Re(U)$ to $\Re(V)$ and a mapping from $\Re(V)$ to $\Re(U)$, that is,

$$\hat{f} : \Re(U) \to \Re(V), R| \to \hat{f}(R) \in \Re(V), \forall R \in \Re(U);$$

$$\hat{f}(R) = \bigcup_{x \in U} \{f(x) \times f(R_s(x))\}.$$

$$\hat{f}^{-1} : \Re(V) \to \Re(U), T| \to \hat{f}^{-1}(T) \in \Re(U), \forall T \in \Re(V);$$

$$\hat{f}^{-1}(T) = \bigcup_{y \in V} \{f^{-1}(y) \times f^{-1}(T_s(y))\}.$$

Then $\hat{f}$ and $\hat{f}^{-1}$ are called relation mapping and inverse relation mapping induced by f respectively; $\hat{f}(R)$ and $\hat{f}^{-1}(T)$ are called binary relations induced by f on V and U respectively. In the subsequent discussion, we simply denote $\hat{f}$ and $\hat{f}^{-1}$ by f and f^{-1} respectively.

Theorem 3.2. Let $f : U \to V, R_1, R_2 \in \Re(U)$. If f is both type-1 and type-2 consistent with respect to R_1 and R_2, then $f(R_1 \cap R_2) = f(R_1) \cap f(R_2)$.

Proof.

$$f(R_1 \cap R_2) = \bigcup_{x \in U} \{f(x) \times f((R_1 \cap R_2)_s(x))\}$$
$$\subseteq \bigcup_{x \in U} \{f(x) \times (f((R_1)_s(x)) \cap f((R_2)_s(x)))\}$$
$$= \bigcup_{x \in U} \{f(x) \times f((R_1)_s(x)) \cap f(x) \times f((R_2)_s(x))\}$$
$$\subseteq \left(\bigcup_{x \in U} \{f(x) \times f((R_1)_s(x))\} \right) \cap \left(\bigcup_{x \in U} \{f(x) \times f((R_2)_s(x))\} \right)$$
$$= f(R_1) \cap f(R_2).$$

Next, we are to prove the inverse inclusion.

Let $(y_1, y_2) \in f(R_1) \cap f(R_2)$. Then $(y_1, y_2) \in f(R_1)$ and $(y_1, y_2) \in f(R_2)$. By the definition of $f(R_1)$, there exists $x_1 \in U$ such that $(y_1, y_2) \in f(x_1) \times f((R_1)_s(x_1))$, which implies $y_1 = f(x_1)$ and $y_2 \in f((R_1)_s(x_1))$. Similarly, there exists $x_2 \in U$ such that $(y_1, y_2) \in f(x_2) \times f((R_2)_s(x_2))$, which implies $y_1 = f(x_2)$ and $y_2 \in f((R_2)_s(x_2))$. Thus $f(x_1) = f(x_2)$ and $y_2 \in f((R_1)_s(x_1)) \cap f((R_2)_s(x_2))$. Since f is a type-2 consistent function with respect to R_1 and R_2, we have $(R_1)_s(x_1) = (R_1)_s(x_2)$ and $(R_2)_s(x_1) = (R_2)_s(x_2)$. Hence by Theorem 2.4, $y_2 \in f((R_1)_s(x_2)) \cap f((R_2)_s(x_2)) = f((R_1)_s(x_2) \cap (R_2)_s(x_2)) = f((R_1 \cap R_2)_s(x_2))$. Then we can conclude that $(y_1, y_2) = (f(x_2), y_2) \in f(x_2) \times f((R_1 \cap R_2)_s(x_2)) \subseteq f(R_1 \cap R_2)$. Thus $f(R_1 \cap R_2) \supseteq f(R_1) \cap f(R_2)$. Therefore $f(R_1 \cap R_2) = f(R_1) \cap f(R_2)$.

Corollary 3.3. Let $f : U \to V, R_1, R_2, \cdots, R_n \in \Re(U)$. If f is both type-1 and type-2 consistent with respect to each relation R_i $(i \le n)$, then $f\left(\bigcap_{i=1}^{n} R_i\right) = \bigcap_{i=1}^{n} f(R_i)$.

Proof. It is similar to the proof of Theorem 3.2.

Theorem 3.4. Let $f : U \to V, R \in \Re(U)$. If f is both type-1 and type-2 consistent with respect to R, then $f^{-1}(f(R)) = R$.

Proof. Let $(x_1, x_2) \in R$, namely, $x_2 \in R_s(x_1)$. Thus $f(x_2) \in f(R_s(x_1))$. By the definition of $f(R)$, we have that $(f(x_1), f(x_2)) \in f(R)$. Let $y_1 = f(x_1)$ and $y_2 = f(x_2)$, then $y_2 \in f(R)_s(y_1)$. Thus $f^{-1}(y_2) \subseteq f^{-1}(f(R)_s(y_1))$. It follows that $f^{-1}(y_1) \times f^{-1}(y_2) \subseteq f^{-1}(y_1) \times f^{-1}(f(R)_s(y_1)) \subseteq f^{-1}(f(R))$, which implies $(x_1, x_2) \in f^{-1}(f(R))$. Therefore $f^{-1}(f(R)) \supseteq R$. Next, we are to prove that $f^{-1}(f(R)) \subseteq R$.

Let $(x_1, x_2) \in f^{-1}(f(R))$, then there exists $y_1 \in V$ such that $(x_1, x_2) \in f^{-1}(y_1) \times f^{-1}((f(R))_s(y_1))$. This implies $x_1 \in f^{-1}(y_1)$ and $x_2 \in f^{-1}((f(R))_s(y_1))$. Hence $y_1 = f(x_1)$ and $f(x_2) \in (f(R))_s(y_1)$. Let $y_2 = f(x_2)$, then $(y_1, y_2) \in f(R)$. By the definition of $f(R)$, there exists $x_3 \in U$ such that $(y_1, y_2) \in f(x_3) \times f(R_s(x_3))$. This implies $y_1 = f(x_3)$ and $y_2 \in f(R_s(x_3))$. Hence $f(x_1) = f(x_3)$ and $f(x_2) \in f(R_s(x_3))$. Since f is type-2 consistent with respect to R, we have $R_s(x_3) = R_s(x_1)$ and $f(x_2) \in f(R_s(x_1))$. Hence $x_2 \in f^{-1}(f(R_s(x_1)))$. Again, since f is type-1 consistent with respect to R, by Theorem 2.2, $x_2 \in f^{-1}(f(R_s(x_1))) = R_s(x_1)$. Thus $(x_1, x_2) \in R$. It follows that $f^{-1}(f(R)) \subseteq R$. Therefore $f^{-1}(f(R)) = R$.

Corollary 3.5. Let $f : U \to V, R_1, R_2, \cdots, R_n \in \Re(U)$. If f is both type-1 and type-2 consistent with respect to each relation R_i $(i \le n)$, then

$$f^{-1}\left(f\left(\bigcap_{i=1}^{n} R_i\right)\right) = \left(\bigcap_{i=1}^{n} R_i\right).$$

Proof. It is similar to the proof of Theorem 3.4.

4 Homomorphism Between Relation Information Systems and Its Properties

By means of the results of the above sections, we introduce the notion of a homomorphism between two information systems and show that reductions of the original system and image system are equivalent to each other.

Definition 4.1. Let U and V be finite universes, $f : U \to V$ a mapping from U to V, and $\mathbf{R} = \{R_1, R_2, \cdots, R_n\}$ a family of binary relations on U, let $f(\mathbf{R}) = \{f(R_1), f(R_2), \cdots, f(R_n)\}$. Then the pair $(U, \mathbf{R})$ is referred to as a relation information system, and the pair $(V, f(\mathbf{R}))$ is referred to as a $f-$induced relation information system of $(U, \mathbf{R})$.

By Theorem 3.2, we can introduce the following concept.

Definition 4.2. Let $(U, \mathbf{R})$ be a relation information system and $(V, f(\mathbf{R}))$ a $f-$ induced relation information system of $(U, \mathbf{R})$. If $\forall R_i \in \mathbf{R}$, f is both type-1 and type-2 consistent with respect to R_i on U, then f is referred to as a homomorphism from $(U, \mathbf{R})$ to $(V, f(\mathbf{R}))$.

Remark. After the notion of homomorphism is introduced, all the theorems and corollaries in the above sections may be seen as the properties of homomorphism.

Definition 4.3. Let $(U, \mathbf{R})$ be a relation information system. The subset $\mathbf{P} \subseteq \mathbf{R}$ is referred to as a reduct of $\mathbf{R}$ if $\mathbf{P}$ satisfies the following conditions:
  (1) $\cap \mathbf{P} = \cap \mathbf{R}$;   (2) $\forall R_i \in \mathbf{P}$, $\cap \mathbf{P} \subset \cap (\mathbf{P} - R_i)$.

Theorem 4.4. Let $(U, \mathbf{R})$ be a relation information system, $(V, f(\mathbf{R}))$ a $f-$ induced relation information system of $(U, \mathbf{R})$, and f a homomorphism from $(U, \mathbf{R})$ to $(V, f(\mathbf{R}))$. Then $\mathbf{P} \subseteq \mathbf{R}$ is a reduct of $\mathbf{R}$ if and only if $f(\mathbf{P})$ is a reduct of $f(\mathbf{R})$.

Proof. $\Rightarrow$ Since $\mathbf{P}$ is a reduct of $\mathbf{R}$, we have $\cap \mathbf{P} = \cap \mathbf{R}$. Hence $f(\cap \mathbf{P}) = f(\cap \mathbf{R})$. Since f is a homomorphism from $(U, \mathbf{R})$ to $(V, f(\mathbf{R}))$, by Definition 4.2 and Corollary 3.3, we have $\cap f(\mathbf{P}) = \cap f(\mathbf{R})$. Assume that $\exists R_i \in \mathbf{P}$ such that $\cap (f(\mathbf{P}) - f(R_i)) = \cap f(\mathbf{P})$. Because $f(\mathbf{P}) - f(R_i) = f(\mathbf{P} - R_i)$, we have that $\cap (f(\mathbf{P}) - f(R_i)) = \cap f(\mathbf{P} - R_i) = \cap f(\mathbf{P}) = \cap f(\mathbf{R})$. Similarly, by Definition 4.2 and Corollary 3.3, it follows that $f(\cap (\mathbf{P} - R_i)) = f(\cap \mathbf{R})$. Thus $f^{-1}(f(\cap (\mathbf{P} - R_i))) = f^{-1}(f(\cap \mathbf{R}))$. By Definition 4.2 and Corollary 3.5, $\cap (\mathbf{P} - R_i) = \cap \mathbf{R}$. This is a contradiction to that $\mathbf{P}$ is a reduct of $\mathbf{R}$.
  $\Leftarrow$ Let $f(\mathbf{P}) \subseteq f(\mathbf{R})$ be a reduct of $f(\mathbf{R})$, then $\cap f(\mathbf{P}) = \cap f(\mathbf{R})$. Since f a homomorphism from $(U, \mathbf{R})$ to $(V, f(\mathbf{R}))$, by Definition 4.2 and Corollary 3.3, we have $f(\cap \mathbf{P}) = f(\cap \mathbf{R})$. Hence $f^{-1}(f(\cap \mathbf{P})) = f^{-1}(f(\cap \mathbf{R}))$. By Definition 4.2 and Corollary 3.5, $\cap \mathbf{P} = \cap \mathbf{R}$. Assume that $\exists R_i \in \mathbf{P}$ such that $\cap (\mathbf{P} - R_i) = \cap \mathbf{R}$, then $f(\cap (\mathbf{P} - R_i)) = f(\cap \mathbf{R})$. Again, by Definition 4.2 and Corollary 3.3, we have $\cap f(\mathbf{P} - R_i) = \cap f(\mathbf{R})$. Hence $\cap (f(\mathbf{P}) - f(R_i)) = \cap f(\mathbf{R})$. This is a contradiction to that $f(\mathbf{P})$ is a reduct of $f(\mathbf{R})$. This completes the proof of this theorem.

By Theorem 4.4, we immediately get the following corollary.

Corollary 4.5. Let $(U, \mathbf{R})$ be a relation information system, $(V, f(\mathbf{R}))$ a $f-$ induced relation information system of $(U, \mathbf{R})$, and f a homomorphism from $(U, \mathbf{R})$ to $(V, f(\mathbf{R}))$. Then $\mathbf{P} \subseteq \mathbf{R}$ is is superfluous in $\mathbf{R}$ if and only if $f(\mathbf{P})$ is superfluous in $f(\mathbf{R})$.

The following example is employed to illustrate our idea in this paper.

Example 4.6. Let $(U, \mathbf{R})$ be a relation information system, where $U = \{x_1, x_2, \cdots, x_{10}\}$, $\mathbf{R} = \{R_1, R_2, R_3\}$,
$R_1 = \{(x_2, x_3), (x_2, x_6), (x_5, x_2), (x_5, x_3), (x_5, x_6), (x_5, x_8), (x_7, x_{12}),$
$(x_7, x_{13}), (x_7, x_{14}), (x_7, x_{15}), (x_8, x_3), (x_8, x_6) (x_9, x_{12}), (x_9, x_{13}),$
$(x_9, x_{14}), (x_9, x_{15}), (x_{10}, x_{12}), (x_{10}, x_{13}), (x_{10}, x_{14}), (x_{10}, x_{15})\},$

$$R_2 = (x_1, x_{12}), (x_1, x_{13}), (x_1, x_{14}), (x_1, x_{15}), (x_2, x_3), (x_2, x_6), (x_4, x_{12}),$$
$$(x_4, x_{13}), (x_4, x_{14}), (x_4, x_{15}), (x_5, x_2), (x_5, x_8), (x_8, x_3), (x_8, x_6),$$
$$(x_{11}, x_{12}), (x_{11}, x_{13}), (x_{11}, x_{14}), (x_{11}, x_{15})\},$$
$$R_3 = (x_1, x_7), (x_1, x_9), (x_1, x_{10}), (x_2, x_3), (x_2, x_6), (x_4, x_7), (x_4, x_9),$$
$$(x_4, x_{10}), (x_5, x_3), (x_5, x_6), (x_8, x_3), (x_8, x_6), (x_{11}, x_7), (x_{11}, x_9),$$
$$(x_{11}, x_{10}), (x_{12}, x_5), (x_{13}, x_5), (x_{14}, x_5), (x_{15}, x_5)\}$$
$$R_1 \cap R_2 \cap R_3 = \{(x_2, x_3), (x_2, x_6), (x_8, x_3), (x_8, x_6)\}.$$

Let $V = \{y_1, y_2, y_3, y_4, y_5, y_6\}$. Define a mapping as follows:

x_1, x_4, x_{11}	x_2, x_8	x_3, x_6	x_5	x_7, x_9, x_{10}	$x_{12}, x_{13}, x_{14}, x_{15}$
y_1	y_2	y_3	y_4	y_5	y_6

Then $f(\mathbf{R}) = \{f(R_1), f(R_2), f(R_3)\}$, where
$f(R_1) = \{(y_2, y_3), (y_4, y_2), (y_4, y_3), (y_5, y_6)\},$
$f(R_2) = \{(y_1, y_6), (y_2, y_3), (y_4, y_2)\},$
$f(R_3) = \{(y_1, y_5), (y_2, y_3), (y_4, y_3), (y_6, y_4)\}.$
And $(V, f(\mathbf{R}))$ is the $f-$ induced relation information system of $(U, \mathbf{R})$. It is very easy to verify that f is a homomorphism from $(U, \mathbf{R})$ to $(V, f(\mathbf{R}))$.

We can see that $f(R_1)$ is superfluous in $f(\mathbf{R}) \Leftrightarrow R_1$ is superfluous in $\mathbf{R}$ and that $\{f(R_2), f(R_3)\}$ is a reduct of $f(\mathbf{R}) \Leftrightarrow \{R_2, R_3\}$ is a reduct of $\mathbf{R}$. Therefore, we can reduce the original system by reducing the image system and reduce the image system by reducing the original system. That is, the reductions of the original system and image system are equivalent to each other.

5   Conclusions

In this paper, we point out that a mapping between two universes can induce a binary relation on one universe according to the given relation on the other universe. For a relation information system, we can consider it as a composition of some generalized approximation spaces on the same universe. The mapping between generalized approximation spaces can be explained as a mapping between the given relation information systems. A homomorphism is a special mapping between two relation information systems. Under the condition of homomorphism, we discuss the characters of relation information systems, and find out that the reductions of the original system and image system are equivalent to each other. These results may have potential applications in knowledge reduction, decision making and reasoning about data, especially for the case of two relation information systems. Our results also illustrate that some characters of a system are guaranteed in explanation system, i.e., a system gain acknowledgement from another system.

Acknowledgement. This research is supported by Natural Science of Foundation of China (Grant No. 10571025).

References

1. J.W. Grzymala-Busse, Algebraic properties of knowledge representation systems, in : Proceedings of the ACM SIGART International Symposium on Methodologies for Intelligent Systems, Knoxville, 1986, pp. 432-440.
2. J.W. Graymala-Busse, W.A.Sedelow Jr., On rough sets, and information system homomorphism, Bull. Pol. Acad. Sci. Tech. Sci. 36 (3,4) (1988) 233-239.
3. M. Kryszkiewicz, Rules in incomplete information systems, Information Sciences 113 (1999) 271-292.
4. D.Y. Li, Y.C. Ma, Invariant characers of information systems under some homomorphisms, Information Sciences 129 (2000) 211-220.
5. T.Y. Lin, Neighborhood systems and approximation in database and knowledge base systems, in: Proceedings of the Fourth International Symposium on Methodologies of Intelligent Systems, Charlotte, 1989, pp. 75-86.
6. Z. Pawlak, Rough sets. Theoretical Aspects of Reasoning about Data, Kluwer Acadedmic Publishers, Dordrecht, 1991.
7. Z. Pawlak, Rough sets. International Journal of Computer and Information Sciences 11 (1982) 341-356.
8. S.K.Pal, A.Skowron, (Eds.), Rough-Fuzzy Hybridization: A New Trend in Decision Making, Springer, Berlin, 1999.
9. P. Pagliani, Transforming information systems, in:Proceedings of the 10th International Conference on Rough Sets, Fuzzy Sets, Data Mining and Granular Computing, Canada, 2005, pp. 660-670.
10. D. Slezak, Searching for dynamic reducts in inconsistent decision tables, in: Proceedings of IPMU' 98, France, Vol.2, 1998, pp. 1362-1369.
11. W. -Z. Wu, M. Zhang, H.-Z. Li, J.-S. Mi, Knowledge reduction in random information systems via Dempster-Shafer theory of evidence, Information Sciences 174 (2005) 143-164. Information Sciences 174 (2005) 143-164.

Dynamic Reduction Based on Rough Sets in Incomplete Decision Systems

Dayong Deng[1,2] and Houkuan Huang[1]

[1] School of Computer and Information Technology, Beijing Jiaotong University,
Beijing, PR China, 100044
dayongd@163.com, hkhuang@center.njtu.edu.cn
[2] Zhejiang Normal University, Jinhua, Zhejiang Province, PR China, 321004

Abstract. In this paper we investigate the dynamic characteristics in an incomplete decision system while information is increasing. We modify the definition of reduction of condition attributes in this case, and present algorithms of reduction in order to deal with increase information.

Keywords: Rough Sets, Incomplete Decision, Increase Information, Attribute Reduction.

1 Introduction

When collecting information about a given topic in a certain moment in time, it may happen that we do not exactly know all the details of the issue in question. This lack of knowledge leads to an incomplete information system. Rough set theory is a valid mathematical tool, which deals with imprecise, vague and incomplete information[10,11,12]. In general, rough set theory deals with information in complete information systems. But in recent years, there are many people, who disposed incomplete information with rough set theory, and presented several methods of dealing with missing attribute values[3]. M.Kryszkiewicz[1,2] proposed a tolerance relation between objects, which is reflexive, symmetric but not transitive. J.Stefanowskis' model[6,7] is based on similarity relation, which is reflexive, transitive, but not symmetric. J.W.Grzymala-Busse's model[4,5] is based on characteristic relation, which is only reflexive. S.Greco's model[8] is based on similarity relation, which is transitive, but not reflexive or symmetric. G.Y.Wang[9] extended M.Kryszkiewicz's model so that his model fits real world more. But all of indiscernibility relations in these models are static, and they could not fit the case of increase information in incomplete information system. In the paper we consider the case of incomplete decision systems with increase information based on M.Kryszkiewicz's model. We investigate its dynamic properties and present new algorithms to get reducts while information is increasing. The method of reduction preserves their positive regions as well as other important information in these incomplete information systems such that the reducts are not put at a disadvantage when information is increasing.

The rest of the paper is organized as follows. In section 2 we introduce the basic concepts of incomplete information systems. In section 3 we investigate

J.T. Yao et al. (Eds.): RSKT 2007, LNAI 4481, pp. 76–83, 2007.
© Springer-Verlag Berlin Heidelberg 2007

their dynamic properties. In section 4 we present new algorithms of dynamic reduction. In section 5 an example is given to show the ideas of new algorithms. At last, we draw a conclusion in section 6.

2   Information Systems

An information system is a pair $IS = (U, A)$, where U is the universe of discourse with a finite number of objects(or entities), A is a set of attributes defined on U. Each $a \in A$ corresponds to the function $a : U \to V_a$, where V_a is called the value set of a. Elements of U are called situation, objects or rows, interpreted as, e.g., cases, states.

With any subset of attributes $B \subseteq A$, we associate the information set for any object $x \in U$ by

$$Inf_B(x) = \{(a, a(x)) : a \in B\}$$

An equivalence relation called B-indiscernible relation is defined by

$$IND(B) = \{(x, y) \in U \times U : Inf_B(x) = Inf_B(y)\}$$

Two objects x, y satisfying the relation $IND(B)$ are indiscernible by attributes from B. $[x]_B$ is referred to as the equivalence class of $IND(B)$ defined by x. The equivalence classes of $IND(B)$ are denoted by

$$U/B = \{[x]_B : x \in U\}.$$

A minimal subset B of A such that $IND(B) = IND(A)$ is called a reduct of IS.

Suppose $IS = (U, A)$ is an information system, $B \subseteq A$ is a subset of attributes, and $X \subseteq U$ is a subset of discourse, the sets

$$\underline{B}(X) = \{x \in U : [x]_B \subseteq X\}, \overline{B}(X) = \{x \in U : [x]_B \bigcap X \neq \phi\}$$

are called B-lower approximation and B-upper approximation respectively. The lower approximation is also called positive region, denoted by $POS_B(X)$.

A special type of information system is called decision system $DS = (U, A \cup \{d\})$, where $\{d\} \cap A = \phi$, A is a set of condition attributes, and d is a distinguished attribute called conclusion attribute. In a decision system the positive region of the decision attribute corresponding to the condition attributes is denoted by $POS_A(d)$:

$$POS_A(d) = \bigcup_{Y_i \in U/\{d\}} (POS_A(Y_i))$$

It may happen that some values of attributes for objects in information systems are missing. These information systems are called incomplete information systems. The missing values are called null values, which are denoted by $*$. Therefore, a similarity relation could be defined as follows[1,2]:

$$SIM(B) = \{(x, y) \in U \times U : \forall a \in B(a(x) = a(y) \ or \ a(x) = * \ or \ a(y) = *)\}$$

Let $S_B(x)$ denotes the object set $\{y \in U : (x, y) \in SIM(B)\}$, where $B \subseteq A$. The lower and upper approximation of a concept $X \subseteq U$ are defined as follows respectively:

$$\underline{B}(X) = \{x \in U : S_B(x) \subseteq X)\}$$

$$\overline{B}(X) = \{x \in U : S_B(x) \bigcap X \neq \emptyset\}$$

If there is not confusion, we will also denote the set of tolerance classes $S_B(x)$ by U/B, and the B-lower approximation of X is also called the positive region, denoted by $POS_B(X)$.

For $B \subseteq A$, $C \subseteq A$, we call the cover of U/B is finer than that of U/C, denoted by $U/B \subseteq U/C$, if for any tolerance class $S_B(x)$ in U/B there exists a tolerance class $S_C(x)$ in U/C such that $S_B(x) \subseteq S_C(x)$.

In incomplete decision systems, we assume that values of conclusion attributes are usually complete in the sequel.

3   Incomplete Systems with a Monotonic Increase of Information

In [13] G.Cattaneo and D.Ciucci defined three ways of increasing the knowledge in incomplete information systems. In this paper we are only dealing with the first case. Its definition is formalized in the following way.

Definition 1. Let $IS^{(t_i)} = (U_i, A_i)$ and $DS^{(t_{i+1})} = (U_{i+1}, A_{i+1})$, with t_i, $t_{i+1} \in R$, $t_i \leq t_{i+1}$ be two incomplete information systems, where $U_i = U_{i+1}$. The attributes in A_i are the same as that in A_{i+1}. We will say that there is a monotonic increase of information in the information system IS: For $\forall x \in U_i$ and $\forall a^{t_i} \in A_i$, $a^{t_i}(x) \neq *$ implies $a^{t_i}(x) = a^{t_{i+1}}(x)$. In such a case, we will denote by $IS^{(t_i)} \preceq_1 IS^{(t_{i+1})}$.

$IS^{(t_i)} \preceq_1 IS^{(t_{i+1})}$ means that, in the information system IS the universe of discourse and the attributes do not change, but the values of attributes may be changed from unknown to known. Because we only investigate this case, we will denote U_i by U in the sequel.

Definition 2. Let $IS^{(t_i)} = (U, A_i)(t_i \in R)$ be a series of incomplete information systems with a monotonic increase of information, i.e. $IS^{(t_i)} \preceq_1 IS^{(t_{i+1})}$. We say the information system IS is a complete information system if it satisfies the condition:

$$IS = \lim_{i \to \infty} IS^{(t_i)}$$

From definition 2, there are two types of complete information systems corresponding to a series of incomplete information systems with a monotonic increase of information(complete information systems, in short): (1) All of values of attributes are known. (2) Some values of attributes will be unknown forever. Without generality we assume that all of values of attributes are known in complete information systems.

We will investigate properties of incomplete information systems with a monotonic increase of information in the sequel.

Proposition 1. Suppose $IS^{(t_i)} = (U, A_i) \preceq_1 IS^{(t_{i+1})} = (U, A_{i+1})$, with $t_i, t_{i+1} \in R$, $t_i \leq t_{i+1}$ be two incomplete information systems, Then for $\forall a \in A_i$ and $\forall x \in U$, we have

$$S_{\{a\}}^{t_{i+1}}(x) \subseteq S_{\{a\}}^{t_i}(x)$$

Proof. In terms of the definition $S_{\{a\}}(x) = \{y \in U : (x, y) \in SIM(\{a\})\}$, we have $\forall y(y \in S_{\{a\}}^{t_{i+1}}(x) \Rightarrow y \in S_{\{a\}}^{t_i}(x))$. Therefore $S_{\{a\}}^{t_{i+1}}(x) \subseteq S_{\{a\}}^{t_i}(x)$.

Corollary 1. Suppose $IS^{(t_i)} = (U, A_i) \preceq_1 IS^{(t_{i+1})} = (U, A_{i+1})$, with $t_i, t_{i+1} \in R$, $t_i \leq t_{i+1}$ be two incomplete information systems, then $U/B_{i+1} \subseteq U/B_i$ for $B \subseteq A$.

Corollary 2. Suppose $IS^{(t_i)} = (U, A_i) \preceq_1 IS^{(t_{i+1})} = (U, A_{i+1})$, with $t_i, t_{i+1} \in R$, $t_i \leq t_{i+1}$ be two incomplete information systems, $IS = (U, A)$ their corresponding complete information system and $X \subseteq U$ a concept. Then

$$\underline{B_i}(X) \subseteq \underline{B_{i+1}}(X) \subseteq \underline{B}(X)$$

$$\overline{B}(X) \subseteq \overline{B}_{i+1}(X) \subseteq \overline{B}_i(X)$$

for $\forall B \subseteq A$.

Theorem 1. Suppose $DS^{(t_i)} = (U, A_i \bigcup \{d\}) \preceq_1 DS^{(t_{i+1})} = (U, A_{i+1} \bigcup \{d\})$, with $t_i, t_{i+1} \in R$, $t_i \leq t_{i+1}$, be two incomplete decision systems, Then

$$POS_{B_i}(\{d\}) \subseteq POS_{B_{i+1}}(\{d\})$$

for $B \subseteq A$.

Proof. It can be got directly from Corollary 2.

Corollary 3. Suppose $DS^{(t_i)} = (U, A_i \bigcup \{d\})$ is an incomplete decision system, $DS = (U, A \bigcup \{d\})$ is its corresponding complete decision system, Then

$$POS_{B_i}(\{d\}) \subseteq POS_B(\{d\})$$

for $B \subseteq A$.

From the above propositions, the positive regions in incomplete decision systems are increasing with increasing information in them. We should not delete any of condition attributes unless these condition attributes are confirmed not to influence the positive regions in the series of incomplete decision systems. In the next section we will investigate reduction of condition attributes in incomplete decision systems.

4   Dynamic Reduction

In decision systems with missing values almost all of existed methods are to get reducts in the criterion of the positive regions preserved, these methods don't consider dynamic increase of information. In this section we will investigate reduction in this case. The criterion of reduction, except for preserved positive region, is to delete condition attributes in which there are no null values in the negative positive region, i.e. all of elements with missing value are in the positive region. It is easy to prove that these deleted condition attributes are irrelative to the positive regions in incomplete decision systems with increase information in terms of above propositions. In terms of the criteria, the algorithm of reduction in an incomplete decision system is presented as follows:

Algorithm 1: Static reduction of incomplete decision system(SRIDS, In short).

> Input: An incomplete decision system $DS^{(t_i)} = (U, A_i \bigcup \{d\})$
> Output: A reduct of $DS^{(t_i)} = (U, A_i \bigcup \{d\})$
> Step1: $U_1 = POS_{A_i}(\{d\})$, $B = A_i$
> Step2: $U_2 = U - U_1$
> Step3: For j=1 to $|A_i|$
>   { $flag = 1$;
>   For k=1 to $|U_2|$
>   If $a_j(x_k) = *$ Then $flag = 0$;
>   If flag and $POS_{B-\{a_j\}}(\{d\}) = U_1$
>   Then $B = B - \{a_j\}$; }
> Step4: Output the reduct B

In algorithm 1, $DS^{(t_i)}$ represents the state of the decision system DS at t_i. The symbol $flag$ is to decide whether any elements in U_2(it stands for the negative positive region) are missing values, if $flag = 1$ then there are no null values in the negative positive region, or else there are some null values. $|\bullet|$ denotes the cardinality of the set, and x_k is an element of U_2.

The difference between algorithm 1 and other classical algorithms is whether the missing values in the negative positive region should be considered when condition attributes are reduced. The former considers the null values of condition attributes in order to avoid a disadvantage for the reduced condition attributes to the positive region in the future. The later only consider the positive region at a moment.

The time complexity of algorithm 1 is decided by that of counting positive region. Suppose we utilize the algorithm in literature [14] to compute positive region, whose time complexity is $O(|A_i||U|log|U|)$. Therefore, the time complexity of algorithm 1 is $O(|A_i|^2|U|log|U|)$.

Suppose the incomplete decision system $DS^{(t_i)} = (U, A_i \bigcup \{d\})$ is dynamically increasing its information, i.e. $DS^{(t_i)} \preceq_1 DS^{(t_{i+1})}$, with $t_i, t_{i+1} \in R$, $t_i \leq t_{i+1}$, and the maximum of i is equal to n. In this case we could call the above algorithm iteratively. The algorithm of reduction with respect to the dynamical incomplete decision system $DS^{(t_i)}$ is presented as follows:

Algorithm 2: Dynamical reduction of an incomplete decision system with increase information

Input: An incomplete decision system $DS^{(t_i)} = (U, A_i \bigcup \{d\})$ with increase information.

Output: A dynamic reduct with respect to the incomplete decision system DS

> For i=1 to n
> { SRIDS($DS^{(t_i)}$);
> $DS^{(t_i)} = (U, B)$;
> }

B denotes a reduct of the incomplete decision system DS at t_i. Because the time complexity of algorithm 1 is $O(|A_i|^2 |U| log |U|)$, the time complexity of algorithm 2 is $O(n|A_i|^2 |U| log |U|)$. The Algorithm 2 counts the positive region iteratively. We could improve it by avoiding the iterative workload. The improved algorithm is presented as follows:

Algorithm 3: Improved dynamic reduction of an incomplete decision system.

Input: An incomplete decision system $DS^{(t_i)} = (U, A_i \bigcup \{d\})$ with increase information.

Output: A dynamic reduct with respect to the incomplete decision system DS

> Step 1: $B = A$; $U_1 = \emptyset$;
> Step 2: $U_2 = U$;
> Step 3: For i=1 to n
> > {$U_3 = POS_B(\{d\})$;
> > $U_1 = U_1 \bigcup U_3$;
> > $U_2 = U_2 - U_3$;
> > $C = B$;
> > For j=1 to $|B|$
> > { $flag = 1$;
> >   For k=1 to $|U_2|$
> >   If $a_j(x_k) = *$ then $flag = 0$;
> >   If flag and $U_1 = POS_{C-\{a_j\}}(\{d\})$ then $C = C - \{a_j\}$;
> >   } //End for j
> > Output C;
> > B=C;
> > } // End for i

In algorithm 3, U_1 denotes positive region of the incomplete decision system, U_2 negative positive region, U_3 the incremental positive region in the rest of the incomplete decision system, $POS_B(\{d\})$ the positive region of incomplete decision system $DS' = (U_2, B \bigcup \{d\})$, and $POS_{C-\{a_j\}}(\{d\})$ the positive region of incomplete decision system $DS'' = (U, (C - \{a_j\}) \bigcup \{d\})$. The output value of C is a reduct at t_i. The rest of symbols are the same as that of algorithm 1.

The algorithm 3 could reduce the the iterative workload of computing positive region. In algorithm 3 we could only compute the additive positive region $POS_B(\{d\})$, not the whole positive region of the decision system, although the time complexity of algorithm 3 is the same as that of algorithm 2.

5   Example

Suppose the incomplete decision system $DS^{(t_i)} = (U, A_i \bigcup \{d\})$ is dynamically increasing its information, where $DS^{(t_1)}$ is denoted by Table 1, $DS^{(t_2)}$ is denoted by Table 2, and $DS^{(t_1)} \preceq_1 DS^{(t_2)}$, $A = \{a_1, a_2, a_3\}$. In $DS^{(t_1)}$ $a_1(x_3) = *$, but $a_1(x_3) = 2$ in $DS^{(t_2)}$. In terms of Algorithm 1 we could get the reduct $\{a_1, a_2\}$ at t_1. In table 1, although the attributes a_1 and a_3 could be deleted if we only preserve the positive region, there are some missing values of a_1 in the negative positive region, while there are no missing values of a_3 in the negative positive region. At t_2 we could get more elements in the positive region. For example, the element x_3 is not in the positive region at t_1, but it is in the positive region at t_2. It is easy to know the reduct of incomplete information DS at t_2 is also $\{a_1, a_2\}$ from table 2 in term of algorithm 1. That is to say, the condition attribute a_1 should not be reduced at t_1 in the incomplete decision system with increase information.

Table 1. Incomplete Decision system DS at t_1

U	a1	a2	a3	d
x1	0	0	1	1
x2	1	1	*	1
x3	*	0	1	0
x4	0	2	1	0
x5	0	0	1	1
x6	3	1	*	1
x7	3	2	1	0
x8	*	2	1	0

Table 2. Incomplete Decision system DS at t_2

U	a1	a2	a3	d
x1	0	0	1	1
x2	1	1	*	1
x3	2	0	1	0
x4	0	2	1	0
x5	0	0	1	1
x6	3	1	*	1
x7	3	2	1	0
x8	*	2	1	0

6   Conclusion

In the paper we investigate some properties of incomplete decision systems with increase information, and the reduction of condition attributes in this case. A new method of reduction is presented, in which we not only consider positive region in an incomplete decision system but also its potential influence on positive region in the future.

References

1. Kryszkiewicz, M.: Rough Set Approach to Incomplete Information Systems. Information Science1 112(1-4) (1998) 39-49.
2. Kryszkiewicz, M.: Rules in Incomplete Information Systems. Information Science 113(3-4) (1999) 271-292.
3. Grzymala-Busse, J.W., Hu, M.: A Comparison of Several Approaches to Missing Attribute Values in Data Mining. In: Proceedings of 2nd International Conference on Rough Sets and Current Trends in Computing(RSCTC2000), Canada (2000) 378-385.
4. Grzymala-Busse, J.W.: Characteristic Relations for Incomplete Data:A Generalization of the Indiscernibility Relation. In:Proceedings of 4th International Conference on Rough Sets and Current Trends in Computing(RSCTC2004), Sweden (2004) 254-263.
5. Grzymala-Busse, J.W.: Incomplete Data and Generalization of Indiscernibility Relation, Definability, and Approximation. In:Proceedings of 10th International Conference on Rough Sets, Fuzzy Sets, Data Mining, and Granular Computing (RSRDGrC2005), Canada (2005) 244-253.
6. Stefanowski, J., Tsoukis, A.: On the Extension of Rough Sets under Incomplete Information. In:Proceedings of 7th International Workshop on Rough Sets, Fuzzy Sets, Data Mining, and Granular Computing(RSRDGrC1999), Japan (1999) 73 - 81.
7. Stefanowski, J., Tsoukis, A.: Incomplete Information Tables and Rough Classification. Computational Intelligence 17(3) (2001) 545-566.
8. Greco, S.,Matarazzo, B., Slowinski, R.: Dealing with Missing Data in Rough Set Analysis of Multi-attribute and Multi-criteria Decision Problems. In Decision Making:Recent developments and Worldwide Applications,ed. by S.H.Zanakis, G.Doukidis, and Z.Zopounidis, Kluwer Academic Publishers, Dordrecht (2000) 295-316.
9. Wang, G.Y.: Extension of Rough Set Under Incomplete Information Systems. Journal of Computer Research and Development(in Chinese) 39(10) (2002) 1238-1243.
10. Wang,G.Y.: Rough Set Theory and Knowledge Discovery(in Chinese). Xi'an Jiaotong University Press (2001).
11. Pawlak, Z.: Rough sets-Theoretical Aspect of Reasoning about Data. Kluwer Academic (1991).
12. Liu, Q.: Rough Sets and Rough Reasoning. Science Press(in Chinese) (2001).
13. Cattaneo, G., Ciucci, D.: Investigation about Time Monotonicity of Similarity and Preclusive Rough Approximations in Incomplete Information Systems. In:Proceedings of 4th International Conference, RSCTC2004, Uppsala, Sweden (2004) 38-48.
14. Liu,S.H.,Sheng,Q.J.,Shi,Z.Z.:A New Method for Fast Computing Positive Region. Journal of Computer Research and Development(in Chinese) 40(5) (2003) 637-642.

Entropies and Co–entropies for Incomplete Information Systems*

Daniela Bianucci, Gianpiero Cattaneo, and Davide Ciucci

Dipartimento di Informatica, Sistemistica e Comunicazione
Università di Milano – Bicocca
Via Bicocca degli Arcimboldi 8, I–20126 Milano, Italia
{bianucci,cattang,ciucci}@disco.unimib.it

Abstract. A partitioning approach to the problem of dealing with the entropy of incomplete information systems is explored. The aim is to keep into account the incompleteness and at the same time to obtain a probabilistic partition of the information system. For the resulting probabilistic partition, measures of entropy and co–entropy are defined, similarly to the entropies and co–entropies defined for the complete case.

Keywords: entropy, co–entropy, incomplete information system.

1 Introduction: Qualitative and Quantitative Valuations of Roughness for Complete Information Systems

In this work, we discuss the entropy of incomplete information systems as an extension of the approach based on partitions from complete information systems. In order to introduce an approach of *probability partition* from an incomplete information system, let us first recall how one gets a partition from a complete information system, and thus how one can apply a measure of rough entropy when dealing with an information system.

Let us recall that the original Pawlak approach to rough sets is essentially based on an *approximation space*, i.e., a pair $\langle X, \pi \rangle$ where X is a (finite) set, called the *universe* of *objects*, and $\pi = \{A_1, A_2, \ldots, A_N\}$ is a partition of X, in general induced by the indistinguishability equivalence relation from a complete information system [1]. The subsets A_j are the *elementary sets* (or also *events*), each of which can be interpreted as a *granule of knowledge* supported by the partition. We denote by $gr_\pi(x)$ the granule (equivalence class) from π which contains the point $x \in X$. In the rough set theory, once fixed a partition π of X, any of its subsets H can be approximated from the bottom and from the top by the two *lower* and *upper approximations* defined respectively as: $l_\pi(H) := \cup\{A_i \in \pi : A_i \subseteq H\}$ and $u_\pi(H) := \cup\{A_j \in \pi : H \cap A_j \neq \emptyset\}$, producing the rough approximation of H defined as the pair $r_\pi(h) = (l_\pi(H), u_\pi(H))$ (with trivially $l_\pi(H) \subseteq H \subseteq u_\pi(H)$), see [2] for a complete discussion. We can also

---

* The author's work has been supported by MIUR\PRIN project "Automata and Formal languages: mathematical and application driven studies."

define the *boundary region* of H as $b_\pi(H) = u(H) \setminus l(H)$, and its *external region* as $e_\pi(H) = X \setminus u(H)$. Obviously, whatever be the starting original partition π, for any subset H the triple $\pi(H) = \{l_\pi(H), b_\pi(H), e_\pi(H)\}$ is a new partition of X, which depends from the choice of the subset H.

These considerations can be applied to the case of a complete Information System (IS), formalized by a triple $IS := \langle X, Att, F \rangle$ consisting of a nonempty finite set X of objects, a nonempty finite set of attribute Att, and a mapping $F : X \times Att \to V$ which assigns to any object $x \in A$ the value $F(x,a)$ assumed by the attribute $a \in Att$ [1,3,4]. Indeed, in this IS case the partition generated by a set of attributes $\mathcal{A}$, denoted by $\pi_{\mathcal{A}}(IS)$, consists of equivalence classes of *indistinguishable* objects A_i, i.e., two objects $x, y \in A_i$ iff for any attribute $a \in \mathcal{A}$, the condition $F(x,a) = F(y,a)$ holds.

In many applications it is of a certain interest to analyze the variations occurring inside two information systems labelled with two parameters t_1 and t_2. In particular, one has to do mainly with the following two cases in both of which the set of objects remains invariant:

(1) *dynamics* (see [5]), in which $IS_{t_1} = (X, Att_1, F_1)$ and $IS_{t_2} = (X, Att_2, F_2)$ are under the conditions that $Att_1 \subset Att_2$ and $\forall x \in X$, $\forall a_1 \in Att_1$: $F_2(x, a_1) = F_1(x, a_1)$. This situation corresponds to a dynamical increase of knowledge (t_1 and t_2 are considered as time parameters, with $t_1 < t_2$) for instance in a medical database the increase corresponds to the fact that during the researches on the disease some symptoms which have been neglected at time t_1 become relevant at time t_2 under some new investigations.

(2) *reduct*, in which $IS_{t_1} = (X, Att_1, F_1)$ and $IS_{t_2} = (X, Att_2, F_2)$ are under the conditions that $Att_2 \subset Att_1$ and $\forall x \in X$, $\forall a_2 \in Att_2$: $F_2(x, a_2) = F_1(x, a_2)$. In this case it is of a certain interest to verify if the corresponding partitions are invariant $\pi_{Att_2}(IS_{t_2}) = \pi_{Att_1}(IS_{t_1})$, or not.

From the point of view of the rough approximations of subsets Y of the universe X, both these cases can be treated under a unified formal framework in which during the time evolution $t_1 \to t_2$ one try to relate the corresponding variation of partitions $\pi_{t_1} \to \pi_{t_2}$ with, for instance, the boundary transformation $b_{t_1}(Y) \to b_{t_2}(Y)$. First of all, as to the partitions of X, whose collection will be denoted by $\Pi(X)$, their more interesting structure is the one of complete lattice (see [6]) with respect to the partially order relation $\pi_1 \preceq \pi_2$, which can be formalized in one of the following mutually equivalent forms: (por1) $\forall A \in \pi_1$, $\exists B \in \pi_2$: $A \subseteq B$; (por2) $\forall B \in \pi_2$, $\exists \{A_{i_1}, A_{i_2}, \ldots, A_{i_h}\} \subseteq \pi_1$: $B = A_{i_1} \cup A_{i_2} \cup \ldots \cup A_{i_h}$; (por3) $\forall x \in X$, $gr_{\pi_1}(x) \subseteq gr_{\pi_2}(x)$ (as shown in [7], an extension of these three formulations to the case of coverings leads to different binary relations of quasi–orderings). The lattice $\Pi(X)$ of all partitions of X is lower bounded by the least element $\pi_d := \{\{x\} : x \in X\}$ (the *discrete* partition) consisting of all singletons from X, and the greatest element $\pi_t := \{X\}$ (the *trivial* partition) whose unique equivalence class is the whole universe. If $\pi_1 \preceq \pi_2$ we say that π_1 (resp., π_2) is *finer* (resp., *coarser*) than π_2 (resp., π_1). The induced *strict ordering* on partition, denoted by $\pi_1 \prec \pi_2$, is defined as $\pi_1 \preceq \pi_2$ and $\pi_1 \neq \pi_2$. This means that it must exists at least an equivalence class $B_i \in \pi_2$

such that its partition with respect to π_1 is formed at least of two subsets, i.e., $\exists\{A_{i_1}, A_{i_2}, \ldots, A_{i_p}\} \subseteq \pi_1$, with $p \geq 2$, s.t. $B_i = A_{i_1} \cup A_{i_2} \cup \ldots \cup A_{i_p}$.

Let us note that if $\pi_1 \preceq \pi_2$, then the two rough approximations of a given subset Y, $r_{\pi_i}(Y) = (l_{\pi_i}(Y), u_{\pi_i}(Y))$, for $i = 1, 2$, are such that $l_{\pi_2}(Y) \subseteq l_{\pi_1}(Y) \subseteq Y \subseteq u_{\pi_1}(Y) \subseteq u_{\pi_2}(Y)$, i.e., the rough approximation of Y with respect to the partition π_1 is *better* than the rough approximation of the same subset with respect to π_2. This leads to a first but only *qualitative* valuation of the *roughness* of a subset Y of the universe expressed by the law: $\pi_1 \preceq \pi_2$ implies that for $\forall Y$, $b_{\pi_1}(Y) \subseteq b_{\pi_2}(Y)$. The delicate point is that the condition of strict ordering $\pi_1 \prec \pi_2$ does not assure that for $\forall Y$, $b_{\pi_1}(Y) \subset b_{\pi_2}(Y)$. It is possible to give some very simple counter–examples in which notwithstanding $\pi_1 \prec \pi_2$ one has that $\exists Y_0$: $b_{\pi_1}(Y_0) = b_{\pi_2}(Y_0)$ [8,7], and this is not a desirable behavior of such a qualitative valuation of roughness. On the other hand, in many practical applications (for instance in the attribute reduction procedure), it is interesting not only to have a possible qualitative valuation of the roughness of a generic subset Y, but also a *quantitative* valuation formalized by a mapping $E : \Pi(X) \times 2^X \to [0, 1]$ assumed to satisfy (at least) the following two minimal requirements:

(re1) the *strict monotonicity condition*: for any $Y \in 2^X$, $\pi_1 \prec \pi_2$ implies $E(\pi_1, Y) < E(\pi_2, Y)$;

(re2) the *boundary conditions*: for $\forall Y \in 2^X$, $E(\pi_d, Y) = 0$ and $E(\pi_t, Y) = 1$.

In the sequel, sometimes we will use the notation $E_\pi : 2^X \to [0, 1]$ to denote the above mapping in which the partition $\pi \in \Pi(X)$ is considered fixed once for all. The interpretation of condition (re2) is possible under the assumption that a quantitative valuation of the roughness $E_\pi(Y)$ should be *directly* related to its boundary by $|b_\pi(Y)|$. From this point of view, the value 0 corresponds to the discrete partition for which the boundary of any subset Y is empty, and so its rough approximation is $r_{\pi_d}(Y) = (Y, Y)$ with $|b_{\pi_d}(Y)| = 0$, i.e., a crisp situation. On the other hand, the value 1 corresponds to the trivial partition in which the boundary of any nontrivial subset $Y(\neq \emptyset, X)$ is the whole universe, and so its rough approximation is $r_{\pi_t}(Y) = (\emptyset, X)$ with $|b_{\pi_t}(Y)| = |X|$, i.e., the minimum of sharpness or maximum of roughness.

This being stated, in literature one can find a lot of quantitative *measures of roughness* of Y relatively to a given partition $\pi \in \Pi(X)$ formalized as mappings $\rho_\pi : 2^X \to [0, 1]$. The *accuracy* of the set Y with respect to the partition is then defined as $\alpha_\pi(Y) = 1 - \rho_\pi(Y)$. Two of the more interesting roughness measures are $\rho_\pi^{(P)}(Y) := \frac{|b_\pi(Y)|}{|u_\pi(Y)|}$ [3] and $\rho_\pi^{(C)}(Y) := \frac{|b_\pi(Y)|}{|X|}$ [7]. These roughness measures satisfy the above "boundary" condition (re2), but their drawback is that the strict condition on partitions $\pi_1 \prec \pi_2$ does not assure a corresponding strict behavior $\forall Y$, $b_{\pi_1}(Y) \subset b_{\pi_2}(Y)$, and so also the strict correlation $\rho_{\pi_1}(Y) < \rho_{\pi_2}(Y)$ cannot be inferred. In other words, in general a rough measure is monotonic, but not strictly monotonic, contrary to the above requirement (re1).

This drawback can be overcome according to at least two strategies: either by some new strictly monotonic roughness measures or maintaining one of the monotonic roughness measures ρ_π and considering a strict monotonic function

$\Omega : \Pi(X) \to [0,1]$ in such a way that the new mapping $E(\pi, Y) := \rho_\pi(Y) \cdot \Omega(\pi)$ turns out to be strictly monotonic. In this paper we explore this second possibility in which, for the sake of simplicity, the required function is not the normalized Ω but it is given by a co–entropy function (also granularity measure) $E : \Pi(X) \to [0, k]$, where k is a suitable constant, and from which it is possible to induce the normalized $\Omega(Y) = E(Y)/k$. This is discussed in the following section.

2   Global and Pointwise Entropies and Co–entropies from Partitions

Given a partition $\pi = (A_1, A_2, \ldots, A_N)$ of the universe X by the *elementary events* A_i, one can construct the σ–algebra $\mathcal{E}(\pi)$ of *events* generated by π consisting of the empty set and all the set theoretic unions of elementary events. In the measurable space $(X, \mathcal{E}(\pi))$ the *counting measure* $m_\pi : \mathcal{E}(\pi) \to \mathbb{R}_+$ assigns to any event E the corresponding measure $m_\pi(E) = |E|$ (its cardinality). In this space with measure $(X, \mathcal{E}(\pi), m_\pi)$ we can introduce the vector $\boldsymbol{m}(\pi) = (m_\pi(A_1), m_\pi(A_2), \ldots, m_\pi(A_N))$, with $m_\pi(A_j) > 0$ for every j and $\sum_{j=1}^N m_\pi(A_j) = |X|$, which is a measure of the granulation, called the *granularity distribution* induced by π. Finally, it is possible to introduce the vector $\boldsymbol{p}(\pi) := (p_\pi(A_1), p_\pi(A_2), \ldots, p_\pi(A_N))$, where each $p_\pi(A_j) := \frac{|A_j|}{|X|}$ represents the probability of occurrence of the granule A_j. Since for each j we have that $p_\pi(A_j) > 0$ and $\sum_{j=1}^N p_\pi(A_j) = 1$, the vector $\boldsymbol{p}(\pi)$ constitutes a *probability distribution* induced by granulation. This being stated, in this section we consider the following two quantities depending from the partition π:

$$E(\pi) = \frac{1}{|X|} \sum_{i=1}^N |A_i| \log |A_i| \qquad (2.1a)$$

$$H(\pi) := - \sum_{j=1}^N p(A_j) \cdot \log p(A_j) = - \frac{1}{|X|} \sum_{i=1}^N |A_i| \log \frac{|A_i|}{|X|}. \qquad (2.1b)$$

Let us note that $E(\pi)$ depends only from the *granularity distribution* $\boldsymbol{m}(\pi)$, whereas $H(\pi)$ depends from the *probability distribution* $\boldsymbol{p}(\pi)$. In our opinion this leads to two different semantical interpretations of these quantities. Indeed, in agreement with the *information theory*, since the granule A_j has probability $p(A_j)$, we shall say that the quantity $I[p(A_j)] := -\log p(A_j)$ is the *uncertainty* associated with the granule A_j. Thus, the quantity $H(\pi)$, as expectation of the discrete random variable $I[p(A_j)]$ with probability $p(A_j)$, is the *average uncertainty* relatively to the probability distribution $\boldsymbol{p}(\pi)$, i.e., it measures the *uncertainty* of the granulation. According to Shannon [9], $H(\pi)$ is called the *entropy* of the partition π. Besides this entropy, the quantity $E(\pi)$ can be defined as *co–entropy* owing to its complementarity role with respect to the entropy $H(\pi)$ formalized by the identity $E(\pi) + H(\pi) = \log |X|$, whatever be the partition π. Let us note that in [10] this quantity has been called *measure of the granularity* since it "is basically an expectation of granularity with respect to all subsets in a partition". The following strict monotonic (resp., anti–monotonic) behavior of co–entropy (resp., entropy) is a standard result (for the entropy see for instance [11]): $\pi_1 \prec \pi_2$ implies $E(\pi_1) < E(\pi_2)$ and $H(\pi_2) < H(\pi_1)$. Since the trivial (resp., discrete) partition π_t (resp., π_d) is the greatest (resp., least) element of the lattice of all partitions, it is easy to see that for any partition π it is $0 = E(\pi_d) \le E(\pi) \le E(\pi_t) = \log|X|$, according to the fact that "the coarsest partition π_t has the maximum granularity value $\log|X|$ and the finest partition π_d has the minimum granularity value 0" [10]. So the required normalized co–entropy (granular entropy) is $\Omega(\pi) := \frac{E(\pi)}{\log|X|}$.

Note that in [12] the entropy $H(\pi)$, with the corresponding anti–monotonic behavior, has been assumed as a "measure of granularity", but this (formally legitimate) choice is in contrast with the strict monotonicity (meta–)requirement subsumed by $E(\pi, Y)$ as "local" (i.e., depending from Y) measure of roughness. Further, this choice suffers also of another drawback. As previously underlined, in information theory the entropy $H(\pi)$ is interpreted as a measure of the uncertainty of the probability distribution generated by the partition π. In conclusion, the different behaviors of $H(\pi)$ and $E(\pi)$ with respect to the variation of the partition π lead to different semantics: $H(\pi)$ can be interpreted as a measure of the *information uncertainty*, $E(\pi)$ as a measure of *partition granularity*, and $\rho_\pi(Y) \cdot E(\pi)$ as a *local* measure of *rough granularity*. The finer is the partition and the greatest (resp., lower) is the uncertainty (resp., the roughness).

In order to appreciate a possible generalization of these arguments to the case of incomplete IS, for instance according to the approach of [13], in [7] it has been introduced also in the partition context the new notions of *pointwise* entropy and co–entropy as the two mappings in which the sum involves the "local" information given by all the equivalence classes $gr(x)$, with corresponding "probabilities" $\mu_\pi(x) = \frac{|gr(x)|}{|X|}$, for the object x ranging on the universe X:

$$E_{LX}(\pi) = \frac{1}{|X|} \sum_{x \in X} |gr(x)| \cdot \log|gr(x)| = \frac{1}{|X|} \sum_{i=1}^{N} |A_i|^2 \cdot \log|A_i| \qquad (2.2a)$$

$$H_{LX}(\pi) = -\sum_{x \in X} \mu_\pi(x) \cdot \log\mu_\pi(x) = -\frac{1}{|X|} \sum_{i=1}^{N} |A_i|^2 \cdot \log\frac{|A_i|}{|X|} \qquad (2.2b)$$

Trivially, $\forall \pi \in \Pi(X)$, $0 \le E(\pi) \le E_{LX}(\pi)$. In the sequel, we refer to $E(\pi)$ as the *global* entropy and to $E_{LX}(\pi)$ as the *pointwise* one. Moreover, setting $\mu(\pi) := \sum_{x \in X} \mu(x)$, one gets that $E_{LX}(\pi) + H_{LX}(\pi) = \log|X| \cdot \mu(\pi)$, with this latter depending on the partition π. Note that the probability vector $\boldsymbol{p}_{LX} := (\mu_\pi(x_1), \mu_\pi(x_2), \ldots, \mu_\pi(x_{|X|}))$ is not a probability distribution since the sum of its components is $\mu(\pi) \ge 1$. Notwithstanding this drawback, from (2.2a) it follows that the strict monotonicity condition holds also for the pointwise co–entropy: $\pi_1 \prec \pi_2$ implies $E_{LX}(\pi_1) < E_{LX}(\pi_2)$. Of course, in this case one has that for any π: $0 \le E_{LX}(\pi) \le |X| \cdot \log|X|$, with corresponding normalized co–entropy (granulation measure) $\Omega_{LX}(\pi) := \frac{E_{LX}(\pi)}{|X| \cdot \log|X|}$. Unfortunately, H_{LX} presents *neither* monotonic *nor* anti–monotonic behavior.

3 Incomplete Information Systems and Definition Domain

An *incomplete* information system is formalized as a triple $\langle X, Att, F \rangle$ where F is a mapping *partially* defined on a subset $\mathcal{D}(F)$ of $X \times Att$ under the following two *non–redundancy* conditions: (1) about objects: for every object $x \in X$ there exists at least an attribute $a \in Att$ such that $(x, a) \in \mathcal{D}(F)$; (2) about attributes: for every attribute $a \in Att$ there exists at least an object $x \in X$ such that $(x, a) \in \mathcal{D}(F)$. In this way also the mapping representation of an attribute a is partially defined on the *definition domain* $X_a := \{x \in X : (x, a) \in \mathcal{D}(F)\}$ of X (which is nonempty owing to the non–redundancy condition (2) about attributes) as the surjective mapping $f_a : X_a \mapsto val(a)$, where $val(a) := \{F(x, a) : x \in X_a\}$ is the set of all *possible values* of the attribute a. The non–redundancy condition (1) about objects assures that $\bigcup_{a \in Att} X_a = X$ (covering condition about attribute definition domains). Adding to $val(a)$ the further *null value* $*$, we obtain the new set $val^*(a)$ and it is possible to extend the partially defined mapping f_a to a global defined one, denoted by $f_a^* : X \mapsto val^*(a)$, which assigns to any object $x \in X$ the value $f_a^*(x) = f_a(x)$ if $x \in X_a$, and the value $f_a^*(x) = *$ otherwise.

Also in the case of incomplete information systems, if one fixes an attribute a and denotes by $\alpha_i \in val(a)$, the subset of the universe $A_i = f_a^{-1}(\alpha_i) = \{x \in X_a : f_a(x) = \alpha_i\}$ is the *elementary event* of all objects for which the attribute a assumes the value α_i. Further, for any family of attributes $\mathcal{A}$ one can construct the "common" definition domain $X_{\mathcal{A}} = \bigcup_{a \in \mathcal{A}} X_a$ and then it is possible to consider the multi–attributes mapping $f_{\mathcal{A}}$ assigning to any object $x \in X_{\mathcal{A}}$ the corresponding collection of values $f_{\mathcal{A}}(x) = (f_a^*(x))_{a \in \mathcal{A}}$, obtaining a mapping $f_{\mathcal{A}} : X_{\mathcal{A}} \mapsto val^*(\mathcal{A})$, with $val^*(\mathcal{A}) \subseteq \Pi_{a \in \mathcal{A}} val^*(a)$ the range of the mapping $f_{\mathcal{A}}$. Note that owing to the non–redundancy conditions for any $a \in Att$ at least one of the $f_a^*(x) \neq *$, and so $val^*(\mathcal{A})$ excludes the string consisting of all $*$. In order to extend to an incomplete information system the properties and considerations about entropy and co–entropy of partitions described at the end of section (2), we have at least two different possibilities [7].

(i)   For any possible "value" $\alpha \in val^*(\mathcal{A})$, one can construct the *granule* $f_{\mathcal{A}}^{-1}(\alpha) = \{x \in X_{\mathcal{A}} : f_{\mathcal{A}}(x) = \alpha\}$ of X labelled by α, also denoted by $[\mathcal{A}, \alpha]$. The family of granules $gr(\mathcal{A}) = \{[\mathcal{A}, \alpha] : \alpha \in val^*(\mathcal{A})\}$ plus the *null granule* $[\mathcal{A}, *] = X \setminus X_{\mathcal{A}}$ (i.e., the collection of the objects in which all the attributes are unknown) constitutes a partition of the universe X, in which $gr(\mathcal{A})$ is a partition of the subset $X_{\mathcal{A}}$ of X (which can be considered as a "partial" partition of X).

(ii) Otherwise, we can consider the *covering* generated by a *similarity* (reflexive and symmetric, but in general nontransitive) relation. In the case of incompleteness it is often used the following relation [14]: two objects $x, y \in X$ are said to be *similar* if and only if $\forall a_i \in \mathcal{A} \subseteq Att$, either $f_{a_i}(x) = f_{a_i}(y)$ or $f_{a_i}(x) = *$ or $f_{a_i}(y) = *$.

The corresponding options are the following two. The first one, related to the above point (i) and investigated in this paper, involves partial partitions (related to "probabilistic partitions", i.e., partitions with respect to a measure m on events from X for which $m(X \setminus X_\mathcal{A}) = 0$). The second one, related to the point (ii), widely treated in literature by almost all the authors devoted to this argument (see for instance [13,15]), is applied to coverings [7]. The main point of difference, which gives to the approach (i) a real content of novelty, is that it is based on a generalization to probability partitions of the more economical *global* co–entropy (2.1a), whereas the approach (ii) generalizes the more complex *pointwise* co–entropy (2.2a) applied to coverings. Let us recall that in [7] different attempts has been investigated in order to give a global notion of co–entropy in the context of coverings, but all these attempts has been failed from the point of view of the monotonicity requirement. Finally, it is important that the results about incomplete ISs are not confused with the intrinsic arguments about complete ISs. These latter (as treated for instance in [16,12,10]), has to do with a narrow situation whose extension to the incomplete case is not trivial, and certainly original. This is what we discuss in the remaining part of the paper.

4   Entropies for Incomplete Information Systems

For any subset $\mathcal{A}$ of attributes of an incomplete information system, for the sake of simplicity, let us denote $val^*(\mathcal{A})$ as $V_\mathcal{A}^*$, for any $\alpha \in V_\mathcal{A}^*$ the corresponding granule $f_A^{-1}(\alpha)$ as A_α and let us set $X_\mathcal{A}^* = X \setminus X_\mathcal{A}$. Let us remark that the following holds: $x \notin X_\mathcal{A}$ iff $\forall a \in \mathcal{A} : f^*(x) = *$. Hence, the complementing (of $X_\mathcal{A}$ with respect to X) domain $X_\mathcal{A}^*$ is the collection of all states in which each attribute f_a of the family $\mathcal{A}$ is not defined (or in the information table the row corresponding to the object $x \in X_\mathcal{A}^*$ assumes the value $*$ in correspondence of any attribute $a \in \mathcal{A}$). From now on, if no confusion is likely, we simply use X^* instead of $X_\mathcal{A}^*$.

Now, we can define the measures $m_\mathcal{A}(A_\alpha) = |A_\alpha|$ and $m_\mathcal{A}(X^*) = 0$, and so $m_\mathcal{A}(X) = m_\mathcal{A}(\bigcup_\alpha A_\alpha \cup X^*) = \sum_\alpha m_\mathcal{A}(A_\alpha) + m_\mathcal{A}(X^*) = |X_\mathcal{A}|$, with the natural extension to the σ–algebra of events $\mathcal{E}_\mathcal{A}(X)$ from X generated by the elementary events $\{A_\alpha : \alpha \in V_\mathcal{A}^*\} \cup \{A^* \in 2^X : A^* \subseteq X^*\}$ (with $m(A^*) = 0$), obtaining in this way a *finite measure* $m_\mathcal{A} : \mathcal{E}_\mathcal{A}(X) \to \mathbb{R}_+$ depending from the set of attributes $\mathcal{A}$. In particular, the measure of the whole universe changes with the choice of $\mathcal{A}$. The corresponding probabilities are then $p(A_\alpha) = \frac{m_\mathcal{A}(A_\alpha)}{m_\mathcal{A}(X)} = \frac{|A_\alpha|}{|X_\mathcal{A}|}$ and $p(X^*) = 0$. According to a widely used terminology, the collection $\pi(\mathcal{A}) = \{A_\alpha : \alpha \in V_\mathcal{A}^*\}$ is a *probability partition* in the sense that the following hold: (1)   each $p(A_\alpha) > 0$; (2) $p(\bigcup_\alpha A_\alpha) = 1$; (3) $p(A_\alpha \cap A_\beta) = 0$ for $\alpha \neq \beta$.

Also in this case it is possible to define the *co–entropy* and the *entropy* of the probability partition generated by $\mathcal{A}$, similarly to (2.1a) and (2.1b), as follows:

$$E(\mathcal{A}) = \frac{1}{m_\mathcal{A}(X)} \sum_{\alpha \in V_\mathcal{A}^*} m(A_\alpha) \log m(A_\alpha) \tag{4.1a}$$

$$H(\mathcal{A}) = - \sum_{\alpha \in V_\mathcal{A}^*} p(A_\alpha) \log p(A_\alpha) = - \sum_{\alpha \in V_\mathcal{A}^*} \frac{m_\mathcal{A}(A_\alpha)}{m_\mathcal{A}(X)} \log \frac{m_\mathcal{A}(A_\alpha)}{m_\mathcal{A}(X)} \tag{4.1b}$$

Trivially, $H(\mathcal{A})+E(\mathcal{A}) = \log|X_{\mathcal{A}}| = \log(m_{\mathcal{A}}(X))$, i.e., the non–negative quantity $E(\mathcal{A})$ "complements" the entropy $H(\mathcal{A})$ with respect to the value $\log(m_{\mathcal{A}}(X))$, which depends on the attribute collection $\mathcal{A}$. Let us remark that under the order condition $\mathcal{A} \subseteq \mathcal{B}$ on attributes we cannot state in general either $H(\mathcal{A}) \leq H(\mathcal{B})$ or $H(\mathcal{A}) \geq H(\mathcal{B})$. If for any collection $\mathcal{A}$ of attributes one defines the (globally normalized) probability $p^*(A_\alpha) = \frac{|A_\alpha|}{|X|}$, then the following definition can be given.

Definition 4.1. *Let $\langle X, Att, F \rangle$ be an incomplete information system, $\mathcal{A} \subseteq Att$ a collection of attributes, $X_{\mathcal{A}} \subseteq X$ the corresponding definition domain, and $\pi(\mathcal{A})$ the related pseudo–probability partition (pseudo since the condition (2) of probability partitions must be substituted by $p^*(\cup_\alpha A_\alpha) = |X_{\mathcal{A}}|/|X| \leq 1$).*
Then, we define the following co–entropy and entropy:

$$\tilde{E}(\mathcal{A}) := \left(\frac{|X| - |X_{\mathcal{A}}|}{|X|} \right) \log|X| + \frac{1}{|X|} \sum_{\alpha \in V_{\mathcal{A}}^*} |A_\alpha| \log|A_\alpha| \tag{4.2a}$$

$$\tilde{H}(\mathcal{A}) := - \sum_{\alpha \in V_{\mathcal{A}}^*} p^*(A_\alpha) \log p^*(A_\alpha) = -\frac{1}{|X|} \sum_{\alpha \in V_{\mathcal{A}}^*} |A_\alpha| \log \frac{|A_\alpha|}{|X|} \tag{4.2b}$$

Also in this case we have that $\tilde{H}(\mathcal{A}) + \tilde{E}(\mathcal{A}) = \log|X|$. The following important result about monotonicity holds.

Theorem 4.1 (Monotonicity of $\tilde{H}(\mathcal{A})$). *Given an incomplete information system, let $\mathcal{A} \subseteq \mathcal{B}$ be two collections of attributes, and $\pi(\mathcal{B})$ and $\pi(\mathcal{A})$ the corresponding probability partitions. Then we have $\tilde{H}(\mathcal{A}) \leq \tilde{H}(\mathcal{B})$.*
Moreover, under the condition $|X_{\mathcal{B}}| > |X_{\mathcal{A}}|$ the following strict monotonicity holds: $\mathcal{A} \subset \mathcal{B}$ implies $\tilde{H}(\mathcal{A}) < \tilde{H}(\mathcal{B})$.

As a direct consequence of theorem 4.1, and making use of $\tilde{H}(\mathcal{A}) + \tilde{E}(\mathcal{A}) = \log|X|$, we have the following corollary regarding the co–entropy $\tilde{E}(\mathcal{A})$.

Corollary 4.1 (Anti–monotonicity of $\tilde{E}(\mathcal{A})$). *Let $\mathcal{A}, \mathcal{B}$ be two collections of attributes such that $\mathcal{A} \subseteq \mathcal{B}$. Then we have $\tilde{E}(\mathcal{B}) \leq \tilde{E}(\mathcal{A})$.*

5   Conclusions and Open Problems

We have illustrated a partitioning approach for incomplete information systems which take into account the incomplete nature producing at the same time a *probability* partition from one side (probability $p(A_\alpha) = |A_\alpha|/|X_{\mathcal{A}}|$) and a pseudo–probability partition on the other side (probability $p^*(A_\alpha) = |A_\alpha|/|X|$). We have then presented a definition of entropy and co–entropy for incomplete information systems based on the described partitioning approach.

We have shown that the entropy behaves monotonically and the co–entropy anti–monotonically, with respect to the collections of attributes. Let us stress that both the here defined co–entropies (4.1a) and (4.2a) result to be a generalization of the co–entropy (2.1a) of complete information systems.

The further step in this research will be the application of our co–entropy to the construction of reducts and rules in "real" information tables and the comparison, also from a computational point of view, with the "pointwise" co–entropy based on coverings considered in [7]. Indeed, even if the procedures to compute the here introduced co-entropy and the "pointwise" one are in the same complexity class, it can be easily seen that the former one always requires less operations than the last one.

References

1. Pawlak, Z.: Information systems - theoretical foundations. Information Systems **6** (1981) 205–218
2. Cattaneo, G.: Abstract approximation spaces for rough theories. [17] 59–98
3. Pawlak, Z.: Rough sets: Theoretical Aspects of Reasoning about Data. Kluwer Academic Publishers, Dordrecht (1991)
4. Komorowski, J., Pawlak, Z., Polkowski, L., Skowron, A.: Rough sets: A tutorial. In Pal, S., Skowron, A., eds.: Rough Fuzzy Hybridization. Springer–Verlag, Singapore (1999) 3–98
5. Cattaneo, G., Ciucci, D.: Investigation about Time Monotonicity of Similarity and Preclusive Rough Approximations in Incomplete Information Systems. Volume 3066 of Lecture Notes in Artificial Intelligence., Springer–Verlag (2004) 38–48
6. Preparata, F.P., Yeh, R.T.: Introduction to Discrete Structures. Addison–Wesley Publ. Comp., Reading, Massachusetts (1973)
7. Bianucci, D., Cattaneo, G., Ciucci, D.: Entropies and co–entropies of coverings with application to incomplete information systems. Fundamenta Informaticae **75** (2007) 77–105
8. Beaubouef, T., Petry, F.E., Arora, G.: Information–theoretic measures of uncertainty for rough sets and rough relational databases. Journal of Information Sciences **109** (1998) 185–195
9. Shannon, C.E.: A mathematical theory of communication. The Bell System Technical Journal **27** (1948) 379–423, 623–656
10. Yao, Y.: Probabilistic approaches to rough sets. Expert Systems **20** (2003) 287–297
11. Reza, F.M.: An Introduction to Information theory. Dover Publications, New York (1994) (originally published by Mc Graw-Hill, New York, 1961).
12. Duntsch, I., Gediga, G.: Roughian: Rough information analysis. International Journal of Intelligent Systems **16** (2001) 121–147
13. Liang, J., Xu, Z.: Uncertainty measure of randomness of knowledge and rough sets in incomplete information systems. Intelligent Control and Automata **4** (2000) 2526–2529 Proc. of the 3rd World Congress on Intelligent Control and Automata.
14. Kryszkiewicz, M.: Rough set approach to incomplete information systems. Information Sciences **112** (1998) 39–49
15. Huang, B., He, X., Zhong, X.: Rough entropy based on generalized rough sets covering reduction. Journal of Software **15** (2004) 215–220
16. Gediga, G., Duntsch, I.: Rough approximation quality revisited. Artificial Intelligence **132** (2001) 219–234
17. Polkowski, L., Skowron, A., eds.: Rough Sets in Knowledge Discovery 1. Physica–Verlag, Heidelberg, New York (1998)

Granular Computing Based on a Generalized Approximation Space

Jian-Min Ma[1], Wen-Xiu Zhang[1], Wei-Zhi Wu[2], and Tong-Jun Li[1,2]

[1] Institute for Information and System Sciences, Faculty of Science,
Xi'an Jiaotong University, Xi'an, Shaan'xi 710049, P.R. China
majm@mail.xjtu.edu.cn, wxzhang@mail.xjtu.edu.cn
[2] Mathematics, Physics and Information College,
Zhejiang Ocean University, Zhoushan, Zhejiang, 316004, P.R. China
wuwz8681@sina.com, litj@zjou.net.cn

Abstract. A family of overlapping granules can be formed by granulating a finite universe under a binary relation in a set-theoretic setting. In this paper, we granulate a universe by a binary relation and obtain a granular universe. And then we define two kinds of operators between these two universes, study properties of them. By combining these two kinds of operators, we get two pairs of approximation operators. It is proved that one kind of combination operators is just the approximation operators under a generalized approximation space defined according to Pawlak's rough set theory.

Keywords: Generalized approximation space, $L-$lower approximation operator, $H-$upper approximation operator, Similarity relation.

1 Introduction

Granular computing is a label of theories, methodologies, techniques, and tools that makes use of granules, i.e., groups, classes, or clusters of a universe, in the process of problem solving [10, 14, 16]. Since Pawlak introduced the theory of rough sets [7,8], it has made granular computing popular . Hobbs [2] introduced the concepts of granularity in 1985. Later the concept "granular computing" was suggested by Zadeh [15, 16] for the first time in 1996. The basic ideas of information granulation have been explored in many fields, such as rough sets, fuzzy sets, cluster analysis, database, machine learning, data -mining, and so on. There is a renewed and fast growing interest in the study of granular computing [3, 4, 10, 13].

As a concrete theory of granular computing, rough set model enables us to precisely define and analyze many notions of granular computing. The results provide an in-depth understanding of granular computing. Many models of granular computing have been proposed and studied [16, 11]. However, there are many fundamental issues in granular computing, such as granulation of the universe, description of granules, relationships between granules, and computing with granules.

J.T. Yao et al. (Eds.): RSKT 2007, LNAI 4481, pp. 93–100, 2007.
© Springer-Verlag Berlin Heidelberg 2007

Yao [12] proposed a concrete model of granular computing based on a simple granulation structure, namely, a partition of a universe. Results from rough sets, quotient space theory, belief functions, and power algebra are reformulated, reinterpreted, and combined for granular computing. For the universe and the coarse-grained universe induced by an equivalence relation, two basic operation called zooming-out and zooming-in operations are introduced. And Computations in these universes can be connected through the two operations.

Because the equivalence relation in [12] is too strong to be obtained in general, we only consider a reflexive relation on a universe which is easy to obtain usually. Then a covering model can be obtained by granulating a finite set of a universe based on the reflexive relation [6]. And we cited definitions of zooming-out and zooming-in operations in [12] and discussed the covering model of granular computing [6]. However, relationships between subsets of a coarse-grained universe would not hold in the universe. Furthermore, although rough set approximations of a classical subset of a universe in a generalized approximation space [17] can be obtained by a combination of these operations, the duality may not hold.

In this paper, we first granulate a finite set of a universe into a family of overlapping granules based on a general binary relation. We introduce two kinds of operators between a universe and the granulated universe, and study their properties. Then we combine them to two pairs of approximation operators, which are used to study connections between computations in the two universes. It is also proved that approximation representations of a generalized approximation space can be obtained by combining them, and the duality always holds for the different combinations.

This paper is organized as follows. Section 2 introduces two kinds of operators between a universe and a granulated universe, and studies their properties. Section 3 shows new operations formed by different combining the two operations, investigates their properties, and discusses connections between computations in the two universes. Finally, Section 4 concludes the paper.

2   Preliminaries

Let U be a finite and nonempty set called a universe, and $r \subseteq U \times U$ a binary relation on the universe U. For any $x \in U$, the set $r(x) = \{y \in U; \ (x,y) \in r\}$ is called the successor neighborhood of x. The relation r is referred to as serial if for any $x \in U$, there exists $y \in U$ such that $y \in r(x)$. r is referred to as reflexive if for all $x \in U$, $x \in r(x)$; r is referred to as symmetric if for all $x, y \in U$, $x \in r(y)$ implies $y \in r(x)$; r is referred to as transitive if for all $x, y, z \in U$, $x \in r(y)$ and $y \in r(z)$ implies $x \in r(z)$; r is referred to as Euclidean if for all $x, y, z \in U$, $y \in r(x)$ and $z \in r(x)$ implies $z \in r(y)$ [9,17]. Furthermore, r is referred to as a similarity relation on U if it is reflexive and transitive; r is referred to as a tolerance relation on U if it is reflexive and symmetric. For any binary relation $r \subseteq U \times U$, the pair (U, r) is referred to as a generalized approximation space.

For a generalized approximation space (U, r), the family of all successor neighborhood, denoted by $\mathcal{A} = \{r(x); \ x \in U\}$, is commonly knows as the granulated set. If r is reflexive, it forms a covering of U, namely, a family of overlapping subsets whose union is U. For any $x \in U$, the successor neighborhood $r(x)$ is considered as a whole granule instead of many individuals [12]. It is a subset of U and an element of $\mathcal{A}$. We use $|r(x)|$ to denote the whole granule $r(x)$, and call $A = \{|r(x)|; \ x \in U\}$ a granulated universe. We denote by 2^U the power set of the universe U, and by ^{c} the set complement operator.

Definition 1. *Let (U, r) be a generalized approximation space. For any $B \subseteq A$, the mapping $f : 2^A \to 2^U$ is given by*

$$f(B) = \{x \in U; \ |r(x)| \in B\}.$$

Then we can get the following properties: for any $B, C \subseteq A$,
  (1) $f(\emptyset) = \emptyset$;
  (1) $f(A) = U$;
  (2) $f(B \cup C) = f(B) \cup f(C)$;
  (3) $f(B \cap C) = f(B) \cap f(C)$;
  (4) $f(B^c) = f(B)^c$;
  (5) $B \subseteq C \Leftrightarrow f(B) \subseteq f(C)$.

Definition 2. *Let (U, r) be a generalized approximation space and $X \subseteq U$. A pair (L, H) of mappings $L, H : 2^U \to 2^A$ is defined as follows:*

$$L(X) = \{|r(x)|; \ r(x) \subseteq X\},$$
$$H(X) = \{|r(x)|; \ r(x) \cap X \neq \emptyset\}.$$

They are called $L-$lower and $H-$upper approximation of X, respectively.

By Definition 2 we can easily get that for a generalized approximation space (U, r), $X, Y \subseteq U$,
  (LH1) $H(\emptyset) = \emptyset$;
  (LH2) $L(U) = A$;
  (LH3) $L(X^c) = H(X)^c, H(X^c) = L(X)^c$;
  (LH4) $L(X \cap Y) = L(X) \cap L(Y), H(X \cap Y) \subseteq H(X) \cap H(Y)$;
  (LH5) $L(X \cup Y) \supseteq L(X) \cup L(Y), H(X \cup Y) = H(X) \cup H(Y)$;
  (LH6) $X \subseteq Y \Rightarrow L(X) \subseteq L(Y), H(X) \subseteq H(Y)$;
  (LH7)   Let $B_n(X) = H(X) - L(X)$, then $B_n(X^c) = B_n(X)$.
  If r is serial, we have $L(X) \subseteq H(X)$, $L(\emptyset) = \emptyset$ and $H(U) = A$.

Remark 1. In fact, there exists $X \subseteq U$ and $X \neq \emptyset$ such that $L(X) = \emptyset$; and there is $X \neq U$ such that $H(X) = A$.

In general, the following formulas may not hold:

$$H(X \cap Y) = H(X) \cap H(Y),$$
$$L(X \cup Y) = L(X) \cup L(Y).$$

Example 1. Suppose $U = \{x_1, x_2, x_3, x_4, x_5\}$, and $r \subseteq U \times U$ be a binary relation on U satisfying: $r(x_1) = \{x_3, x_4\}$, $r(x_2) = \{x_1, x_2, x_4\}$, $r(x_3) = \{x_3\}$, $r(x_4) = \{x_4\}$, $r(x_5) = \{x_2, x_5\}$.

(1) Take $X = \{x_3, x_4\}$ and $Y = \{x_2, x_3\}$. Then $X \cap Y = \{x_3\}$ and $H(X \cap Y) = \{|r(x_1)|, |r(x_3)|\}$, but $H(X) \cap H(Y) = \{|r(x_1)|, |r(x_2)|, |r(x_3)|, |r(x_4)|\} \cap \{|r(x_1)|, |r(x_2)|, |r(x_3)|, |r(x_5)|\} = \{|r(x_1)|, |r(x_2)|, |r(x_3)|\}$. Hence $H(X \cap Y) \subset H(X) \cap H(Y)$.

(2) Take $X = \{x_1, x_2\}$ and $Y = \{x_3\}$. Then $X \cup Y = \{x_1, x_2, x_3\}$ and $L(X \cup Y) = \{|r(x_3)|, |r(x_5)|\}$, however $L(X) \cup L(Y) = \emptyset \cup \{|r(x_3)|\} = \{|r(x_3)|\}$. Hence $L(X) \cup L(Y) \subset L(X \cup Y)$.

Proposition 1. *Let (U, r) be a generalized approximation space and $X, Y \subseteq U$. Note that $\underline{Z}(X, Y) = \{|r(x)|;\ r(x) \subseteq X \cup Y, |r(x)| \in B_n(X) \cap B_n(Y)\}$. Then*

$$L(X \cup Y) = L(X) \cup L(Y) \cup \underline{Z}(X, Y).$$

Proof. It is easy to see that $L(X) \cup L(Y) \subseteq L(X \cup Y)$. Then $|r(x)| \in L(X \cup Y) - L(X) \cup L(Y)$ if and only if $r(x) \subseteq X \cup Y$, $r(x) \not\subseteq X$ and $r(x) \not\subseteq Y$. Then $|r(x)| \in L(X \cup Y) - L(X) \cup L(Y)$ if and only if $r(x) \subseteq X \cup Y$, and $|r(x)| \in B_n(X) \cap B_n(Y)$. That is $L(X \cup Y) - L(X) \cup L(Y) = \underline{Z}(X, Y)$. Therefore $L(X \cup Y) = L(X) \cup L(Y) \cup \underline{Z}(X, Y)$.

Proposition 2. *Let (U, r) be a generalized approximation space and $X, Y \subseteq U$. Note that $\overline{Z}(X, Y) = \{|r(x)|;\ r(x) \cap (X \cap Y) = \emptyset, |r(x)| \in B_n(X) \cap B_n(Y)\}$. Then*

$$H(X \cap Y) = H(X) \cap H(Y) - \overline{Z}(X, Y).$$

Proof. By (LH5) we can get that $H(X \cap Y) \subseteq H(X) \cap H(Y)$. Then $|r(x)| \in H(X) \cap H(Y) - H(X \cap Y)$ if and only if $r(x) \cap X \neq \emptyset, r(x) \cap Y \neq \emptyset$ and $r(x) \cap (X \cap Y) = \emptyset$. Then $|r(x)| \in H(X) \cap H(Y) - H(X \cap Y)$ if and only if $|r(x)| \in B_n(X) \cap B_n(Y)$, and $r(x) \cap (X \cap Y) = \emptyset$. Therefore $H(X) \cap H(Y) - H(X \cap Y) = \overline{Z}(X, Y)$. Thus $H(X \cap Y) = H(X) \cap H(Y) - \overline{Z}(X, Y)$.

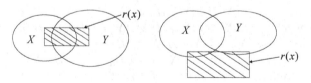

Fig. 1. $\underline{Z}(X, Y)$ Fig. 2. $\overline{Z}(X, Y)$

3 Rough Approximation Representations on U and A

Pawlak's classical rough set theory shows that lower and upper approximations of a classical set are also subsets of the same universe. From Definition 1 and

Definition 2 we can combine these operators f and (L, H), and get some new operators on a same universe.

By Definitions 1 and 2, one can easily obtain lower and upper approximations of a classical subset of the generalized approximation space by performing a combination of (L, H) and f as follows:

$$fL(X) = f(\{|r(x)|; \ r(x) \subseteq X\}) = \{x; \ r(x) \subseteq X\},$$
$$fH(X) = f(\{|r(x)|; \ r(x) \cap X \neq \emptyset\}) = \{x; \ r(x) \cap X \neq \emptyset\}.$$

Then $fL, fH : 2^U \to 2^U$ are called $fL-$lower and $fH-$upper approximation operators, respectively.

Since we have studied properties of f and (L, H), we can easily get the following properties for any $X, Y \subseteq U$:

(fLH1) $fH(\emptyset) = \emptyset$;

(fLH2) $fL(U) = U$;

(fLH3) $fL(X^c) = fH(X)^c, \ fH(X^c) = fL(X)^c$;

(fLH4) $fL(X \cap Y) = fL(X) \cap fL(Y),$
$fH(X \cap Y) \subseteq fH(X) \cap fH(Y)$;

(fLH5) $fL(X \cup Y) \supseteq fL(X) \cup fL(Y),$
$fH(X \cup Y) = fH(X) \cup fH(Y)$;

(fLH6) $X \subseteq Y \Rightarrow fL(X) \subseteq fL(Y), \ fH(X) \subseteq fH(Y)$;

(fLH7) Let $B_1(X) = fH(X) - fL(X)$, then $B_1(X^c) = B_1(X)$.

Note that (fLH7) means simply that if we cannot decide when an object is in X, we obviously cannot decide whether it is in X^c either. (fLH3) shows that fL and fH are the dual approximation operators.

If r is reflexive, then $fL(X) \subseteq X \subseteq f(H(X), \ fL(\emptyset) = \emptyset, fH(U) = U$.

Proposition 3. *Let (U, r) be a generalized approximation space. If r is reflexive and Euclidean, for any $X \subseteq U$, we have*

(1) $fH(X) = fL(fH(X))$;

(2) $fL(X) = fH(fL(X))$.

Proof. Since r is reflexive, we have $fL(fH(X)) \subseteq fH(X)$. Take $x \in fH(X)$. Then $r(x) \cap X \neq \emptyset$. For any $y \in r(x)$, we have $r(x) \subseteq r(y)$ because r is Euclidean. Hence $r(y) \cap X \neq \emptyset$ and $y \in fH(X)$. By the arbitrariness of y we can get $r(x) \subseteq fH(X)$ which leads to $x \in fL(fH(X))$. Therefore $fL(fH(X)) = fH(X)$. By (1) and (fLH3) we can easily get (2).

According to Propositions 1 and 2 we can get:

Proposition 4. *Let (U, r) be a generalized approximation space. Then for any $X, Y \subseteq U$, we have*

(1) $fL(X \cup Y) = fL(X) \cup fL(Y) \cup \underline{Z}_1(X, Y)$, where
$\underline{Z}_1(X, Y) = \{x \in U; \ r(x) \subseteq X \cup Y, \ r(x) \subseteq B_1(X) \cap B_1(Y)\}.$

(2) $fH(X \cap Y) = fH(X) \cap fH(Y) - \overline{Z}_1(X, Y)$, where
$\overline{Z}_1(X, Y) = \{x \in U; \ r(x) \cap (X \cap Y) = \emptyset, \ r(x) \subseteq B_1(X) \cap B_1(Y)\}.$

For a subset $B \subseteq A$ we can obtain a subset $f(B) \subseteq U$, and then obtain a pair of subsets $Lf(B)$ and $Hf(B)$ as follows:

$$Lf(B) = \{|r(x)|; \ r(x) \subseteq f(B)\} = \{|r(x)|; \ r(x) \subseteq \{y; |r(y)| \in B\}\},$$

$$Hf(B)) = \{|r(x)|; \ r(x) \cap f(B) \neq \emptyset\} = \{|r(x)|; \ \exists y \in r(x), |r(y)| \in B\}.$$

Then $Lf, Hf : 2^A \to 2^A$ are called Lf–lower and Hf–upper approximation operators, respectively. And these two approximation operators give out the approximation representations of B in the granulated universe A.

So for any $B, C \subseteq A$ we have the following properties:

(LHf1) $Hf(\emptyset) = \emptyset$;

(LHf2) $Lf(A) = A$;

(LHf3) $Lf(B^c) = (Hf(B))^c$, $Hf(B^c) = (Lf(B))^c$;

(LHf4) $Lf(B \cap C) = Lf(B) \cap Lf(C)$,
        $Hf(B \cap C) \subseteq Hf(B) \cap Hf(C)$;

(LHf5) $Lf(B \cup C) \supseteq Lf(B) \cup Lf(C)$,
        $Hf(B \cup C) = Hf(B) \cup Hf(C)$;

(LHf6) $B \subseteq C \Rightarrow Lf(B) \subseteq Lf(C)$, $Hf(B) \subseteq Hf(C)$;

(LHf7) Let $B_2(B) = Hf(B) - Lf(B)$, then $B_2(B^c) = B_2(B)$.

If r is serial, $Lf(\emptyset) = \emptyset$ and $Hf(A) = A$; if r is reflexive, we have $Lf(B) \subseteq B \subseteq Hf(B)$.

For any $X \subseteq U$ and $B \subseteq A$, by the different combinations of f and (L, H) we can get

$$fL(f(B)) = f(Lf(B)),$$

$$Lf(L(X)) = L(fL(X)).$$

Therefore, we can easily get

Lemma 1. *Let (U, r) be a generalized approximation space. If r is a similarity relation, for any $X \subseteq U$ we have*

   (1) $L(fL(X)) = L(X)$;
   (2) $H(fH(X)) = H(X)$.

Proof. Since r is reflexive, $L(fL(X)) \subseteq L(X)$. Conversely, take $|r(x)| \in L(X)$. Then $r(x) \subseteq X$. Since r is transitive, for any $y \in r(x)$ we have $r(y) \subseteq r(x)$. Therefore $|r(y)| \in L(X)$ and $y \in fL(X)$. By the arbitrariness of y we can get $r(x) \subseteq fL(X)$. Hence $|r(x)| \in L(fL(X))$, and $L(X) \subseteq L(fL(X))$. From which we get $L(fL(X)) = L(X)$. By (LHf3) we can prove (2). $\square$

However, the following formulae may not hold:

$$fL(f(B)) = f(B),$$

$$fH(f(B)) = f(B).$$

Example 2. Suppose $U = \{x_1, x_2, x_3, x_4, x_5\}$, and r is a similarity relation on U satisfying: $r(x_1) = \{x_1, x_3, x_4\}$, $r(x_2) = \{x_2\}$, $r(x_3) = \{x_3, x_4\}$, $r(x_4) = \{x_4\}$, $r(x_5) = \{x_2, x_5\}$. Then $A = \{|r(x_1)|, |r(x_2)|, |r(x_3)|, |r(x_4)|, |r(x_5)|\}$.

(1)   Take $B = \{|r(x_1)|, |r(x_3)|\}$, then $f(B) = \{x_1, x_3\}$, and $L(f(B)) = L(\{x_1, x_3\}) = \emptyset$. But $fL(f(B)) = f(\emptyset) = \emptyset \neq f(B)$.

(2)   Take $B = \{|r(x_4)|, |r(x_5)|\}$, then $f(B) = \{x_4, x_5\}$, and $H(f(B)) = \{|r(x_1)|, |r(x_3)|, |r(x_4)|, |r(x_5)|\}$. However $fH(f(B)) = f(\{|r(x_1)|, |r(x_3)|, |r(x_4)|, |r(x_5)|\}) = \{x_1, x_3, x_4, x_5\} \neq f(B)$.

Proposition 5. *Let (U, r) be a generalized approximation space. If r is a similarity relation, then for any $B \subseteq A$ we have*

(1) $Lf(Lf(B)) = Lf(B)$;

(2) $Hf(Hf(B)) = Hf(B)$.

In addition, according to Propositions 1, 2 and properties of f we can get:

Proposition 6. *Let (U, r) be a generalized approximation space. If r is reflexive, then for any $B, C \subseteq A$,*

(1) $Lf(B \cup C) = Lf(B) \cup Lf(C) \cup \underline{Z}_2(B, C)$; where
  $\underline{Z}_2(B, C) = \{|r(x)|;\ r(x) \subseteq f(B \cup C), r(x) \subseteq B_2(B) \cap B_2(C)\}$.

(2) $Hf(B \cap C) = Hf(B) \cap Hf(C) - \overline{Z}_2(B, C)$; where
  $\overline{Z}_2(B, C) = \{|r(x)|;\ r(x) \cap f(B \cap C) = \emptyset, r(x) \subseteq B_2(B) \cap B_2(C)\}$.

Literatures [7, 17] define lower and upper approximation operators for a generalized approximation space (U, R) with R being a binary relation on U as follows:

$$\underline{R}(X) = \{x \in U; R_s(x) \subseteq X\},$$
$$\overline{R}(X) = \{x \in U; R_s(x) \cap X \neq \emptyset\},$$

where $R_s(x)$ denotes the successor neighborhood of x. Obviously, for a generalized approximation space (U, r), $fL(X) = \underline{r}(X)$ and $fH(X) = \overline{r}(X)$. Since we have studied properties of operators f and (L, H), we can easily get properties of $(\underline{r}(X), \overline{r}(X))$.

4   Conclusion

Granular computing is a way of thinking that relies in our ability to perceive the real world under various grain sizes, to abstract and consider only those things that serve our present interest, and to switch among different granularities. In this paper, two kinds of operators have been introduced between a universe and a granulated universe based on a generalized binary relation. Connections between the elements of a universe and the elements of a granulated universe, as well as connections between computations in the two universes are investigated by two pairs of combination operators.

Acknowledgement

This work was supported by a grant from the National Natural Science Foundation of China (No. 60673096), the National 973 Program of China (No. 2002CB 312200) and the Scientific Research Project of the Education Department of Zhejiang Province in China (No. 20061126).

References

1. Dubois, D., Prade, H.: Rough fuzzy sets and fuzzy rough sets, International Journal of General Systems, 17 (1990) 191-208
2. Hobbs, J.R.: Granularity. In: Proc. of IJCAI, Los Angeles, (1985) 432-435
3. Lin, T.Y.: Granular computing on binary relations I: data mining and neighborhood systems, II: rough set representations and belief functions. In: Rough Sets in Knowledge Discovery 1. Polkowski, L., Skowron, A(Eds.) Physica-Verlag, Heidelberg (1998) 107-140
4. Lin, T.Y.: Generating concept hierarchies/networks:mining additional semantics in relational data. Advances in Knowledge Discovery and Data Mining, Proceedings of the 5th Pacifi-Asia Conference, Lecture Notes on Artificial Intelligence 2035 (2001) 174-185
5. Lin, T.Y.: Granular computing, LNAI 2639 (2003) 16-24
6. Ma, J.-M., Zhang, W.-X., Li,T.-J.: A covering model of granular computing, Proceedings of 2005 International Conference in Machine Learning and Cybernetics (2005) 1625-1630
7. Pawlak, Z.: Rough sets, International Journal of Computer and Information Science 11 (1982) 341–356
8. Pawlak, Z.: Rough sets, Kluwer Academic Publishers. Dordrecht Boston London (1991)
9. Skowron, A., Stepaniuk, J.: Information granules: towards foundations of granular computing, International Journal of Intelligent Systems 16 (2001) 57–85
10. Yao, Y.Y.: Granular computing: basic issues and possible solutions, Proceedings the 5th joint conference on information sciences (2000) 186-189
11. Yao, Y.Y.: Information granulation and rough set approximation, International Journal of Intelligent systems, 16 (2001) 87-104
12. Yao, Y.Y.: A partition model of granular computing, LNCS 3100 (2004) 232-253
13. Yao, J.T., Yao, Y.Y.: Introduction of classification rules by granular computing, LNAI 2475 (2002) 331-338
14. Zadeh, L.A.: Fuzzy sets and information granularity. In: Advances in Fuzzy Set Theory and Applications, Gupta, N, Ragade, R and Yager. eds. NorthHolland, Amsterdam (1979) 3-18
15. Zadeh, L.A.: Fuzzy logic-computing with words. IEEE Transactions on Fuzzy Systems, 4 (1996) 103-111
16. Zadeh, L.A.: Towards a theory of fuzzy information granulation and its centrality in human reasoning and fuzzy logic, Fuzzy Sets and Systems **19** (1997) 111-127
17. Zhang, W.-X., Wu, W.-Z., Liang, J.-Y., Li, D.-Y.: Theory and Methods of Rough Sets, Science Press, Beijing (2001)

A General Definition of an Attribute Reduct

Yan Zhao, Feng Luo, S.K.M. Wong, and Yiyu Yao

Department of Computer Science, University of Regina
Regina, Saskatchewan, Canada S4S 0A2
{yanzhao,luo202,yyao}@cs.uregina.ca, skmwong@rogers.com

Abstract. A reduct is a subset of attributes that are jointly sufficient and individually necessary for preserving a particular property of a given information table. A general definition of an attribute reduct is presented. Specifically, we discuss the following issues: First, there are a variety of properties that can be observed in an information table. Second, the preservation of a certain property by an attribute set can be evaluated by different measures, defined as different fitness functions. Third, by considering the monotonicity property of a particular fitness function, the reduct construction method needs to be carefully examined. By adopting different heuristics or fitness functions for preserving a certain property, one is able to derive most of the existing definitions of a reduct. The analysis brings new insight into the problem of reduct construction, and provides guidelines for the design of new algorithms.

Keywords: attribute reducts, property preservation functions, monotonicity of evaluation function.

1 Introduction

In many data analysis applications, information and knowledge are stored and represented in an information table, where a set of objects is described by a set of attributes. We are faced with one practical problem: for a particular property, whether all the attributes in the attribute set are always necessary to preserve this property. Using the entire attribute set for describing the property is time-consuming, and the constructed rules may be difficult to understand, to apply or to verify. In order to deal with this problem, attribute selection is required. The theory of rough sets has been applied to data analysis, data mining and knowledge discovery. A fundamental notion supporting such applications is the concept of reducts [4]. The objective of reduct construction is to reduce the number of attributes, and at the same time, preserve the property that we want.

In the literature of rough set theory, there are many definitions of a reduct, and each focuses on preserving one specific type of property. This results in two problems: first, all these existing definitions have the same structure, and there is a lack of a higher level of abstraction. Second, along with the increasing requirements of data analysis, we need to find more properties of an information table. This naturally leads to more definitions in different forms. For these two reasons, a general definition of an attribute reduct is necessary and useful [5]. A general definition is suggested in Section 2. After that, three issues are discussed in detail.

J.T. Yao et al. (Eds.): RSKT 2007, LNAI 4481, pp. 101–108, 2007.
© Springer-Verlag Berlin Heidelberg 2007

2   A Definition of a Reduct

An information table provides a convenient way to describe a finite set of objects called a universe by a finite set of attributes [4]. It represents all available information and knowledge. That is, objects are only perceived, observed, or measured by using a finite number of attributes.

Definition 1. *An information table is the following tuple:*

$$S = (U, At, \{V_a \mid a \in At\}, \{I_a \mid a \in At\}),$$

where U is a finite nonempty set of objects, At is a finite nonempty set of attributes, V_a is a nonempty set of values of $a \in At$, and $I_a : U \rightarrow V_a$ is an information function that maps an object of U to exactly one value in V_a.

A general definition of an attribute reduct is given as follows.

Definition 2. *Given an information table $S = (U, At, \{V_a \mid a \in At\}, \{I_a \mid a \in At\})$, consider a certain property $\mathbb{P}$ of S and $R \subseteq A \subseteq At$. An attribute set R is called a reduct of $A \subseteq At$ if it satisfies the following three conditions:*

(1.) *Evaluability condition: the property can be represented by an evaluation function $e : 2^{At} \longrightarrow (L, \preceq)$;*
(2.) *Jointly sufficient condition: $e(A) \preceq e(R)$;*
(3.) *Individually necessary condition: for any $R' \subset R$, $\neg(e(A) \preceq e(R'))$.*

An evaluation or fitness function, $e : 2^{At} \longrightarrow (L, \preceq)$, maps an attribute set to an element of a poset L equipped with the partial order relation $\preceq$, i.e., $\preceq$ is reflexive, anti-symmetric and transitive. For each property, we can use an evaluation function as its indicator. Normally, the fitness function is not unique. By applying the function e, we are able to pick the attribute set that preserves the property $\mathbb{P}$. Suppose we target the attribute set A, then the evaluation of a candidate reduct R ($e(R)$) should be the same or superior to $e(A)$. In many cases, we have $e(R) = e(A)$.

There are many properties that can be observed in an information table. The discovery of a certain property allows us to describe the information of the universe, or to predict the unseen data in the future. A property can be well-defined and easy to be observed, for example, the size of the dataset and the dimension of the description space. Alternatively, a property can be understood as a previously unknown pattern to be discovered by a data analysis task, for example, an association of attributes, a cluster of objects, a set of classification rules, a preference ordering of objects, or the similarities or differences among objects.

3   Interpretations of Properties

To classify the properties of an information table is not an easy task, as properties have internal relationships, and there is no clear cut between different properties.

In addition, the number of properties is huge. Therefore, we only list some of the well-known properties.

3.1 Property $\mathbb{P}_1$: Descriptions of Object Relations

A binary object relation (i.e., a subset of $U \times U$) represents associations of one object with other objects (perhaps the same one). Pawlak defines the indiscernibility relation to summarize all indiscernible object pairs [4]. Given an attribute set $A \subseteq At$, the indiscernibility relation is defined as:

$$IND(A) = \{(x, y) \in U \times U \mid \forall a \in A, I_a(x) = I_a(y)\}. \tag{1}$$

If $(x, y) \in IND(A)$, then x and y are indiscernible with respect to A. We can also define a discernibility relation as:

$$DIS(A) = \{(x, y) \in U \times U \mid \forall a \in A, I_a(x) \neq I_a(y)\}. \tag{2}$$

If $(x, y) \in DIS(A)$, then x and y are different, and are discernible by any attribute in A. It is easy to relax the indiscernibility or the discernibility relation to define a similarity relation. For the indiscernibility relation IND, $IND(At)$ is finest, and $IND(\emptyset)$ is the coarsest. All relations form a poset under the set inclusion relation, which is embedded in $(2^{U \times U}, \subseteq)$. For the discernibility relation DIS, the order is reversed.

Skowron and Rauszer suggest a discernibility matrix that stores all the attributes that differentiate between any two objects of the universe [6]. Given an information table S, its discernibility matrix $\mathbf{dm}$ is a $|U| \times |U|$ matrix with each element $\mathbf{dm}(x, y)$ defined as:

$$\mathbf{dm}(x, y) = \{a \in At \mid I_a(x) \neq I_a(y), x, y \in U\},$$

where $|.|$ indicates cardinality of a set. The discernibility matrix $\mathbf{dm}$ is symmetric and $\mathbf{dm}(x, x) = \emptyset$. It is easy to verify that:

$$\forall (x, y) \in IND(A), A \cap \mathbf{dm}(x, y) = \emptyset;$$
$$\forall (x, y) \in DIS(A), A \subseteq \mathbf{dm}(x, y).$$

3.2 Property $\mathbb{P}_2$: Descriptions of Relative Object Relations

The indiscernibility, discernibility relations can be defined regarding to the labels of the objects. That is, we concern the indiscernibility relation of two objects if and only if they have the same label, and we concern the discernibility relation of two objects if and only if their labels are different.

Given an attribute that labels objects, an information table can be written as $S = (U, At = C \cup D, \{V_a \mid a \in At\}, \{I_a \mid a \in At\})$, where D is called the set of *decision attributes*, and C is called the set of *conditional attributes*. The D-relative indiscernibility and discernibility relations can be defined as:

$$IND_D(A) = \{(x, y) \in U \times U \mid \forall a \in A, I_a(x) = I_a(y) \wedge I_D(x) = I_D(y)\},$$
$$DIS_D(A) = \{(x, y) \in U \times U \mid \forall a \in A, I_a(x) \neq I_a(y) \wedge I_D(x) \neq I_D(y)\}.$$

Skowron and Rauszer's discernibility matrix can be used to store all the D-relative discernibility relations.

3.3   Property $\mathbb{P}_3$: Partitions of an Information Table

An indiscernibility relation induces a partition of the universe, denoted as π_A or $U/IND(A)$. Each block of the partition,

$$[x]_A = \{y \in U \mid \forall a \in A, I_a(x) = I_a(y)\}, \tag{3}$$

is an equivalence class containing x. For any two objects $x, y \in [x]_A$, $(x, y) \in IND(A)$.

One can obtain a finer partition by further dividing the equivalence classes of a partition. A partition π_1 is a refinement of another partition π_2, or equivalently, π_2 is a coarsening of π_1, denoted by $\pi_1 \preceq \pi_2$, if every block of π_1 is contained in some block of π_2. The partition $U/IND(At)$ is the finest partition and the partition $U/IND(\emptyset)$ is the coarsest partition. All partitions form a poset under the refinement relation, denoted as $(\Pi(U), \preceq)$.

3.4   Property $\mathbb{P}_4$: Descriptions of Concepts

To describe a concept, rough set theory introduces a pair of lower approximation $(\underline{apr})$ and upper approximation $(\overline{apr})$. Given an attribute set $A \subseteq At$, the lower and upper approximations of $X \subseteq U$ induced by A are defined by:

$$\underline{apr}_A(X) = \bigcup\{[x]_A \mid [x]_A \subseteq X\} = \bigcup\{[x]_A \mid \frac{|[x]_A \cap X|}{|[x]_A|} = 1\}; \tag{4}$$

$$\overline{apr}_A(X) = \bigcup\{[x]_A \mid [x]_A \cap X \neq \emptyset\} = \bigcup\{[x]_A \mid 0 < \frac{|[x]_A \cap X|}{|[x]_A|} \leq 1\}, \tag{5}$$

Probabilistic rough set models [9,11] relax the precision threshold from 1 to $\beta \in (0.5, 1]$. The β-level lower and upper approximations are defined as:

$$\underline{apr}_A^\beta(X) = \bigcup\{[x]_A \mid \frac{|[x]_A \cap X|}{|[x]_A|} \geq \beta\},$$

$$\overline{apr}_A^\beta(X) = \bigcup\{[x]_A \mid 0 < \frac{|[x]_A \cap X|}{|[x]_A|} < \beta\}.$$

A pair of approximation operators $(\underline{apr}_1^\beta, \overline{apr}_1^\beta)$ is larger than another pair of approximation operators $(\underline{apr}_2^\beta, \overline{apr}_2^\beta)$, or equivalently, $(\underline{apr}_2^\beta, \overline{apr}_2^\beta)$ is smaller than $(\underline{apr}_1^\beta, \overline{apr}_1^\beta)$, denoted by $(\underline{apr}_1^\beta, \overline{apr}_1^\beta) \preceq (\underline{apr}_2^\beta, \overline{apr}_2^\beta)$, if $\underline{apr}_2^\beta(X) \subseteq \underline{apr}_1^\beta(X)$ for all $X \subseteq U$. The approximation operator pair $(\underline{apr}_{At}^\beta, \overline{apr}_{At}^\beta)$ is the largest one, and the approximation pair $(\underline{apr}_\emptyset^\beta, \overline{apr}_\emptyset^\beta)$ is the smallest one. All approximation operators form a poset under the set inclusion relation, which is embedded in $((\underline{apr} : 2^U \longrightarrow 2^U, \overline{apr} : 2^U \longrightarrow 2^U), \preceq)$.

3.5   Property $\mathbb{P}_5$: Classification of a Set of Concepts

The decision attribute of an information table classifies the universe into a family of classes $U/IND(D)$. The union of all the lower approximations of those classes can be defined as the positive region, and the rest is called the boundary region. That is:

$$POS_A(D) = \bigcup_{X_i \in U/IND(D)} \underline{apr}_A(X_i); \tag{6}$$

$$BND_A(D) = U - POS_A(D). \tag{7}$$

Based on the β-lower and upper approximations, the β-positive and boundary regions can be defined. For example, $POS_A^\beta(D) = \bigcup_{X_i \in U/IND(D)} \underline{apr}_A^\beta(X_i)$. If $\beta \in (0.5, 1]$, we have $POS_A^\beta(D) \geq POS_A(D)$.

A positive region $POS_1(D)$ is larger than another positive region $POS_2(D)$, or equivalently, $POS_2(D)$ is smaller than $POS_1(D)$, if $POS_2(D) \subseteq POS_1(D)$. The positive region $POS_{At}(D)$ is the largest positive region, and the positive region $POS_\emptyset(D)$ is the smallest one. All positive regions form a poset under the subset relation, which is embedded in $(2^U, \subseteq)$.

4   Evaluation Functions for Property Preservation

For a certain property $\mathbb{P}$, we can use various fitness functions to evaluate the degree of satisfiability of the property by an attribute set. Some functions reflect the definition of the property directly; some reflect the definition of the property indirectly.

4.1   Evaluate the Description of an Object Relation

According to properties $\mathbb{P}_1$ and $\mathbb{P}_2$, we can directly use the following function to evaluate the property preservation by a set of attributes:

$$e_\mathbb{P} : 2^{At} \longrightarrow (2^{U \times U}, \subseteq).$$

A property $\mathbb{P} \in \{\mathbb{P}_1, \mathbb{P}_2\}$ can be one of the $IND(A), DIS(A)$ and $SIM(A)$ relations and the D-relative relations. The standard reduct construction method implements the IND relation as the evaluation function [4]. Yao and Zhao explore the DIS relations and the IND–DIS relations for reduct construction [10].

A property $\mathbb{P} \in \{\mathbb{P}_1, \mathbb{P}_2\}$ can also be quantified using the function:

$$e_\mathbb{P} : 2^{At} \longrightarrow (\Re, \leq),$$

where $\Re$ is the set of real numbers. We use the cardinality of the set of object pairs satisfying a certain relation. Owing to the fact that the relations can be represented as a discernibility matrix, to count the number of $\mathbf{dm}(x, y) \in \mathbf{dm}$ such that $A \cap \mathbf{dm}(x, y) = \emptyset$ is equivalent to counting the cardinality of $IND(A)$. At the meantime, to count the number of $\mathbf{dm}(x, y) \in \mathbf{dm}$ such that $A \subseteq \mathbf{dm}(x, y)$ is equivalent to counting the cardinality of $DIS(A)$.

4.2   Evaluate a Partition of the Universe

According to this property, we can use the following function to evaluate the property preservation by a set of attributes directly:

$$e_{\mathbb{P}_3} : 2^{At} \longrightarrow (\Pi(U), \preceq).$$

Since only the indiscernibility relation is an equivalence relation and be able to partition the universe, this type of property can be considered as a variation of property $\mathbb{P}_1$.

A partition of the universe changes the information entropy of the configuration. That means, we can evaluate the partition by calculating the information entropy. The evaluation function can be defined as:

$$e_{\mathbb{P}_3} : 2^{At} \longrightarrow (\Re, \leq).$$

For $A \subseteq At$, the information entropy is defined as $H(A) = -\sum p(\phi_A) \log p(\phi_A)$ where ϕ_A is a configuration defined by an attribute set A, and $p(\phi_A)$ is the probability of a configuration in the information table. The entire information table contains $H(At)$ bits of information.

4.3   Evaluate the Description of a Concept

According to this property, we can use the following function to evaluate the property preservation by a set of attributes directly:

$$e_{\mathbb{P}_4} : 2^{At} \longrightarrow ((\underline{apr} : 2^U \longrightarrow 2^U, \overline{apr} : 2^U \longrightarrow 2^U), \preceq).$$

This type of functions map a set A of attributes to a pair of approximation operators.

A function representing this property can also be defined as a mapping from an attribute set A to the lower approximation operator $\underline{apr}_A$. In the probabilistic cases, it can be defined as a mapping to $\underline{apr}_A^\beta$. Therefore, the function is written as:

$$e_{\mathbb{P}_4} : 2^{At} \longrightarrow (\underline{apr} : 2^U \longrightarrow 2^U, \preceq).$$

4.4   Evaluate a Classification

According to this property, we can use the following function to evaluate the property preservation by a set of attributes directly:

$$e_{\mathbb{P}_5} : 2^{At} \longrightarrow (2^U, \subseteq).$$

The positive and the boundary regions can be directly used. The positive region is defined for reduct construction by Pawlak [4]. Practically, $POS_A^\beta(D)$ has been applied for constructing reducts by many researchers [2,7,11].

It is natural to extend the above function to the following form:

$$e_{\mathbb{P}_5} : 2^{At} \longrightarrow (\Re, \leq).$$

The function e can be interpreted as the counting of $POS_A(D)$ or $BND_A(D)$, or its extension. For example, a classification accuracy measure $\gamma(A, D)$ has been studied to evaluate the ratio of the positive region with respect to the cardinality of the universe:

$$\gamma^\beta(A, D) = \frac{|POS_A^\beta(D)|}{|U|}.$$

The γ criterion is widely applied for reduct construction. The γ^β criterion is also applied for computing reducts by many authors [2,3,11].

The conditional entropy reflects the classification accuracy from the information-theoretic viewpoint. The conditional entropy of D given an attribute set $A \subseteq C$ is defined as:

$$H(D|A) = - \sum_{X_i \in U/IND(D)} p(X_i)p(X_i|\phi_A) \log p(X_i|\phi_A),$$

The conditional entropy can be used as a quantitative measure of this property, and is applied for reduct construction [3]. Other information theoretic approach has been studied by many researchers [1,8].

5   The Monotonicity of Property Evaluation Functions

It is important to note that some functions are monotonic with respect to the set inclusion, while some are not. For example, the relations IND, DIS and the information entropy H have the monotonicity property with respect to the set inclusion, however, the γ^β measure does not have the monotonicity property.

If the function e is monotonic with respect to the set inclusion of attribute sets, according to the definition, we need to check all the subsets of a candidate reduct, in order to confirm that a candidate reduct is a reduct. On the other hand, if e is not monotonic regarding the set inclusion, we need to search more attribute sets, and the situation is more complicated.

The non-monotonicity property of the fitness function has not received enough attention by the rough set community. Due to a lack of consideration of this issue, some of the reduct construction strategies are not entirely reasonable. For example, the measure $\gamma^\beta(P, D) = \gamma^\beta(C, D)$ has been inappropriately used by many researchers [2,3,11]. By emphasizing the equality relation, one might miss some attribute sets that also are reducts, and with the γ^β value greater than or equal to $\gamma^\beta(C, D)$.

6   Conclusion

This paper introduces a general definition of an attribute reduct, and presents a critical review of the existing reduct construction algorithms. It is found that the differences among different definitions of a reduct, and associated reduct construction algorithms, lie in the properties they try to preserve. Various qualitative and quantitative functions can be used to evaluate the degree of preservation of a certain property. The monotonicity property of an evaluation function needs to be emphasized. When the monotonicity property holds, the equality relation can be simply used to verify a candidate reduct; otherwise, a partial order relation $\preceq$ needs to be used. The analysis provides new insight of the existing studies, points out common insufficient consideration of monotonicity in some of the existing algorithms, and gives guidelines for the design of new algorithms.

References

1. Beaubouef, T., Petry, F.E., Arora, G.: Information-theoretic measures of uncertainty for rough sets and rough relational databases. Information Sciences 109 (1998) 185-195
2. Beynon, M.: Reducts within the variable precision rough sets model: a further investigation. European Journal of Operational Research 134 (2001) 592-605
3. Hu, Q., Yu D., Xie, Z.: Information-preserving hybrid data reduction based on fuzzy-rough techniques. Pattern Recognition Letters 27 (2006) 414-423
4. Pawlak, Z.: Rough sets. International Journal of Computer Information and Science 11 (1982) 341-356
5. Qiu, G.F., Zhang, W.X., Wu, W.Z.: Charaterization of attributes in generalized approximation representation spaces. LNAI 3461 (2005) 84-93
6. Skowron, A., Rauszer, C.: The discernibility matrices and functions in information systems. In: Slowiński, R. (ed.): Intelligent Decision Support, Handbook of Applications and Advances of the Rough Sets Theory. Dordrecht, Kluwer (1992)
7. Swiniarski, R.W.: Rough sets methods in feature reduction and classification. International Journal of Applied Mathematics and Computer Science 11 (2001) 565-582
8. Wang, G.Y., Zhao, J., Wu, J.: A comparitive study of algebra viewpoint and information viewpoint in attribute reduction. Foundamenta Informaticae 68 (2005) 1-13
9. Yao, Y.Y., Wong, S.K.M.: A decision theoretic framework for approximating concepts. International Journal of Man-machine Studies 37 (1992) 793-809
10. Yao, Y.Y., Zhao, Y.: Conflict analysis based on discernibility and indiscernibility. Proceedings of 2007 IEEE Symposium on Foundations of Computational Intelligence (2007)
11. Ziarko W.: Variable precision rough set model. Journal of Computer and System Sciences 46 (1993) 39-59

Mining Associations for Interface Design

Timothy Maciag[1], Daryl H. Hepting[1], Dominik Ślęzak[2],
and Robert J. Hilderman[1]

[1] Department of Computer Science, University of Regina
3737 Wascana Parkway, Regina, SK, S4S 0A2 Canada
[2] Infobright Inc.
218 Adelaide St. W, Toronto, ON, M5H 1W8 Canada
maciagt@cs.uregina.ca, dhh@cs.uregina.ca, slezak@infobright.com,
robert.hilderman@uregina.ca

Abstract. Consumer research has indicated that consumers use compensatory and non-compensatory decision strategies when formulating their purchasing decisions. Compensatory decision-making strategies are used when the consumer fully rationalizes their decision outcome whereas non-compensatory decision-making strategies are used when the consumer considers only that information which has most meaning to them at the time of decision. When designing online shopping support tools, incorporating these decision-making strategies with the goal of personalizing the design of the user interface may enhance the overall quality and satisfaction of the consumer's shopping experiences. This paper presents work towards this goal. The authors describe research that refines a previously developed procedure, using techniques in cluster analysis and rough sets, to obtain consumer information needed in support of designing customizable and personalized user interface enhancements. The authors further refine their procedure by examining and evaluating techniques in traditional association mining, specifically conducting experimentation using the Eclat algorithm for use with the authors' previous work. A summary discussing previous work in relation to the new evaluation is provided. Results are analyzed and opportunities for future work are described.

Keywords: Association mining, clustering, rough sets, usability, personalization.

1 Introduction

The world wide web is increasingly changing the way consumers browse for and purchase items. Millions of consumers engage in purchasing and consuming goods and services from online stores each day. Given this rapid increase in e-market activities there has been an increased demand for more usable tools that more effectively support the online consumer in formulating satisfying decision outcomes. Design of these systems could incorporate functionality that enables consumers to quickly and easily browse for and retrieve items in which they are

J.T. Yao et al. (Eds.): RSKT 2007, LNAI 4481, pp. 109–117, 2007.

interested. Providing enhanced options to customize and personalize the support interface could greatly enrich the consumers online shopping experiences [1]. Modelling consumer decision-making strategies in the design of the user interface may aid in achieving this end.

1.1   Consumer Decision-Making

Consumer research has indicated that consumers generally employ two types of decision-making strategies in their purchasing decisions: compensatory and non-compensatory [2].

Compensatory decision strategies are used when the decision maker applies a strict, fully rationalized thought process based on pre-defined preferences, ratings, or rankings to formulate a final decision [3]. The decision-maker will systematically weigh all possible alternatives in order to form the best possible decision outcome. Compensatory decision strategies have the potential to be quite complex in that: the consumer may not always be an expert in the decision domain, the consumer may not value certain attributes yet need to consider them when they formulate their decisions, the decision outcome may consist of an overabundance of information forcing the consumer to filter through results for wanted information, and/or the consumer may have criteria present in every decision yet they still must specify these value(s) in each decision formulation.

Non-compensatory decision strategies are used when the decision-maker applies *bounded rationality* [4]. Bounded rationality refers to the limitations in the human capacity for reaching fully rationalized decision outcomes (i.e. those decision outcomes that consider all facets of available information as in compensatory strategies). Decision makers will often arrive at a final decision based on *ad hoc* decision strategies using a variety of factors, which include: pre-defined and developing preferences, ratings and rankings (total or subset), the interface design, in addition to others [2,5].

1.2   Usability in Online Shopping Environments

There has been considerable research into understanding what constitutes a satisfying user interface for online shopping environments. Jedetski et al. [6] discuss that the design of the user interface is paramount in whether or not users have a satisfying experience in such support tools. In terms of developing satisfying user interfaces, providing consumers with enhanced options such as the ability to customize and personalize their user interface will ensure that they have a satisfying shopping experience [1,7,8]. Holland et al. [9], describe a technique utilizing methods in association mining to gather user preferences in support of developing personalized user interfaces from online user logs. As well, Li and Kit [10] describe a method to enhance the usability of online support tools by utilizing data mining techniques to mine associated information to design and develop a better link navigational structure of a website. Depending on the amount of data and information that consumers must provide, this task has potential to be a highly complex and time consuming. Maciag et al. [8,3]

describe a technique to reduce the complexity of this task by reducing the amount of consumer information required. The primary idea of their research was to formalize the foundations of a personalization procedure aimed at clustering consumers into groups bearing similar attribute values and product preferences.

2   A Review of Previous Work

In Maciag et al. [8,3], web-based shopping support tools were developed to conduct a usability evaluation. The authors chose to base their evaluation using a software support tool designed by the United States Environment Protection Agency (US-EPA) that enabled product comparisons between 29 environmentally preferable cleaning products using eight product attributes. Table 1 provides a listing of these attributes and their corresponding values.

Table 1. US-EPA attributes (with abbreviations) and corresponding values

Attribute (abbreviation)	Values
Skin Irritation (skin)	exempt, negligible-slight, slight, medium, strong, not reported
Food chain exposure (fce)	exempt, $\leq$ 5000, $\leq$ 10000, $\leq$ 15000, > 15000, not reported
Air pollution potential (air)	N/A, 0%, $\leq$ 1%, $\leq$5%, $\leq$ 15%, $\leq$ 30%, > 30%, not reported
Product contains fragrance (frag)	yes, no
Product contains dye (dye)	yes, no
Product is a concentrate (con)	yes, no
Product packaging made of recyclable paper (rec)	N/A, yes
Product minimizes exposure to concentrate (exp)	N/A, yes, no/small sizes, no

56 participants were recruited to complete a series of tasks on the support tools, which included completion of a questionnaire that asked participants to rank the eight attributes described in Table 1 using a four point scale: *unimportant, somewhat important, important, very important*. In addition participants were asked to select which of the 29 cleaning products they would consider purchasing for personal use. This information was used to develop a procedure to gather consumer information in support of clustering the participants into groups having similar attribute and product preferences.

Figure 1 illustrates the authors' procedure in Maciag et al. [3]. First, the 29 cleaning products were clustered, generating four product clusters. A decision system was constructed, comprised of 16 attributes (based on participant rankings) and one decision attribute (based on the participant product selections and product cluster values). Using the Rough Set Exploration System (RSES) [11], rough set reduction techniques were performed on the decision system. Utilizing

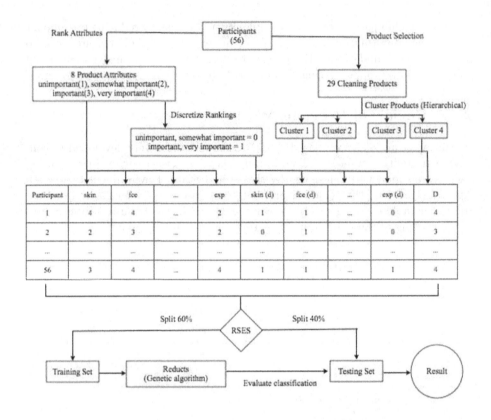

Fig. 1. Diagram illustrating the procedure in Maciag et al. [3]

the genetic algorithm functionality in RSES, the authors formulated the top ten *reducts* (the set(s) of attributes needed to discern objects (e.g. the 56 participants) in a decision system [12]) for the training set and tested the results using participant data in the testing set. Three of the top ten reducts generated consisted of only two of the 16 attributes and had classification accuracy of 100% and total coverage of 88%. It is important to note that the reason why the total coverage was slightly reduced is that some participants in the testing set could not be classified accordingly based on the reduct attributes generated by the genetic algorithm procedure provided by RSES. Thus, the authors proposed a system design where, upon initialization (i.e. a consumer's first use of the support tool), consumers could be given a reduced questionnaire that would elicit their preferences with respect to the attributes represented in the reducts. Consumers would be placed in the appropriate cluster based on their response and a customizable and personalized user interface would be displayed specific to the cluster group's preferences.

The work described in this paper will further refine the procedure described in Maciag et al. [3] (Figure 1). Specifically, the research described here will build on previous work by examining and evaluating techniques in association mining to aid in the task of designing the personalized aspects of the user interface after the initial clustering procedure, as described in Maciag et al. [3], is performed. The concepts of compensatory and non-compensatory decision strategies are used to provide the basis for design.

3   Experiment Design and Results

The authors examined and evaluated the Eclat algorithm [13,14,15], to be used in conjunction with the work described in Maciag et al. [3], as a means to gather useful consumer data and information in support of personalizing aspects of a user interface. Eclat is a data mining algorithm that is used for mining frequent item sets, i.e. sets of *transactions* containing associated values meeting minimum *support* and *confidence* thresholds. These thresholds are described in Equations 1 and 2.

$$Support(X \to Y) = P(X \cup Y) \tag{1}$$

$$Confidence(X \to Y) = \frac{P(X \cup Y)}{P(X)} \tag{2}$$

The Eclat algorithm can be used to determine whether certain items, e.g. item X and item Y, are associated in some fashion [16,14]. For instance, *what is the percentage items X and Y are purchased together?* [14]. Eclat functions by performing a depth-first traversal of a prefix tree to formulate *association rules*, i.e. the set(s) of rules that could be used to describe relationships among data. Figure 2 provides a classic illustration of the Eclat algorithm [16,14,15].

The authors utilized the Eclat algorithm to examine and evaluate the associations among certain aspects of user and product data obtained from the usability evaluation described previously. Figure 3 illustrates the steps taken in the authors' analysis. Eclat software[1] was used to analyze the associations between the total set of 29 cleaning products, the sets of products belonging to each of the four clusters generated in Maciag et al. [3], as well as the stated attribute rankings of those participants assigned within each cluster as per the procedure described in Maciag et al. [3]. A minimum support/confidence threshold of 75% was used in the examination. This support/confidence threshold was chosen since lower thresholds would provide a more loosely bound collection of associated attributes (personalized aspects would potentially loose meaning) whereas as higher thresholds would yield a more tightly bound collection of associated attributes (reduction of design possibilities).

4   Discussion

Tables 2 and 3 provide the results of the authors' examination. The first column of Table 2 represents the cluster value (the total set of products and clusters

---

[1] The authors used the Eclat software developed by Borgelt, http://fuzzy.cs. uni-magdeburg.de/ borgelt/eclat.html, (Fall 2006) [16].

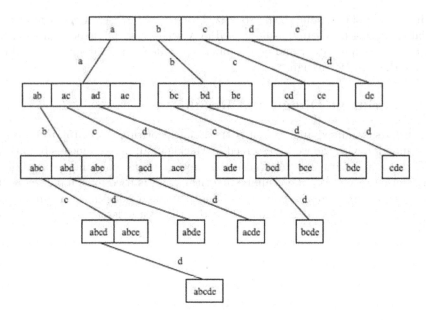

Fig. 2. Classic illustration of Eclat [16,14]. The empty root is omitted from the illustration. The depth-first traversal begins at the left-most item, *a*, and traverses the tree structure (backtracking when necessary) until all items are analyzed.

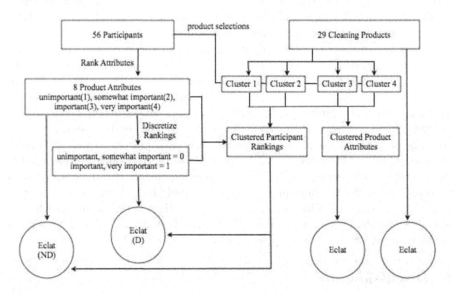

Fig. 3. Diagram illustrating the Eclat procedure used by the authors. Note, on the bottom left hand side, non-discretized attribute rankings (four point ranking scale) are denoted as *ND* and discretized rankings (binary scale) are denoted as *D*.

Table 2. Eclat results for the product analysis (total set and four product clusters). Please refer to Section 2 for definitions and attribute abbreviations. Note there were interesting results below the minimum support/confidence thresholds. These associations are italicized.

Cluster	Associated Attributes	Confidence
All Products	frag=no	86%
1	rec=N/A, dye=no, frag=no, fce=exempt	78%
	dye=no, frag=no, fce=exempt	89%
	fce=exempt	100%
	frag=no	89%
	dye=no	89%
	rec=N/A	78%
2	frag=no, fce=not reported	83%
3	rec=yes, exp=no, con=yes	83%
	exp=no, con=yes	100%
	rec=yes	83%
4	dye=no, frag=no	78%
	fce=exempt, frag=no	78%
	frag=no	89%
	fce=exempt	78%
	dye=no	78%
	dye=no, fce=exempt, frag=no	*67%*

1 to 4) as generated by the procedure in Maciag et al. [3], the second column represents the set(s) of associated product attributes generated by Eclat that met the 75% support threshold, and the final column represents the confidence threshold (as %) of the set(s) of associated product attributes. The first column of Table 3 represents the cluster value similarly seen in Table 2, the second column indicates whether the attribute rankings were discretized or not (as illustrated in Figures 1 and 3), the third column represents the set(s) of associated attribute rankings generated by Eclat that met the 75% support threshold, and the final column represents the confidence threshold (as %) of the set(s) of associated attribute rankings.

The results described in Table 2 could be used to indicate how to incorporate compensatory decision strategies in the design of a personalized user interface. Using this information, the interface design could highlight those attributes that are highly associated. For example, products in cluster 1 are strongly associated by the following: they contain no dye or fragrance, are not made of recyclable paper, and are mostly exempt from food chain exposure. Since consumers who are assigned to this cluster would normally select products that have these attribute values, they could initially be included in their product comparisons.

The results described in Table 2 could be used to indicate how to incorporate non-compensatory decision strategies in the design of a personalized user interface. Here, the attributes that are highly associated and favourably ranked could

Table 3. Eclat results for the participant rankings. Note, non-discretized (four point ranking scale) are denoted as *ND* under the Type label and discretized rankings (binary scale) are denoted as *D*. Refer to Section 2 for definitions and attribute abbreviations. Note, interesting results below the minimum support/confidence thresholds are italicized.

Cluster	Type	Associated Rankings	Confidence
All Products	D	air=important, skin=important	88%
	D	skin=important	95%
	D	air=important	93%
1	–	*no participants assigned*	–
2	–	*only 1 participant assigned*	–
3	D	exp=not important, rec=not important, con=not important, dye=not important, frag=important, air=important, skin=important	100%
	ND	rec=somewhat important, con=somewhat important, air=very important	100%
4	*D*	*rec=important, air=important, skin=important*	*74%*
	D	air=important, skin=important	93%
	D	skin=important	97%
	D	air=important	95%
	D	rec=important	77%
	ND	*skin=very important*	*72%*

be highlighted on the interface display, while omitting all other non-associated and non-favourably ranked attributes. This information could be used to design and deploy a personalized user interface specific to each cluster's attribute and product preferences.

5   Conclusion and Future Work

This paper described refinement of a procedure to design personalized user interfaces for online shopping support tools. The emphasis of the research described was to formalize a procedure to provide information in support of enhancing the functionality of these types of support tools by accommodating the diversity in consumer decision-making. The authors refined previous research by examining algorithms in association mining, specifically examining the Eclat algorithm. The authors illustrated how Eclat could be used, in conjunction with the authors' previous work, to obtain consumer information in support of designing personalized support tools to enhance the user interface design and potentially increase consumer satisfaction while using such tools.

Future work will include software implementation of the procedure described and further examination through usability evaluation. The authors also plan to examine and evaluate the procedure described in this paper in similar application domains that provide consumers with options to compare items using additional product attributes. As well, techniques to develop metrics for evaluating consumer decision accuracy using concepts described in this paper are currently being designed and evaluated.

References

1. Ha, S.H.: Helping Online Customers Decide Through Web Personalization. IEEE Intelligent Systems (2002) 34–43
2. Bettman, J.R., Luce, M.F., Payne, J.W.: Constructive Consumer Choice Processes. Journal of Consumer Research **25** (1998) 187–217
3. Maciag, T.J., Hepting, D.H., Slezak, D.: Consumer Modelling in Support of Interface Design. In Proc. IEEE International Conference on Hybrid Information Technology **2** (2006) 153–160
4. Simon, H.A.: A Behavioral Model of Rational Choice. Economics **69** (1955) 99–118
5. Hsee, C.K., Leclerc, F.: Will Products Look More Attractive When Presented Separately or Together? Journal of Consumer Research **25** (1998) 175–186
6. Jedetski, J., Adelman, L., Yeo, C.: How Web Site Decision Technology Affects Consumers. IEEE Internet Computing (2002) 72–79
7. Eirinaki, M., Vazirgiannis, M.: Web Mining for Web Personalization. ACM Transaction on Internet Technology **3** (2003) 1–27
8. Maciag, T., Hepting, D.H., Slezak, D.: Personalizing User Interfaces for Environmental Decision Support Systems. In Proc. Rough Sets and Soft Computing in Intelligent Agent and Web Technology (2005)
9. Holland, S., Ester, M., Kiessling, W.: Preference Mining: A Novel Approach on Mining User Preferences for Personalized Applications. In Proc. Principles and Practice of Knowledge Discovery in Databases (PKDD) (2003)
10. Li, C.H., Kit, C.C.: Web Structure Mining for Usability Analysis. In Proc. IEEE/WIC/ACM Web Intelligence (2005)
11. Bazan, J., Szczuka, M.: The Rough Set Exploration System. Transactions on Rough Sets 3 (2005) 37–56
12. Pawlak, Z.: Rough Sets, Theoretical Aspects of Reasoning About Data. Kluwer Academic Publishers (1991)
13. Zaki, M.J.: Scalable Algorithms for Association Mining. In Proc. IEEE Transactions on Knowledge and Data Engineering **12** (2000)
14. Mahanti, A., Alhajj, R.: Visual Interface for Online Watching of Frequent Itemset Generation in Apriori and Eclat. In Proc. IEEE International Conference on Machine Learning and Applications (2005)
15. Ceglar, A., Roddick, J.F.: Association Mining. ACM Computing Surveys **38** (2006)
16. Borgelt, C.: Implementations of Apriori and Eclat. In Proc. Workshop on Frequent Item Set Mining Implementations (FIMI) (2003)

Optimized Generalized Decision in Dominance-Based Rough Set Approach

Krzysztof Dembczyński[1], Salvatore Greco[2], and Wojciech Kotłowski[1], and Roman Słowiński[1,3]

[1] Institute of Computing Science, Poznań University of Technology,
60-965 Poznań, Poland
{kdembczynski,wkotlowski,rslowinski}@cs.put.poznan.pl
[2] Faculty of Economics, University of Catania, 95129 Catania, Italy
salgreco@unict.it
[3] Institute for Systems Research, Polish Academy of Sciences, 01-447 Warsaw, Poland

Abstract. Dominance-based Rough Set Approach (DRSA) has been proposed to deal with multi-criteria classification problems, where data may be inconsistent with respect to the dominance principle. However, in real-life datasets, in the presence of noise, the notions of lower and upper approximations handling inconsistencies were found to be excessively restrictive which led to the proposal of the variable consistency variant of the theory. In this paper, we deal with a new approach based on DRSA, whose main idea is based on the error corrections. A new definition of the rough set concept known as generalized decision is introduced, *the optimized generalized decision*. We show its connections with statistical inference and dominance-based rough set theory.

1 Introduction

The multicriteria classification problem [10] consists in assignment of objects to pre-defined *decision classes* Cl_t, $t \in T = \{1, \dots, n\}$. It is assumed that the classes are preference-ordered according to an increasing order of class indices, i.e. for all $r, s \in T$, such that $r > s$, the objects from Cl_r are strictly preferred to the objects from Cl_s. The objects are evaluated on a set of *condition criteria* (i.e., attributes with preference-ordered domains). It is assumed that a better evaluation of an object on a criterion, with other evaluations being fixed, should not worsen its assignment to a decision class. The problem of multicriteria classification can also be seen as a data analysis problem, under assumption of *monotone relationship* between the decision attribute and particular condition attributes, i.e. that the expected decision value increases (or decreases) with increasing (or decreasing) values on condition attributes. This definition is valid only in the probabilistic sense, so it may happen that there exists in the dataset X an object x_i not worse than another object x_k on all condition attributes, however x_i is assigned to a worse class than x_k; such a situation violates the monotone nature of data, so we shall call objects x_i and x_k *inconsistent with respect to dominance principle*.

J.T. Yao et al. (Eds.): RSKT 2007, LNAI 4481, pp. 118–125, 2007.

Rough set theory [9] has been adapted to deal with this kind of inconsistency and the resulting methodology has been called *Dominance-based Rough Set Approach* (DRSA) [5,6]. In DRSA, the classical indiscernibility relation has been replaced by a dominance relation. Using the rough set approach to the analysis of multicriteria classification data, we obtain lower and the upper (rough) approximations of unions of decision classes. The difference between upper and lower approximations shows inconsistent objects with respect to the dominance principle. Another, equivalent picture of this problem can be expressed in terms of the *generalized decision* concept [3,4].

Unfortunately, it can happen, that due to the random nature of data and due to the presence of noise, we loose too much information, thus making the DRSA inference model not accurate. In this paper, a new approach is proposed, based on combinatorial optimization for dealing with inconsistency, which can be viewed as a slightly different way of introducing variable precision in the DRSA. The new approach is strictly based on the generalized decision concept. It is an invasive method (contrary to DRSA), which reassigns the objects to different classes when they are traced to be inconsistent. We show, that this approach has statistical foundations and is strictly connected with the standard dominance-based rough set theory.

We assume that we are given a set $X = \{x_1, \ldots, x_\ell\}$, consisting of ℓ objects, with their decision values (class assignments) $Y = \{y_1, \ldots, y_\ell\}$, where each $y_i \in T$. Each object is described by a set of m condition attributes $Q = \{q_1, \ldots, q_m\}$ and by $\mathrm{dom}q_i$ we mean the set of values of attribute q_i. By the *attribute space* we mean the set $V = \mathrm{dom}q_1 \times \ldots \times \mathrm{dom}q_m$. Moreover, we denote the evaluation of object x_i on attribute q_j by $q_j(x_i)$. Later on we will abuse the notation a little bit, identifying each object x with its evaluations on all the condition attributes, $x \equiv (q_1(x), \ldots q_m(x))$. By a *class* $Cl_t \subset X$, we mean the set of objects, such that $y_i = t$, i.e. $Cl_t = \{x_i \in X : y_i = t, 1 \le i \le \ell\}$.

The article is organized in the following way. Section 2 describes main elements of DRSA. Section 3 presents an algorithmic background for new approach. The concept of optimized generalized decision is introduced in Section 4. In Section 5 the connection with statistical inference is shown. The paper ends with conclusions. Proofs of the theorems are omitted due to the space limit.

2 Dominance-Based Rough Set Approach

Within DRSA [5,6], we define the *dominance* relation D as a binary relation on X in the following way: for any $x_i, x_k \in X$ we say that x_i *dominates* x_k, $x_i D x_k$, if on every condition attribute, x_i has evaluation not worse than x_k, $q_j(x_i) \ge q_j(x_k)$, for all $1 \le j \le m$. The dominance relation D is a partial preorder on X, i.e. it is reflexive and transitive. The *dominance principle* can be expressed as follows. For all $x_i, x_j \in X$ it holds:

$$x_i D x_j \implies y_i \ge y_j \tag{1}$$

The rough approximations concern granules resulting from information carried out by the decisions. The *decision granules* can be expressed by unions of decision classes: for all $t \in T$

$$Cl_t^{\geq} = \{x_i \in X : y_i \geq t\}, \qquad Cl_t^{\leq} = \{x_i \in X : y_i \leq t\}. \tag{2}$$

The *condition granules* are dominating and dominated sets defined as:

$$D^+(x) = \{x_i \in X : x_i D x\}, \qquad D^-(x) = \{x_i \in X : x D x_i\}. \tag{3}$$

Lower rough approximations of $Cl_t^{\geq}$ and $Cl_t^{\leq}$, $t \in T$, are defined as follows:

$$\underline{Cl_t^{\geq}} = \{x_i \in X : D^+(x_i) \subseteq Cl_t^{\geq}\}, \qquad \underline{Cl_t^{\leq}} = \{x_i \in X : D^-(x_i) \subseteq Cl_t^{\leq}\}. \tag{4}$$

Upper rough approximations of $Cl_t^{\geq}$ and $Cl_t^{\leq}$, $t \in T$, are defined as follows:

$$\overline{Cl_t^{\geq}} = \{x_i \in X : D^-(x_i) \cap Cl_t^{\geq} \neq \emptyset\}, \quad \overline{Cl_t^{\leq}} = \{x_i \in X : D^+(x_i) \cap Cl_t^{\leq} \neq \emptyset\}. \tag{5}$$

In the rest of this section we focus our attention on the *generalized decision* [3]. Consider the following definition of generalized decision $\delta_i = [l_i, u_i]$ for object $x_i \in X$, where:

$$l_i = \min\{y_j : x_j D x_i, x_j \in X\}, \tag{6}$$

$$u_i = \max\{y_j : x_i D x_j, x_j \in X\}. \tag{7}$$

In other words, the generalized decision reflects an interval of decision classes to which an object may belong due to the inconsistencies with the dominance principle caused by this object. Obviously, $l_i \leq y_i \leq u_i$ for every $x_i \in X$ and if $l_i = u_i$, then object x_i is consistent with respect to the dominance principle with every other object $x_k \in X$.

Let us remark that the dominance-based rough approximations may be expressed using generalized decision:

$$\begin{aligned} \underline{Cl_t^{\geq}} &= \{x_i \in X : l_i \geq t\} & \overline{Cl_t^{\geq}} &= \{x_i \in X : u_i \geq t\} \\ \underline{Cl_t^{\leq}} &= \{x_i \in X : u_i \leq t\} & \overline{Cl_t^{\leq}} &= \{x_i \in X : l_i \leq t\}. \end{aligned} \tag{8}$$

It is also possible to obtain generalized decision using the rough approximation:

$$l_i = \max\left\{t : x_i \in \underline{Cl_t^{\geq}}\right\} = \min\left\{t : x_i \in \overline{Cl_t^{\leq}}\right\} \tag{9}$$

$$u_i = \min\left\{t : x_i \in \underline{Cl_t^{\leq}}\right\} = \max\left\{t : x_i \in \overline{Cl_t^{\geq}}\right\} \tag{10}$$

Those two descriptions are fully equivalent. For the purpose of this text we will look at the concept of generalized decision from a different point of view. Let us define the following relation: the decision range $\alpha = [l^\alpha, u^\alpha]$ is *more informative* than $\beta = [l^\beta, u^\beta]$ if $\alpha \subseteq \beta$. We show now that the generalized decision concept (thus also DRSA rough approximations) is in fact the unique optimal non-invasive approach that holds the maximum possible amount of information which can be obtained from given data:

Theorem 1. *The generalized decisions* $\delta_i = [l_i, u_i]$, *for* $x_i \in X$, *are most informative ranges from any set of decisions ranges of the form* $\alpha_i = [l_i^\alpha, u_i^\alpha]$ *that have the following properties:*

1. *The sets* $\{(x_i, l_i^\alpha) \colon x_i \in X\}$ *and* $\{(x_i, u_i^\alpha) \colon x_i \in X\}$, *composed of objects with, respectively, decisions* l_i^α *and* u_i^α *assigned instead of* y_i *are consistent with the dominance principle.*
2. *For each* $x_i \in X$ *it holds* $l_i^\alpha \le y_i \le u_i^\alpha$.

3 Minimal Reassignment

A new proposal of the definitions of lower and upper approximations of unions of classes is based on the concept of *minimal reassignment*. At first, we define the reassignment of an object $x_i \in X$ as changing its decision value y_i. Moreover, by minimal reassignment we mean reassigning the smallest possible number of objects to make the set X consistent (with respect to the dominance principle). One can see, that such a reassignment of objects corresponds to indicating and correcting possible errors in the dataset, i.e. it is an invasive approach.

We denote the minimal number of reassigned objects from X by R. To compute R, one can formulate a linear programming problem. Such problems were already considered in [2] (in the context of binary and multi-class classification) and also in [1] (in the context of boolean regression). Here we formulate a similar problem, but with a different aim.

For each object $x_i \in X$ we introduce $n-1$ binary variables d_{it}, $t \in \{2, \dots, n\}$, having the following interpretation: $d_{it} = 1$ iff object $x_i \in Cl_t^\ge$ (note that always $d_{i1} = 1$, since $Cl_1^\ge = X$). Such interpretation implies the following conditions:

$$\text{if } t' \ge t \text{ then } d_{it'} \le d_{it} \tag{11}$$

for all $i \in \{1, \dots, \ell\}$ (otherwise it would be possible that there exists object x_i belonging to $Cl_{t'}^\ge$, but not belonging to $Cl_t^\ge$, where $t' > t$). Moreover, we give a new value of decision y_i' to object x_i according to the rule: $y_i' = 1 + \sum_{t=2}^n d_{it}$ (the highest t such that x_i belongs to $Cl_t^\ge$). So, for each object $x_i \in U$ the cost function of the problem can be formulated as $R_i = (1 - d_{i,y_i}) + d_{i,y_i+1}$. Indeed, the value of decision for x_i changes iff $R_i = 1$ [4].

The following conditions must be satisfied for X to be consistent according to (1):

$$d_{it} \ge d_{jt} \qquad \forall i, j \colon x_i D x_j \quad 2 \le t \le n \tag{12}$$

Finally, we can formulate the problem in terms of integer linear programming:

$$\text{minimize} \quad R = \sum_{i=1}^\ell R_i = \sum_{i=1}^\ell \left((1 - d_{i,y_i}) + d_{i,y_i+1} \right) \tag{13}$$

subject to $d_{it'} \leq d_{it}$ $1 \leq i \leq \ell,$ $2 \leq t < t' \leq n$

$d_{it} \geq d_{jt}$ $1 \leq i, j \leq \ell,$ $x_i D x_j,$ $2 \leq t \leq n$

$d_{it} \in \{0, 1\}$ $1 \leq i \leq \ell,$ $2 \leq t \leq n$

The matrix of constraints in this case is totally unimodular [2,8], because it contains in each row either two values 1 and -1 or one value 1, and the right hand sides of the constraints are integer. Thus, we can relax the integer condition reformulating it as $0 \leq d_{it} \leq 1$, and get a linear programming problem. In [2], the authors give also a way for further reduction of the problem size. Here, we prove a more general result using the language of DRSA.

Theorem 2. *There always exists an optimal solution of (13),* $y_i^* = 1 + \sum_{t=2}^{n} d_{it}^*$, *for which the following condition holds:* $l_i \leq y_i^* \leq u_i,\ 1 \leq i \leq \ell$.

Theorem 2 enables a strong reduction of the number of variables. For each object x_i, variables d_{it} can be set to 1 for $t \leq l_i$, and to 0 for $t > u_i$, since there exists an optimal solution to (13) with such values of the variables. In particular, if an object x_i is consistent (i.e. $l_i = u_i$), the class assignment for this object remains the same.

4   Construction of the Optimized Generalized Decisions

The reassignment cannot be directly applied to the objects from X, since the optimal solution may not be unique. Indeed, in some cases one can find different subsets of X, for which the change of decision values leads to the same value of cost function R. It would mean that the reassignment of class labels for some inconsistent objects depends on the algorithm used, which is definitely undesirable. To avoid that problem, we must investigate the properties of the set of optimal feasible solutions of the problem (13).

Let us remark the set of all feasible solutions to the problem (13) by F, where by solution f we mean a vector of new decision values assigned to objects from X, i.e. $f = (f_1, \ldots, f_\ell)$, where f_i is the decision value assigned by solution f to object x_i. We also denote the set of optimal feasible solutions by OF. Obviously, $OF \subset F$ and $OF \neq \emptyset$, since there exist feasible solutions, e.g. $f = (1, \ldots, 1)$.

Assume that we have two optimal feasible solutions $f = (f_1, \ldots, f_\ell)$ and $g = (g_1, \ldots, g_\ell)$. We define "min" and "max" operators on F as $\min\{f, g\} = (\min\{f_1, g_1\}, \ldots, \min\{f_\ell, g_\ell\})$ and $\max\{f, g\} = (\max\{f_1, g_1\}, \ldots, \max\{f_\ell, g_\ell\})$. The question arises, whether if $f, g \in OF$ then $\min\{f, g\}$ and $\max\{f, g\}$ also belong to OF? The following lemma gives the answer:

Lemma 1. *Assume* $f, g \in OF$. *Then* $\min\{f, g\}, \max\{f, g\} \in OF$.

Having the lemma, we can start to investigate the properties of the order in OF. We define a binary relation $\succeq$ on OF as follows:

$$\forall_{f, g \in OF}\quad (f \succeq g \Leftrightarrow \forall_{1 \leq i \leq \ell} f_i \geq g_i) \tag{14}$$

It can be easily verified that it is a partial order relation. We now state the following theorem:

Theorem 3. *There exist the greatest and the smallest element in the ordered set $(OF, \succeq)$*

Theorem 3 provides the way to define for all $x_i \in X$ *the optimized generalized decisions* $\delta_i^* = [l_i^*, u_i^*]$ as follows:

$$l_i^* = y_{*i} = \min\{f_i \colon f \in OF\} \tag{15}$$
$$u_i^* = y_i^* = \max\{f_i \colon f \in OF\} \tag{16}$$

Of course, both l^* and u^* are consistent with respect to dominance principle (since they belong to OF). The definitions are more resistant to noisy data, since they appear as solutions with minimal number of reassigned objects. It can be shown that using the classical generalized decision, for any consistent set X we can add two "nasty" objects to X (one, which dominates every object in X, but has the lowest possible class, and another which is dominated by every object in X, but has the highest possible class) to make the generalized decisions completely noninformative, i.e. for every object $x_i \in X$, l_i equals to the lowest possible class and u_i equals to the highest possible class. If we use the optimized generalized decisions to this problem, two "nasty" objects will be relabeled (properly recognized as errors) and nothing else will change.

Optimized generalized decision is a direct consequence of the non-uniqueness of optimal solution to the minimal reassignment problem (so also to the problems considered in [2,1]). Also note that using (15) and (16) and reversing the transformation with 8 we end up with new definitions of *optimized lower and upper approximations*.

The problem which is still not solved is how to find the smallest and the greatest solutions in an efficient way. We propose to do this as follows: we modify the objective function of (13) by introducing the additional term:

$$R' = \epsilon \sum_{i=1}^{\ell} \sum_{t=l_i}^{u_i} d_{it} \tag{17}$$

and when we seek the greatest solution we subtract R' from the original objective function, while when we seek the smallest solution we add R' to the original objective function, so we solve two linear programs with the following objective functions:

$$R_\pm = \sum_{i=1}^{\ell} R_i \pm R' = R \pm R' \tag{18}$$

To prove, that by minimizing the new objective function we indeed find what we require, we define $I = \sum_{i=1}^{\ell}(u_i - l_i)$. The following theorem holds:

Theorem 4. *When minimizing objective functions (18) one finds the smallest and the greatest solution provided $\epsilon < I^{-1}$.*

Note that the solutions to the modified problem are unique.

5   Statistical Base of Minimal Reassignment

In this section we introduce a statistical justification for the described approach. We consider here only the binary (two-class) problem, however this approach can be extended to the multi-class case. We state the following assumptions: each pair $(x_i, y_i) \in X \times Y$ is a realization of random vector $(\mathcal{X}, \mathcal{Y})$, independent and identically distributed (i.i.d.) [7,12]. Moreover, we assume that the statistical model is of the form $\mathcal{Y} = b(\mathcal{X}) \oplus \epsilon$, where $b(\cdot)$ is some function, such that $b(x) \in \{0, 1\}$ for all $x \in V$ and $b(x)$ is isotonic (monotone and decreasing) for all $x \in V$. We observe y which is the composition ($\oplus$ is binary addition) of $b(x)$ and some variable ϵ which is the random noise. If $\epsilon(x) = 1$, then we say, that the decision value was *misclassified*, while if $\epsilon(x) = 0$, than we say that the decision value was correct. We assume that $\Pr(\epsilon = 1) < \frac{1}{2} \equiv p$ and it is independent of x, so each object is misclassified with the same probability p.

We now use the maximum likelihood estimate (MLE). We do not know the real decision values $b(x_i) \equiv b_i$ for all $x_i \in X$ and we treat them as parameters. We fix all x_i and treat only y_i as random. Finally, considering $B = \{b_1, \ldots, b_\ell\}$ and denoting by ϵ_i the value of variable ϵ for object x_i, the MLE is as follows:

$$L(B; Y) = \Pr(Y|B) = \prod_{i=1}^{\ell} \Pr(y_i|b_i) = \prod_{i=1}^{\ell} p^{\epsilon_i}(1-p)^{1-\epsilon_i} \qquad (19)$$

Taking minus logarithm of (19) (the negative log-likelihood) we equivalently minimize:

$$-\ln L(B; Y) = -\sum_{i=1}^{\ell} (\epsilon_i \ln p - (1 - \epsilon_i) \ln(1-p)) = \ln \frac{1-p}{p} \sum_{i=1}^{\ell} \epsilon_i + \ell \ln(1-p)$$
$$(20)$$

We see, that for any fixed value of p, the negative log-likelihood reaches its minimum when the sum $\sum_{i=1}^{\ell} \epsilon_i$ is minimal. Thus, for any p, to maximize the likelihood, we must minimize the number of misclassifications. This is equivalent to finding values $b_i, 1 \leq i \leq \ell$, which are monotone, i.e. consistent with the dominance principle, and such that the number of misclassifications $\sum_{i=1}^{\ell} \epsilon_i = \sum_{i=1}^{\ell} |b_i - y_i|$ is minimal. Precisely, this is the two-class problem of minimal reassignment.

Finally, we should notice that for each $x \in X$, $b(x)$ is the most probable value of y (decision) for given x, since $p < \frac{1}{2}$. Therefore, we estimate the decision values, that would be assigned to the object by the *optimal Bayes classifier* [7], i.e. the classifier which has the smallest expected error.

6   Conclusions

We propose a new extension of the Dominance-based Rough Set Approach (DRSA), which involves a combinatorial optimization problem concerning minimal reassignment of objects. As it is strongly related to the standard DRSA, we describe our approach in terms of the generalized decision concept. By reassigning the minimal number of objects we end up with a non-univocal optimal solution. However, by considering the whole set of optimal solution, we can optimize the generalized decision, so as to make it more robust in the presence of noisy data.

On the other hand, reassigning the objects to different classes in view of making the dataset consistent, has a statistical justification. Under assumption of common misclassification probability for all of the objects, it is nothing else than a maximum likelihood estimate of the optimal Bayes classifier.

References

1. Boros, E., Hammer, P. L., Hooker, J. N.: Boolean regression. Annals of Operations Research, **58** (1995) 201–226
2. Chandrasekaran, R., Ryu, Y. U., Jacob, V., Hong, S.: Isotonic separation. IN-FORMS J. Comput., **17** (2005) 462–474
3. Dembczyński, K., Greco, S., Słowinski, R.: Second-order Rough Approximations in Multi-criteria Classification with Imprecise Evaluations and Assignments, LNAI, **3641** (2005) 54–63
4. Dembczyński, K., Greco, S., Kotłowski, W., Słowiński, R.: Quality of Rough Approximation in Multi-Criteria Classification Problems. LNCS **4259** (2006) 318-327.
5. Greco S., Matarazzo, B.: Słowiński, R.: Rough approximation of a preference relation by dominance relations. European Journal of Operational Research, **117** (1999) 63–83
6. Greco S., Matarazzo, B., Słowiński, R.: Rough sets theory for multicriteria decision analysis. European Journal of Operational Research, **129** (2001) 1–47
7. Hastie, T., Tibshirani, R., Friedman J.: The Elements of Statistical Learning. Springer, Berlin 2003.
8. Papadimitriou, C. H., Steiglitz, K.: Combinatorial Optimization. Dover Publications, New York (1998).
9. Pawlak, Z.: Rough sets. International Journal of Information & Computer Sciences, **11** (1982) 341–356
10. Roy, B., Decision science or decision aid science?. European Journal of Operational Research, **66** (1993) 184–203.
11. Słowiński, R., Greco S., Matarazzo, B.: Rough Set Based Decision Support. Chapter 16 [in]: Burke, E. K., Kendall, G. (eds.): Search Methodologies: Introductory Tutorials on Optimization and Decision Support Techniques. Springer-Verlag, New York (2005) 475–527.
12. Vapnik, V.: Statistical Learning Theory. Wiley-Interscience, New York (1998)

Monotonic Variable Consistency Rough Set Approaches

Jerzy Błaszczyński[1], Salvatore Greco[2], Roman Słowiński[1,3],
and Marcin Szeląg[1]

[1] Institute of Computing Science, Poznań University of Technology,
60-965 Poznań, Poland
{jblaszczynski,rslowinski,mszelag}@cs.put.poznan.pl
[2] Faculty of Economics, University of Catania,
Corso Italia, 55, 95129 Catania, Italy
salgreco@unict.it
[3] Institute for Systems Research, Polish Academy of Sciences,
01-447 Warsaw, Poland

Abstract. We consider new definitions of Variable Consistency Rough
Set Approaches that employ monotonic measures of membership to the
approximated set. The monotonicity is understood with respect to the
set of considered attributes. This kind of monotonicity is related to the
monotonicity of the quality of approximation, considered among basic
properties of rough sets. Measures that were employed by approaches
proposed so far lack this property. New monotonic measures are considered in two contexts. In the first context, we define Variable Consistency
Indiscernibility-based Rough Set Approach (VC-IRSA). In the second
context, new measures are applied to Variable Consistency Dominance-
based Rough Set Approaches (VC-DRSA). Properties of new definitions
are investigated and compared to previously proposed Variable Precision
Rough Set (VPRS) model, Rough Bayesian (RB) model and VC-DRSA.

1 Introduction to Monotonicity Problem in Rough Sets

One of the basic concepts in rough set approaches is the decision table, defined
as an universe of objects U, described by a set $A = \{a_1, a_2, \ldots, a_n\}$ of criteria
and/or regular attributes. The set of attributes is divided into a set of condition attributes C and a set of decision attributes D, where $C \cup D = A$. We
use indiscernibility relation for regular attributes and dominance relation for
criteria to define elementary sets and dominance cones in the space of condition
attributes, respectively. When we consider Indiscernibility-based Rough Set Approach (IRSA), the set of attributes is composed of regular attributes only. In
order to analyze a decision table within Dominance-based Rough Set Approach
(DRSA), we need at least one condition criterion and at least one decision criterion. Decision attributes introduce partition of U into decision classes. In DRSA,
decision classes X_i are ordered. If $i < j$, then class X_i is considered to be worse
than X_j. The ordered classes form upward unions $X_i^{\geq} = \bigcup_{t \geq i} X_t$ and downward

J.T. Yao et al. (Eds.): RSKT 2007, LNAI 4481, pp. 126–133, 2007.

unions $X_i^{\leq} = \bigcup_{t \leq i} X_t$. Further, we define P-lower approximations of decision classes or unions of decisions classes, where P is a non-empty subset of C.

For $X \subseteq U, P \subseteq C, y \in U$, given lower threshold α_X, let $f_X^P(y)$ be a measure used in the definition of P-lower approximation of set X in one of two ways:

$$\underline{P}(X) = \{y \in U : f_X^P(y) \geq \alpha_X\} \tag{1}$$

$$or \quad \underline{P}(X) = \{y \in X : f_X^P(y) \geq \alpha_X\}. \tag{2}$$

Analogically, for upper threshold β_X, we can define (1) and (2) using $g_X^P(y)$ as

$$\underline{P}(X) = \{y \in U : g_X^P(y) \leq \beta_X\} \tag{3}$$

$$or \quad \underline{P}(X) = \{y \in X : g_X^P(y) \leq \beta_X\}. \tag{4}$$

It is reasonable to require that measure $f_X^P(y)$ $(g_X^P(y))$ is monotonically non-decreasing (non-increasing) with respect to (w.r.t.) P. Formally, $f_X^P(y)$ $(g_X^P(y))$ is monotonically non-decreasing (non-increasing) if and only if (iff) for all $y \in U$, $P \subseteq R \subseteq C$, $X \subseteq U$, $f_X^P(y) \leq f_X^R(y)$ $(g_X^P(y) \geq g_X^R(y))$.

Monotonicity of measure $f_X^P(y)$ or $g_X^P(y)$ ensures monotonicity w.r.t. set of attributes of P-lower approximation defined according to (2) or (4), respectively. A rough set approach is called monotonic if its P-lower approximation has this property. This kind of monotonicity is one of the basic properties of rough sets. It is important also because it simplifies rule induction, in which P-lower (and P-upper) approximations are generalized by rules.

Rough membership measure is used to control positive regions in Variable Precision Rough Set (VPRS) model [6] and in Variable Consistency Dominance-based Rough Set Approaches (VC-DRSA) [1]. Originally, rough membership function was introduced in [4].

In IRSA, rough membership of $y \in U$ to $X \subseteq U$ w.r.t. $P \subseteq C$, is defined as

$$\mu_X^P(y) = \frac{|I_P(y) \cap X|}{|I_P(y)|},$$

where $I_p(y)$ denotes a set of objects indiscernible with object y when considering set of attributes P (i.e., elementary set composed of all objects having the same description as object y, which is also called a granule). Rough membership captures a ratio of objects that belong to granule $I_P(y)$ and to considered set X, among all objects belonging to granule $I_P(y)$. This measure can also be treated as an estimate of conditional probability $Pr(x \in X | x \in I_P(y))$. In IRSA, the rough membership is used in definition (1), so it is expected to be monotonically non-decreasing. Unfortunately, this property does not hold. Let us consider the following example: U composed of objects y_1, y_2, y_3 and two sets $X_1 = \{y_1\}$, $X_2 = \{y_2, y_3\}$. First, we consider only one attribute, thus $P = \{a_1\}$. All objects have the same value on a_1 (i.e., they all belong to the same granule). We can observe that $\mu_{X_2}^P(y_1) = \mu_{X_2}^P(y_2) = \mu_{X_2}^P(y_3) = 0.66$. We extend set P to set $R = \{a_1, a_2\}$. It may happen that on a_2 objects y_1 and y_2 have the same value while y_3 has different value. Thus, we have two granules. The first one consisting of objects y_1, y_2 and the other one composed of object y_3. The value of rough membership in the first granule drops to 0.5.

Other measures than rough membership are also used in rough set approaches. For example, confirmation measures [2] are used together with rough membership in Parameterized Rough Sets [3]. Confirmation measures quantify the degree to which membership of object y to given elementary set $I_p(y)$ provides "evidence for or against" or "support for or against" assignment to considered set X. According to [3], confirmation measures are used within definition (1), so they should be monotonically non-decreasing. Unfortunately, the well-known confirmation measures do not have this property. Bayes factor is an interestingness measure that has similar properties as confirmation measures (its formulation is very close to the confirmation measure l [2]). It is used in Rough Bayesian (RB) model [5]. Bayes factor for $y \in U$ and $X \subseteq U$ w.r.t. $P \subseteq C$ is defined as

$$B_X^P(y) = \frac{|I_P(y) \cap X||\neg X|}{|I_P(y) \cap \neg X||X|}.$$

The Bayes factor can also be seen as a ratio of two inverse probabilities $Pr(x \in I_P(y)|x \in X)$ and $Pr(x \in I_P(y)|x \in \neg X)$. This measure is also used in definition (1), so it is expected to be monotonically non-decreasing w.r.t. set of attributes. Unfortunately, this is not the case. In the example considered above, Bayes factor for object y_2 drops from $B_{X_2}^P(y_2) = 1$ to $B_{X_2}^R(y_2) = 0.5$.

In DRSA, rough membership is defined for $y \in U$, $P \subseteq C$, $X^{\geq}, X^{\leq} \subseteq U$, as

$$\mu_{X^{\geq}}^P(y) = \frac{|D_P^+(y) \cap X^{\geq}|}{|D_P^+(y)|}, \quad \mu_{X^{\leq}}^P(y) = \frac{|D_P^-(y) \cap X^{\leq}|}{|D_P^-(y)|},$$

where $D_p^+(y)$, $D_p^-(y)$ denotes P-positive and P-negative dominance cone of object y (i.e., granule that is composed of objects dominating y, and granule composed of objects dominated by y, both w.r.t. P) and $X^{\geq}$, $X^{\leq}$ denotes upward and downward union of sets, respectively. Moreover, $\mu_{X^{\geq}}^P(y)$ and $\mu_{X^{\leq}}^P(y)$ can be interpreted as estimates of probability $Pr(z \in X^{\geq}|zD_Py)$ and $Pr(z \in X^{\leq}|yD_Pz)$, respectively (zD_Py denotes object z dominates object y w.r.t. P).

Formulation of Bayes factor for $y \in U$, $X^{\geq}, X^{\leq} \subseteq U$, $P \subseteq C$, is as follows:

$$B_{X^{\geq}}^P(y) = \frac{|D_P^+(y) \cap X^{\geq}||\neg X^{\geq}|}{|D_P^+(y) \cap \neg X^{\geq}||X^{\geq}|}, \quad B_{X^{\leq}}^P(y) = \frac{|D_P^-(y) \cap X^{\leq}||\neg X^{\leq}|}{|D_P^-(y) \cap \neg X^{\leq}||X^{\leq}|}.$$

Both these measures, used within DRSA in definition (2), are not monotonically non-decreasing w.r.t. set of attributes. To show this lack, it suffices to add to the example considered above that object y_3 has worse evaluation on attribute a_2 than objects y_1, y_2 and we consider union $X_2^{\geq}$.

2   Monotonic Variable Consistency Indiscernibility-Based Rough Set Approaches

Our motivation for proposing Variable Consistency Indiscernibility-based Rough Set Approaches (VC-IRSA) comes from the need of ensuring monotonicity of lower and upper approximations of sets w.r.t. set of attributes. The main difference between VC-IRSA and VPRS and RB model is that we include to P-lower approximation only objects that belong to the approximated set (def. (2) and (4)). In the other approaches, whole granules are included to lower approximation (def. (1) and (3)). Remark that a granule included in P-lower approximation may be composed of some inconsistent objects. Enlarging set P of attributes to $R \supset P$, the P-inconsistent objects may become R-discernible and thus, if we would like to preserve monotonicity of lower approximations, then we should keep in the R-lower approximation the R-discernible objects that do not belong to the approximated set. This can be considered controversial, however.

We propose definitions of measures that ensure monotonicity of VC-IRSA. In the first definition, the number of acceptable objects from outside of the approximated set is controlled by an inconsistency measure $\epsilon_{X_i}^P(y)$, that for $y \in U, P \subseteq R \subseteq C, X_i, \neg X_i \subseteq U$, where $\neg X_i = U - X_i$, is defined as

$$\epsilon_{X_i}^P(y) = \frac{|I_P(y) \cap \neg X_i|}{|\neg X_i|}.$$

Theorem. For all $P' \subseteq P'' \subseteq C, X_i \subseteq U, y \in U$, inconsistency measure $\epsilon_{X_i}^P(y)$ is monotonically non-increasing w.r.t. sets of attributes P' and P'':

$$\epsilon_{X_i}^{P'}(y) \geq \epsilon_{X_i}^{P''}(y). \qquad \square$$

In this case, definition (4) takes the form:

$$\underline{P}^{\beta_{X_i}}(X_i) = \{y \in X_i : \epsilon_{X_i}^P(y) \leq \beta_{X_i}\}, \qquad (5)$$

where parameter $\beta_{X_i} \in [0, 1]$ reflects the greatest degree of inconsistency acceptable to include object y to the lower approximation of set X_i. Let us observe that $\epsilon_{X_i}^P(y)$ is an estimate of inverse probability $Pr(x \in I_P(y) | x \in \neg X_i)$.

Other two measures that ensure monotonicity of VC-IRSA are as follows:

$$\underline{\mu}_{X_i}^P(y) = min_{R \supseteq P} \frac{|I_R(y) \cap X_i|}{|I_R(y)|}, \quad \overline{\mu}_{X_i}^P(y) = max_{R \subseteq P} \frac{|I_R(y) \cap X_i|}{|I_R(y)|}.$$

Theorem. For all $P' \subseteq P'' \subseteq C, X_i \subseteq U, y \in U$:

$$\underline{\mu}_{X_i}^{P'}(y) \leq \underline{\mu}_{X_i}^{P''}(y), \quad \overline{\mu}_{X_i}^{P'}(y) \leq \overline{\mu}_{X_i}^{P''}(y),$$

that is, measures $\underline{\mu}_{X_i}^P(y)$ and $\overline{\mu}_{X_i}^P(y)$ are monotonically non-decreasing w.r.t. sets of attributes P' and P''. Moreover, for all $P \subseteq C, X_i \subseteq U, y \in U$:

$$\underline{\mu}_{X_i}^P(y) \leq \overline{\mu}_{X_i}^P(y). \qquad \square$$

Since $\underline{\mu}_{X_i}^P(y) \leq \overline{\mu}_{X_i}^P(y)$, we define rough membership interval $[\underline{\mu}_{X_i}^P(y), \overline{\mu}_{X_i}^P(y)]$.

We define P-lower approximation of X_i by means of $\underline{\mu}_{X_i}^P(y), \overline{\mu}_{X_i}^P(y)$ and lower thresholds $\underline{\alpha}_{X_i}, \overline{\alpha}_{X_i} \in [0, 1]$, as

$$\underline{P}^{\underline{\alpha}_{X_i}, \overline{\alpha}_{X_i}}(X_i) = \{y \in X_i : \underline{\mu}_{X_i}^P(y) \geq \underline{\alpha}_{X_i} \wedge \overline{\mu}_{X_i}^P(y) \geq \overline{\alpha}_{X_i}\}. \qquad (6)$$

The notion of rough membership interval can be further extended to multi-dimensional VC-IRSA. We define multi-dimensional inconsistency measure $\underline{\gamma}_{X_i}^P(y, \underline{\alpha}_{X_i})$ and multi-dimensional consistency measure $\overline{\gamma}_{X_i}^P(y, \overline{\alpha}_{X_i})$ as

$$\underline{\gamma}_{X_i}^P(y, \underline{\alpha}_{X_i}) = |(R, z) : R \supseteq P, z \in I_R(y), \ \mu_{X_i}^R(z) < \underline{\alpha}_{X_i}|,$$
$$\overline{\gamma}_{X_i}^P(y, \overline{\alpha}_{X_i}) = |(R, z) : R \subseteq P, z \in I_R(y), \ \mu_{X_i}^R(z) \geq \overline{\alpha}_{X_i}|.$$

Measures $\underline{\gamma}_{X_i}^P$ and $\overline{\gamma}_{X_i}^P$ are monotonic w.r.t. set of attributes.

Theorem. For all $P' \subseteq P'' \subseteq C$, $\underline{\alpha}_{X_i}, \overline{\alpha}_{X_i} \in [0, 1]$, $X_i \subseteq U$, $y \in U$:

$$\underline{\gamma}_{X_i}^{P'}(y, \underline{\alpha}_{X_i}) \geq \underline{\gamma}_{X_i}^{P''}(y, \underline{\alpha}_{X_i}), \quad \overline{\gamma}_{X_i}^{P'}(y, \overline{\alpha}_{X_i}) \leq \overline{\gamma}_{X_i}^{P''}(y, \overline{\alpha}_{X_i}). \qquad \square$$

We can define P-lower approximation of X_i by $\underline{\gamma}_{X_i}^P(y, \underline{\alpha}_{X_i}), \overline{\gamma}_{X_i}^P(y, \overline{\alpha}_{X_i})$ as

$$\underline{P}^{\underline{\alpha}_{X_i}, \overline{\alpha}_{X_i}, \underline{\psi}_{X_i}, \overline{\psi}_{X_i}}(X_i) = \{y \in X_i : \underline{\gamma}_{X_i}^P(y, \underline{\alpha}_{X_i}) \leq \underline{\psi}_{X_i} \wedge \overline{\gamma}_{X_i}^P(y, \overline{\alpha}_{X_i}) \geq \overline{\psi}_{X_i}\}, \tag{7}$$

where $\underline{\psi}_{X_i} \in \{0, 1, \ldots\}$, $\overline{\psi}_{X_i} \in \{1, 2, \ldots\}$ is a threshold of inconsistency and consistency, respectively. Let us observe that in case $\underline{\psi}_{X_i} = 0$ and $\overline{\psi}_{X_i} = 1$, the lower approximation based on $\underline{\gamma}_{X_i}^P(y, \underline{\alpha}_{X_i})$ and $\overline{\gamma}_{X_i}^P(y, \overline{\alpha}_{X_i})$ coincide with the lower approximation based on $\underline{\mu}_{X_i}^P(y)$ and $\overline{\mu}_{X_i}^P(y)$.

3   Monotonic Variable Consistency Dominance-Based Rough Set Approaches

We reformulate definitions of monotonic approaches presented in section 2 to introduce monotonic definitions of VC-DRSA. Instead of granule $I_P(y)$, we use positive dominance cone $D_P^+(y)$ or negative dominance cone $D_P^-(y)$. Instead of set X_i, we consider upward union $X_i^{\geq}$ or downward union $X_i^{\leq}$.

The inconsistency measures $\epsilon_{X_i^{\geq}}^P(x)$ and $\epsilon_{X_i^{\leq}}^P(y)$ for $x \in X_i^{\geq}, y \in X_i^{\leq}, X_i^{\geq}, X_i^{\leq}, X_{i-1}^{\leq}, X_{i+1}^{\geq} \subseteq U$, $P \subseteq R \subseteq C$, are defined as

$$\epsilon_{X_i^{\geq}}^P(x) = \frac{|D_P^+(x) \cap X_{i-1}^{\leq}|}{|X_{i-1}^{\leq}|}, \quad \epsilon_{X_i^{\leq}}^P(y) = \frac{|D_P^-(y) \cap X_{i+1}^{\geq}|}{|X_{i+1}^{\geq}|}.$$

Theorem. For all $P' \subseteq P'' \subseteq C$, $x \in X_i^{\geq}, y \in X_i^{\leq}, X_i^{\geq}, X_i^{\leq} \subseteq U$, measures $\epsilon_{X_i^{\geq}}^P(x)$ and $\epsilon_{X_i^{\leq}}^P(y)$ are monotonically non-increasing w.r.t. P' and P'':

$$\epsilon_{X_i^{\geq}}^{P'}(x) \geq \epsilon_{X_i^{\geq}}^{P''}(x), \quad \epsilon_{X_i^{\leq}}^{P'}(y) \geq \epsilon_{X_i^{\leq}}^{P''}(y). \qquad \square$$

We define P-lower approximation of $X_i^{\geq}$ and $X_i^{\leq}$, for $\beta_{X_i^{\geq}}, \beta_{X_i^{\leq}} \in [0, 1]$, as

$$\underline{P}^{\beta_{X_i^{\geq}}}(X_i^{\geq}) = \{x \in X_i^{\geq} : \epsilon_{X_i^{\geq}}^P(x) \leq \beta_{X_i^{\geq}}\}, \tag{8}$$

$$\underline{P}^{\beta_{X_i^{\leq}}}(X_i^{\leq}) = \{y \in X_i^{\leq} : \epsilon_{X_i^{\leq}}^P(y) \leq \beta_{X_i^{\leq}}\}. \tag{9}$$

Other four measures that ensure monotonicity of VC-DRSA are as follows:

$$\underline{\mu}^P_{X^\geq_i}(x) = \min_{\substack{R \supseteq P, \\ z \in D^+_R(x) \cap X^\geq_i}} \frac{|D^+_R(z) \cap X^\geq_i|}{|D^+_R(z)|}, \quad \overline{\mu}^P_{X^\geq_i}(x) = \max_{\substack{R \subseteq P, \\ z \in D^-_R(x) \cap X^\geq_i}} \frac{|D^+_R(z) \cap X^\geq_i|}{|D^+_R(z)|},$$

$$\underline{\mu}^P_{X^\leq_i}(y) = \min_{\substack{R \supseteq P, \\ z \in D^-_R(y) \cap X^\leq_i}} \frac{|D^-_R(z) \cap X^\leq_i|}{|D^-_R(z)|}, \quad \overline{\mu}^P_{X^\leq_i}(y) = \max_{\substack{R \subseteq P, \\ z \in D^+_R(y) \cap X^\leq_i}} \frac{|D^-_R(z) \cap X^\leq_i|}{|D^-_R(z)|}.$$

Theorem. For all $P' \subseteq P'' \subseteq C$, $x \in X^\geq_i, y \in X^\leq_i$, $X^\geq_i, X^\leq_i \subseteq U$:

$$\underline{\mu}^{P'}_{X^\geq_i}(x) \leq \underline{\mu}^{P''}_{X^\geq_i}(x), \quad \overline{\mu}^{P'}_{X^\geq_i}(x) \leq \overline{\mu}^{P''}_{X^\geq_i}(x),$$

$$\underline{\mu}^{P'}_{X^\leq_i}(y) \leq \underline{\mu}^{P''}_{X^\leq_i}(y), \quad \overline{\mu}^{P'}_{X^\leq_i}(y) \leq \overline{\mu}^{P''}_{X^\leq_i}(y),$$

that is, measures $\underline{\mu}^P_{X^\geq_i}(x)$, $\overline{\mu}^P_{X^\geq_i}(x)$, $\underline{\mu}^P_{X^\leq_i}(y)$ and $\overline{\mu}^P_{X^\leq_i}(y)$ are monotonically non-decreasing w.r.t. sets of attributes P' and P''. Moreover, for all $P \subseteq C$:

$$\underline{\mu}^P_{X^\geq_i}(x) \leq \overline{\mu}^P_{X^\geq_i}(x), \quad \underline{\mu}^P_{X^\leq_i}(y) \leq \overline{\mu}^P_{X^\leq_i}(y). \qquad \square$$

Consequently, we introduce rough membership intervals $[\underline{\mu}^P_{X^\geq_i}(x), \overline{\mu}^P_{X^\geq_i}(x)]$ and $[\underline{\mu}^P_{X^\leq_i}(y), \overline{\mu}^P_{X^\leq_i}(y)]$. We define P-lower approximation of union $X^\geq_i$ and $X^\leq_i$, for lower thresholds $\underline{\alpha}_{X^\geq_i}, \overline{\alpha}_{X^\geq_i}, \underline{\alpha}_{X^\leq_i}, \overline{\alpha}_{X^\leq_i} \in [0,1]$, as

$$\underline{P}^{\underline{\alpha}_{X^\geq_i}, \overline{\alpha}_{X^\geq_i}}(X^\geq_i) = \{x \in X^\geq_i : \underline{\mu}^P_{X^\geq_i}(x) \geq \underline{\alpha}_{X^\geq_i} \wedge \overline{\mu}^P_{X^\geq_i}(x) \geq \overline{\alpha}_{X^\geq_i}\}, \qquad (10)$$

$$\underline{P}^{\underline{\alpha}_{X^\leq_i}, \overline{\alpha}_{X^\leq_i}}(X^\leq_i) = \{y \in X^\leq_i : \underline{\mu}^P_{X^\leq_i}(y) \geq \underline{\alpha}_{X^\leq_i} \wedge \overline{\mu}^P_{X^\leq_i}(y) \geq \overline{\alpha}_{X^\leq_i}\}. \qquad (11)$$

The notion of rough membership interval can be further extended to multi-dimensional VC-DRSA. We define multi-dimensional inconsistency (consistency) measures $\underline{\gamma}^P_{X^\geq_i}(x, \underline{\alpha}_{X^\geq_i})$ and $\underline{\gamma}^P_{X^\leq_i}(y, \underline{\alpha}_{X^\leq_i})$ $(\overline{\gamma}^P_{X^\geq_i}(x, \overline{\alpha}_{X^\geq_i})$ and $\overline{\gamma}^P_{X^\leq_i}(y, \overline{\alpha}_{X^\leq_i}))$ as

$$\underline{\gamma}^P_{X^\geq_i}(x, \underline{\alpha}_{X^\geq_i}) = |(R, z) : R \supseteq P, z \in D^+_R(x) \cap X^\geq_i, \mu^R_{X^\geq_i}(z) < \underline{\alpha}_{X^\geq_i}|,$$

$$\overline{\gamma}^P_{X^\geq_i}(x, \overline{\alpha}_{X^\geq_i}) = |(R, z) : R \subseteq P, z \in D^-_R(x) \cap X^\geq_i, \mu^R_{X^\geq_i}(z) \geq \overline{\alpha}_{X^\geq_i}|,$$

$$\underline{\gamma}^P_{X^\leq_i}(y, \underline{\alpha}_{X^\leq_i}) = |(R, z) : R \supseteq P, z \in D^-_R(y) \cap X^\leq_i, \mu^R_{X^\leq_i}(z) < \underline{\alpha}_{X^\leq_i}|,$$

$$\overline{\gamma}^P_{X^\leq_i}(y, \overline{\alpha}_{X^\leq_i}) = |(R, z) : R \subseteq P, z \in D^+_R(y) \cap X^\leq_i, \mu^R_{X^\leq_i}(z) \geq \overline{\alpha}_{X^\leq_i}|.$$

Measures $\underline{\gamma}^P_{X^\geq_i}(x, \underline{\alpha}_{X^\geq_i})$, $\overline{\gamma}^P_{X^\geq_i}(x, \overline{\alpha}_{X^\geq_i})$, $\underline{\gamma}^P_{X^\leq_i}(y, \underline{\alpha}_{X^\leq_i})$ and $\overline{\gamma}^P_{X^\leq_i}(y, \overline{\alpha}_{X^\leq_i})$ are monotonic w.r.t. set of attributes, as stated by the following theorem.

Theorem. For all $P' \subseteq P'' \subseteq C$, $\underline{\alpha}_{X^\geq_i}, \overline{\alpha}_{X^\geq_i}, \underline{\alpha}_{X^\leq_i}, \overline{\alpha}_{X^\leq_i} \in [0,1]$, $x \in X^\geq_i, y \in X^\leq_i$, $X^\geq_i, X^\leq_i \subseteq U$:

$$\underline{\gamma}^{P'}_{X^\geq_i}(x, \underline{\alpha}_{X^\geq_i}) \geq \underline{\gamma}^{P''}_{X^\geq_i}(x, \underline{\alpha}_{X^\geq_i}), \quad \overline{\gamma}^{P'}_{X^\geq_i}(x, \overline{\alpha}_{X^\geq_i}) \leq \overline{\gamma}^{P''}_{X^\geq_i}(x, \overline{\alpha}_{X^\geq_i}),$$

$$\underline{\gamma}^{P'}_{X^\leq_i}(y, \underline{\alpha}_{X^\leq_i}) \geq \underline{\gamma}^{P''}_{X^\leq_i}(y, \underline{\alpha}_{X^\leq_i}), \quad \overline{\gamma}^{P'}_{X^\leq_i}(y, \overline{\alpha}_{X^\leq_i}) \leq \overline{\gamma}^{P''}_{X^\leq_i}(y, \overline{\alpha}_{X^\leq_i}). \qquad \square$$

We define P-lower approximation of union $X_i^{\geq}$, $X_i^{\leq}$ by measures $\underline{\gamma}_{X_i^{\geq}}^{P}(x, \underline{\alpha}_{X_i^{\geq}})$, $\overline{\gamma}_{X_i^{\geq}}^{P}(x, \overline{\alpha}_{X_i^{\geq}})$, $\underline{\gamma}_{X_i^{\leq}}^{P}(y, \underline{\alpha}_{X_i^{\leq}})$ and $\overline{\gamma}_{X_i^{\leq}}^{P}(y, \overline{\alpha}_{X_i^{\leq}})$, for inconsistency thresholds $\underline{\psi}_{X_i^{\geq}}$, $\underline{\psi}_{X_i^{\leq}} \in \{0, 1, \ldots\}$ and consistency thresholds $\overline{\psi}_{X_i^{\geq}}, \overline{\psi}_{X_i^{\leq}} \in \{1, 2, \ldots\}$, as

$$\underline{P}^{\underline{\alpha}_{X_i^{\geq}}, \overline{\alpha}_{X_i^{\geq}}, \underline{\psi}_{X_i^{\geq}}, \overline{\psi}_{X_i^{\geq}}}(X_i^{\geq}) = \{x \in X_i^{\geq} : \underline{\gamma}_{X_i^{\geq}}^{P}(x, \underline{\alpha}_{X_i^{\geq}}) \leq \underline{\psi}_{X_i^{\geq}} \wedge \overline{\gamma}_{X_i^{\geq}}^{P}(x, \overline{\alpha}_{X_i^{\geq}}) \geq \overline{\psi}_{X_i^{\geq}}\},$$
(12)

$$\underline{P}^{\underline{\alpha}_{X_i^{\leq}}, \overline{\alpha}_{X_i^{\leq}}, \underline{\psi}_{X_i^{\leq}}, \overline{\psi}_{X_i^{\leq}}}(X_i^{\leq}) = \{y \in X_i^{\leq} : \underline{\gamma}_{X_i^{\leq}}^{P}(y, \underline{\alpha}_{X_i^{\leq}}) \leq \underline{\psi}_{X_i^{\leq}} \wedge \overline{\gamma}_{X_i^{\leq}}^{P}(y, \overline{\alpha}_{X_i^{\leq}}) \geq \overline{\psi}_{X_i^{\leq}}\}.$$
(13)

4   Illustrative Example

Let us consider VC-IRSA and the set of objects shown in Fig. 1a. First, let us observe that according to def. (5), $\underline{P}^{0}(X_2) = \{y_2, y_3\}$. Object $y_1 \notin \underline{P}^{0}(X_2)$ because $\epsilon_{X_2}^{P}(y_1) = \frac{1}{3}$. Second, applying def. (6), we have $\underline{P}^{\frac{1}{2}, \frac{2}{3}}(X_2) = \{y_1, y_2, y_3\}$. It is worth noting that because $\mu_{X_2}^{\{a_1\}}(y_1) = \frac{2}{3}$, $\mu_{X_2}^{\{a_2\}}(y_1) = \frac{1}{2}$ and $\mu_{X_2}^{P}(y_1) = \frac{1}{2}$, then $\overline{\mu}_{X_2}^{P}(y_1) = \frac{2}{3}$. Thus, $\overline{\mu}_{X_i}^{P}(y)$ is monotonically non-decreasing. The same can be shown for $\underline{\mu}_{X_i}^{P}(y)$. Third, according to def. (7), we have $\underline{P}^{\frac{1}{2}, \frac{2}{3}, 0, 1}(X_2) = \{y_1, y_2, y_3\}$. Let us show the monotonicity of $\overline{\gamma}_{X_i}^{P}(y, \alpha_{X_i})$. We have $\overline{\gamma}_{X_2}^{\{a_1\}}(y_3, \frac{2}{3}) = 1$, $\overline{\gamma}_{X_2}^{\{a_2\}}(y_3, \frac{2}{3}) = 3$ and $\overline{\gamma}_{X_2}^{\{a_1, a_2\}}(y_3, \frac{2}{3}) = 5$.

We present properties of VC-DRSA by example in Fig. 1b. We distinguish two unions: $X_1^{\leq}$ and $X_2^{\geq}$. First, we can observe that $\underline{P}^{\frac{1}{3}}(X_2^{\geq}) = \{y_1, y_2\}$ (def. (8)). Object $y_3 \notin \underline{P}^{\frac{1}{3}}(X_2^{\geq})$ because $\epsilon_{X_2^{\geq}}^{P}(y_3) = \frac{2}{3}$. We can notice that $\epsilon_{X_2^{\geq}}^{\{a_1\}}(y_2) = \frac{1}{3}$, $\epsilon_{X_2^{\geq}}^{\{a_2\}}(y_2) = \frac{2}{3}$, while $\epsilon_{X_2^{\geq}}^{P}(y_2) = \frac{1}{3}$. This shows monotonically non-increasing behavior of $\epsilon_{X_2^{\geq}}^{P}(x)$ w.r.t. set of attributes. Second, applying def. (10), we obtain $\underline{P}^{\frac{1}{2}, \frac{2}{3}}(X_2^{\geq}) = \{y_1, y_2\}$. It is worth noting that because $\mu_{X_2^{\geq}}^{\{a_1\}}(y_2) = \frac{2}{3}$,

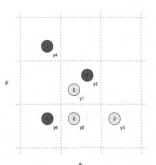

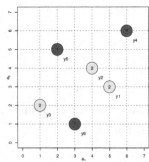

a) Exemplary set of objects described by means of set P of two condition attributes a_1 and a_2.

b) Exemplary set of objects described by means of set P of two condition gain criteria a_1 and a_2.

Fig. 1. Exemplary data sets. Objects marked with 1 and 2 belong to class X_1 and X_2.

$\mu_{X_2^\geq}^{\{a_2\}}(y_3) = \frac{3}{5}$, $\mu_{X_2^\geq}^P(y_3) = \frac{3}{5}$, then $\overline{\mu}_{X_2^\geq}^P(y_1) = \frac{2}{3}$. Thus, $\overline{\mu}_{X_i^\geq}^P(x)$ is monotonically non-decreasing. The same can be shown for $\underline{\mu}_{X_i^\geq}^P(x)$. Third, by def. (12), $\underline{P}^{\frac{1}{2},\frac{2}{3},0,1}(X_2^\geq) = \{y_1, y_2\}$. Let us show monotonicity of $\overline{\gamma}_{X_i^\geq}^P(x, \overline{\alpha}_{X_i^\geq})$. We have $\overline{\gamma}_{X_2^\geq}^{\{a_1\}}(y_2, \frac{2}{3}) = 1$, $\overline{\gamma}_{X_2^\geq}^{\{a_2\}}(y_2, \frac{2}{3}) = 0$ and $\overline{\gamma}_{X_2^\geq}^{\{a_1, a_2\}}(y_2, \frac{2}{3}) = 1$.

5 Conclusions

In this paper, we have presented several definitions of monotonic Variable Consistency Rough Set Approaches that employ indiscernibility and dominance relations. The monotonicity of lower (and upper) approximations w.r.t. set of attributes is an important property from the perspective of rule induction, because it permits incremental search of conditions. We have proposed two types of definitions of monotonic approaches. The first one involves inconsistency measure, which formulation is similar to Bayes factor and confirmation measure l. The second one consists in using monotonic measures based on the rough membership measure, which itself is not monotonic. Computation of lower approximations according to definitions of the second type is an NP-hard problem, equivalent to induction of a set of all rules. On the other hand, computation of approximations and rules induction can be combined.

Acknowledgments

The first and the third author wish to acknowledge financial support from the Ministry of Science and Higher Education (grant 3T11F 02127). The research of the second author is supported by Ministry of Education, University and Scientific Research.

References

1. J. Błaszczyński, S. Greco, R. Słowiński, M. Szeląg, *On Variable Consistency Dominance-based Rough Set Approaches*. In: LNAI, vol. 4259, Springer-Verlag, Berlin 2006, pp. 191-202.
2. B. Fitelson, *Studies in Bayesian Confirmation Theory*. Ph. D. thesis, University of Wisconsin-Madison, 2001.
3. S. Greco, B. Matarazzo, R. Słowiński, *Rough Membership and Bayesian Confirmation Measures for Parameterized Rough Sets*. In: LNAI, vol. 3641, Springer-Verlag, Berlin 2005, pp. 314-324.
4. Z. Pawlak, A. Skowron, *Rough Membership Functions*. In: R. R. Yager, M. Fedrizzi and J. Kacprzyk (eds.): *Advances in the Dampster-Shafer Theory of Evidence*, Wiley, New York 1994, pp. 251-271.
5. D. Ślęzak, *Rough Sets and Bayes Factor*. Trans. Rough Sets III, 2005, pp. 202-229.
6. W. Ziarko, *Variable Precision Rough Set Model*. J. Comput. Syst. Sci. 46(1), 1993, pp. 39-59.

Bayesian Decision Theory
for Dominance-Based Rough Set Approach

Salvatore Greco[1], Roman Słowiński[2], and Yiyu Yao[3]

[1] Faculty of Economics, University of Catania,
Corso Italia, 55, 95129 Catania, Italy
salgreco@unict.it
[2] Institute of Computing Science, Poznań University of Technology,
60-965 Poznań, and Institute for Systems Research,
Polish Academy of Sciences, 01-447 Warsaw, Poland
roman.slowinski@cs.put.poznan.pl
[3] Department of Computer Science, University of Regina,
S4S0A2 Regina, Saskatchewan, Canada
yyao@cs.uregina.ca

Abstract. Dominance-based Rough Set Approach (DRSA) has been proposed to generalize classical rough set approach when consideration of monotonicity between degrees of membership to considered concepts has to be taken into account. This is typical for data describing various phenomena, e.g., "the larger the mass and the smaller the distance, the larger the gravity", or "the more a tomato is red, the more it is ripe". These monotonicity relationships are fundamental in rough set approach to multiple criteria decision analysis. In this paper, we propose a Bayesian decision procedure for DRSA. Our approach permits to take into account costs of misclassification in fixing parameters of the Variable Consistency DRSA (VC-DRSA), being a probabilistic model of DRSA.

Keywords: Bayesian Decision Theory, Dominance, Rough Set Theory, Variable Consistency, Cost of Misclassification.

1 Introduction

Rough set theory has been proposed by Pawlak in the early 80s [5,6] as a tool for reasoning about data in terms of granules of knowledge. While the original rough set idea is very useful for classification support, it is not handling a background knowledge about monotonic relationship between evaluation of objects on condition attributes and their evaluation on decision attributes. Such a knowledge is typical for data describing various phenomena and for data describing multiple criteria decision problems. E.g., "the larger the mass and the smaller the distance, the larger the gravity", "the more a tomato is red, the more it is ripe" or "the better the school marks of a pupil, the better his overall classification". The monotonic relationships within multiple criteria decision problems follow from preferential ordering of value sets of attributes (scales of criteria), as well as preferential ordering of decision classes. In order to handle these monotonic

J.T. Yao et al. (Eds.): RSKT 2007, LNAI 4481, pp. 134–141, 2007.

relationships between conditions and decisions, Greco, Matarazzo and Słowiński [2,3,7] proposed to substitute the indiscernibility relation for a dominance relation. Dominance-based Rough Set Approach (DRSA) permits approximation of ordered sets. When dealing with preferences, monotonicity is expressed through the following relationship: "the better is an object with respect to (w.r.t.) considered points of view (criteria), the more it is appreciated". The definitions of rough approximations originally introduced in DRSA are based on a strict application of the dominance principle. However, when defining non-ambiguous objects, it is reasonable to accept a limited proportion of negative examples, particularly for large data tables. Such extended version of DRSA is called Variable Consistency DRSA model (VC-DRSA) [4] being a probabilistic model of DRSA. The focus of this paper is on extending the Bayesian decision theoretic framework [1], already introduced in case of classical rough set approach [8], to the VC-DRSA model. The paper is organized as follows. In the next section, the general principle of DRSA are recalled, together with a presentation of VC-DRSA. In the third section, a Bayesian decision procedure for DRSA is presented. The last sections contains conclusions.

2   Dominance-Based Rough Set Approach

In data analysis, information about objects can be represented in the form of an information table. The rows of the table are labelled by objects, whereas columns are labelled by attributes and entries of the table are attribute-values. Formally, by an information table we understand the 4-tuple $S = <U, Q, V, f>$, where U is a finite set of objects, Q is a finite set of attributes, $V = \bigcup_{q \in Q} V_q$, where V_q is a value set of the attribute q, and $f : U \times Q \to V$ is a total function such that $f(x, q) \to V_q$ for every $q \in Q$, $x \in U$, called an information function [6]. The set Q is, in general, divided into set C of condition attributes and set D of decision attributes. Assuming that all condition attributes $q \in C$ are criteria, let $\succeq_q$ be a weak preference relation on U w.r.t. criterion q such that $x \succeq_q y$ means "x is at least as good as y w.r.t. criterion q". We suppose that $\succeq_q$ is a complete preorder, i.e. a strongly complete and transitive binary relation, defined on U on the basis of evaluations $f(\cdot, q)$. Without loss of generality, we can assume that for all $x, y \in U$, $x \succeq_q y$ iff $f(x, q) \geq f(y, q)$. Furthermore, let us assume that the set of decision attributes D (possibly a singleton $\{d\}$) makes a partition of U into a finite number of decision classes $Cl = \{Cl_t, t \in T\}$, $T = \{1, ..., n\}$, such that each $x \in U$ belongs to one and only one class $Cl_t \in Cl$. We suppose that the classes are preference-ordered, i.e. for all $r, s \in T$, such that $r > s$, the objects from Cl_r are preferred to the objects from Cl_s. More formally, if $\succeq$ is a comprehensive weak preference relation on U, i.e. if for all $x, y \in U$, $x \succeq y$ means "x is at least as good as y", we suppose:

$$[x \in Cl_r, y \in Cl_s, r > s] \Rightarrow [x \succeq y \text{ and not } y \succeq x].$$

The above assumptions are typical for consideration of a multiple-criteria sorting problem. The sets to be approximated are called *upward union* and *downward union* of classes, respectively:

$$Cl_t^{\geq} = \bigcup_{s \geq t} Cl_s, \quad Cl_t^{\leq} = \bigcup_{s \leq t} Cl_s, \ t = 1, ..., n.$$

The statement $x \in Cl_t^{\geq}$ means "x belongs to at least class Cl_t", while $x \in Cl_t^{\leq}$ means "x belongs to at most class Cl_t". Let us remark that $Cl_1^{\geq} = Cl_n^{\leq} = U$, $Cl_n^{\geq} = Cl_n$ and $Cl_1^{\leq} = Cl_1$. Furthermore, for $t = 2, ..., n$, we have:

$$Cl_{t-1}^{\leq} = U - Cl_t^{\geq} \text{ and } Cl_t^{\geq} = U - Cl_{t-1}^{\leq}.$$

The key idea of rough sets is approximation of one knowledge by another knowledge. In classical rough set approach (CRSA) [6], the knowledge approximated is a partition of U into classes generated by a set of decision attributes; the knowledge used for approximation is a partition of U into elementary sets of objects that are indiscernible with respect to a set of condition attributes. The elementary sets are seen as "granules of knowledge". In DRSA [2,3,7], where condition attributes are criteria and classes are preference-ordered, the knowledge approximated is a collection of upward and downward unions of classes and the "granules of knowledge" are sets of objects defined using a dominance relation, instead of an indiscernibility relation used in CRSA. This is the main difference between CRSA and DRSA. In the following, in order to gain some more flexibility, we use the *variable consistency* DRSA model [4] which has its counterpart within the CRSA in the variable precision rough set approach [9,10]. Let us define now the dominance relation. We say that "x dominates y w.r.t. $P \subseteq C$, denoted by xD_Py, if $x \succeq_q y$ for all $q \in P$.

Given a set of criteria $P \subseteq C$ and $x \in U$, the "granules of knowledge" used for approximation in DRSA are:

– a set of objects dominating x, called P-*dominating set*,

$$D_P^+(x) = \{y \in U : yD_Px\},$$

– a set of objects dominated by x, called P-*dominated set*,

$$D_P^-(x) = \{y \in U : xD_Py\}.$$

For any $P \subseteq C$ we say that $x \in U$ belongs to $Cl_t^{\geq}$ with no ambiguity at consistency level $l \in (0, 1]$, if $x \in Cl_t^{\geq}$ and at least $l \times 100\%$ of all objects $y \in U$ dominating x w.r.t. P also belong to $Cl_t^{\geq}$, i.e.

$$\frac{|D_P^+(x) \cap Cl_t^{\geq}|}{|D_P^+(x)|} \geq l \quad (i)$$

where, for any set A, $|A|$ denotes its cardinality.

In this case, we say that x is a non-ambiguous object at consistency level l w.r.t. the upward union $Cl_t^{\geq}$ ($t = 2, ..., n$). Otherwise, we say that x is an ambiguous object at consistency level l w.r.t. the upward union $Cl_t^{\geq}$ ($t = 2, ..., n$).

Let us remark that $\frac{|D_P^+(x) \cap Cl_t^\geq|}{|D_P^+(x)|}$ can be interpreted as an estimation of the probability $P(y \in Cl_t^\geq | yD_Px)$ in the data table and thus (i) can be rewritten as

$$P(y \in Cl_t^\geq | yD_Px) \geq l.$$

The level l is called consistency level because it controls the degree of consistency between objects qualified as belonging to $Cl_t^\geq$ without any ambiguity. In other words, if $l < 1$, then no more than $(1-l) \times 100\%$ of all objects $y \in U$ dominating x w.r.t. P do not belong to $Cl_t^\geq$ and thus contradict the inclusion of x in $Cl_t^\geq$. Analogously, for any $P \subseteq C$ we say that $x \in U$ belongs to $Cl_t^\leq$ with no ambiguity at consistency level $l \in (0,1]$, if $x \in Cl_t^\leq$ and at least $l \times 100\%$ of all objects $y \in U$ dominated by x w.r.t. P also belong to $Cl_t^\leq$, i.e.

$$\frac{|D_P^-(x) \cap Cl_t^\leq|}{|D_P^-(x)|} \geq l. \qquad (ii)$$

In this case, we say that x is a non-ambiguous object at consistency level l w.r.t. the downward union $Cl_t^\leq$ $(t = 1, ..., n-1)$. Otherwise, we say that x is an ambiguous object at consistency level l w.r.t. the downward union $Cl_t^\leq$ $(t = 1, ..., n-1)$. Let us remark that $\frac{|D_P^-(x) \cap Cl_t^\leq|}{|D_P^-(x)|}$ can be interpreted as an estimation of the probability $P(y \in Cl_t^\leq | xD_Py)$ in the data table and thus (ii) can be rewritten as

$$P(y \in Cl_t^\leq | xD_Py) \geq l.$$

The concept of non-ambiguous objects at some consistency level l leads naturally to the definition of P-*lower approximations* of the unions of classes $Cl_t^\geq$ and $Cl_t^\leq$, denoted by $\underline{P}^l(Cl_t^\geq)$ and $\underline{P}^l(Cl_t^\leq)$, respectively:

$$\underline{P}^l(Cl_t^\geq) = \left\{ x \in Cl_t^\geq : \frac{|D_P^+(x) \cap Cl_t^\geq|}{|D_P^+(x)|} \geq l \right\},$$

$$\underline{P}^l(Cl_t^\leq) = \left\{ x \in Cl_t^\leq : \frac{|D_P^-(x) \cap Cl_t^\leq|}{|D_P^-(x)|} \geq l \right\}.$$

P-*lower approximations* of the unions of classes $Cl_t^\geq$ and $Cl_t^\leq$ can also be formulated in terms of conditional probabilities as follows:

$$\underline{P}^l(Cl_t^\geq) = \left\{ x \in Cl_t^\geq : P(y \in Cl_t^\geq | yD_Px) \geq l \right\},$$

$$\underline{P}^l(Cl_t^\leq) = \left\{ x \in Cl_t^\leq : P(y \in Cl_t^\leq | xD_Py) \geq l \right\}.$$

Given $P \subseteq C$ and consistency level $l \in (0,1]$, we can define the P-*upper approximations* of $Cl_t^\geq$ and $Cl_t^\leq$, denoted by $\overline{P}^l(Cl_t^\geq)$ and $\overline{P}^l(Cl_t^\leq)$, by complementarity of $\underline{P}^l(Cl_t^\geq)$ and $\underline{P}^l(Cl_t^\leq)$ w.r.t. U:

$$\overline{P}^l(Cl_t^\geq) = U - \underline{P}^l(Cl_{t-1}^\leq), \quad t = 2, ..., n,$$

$$\overline{P}^l(Cl_t^{\leq}) = U - \underline{P}^l(Cl_{t+1}^{\geq}), \quad t = 1, ..., n - 1.$$

$\overline{P}^l(Cl_t^{\geq})$ can be interpreted as the set of all the objects belonging to $Cl_t^{\geq}$, possibly ambiguous at consistency level l. Analogously, $\overline{P}^l(Cl_t^{\leq})$ can be interpreted as the set of all the objects belonging to $Cl_t^{\leq}$, possibly ambiguous at consistency level l. The P-boundaries (P-doubtful regions) of $Cl_t^{\geq}$ and $Cl_t^{\leq}$ are defined as:

$$Bn_P^l(Cl_t^{\geq}) = \overline{P}^l(Cl_t^{\geq}) - \underline{P}^l(Cl_t^{\geq}),$$

$$Bn_P^l(Cl_t^{\leq}) = \overline{P}^l(Cl_t^{\leq}) - \underline{P}^l(Cl_t^{\leq}).$$

The variable consistency model of the dominance-based rough set approach provides some degree of flexibility in assigning objects to lower and upper approximations of the unions of decision classes. It can easily be shown that for $0 < l' < l \leq 1$,

$$\underline{P}^l(Cl_t^{\geq}) \subseteq \underline{P}^{l'}(Cl_t^{\geq}), \quad \overline{P}^l(Cl_t^{\geq}) \supseteq \overline{P}^{l'}(Cl_t^{\geq}), \quad t = 2, ..., n,$$

$$\underline{P}^l(Cl_t^{\leq}) \subseteq \underline{P}^{l'}(Cl_t^{\leq}), \quad \overline{P}^l(Cl_t^{\leq}) \supseteq \overline{P}^{l'}(Cl_t^{\leq}), \quad t = 1, ..., n - 1.$$

The dominance-based rough approximations of upward and downward unions of classes can serve to induce a generalized description of objects contained in the information table in terms of "if..., then..." decision rules. The following two basic types of variable-consistency decision rules can be induced from lower approximations of upward and downward unions of classes:

1. $D_{\geq}$-decision rules with the following syntax:
   "if $f(x, q1) \geq r_{q1}$ and $f(x, q2) \geq r_{q2}$ and ... $f(x, qp) \geq r_{qp}$, then $x \in Cl_t^{\geq}$"
   in $\alpha\%$ of cases, where $t = 2, ..., n$, $P = \{q1, ..., qp\} \subseteq C$,
   $(r_{q1},...,r_{qp}) \in V_{q1} \times V_{q2} \times ... \times V_{qp}$.

2. $D_{\leq}$-decision rules with the following syntax:
   "if $f(x, q1) \leq r_{q1}$ and $f(x, q2) \leq r_{q2}$ and ... $f(x, qp) \leq r_{qp}$, then $x \in Cl_t^{\leq}$"
   in $\alpha\%$ of cases, where $t = 1, ..., n - 1$, $P = \{q1, ..., qp\} \subseteq C$,
   $(r_{q1},...,r_{qp}) \in V_{q1} \times V_{q2} \times ... \times V_{qp}$.

3 The Bayesian Decision Procedure for DRSA

Let $P(y \in Cl_t^{\geq}|yD_Px)$ be the probability of an object $y \in U$ to belong to $Cl_t^{\geq}$ given yD_Px, that is the probability that y belongs to a class of at least level t, given that y dominates x w.r.t. set of criteria $P \subseteq C$. Analogously, let $P(y \in Cl_t^{\leq}|xD_Py)$ be the probability of an object $y \in U$ to belong to $Cl_t^{\leq}$ given xD_Py, that is the probability that x belongs to a class of at most level t, given that y is dominated by x w.r.t. set of criteria $P \subseteq C$. One can also consider probabilities $P(y \in Cl_{t-1}^{\leq}|yD_Px)$ and $P(y \in Cl_{t+1}^{\geq}|xD_Py)$. Obviously, we have that $P(y \in Cl_{t-1}^{\leq}|yD_Px) = 1 - P(y \in Cl_t^{\geq}|yD_Px)$, $t = 2, ..., n$, and $P(y \in Cl_{t+1}^{\geq}|xD_Py) = 1 - P(y \in Cl_t^{\leq}|xD_Py)$, $t = 1, ..., n - 1$.

Let $\lambda(z \in Cl_t^{\geq} | z \in Cl_t^{\geq})$ denote the loss for assigning an object $z \in U$ to $Cl_t^{\geq}$ when this is true, i.e. when condition $z \in Cl_t^{\geq}$ holds, $t = 2, ..., n$. Analogously,

- $\lambda(z \in Cl_t^{\geq} | z \in Cl_{t-1}^{\leq})$ denotes the loss for assigning an object $z \in U$ to $Cl_t^{\geq}$ when this is false, i.e. when condition $z \in Cl_{t-1}^{\leq}$ holds, $t = 2, ..., n$,
- $\lambda(z \in Cl_t^{\leq} | z \in Cl_t^{\leq})$ denotes the loss for assigning an object $z \in U$ to $Cl_t^{\leq}$ when this is true, i.e. when condition $z \in Cl_t^{\leq}$ holds, $t = 1, ..., n - 1$,
- $\lambda(z \in Cl_t^{\leq} | z \in Cl_{t+1}^{\geq})$ denotes the loss for assigning an object $z \in U$ to $Cl_t^{\leq}$ when this is false, i.e. when condition $z \in Cl_{t+1}^{\geq}$ holds, $t = 1, ..., n - 1$.

In the following, we suppose for simplicity that the above losses are independent from object z.

Given an object $y \in U$, such that yD_Px, the expected losses $R(y \in Cl_t^{\geq} | yD_Px)$ and $R(y \in Cl_{t-1}^{\leq} | yD_Px)$ associated with assigning y to $Cl_t^{\geq}$ and $Cl_{t-1}^{\leq}$, $t = 2, ..., n$, respectively, can be expressed as:

$$R(y \in Cl_t^{\geq} | yD_Px) = \lambda(y \in Cl_t^{\geq} | y \in Cl_t^{\geq}) P(y \in Cl_t^{\geq} | yD_Px) +$$

$$\lambda(y \in Cl_t^{\geq} | y \in Cl_{t-1}^{\leq}) P(y \in Cl_{t-1}^{\leq} | yD_Px),$$

$$R(y \in Cl_{t-1}^{\leq} | yD_Px) = \lambda(y \in Cl_{t-1}^{\leq} | y \in Cl_t^{\geq}) P(y \in Cl_t^{\geq} | yD_Px) +$$

$$\lambda(y \in Cl_{t-1}^{\leq} | y \in Cl_{t-1}^{\leq}) P(y \in Cl_{t-1}^{\leq} | yD_Px).$$

By applying the Bayesian decision procedure, we obtain the following minimum-risk decision rules:

- assign y to $Cl_t^{\geq}$ if $R(y \in Cl_t^{\geq} | yD_Px) \geq R(y \in Cl_{t-1}^{\leq} | yD_Px)$,

- assign y to $Cl_{t-1}^{\leq}$ if $R(y \in Cl_t^{\geq} | yD_Px) < R(y \in Cl_{t-1}^{\leq} | yD_Px)$.

It is quite natural to assume that

$$\lambda(z \in Cl_t^{\geq} | z \in Cl_t^{\geq}) < \lambda(z \in Cl_{t-1}^{\leq} | z \in Cl_t^{\geq}) \quad \text{and}$$

$$\lambda(z \in Cl_{t-1}^{\leq} | z \in Cl_{t-1}^{\leq}) < \lambda(z \in Cl_t^{\geq} | z \in Cl_{t-1}^{\leq}).$$

That is, the loss of classifying an object belonging to $Cl_t^{\geq}$ into the correct class $Cl_t^{\geq}$ is smaller than the loss of classifying it into the incorrect class $Cl_{t-1}^{\leq}$; whereas the loss of classifying an object not belonging to $Cl_t^{\geq}$ into the class $Cl_t^{\geq}$ is greater than the loss of classifying it into the class $Cl_{t-1}^{\leq}$. With this loss function and the fact that $P(y \in Cl_t^{\geq} | yD_Px) + P(y \in Cl_{t-1}^{\leq} | yD_Px) = 1$, the above decision rules can be expressed as:

- assign y to $Cl_t^{\geq}$ if $P(y \in Cl_t^{\geq} | yD_Px) \geq \alpha_t$,

- assign y to $Cl_{t-1}^{\leq}$ if $P(y \in Cl_t^{\geq} | yD_Px) < \alpha_t$,

$$\text{where} \quad \alpha_t = \frac{\lambda(z \in Cl_t^{\geq}|z \in Cl_{t-1}^{\leq}) - \lambda(z \in Cl_{t-1}^{\leq}|z \in Cl_{t-1}^{\leq})}{\Lambda^{\geq}}, \text{ and}$$

$$\Lambda^{\geq} = \lambda(z \in Cl_t^{\geq}|z \in Cl_{t-1}^{\leq}) + \lambda(z \in Cl_{t-1}^{\leq}|z \in Cl_t^{\geq}) -$$

$$\lambda(z \in Cl_{t-1}^{\leq}|z \in Cl_{t-1}^{\leq}) - \lambda(z \in Cl_t^{\geq}|z \in Cl_t^{\geq}).$$

Given an object $y \in U$, such that xD_Py, the expected losses $R(y \in Cl_t^{\leq}|xD_Py)$ and $R(y \in Cl_{t+1}^{\geq}|xD_Py)$ associated with assigning y to $Cl_t^{\leq}$ and $Cl_{t+1}^{\geq}$, respectively, can be expressed as:

$$R(y \in Cl_t^{\leq}|xD_Py) = \lambda(y \in Cl_t^{\leq}|y \in Cl_t^{\leq})P(y \in Cl_t^{\leq}|xD_Py) +$$

$$\lambda(y \in Cl_t^{\leq}|y \in Cl_{t+1}^{\geq})P(y \in Cl_{t+1}^{\geq}|xD_Py),$$

$$R(y \in Cl_{t+1}^{\geq}|xD_Py) = \lambda(y \in Cl_{t+1}^{\geq}|y \in Cl_t^{\leq})P(y \in Cl_t^{\leq}|xD_Py) +$$

$$\lambda(y \in Cl_{t+1}^{\geq}|y \in Cl_{t+1}^{\geq})P(y \in Cl_{t+1}^{\geq}|xD_Py).$$

By applying the Bayesian decision procedure, we obtain the following minimum-risk decision rules:

- assign y to $Cl_t^{\leq}$ if $R(y \in Cl_t^{\leq}|xD_Py) \geq R(y \in Cl_{t+1}^{\geq}|xD_Py)$,

- assign y to $Cl_{t+1}^{\geq}$ if $R(y \in Cl_t^{\leq}|xD_Py) < R(y \in Cl_{t+1}^{\geq}|xD_Py)$.

It is quite natural to assume that

$$\lambda(z \in Cl_t^{\leq}|z \in Cl_t^{\leq}) < \lambda(z \in Cl_{t+1}^{\geq}|z \in Cl_t^{\leq}) \quad \text{and}$$

$$\lambda(z \in Cl_{t+1}^{\geq}|z \in Cl_{t+1}^{\geq}) < \lambda(z \in Cl_t^{\leq}|z \in Cl_{t+1}^{\geq}).$$

That is, the loss of classifying an object belonging to $Cl_t^{\leq}$ into the correct class $Cl_t^{\leq}$ is smaller than the loss of classifying it into the incorrect class $Cl_{t+1}^{\geq}$; whereas the loss of classifying an object not belonging to $Cl_t^{\leq}$ into the class $Cl_t^{\leq}$ is greater than the loss of classifying it into the class $Cl_{t+1}^{\geq}$. With this loss function and the fact that $P(y \in Cl_t^{\leq}|xD_Py) + P(y \in Cl_{t+1}^{\geq}|xD_Py) = 1$, the above decision rules can be expressed as:

- assign y to $Cl_t^{\leq}$ if $P(y \in Cl_t^{\leq}|xD_Py) \geq \beta_t$,

- assign y to $Cl_{t+1}^{\leq}$ if $P(y \in Cl_t^{\leq}|xD_Py) < \beta_t$,

$$\text{where} \quad \beta_t = \frac{\lambda(z \in Cl_{t+1}^{\geq}|z \in Cl_{t+1}^{\geq}) - \lambda(z \in Cl_{t+1}^{\geq}|z \in Cl_t^{\leq})}{\Lambda^{\leq}}, \text{ and}$$

$$\Lambda^{\leq} = \lambda(z \in Cl_t^{\leq}|z \in Cl_{t+1}^{\geq}) + \lambda(z \in Cl_{t+1}^{\geq}|z \in Cl_t^{\leq}) -$$

$$\lambda(z \in Cl_{t+1}^{\geq}|z \in Cl_{t+1}^{\geq}) - \lambda(z \in Cl_t^{\leq}|z \in Cl_t^{\leq}).$$

Using the values of parameters α_t and β_t obtained using the Bayesian decision procedure, we can redefine the P-*lower approximations* of the unions of classes $Cl_t^{\geq}$ and $Cl_t^{\leq}$, denoted by $\underline{P}^{\alpha_t}(Cl_t^{\geq})$ and $\underline{P}^{\beta_t}(Cl_t^{\leq})$, as follows:

$$\underline{P}^{\alpha_t}(Cl_t^{\geq}) = \left\{ x \in Cl_t^{\geq} : \frac{|D_P^+(x) \cap Cl_t^{\geq}|}{|D_P^+(x)|} \geq \alpha_t \right\},$$

$$\underline{P}^{\beta_t}(Cl_t^{\leq}) = \left\{ x \in Cl_t^{\leq} : \frac{|D_P^-(x) \cap Cl_t^{\leq}|}{|D_P^-(x)|} \geq \beta_t \right\}.$$

4 Conclusions

In this paper, we proposed a Bayesian decision procedure for DRSA that permits to take into account costs of misclassification in fixing parameters of the probabilistic model of DRSA, i.e. VC-DRSA. Future research will focus on investigation of the formal properties of the proposed model and on comparison of its performance with competitive models in data analysis.

Acknowledgement. The second author wishes to acknowledge financial support from the Polish Ministry of Science and Higher Education (3T11F 02127).

References

1. Duda, R. O., Hart., P. E.: Pattern Classification and Scene Analysis, Wiley (1973).
2. Greco, S., Matarazzo, B., Słowiński, R.: Rough set theory for multicriteria decision analysis, European Journal of Operational Research, **129** (2001) 1-47.
3. Greco, S., Matarazzo, B., Słowiński R.: Decision rule approach. In: J. Figueira, S. Greco, M. Erghott (Eds.), Multiple Criteria Decision Analysis: State of the Art Surveys, Springer (2005) 507-563.
4. Greco, S., Matarazzo, B., Słowiński, R., Stefanowski, J.: Variable consistency model of dominance-based rough sets approach. In: W. Ziarko, Y. Yao (Eds.), Rough Sets and Current Trends in Computing, LNAI **2005**. Springer, Berlin (2001) 170-181.
5. Pawlak, Z.: Rough Sets, International Journal of Computer and Information Sciences **11** (1982) 341-356.
6. Pawlak, Z.: Rough Sets, Kluwer (1991).
7. Słowiński, R., Greco, S., Matarazzo, B.: Rough set based decision support. In: E. Burke, G. Kendall (Eds.), Search Methodologies: Introductory Tutorials in Optimization and Decision Support Techniques, Springer (2005) 475-527.
8. Yao, Y.Y., Wong, S.K.M.: A decision theoretic framwork for approximating concepts, International Journal of Man-machine Studies **37** (1992) 793-809.
9. Ziarko W.: Variable precision rough sets model, Journal of Computer and Systems Sciences **46** (1993) 39-59.
10. Ziarko W.: Rough sets as a methodology for data mining. In: L. Polkowski, A. Skowron (Eds.), Rough Sets in Data Mining and Knowledge Discovery, **1**. Physica (1998) 554-576.

Ranking by Rough Approximation of Preferences for Decision Engineering Applications

Kazimierz Zaras[1] and Jules Thibault[2]

[1] Département des Sciences de la Gestion
Université du Québec en Abitibi-Témiscamingue Rouyn-Noranda
Québec, Canada J9X 5E4
Kazimierz.Zaras@uqat.ca
[2] Department of Chemical Engineering, University of Ottawa
Ottawa, Ontario, Canada K1N 6N5
Jules.Thibault@uottawa.ca

Abstract. A pulping process is studied to illustrate a new methodology in the field of decision engineering, which relies on the Dominance Rough-Set-based Approach (DRSA) to determine the optimal operating region. The DRSA performs a rough approximation of preferences on a small set of Pareto-optimal experimental points to infer the decision rules with and without considering thresholds of indifference with respect each attribute in the decision table. With thresholds of indifference, each rule can be represented by three discrete values (i.e. 0; 0.5; 1). A value of (1) indicates the first point, in a pair wise comparison, is strictly preferred to the second point from the Pareto domain. A value of (0) indicates the opposite relation whereas a value of (0.5) indicates that the two points are equivalent from an engineering point of view. These decision rules are then applied to the entire set of points representing the Pareto domain. The results show that the rules obtained with the indifference thresholds improve the quality of approximation.

Keywords: Neutral network, Genetic algorithm, Pareto domain, Preferences, Multicriteria analysis, Dominance-based Rough Set Approach.

1 Introduction

During the operation of an industrial process, the operator should ideally select values of input parameters/variables from a performance criterion point of view. The main problem facing the decision maker is that the range of parameter/variable values is usually very large and the number of their combinations is even larger such that a decision aid methodology is required to assist the decision maker in the judicious selection of all values of the process parameter/variables that lead to the best compromise solution in the eyes of the expert that has a profound knowledge of the process. This is at the core of a new decision engineering methodology, which mainly consists of three steps:

1. Process modelling,
2. Determination of the Pareto domain defined in terms of input parameters, and
3. Pareto Set ranking by the Rough Set Method.

J.T. Yao et al. (Eds.): RSKT 2007, LNAI 4481, pp. 142–148, 2007.

The rough set method allows capturing relatively easily valuable, at time unconscious, information, from an expert that a profound knowledge about the operation for the process in other to establish a ranking method that will be used to rank the entire Pareto domain.

2 Process Modelling

This methodology will be illustrated using a pulping process example. In the pulping process it is necessary to choose, using an appropriate multicriteria methodology, the set of operating conditions that will give the optimal quality of the pulp and the resulting paper sheet. The pulping process is a very complex nonlinear process for which a model is not readily available. It is therefore desired to have a model that can predict the various quality characteristics of the final product. To derive this model, a series of experiments were conducted by Lanouette et al. [4] in a pilot-scale pulp processing plant located in the Pulp and Paper Research Centre at Université du Québec à Trois-Rivières.

Among the numerous performance criteria, four objective criteria were retained as the most important ones for this process (see Thibault et al. [7] and Renaud et al. [6] for a more complete description of the process). The aim in this process is to maximize both the ISO brightness (Y_1) and the rupture length (Y_4) of the resulting paper sheet, while reducing the specific refining energy requirement (Y_2) and the extractive contents (Y_3). The experimental design that was used to perform the experiment is a D-Optimal design where seven input variables were considered. A D-Optimal design consists of a group of design points chosen to maximize the determinant of the Fisher information matrix ($X'X$). To model each of the four performance criteria of the process, stacked feedforward neural networks were used. Each neural network used the seven input process variables.

3 Determination of the Pareto Domain

The next step of the methodology consists of determining the region circumscribing all feasible solutions of the input variables represented by a large number of data points. An extension of the traditional genetic algorithm is suggested to deal with discretized data by introducing the dominance concept (see [3]). The procedure to obtain a good approximation of the Pareto domain is relatively simple. The n points randomly chosen initialize the search algorithm. For each point, the performance criteria are evaluated. A pair wise comparison of all points approximating the Pareto domain is performed. Then a dominance function, consisting of counting the number of times a given point is dominated by the other points, is calculated. A fraction of the dominated points corresponding to those most dominated is discarded. The non-dominated and the least dominated points are retained and recombined to replace the most dominated ones that were discarded. The recombination procedure is applied until all points are non-dominated. In the case of the pulping process, the Pareto domain defined in terms of input variables is represented by 6000 points. This number of points is too numerous to allow the decision-maker to easily select the zone of optimal conditions. For this reason it is necessary to use a ranking algorithm to establish the optimal region of operation. The next step of this overall methodology deals with this problem. The particular method used in this investigation is the Rough Set Method which is based on the Dominance Rough-Set-Based Approach (DRSA) (see [1]).

4   Ranking the Entire Pareto Set Using the Rough Set Method

The Rough Set Method is used to rank a large number of non-dominated points approximating the Pareto domain. The implementation of this ranking scheme is based on the Rough Set theory suggested by Pawlak [5], and developed by Greco et al. [1-2,7] and Zaras [9], a method known as the Dominance Rough-Set–based Approach (DRSA).

The procedure of this ranking method can be summarized as follows (see Thibault et al. [8]). First, a handful of points, usually (4-7), from different regions of the Pareto domain are selected and presented to a human expert who has a profound knowledge of the process. The expert is given the task of ordering the subset of points from the most preferred to the least preferred (Table 1). After creating the ranked subset, the expert specifies the indifference threshold for each criterion. The indifference threshold corresponds to measurement error as well as possible limits in the human detection of differences in a given criterion. Especially, the indifference threshold for a particular criterion is defined as the difference between two values of that criterion that is not considered significant enough to rank one value as preferred over another (Table 2).

Table 1. Subset of points from the Pareto domain ranked by the expert

Point	Y_1	Y_2	Y_3	Y_4
16	66.95	7.04	0.311	4.07
271	69.49	7.64	0.218	3.76
223	68.82	7.26	0.166	3.54
4671	69.46	7.91	0.222	3.89
12	66.99	7.25	0.526	4.14
2	68.68	6.29	0.469	2.55
1	67.85	6.53	0.273	1.94

The next step is to establish a set of rules that are based on the expert's ranked subset and indifference thresholds. Here, each point in the ranked subset is compared to every other point within that set in order to define "rules of preference" and "rules of non-preference". Each rule can be represented by a vector containing two (i.e. 0; 1, see Table 3) or three values (i.e. 0; 0,5; 1, see Table 4) depending if the comparison is performed without or with indifference thresholds. The dimension of each vector is equal to the number of attributes. In the case without thresholds a value of (1) for a given criterion indicates the first point of the compared pair is preferred to the second point whereas a value of (0) indicates the criterion of the second point is preferred to the first point. However, it is not known if the point is weakly or strongly preferred for that particular criterion. In the case with thresholds of indifference, a value of (1) indicates the first point of the compared pair is strictly preferred to the second point from the Pareto domain, a value of (0) indicates the opposite relation and (0.5) indicates the indifference because the gap between the two values of the criterion is not sufficient to allow choosing one point over the other. The conjunctions $(0.5 \vee 1)$ and $(0 \vee 0.5)$ indicate the weak preference and the weak non-preference, respectively. These decision rules are then applied to the whole set of points approximating the Pareto domain where all pairs of points are compared to determine if they satisfy a preference or a non-preference rule.

Table 2. Indifference thresholds for each criterion

Criterion	Description	Threshold
Y_1	ISO Brightness	0.50
Y_2	Refining Energy	0.40
Y_3	Extractives Content	0.05
Y_4	Rupture Length	0.30

Table 3. Set of rules for the ranked set without indifference thresholds

Preference rules				Non-preference rules			
Y_1	Y_2	Y_3	Y_4	Y_1	Y_2	Y_3	Y_4
0	0	1	1	1	1	0	0
1	1	0	1	0	0	1	0

The quality of the approximation expresses the ratio of all pairs of points in the ranked subset correctly ordered by "rules of preference" and "rules of non-preference" to the number of all the pairs of points in the ranked subset. The quality of approximation in the case without thresholds is equal to 0.38.

The quality of the approximation in the case with thresholds is equal to 0.57, which indicates that the quality of the approximation with the thresholds is significantly improved.

The last step of the Rough Set method is to perform a pair wise comparison of all 6000 points of the Pareto domain to determine if a preference or non-preference rule applies. If a preference rule is determined, the score of the first point is incremented by one and the score of the second point is decreased by one. The opposite operation is performed if a non-preference rule is identified. The scores of all Pareto-optimal points were initially set to zero. When all points are compared, the point that has the highest score is considered to be the optimal point. It is however preferable to examine the zone of the Pareto domain where a given percentage of the best points are located rather than considering an individual point. The Rough Set Method provides a clear recommendation as to the optimal zone of operation.

Table 4. Set of rules for the ranked set with indifference thresholds

Preference rules				Non-preference rules			
Y_1	Y_2	Y_3	Y_4	Y_1	Y_2	Y_3	Y_4
-	-	0.5	1	-	-	0.5	0
-	1	-	$0.5 \vee 1$	-	0	-	$0.5 \vee 0$
0.5	0.5	1	0.5	0.5	0.5	0	0.5
1	0.5	0	0.5	1	0	0	0

5 Results and Conclusions

Results obtained using the Rough Set Method (RSM) are presented in Fig. 1 without thresholds and in Fig. 2 with thresholds. Two-dimensional graphical projections show the results for both cases of the ranking the 6000 points of the Pareto front. The first 10% corresponding to the highly-ranked points are plotted using dark points.

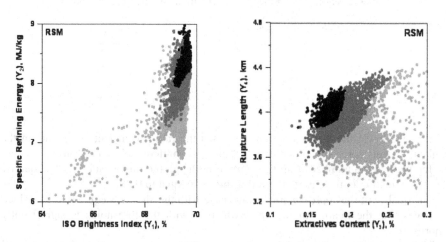

Fig. 1. Graph of the Pareto Front ranked by RSM without thresholds

The optimal region satisfies very well three of the four criteria. The choice of the expert is very clear, he can sacrifice having a higher specific refining energy (Y_2 being highest in the optimal region when it should be lowest) to have all the other criteria (Y_1, Y_3 and Y_4) being satisfied extremely well. In RSM, there is always at least one criterion that has to be sacrificed because a preference rule cannot contain all ones. Indeed, if a rule contained all ones this would mean that one of the two points that led to that rule would dominate the other point.

In this paper, two Dominance Rough–Set-based Approaches have been compared based on the extraction of rules from a subset of Pareto-optimal points ranked by a DM with and without indifference thresholds. The comparison of the quality of approximation indicates that the quality performance is improved with using the thresholds. However, the improved quality doesn't reduce the region of highly-ranked points. On the contrary, we can see on the graphical projections of the Pareto Front that this region is getting larger. There seems to have more nuance in the choice of the decision maker.

The introduction of indifference thresholds to the Dominance Rough-Set-based Approach (DRSA) also allows to make a difference between weaker and strict partial preferences with respect to each criterion for each decision rule.

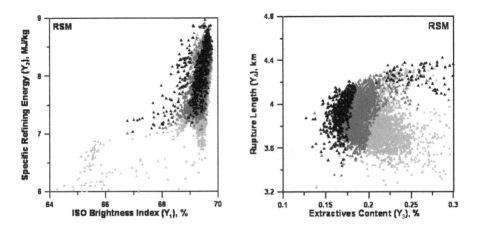

Fig. 2. Graph of the Pareto Front ranked by RSM with thresholds

References

1. Greco, S., Matarazzo B., Slowinski. R.: Rough approximation of a preference relation by dominance relations. EJOR, 117, 2, (1999) 63-83.
2. Greco, S., Matarazzo B., Slowinski. R.: Rough Set methodology for sorting problems in presence of multiple attributes and criteria, EJOR 138, 2, (2002) 247-259.
3. Fonteix, C., Bicking, F. Perrin, E., Marc, I.: Haploïd and diploïd algorithms, a new approach for global optimization: compared performances, Int. J. Sys. Sci. 26, 10, (1995), 1919-1933.
4. Lanouette, R., Thibault, J., Valade, J.L.: Process modeling with neural networks using small experimental datasets, Comp. Chem. Engng, (1999) 23, 1167-1176.
5. Pawlak, Z.: Rough Sets. Theoretical aspects of reasoning about data. Kluwer Academic Publishers, Dordrecht (1991).
6. Renaud, J., Thibault, J., Lanouette, R., Kiss, L.N., Zaras, K., Fonteix., C.: Comparison of two multicriteria decision aid methods: Net Flow and Rough Set Methods in a high yield pulping process, , EJOR 177, 3, (2007) 1418-1432.

7. Slowinski, R., Greco, S., Matarazzo, B.: Rough Set based decision support. Chapter 16 [in] Burke, E.K., and Kendall, G. (eds.), Search Methodologies: Introductory Tutorials in Optimization and Decision Support Techniques, Springer-Verlag, New York, 2005, pp. 475-527.
8. Thibault, J., Taylor, D., Yanofsky, C., Lanouette, R., Fonteix, C., Zaras, K.: Multicriteria optimization of a high yield pulping process with Rough Sets. Chem. Eng. Sci. 58 (2003) 203-213.
9. Zaras, K.: Rough approximation of preference relation by a multiattribute dominance for deterministic, stochastic and fuzzy decision problems, EJOR, 159, 1, (2004), 196-206.

Applying a Decision Making Model in the Early Diagnosis of Alzheimer's Disease

Ana Karoline Araújo de Castro and Plácido Rogério Pinheiro,
and Mirian Caliope Dantas Pinheiro

Master Degree in Applied Computer Sciences
University of Fortaleza
Av. Washington Soares, 1321 - Bloco J sala 30,
CEP: 60811-905, Fortaleza, Ceará, Brazil
akcastro@gmail.com
{placido,caliope}@unifor.br

Abstract. This study considers the construction of a multicriteria model to assist in the early diagnosis of Alzheimer's disease. Alzheimer's disease is considered the most frequent of the dementias and it is responsible for about 50% of the cases. Due to this fact and the therapeutical limitations in the most advanced stage of the disease, early diagnosis of Alzheimer's disease is extremely important and it can provide better life conditions to patients and their families.

Keywords: Early diagnosis, alzheimer's, multicriteria.

1 Introduction

Demographic studies in developed and developing countries have showed a progressive and significant increase in the elderly population in the last years [14]. According to the [11], chronic disease is a main cause of incapacity and cardiopathy, cancer and vascular accidents continue to be the main cause of death in 65 year olds. The Alzheimer's disease contributed to almost 44,000 deaths in 1999.

Alzheimer's disease is the most frequent cause of dementia and is responsible (alone or in association with other diseases) for 50% of the cases in western countries [14]. The dementias are syndromes described by a decline in memory and other neuropsychological changes especially occurring in the elderly and increasing exponentially in function of age. According to [8], despite that high incidence, doctors fail to detect dementia in 21 to 72% of their patient.

Considering the few alternative therapies and greater effectiveness of treatments after early diagnosis, identifying the cases that are high-risk for becoming dementia take on capital importance [10].

For this reason, systems of classification with specific criteria of inclusion and exclusion allow the definition of the diagnosis the Alzheimer's disease to be clinically viable. The criteria were based on clinical information and in laboratory tests.

J.T. Yao et al. (Eds.): RSKT 2007, LNAI 4481, pp. 149–156, 2007.

The main focus of this paper is to develop a multicriteria model for aiding in decision making of the early diagnosis of Alzheimer's disease. This disease is difficult to diagnose due to its initial subtle symptoms, which progress slowly until it becomes obvious and destructive. Therefore a study was developed with a patient that showed symptoms of dementia. It was necessary to construct value scales originating from semantic judgments of value with the objective of defining a ranking with the classification of the dementia in patient analysis. Finally conclusions and futures works are shown.

2   Importance of Early Diagnosis of Alzheimer's Disease

Alzheimer's disease is the most common type of dementia. People with any type of cognitive loss represent 5-10% of the population of 65 years or more, and more than 50% of the cases are due to Alzheimer's disease [7].

For this reason, the early diagnosis of this pathology is of great importance; mainly if we take in consideration the therapeutic advances and the relative optimism in relation to new active drugs in Central Nervous System.

Nowadays, there is an attempt to diagnose as soon as possible. It is necessary to carefully evaluate Mild Cognitive Damage (MCD), that represents a deficit in episodic memory tasks (relevant to day to day), but that still is insufficient for the diagnosis of dementia. This occurs because it is increasingly evident that many of these people with MCD can be found in a pre-clinical stage of Alzheimer's disease [12] and that this progression up to full dementia can take several years.

2.1   Criteria for Diagnosis

Systems of classification with specific inclusion and exclusion criteria allow for the clinically probable definition and diagnosis of Alzheimer's disease. These criteria were based on clinical information and laboratory tests.

There are at least four different systems that were proposed for the clinical diagnosis of Alzheimer's disease:

1. ICD-10 (International Classification of Diseases)[15].
2. DSM-III-R and DSM-IV (Diagnostic and Statistical Manual of Mental Disorders)[6].
3. NINCDS-ADRDA (National Institute of Neurological and Communicative Disorders and Stroke/Alzheimer's Disease and Related Disorders Association)[9].
4. CDR (Clinical Dementia Rating)[10].

[6] proposed a global criterion for evaluating the gravity of the dementias, especially those motived by Alzheimer's disease. This criterion was denominated "Clinical Dementia Rating" (CDR). CDR evaluates six behavioral and cognitive categories and can be used by neurologists, psychiatrists, psychologists and others professionals that study cognitive functions in the elderly. Another advantage is that it is highly reliable. CDR emphasizes cognitive operations without being

bound to the medical, social and emotional aspects of the patient. Thus it is applied in the light cases as well as the more severe cases of dementia and even for patients with questionable diagnosis.

In spite of the fact that CDR does not meet all criteria necessary for the clinical judgment of the disease, initially, this will not affect the initial application of the model. Because at this moment, we are carrying out the initial tests in order to subsequently apply new judgments with more patients to improve the initially proposed model.

Two categories of the CDR are described in table 1.

Table 1. Classification of the categories memory and orientation evaluated by Clinical Dementia Rating [10]

Impairment	None (0)	Questionable (0,5)	Mild (1)	Moderate (2)	Severe (3)
Memory	No memory loss or slight inconstant forgetfulness	Consistent slight forgetfulness; partial recollection of events; "benign" forgetfulness	Moderate memory loss; more marked for recent events; defect interferes with everyday activities	Severe memory loss; only highly learned material retained; new material rapidly lost	Severe memory loss; only fragments remain
Orientation	Fully oriented	Fully oriented except for slight difficulty with time relationships	Moderate difficulty with time relationships; oriented for place at examination; may have geographic disorientation elsewhere	Severe difficulty with time relationships; usually disoriented to time, often to place	Oriented to person only

In this study only the CDR criteria will used, therefore a multicriteria approach for aiding in early diagnosis of Alzheimer's disease will be used.

3   Decision Making Tool

The evaluation of early diagnosis of Alzheimer's disease is complex, and requires a high level of expertise. In order to help in decision making for early diagnosis of Alzheimer's disease we use a multicriteria methodology, which is based on two main characteristics:

1. The existence of several criteria.
2. Solutions for attending to the needs of the actors involved.

According to [2], in decision making it is necessary to look for elements that can answer the questions raised in order to clarify and make recommendations or increase the coherency between the evolution of the process and the objectives and values considered in the environment.

A substantial reading on MCDA methods can be found in [1, 2, 3, 4, 13], where the authors address the definitions and the problems that are involved in the decision making process.

The evaluation process is composed of the construction of judgment matrixes and constructing value scales for each Fundamental point of view (FPV) already

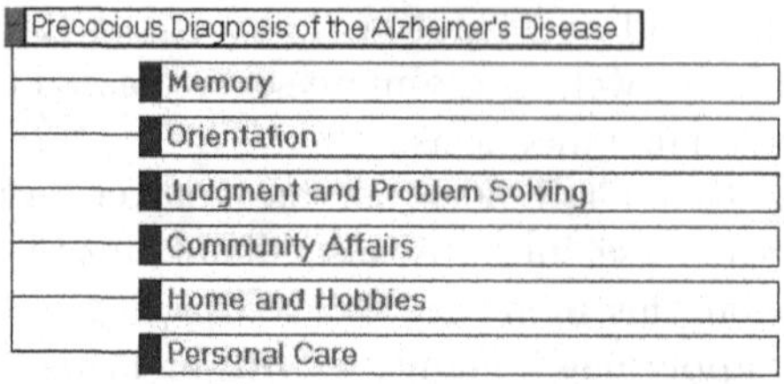

Fig. 1. Problem value tree

defined. The construction of cardinal value scales will be implemented through the MACBETH methodology developed by [2].

Figure 1 shows the tree corresponding to the FPV's that are used in evaluation of early diagnosis of Alzheimer's disease in a given patient.

Each Fundamental Point of View (FPV) is constituted by actions, which will be the alternatives used for the classification of each dementia (the acronyms in parentheses correspond to the names of the alternatives used in Macbeth): No Dementia (ND), Questionable Dementia (QD), Mild Dementia (MD), Moderate Dementia (MOD) and Severe Dementia (SD). The order that alternatives have in the judgment matrix was determined by everyday facts of the patient in comparasion with his life history or his performance in the past.

After the judgment of each FPV according to the each patient's information the value scales will be generated, which correspond to the Current Scale field in the judgment matrix. The resulting values will be used in the final judgment of the patient's diagnosis.

4   Application of the Model in the Decision Making System

The application of the model in multicriteria together with CDR was done by means of unstructured interviews carried out by the examiner who did not know the clinical diagnosis. A family member was also interviewed, preferably the spouse or a son or daughter who lives together on a daily basis with the individual. In rare cases, when the elderly patient lived alone, we talked with a close relative or a friendly neighbor. The coherence of this information was evaluated. In a few cases the interview was done with another type of informant [10].

As an example of a practical case, we will describe two case histories of patients that were used in this decision making process.

4.1   Case History 1

Mrs. Y is 80 years old. She was a bilingual secretary in a big firm. According to her friend and neighbor for many years (informant selected because the patient is single and lives alone). She presented difficult in remembering appointments, even those as important as doctors' appointments. Many times she would call the doctor without realizing that she had already called [10].

All things considered, we constructed the matrixes in Macbeth software and the evaluations were done. After the judgment of the each FPV, a scale of values that correspond to the Current Scale field in the judgment matrix will be generated. This can be observed in figure 2 where the judgment matrix of the FPV Memory is shown. For the evaluation, the history of the patient was considered together with the classifications defined by the CDR showed previously in table 1.

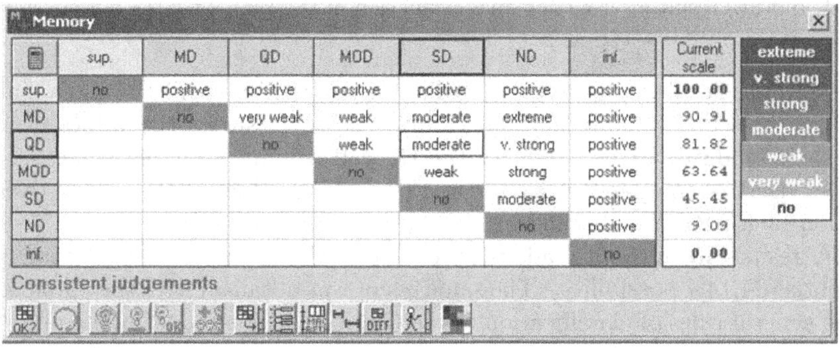

	sup.	MD	QD	MOD	SD	ND	inf.	Current scale
sup.	no	positive	positive	positive	positive	positive	positive	100.00
MD		no	very weak	weak	moderate	extreme	positive	90.91
QD			no	weak	moderate	v. strong	positive	81.82
MOD				no	weak	strong	positive	63.64
SD					no	moderate	positive	45.45
ND						no	positive	9.09
inf.							no	0.00

Fig. 2. Judgment of the Memory FPV for the patient Mrs. Y

After evaluating the alternatives of all the FPV's individually, an evaluation of the FPV's in one matrix only was carried out. For this, a judgment matrix was created in which the decision maker's orders are defined according to the patient's characteristics. In summary, the order of the FPV's in the judgment matrix was defined in the following manner: Orientation, Memory, Judgment and Problem Solving, Community Affairs, Personal Care and Home and Hobbies. Figure 3 presents the judgment matrix of the FPV's.

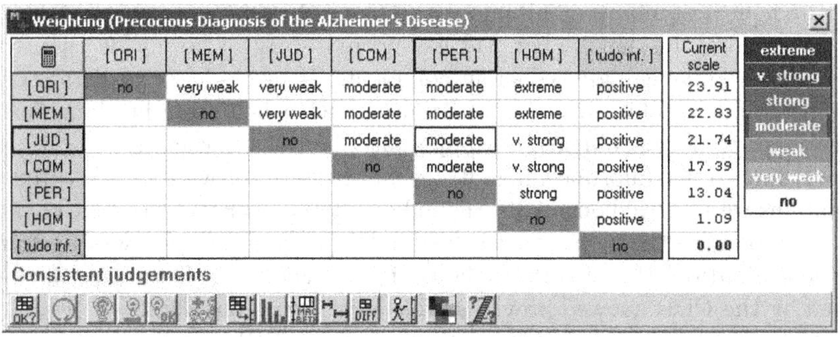

	[ORI]	[MEM]	[JUD]	[COM]	[PER]	[HOM]	[tudo inf.]	Current scale
[ORI]	no	very weak	very weak	moderate	moderate	extreme	positive	23.91
[MEM]		no	very weak	moderate	moderate	extreme	positive	22.83
[JUD]			no	moderate	moderate	v. strong	positive	21.74
[COM]				no	moderate	v. strong	positive	17.39
[PER]					no	strong	positive	13.04
[HOM]						no	positive	1.09
[tudo inf.]							no	0.00

Fig. 3. Judgment of all the FPV's for Mrs. Y

Table of scores							☒
Options	Overall	MEM	ORI	JUD	COM	HOM	PER
ND	10.59	9.09	12.50	7.69	9.09	88.89	10.00
QD	77.42	81.82	50.00	84.62	90.91	77.78	90.00
MD	84.14	90.91	75.00	92.31	81.82	66.67	80.00
MOD	71.18	63.64	87.50	69.23	63.64	44.44	70.00
SD	37.96	45.45	25.00	38.46	45.45	11.11	40.00
[tudo sup.]	100.00	100.00	100.00	100.00	100.00	100.00	100.00
[tudo inf.]	0.00	0.00	0.00	0.00	0.00	0.00	0.00
Weights :		0.2283	0.2391	0.2174	0.1739	0.0109	0.1304

Fig. 4. Indexes resulting from the final evaluation of the FPV's for the patient Mrs. Y

After all the evaluations were carried out, a matrix for each FPV and a matrix with all the FPV's, show the final result. The table shows the values of each alternative judged. Figure 4 shows the Overall column with the values of the each alternative compared with all the FPV's. The values of the alternatives in relation each FPV are also shown.

The results in figure 4 lead us to conclude that Mrs. Y suffers from mild dementia (84.14% possibility). The same result was obtained with the application of the general rules of classification of the Clinical Dementia Rating [6] and [10]. An advantage found in the application of these multicriteria models was that achieved a ranking of all the alternatives, so it was possible to determine the chance that a given patient has of acquiring Alzheimer's disease in each one of the categories of memory. The result of the application of this model concluded that Mrs. Y has an 84.14% possibility of acquiring mild dementia, 77.42% possibility of acquiring questionable dementia, 71.18% possibility of acquiring moderate dementia, 37.96% possibility of acquiring severe dementia and 10.59% possibility of acquiring no dementia.

4.2   Case History 2

Mr. X is 77 years old. He was an accountant. According to his daughter, after he retired, he lost many of his more complex interests and preferred to stay at home reading his newspaper. He has no more say in family decisions, but when requested, he can give coherent comments. He attends to the details of his retirement and pays his bills, but sometimes requests help from his daughter. He drives a car without problems, but apparently he was distruct [10].

According to this information, matrixes in Macbeth software were created and evaluations were carried out. After the judgment of each FPV, the scale of values that correspond to the Current Scale field in the judgment matrix was generated. For the evaluation, the history of the patient, together with the classifications defined by the CDR showed previously in table 1 was considered.

After the evaluation of the alternatives of all the FPV's individually an evaluation of the FPV's in only one matrix was carried out. For this, a judgment matrix was created in which the decision maker's orders were defined to come from the patient's characteristics. In summary, the order of the FPV's in the judgment matrix was defined in the following way: Home and Hobbies, Community Affairs, Judgment and Problem Solving, Orientation, Memory and Personal Care.

After all the evaluations were carried out, a matrix for each FPV and a matrix with all the FPV's, show the final result. The table shows the values of each alternative judgment.

The result conclude that Mr. X suffers from mild dementia (87.92% of possibility). The result obtained with the application the general rules of classification of the Clinical Dementia Rating was different [6] and [10]. The result found was that the patient has questionable dementia. This is a stage of dementia a little more progressed than that found with the application of the model in multicriteria. An advantage found in application of these multicriteria models was that a ranking of all the alternatives allowed for the possibilities of determining the chances that a given patient had of acquiring Alzheimer's disease in each one the memory categories. The result of the application of the model concluded that Mr. X has an 87.92% possibility of acquiring mild dementia, 79.48% possibility of acquiring moderate dementia, 73.23% possibility of acquiring questionable dementia, 44.24% possibility of acquiring severe dementia and 42.62% possibility of acquiring no dementia.

5   Conclusion

As a consequence of the progressive increase in the elderly population in the last years, there has been an increase in the prevalence of dementias. There are on average 70 diseases that can cause dementia, but Alzheimer's disease is the most frequent and is responsible (separately or in association) for 50% of cases in western countries [14].

Several systems of classification with specific criteria of inclusion and exclusion permit the definition of early diagnosis of Alzheimer's disease. A multicriteria analysis that aids in decision making for these diagnoses was applied to one of these systems, the Clinical Dementia Rating (CDR).

After evaluating the judgment matrixes that FPV's were generated based on cognitive - behavior categories of the CDR, a case study was carried out. The history of the behavior of the three patients was analyzed and judgment matrixes were generated from the information obtained.

The result these matrixes was satisfactory because it confirmed the result obtained in the application of the same case study applied with the general rules for the classification of the CDR. The patient Mr. X yielded a different result than that which was found with the application of the general rules of the CDR. However, the application of multicriteria for this case study revealed the percentage of all the possibilities of memory deficiency that are possible in the patient.

As a future project, this model can be extended with the inclusion of new criteria or new models which can be developed that use other systems of classification. Another project that can be developed could be the application of these models in a given group of people that would serve to verify its validity.

Another interesting project could be the application of the model to cases of patients with suspected dementia of unknown causes but that are difficult to diagnose.

The application of the Dominance-based Rough Set Approach method [5] that is based on ordinal properties of evaluations is being analyzed now.

Acknowledgments. Ana Karoline Araújo de Castro is thankful to FUNCAP (Fundação Cearense de Apoio ao Desenvolvimento Científico e Tecnológico) for the support she has received for this project.

References

1. Costa, C.A.B., Beinat, E., Vickerman, R.: Introduction and Problem Definition, CEG-IST Working Paper (2001).
2. Costa, C.A.B., Corte, J.M.D., Vansnick, J.C.: Macbeth, LSE-OR Working Paper (2003).
3. Drucker, P.F.: The Effective Decision. Harvard Business Review on Decision Making, Harvard Business School Press, Boston (2001).
4. Goodwin, P., Wright, G.: Decision Analysis for Management Judgment, John Wiley and Sons, Chicester (1998).
5. Greco, S., Matarazzo, B., Slowinski, R.: Rough Sets Theory for Multicriteria Decision Analysis, European J. of Operational Research **129** (2001) 1–47.
6. Hughes, C.P., Berg, L., Danzinger, W.L., Coben, L.A., Martin, R.L.: A New Clinical Scale for the Staging of Dementia, British Journal of Psychiatry **140** (1982) 566–572.
7. Jellinger, K., Oanielczyk, W., Fischer P. et al.: Clinicopathological Analysis of Dementia Disorders in the Elderly, Journal Neurology **5ci** (1990) 95:239–258.
8. Mayo Fundation for Medical Education and Research. Homepage at http://www.mayo.edu/geriatrics-rst/Dementia.I.html
9. Mckhann, G.D., Drachman, D., Folstein, M., Katzman, R., Price, D., Stadlan, E.M.: Clinical Diagnosis of Alzheimer's Disease: Report of the NINCDS-ADRDA Work Group under the Auspices of the Department of Health and Human Services Task Force on Alzheimer's Disease, Neurology **34** (1984) 939–944.
10. Morris, J.: The Clinical Dementia Rating (CDR): Current Version and Scoring Rules, Neurology **43(11)** (1993) 2412–2414.
11. National Center for Health Statistics, United States, 2000 with Adolescent Health Chartbook, Hyattsville, Maryland (2000).
12. Petersen, R.C., Smith, G.E., Waring, S.C. et al.: Mild Cognitive Impairment: Clinical Characterization and Outcome, Archives Neurology **56** (1999) 303–308.
13. Pinheiro, P.R., Souza, G.G.C.: A Multicriteria Model for Production of a Newspaper. Proc: The 17th International Conference on Multiple Criteria Decision Analysis, Canada (2004) 315–325.
14. Porto, C.S., Fichman, H.C., Caramelli, P., Bahia, V. S., Nitrini, R.: Brazilian Version of the Mattis Dementia Rating Scale Diagnosis of Mild Dementia in Alzheimer.s Disease, Arq Neuropsiquiatr **61(2-B)** (2003) 339–345.
15. World Health Organization. Homepage at http://www.who.int/classifications/ icd/en/

Singular and Principal Subspace of Signal Information System by BROM Algorithm

Władysław Skarbek

Warsaw University of Technology, Faculty of Electronics and Information Technology,
00-665 Warszawa, Nowowiejska 15/19, Poland
W.Skarbek@ire.pw.edu.pl

Abstract. A novel algorithm for finding algebraic base of singular subspace for signal information system is presented. It is based on Best Rank One Matrix (BROM) approximation for matrix representation of information system and on its subsequent matrix residua. From algebraic point of view BROM is a kind of power method for singular value problem. By attribute centering it can be used to determine principal subspace of signal information system and for this goal it is more accurate and faster than Oja's neural algorithm for PCA while preserving its adaptivity to signal change in time and space. The concept is illustrated by an exemplary application from image processing area: adaptive computing of image energy singular trajectory which could be used for image replicas detection.

Keywords: Principal Component Analysis, Singular Value Decomposition, JPEG compression, image replica detection, image energy singular trajectory.

1 Introduction

Signal information system is a special kind of information system in Pawlak's sense [1] in which objects are certain signal (multimedia) objects such as images, audio tracks, video sequences while attributes are determined by certain discrete elements drawn from spatial, temporal, or transform domain of signal objects.

For instance the Discrete Cosine Transform frequency channel represents a DCT coefficient which specifies a share of such frequency in the whole signal object.

Having n signal objects and m DCT frequency channels we get an information system which can be represented by a matrix $A \in \mathbb{R}^{m \times n}$. In case of Discrete Fourier Transform the matrix has complex elements and $A \in \mathbb{C}^{m \times n}$.

The columns of $A \in \mathbb{R}^{m \times n}$ define n attributes $A = [a_1, \ldots, a_n]$ and they can be considered as elements of m dimensional vector space: $a_i \in \mathbb{R}^m$, $i = 1, \ldots, n$.

In order to use algebraic properties of the vectorial space the attributes should have common physical units or they should be made unit-less, for instance by an affine transform, such as attribute centering and scaling.

Having attributes in the vectorial space we can define their dependence by the concept of linear combinations and by the related concepts of linear independence and linear subspace.

J.T. Yao et al. (Eds.): RSKT 2007, LNAI 4481, pp. 157–165, 2007.

In such approach we start from the subspace span(A) which includes all finite linear combinations of columns of A, i.e. attributes $a_1, \ldots, a_n$. If the dimension of span(A) equals to r, i.e. if rank(A) = r then we can find a nested sequence of $r+1$ linear subspaces of increasing dimensionality starting from the null subspace $\mathcal{S}_0 := \{0_m\}$ and ending at $\mathcal{S}_r :=$span(A) :

$$\mathcal{S}_0 \subset \mathcal{S}_1 \subset \cdots \subset \mathcal{S}_{r-1} \subset \mathcal{S}_r \ .$$

In the infinite number of nested subspace sequences for the given matrix A, there is a specific class of singular subspaces defined for A by the condition of minimum projection error. Namely, let $P_{\mathcal{S}}a$ be the orthogonal projection of a onto the subspace $\mathcal{S}$. Then the *singular subspace* $\mathcal{S}_q$ of dimension $q <$rank(A) minimizes the following squared projection error in norm l_2 :

$$\mathcal{S}_q := \arg \min_{\dim(\mathcal{S})=q} \sum_{i=1}^{n} \|a_i - P_{\mathcal{S}}a_i\|_2^2 \ .$$

The unit vector u spanning $\mathcal{S}_1$ is called the *singular direction*. We say that the attributes $a_1, \ldots, a_n$ are centered if their mean is zero vector, i.e. $\sum_{i=1}^{n} a_i = 0_m$. In case of centered attributes the singular subspace is called the *principal subspace* and the singular direction is called the *principal direction*.

In practice the singular subspace of matrix A is found from Singular Value Decomposition (SVD) of matrix A to orthogonal matrices U, V and diagonal matrix Σ :

$$A = U\Sigma V^t, \ U \in \mathbb{R}^{m \times r}, \ U^t U = I_{r \times r}$$
$$\Sigma = \mathrm{diag}(\sigma_1, \ldots, \sigma_r), \ V \in \mathbb{R}^{n \times r}, \ V^t V = I_{r \times r} \ .$$

Namely, the first q columns of $U = [u_1, \ldots, u_r]$ span the singular subspace $\mathcal{S}_q = $span($u_1, \ldots, u_r$).

Traditionally principal subspaces are obtained from Eigenvector Decomposition (EVD) of the outer product of centered matrix A :

$$AA^t = U\Lambda U^t$$

This procedure is known as Principal Component Analysis (PCA) – one of the most famous transformations in signal theory [2]. The traditional approach, though very efficient, is not adaptive to change of matrix A. In case of PCA there is also well known Oja neural scheme [3] which stochastically approximates the principal direction. However, it can be used to centered data only and therefore is not applicable to the general case of singular direction. In this paper another adaptive scheme is presented. It is based on analysis of rank one matrix approximations of the information system.

2   BROM Algorithm for Singular Subspace

Let us consider a signal information system with m objects and n real valued attributes. Then it is represented by a matrix $A = [a_1, \ldots, a_n] \in \mathbb{R}^{m \times n}, \ a_i \in \mathbb{R}^m$.

If the linear subspace spanned by attributes a_i has the dimension r, i.e. if rank$(A) = r$, then for any $q < r$ we consider the following problem: *Find a matrix $X \in \mathbb{R}^{m \times n}$ of rank q which is the best approximation of A in Frobenius norm, i.e. it minimizes $\|A - X\|_F$*. We call this problem as *best rank q matrix* and in particular for $q = 1$ we have BROM problem, i.e. *best rank one matrix* problem.

It appears that it is enough to know an algorithm for BROM problem, in order to get incrementally the solution X_q for any $q < r$:

$$X_0 = 0_{m \times n}; \text{ for } q = 1, \ldots, \text{rank}(A) - 1 : \ X_q := X_{q-1} + \text{BROM}(A - X_{q-1}); \quad (1)$$

On the other hand the matrix X of rank one has a shorter nonlinear parametrization with $m + n$ variables. Namely the following property is true: *The matrix $X \in \mathbb{R}^{m \times n}$ is of rank one if and only if there exist vectors $u \in \mathbb{R}^m$, $v \in \mathbb{R}^n$, $u \neq 0, v \neq 0$, such that $X = uv^t$*. Therefore we can state the following optimization goal function e of two vectorial parameters u, v :

$$e(u, v) := \|A - uv^t\|_F^2, \ u \in \mathbb{R}^m, \ v \in \mathbb{R}^n \quad (2)$$

It is easy to find a necessary and sufficient condition for the stationary points of e, i.e. the zero gradient points of $e(u, v)$:

$$v = \frac{A^t u}{\|u\|^2}, \ u = \frac{Av}{\|v\|^2} \quad (3)$$

Moreover, at fixed u (v) the actual minimum of e can be explicitly found:

- At fixed $u \in \mathbb{R}^m, u \neq 0$ the optimal $v_{opt} \in \mathbb{R}^n$ minimizing $e(u, v)$ has the form:

$$v_{opt} = \frac{A^t u}{\|u\|^2}, \ e(u, v_{opt}) \leq e(u, v), \ \forall v \in \mathbb{R}^n$$
$$e(u, v_{opt}) = \|A\|_F^2 - \|u\|_2^2 \|v_{opt}\|_2^2 \quad (4)$$

- At fixed $v \in \mathbb{R}^n, v \neq 0$ the optimal $u_{opt} \in \mathbb{R}^m$ minimizing $e(u, v)$ is of the form:

$$u_{opt} = \frac{Av}{\|v\|^2}, \ e(u_{opt}, v) \leq e(u, v), \ \forall u \in \mathbb{R}^m$$
$$e(u_{opt}, v) = \|A\|_F^2 - \|u_{opt}\|_2^2 \|v\|_2^2 \quad (5)$$

The algorithm BROM looks for the best rank one matrix uv^t of the matrix A by the following locally optimal steps for $i = 0, 1, \cdots$:

1. for u_i determine the optimal v_{i+1};
2. for v_{i+1} determine the optimal u_{i+1}.

Using the explicit formulas for u_{opt} and v_{opt} we get the following iterative scheme:

$$u_0 := \text{ nonzero column of } A$$
$$v_{i+1} := \frac{A^t u_i}{\|u_i\|^2}, \ u_{i+1} := \frac{Av_{i+1}}{\|v_{i+1}\|^2}, \ i = 0, 1, 2, \ldots \quad (6)$$

For practical use the following form of BROM has been elaborated which returns the singular direction in vector u.

algorithm $[u, v] := brom(A)$
    $u :=$ the first nonzero column of A
    if $u = 0$ **then** *return* **endif**
    do
$$v := \frac{A^t u}{\|u\|^2}; \ u := \frac{Av}{\|v\|^2}$$
    until $(\|u\| \cdot \|v\|$ stabilizes$)$
    $v := v * \|u\|; \ u := u/\|u\|$
endalgorithm

Since at the exit of BROM we have $v = A^t u$, the coordinate of column a_i w.r.t. to the vector u is v_i, $i = 1, \ldots, n$. It is interesting that when a symmetric matrix A is input of the BROM, the algorithm produces as u the eigenvector corresponding to the eigenvalue λ_{max} of maximum absolute value and $\lambda_{max} = u^t A u$. This follows from the observation that modulo a scaling factor, the BROM algorithm performs iterations of *the power method* for the matrix A^2 which has the same eigenvectors as A, but for squared eigenvalues. Since the power method computes the maximal eigenvalue for A^2 then for A it corresponds to the maximum absolute eigenvalue.

3   Outline of BROM's Convergence Analysis

The strict proof of convergence for BROM has been recently developed by the author, but the limit of pages for this paper allows only for an outline of BROM's convergence analysis.

We analyze the sequences defined by (6). The first observation concerns the behavior of norms for the sequences. Namely, the norms of vectorial sequences u_i and v_i satisfy the following inequalities for $i = 1, 2, \cdots$:

$$\|u_i\| \|v_i\| \leq \|A\|_F$$
$$\|u_i\| \|v_i\| \leq \|u_i\| \|v_{i+1}\| \leq \|u_{i+1}\| \|v_{i+1}\|$$
$$\|u_i\| \leq \|u_{i+1}\|, \ 1 \leq \|v_i\| \leq \|v_{i+1}\|$$

Hence, the norms of the sequences are bounded and monotonic. Thus the sequences of norms $\|u_i\|, \|v_i\|$ are convergent.

Let λ be an eigenvalue of the matrix B. Then we denote by $W(B, \lambda)$ the subspace of all eigenvectors defined by eigenvalue λ :

$$W(B, \lambda) := \{u : \ Bu = \lambda u\}$$

The remaining convergence analysis can be summarized in six properties which are stated in the following theorem.

Theorem 1 (on convergence of BROM).

1. *The vectorial sequence u_i is convergent in l_2 to an eigenvector u_* of the matrix AA^t corresponding to the largest eigenvalue λ for which the initial vector u_0 is not perpendicular to the eigenvector subspace $W(AA^t, \lambda)$.*

2. *The vectorial sequence v_i is convergent in l_2 to the vector $v_* = A^t u_* / \|u_*\|^2$.*
3. *The vector $u_* / \|u_*\|$ is the singular direction with the singular value $\sqrt{\lambda}$ and the singular coordinates $\sqrt{\lambda} v_* / \|v_*\|$.*
4. *The matrix sequence $u_i v_i^t$ is convergent in Frobenius norm to the matrix $u_* v_*^t$ which is the stationary point of the objective function $e(u, v)$.*
5. *If u_0 is not perpendicular to the eigenvector subspace $W(AA^t, \lambda_{max})$ for the maximal eigenvalue of the matrix AA^t then the matrix sequence $u_i v_i^t$ is convergent w.r.t. Frobenius norm to the matrix $u_* v_*^t$ which is the global minimum of the objective function $e(u, v)$ and $e(u_*, v_*) = \|A\|_F^2 - \lambda_{max}$.*
6. *The stop condition for BROM algorithm selected in the form:*

$$\|u_{i+1}\| \|v_{i+1}\| - \|u_i\| \|v_i\| < \epsilon$$

implies the stabilization of the objective function: $e(u_i, v_i) - e(u_{i+1}, v_{i+1}) < \epsilon$.

The above theorem explains why in the stop condition we can replace the original requirement for $u_i v_i^t$ stabilization by less costly condition of stabilization for the norm product $\|u_i\| \|v_i\|$. Namely, the convergence of the matrix sequence $u_i v_i^t$ enables observing of this convergence indirectly in the range of the error function $e(u_i, v_i)$. But from the last property of the theorem we have seen that the convergence of $e(u_i, v_i)$ can be detected from the convergence of the sequence $\|u_i\| \|v_i\|$.

4   Application: Image Energy Singular Trajectory

In many applications images are decomposed into small size blocks in order to make local analysis which is more efficient or more problem relevant. The blocks of the decomposition can be disjoint or overlapping. For instance in JPEG compression [4] the blocks are of size 8×8 and they are not overlapping. In some applications the order of blocks is irrelevant while in some their relative locations in the sequence is important.

Having a fixed ordering of blocks, for instance according the raster scan or along the Hilbert curve, we can introduce a concept of signal energy trajectory for the given image. Namely, each image block is characterized by the signal energy measured by the sum of squared pixel intensities.

The signal energy as a block feature is not invariant to most image processing operations. However, if we consider the fractional distribution of the energy in singular channels defined by singular directions of image blocks, the situation is much better and such features can be used for instance to image replica detection [5] even if replica have been *invisibly processed* to cheat web robots.

Let us define the *image energy singular trajectory* of rank r more formally. Let $f_1, \ldots, f_L$ be the sequence of pixel blocks drawn from the image f. It means that $f_i := f|_{D_i}$ for a rectangular sub-domain D_i of the image domain D. Performing the singular decomposition of the matrix f_i, we consider only r dominant singular values $\sigma_i(1), \ldots, \sigma_i(r)$.

It is well known that the signal energy of block f_i is decomposed into the sum of all squared singular values of f_i :

$$\|f_i\|_F^2 = \sum_k \sigma_i^2(k), \quad \sum_k \frac{\sigma_i^2(k)}{\|f_i\|_F^2} = 1$$

The image energy singular trajectory of rank r is defined as the sequence of points in r dimensional unit cube $[0,1]^r$:

$$\left(\frac{\sigma_i^2(1)}{\|f_i\|_F^2}, \ldots \frac{\sigma_i^2(r)}{\|f_i\|_F^2} \right), \ i = 1, \ldots, L$$

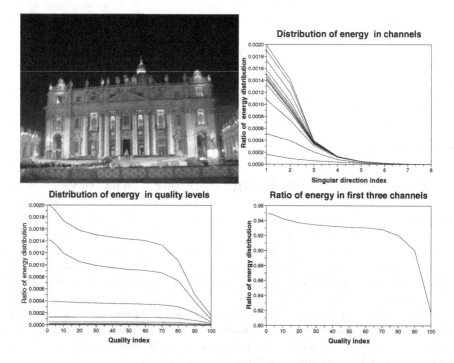

Fig. 1. Dependence of energy distribution in singular channels for various JPEG compression quality. For bottom graphs lower curves correspond to higher index of energy channel while for the right upper graph to the higher quality index.

The rationale for this novel concept is getting a fine characterization of signal energy distribution with its spatial coherency to be represented by trajectory concept. Since trajectories can be normalized by its re-sampling to a standard discrete interval, changes of image resolutions, cutting of windows, and local affine image transformations could be detected in trajectory segments by point proximity analysis for trajectories in time-energy space.

If small rank trajectories are enough for a particular application then we expect that BROM algorithm can be recommended in place of standard SVD algorithm. In the remaining part of this section few such practical cases are analyzed.

JPEG quality measure. In image compression the quality is usually measured by error image global analysis. The error image is the difference between decoder's output image and encoder's input image computed pixel-wise.

The most popular image fidelity measures are based on mean squared error (MSE) which is a scaled version of squared Frobenius norm for the error image. More subtle fidelity measures have vectorial character such as SVD based measures [6].

We analyze the distribution of compression error energy for an image of architectural scenes. Let JPEG quality index be from the set:

$$Q = \{1, 2, 5, 10, 20, 30, 40, 50, 60, 70, 80, 100\} .$$

Then for the image of Fig. 1 the average distribution of energy in error images for those quality indexes w.r.t. all eight singular directions is presented. We see that most of energy is included in the first three singular directions (so called the energy channels) – from 95% for low quality images down to 80% for high quality image. We observe that the proposed measure is uncorrelated with human eye sensitivity: the highest drop of energy corresponds to best visual quality interval [70, 100]. This is very desirable property for image replica detection.

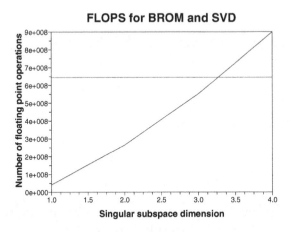

FLOPS for BROM and SVD

Fig. 2. Number of floating point operation at computing image energy singular trajectory by BROM and SVD (horizontal line) as function of singular subspace dimension

Trajectory complexity for BROM and SVD. It is interesting to compare the computational efficiency of using BROM to find the image energy singular trajectory w.r.t. the classical approach using Singular Value Decomposition for image blocks. Since SVD returns all singular values of image blocks it seems that

BROM should be much faster. The expectation is confirmed up to rank $r = 3$ (cf. Fig. 2) at the arithmetic precision 10^{-10} for singular values.

For this kind of applications when *signal information systems* are relatively small the adaptivity of BROM results in less than 5% reducing of complexity.

At comparison, the SVD complexity has been evaluated on the basis of the formulas given in [7]. Other experiments show that BROM algorithm is more accurate and faster than Oja algorithm [3] when applied for centered data to find the principal direction.

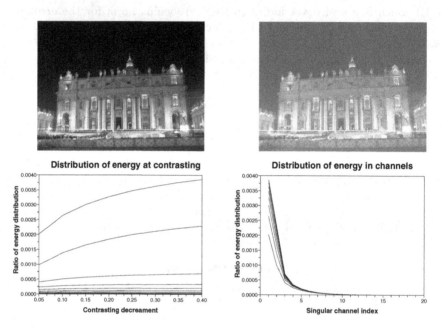

Fig. 3. Dependence of energy distribution in singular channels for various image contrasting and brightening operations

Energy distribution at image contrasting and brightening. In this experiment we change the image contrast and its brightness using scaling parameter $s \in (0, 1)$:

$$f_s := (1 - s) \cdot f + s \cdot f_{max}$$

where f_{max} is the maximum value of image f. This kind of image processing operations in the same time reduces contrast by factor $(1 - s)$ and in a sense compensates this by increase of brightness to preserve the maximum value of image intensity. For small s the processing effect is invisible, but the change in the signal energy is significant.

Figure 3 shows two processed versions of image from Fig. 1. The left image is processed with $s = 0.05$ and the right one with $s = 0.4$.

Experiments show that the error in image energy distribution slightly depends on s and for singular channels with indices higher than three (lower curves on the left graph of Fig. 3) is marginal (less than 0.04%). The maximum change is observed for the dominant singular subspace (top curve) but even for maximal $s = 0.4$ the change of energy share for this channel is less than 0.4%. It means that the energy distribution is a good invariant for image contrasting and brightening operations and it in this context is useful for image replica detection.

The right graph on Fig. 3 confirms *the rule of three* observed in context of JPEG image quality: *the most of change in image energy distribution is observed in the first three singular channels.* The curves in this graph are indexed by the parameter s with $s = 0.05$ for the bottom curve and $s = 0.4$ for the top one.

5 Conclusions

BROM algorithm is a practical alternative for SVD in finding algebraic base of singular subspace for signal information systems.

From algebraic point of view BROM is a kind of power method applied for singular value problem and it can be specialized to be a novel power method for eigenvalue problem.

By attribute centering it can be used to determine principal subspace of signal information system and for this goal it is more accurate and faster than Oja's neural algorithm for PCA while preserving its adaptivity to signal change in time and space.

In exemplary application computing image energy singular trajectory it appears faster than SVD for rank less than four. The trajectory approach is the useful tool applicable to image replica detection.

Acknowledgment. The work presented was developed within VISNET 2, a European Network of Excellence (http://www.visnet-noe.org), funded under the European Commission IST FP6 Programme.

References

1. Pawlak, Z.: Information Systems Theoretical Foundations. Information Systems **3** (1981) 205–218
2. Jolliffe, I.T.: Principal Component Analysis. Springer-Verlag (2002)
3. Skarbek, W., Pietrowcew, A., Sikora, R.: The modified Oja-RLS algorithm, stochastic convergence analysis and application for image compression. Fundamenta Informaticae **36** (1998) 345–365
4. Pennebaker, W.B., Mitchell, J.L.: JPEG: Still Image Data Compression Standard. Springer (2006)
5. Maret, Y., DuFaux, F., Ebrahimi, T.: Adaptive image replica detection based on support vector classifiers. Signal Processing: Image Communication **21** (2006) 688–703
6. Shnayderman, A., Gusev, A., Eskicioglu, A.M.: An SVD-Based Grayscale Image Quality Measure for Local and Global Assessment. IEEE Trans. Image Process **15** (2006) 422–429
7. Golub, G., Loan, C.: Matrix Computations. The Johns Hopkins University Press (1989)

Biometric Verification by Projections in Error Subspaces

Mariusz Leszczynski and Władysław Skarbek

Warsaw University of Technology, Faculty of Electronics and Information Technology,
00-665 Warszawa, Nowowiejska 15/19, Poland
M.Leszczynski@ire.pw.edu.pl

Abstract. A general methodology for design of biometric verification system is presented. It is based on linear feature discrimination using sequential compositions of several types of feature vector transformations: data centering , orthogonal projection onto linear subspace, vector component scaling, and orthogonal projection onto unit sphere. Projections refer to subspaces in global, within-class, and between-class error spaces. Twelve basic discrimination schemes are identified by compositions of subspace projections interleaved by scaling operations and single projection onto unit sphere. For the proposed discriminant features, the Euclidean norm of difference between query and average personal feature vectors is compared with the threshold corresponding to the required false acceptance rate. Moreover, the aggregation by geometric mean of distances in two schemes leads to better verification results. The methodology is tested and illustrated for the verification system based on facial 2D images.

Keywords: biometrics, face verification, discriminant analysis, singular subspace, within-class errors.

1 Introduction

Biometrics is a research field with a practical goal: create applications for uniquely recognizing humans based upon one or more intrinsic physical and/or behavioral traits including facial 2D/3D image, voice, fingerprints, eye retinas and irises, hand measurements, signature, gait and typing patterns. Biometric verification is one of three tasks which are usually attributed to pattern recognition: object identification, object verification, and similar object searching. However, biometric pattern verification is conceptually different from traditional class membership verification. To understand this point let us consider two pattern verification queries:

1. Given an image of a digit, verify whether the digit is *five*.
2. Given a facial image and a person identifier, verify whether the image matches to this id.

To solve the first problem a model for image class *five* is designed and used to verify the membership of the input image to the queried class. For instance symbol images x are mapped into a space of features $y = \mathcal{M}(x)$ in which memberships to symbol classes are represented by class c probability distributions

J.T. Yao et al. (Eds.): RSKT 2007, LNAI 4481, pp. 166–173, 2007.

$p_c(y)$ Then the predicate $\forall c \neq 5, p_c(y) < p_5(y)$ could be the basis of the verification. Moreover, such verification is optimal since it results in minimum of verification error = false acceptance rate + false rejection rate.

To solve the second problem we may follow the above approach. But then each new human being h in the system should have a new model p_h for his/her facial images in certain feature space. It means that models built for facial databases in training stage cannot be directly used in testing and exploiting stages of such verification system since in practice the sets of *training persons* and *exploiting persons* are different.

From the above examples we see that for the biometric verification we need such a model training procedure which builds a model with parameters to be used by testing and exploiting procedures.

Since natural human centered pattern classes cannot be used in person verification biometric systems, another categorization has to be sought. It appears that differences of human features for the biometric measurements of the same person (within-class differences) and for different persons (between-class features) create a consistent categorization including two specific classes. The specificity of this two classes follows from the fact that means of these two classes are both equal to zero. Moreover, for the within-class feature variation could be sometimes greater than between-class feature variation, i.e. usually the squared within-class errors are of the same magnitude as squared between-class errors.

Therefore, it is natural to look for such a linear transformation $W : \mathbb{R}^N \to \mathbb{R}^n$ of original biometric measurements $x \in \mathbb{R}^N$ (e.g. vectorized pixel matrix of face image or its 2D frequency representation) into a target feature vector $y = W^t x$ for which within-class differences are decreased while between-class differences are increased.

To this goal the class separation measure is defined as the ratio of between-class variation to within-class variation for vectorial data set $\{x_1, \ldots, x_L\}$ represented in columns of matrix $X \in \mathbb{R}^{N \times L}$:

$$v_X := \frac{\text{variation}_b(X)}{\text{variation}_w(X)} \tag{1}$$

where the within and between class variations are defined together with total variation via squared Euclidean distance:

$$\text{variation}_w(X) := \frac{1}{J^2} \sum_{j=1}^{J} \frac{1}{L_j^2} \sum_{i_1, i_2 \in I_j} \|y_{i_1} - y_{i_2}\|^2$$

$$\text{variation}_b(X) := \frac{1}{J^2} \sum_{j_1 \neq j_2} \frac{1}{L_{j_1} L_{j_2}} \sum_{i_1 \in I_{j_1}, i_2 \in I_{j_2}} \|y_{i_1} - y_{i_2}\|^2$$

$$\text{variation}_t(X) := \text{variation}_w(X) + \text{variation}_b(X) =$$

$$\frac{1}{J^2} \sum_{j_1=1}^{J} \sum_{j_2=1}^{J} \frac{1}{L_{j_1} L_{j_2}} \sum_{i_1 \in I_{j_1}, i_2 \in I_{j_2}} \|y_{i_1} - y_{i_2}\|^2 \tag{2}$$

where J is the number of classes, L_j is the number of j-th class samples ($L = L_1 + \cdots + L_J$) whose index set is denoted by I_j, $j = 1, \ldots, J$.

It appears that the class variations are not new concepts as they are scaled forms of class variances which were introduced by Fisher already in thirties of twentieth century [1]:

$$\text{var}_w(X) := \frac{1}{J} \sum_{j=1}^{J} \frac{1}{L_j} \sum_{i \in I_j} \| x_i - \overline{x}^j \|^2$$

$$\text{var}_b(X) := \frac{1}{J} \sum_{j=1}^{J} \| \overline{x}^j - \overline{x} \|^2 \tag{3}$$

$$\text{var}_t(X) := \text{var}_w(X) + \text{var}_b(X) = \frac{1}{J} \sum_{j=1}^{J} \frac{1}{L_j} \sum_{i \in I_j} \| x_i - \overline{x} \|^2$$

where $\overline{x}^j$ is the class mean of all j-th class samples in X and $\overline{x}$ is the grand mean of all samples in X.

Namely, the following relations are true for class variations and class variances:

$$\text{variation}_w(X) = \frac{2\text{var}_t(X)}{J}$$

$$\text{variation}_b(X) = 2 \left(\text{var}_b(X) + \frac{J-1}{J} \text{var}_w(X) \right) \tag{4}$$

$$\text{variation}_t(X) = 2\text{var}_t(X)$$

Hence, the class separation measure v_X is the affine form of Fisher separation measure with coefficients solely dependent on the number of classes J :

$$v_X = J f_X + J - 1 \tag{5}$$

2   Classical Optimization of Fisher Measure

The Fisher class separation measure becomes a goal function w.r.t. transformation matrix $W \in \mathbb{R}^{N \times n}$ when the source data matrix X is replaced by feature data matrix $Y := W^t X$.

The standard approach in optimizing (maximizing) $f(W) := f_{W^t X}$ is replacing the scalar product of two vectors by the trace of their outer product:

$$a^t b = \text{tr}(ab^t), \ \| a \|^2 = \text{tr}(aa^t)$$

Then we observe that the within and between-class variances are traces of within and between-class covariance matrices, respectively:

$$R_w(Y) := \frac{1}{L} \sum_{j=1}^{J} \frac{1}{L_j} \sum_{i \in I_j} (y_i - \overline{y}^j)(y_i - \overline{y}^j)^t = W^t R_w(X) W$$

$$R_b(Y) = \frac{1}{J} \sum_{j=1}^{J} (\overline{y}^j - \overline{y})(\overline{y}^j - \overline{y})^t = W^t R_b(X) W \tag{6}$$

$$f(W) = f_Y = \frac{\text{tr}(R_b(Y))}{\text{tr}(R_w(Y))} = \frac{\text{tr}(W^t R_b(X) W)}{\text{tr}(W^t R_w(X) W))}$$

The results of optimization for $f(W)$ are traditionally called Linear Discriminant Analysis (LDA). Fisher considered the scalar LDA features, i.e. the case of $n = 1$ in which $W = w \in \mathbb{R}^{n \times 1}$, $y = w^t x$ is the scalar and the within and between-class variances are quadratic forms of vectorial variable w. Then the Fisher measure transforms to Rayleigh quotient w.r.t. matrices R_b and R_w :

$$f(w) = f_y = f_{w^t X} = \frac{w^t R_b(X) w}{w^t R_w(X) w} \tag{7}$$

The standard analysis of stationary points for $f(W)$, $W = [w_1, \ldots, w_n]$, $w_i \in \mathbb{R}^N$, $i = 1, \ldots, n$, leads to conclusion that the maximum is achieved by eigenvectors w_i corresponding to the maximal eigenvalue λ_{max} of the following generalized eigenvalue problem:

$$R_b(X)W = \lambda R_w(X)W \tag{8}$$

Therefore the rank of matrix W cannot be higher than the rank of eigenvalue λ_{max}. In practice this rank equals to one and we get the result equivalent to scalar case with $n = 1$. Therefore, the additional requirement should be imposed onto $W : \text{rank}(W) = n$.

If the matrix R_w is not singular then the optimal solution (Fukunaga [2]) at this requirement is achieved from Eigenvalue Decomposition (EVD) of symmetric, semi-definite matrix $R_b' := C_w^{-1} R_b C_w^{-t}$, where C_w is the Cholesky matrix ([3]) for R_w. Firstly, we look for W of rank N as follows:

$$\begin{aligned} R_b W &= \lambda R_w W, \ R_w = C_w C_w^t, \ W' = C_w^t W \\ R_b' &= W' \Lambda (W')^t \\ W &= C_w^{-t} W' \end{aligned} \tag{9}$$

If columns of W' are sorted by decreasing eigenvalues λ_i then we select from W the first n columns as the solution. This procedure works only if $\text{rank}(R_w) = N$ and $\text{rank}(R_b) \geq n$.

In Section 3 we discuss the important case of $\text{rank}(R_w) < N$.

3 Optimization of Fisher Measures by Projections in Error Spaces

In case of singular matrix R_w a sort of regularization is necessary. There are known two general approaches to this problem:

1. Regularization of data by mapping to $Y = \mathcal{P}(X)$ in order to get nonsingular $R_w(Y)$.
2. Regularization of LDA model by imposing an additional constraint on full rank LDA matrix $W = [w_1, \ldots, w_n]$ – for instance orthogonality to kernel space of R_w :

$$w_i \perp \text{ker}(R_w), \ i = 1, \ldots, n \tag{10}$$

In this section a novel point of view on LDA regularization is presented which uses the concept of projections in error subspaces. It unifies in one consistent scheme both approaches and integrates also with Dual Linear Discriminant Analysis (DLDA) [5].

In the presented discriminant analysis the source data matrix $Y_0 := X$ undergoes up to seven linear transformations before reaching the final matrix of features:

$$Y_{t-1} \longrightarrow Y_t, \ t = 1, \ldots, T \leq 7$$

The j-th class indexes I_j are identified by column indexes of data matrix Y_t and they are not changed at data matrix transformations.

There are three types of errors in our approach. They are defined w.r.t. any data matrix $Y = [y_1, \ldots, y_L]$ and with the fixed class assignments I_j, $j = 1, \ldots, J$:

1. *Grand error:* the difference of data vector y_k and the grand mean vector of Y. This can be modelled by the *global centering* operation C_g:

$$\bar{y} = \frac{1}{L} \sum_{i=1}^{L} y_i, \ C_g(y_k) := y_k - \bar{y}, \ k = 1, \ldots, L \qquad (11)$$

2. *Within-class error:* the difference of data vector $y_k, k \in I_j$ and its class mean $\bar{y}^{(j)}$:

$$\bar{y}^{(j)} = \frac{1}{L_j} \sum_{i \in I_j} y_i, \ C_w(y_k) := y_k - \bar{y}^{(j)}, \ k \in I_j, \ j = 1, \ldots, J \qquad (12)$$

3. *Between-class error:* the difference of class mean $\bar{y}^{(j)}$ and grand mean $\bar{y}$:

$$C_b(\bar{y}^{(j)}) := \bar{y}^{(j)} - \bar{y}, \ j = 1, \ldots, J \qquad (13)$$

The error vectors span error linear subspaces denoted as follows:

$$\mathcal{E}_g(Y) := \mathrm{span}(C_g(Y)), \ \mathcal{E}_w(Y) := \mathrm{span}(C_w(Y)), \ \mathcal{E}_b(\overline{Y}) := \mathrm{span}(C_b(\overline{Y})) \qquad (14)$$

Note that $\mathcal{E}_g(Y)$ is related to famous PCA approach recently linked to rough set [4] verification, too.

The singular bases $U^{(g)}$, $U^{(w)}$, $U^{(b)}$ of the error linear subspaces are obtained from Singular Value Decomposition (SVD [3]) for matrices $C_g(Y), C_w(Y), C_b(\overline{Y})$, respectively:

$$C_g(Y) = U^{(g)} \Sigma^{(g)} (V^{(g)})^t, \ C_w(Y) = U^{(w)} \Sigma^{(w)} (V^{(w)})^t, \ C_b(\overline{Y}) = U^{(b)} \Sigma^{(b)} (V^{(b)})^t$$

where diagonal squared matrices $\Sigma^{(\cdot)}$ are of size equal to the rank of centered data matrix $C.()$. In case of grand and within-class centering singular values are ordered from maximal to minimal value while in case of between-class centering the standard SVD order is inverse – the first element on the diagonal is minimal.

Let $Y \in \mathbb{R}^{a \times L}$. Then we identify all singular subspaces of dimension $a' \leq \dim(\mathcal{E}(Y))$ of error spaces by the projection operators which map the space $\mathbb{R}^a$ onto $\mathbb{R}^{a'}$ – the space of projection coefficients w.r.t. to the singular base $U_{a'}$ restricted to the first a' vectors:

1. $\mathcal{P}_{a,a'}^{(g)}$: projection onto grand error singular subspace of dimension a';
2. $\mathcal{P}_{a,a'}^{(w)}$: projection onto within-class error singular subspace of dimension a';
3. $\mathcal{P}_{a,a'}^{(b)}$: projection onto between-class error singular subspace of dimension a'.

Additional operation required after projection is component-wise scaling by the first inverse singular values which create the diagonal matrix $\Sigma_{a'}^{-1}$:

1. $\mathcal{S}_{a'}^{(g)}$: scaling of projected vector in grand error singular subspace of dimension a';
2. $\mathcal{S}_{a'}^{(w)}$: scaling of projected vector in within-class error singular subspace of dimension a';
3. $\mathcal{S}_{a'}^{(b)}$: scaling of projected vector in between-class error singular subspace of dimension a'.

In matrix composition terms the projection and scaling operations have the form:

$$\mathcal{P}_{a,a'}(x) = U_{a'}^t x, \ \mathcal{S}_{a'}(y) = \Sigma_{a'}^{-1} y \tag{15}$$

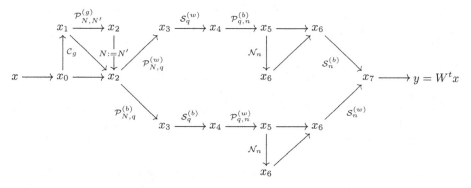

Fig. 1. Diagram of LDA type transformations based on projections onto error singular subspaces

The last operation we use in definition of LDA transformation is the vector length normalization $\mathcal{N}_a$ which can be geometrically interpreted as the projection on the unit sphere in $\mathbb{R}^a$.

$$\mathcal{N}_a(x) := \frac{x}{\|x\|} \tag{16}$$

Using the above notation the all known to authors LDA transformations can be defined via the diagram in Fig.1. It defines altogether 12 LDA type transformations. For instance in face verification system the following transformation path on the diagram gives best results:

In terms of operation compositions we get the following sequence (denoted here by DLDA) which includes also optimal weighting for matching:

$$\mathcal{W}_{DLDA} := \mathcal{S}_n^{(w)} \mathcal{N}_n \mathcal{P}_{q,n}^{(w)} \mathcal{S}_q^{(b)} \mathcal{P}_{N,q}^{(b)} \mathcal{C}_{(g)} \tag{17}$$

It is better than more popular LDA scheme improved by centering and normalization operations:

$$\mathcal{W}_{LDA} := \mathcal{S}_n^{(b)} \mathcal{N}_n \mathcal{P}_{q,n}^{(b)} \mathcal{S}_q^{(w)} \mathcal{P}_{N,q}^{(w)} \mathcal{C}_{(g)} \tag{18}$$

As a matter of fact the discriminant operation maximizing DLDA class separation measure is restricted to the composition $\mathcal{P}_{q,n}^{(w)} \mathcal{S}_q^{(b)} \mathcal{P}_{N,q}^{(b)} \mathcal{C}_{(g)}$ while the final two operations $\mathcal{S}_n^{(w)} \mathcal{N}_n$ are responsible for the optimal thresholding of within-class error which is selected as the distance function for the person id verification.

4   Experiments for Face Verification

From the previous works described in [5] it is already known that in case of face verification the optimization of inverse Fisher ratio (DLDA) leads to better results than the optimization of Fisher ratio (LDA). The final weighting of LDA or DLDA vector components had been also applied since they follow from Gaussian model of class errors.

Moreover, it was also observed that the normalization operation $\mathcal{N}_n$ improves significantly the equal error rate and ROC function face verification based on LDA or DLDA. The reason is explained by weak correlation between within-class error and between-class error. Therefore despite comparable norm magnitude of

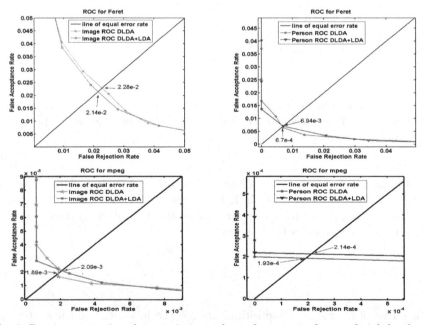

Fig. 2. Receiver operating characteristics and equal error rate for two facial databases *Feret* and *Mpeg* for DLDA and combined LDA+DLDA in single and multi-image scenarios

those errors the projection onto the unit sphere separates them while the final scaling respects the probability of all errors which are projected onto the same point of the unit sphere.

In the experiments described here we analyze two problems for face verification:

1. Is there any combination of LDA and DLDA class errors which improves DLDA?
2. What is the degree of improvement if the verification is based on the several facial images of the same person instead of the single one?

For the first problem we have found (cf. Fig. 2):

- LDA and DLDA class errors are of comparable magnitude.
- Geometric mean of both errors leads to slight improvements of EER and ROC w.r.t. DLDA error alone.
- The maximum, the arithmetic mean, and the harmonic mean of LDA and DLDA class errors give intermediate results between the best DLDA results and significantly worse LDA results.

For the second problem, experiments prove the significant advantage of the multi-image approach. Confront Fig. 2 where by *Person ROC* we mean the id acceptance if at least half of the query images are accepted. The acceptance by single image is described by *Image ROC*.

Conclusions. The general methodology presented for design of biometric verification system based on linear feature discrimination using sequential compositions of several types of feature vector transformations identifies the 12 basic discrimination schemes. The methodology has been tested for the verification system based on facial 2D images allowing for the choice of two best schemes which aggregated by geometric mean of distances leads to the best face verification results.

Acknowledgment. The work presented was developed within VISNET 2, a European Network of Excellence (http://www.visnet-noe.org), funded under the European Commission IST FP6 Programme.

References

1. Fisher, R. A.: The use of multiple measurements in taxonomic problems. Annals of Eugenics **7** (1936) 179–188
2. Fukunaga, K.: Introduction to Statistical Pattern Recognition. Academic Press (1992)
3. Golub, G., Loan, C.: Matrix Computations. The Johns Hopkins University Press (1989)
4. Swiniarski, R.W., Skowron A.: Independent Component Analysis, Principal Component Analysis and Rough Sets in Face Recognition. In Transactions on Rough Sets I. Springer, LNCS **3100** (2004) 392-404
5. Skarbek, W., Kucharski, K., Bober, M.: Dual LDA for Face Recognition. Fundamenta Informaticae **61** (2004) 303–334

Absolute Contrasts in Face Detection with AdaBoost Cascade

Marcin Wojnarski

Warsaw University, Faculty of Mathematics, Informatics and Mechanics
ul. Banacha 2, 02-097 Warszawa, Poland
mwojnars@ns.onet.pl

Abstract. Object detection using AdaBoost cascade classifier was introduced by Viola and Jones in December 2001. This paper presents a modification of their method which allows to obtain even 4-fold decrease in false rejection rate, keeping false acceptance rate – as well as the classifier size and training time – at the same level. Such an improvement is achieved by extending original family of weak classifiers, which is searched through in every step of AdaBoost algorithm, with classifiers calculating *absolute* value of contrast.

Test results given in the paper come from a face localization problem, but the idea of absolute contrasts can be applied to detection of other types of objects, as well.

1 Introduction

The original AdaBoost method is known from the late 1980s as a multiclassifier and a training procedure for a collection of *weak classifiers* (called also *features*), having success rate of about 0.5, to boost them by suitable voting process to very high level of performance [1,2]. Although this training scheme gives a classifier of very good accuracy, the number of simple features used is far too big, making real-time applications impossible [3].

The AdaBoost cascade method proposed by Viola and Jones [3] solves this problem. Viola and Jones connected a number of strong classifiers built with standard AdaBoost algorithm in a sequence, to form a *cascade* of classifiers of increasing complexity. Every stage of the cascade either rejects the analyzed window or passes it to the next stage. Only the last stage may finally accept the window. Thus, to be accepted, a window must pass through the whole cascade, but rejection may happen at any stage.

During detection, most sub-windows of the analyzed image are very easy to reject, so they are rejected at very early stage and do not have to pass the whole cascade. In this way, the average processing time of a single sub-window can be even thousands times lower than in the case of a standard AdaBoost classifier, particularly because the first stages can be very small and fast, and only the last ones have to be large.

Viola and Jones [3] proposed also a set of features for use particularly in face detection (but easily applicable to detection of other objects). This paper

J.T. Yao et al. (Eds.): RSKT 2007, LNAI 4481, pp. 174–180, 2007.
© Springer-Verlag Berlin Heidelberg 2007

presents an extension of that family of features, after which the AdaBoost algorithm – both the original and the cascade one – yields a classifier of much better accuracy. Original features given by Viola and Jones [3] are described in section 2.1, and the extension is presented in section 3.

2 AdaBoost Cascade Algorithm

2.1 The Weak Classifier

For each detection window o of the image being processed, a weak classifier gives a decision $\delta_\omega(o) \in \{-1, +1\}$ indicating membership of the image o to one of two classes, labelled by -1 (negative, e.g. a non-face) and +1 (positive, e.g. a face).

The classifier first calculates a *region contrast* $c(R)$ of the window o:

$$c(R) = \sum_{(x,y) \in R^+} o(x,y) - \sum_{(x,y) \in R^-} o(x,y) \, , \qquad (1)$$

where R is a sub-window of the window o, composed of a *positive sub-region* R^+ and a *negative sub-region* R^-. After computing the contrast, the classifier gives a response of $+1$ if $c(R) >= \theta$ or -1 otherwise. Here, R and θ are parameters of the weak classifier.

There are four types of regions R, presented in Figure 1. Positive sub-region R^+ is drawn in white and negative sub-region R^- is drawn in black. The main advantage of these sub-regions is that they have rectangular shape, so they can be computed very rapidly (in constant time) using an *integral image* representation of the original image. The integral image is computed only once, at the beginning of object detection in a specified image.

Fig. 1. Types of regions used in weak classifiers. White indicates positive sub-region and black negative sub-region.

The contrast region R is parameterized not only by its type $t (t \in \{A, B, C, D\})$, but also by four integral parameters: position (x, y) of its upper-left corner relative to the detection window, width a and height b of sub-regions (all sub-regions of a given region have the same size). Note that the family of all possible features is very large, e.g. in the experiments described further in the paper this family was composed of about 160000 elements. Certainly, only small subset of this family is used in a final strong classifier – AdaBoost algorithm is used to choose the best subset.

2.2   The Strong Classifier

A strong classifier is composed of a number of weak classifiers. Its decision is made by weighted voting: decision of a t-th weak classifier is multiplied by a weight α_t:

$$\gamma_t(o) = \delta_t(o) * \alpha_t$$

and all values $\gamma_t(o)$ are summed up and compared with a threshold Θ to form a final decision. Usually $\Theta = 0$, but when strong classifiers are further connected into a cascade, their thresholds can be different, in order to obtain required rate of false acceptance/rejection at every stage.

The AdaBoost algorithm is used to choose the most discriminative subset of all possible features and to set values of α_t. The algorithm works in an incremental manner, finding consecutive weak classifiers one by one. In every step, the algorithm looks through the whole family of contrast regions R to choose the best one – this is a simple exhaustive search method. However, to evaluate a weak classifier, its threshold θ has to be set, as well, not only the region R. Thus, for *every* possible R the best θ has to be found, which takes $N \log N$ time, where N is the number of training images. Certainly, this procedure is very time-consuming, usually it takes several minutes to choose a next weak classifier.

Every image in the training set has an associated weight, which is a positive real number. The weights are used during evaluation of weak classifiers: images with bigger weights are more important and have more influence on which classifier will be chosen next. The weights get changed after every step of AdaBoost procedure – in this way the successive classifiers found by the algorithm can be different (but not necessarily have to be).

Initially the weights $w_{i,t}$ are equal and sums up to 1:

$$w_{i,1} = \frac{1}{N}$$

for every image $i = 1, \ldots, N$. When the next weak classifier is found, the weights are modified and normalized to sum up to 1:

$$v_{i,t+1} = w_{i,t} e^{-\gamma_t(o_i) y_i} \, ,$$

$$w_{i,t+1} = \frac{v_{i,t+1}}{\sum_{i=1}^{N} v_{i,t+1}} \, ,$$

where $\gamma_t(o_i)$ is a decision value of the recently found classifier and y_i is a true classification ($+1$ or -1) of the i-th image.

2.3   The Cascade

Every stage of a cascade is built using the simple AdaBoost algorithm. When the next stage is created, its threshold Θ is set to the biggest value which still guarantees that the false rejection rate is below a predefined level – this is evaluated on a separate data set (evaluation set), not the training one.

Before the next stage can be created, both the training and the evaluation set must be filtered: images rejected by the new stage have to be removed. In consequence, new negative images have to be generated, as their number would drop roughly by half at every stage.

3   New Type of Weak Classifiers

The family of weak classifiers used in the algorithm proposed by Viola and Jones has a disadvantage. Namely, the contrast computed by a weak classifier, eq. (1), depends on *which* exactly sub-region (R^+ or R^-) is darker, so the weak classifier discriminates between windows with R^+ darker than R^- and R^- darker than R^+.

However, in many cases we would like to discriminate between windows with *the same* or *different* intensity in R^+ and R^-, ignoring information of *which* exactly sub-region is darker. That is because in real images what is important is the existence or lack of an intensity difference, and not the exact sign of the difference. For example, in face localization one may encounter a dark face on a bright background as well as a bright face on a dark background, so a classifier detecting whether there is a difference in intensity would be more discriminative than the one detecting a sign of the difference.

Table 1. Test error rates of face classifiers. Every classifier is characterized by its size (number of weak classifiers at each stage of a cascade) and number of training images (positive+negative examples). The first four classifiers are strong ones, the latter are cascades. FA – false acceptance rate, FR - false rejection rate. ABS – a classifier was built using extended family of weak classifiers, comprising both standard classifiers and the ones computing absolute contrast. In this case, the number of weak classifiers chosen from the absolute-contrast family is given in parentheses.

Classifier	FA	FR
20 weak, 500+500 images	1.4%	9.0%
20 weak, 500+500 images, ABS (4)	0.6%	8.2%
100 weak, 1000+1000 images	0.3%	3.2%
100 weak, 1000+1000 images, ABS (49)	0.4%	0.9%
4+4+10+20, 1000+1000	0.086%	13.1%
4+4+10+20, 1000+1000, ABS (1+2+5+12)	0.024%	14.2%
5+15+30+50+100+200, 1500+1000	0.00120%	4.4%
5+15+30+50+100+200, 1500+1000, ABS (1+6+11+25+50+120)	0.00028%	4.5%

For this reason, we extended the family of weak classifiers with the ones computing *absolute* values of contrasts:

$$c(R) = \left| \sum_{(x,y)\in R^+} o(x,y) - \sum_{(x,y)\in R^-} o(x,y) \right| \qquad (2)$$

3.1   Test Results

The results of strong and cascaded classifiers of different size built of the extended family of weak classifiers, compared to the ones using original family alone, are shown in Table 1. The results come from a face localization problem, but the idea of absolute contrasts can be applied in detection of other types of objects, as well. Two types of errors – false acceptance and false rejection – are considered separately, as in practical applications the former should be hundreds times lower than the latter.

Images of faces came from MPEG-7 and Altkom databases. MPEG-7 contained images of 635 persons, 5 for each subject (3175 in total). Altkom was composed of images of 80 persons, 15 for each subject (1200 in total). The images had 46×56 pixels and contained frontal or slightly rotated faces with fixed eye positions, with varying facial expressions and under different lighting conditions.

The number of positive training examples in each experiment is given in Table 1. Test sets contained the same number of faces as training ones. A validation set was used to find thresholds of strong classifiers in a cascade. All these sets were disjoint in each experiment (i.e. contained images of different persons) and included the same proportion of images from MPEG-7 and Altkom databases.

Negative examples were generated randomly from 10000 large images not containing faces, in on-line fashion. They had to be generated after creation of every new stage of a cascade, so as to compensate for examples correctly rejected by the last stage.

Exemplary face images used in the experiments are shown in Figure 2.

In order to speed up the training, only weak classifiers of even positions and sizes were considered (this applied both to standard and absolute classifiers). This is almost equivalent to scaling down the images by a factor of 2, but has the advantage of not introducing rounding errors.

Results from Table 1 show that using the extended family allows to achieve over 4-fold decrease in one type of error rate (e.g., false rejection), keeping the other one at a similar level, and without a need of increasing classifier size.

It is worth mentioning that using the absolute-contrast classifiers alone, without the standard ones, gives worse results than using the original family alone, so it is good to *extend* the original family, but not to *replace* it with the absolute-contrast one.

Table 1 (first column, in parentheses) contains also information about the number of weak classifiers which were chosen by AdaBoost algorithm from the absolute-contrast family, at every stage of a cascade. Comparison of this information with the total size of each stage shows that the contribution of absolute-contrast classifiers rises with the size of a stage, from 20 or 25% at the first stage to 60% at the last one. This suggests that absolute contrasts are more useful when the problem becomes more difficult.

Fig. 2. Positive examples (faces) used in the training process

3.2 Efficiency of the Cascade and the Training Process

Calculation of a decision of an absolute-contrast weak classifier takes the same amount of time as in the case of a standard classifier, since the only additional operation is a computation of an absolute value. Therefore, the final strong or cascaded classifier is as fast as the one built of original weak classifiers alone.

The use of the extended family does not slow down significantly the training algorithm either. Although the family is twice bigger, searching through it takes (if properly implemented) at most 10% longer. That is because the most time-consuming part of the search through original family is a quest for the optimal threshold θ, given a type, a size and a position of a sub-window (t, a, b, x, y). This requires calculating and sorting of contrasts of a given sub-window on all training images (say N), which takes time of the order of $N \log N$.

When absolute contrasts are also considered, additionally a sorted sequence of absolute values of contrasts is needed. However, this does not require computations of $N \log N$ complexity again, because a sorted sequence of contrasts is already available, and after transformation by absolute-value function this sequence turns into two sorted sequences, which can be merged in linear time. It should be noted here that construction of a cascade is a very time-consuming process, which takes from several hours to several days when executed on a personal computer (CPU 2.0 GHz), so time efficiency of the presented modification is an important feature.

4 Conclusion

The paper presented a modification of Viola and Jones' object detection algorithm. The modified algorithm utilizes an extended family of features which is searched through during construction of strong classifiers. This extension enables 4-fold decrease in false rejection rate without increase in false acceptance rate or classifier size. Moreover, the extension does not influence significantly training time, despite the fact that the family of features is twice bigger. Obviously, resolution of training images does not have to be increased either.

Acknowledgements

The research has been supported by the grant 3T11C00226 from Ministry of Scientific Research and Information Technology of the Republic of Poland.

References

1. Freund, Y., Schapire, R.E.: A decision-theoretic generalization of on-line learning and an application to boosting. Journal of Computer and System Sciences **55** (1997) 119–139
2. Schapire, R.E., Freund, Y., Bartlett, P., Lee, W.S.: Boosting the margin: a new explanation for the effectiveness of voting methods. In: Proc. 14th International Conference on Machine Learning, Morgan Kaufmann (1997) 322–330
3. Viola, P., Jones, M.: Rapid object detection using a boosted cascade of simple features. In: IEEE Computer Vision and Pattern Recognition. Volume 1. (2001) 511–518
4. Xiao, R., Li, M.J., Zhang, H.J.: Robust multipose face detection in images. IEEE Trans. Circuits and Systems for Video Technology **14** (2004) 31–41
5. Skarbek, W., Kucharski, K.: Image object localization by adaboost classifier. In Campilho, A.C., Kamel, M.S., eds.: ICIAR (2). Volume 3212 of Lecture Notes in Computer Science., Springer (2004) 511–518
6. Papageorgiou, C.P., Oren, M., Poggio, T.: A general framework for object detection. In: International Conference on Computer Vision. (1998) 555–562

Voice Activity Detection
for Speaker Verification Systems

Jaroslaw Baszun

Bialystok University of Technology
Faculty of Computer Science, Department of Real Time Systems
Wiejska Str. 45A, 15-351 Bialystok, Poland
jb@ii.pb.bialystok.pl

Abstract. This paper describes voice activity detection algorithm for speaker verification systems based on properties of human speech modulation spectrum i.e. rate of power distribution in modulation frequency domain. Based on the fact that power of modulation components of speech is concentrated in a range from 1 to 16 Hz and depends on rate of syllables uttering by a person, a new effective algorithm was proposed and compared to standard energy and cepstral based detectors. Experiments confirmed reliability of proposed detector and its better ability to detect speech in real street noise with high probability for signals with very low SNR in compare to other methods.

Keywords: voice activity detector, speech modulation spectrum.

1 Introduction

This paper describes the Voice Activity Detector (VAD) for speaker verification system capable of working efficiently in noisy street conditions. Presented here algorithm can also be applied to many other speech-processing applications especially when speech is corrupted by strong even nonstationary noise.

Detection of voice activity in noisy environments is a challenging task and many algorithms have been proposed. They are usually based on energy, spectral or cepstral features [1][2], higher statistics [3] or Hidden Markov Models are also utilized. All these solutions work well when signal to noise ratio (SNR) is high and noise level changes slowly in compare to speech. But in many real situations, where speech processing systems are used speech signal is corrupted by nonstationary noise, e.g. street noise, and these algorithms give poor results.

In this paper psyhoacoustically motivated voice activity detector is proposed which exploits properties of modulation spectrum of human speech [4]. The speech to noise estimate is computed in modulation frequency domain and used as a feature for speech/pause detection.

J.T. Yao et al. (Eds.): RSKT 2007, LNAI 4481, pp. 181–186, 2007.

2 Voice Activity Detector Based on Filtering of Spectral Envelopes

2.1 Modulation Spectrum of Speech

It is know that low-frequency modulations of sound are the carrier of information in speech [5][6]. In the past many studies were made on the effect of noise and reverberation on the human modulation spectrum [4][7] usually described through modulation index (MI) as a measure of the energy distribution in modulation frequency domain i.e. normalized power over modulation for a given frequency band at dominant modulation frequencies of speech. MI vary between analysis frequency bands. The corrupting background noise encountered in real environments can be stationary or changing usually different in compare to the rate of change of speech. Relevant modulation frequency components of speech are mainly concentrated between 1 and 16 Hz with higher energies around $3 - 5$ Hz what corresponding to the number of syllables pronounced per second [4]. Slowly-varying or fast-varying noises will have components outside the speech range. Further, steady tones will only have MI constant component. Additive noise reduces the modulation index. System capable of tracking speech components in modulation domain are very tempting perspective for such fields like speech coders and speech compression [8], speech enhancement [9][10] and voice activity detectors [11].

2.2 Voice Activity Detector Based on Modulation Spectrum of Speech

The idea of the system comes from previous work on speech enhancement systems [10] based on modulation of speech. The block diagram of the system was shown in Fig. 1. Signal from microphone with sampling frequency 16 kHz is split into $M = 512$ frequency bands using Short Time Fourier Transform (STFT) with Hamming window and 50 % overlapping. Next amplitude envelope is calculated for first 256 bands except first eight bands corresponding to frequencies below 250 Hz in speech signal:

$$y_k(nM) = \sqrt{Re^2[x_k(nM)] + Im^2[x_k(nM)]} \tag{1}$$

Amplitude envelope is summed for all bands and filtered by passband IIR filter with center frequency 3.5 Hz and frequency response shown in Fig. 2. The output of the filter is half-wave filtered to remove negative values from output of the filter. The following computation is carried out on the filtered and not filtered envelope:

$$S(nM) = \frac{Y'}{Y - mean(Y) - Y' - mean(Y')} \tag{2}$$

Above parameter is an estimate of speech to noise ratio of analyzed signal. Mean value of filtered and nonfiltered envelope is computed based on exponential averaging with time constant approximately 1 s. The square of this estimate is used as a classification parameter for voce activity detector. Speech decision is based on comparison between classification parameter and the threshold computed based on the following statistics [1]:

$$Thr = mean(d) + \alpha \cdot std(d) \qquad (3)$$

where d is a classification parameter and α controls confidence limits and is usually in the range 1 to 2, here was set to be equal 2. Both mean value and standard deviation is estimated by exponential averaging in pauses with time constant $a = 0.05$. Frame is considered to be active if value of classifier is greater than threshold. To avoid isolated errors on output caused by short silence periods in speech or short interferences correction mechanism described in [12] was implementing. If current state generating by the VAD algorithm does not differ from n previous states then current decision is passed to detector output otherwise the state is treated as a accidental error and output stays unchanged.

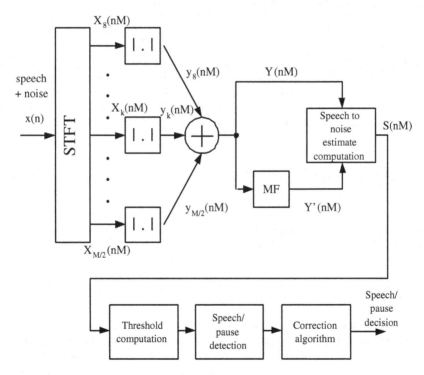

M=512
| . | - amplitude envelope computation
MF - modulation filter

Fig. 1. Voice activity detector block diagram

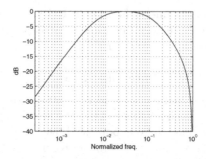

Fig. 2. Magnitude frequency response of modulation filter (MF)

3   Simulation Results

3.1   Simulation Parameters and Classification

Described algorithm was implemented in MATLAB environment. All experiments were realized with speech signals with digitally added street noise and collected in the database with manually marked speech sequences. The algorithm was tested for four different SNRs: 20 dB, 5 dB, 0 dB and −5 dB. Sampling frequency was 16 kHz. The following objective criteria for detector comparison with other algorithms were used [2]: correct detection rate - $P(A)$ and speech/non-speech resolution factor $P(B)$.

$$
\begin{aligned}
P(A) &= P(A/S)P(S) + P(A/N)P(N), \\
P(B) &= P(A/S)P(A/N)
\end{aligned}
\tag{4}
$$

where $P(S)$ and $P(N)$ are rates of speech and pauses in the processed signal.

Proposed algorithm (MDET) was compared to the following voice activity detectors: EDET - energy detector, CEP - one step integral cepstral detector and CEP2 - two step integral cepstral detector [2]. Each algorithm have implemented identical isolated errors correction mechanism according to [12].

3.2   Performance of the Proposed Algorithm

In Table 1 results of carried out experiments was shown. Detectors were tested with signals corrupted by nonstationary street noise because the aim of developed detector is to work as a part of speaker verification system of device localized in noisy streets. Experiments show that all detectors give good results with $SNR = 20$ dB. The difference became from $SNR = 5$ dB and is clear for $SNR = 0$ dB. Cepstral detector is better than energy one but for nonstationary noise also fails for low SNR. Voice activity detector based on modulation spectrum of speech have higher correct detection rate for heavy noise (SNR below 0 dB) in compare to other methods. In Fig. 3 behavior of the presented system with heavy street noise was shown.

Table 1. Computed parameters for tested VADs

Detector	$SNR = 20$ dB		$SNR = 5$ dB		$SNR = 0$ dB		$SNR = -5$ dB	
	$P(A)$	$P(B)$	$P(A)$	$P(B)$	$P(A)$	$P(B)$	$P(A)$	$P(B)$
EDET	0.893	0.786	0.818	0.613	0.657	0.384	0.583	0.154
CEP	0.865	0.693	0.823	0.658	0.811	0.673	0.622	0.210
CEP2	0.914	0.713	0.833	0.674	0.852	0.745	0.632	0.272
MDET	0.982	0.941	0.967	0.912	0.915	0.824	0.863	0.702

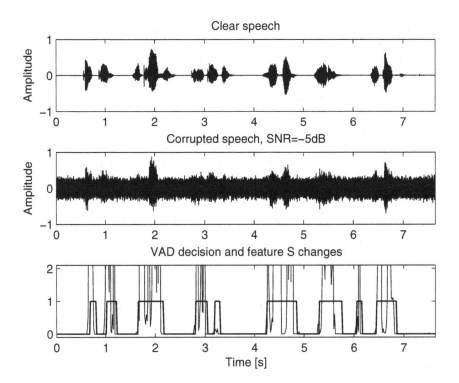

Fig. 3. Example of VAD decision for speech signal corrupted by street noise $SNR = -5$ dB

4 Conclusions

Voice activity detector based on modulation properties of human speech was developed and tested. Tests with real noise sources shown the advantage of presented algorithm over classical solutions based on energy o cepstral features in case of low SNR signals. Carried out experiments confirm that this solution is a very good candidate for VAD in real noisy conditions.

Acknowledgments. The work presented was developed within VISNET 2, a European Network of Excellence (http://www.visnet-noe.org), funded under the European Commission IST FP6 Programme.

References

1. Sovka, P., Pollak, P.: The Study of Speech/Pause Detectors for Speech Enhancement Methods. Proceedings of the 4th European Conference on Speech Communication and Technology, Madrid, Spain (Sep. 1994) 1575–1578
2. Sovka, P., Pollak, P.: Cepstral Speech/Pause Detectors. Proceedings of IEEE Workshop on Non-linear Signal and Image Processing, Neos Marmaras, Halkidiki, Greece, (20-22 Jun. 1995) 388–391
3. Rangoussi, M., Garayannis, G.: Adaptive Detection of Noisy Speech Using Third-Order Statistics. International J. of Adaptive Control and Signal Processing, Vol. 10, (1996) 113–136
4. Houtgast, T., Steeneken, H.J.M.: A Review of the MTF Concept in Room Acoustics and its Use for Estimating Speech Intelligibility in Auditoria. J. Acoust. Soc. Am., Vol. 77, No. 3 (Mar. 1985) 1069–1077
5. Drullman, R., Festen, J.M., Plomp, R.: Effect of Temporal Envelope Smearing on Speech Reception. J. Acoust. Soc. Am. No. 2. (1994) 1053–1064
6. Elhilali, M., Chi, T., Shamma, S.: A Spectro-temporal Modulation Index (STMI) for Assesment of Speech Intelligibility. Speech Communication, Vol. 41. (2003) 331–348
7. Houtgast, T., Steeneken, H.J.M.: The Modulation Transfer Function in Room Acoustics as a Predictor of Speech Intelligibility. Acoustica, Vol. 28, (1973) 66
8. Thompson, J., Atlas, L.: A Non-Uniform Modulation Transform for Audio Coding with Increased Time Resolution. ICASSP (2003), 397–400
9. Hermansky, H., Morgan, N.: RASTA Processing of Speech. IEEE Transaction on Speech and Audio Processing, Vol. 1, No. 4 (Oct. 1994), 587–589
10. Baszun, J., Petrovsky, A.: Flexible Cochlear System Based on Digital Model of Cochlea: Structure, Algorithms and Testing. EUSIPCO (2000), Tampere, Finland, Vol. III, 1863–1866
11. Mesgarani, N., Shamma, S., Slaney, M.: Speech Discrimination Based on Multiscale Spectro-Temporal Modulations. ICASSP (2004), 601–604
12. El-Maleh, K., Kabal, P.: Comparision of Voice Activity Detection Algorithms for Wireless Personal Communications Systems. Proceedings IEEE Canadian Conference Electrical and Computer Engineering, (May 1997) 470–473

Face Detection by Discrete Gabor Jets and Reference Graph of Fiducial Points

Jacek Naruniec and Władysław Skarbek

Warsaw University of Technology, Faculty of Electronics and Information Technology,
00-665 Warszawa, Nowowiejska 15/19, Poland
J.Naruniec@ire.pw.edu.pl, W.Skarbek@ire.pw.edu.pl

Abstract. A novel face detection scheme is described. The facial feature extraction algorithm is based on discrete approximation of Gabor Transform, called Discrete Gabor Jets (DGJ), evaluated in fiducial face points. DGJ is computed using integral image for fast summations in arbitrary windows, and by FFT operations on short contrast signals. Contrasting is performed along radial directions while frequency analysis along angular direction. Fourier coefficients for a small number rings create a long vector which is next reduced to few LDA components. Four fiducial points are only considered: two eye corners and two nose corners. Fiducial points detection is based on face/nonface classifier using distance to point dependent LDA center and threshold corresponding to equal error rate on ROC. Finally, the reference graph is used to detect the whole face. The proposed method is compared with the popular AdaBoost technique and its advantages and disadvantages are discussed.

Keywords: face detection, Gabor filter, Linear Discriminant Analysis, reference graph.

1 Introduction

Face detection is important preprocessing task in biometric systems based on facial images. The result of detection gives the localization parameters and it could be required in various forms, for instance:

- a rectangle covering the central part of face;
- a larger rectangle including forehead and chin;
- eyes centers (the choice of MPEG-7 [1]);
- two eyes inner corners and two nose corners (the choice of this paper).

While from human point of view the area parameters are more convincing (cf. Fig. 1), for face recognition system fiducial points are more important since they allow to perform facial image normalization – the crucial task before facial features extraction and face matching.

We may observe for each face detector the following design scheme [2]:

1. *Define admissible pixels and their local neighborhoods of analysis.* Admissible pixels could cover the whole image area or its sparse subset, for instance edge

J.T. Yao et al. (Eds.): RSKT 2007, LNAI 4481, pp. 187–194, 2007.

Fig. 1. Area versus fiducial points parameters for face localization

or corner points. The neighborhoods could consist of one pixel only or even thousands of them forming rectangular windows of analysis or for instance rings of small rectangles (cf. Fig. 2).

2. *Design a feature extractor* which produces a collection of features for each admissible local neighborhood. It may be as simple as admissible pixel color frequency or as complex as long vector of 100 Angular Radial Transformation (ART) coefficients.

3. *Design a classifier* which decides whether the collection of features extracted from the given neighborhood of analysis could be face relevant. If so the admissible pixel becomes *face relevant point*. It could be a simple classifier based on comparison of feature with a threshold or more complex Support Vector Machine (SVM classifier) using Gaussian kernel.

4. *Define a postprocessing scheme* which selects representative face relevant points defining face locations. The representatives could be obtained as centroids of connected components in the set of all face relevant points or results or more complex clustering scheme combined with graph matching to reject inconsistent ensembles of face relevant points.

On top of the above scheme each detector includes a multi-resolution mechanism to deal with face size. It is implemented either through analysis in image pyramid or by scaling the local neighborhood of analysis together with relevant parameters.

One of the most known face detectors is based on AdaBoost classifier. It was introduced by Viola and Jones in 2001 [3]. Let us trace the design scheme of this prominent method:

1. The local neighborhood of analysis is a small window of size 20×20 scaled up by 20%. The pixel is admissible if and only if it is upper left corner of the window analysis completely included in image domain.

2. In window analysis at fixed positions small contrasting filters are defined of specific size and type. The filter returns the contrast between white and black region defined as the difference between total intensities in the regions.

3. The regional contrast is compared with filter specific threshold giving a weak classifier. The weak decisions are linearly combined using cost coefficients

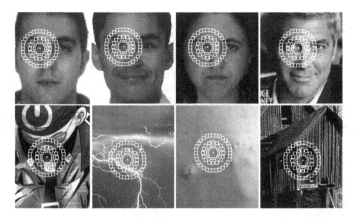

Fig. 2. Rings of small squares as neighborhoods of analysis in our method

elaborated according the AdaBoost machine learning scheme. The AdaBoost is a multi-classifier well known from the late 1980s which due a special weighting scheme of training examples ensures the high performance of strong classifier providing that weak classifiers have the success rate about 0.5. Authors of [3] applied an early and suboptimal heuristics given for AdaBoost training algorithm in [4]. However, their face recognition system described in [5] which also used the AdaBoost concept, contained the optimal training procedure which is methodologically sound. The algorithm proposed by them is a generalization of one described in [6].

4. In postprocessing stage the centroid of enough large connected components of face relevant window corners represents the detected face window.

While AdaBoost is satisfactory solution for facial window detection, its extensions to detect fiducial points, for instance eye centers, are not equally effective. The normalization of facial image based on AdaBoost is not accurate and it results in poor face recognition and verification. In this paper we develop a novel method for detection of face fiducial points which is based on very rough discrete approximation of Gabor transform called here Discrete Gabor Jet (DGJ). The method gives very good results for detection of frontal face views with almost perfect false acceptance rate.

2   DGJ Face Detector

The whole process of face detection consists of several steps illustrated in Fig. 3. The fiducial points are searched only within edge points in the image.

The feature extraction in local neighborhood of analysis is performed in two stages. Firstly the frequency analysis is performed on selected rings (first type coefficients) and on contrasts between pairs of rings (second type coefficients). In the second stage a modified LDA analysis produces 2D features discriminating face and non-face fiducial points. Each type of fiducial point has its specific LDA matrix.

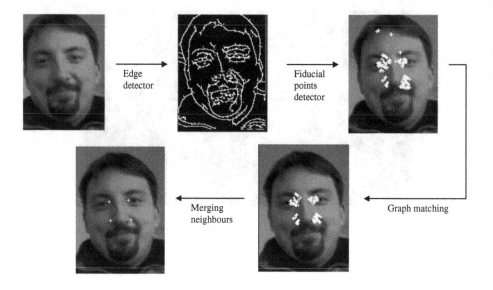

Fig. 3. Illustration of all steps for our face detector

The classifier for the particular fiducial point is based on distance to the centroid of fiducial feature vectors. The standard ROC is built up to tune the distance threshold to the required false acceptance rate.

2.1   Extractor I: Discrete Gabor Jet

The kernel of Gabor filter [7] in spatial image domain is a Gaussian modulated 2D sine wave grating with parameters controlling wave front spatial orientation, wave frequency and rate of attenuation. While Gabor wavelet is accurate tool to represent local forms with complex textures its computation excludes real time applications in pure software implementations.

Therefore we seek for the representation which can describe changes of local image contrasts around the given pixel in both angular and radial direction. To this goal we design rings of small squares and evaluate frequency of luminance changes on such rings (cf. Fig. 2).

There are two types of Gabor jets. The first type detects angular frequency on selected rings while the second type represents angular frequencies for radial contrast between two selected rings.

First type jets. The concept is illustrated in Fig. 4 in upper part.

Each jet of the first type is characterized by radius r of the ring, the number of squares $n = 2^k$, the center (anchoring) point (x, y). The sizes of all squares on the ring are equal. The size is the maximum possible providing that squares do not intersect each other (except of intersection perhaps by one pair).

The sum of pixel values in each square is computed using the integral image like in AdaBoost detector [3]. The sequence of n such values is normalized to be

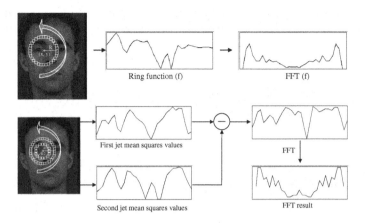

Fig. 4. First type jet (up) – frequency features on single ring of squares, and second type jet – frequency features of double ring of squares

included in the unit interval $[0, 1]$. Finally the obtained sequence f is transformed by FFT. Only the first $n/2$ of DFT complex coefficients are joined to the output feature vector.

Second type jets. The idea is shown in Fig. 4 in bottom part.

Now the jet consists of two rings with radii $r_1 < r_2$ with the same center (x, y) and with the same number $n = 2^k$ of equal size squares.

Like for the first type jets the sum of pixel values in each square is computed using the integral image, but now the mean value of each square is computed. Differences between mean values of each square in the first ring and the corresponding mean values in second ring are taken. Next the obtained differential signal is normalized to the unit interval and then transformed by FFT. Again only the first $n/2$ of DFT complex coefficients are joined to the output feature vector.

In the final design of Gabor jets we take six jets of the first kind and three jets of the second with parameters defined in the following table:

type	1	1	1	1	1	1	2	2	2
n	16	16	16	32	32	64	16	16	32
r_1	8	13	20	16	20	24	10	13	16
r_2	–	–	–	–	–	–	16	20	21

After the first stage of feature extraction, the feature vector has $5 * 16 + 3 * 32 + 64 = 240$ of real components.

2.2   Extractor II: Modified Linear Discriminant Analysis

Having DFT description over local rings as feature vector of length 240 we discriminate them using a Modified Linear Discriminant Analysis (MLDA).

In case of face detection when we deal with two classes only, i.e. with *facial descriptions* and *non-facial descriptions*, the classical LDA enables only scalar discriminative feature. It makes harder separation of two classes by linear approach. Therefore we modify the concepts of within and between-variances and related scatter matrices, in order to get vectorial discriminative features.

Namely, the classical LDA maximizes the Fisher ratio of between-class variance over within-class variance defined as follows ([8],[9]):

$$f_X := \frac{\text{var}_b(X)}{\text{var}_w(X)}$$

$$\text{var}_w(X) := \frac{1}{|I_f|} \sum_{i \in I_f} \|x_i - \overline{x}^f\|^2 + \frac{1}{|I_{\bar{f}}|} \sum_{i \in I_{\bar{f}}} \|x_i - \overline{x}^{\bar{f}}\|^2 \tag{1}$$

$$\text{var}_b(X) := \|\overline{x}^f - \overline{x}\|^2 + \|\overline{x}^{\bar{f}} - \overline{x}\|^2$$

where the training set X of feature vectors is divided into the facial part indexed by I_f and the non-facial part with remaining indices $I_{\bar{f}}$.

It appears that we obtain better discrimination results with the following class separation measure:

$$m_X := \frac{\text{mvar}_b(X)}{\text{mvar}_w(X)}$$

$$\text{mvar}_w(X) := \frac{1}{|I_f|} \sum_{i \in I_f} \|x_i - \overline{x}^f\|^2 \tag{2}$$

$$\text{mvar}_b(X) := \|\overline{x}^f - \overline{x}\|^2 + \frac{1}{|I_{\bar{f}}|} \sum_{i \in I_{\bar{f}}} \|x_i - \overline{x}^f\|^2$$

Like in classical case, the optimization procedure requires replacing variances by traces of scatter matrices:

$$m_X := \frac{\text{trace}(S_{mb}(X))}{\text{trace}(S_{mw}(X))}$$

$$S_{mw}(X) := \frac{1}{|I_f|} \sum_{i \in I_f} (x_i - \overline{x}^f)(x_i - \overline{x}^f)^t \tag{3}$$

$$S_{mb}(X) := (\overline{x}^f - \overline{x})(\overline{x}^f - \overline{x})^t + \frac{1}{|I_{\bar{f}}|} \sum_{i \in I_{\bar{f}}} (x_i - \overline{x}^f)(x_i - \overline{x}^f)^t$$

Since for the large number of positive training examples the scatter matrix S_{mw} is of full rank then we can optimize $m(W) := m_{W^t X}$ w.r.t. the training set X by Cholesky factorization ([10]) of $S_{mw} = C_{mw} C_{mw}^t$ and solving the following EVD problem for the symmetric, semi-definite matrix $S'_{mb} := C_{mw}^{-1} S_{mb} C_{mw}^{-t}$:

$$S_{mb}W = \lambda S_{mw}W, \ S_{mw} = C_{mw} C_{mw}^t, \ W' = C_{mw}^t W$$

$$S'_{mb} = W'\Lambda(W')^t \tag{4}$$

$$W = C_{mw}^{-t} W'$$

If columns of W' are sorted by decreasing eigenvalues λ_i and we want to have $n \geq 2$ LDA features then we select from W the first n columns as the optimal solution.

2.3   Postprocessing: Graph Matching

Let fiducial facial points be depicted according the notation from the Fig. 5. All detected fiducial points to be preserved must pass, for at least one scale, through the graph matching procedure with the following pseudocode.

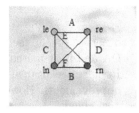

edge	length
A	1.00
B	1.05
C	0.99
D	0.98
E	1.41
F	1.43

Fig. 5. Fiducial points and their symbols used in graph matching algorithm: le - left eye, re - right eye, ln - left nostril, rn - right nostrils corners

```
forall left eyes le do
  forall right eyes re do
    if distance(le.Y,re.Y)<15*scale
      and 30*scale>(re.X-le.X)>20*scale
      forall left nostrils ln do
        if distance(le.X,ln.X)<30*scale
        and 35*scale>(ln.Y-le.Y)>20*scale
          forall right nostrils rn do
            if distance(re.X,rn.X)<30*scale
            and 35*scale>(rn.Y-re.Y)>20*scale
              set distance(le,re):=norm;
              normalize other distances:
                distancen:=distance/norm;
              get total_distance as the sum of distances
                between actual graph and reference graph;
              if total_distance<threshold
                consider points as detected face;
          endfor
        endfor
    endfor
endfor
```

It is interesting that averaged normalized distances between fiducial points indicate their displacement in the corners of a square.

3   Experiments

We have compared DGJ and AdaBoost (AB)methods for several facial databases: Mpeg, Banca, BioID. To experiments only near frontal pose images were selected but with varying lighting conditions. In the following table the false rejection rate (frr in percent) and false acceptance (fa in items) are given. From this comparison we observe that DGJ is a very promising approach for face detection.

Face Base [#]	DGJ frr [%]	AB frr [%]	DGJ fa [#]	AB fa [#]
Mpeg normal. (500)	3.5	11.6	0	0
Banca original (500)	36.0	19.4	4	18
Banca normal. (500)	23.0	39.0	0	0
BioID original (1500)	9.8	17.5	2	9

4   Conclusion

The proposed face detector looking for corners of special facial square using combined feature extraction by DGJ and LDA and next matching to face reference graph has very promising performance. Its false acceptance rate is very low, but false rejection rate still has a room for improvements. However, in experiments it is consistently better than AdaBoost technique.

Acknowledgment. The work presented was developed within VISNET 2, a European Network of Excellence (http://www.visnet-noe.org), funded under the European Commission IST FP6 Programme.

References

1. ISO/IEC 15938-3:2002/Amd.1:2004 Multimedia content description interface. Visual Descriptor Extensions. ISO/IEC (2004)
2. Yang, M. H., Kriegman, D. J., Ahuja, N.: Detecting Faces in Images: A Survey. IEEE Trans. PAMI. **24:1** (2002) 34–58
3. Viola, P., Jones, M.: Robust Real-time Object Detection. Second Int'l Workshop on Statistical and Computational Theories of Vision – Modeling, Learning, Computing and Sampling (2001)
4. Freund, Y., Schapire, R. E.: A decision theoretic generalization of on-line learning and an application to boosting. Journal of Computer and Systems Sciences **55:1** (1997) 119–139
5. Jones, M., Viola, P.: Face Recognition Using Boosted Local Features. Mitsubishi Electric Research Laboratories **TR20003-25** (2003)
6. Shapire, R. E.: The Boosting Approach to Machine Learning – An Overview. MSRI Workshop on Nonlinear Estimation and Classification (2002)
7. Gabor, D.: Theory of communication. Proc. IEE **93:26** (1946) 429–441
8. Fisher, R. A.: The use of multiple measurements in taxonomic problems. Annals of Eugenics **7** (1936) 179–188
9. Fukunaga, K.: Introduction to Statistical Pattern Recognition. Academic Press (1992)
10. Golub, G., Loan, C.: Matrix Computations. The Johns Hopkins University Press (1989)
11. Skarbek, W., Kucharski, K.: Image Object Localization by Adaboost Classifier. Int. Conf. on Image Analysis and Recognition, Lecture Notes on Computer Science **3211** (2004) 511–518

Iris Recognition with Adaptive Coding

Adam Czajka and Andrzej Pacut

Institute of Control and Computation Engineering
Warsaw University of Technology
Nowowiejska 15/19, 00-665 Warsaw, Poland
Biometric Laboratories
Research and Academic Computer Network NASK
Wawozowa 18, 02-796 Warsaw, Poland
{A.Czajka,A.Pacut}@ia.pw.edu.pl, www.BiometricLabs.pl

Abstract. The paper proposes a new iris coding method based on Zak-Gabor wavelet packets. Details of the Zak-Gabor-based coding are presented in the paper, and the method of adaptation the transformation parameters is described. The methodology may be of particular help in development mobile iris systems, where the iris capture devices may present a limited quality. The method was evaluated and presents very favorable results.

Keywords: Iris recognition, biometrics.

1 Iris Measurement and Preprocessing

Biometric authentication starts from acquisition of appropriate biological data characteristic of an individual. We use a dedicated hardware designed and constructed to capture the iris from a convenient distance, with the desired speed and a minimal user cooperation. To illuminate iris we apply a near infrared 850 nm light that meets the ISO recommendations [1]. The system uses the pupil position estimated in real time to guide a person to position the eye, and to release the image capturing process. In this process several frames are captured at varying focal lengths, and the sharpest frame is selected for further analysis. The latter procedure compensates a small depths-of-field typical in iris imaging.

The raw images contain the iris and its surroundings. The iris must be first localized. To detect a boundary between the pupil and the iris, we propose a method which is sensitive to circular dark shapes, and unresponsive to other dark areas as well as light circles, such as specular reflections. This may be achieved by a modified Hough transform that uses the *directional image* to employ the image gradient, rather than the *edge image*, which neglects the gradient direction. A boundary between the iris and the sclera may be approximated by a circle. To determine this boundary, we independently apply Daugman's integro-differential operator [2] to two opposite horizontally placed angular sectors, 45° each, since the entire circular iris boundary may be partially disturbed by eyelids. The two radii of the resulting arcs are averaged to construct a circle approximating the outer iris boundary.

J.T. Yao et al. (Eds.): RSKT 2007, LNAI 4481, pp. 195–202, 2007.

The iris ring limited by the two circular boundaries may still be disrupted by irregular objects like reflections or eyelashes. It is desirable to use occlusion detection that does not assume any particular occlusion shape. We localize non-uniformity points within the iris ring and then construct an occlusion map. First, we calculate the sample variances of the iris image intensity for a set of radial sectors. These variances are compared to the maximum allowed variance obtained for directions in which the probability of iris occlusion is minimal. Those directions in which the calculated variance exceeds the threshold value is marked as an occlusion direction, and the appropriate occlusion radius is stored.

Based on the localized occlusions, we select two opposite 90° wide angular iris sectors. Each iris sector is then transformed by resampling and smoothing to a $P \times R$ rectangle, where $P = 512$ and $R = 16$. The rows f_ℓ of these two rectangles will be further referred to as the *iris stripes*. The experiments (see also [2]) revealed much higher correlation of the iris image in the radial direction, i.e. along the iris stripes, as compared to the angular direction, across the stripes. Figure 1 illustrates the preprocessed iris image and the corresponding iris stripes.

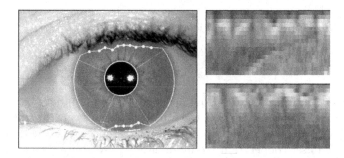

Fig. 1. *Left:* raw camera image processed by our system. The eyelids were automatically detected, and the sectors free of occlusions (marked as white full circles) are selected. Star-like shapes on the pupil are reflections of the illumination NIR diodes, and the '+' marks represent the pupil and the iris centers. *Right:* iris stripes automatically determined for the image shown on the left.

2   Database of Iris Images

Calculations in this work are employing our proprietary database of 720 iris images. The data was collected for 180 different eyes, with 4 images of each eye. We used 3 images of each eye in the estimation stage to calculate the iris templates, and the remaining single image of each eye in the verification stage. Typically, in iris capturing systems with one-eye capture optics, the images taken may be mutually rotated. Thus the mutual rotation of images used in the estimation stage was corrected using the correlation between images. The remaining fourth image was not altered.

3   Iris Features

3.1   Choice of Features

It is often convenient to characterize a discrete-time signal in the frequency domain, thus describing stationary energy distribution. For non-stationary signals, it might be worthwhile to characterize the frequencies locally, and to find the distribution of signal energy in local (possibly overlapping) time segments by application of *time-frequency* or *time-scale* analysis. Similarly, any constant (time-independent) space-homogeneous 1D or 2D pattern can be characterized in a 1D or 2D frequency domain. If a pattern is not space-homogeneous, its spatial frequency contents may be analyzed locally, with the use of *space-frequency* or *space-scale analysis*. Although the iris texture makes a 2D pattern, we simplify it to a set of 1D patterns with a certain loss of information and apply the space-frequency analysis locally to the iris circular sectors to describe their local features and to construct a compact iris features set.

There exist various tools to represent the signal in the mixed space-frequency domain. A family of Windowed Fourier Transforms apply Fourier Transform to windowed signals in time or space. The Gabor transform belongs to this family, and uses a Gaussian window characterized by its width. The window width significantly influences the resulting iris features and must be carefully chosen. We use the space-frequency analysis that employs waveforms indexed by space, scale and frequency simultaneously, what results in a larger set of possible tilling in the space-frequency plane, possibly redundant. This directs our methodology toward a *wavelet packet* analysis. There is a need to select appropriate frequencies and scales simultaneously to make the transformation sensitive to individual features existing in the iris image. In this paper we propose a systematic selection of appropriate scales and frequencies of the iris coding. This approach enables our method to be applied for databases of images of various resolution.

3.2   Application of Zak's Transform

Gaussian-shaped windows are not orthogonal, i.e., the inner product of any two windows is nonzero, therefore Gabor's expansion coefficients cannot be determined in a simple way. The fastest method of Gabor's expansion coefficients determination consists of application of Zak's transform [3] and is often referred to as Zak-Gabor's transform. We outline briefly Zak-Gabor's transform for a single iris stripe and a fixed window width.

Denote by g_s a one-dimensional Gaussian elementary function of the width index s, sampled at points $0, \ldots, P-1$, namely

$$g_s(p) = e^{-\pi\left((p+\frac{1}{2})/2^s\right)^2}, \quad p = 0, \ldots, P-1 \tag{1}$$

where $s = 2, \ldots, S$, and for the stripe length $P = 512$ we set $S = 8$. If P is (typically) chosen to be even, the $\frac{1}{2}$ term in (1) makes g_s to be an even function.

Let M be the number of possible translations of g_s, and K be the number of frequency shifts, where, following Bastiaans [3], we always take $M = P/K$. A shifted and modulated version $g_{mk;s}$ of the elementary function g_s can be constructed, namely

$$g_{mk;s}(p) = g_s(p - mK)e^{ikp2\pi/K}, \quad p = 0 \ldots P - 1 \tag{2}$$

where $m = 0, \ldots, M - 1$ and $k = 0, \ldots, K - 1$ denote the space and frequency shifts, respectively, and g_s is wrapped around in the P-point domain. The finite discrete Gabor transform of the iris stripe f_ℓ is defined as a set of complex coefficients $a_{mk;s\ell}$ that satisfy the Gabor signal expansion relationship, namely

$$f_\ell(p) = \sum_{m=0}^{M-1} \sum_{k=0}^{K-1} a_{mk;s\ell} g_{mk;s}(p), \quad p = 0 \ldots P - 1 \tag{3}$$

Following Bastiaans [3], we further set $K = 2^s$. Note that once the frequency index k is kept constant, $g_{mk;s}$ may be localized in frequency by a modification of s. This is done identically as the scaling in a wavelet analysis, hence we call s the *scale* index. The number of Gabor expansion coefficients $a_{mk;s\ell}$ may be interpreted as the signal's number of degrees of freedom. Note that the number S of scales together with the stripe size P determine both M and K.

The discrete finite Zak transform $\mathscr{Z}f_\ell(\rho, \phi; K, M)$ of a signal f_ℓ sampled equidistantly at P points is defined as the one-dimensional discrete Fourier transform of the sequence $f_\ell(\rho + jK)$, $j = 0, \ldots, M - 1$, namely [3]

$$\mathscr{Z}f_\ell(\rho, \phi; K, M) = \sum_{j=0}^{M-1} f_\ell(\rho + jK)e^{-ij\phi2\pi/M} \tag{4}$$

where $M = P/K$. Discrete Zak's transform is periodic both in frequency ϕ (with the period $2\pi/M$) and location ρ (with the period K). We choose ϕ and ρ within the fundamental Zak interval [3], namely $\phi = 0, 1, \ldots, M - 1$ and $\rho = 0, 1, \ldots, K - 1$.

Application of the discrete Zak transform to both sides of (3) and rearranging the factors yields

$$\mathscr{Z}f_\ell(\rho, \phi; K, M) = \sum_{j}^{M-1} \left[\sum_{m}^{M-1} \sum_{k}^{K-1} a_{mk;s\ell} g_s(\rho + jK - mK)e^{ik\rho2\pi/K} \right] e^{-ij\phi2\pi/M}$$

$$= \left[\sum_{m=0}^{M-1} \sum_{k=0}^{K-1} a_{mk;s\ell} e^{-i2\pi(m\phi/M - k\rho/K)} \right] \left[\sum_{j=0}^{M-1} g_s(\rho + jK)e^{-i2\pi j\phi/M} \right] \tag{5}$$

$$= \mathscr{F}a_{s\ell}(\rho, \phi; K, M) \mathscr{Z}g_s(\rho, \phi; K, M)$$

where $\mathscr{F}a_{s\ell}[\rho, \phi; K, M]$ denotes the discrete 2D Fourier transform of an array of $a_{s\ell}$ that represents Gabor's expansion coefficients determined for the iris stripe

f_ℓ and scale s, and $\mathscr{Z}g_s[\rho, \phi; K, M]$ is discrete Zak's transform of the elementary function g_s. This shows that Gabor's expansion coefficients can be recovered from the product form (5). Once K and M are chosen to be powers of 2 (making also the signal length P to be a power of 2), the calculation of both $\mathscr{Z}f[\rho, \phi; K, M]$ and $\mathscr{Z}g[\rho, \phi; K, M]$, and inversion of 2D Fourier series can employ Fast Fourier Transform thus yielding computation times proportional to those in the FFT.

3.3   Definition of Iris Features

Calculation of Gabor's transform for all iris stripes and for all scales results in a set of coefficients a indexed by the quadruple: within-stripe position, frequency index, scale and stripe index (m, k, s, ℓ). Inspired by Daugman's work [2], we define the signs of the real and imaginary parts of Zak-Gabor coefficients as the feature set $\mathbb{B}$, namely

$$\mathbb{B} = \{\mathrm{sgn}(\Re(a_{mk;s\ell})), \, \mathrm{sgn}(\Im(a_{mk;s\ell}))\} \tag{6}$$

where $m = 0, \ldots, M - 1$, $k = 0, \ldots, K - 1$, $\ell = 0, \ldots, 2R - 1$ and $s = 2, \ldots, S$. Since Fourier's transform is symmetrical for real signals, for each position m the coefficients with the frequency index $k > K/2$ can be ignored. Since $M = P/K$, for each s there are $(N-1)P/2$ coefficients to be determined. Taking into account that this analysis is carried out for all iris stripes, and remembering that $R = 16, S = 8$ and $P = 512$, the total number of coefficients calculated for the iris image is $R(S-1)P = 57,344$. Both real and imaginary parts are coded separately by one bit, hence $N = |\mathbb{B}| = 114,688$ features may be achieved, where $|\cdot|$ denotes the number of elements in a finite set. The features, positioned identically for each iris, may thus form a binary vector. Thus, matching two features requires only a single XOR operation, and the Hamming distance can be applied to calculate the score.

We stress that $\mathbb{B}$ should not be confused with the so called $iriscode^{\mathrm{TM}}$ invented by Daugman. The latter one is a result of an iris image filtering, while $\mathbb{B}$ is constructed with Gabor expansion coefficients.

3.4   Features Selection

The feature set $\mathbb{B}$ selected so far is oversized and only its certain subset will be included into the final feature set. All elements of $\mathbb{B}$ will thus be considered the *candidate features*. We propose a two-stage method that selects Zak-Gabor coefficients. We further consider partitions of all candidate features $\mathbb{B}$ onto *candidate feature families* $\mathbb{B}_{k,s}$, which represent all candidate features that are labeled by the same scale k and frequency s, and differ by space indices m and ℓ, namely

$$\mathbb{B}_{k,s} = \{\mathrm{sgn}(\Re(a_{mk;s\ell})), \mathrm{sgn}(\Im(a_{mk;s\ell})) : m = 0, \ldots, M - 1, \ell = 0, \ldots, 2R - 1\}$$

Stage one: selection of useful features. The first selection stage consists of choosing a subset $\mathbb{B}^0$ of candidate features $\mathbb{B}$, called here the *useful features*. To determine $\mathbb{B}^0$, we analyze a variability of candidate features. For each feature b

we calculate the *within-eye sum of squares* $SS^W(b)$, and the *between-eye sum of squares* $SS^B(b)$. We categorize the features to maximize SS^B and minimize SS^W. We tried several methods to solve this multicriteria maximization problem. The best results were obtained when we minimized the distance from the most desired point on $SS^W \times SS^B$ plane. This point was set as $\left(\min_{b \in \mathbb{B}} SS^W(b), \max_{b \in \mathbb{B}} SS^B(b)\right)$, Fig. 2 (left).

We use the order introduced by the above procedure in the set of candidate features $\mathbb{B}$ in a procedure removing a high correlation of candidate features to increase an 'information density'. We include k-th candidate feature into the set $\mathbb{B}^0$ only if it is *not strongly correlated* with all the features already selected.

We base our useful feature definition on the *decidability coefficient d'* [2] calculated for a given feature subset. We calculate the decidability coefficient for each set of candidate features included into $\mathbb{B}^0$. The decidability varies with the number of candidate features included: it first grows to reach the maximum and then decreases. Experiments show that the decidability d' is highest for the correlation threshold around 0.3, Fig. 2 (right). For this solution there is no between-eye – within-eye overlap of sample distributions, i.e., there are no false matches and no false non-match examples in the estimation data set. The resulting 324 useful features pass to the second feature selection stage. We may add that our procedure included only such features for which $SS^W < SS^B$.

The higher is the number $\nu(k, s)$ of useful features in the candidate features family $\mathbb{B}_{k,s}$, the more important is (k, s) in iris recognition. This enables to categorize these families in the next stage of our selection procedure.

Stage two: selection of feature families $\mathbb{B}_{k,s}$. To finally decide for the best frequencies k and scales s, independently for real or imaginary parts of the Zak-Gabor coefficients, we sort $\mathbb{B}_{k,s}$ by decreasing $\nu(k, s)$ separately for real and imaginary parts of coefficients. This procedure prioritizes the families that are

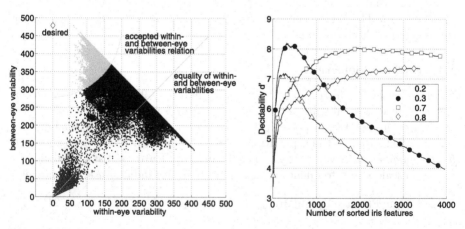

Fig. 2. *Left:* within-eye sum of squares vs. between-eye sum of squares and the area of *useful features*. *Right:* the decidability coefficient vs. number of useful features selected for a few correlation thresholds allowed within the useful features set.

most frequently 'populated' by the useful features. Such candidate feature families resulting in maximum d' are selected to the *final feature set*. This finally produces the iris feature set of 1024 bits (128 bytes), containing only four families. For this final feature set, we achieved the maximum decidability d' and no sample verification errors.

3.5   Features Personalization

Once the optimal feature families, namely the best scale-frequency pairs indexed by s and k, are selected, the iris features set is calculated for those chosen s and k and all $m = 1, \ldots, M - 1$, and $\ell = 0, \ldots, 2R - 1$. Each Zak-Gabor coefficient can 'measure' the correlation between the modulated Gaussian elementary function $g_{mk;s}$ and the corresponding stripe. The question arises how 'robust' are the consecutive Zak-Gabor coefficients against noise, and iris tissue elastic constrictions and dilations.

Due to a significant variability of the iris tissue, some $g_{mk;s}$ may not conform with the iris body, resulting in small coefficients. Such a situation is dangerous, since once the coefficients are close to zero, their signs may depend mostly on a camera noise, and consequently may weaken the final code. This motivates personalization of the iris feature sets that employ only those Zak-Gabor coefficients that exceed experimentally determined threshold, for which the decidability was maximal. Experiments show a far better discrimination between the irises if the personalized coding is employed.

3.6   Template Creation and Verification

Typically, more than one iris image is available for enrollment. For a given eye, a distance is calculated between a feature set of each image and the feature sets calculated for the remaining enrollment images. As the iris template we select this feature set, for which this distance is minimal.

Small eyeball rotations in consecutive images may lead to considerable deterioration of within-eye comparison scores. This rotation can be corrected by maximizing the correlation within the images enrolled. Since during verification the iris image corresponding to the template is unavailable, another methodology must be applied. We use an iterative minimization of the comparison score between Zak-Gabor-based features determined for a set of small artificial shifts of the iris stripes being verified.

4   System Evaluation and Summary

Figure 3 shows a sample distributions of genuine and impostor comparison scores achieved for the proposed coding. No sample errors were observed (FRR=FAR=0%) for the database used. However, the results must be taken with care since we used various images of the same eyes for estimation as well as for verification. Considering statistical guarantees and assuming 95% confidence level for the results obtained we expect FRR < 0.017 and FAR $\in \langle 0.000093, 0.03 \rangle$ in this approach.

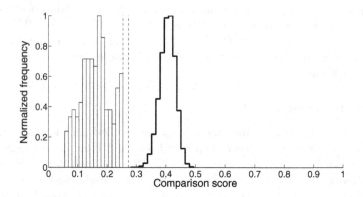

Fig. 3. Sample distributions of genuine and impostor comparison scores in the verification stage achieved for Zak-Gabor-based personalized coding with iterative eye rotation correction. No sample errors were encountered.

The Zak-Gabor coding was used in a number of applications, for instance in remote access scenario [4], in BioSec European project for the purpose of the biometric smart card development and it is also an element of the original iris recognition system prototype with eye aliveness detection [5]. Our feature selection procedure can be applied also to other iris coding methods.

References

1. American National Standards Institute: Information technology — Biometric data interchange formats — Part 6: Iris image data. ISO/IEC 19794-6:2005. ANSI (2005)
2. Daugman, J.: How iris recognition works. IEEE Transactions on circuits and systems for video technology **14** (2004)
3. Bastiaans, M.J.: Gabor's expansion and the zak transform for continuous-time and discrete-time signals. In Zeevi, J., Coifman, R., eds.: Signal and Image Representation in Combined Spaces. Academic Press, Inc. (1995) 1–43
4. Pacut, A., Czajka, A., Strzelczyk, P.: Iris biometrics for secure remote access. In Kowalik, J., ed.: Cyberspace Security and Defense: Research Issues. Springer (2005) 259–278
5. Pacut, A., Czajka, A.: Aliveness detection for iris biometrics. In: 2006 IEEE International Carnahan Conference on Security Technology, 40th Annual Conference, October 17-19, Lexington, Kentucky, USA, IEEE (2006) 122–129

Overview of *Kansei* System and Related Problems

Hisao Shiizuka

Department of Information Design, Kogakuin University
Nishishinjuku 1-24-2, Shinjuku-ku, Tokyo 163-8677, Japan
shiizuka@cc.kogakuin.ac.jp

Abstract. This paper overviews several problems in Kansei engineering such as Comparison of Intelligent systems with Kansei systems, Kansei dialog, family and individuals, framework of a Kansei system, non-verbalized area, and exchange from tacit knowledge to explicit knowledge. Then emphasis is put on importance of the two dimensional space defined by natural and artificial Kansei.

Keywords: Kansei engineering, Natural and artificial Kansei, Kansei systems, Kansei dialog, Non-verbalized area, Tacit knowledge, Explicit knowledge.

1 Introduction

The 19th and 20th Centuries were machine-centered eras. However the 21st Century will be centered on various human senses, where science and technology will be appraised on being harmonious to human beings, and to natural and social environments. Therefore, it is supposed that research and development of advanced technologies cannot be decided from only a technological point. Before the beginning of the 21st century, stand-alone technologies could not solved many issues. For example, it is extremely important that people in the technological and science fields cooperate with people in the human and social science areas. In our era, it is impossible to solve many issues only by technology.

Today stress should be placed on solving issues through a combination of the science and humanities fields. Therefore, conventional problem-solving methods or approaches must be adjusted to properly deal with such issues. In order for technology to aid providing happiness it cannot stand alone. It must cooperate with various fields related to the issues that are being dealt with by the human and social sciences.

Issues which have remained unsolved until the beginning of the 21st century cannot be understood from the perspective of conventional technological views. One of the most important aspects within such issues is to place stress on our senses, consciousness, and feeling. In Japanese these are generally named "*Kansei*". *Kansei* is the total concept of senses, consciousness, and feelings, that relate to human behavior in social living. Especially, *Kansei* relates to the research and developments of products that foster communication and information transfer among people. There are various approaches to achieve this type of communication. However, *Kansei* has not been considered, nor studied sufficiently to date. The objective of this paper is to

J.T. Yao et al. (Eds.): RSKT 2007, LNAI 4481, pp. 203–210, 2007.

discuss the basic concepts and methods of the system and provide a framework for future *Kansei* oriented research.

2 Framework of Intelligent Systems

First of all, let us describe the framework of intelligent systems. As Fig.1 shows, the horizontal axis is expression and learning and the vertical axis is *Kansei*. There are four quadrants. The 1st to 4th quadrants correspond to learning engineering, knowledge engineer- ing, fuzzy engineering and related work, respectively. Although these terms cannot sufficiently express all the fields include-ed in the framework, an intelligent system is understood to exist in the framework. The 4th quadrant includes neural networks, genetic algorithms, genetic programming and chaos systems and is included as a category of the intelligent system. These fields are termed soft computing. Recently, the field of soft computing has been very active.

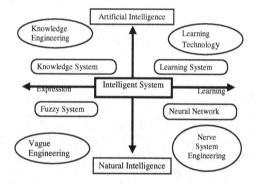

Fig. 1. Framework of Intelligent System

3 Requirement in *Kansei* Dialog

Let us discuss a two dimensional space in which elements required for *Kansei* dialog are mapped. When we build a structure for *Kansei* that has been drawn from then intelligent system, what kind of elements should be placed in the two dimensional space? As shown in Fig.2, it is easier to understand the vertical axis using "natural *Kansei*" of real hu man beings and "artificial *Kansei*," which realizes or expresses "artificially" natural *Kansei*.

The horizontal axis in Fig.2 has its left side as corresponding to "measurement," which means the perception and cognition that is necessary to understand a concept, in other words the thoughts and intentions of humans. Let us name this space "Framework of *Kansei* Systems." In the framework, research is conducted in the area of "natural *Kanei*" so as to build a viewpoint of natural *Kansei* that can be analyzed so as to explain the *Kansei* phenomena in nature with the goal of building a mathematical model of natural *Kansei*. research in the 3rd and 4th quadrants can be understood from such a perspective.

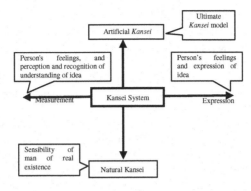

Fig. 2. Elements required in *Kansei* Dialog

Let us name this space "Frame- work of *Kansei* Systems." In the framework, research is conducted in the area of "natural *Kanei*" so as to build a viewpoint of natural *Kansei* that can be analyzed so as to explain the *Kansei* phenomena in nature with the goal of building a mathematical model of natural *Kansei*. Research in the 3rd and 4th quadrants can be understood from such a perspective.

4   Family and Individuals from *Kansei*

The research conducted to analyze and explain the *Kansei* held by we humans naturally is included in the field of natural *Kansei*. The aim is to explain the nature of *Kansei* from a family point of view by aggregating various statistical data and analyzing them. The area of natural *Kansei* overlaps the 3rd and 4th quadrants in Figure 3, so that natural *Kansei* can be closely modeled and thus ex- plained by using mathematical equations. The research points included in the area of natural *Kansei* were chosen to clarify the nature of the family of human *Kansei*.

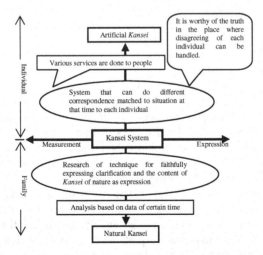

Fig. 3. Family and Individuals from *Kansei*

In contrast, the field of artificial *Kansei* overlaps the 1st and 2nd quadrants and is used to build a system that can properly respond to individual situations. In other words, the proposed system can provide different services for varied individuals. The most important asset of the system is that it can deal with differences among individuals. Therefore, artificial *Kansei* can allow for individualization within mass-production. It is in this area that a company can then foster flexible decisions, strategies and policies.

Hitherto, research on *Kansei* was pursued on natural *Kansei*. The methods employed and applications were, as shown by the 3rd and 4th quadrants in Fig.4, about cognitive science and modeling techniques. For example, questionnaires analyzed by the semantic differential method and other multivariate analyses were used to clarify the nature of human *Kansei* from the viewpoint of their forming a group, a related family. Such results can structure a database. Much of the research conducted, which was on measurements and questionnaires, can be included in this database. The 4th quadrant is mainly relating to modeling and methodology. Adjective words and *Kansei* words from the early stages of *Kansei* research have been evaluated so as to determine their validity for inclusion in the database. Multivariate analyses were widely employed to analyze statistical data. Since then, fuzzy systems, neural networks, genetic algorithms and so on have been used as effective tools. Recently, the method of rough sets has been widely utilized in re- search regarding *Kansei*. Scientists using rough sets intend to discover knowledge or rules that are latent within the data. Understanding the concept of rough sets is vital to insure its value as an effective analytical devise for *Kansei*. Recently, a research group on rough sets published "Rough sets and *Kansei*"[1], justifying the rough set method.

5 Framework of a *Kansei* System

As mentioned above research on natural *Kansei* intends to study the real subject of human *Kansei*. Research on the area of artificial *Kansei* is seen in the 1st and 2nd quadrants. The most distinguishing feature of this research is to construct a system that will behave differently according to each situation and each individual. That is, the most valuable aspects in the field are that various services can be provided to various people and that such a system can deal with differences between individuals.

It is important to hold *Kansei* as a system, to understand *Kansei* from the point of a systematic view. In the field of natural *Kansei*, cognitive science and modeling are studied from this perspective in the 3rd and 4th quadrants. However, the field of artificial *Kansei*, placed in the upper portion of the two dimensional space, has *Kansei* expressions in the 1st quadrant and *Kansei* recognition in the 2nd quadrant. This design represents *Kansei* expression. This design is based upon a wide range of research. In particular, by employing *Kansei* methods, new perspectives and new methodologies can be brought to universal and information design.

The philosophy of universal design(UD) is to design products or provide urban designs where various people can use a product or a city with the least amount of difficulty. As universal design relates to various places and situations in daily living, many experts from myriad fields participate. These fields include archit-ecture, housing, transportation, urban design, industrial products, daily goods, and information techno- logy. As mentioned above, universal design is used for making products, buildings and urban environments more user-friendly.

The following are the 7 rules of universal design. The 7 rules clarify the direction and method of design in relation to wide areas for environments, products and communications. The 7 rules are employed to not only direct design evaluation and the design process, but also to enlighten consumers, as well as designers, about which products and environments are user-friendly. The 7 rules of universal design are specifically written as follows:

(Principle 1) Fairly usable for any person
(Principle 2) Freedom in usage is high.
(Principle 3) Usage is simple.
(Principle 4) The required information is understandable.
(Principle 5) The design is free from careless mistakes and dangers.
(Principle 6) The design is easy to use: there are no difficult physical requirements
    and it can be used with little power.
(Principle 7) There is sufficient space and size for easy access.

It is expected that the methodology based on *Kansei* can be utilized in information design. The rapid development of technology and media require drastic and more qualitative changes in the handling of information. In particular, "learning" is being termed a business in the 21st century. Various learning methods or chances are being provided in parallel and concurrently for all generations. For instance, there are real time conversations, gestures and physical interaction, facial expressions, new reading methods, smart TV programs, entertainment and so on. This is a mode of entertainment that makes people fully delighted after giving information, and that changes the learned information into intelligence[2].

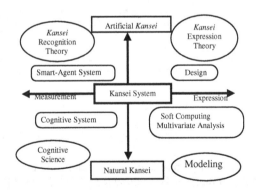

Fig. 4. Framework of a Kansei system

The above has been named entertainment computing[3]: entertainment connected with technology and information. Entertainment computing is included, regarding *Kansei* expressions, in the 1st quadrant; thus, a connection between information engineering and the entertainment industry can be developed through *Kansei*.

Kansei recognition within the 2nd quadrant, shown in Fig.4, is an undeveloped re-search field. While it is studied to make a robot mimic human Kansei, this is an approach aimed at hard robots. It is much more important to develop Kansei recognition within the robot. This places stress on developing software robots. For example, this would be a non-hardware robot (software robot) that moves freely in an Internet (cyberspace) envi- ronment and gathers data. A robot based on *Kansei* would have wider applications than one based on hardware.

6 *Kansei* in Non-verbalized Area

The phrase "Today is an information society" is heard frequently. Even if this so, it is important to show whether this statement is true or not.   What does it state? Knowledge is a reward obtained from understanding through experience. Even if information is presented, it may not be transferred accurately. When relations are constructed so as to mutually act with other persons or systems, and when patterns and meanings of information are assimilated with each other, then the transfer of knowledge will occur.

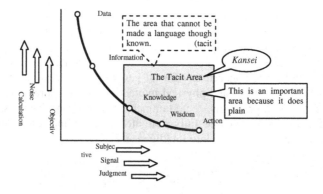

Fig. 5. *Kansei* in non-verbalized area (modified from [2])

When information is presented after being integrated and put in order, knowledge cquisition will be realized. The stimulation of this information creates experience. What is obtained from such experiences is knowledge. It is important that knowledge becomes wisdom after the understanding of knowledge is deepened. Fig.5 shows this flow.

Wisdom is the vaguest, the most individual understanding, and it is more abstract and philosophical than other levels of knowledge or understanding. There are many aspects as to how wisdom is acquired and developed that are unknown.

Wisdom is thought of as a kind of "meta-knowledge," which is based on the mixture and the relations between all processes understood from experiences. Wisdom is also understood as the result of deep thought, evaluation, reflection and interpretation: these are very individual processes. Wisdom cannot be created by a process such as data and information. It is also impossible to share wisdom with another person in the same way that knowledge can be shared. It is understood that each individual obtains wisdom by themselves. As Fig.6 shows, *Kansei* is thought to be the bridge from knowledge and intelligence to wisdom, in the flow of the understanding process that starts from data. That is, since the contents understood (impression) are different among individuals, *Kansei* bridges them and such understanding occurs at an extremely individual level.

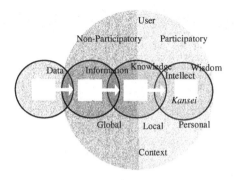

Fig. 6. *Kansei* in the process of context understanding (modified from [2])

7 *Kansei* to Exchange Tacit Knowledge to Explicit Knowledge

In this section the most important point to be addressed is how to treat the area of sense that cannot be verbalized even if it is known. This includes implicit knowledge. Generally knowledge can be classified into tacit knowledge and explicit knowledge. It is said that human knowledge is mostly tacit knowledge. Nevertheless, it is necessary to verbalize and objectify senses and concepts in order to understand them and to transfer the related ideas coupled with them. That is, it is essential to change tacit knowledge into explicit knowledge. However, it is said that it is difficult to execute a rapid evaluation and transfer of knowledge (tacit knowledge) if it is possessed with a body or as an intuition. Yet, it is essential that objective knowledge (explicit knowledge) be employed in the process.

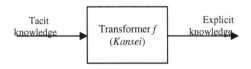

Fig. 7. Explicit knowledge from tacit knowledge

Knowledge of a field, the skill of a worker or researcher, sense toward markets and customers and the ability to find qualities are tacit knowledge. Manuals and guidelines of operational procedures, systematic analysis and verbalized questions and methods, and specifications of a product and design are explicit knowledge. Since Kansei is naturally different for individuals, it is a problem that knowledge transfer does not occur as well as is ex- pected. Therefore, we need a transfer device between people and systems, that is, a Kansei transferring device is required: one like that which is shown in Fig.7. This is also equivalence to that it is variously in how to feel the person, shown in Fig.8.

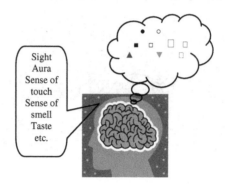

Fig. 8. Kansei transform: It is variously in how to feel the person

8   Concluding Remarks

As explained at the beginning of this paper, the framework of a *Kansei* system shown in Figure 4 is proposed and related topics are discussed in order to answer the questions of how human *Kansei* (senses) functions, how these relate with each other, and which method is most appropriate for *Kansei*. It should be noted that the two dimensional space can be divided by defining the vertical axis of natural *Kansei* and artificial *Kansei* and by the horizontal axis of measurement and expression. Research on artificial *Kansei* has not started yet. Such research is important in the development of software robots (smart agents). Also, it should be emphasized that *Kansei* plays an important role in the process of transferring tacit knowledge into explicit knowledge in the non-verbalized area as Figure 5 shows. As Figure 3 shows, this area is shared with the area of artificial *Kansei* corresponding to an individual. *Kansei* is extremely personal and user- participatory, that is, *Kansei* is different among individuals. It is also important to decide, in realization of artificial *Kansei*, what should be taken into consideration, what should be removed from specification and how much vagueness can be accepted.

References

1. Mori, N., Tanaka, H., Inoue, K.: Rough Sets and Kansei, Kaibundo (2004) (in Japanese)
2. Jacobson, R. (ed.): Information Design, MIT Press (2000)
3. Special Issue on Entertainment Computing, Information Processing, 44 (2003) (in Japanese)
4. Imamura, K. and Shiizuka, H.: Framework of a Kansei System for Artificial Kansei, Transaction of Kansei Engineering, 4 (2004) 13-16 (in Japanese)

Reduction of Categorical and Numerical Attribute Values for Understandability of Data and Rules

Yuji Muto, Mineichi Kudo, and Yohji Shidara

Division of Computer Science
Graduate School of Information Science and Technology,
Hokkaido University, Sapporo 060-0814, Japan
{muto,mine,shidara}@main.ist.hokudai.ac.jp

Abstract. In this paper, we discuss attribute-value reduction for raising up the understandability of data and rules. In the traditional "reduction" sense, the goal is to find the smallest number of attributes such that they enable us to discern each tuple or each decision class. However, once we pay attention also to the number of attribute values, that is, the size/resolution of each attribute domain, another goal appears.

An interesting question is like, which one is better in the following two situations 1) we can discern individual tuples with a single attribute described in fine granularity, and 2) we can do this with a few attributes described in rough granularity. Such a question is related to understandability and Kansei expression of data as well as rules. We propose a criterion and an algorithm to find near-optimal solutions for the criterion. In addition, we show some illustrative results for some databases in UCI repository of machine learning databases.

Keywords: Attribute Values, Reduction, Grouping, Granularity, Understandability.

1 Introduction

In general, the data handled in the fields of pattern recognition and data mining are expressed by a set of measurement values. These measurement values are strongly affected by the intention of the observer and the precision of measurement. Therefore, there is no deterministic evidence that the given representation is essential. There might exist more appropriate representation of data in evoking our "Kansei". If so, it would bring easiness/intuition for the analyst to interpret them. Here "Kansei" is a Japanese word which refers to the psychological image evoked by the competing sensations of external stimuli, and affected by emotions and personal sense of values. This word is often used in the various fields which are derived from "Kansei Engineering" proposed by Nagamachi [1], such as Kansei Information Processing, Kansei Information Retrieval, Kansei mining, and so on [2,3].

J.T. Yao et al. (Eds.): RSKT 2007, LNAI 4481, pp. 211–218, 2007.

Obviously, too rough description is not enough for distinguishing an object from the others. While, too fine description is redundant. That is, it is desired to find an appropriate roundness of representation on that problem domain and the data available. Such a trial is seen as "reduction" in rough sets [4,5]. In reduction, we seek the minimum number of attributes enough for discerning individual objects. However, such a reduction is not always sufficient in roundness viewpoint. To find an appropriate Kansei representation, the attribute values are also required to be a few or succinct. To achieve this goal, we consider reduction of attribute values in addition to reduction of attributes. We do this according to several criteria on the basis of discernibility. This way includes the traditional reduction, as will be described later on. We have already started discussion on this issue [6,7]. In this paper, we describe the methodology in a more general scheme and show its application to rule generation.

2   Reduction and Partition

Rough sets theory [4,5] is known as one of the theories which give mathematical frameworks to describe objects and to understand them. One of purposes of rough sets theory is to find the minimal subsets of attributes keeping some discernibility on a given equivalence relation. This goal is referred to as "reduction". However, the number of attribute values, not the number of attributes, has not been considered so far. Therefore, it can happen that the number of attributes is small but those attributes are described in a very fine way, sometimes with infinitely many values. Such a fine description is not useful in the light of understandability of data or rules or both.

A given equivalence relation specifies "granularity" which gives a basic level of expression of any concept, and each equivalence class gives a granule [8,9,10]. We can see that the traditional reduction is the procedure to make granules expand maximally in some attributes. However, it does not simplify the representation in the selected attributes. For example, in Fig. 1, in reduction we can discern all the data in the first attribute only (Fig. 1 (b)), but, in fact, two attribute values are sufficient to separate two decision classes (Fig. 1 (c)). With this rough representation, we ca use some linguistic representation (Kansei representation), say "low or high" or "small or large."

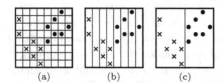   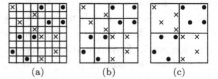

Fig. 1. Reduction and partition: (a) base relation, (b) reduction, and (c) the simplest partition

Fig. 2. Several types of granularity: (a) Measurement granularity R_M , (b) Individual-discernible granularity R_I, and (c) Class-discernible granularity R_D

3 Reduction of Attribute Values Using Granularity

Let us describe formally the reduction of attribute values or granularity. First, let us consider m different attributes $A_1, A_2, \ldots, A_m$, and denote their domains $D_i = \{a_{i1}, a_{i2}, \ldots, a_{in_i}\}(i = 1, 2, \ldots, m)$. In addition, let us consider a class attribute C with its domain $D_C = \{c_1, c_2, \ldots, c_{n_C}\}$. Here, we denote the universe by $U = D_1 \times D_2 \times \cdots \times D_m$. An information table T is given as a subset of U. A tuple of T is expressed by (x, y) with $x = (x_1, x_2, \ldots, x_m)$, $x_i \in D_i$, $y \in D_C$. By (x_i, y_i), we denote ith tuple. We introduce several types of granularity as follows.

3.1 Discernible Granularity

First, we call the granularity of given data itself the "measurement granularity" (R_M). We give it on account of the measurement precision, the description precision, and computational expression on memory. Here we assume that the measurement granularity is fine enough for discerning individuals. However, it can happen that the measurement granularity is not enough for this goal. Then we require a finer granularity in measurement by adopting higher resolution devices or manners. Next, we define two types of granularity on the basis of two kinds of discernibility: 1) "individual-discernible granularity" (R_I) requiring

$$x_i \neq x_j \Rightarrow [x_i]_{R_I} \neq [x_j]_{R_I} \text{ for } \forall i \neq j,$$

and 2) "class-discernible granularity" (R_D) requiring

$$y_i \neq y_j \Rightarrow [x_i]_{R_D} \neq [x_j]_{R_D} \text{ for } \forall i \neq j,$$

where $[x]_R$ is an equivalence class including x in equivalence relation R. An example is shown in Fig. 2. These types of granularity are assumed to be refinements of the successors, that is, $R_M \leq R_I \leq R_D$.

3.2 Evaluation of Granularity

From the viewpoint of understandability of data or rules, it is preferable to minimize the number of attributes and the number of attribute values as long as it ensures some discernibility. Let n_i be the number of attribute values in ith attribute and n be the total sum of them. Then, we propose the following criterion to minimize:

$$J = \prod_i^m n_i + kH\left(\frac{n_1}{n}, \frac{n_2}{n}, \ldots, \frac{n_m}{n}\right) / \log_2 m,$$

$$\text{where, } H\left(\frac{n_1}{n}, \frac{n_2}{n}, \ldots, \frac{n_m}{n}\right) = -\sum_i^m \frac{n_i}{n} \log_2 \frac{n_i}{n}, \quad n = \sum_i^m n_i.$$

Here, k takes a real value in $[-1, 1]$. Especially we use $k = \pm 1$. For $k = -1$, it enhances the equality $n_1 \cong n_2 \cong \cdots \cong n_m$. While, $k = 1$ enhances the reverse, that is, a concentration on a few attributes such as $n_1 = n - m + 1, n_2 = \cdots = n_m = 1$. In either case, the first term is dominant, so it requires the minimal number of attribute values. It is also noted that $n_i = 1$ does not increase the value of the first term.

3.3 Algorithm

To find the granularity that minimizes the criterion, an exhaustive search is not practical. Therefore, we use a greedy algorithm to find a near-optimal granularity for the criterion. The algorithm is shown in Fig. 3. The method starts with the measurement granularity and searches a more rough partition of attribute values. This partition is performed as long as the criterion decreases and the granularity still keeps discernibility. Here, we only consider the equally sized partition (granules). Numerical (ordered) attributes and categorical (non-ordered) attributes are differently dealt with.

- **Initialization**
    1. Set measurement granularity R_M for given attributes. Mark all attributes by 'reducible.' Let $n_i^{(0)}$ be the number of attribute values in ith attribute for all i. Let $t = 1$.
    2. Compute the evaluation value J with $n_1^{(0)}, n_2^{(0)}, \ldots, n_m^{(0)}$.
- **Repeat the following steps as long as discernibility is kept:**
    1. For all reducible attributes, compute the minimum prime number $q_i^{(t)}$ of $n_i^{(t-1)}$ to derived $n_i^{(t)} = n_i^{(t-1)}/q_i^{(t)}$ for $i = 1, 2, \ldots, m$.
    2. Compute the decrease amount $\Delta J_i^{(t)}$ with $n_1^{(t-1)}, n_2^{(t-1)}, \ldots, n_i^{(t)}, \ldots, n_m^{(t-1)}$.
    3. Choose j such as $j = \arg\max_i \Delta J_i^{(t)}$.
    4. (a) Compute the discretization with $n_j^{(t)}$ equally-sized intervals if jth attribute is numerical.
       (b) Compute the partition with $n_j^{(t)}$ equally-sized groups which consist of q_j attribute values if jth attribute is categorical. If there are several possible partitions, compute all the possible partitions.
    5. (a) If discernibility is kept after the discretization/partition, perform the discretization/partition.
          Let $n_i^{(t)} = n_i^{(t-1)}$ for all $i \neq j$, $n_j^{(t)} = n_j^{(t-1)}/q_j^{(t)}$, $J^{(t)} = J^{(t-1)} - \Delta J_j^{(t)}$.
       (b) Otherwise, let $n_i^{(t)} = n_i^{(t-1)}$ for all i, and $J^{(t)} = J^{(t-1)}$.
          The jth attribute will not be reduced with 'irreducible' mark.
    6. If there is no reducible attribute, then terminate the process. Otherwise, let $t = t + 1$.

Fig. 3. Greedy bottom-up partition algorithm

Table 1. Statistics of Datasets

Dataset	#Attribute	#Attribute (categorical)	#Attribute (numerical)	#Object	#Class	Majority Class
Hepatitis	19	13	6	80	2	0.84
Mushroom	22	22	-	5644	2	0.62

4 Experiments and Discussion

We dealt with two datasets of Hepatitis and Mushroom from the UCI repository of Machine Learning databases [12] to evaluate the proposed criteria. Table 1 shows the statistics of those datasets. Here, we removed the instances which include missing values.

We found in order the measurement granularity, individual-discernible granularity and class-discernible granularity for parameter $k = \pm 1$ in J.

The results are shown in Tables 2 and 3. In individual-discernible granularity we succeeded to reduce the attribute size largely from their original size in measurement granularity. In addition, those sizes were further reduced in class discernibility. As for the difference of granularity with two types of k. As expected, the number of attributes is largely reduced, while some attributes have large domain sizes in the results of $k = 1$. In the contrary, in $k = -1$, the number of attribute values are largely reduced. If we focus on the understandability of data or rules, the latter is better.

Table 2. Granularity of Hepatitis (2 classes). In type, N means numerical attribute and C means categorical attribute.

			$k = 1$		$k = -1$	
Attribute	Type	R_M	R_I	R_D	R_I	R_D
Age	N	54	1	1	1	1
Sex	C	2	1	1	1	1
Steroid	C	2	1	1	2	2
Antivirals	C	2	1	1	2	1
Fatigue	C	2	2	1	2	1
Malaise	C	2	1	1	1	1
Anorexia	C	2	1	1	1	1
Liver_Big	C	2	1	1	2	1
Liver_Firm	C	2	2	1	1	1
Spleen_Palpable	C	2	2	2	2	1
Spiders	C	2	1	1	1	1
Ascites	C	2	2	2	1	1
Varices	C	2	1	1	2	2
Bilirubin	N	45	1	1	1	1
Alk_Phosphate	N	256	1	1	32	32
Sgot	N	432	432	216	1	1
Albumin	N	30	1	1	5	5
Protime	N	100	1	1	1	1
Histology	C	2	1	1	1	1
#Attribute		19	5	3	8	4

Table 3. Granularity of Mushroom (2 classes)

		$k = 1$		$k = -1$	
Attribute	R_M	R_I	R_D	R_I	R_D
cap-shape	6	6	1	6	1
cap-surface	4	4	2	4	2
cap-color	10	5	1	5	1
bruises?	2	1	1	1	1
odor	9	3	1	3	1
gill-attachment	4	2	1	2	1
gill-spacing	3	3	1	3	1
gill-size	2	1	1	2	2
gill-color	12	6	3	6	1
stalk-shape	2	2	2	2	2
stalk-root	6	3	3	1	1
stalk-surface-above-ring	4	2	1	2	2
stalk-surface-below-ring	4	2	1	2	1
stalk-color-above-ring	9	3	1	3	1
stalk-color-below-ring	9	3	1	3	1
veil-type	2	1	1	1	1
veil-color	4	1	1	1	1
ring-number	3	1	1	1	1
ring-type	8	1	1	1	1
spore-print-color	9	3	3	3	3
population	6	3	3	3	3
habitat	7	4	4	4	2
#Attribute	22	16	7	16	7

Table 4. Abstract words which correspond to merged values

odor	
merged values	abstract word
{foul, anise, musty}	unpleasant odor
{none, spicy, fishy}	pleasing odor
{pungent, creosote, almond}	irritating odor

stalk-surface-above-ring	
merged values	abstract word
{fibrous, silky}	fibrous
{scaly, smooth}	rough or smooth

spore-print-color	
merged values	abstract word
{brown,orange,yellow}	brownish yellow
{buff,black,purple}	purplish brown
{chocolate,green,white}	barely colored

habitat	
merged values	abstract word
{woods, leaves, paths}	in forest
{grasses,waste, meadows,urban}	plain

In Hepatitis, in case of $k = 1$, the number of attributes were reduced from 19 to 5 in R_I, and furthermore were reduced from 5 to 3 in R_D (Table 2). In addition, the number of attribute values were reduced from 100 to 1 at maximum. In Mushroom, the number of attributes were also reduced from 22 to 16 (R_I), and from 16 to 7 (R_D) (Table 3). These results show a success both in the traditional "reduction," and in our "further reduction."

Once we have a small number of attribute values, then it is possible to assign more "appropriate" words for them instead of their original names or resolution. So, we translated the attribute values to a small set of "abstract words" in

---

Rule 1: Ascites = yes, Protime $\leq$ 51, Anorexia = no $\Rightarrow$ **DIE** (TPrate = 0.54, FPrate = 0.00)
Rule 2: Spiders = yes, Protime $\leq$ 39, Histology = yes
$\Rightarrow$ **DIE** (TPrate = 0.46, FPrate = 0.02)
Rule 3: Histology = no $\Rightarrow$ **LIVE** (TPrate = 0.69, FPrate = 0.08)
Rule 4: Protime > 51 $\Rightarrow$ **LIVE** (TPrate = 0.78, FPrate = 0.15)
Rule 5: Anorexia = yes $\Rightarrow$ **LIVE** (TPrate = 0.18, FPrate = 0.00)

(a) Extracted rules in R_M.

Rule 1: Albumin $\leq$ **1/5** $\Rightarrow$ **DIE** (TPrate = 0.23, FPrate = 0.00)
Rule 2: Spleen_Palpable = yes, Liver_Big = yes, Albumin $\leq$ **3/5**
$\Rightarrow$ **DIE** (TPrate = 0.31, FPrate = 0.02)
Rule 3: Antivirals = yes, Steroid = no and Liver_Big = yes, Albumin $\leq$ **3/5**
$\Rightarrow$ **DIE** (TPrate = 0.15, FPrate = 0.00)
Rule 4: Albumin > **3/5** $\Rightarrow$ **LIVE** (TPrate = 0.73, FPrate = 0.08)
Rule 5: Steroid = yes, Albumin > **1/5** $\Rightarrow$ **LIVE** (TPrate = 0.55, FPrate = 0.15)
Rule 6: LIVER_BIG = no $\Rightarrow$ **LIVE** (TPrate = 0.19, FPrate = 0.00)

(b) Extracted rules in R_I.

Rule 1: Albumin $\leq$ **1/5** $\Rightarrow$ **DIE** (TPrate = 0.23, FPrate = 0.00)
Rule 2: Steroid = no, Albumin $\leq$ **3/5** $\Rightarrow$ **DIE** (TPrate = 0.62, FPrate = 0.15)
Rule 3: Albumin > **3/5** $\Rightarrow$ **LIVE** (TPrate = 0.73, FPrate = 0.08)
Rule 4: Steroid = yes, Albumin > **1/5** $\Rightarrow$ **LIVE** (TPrate = 0.55, FPrate = 0.15)

(c) Extracted rules in R_D.

Fig. 4. Extracted rules for Hepatitis ($k = -1$). In the condition, i/j means ith-level of j levels.

Rule 1: spore-print-color in {black, brown, purple, white},
population in {abundant, numerous, scattered, several, solitary}
odor in {almond, anise, none}   ⇒ **Edible** (TPrate = 1.000, FPrate = 0.000)
Rule 2: odor in {creosote, foul, musty, pungent} ⇒ **Poisonous** (TPrate = 0.96, FPrate = 0.00)
Rule 3: spore-print-color = green   ⇒ **Poisonous** (TPrate = 0.03, FPrate = 0.00)
Rule 4: population = clustered   ⇒ **Poisonous** (TPrate = 0.02, FPrate = 0.00)

(a) Extracted rules in R_M (4/4).

Rule 1: spore-print-color in {*purplish brown, brownish yellow*},
gill-size = broad   ⇒ **Edible** (TPrate = 0.92, FPrate = 0.00)
Rule 2: spore-print-color in {*purplish brown, brownish yellow*},
odor in {*unpleasant odor, pleasing odor*} ⇒ **Edible** (TPrate = 0.86, FPrate = 0.00)
Rule 3: odor = *unpleasant odor*,
spore-print-color = *barely colored* ⇒ **Poisonous** (TPrate = 0.75, FPrate = 0.00)
Rule 4: habitat = *plain*,
spore-print-color = *barely colored* ⇒ **Poisonous** (TPrate = 0.37, FPrate = 0.00)

(b) Extracted rules in R_I (top 4/9).

Rule 1: spore-print-color in {*purplish brown, brownish yellow*},
gill-size = broad   ⇒ **Edible** (TPrate = 0.92, FPrate = 0.00)
Rule 2: spore-print-color in {*purplish brown, brownish yellow*},
stalk-shape = tapering   ⇒ **Edible** (TPrate = 0.74, FPrate = 0.00)
Rule 3: stalk-surface-above-ring = *fibrous*, gill-size = broad,
spore-print-color = *barely colored*, ⇒ **Poisonous** (TPrate = 0.69, FPrate = 0.00)
Rule 4: habitat = *plain*,
spore-print-color = *barely colored* ⇒ **Poisonous** (TPrate = 0.37, FPrate = 0.00)

(c) Extracted rules in R_D (top 4/10).

Fig. 5. Extracted rules for Mushroom ($k = -1$). The number in parentheses shows number of rules selected from all extracted rules.

R_I and R_D. It would help us to understand the data and rules derived from them (Table 4). For numerical values we used a word 'i/j' for showing ith of possible j levels. We compared the rules in measurement granularity and those in discernible granularity. Here, we used C4.5 [11] as the rule extraction method. We calculated the true positive (TP) rate (the ratio of positive samples correctly classified to all positive samples) and the false positive (FP) rate (the ratio of negative samples incorrectly classified to all negative samples).

The results are shown in Figs. 4 and 5. The value of TP rate decreases in the order of R_M,R_I, and R_D. While the easiness of interpretation increase reversely in the order R_M, R_I, and R_D. It is also noted that a larger variety of rules obtained in R_I or R_D or both compared with in R_M.

5   Conclusions

We have discussed on appropriate roughness of description using "granularity", for raising up the understandability of data and rules. We defined some types of granularity on the basis of discernibility criteria, and proposed an algorithm to find near-optimal solutions. In experiments, it has shown that reduced granularity helped us to read/understand the data and the derived rules. In addition, such a way enables us to give some "abstract words" for data, which evokes even our "Kansei."

References

1. Nagamachi, M.: Kansei Engineering: A new ergonomic consumer-oriented technology for product development. International Journal of Industrial Ergonomics **15** (1995) 3–11.
2. Tsutsumi, K., Ito, N., and Hashimoto, H.: A Development of the Building Kansei Information Retrieval System. Xth International Conference on Computing in Civil and Building Engineering, June 02-04 (2004) 174–175.
3. Yanagisawa, H., Fukuda, S.: Interactive Reduct Evolutional Computation for Aesthetic Design. Journal of Computing and Information Science in Engneering **5**(1) (2005) 1–7.
4. Pawlak, Z.: Rough Sets: Theoretical Aspects of Reasoning about Data. Kluwer Academic Publishers (1991).
5. Ziarko, W.: Variable Precision Rough Set Model. Journal of Computer and System Sciences **46** (1993) 39–59.
6. Muto, Y., Kudo, M.: Discernibility-Based Variable Granularity and Kansei Representations. Rough Sets, Fuzzy Sets, Data Mining, and Granular Computing (RSFDGrC 2005), Lecture Notes in Artificial Intelligence, **3641** (2005) 692–700.
7. Muto, Y., Kudo, M., and Murai, T.: Reduction of Attribute Values for Kansei Representation. Journal of Advanced Computational Intelligence and Intelligent Informatics **10**(5) (2006) 666–672.
8. Zadeh, L. A.: Fuzzy Sets and Information Granularity. Advances in Fuzzy Set Theory and Applications, (1979) 3–18.
9. Lin, T. Y.: Granular Computing on Binary Relation I : Data Mining and Neighborhood Systems. II : Rough Set Representations and Belief Functions. In L. Polkowski and A. Skowron (eds.), Rough Sets in Knowledge Discovery 1 : Methodology and Applications, Physica-Verlag, (1998) 107–121, 122–140.
10. Murai, T., Resconi, G., Nakata, M., and Sato, Y.: Granular Reasoning Using Zooming In & Out: Part 1. Propositional Reasoning. G.Wang et al.(eds.), Rough sets, Fuzzy sets, Data mining, and Granular Computing, LNAI **2639** (2003) 421–424.
11. Quinlan, J. R. : C4.5: Programs for Machine Learning. MK, San Mateo, CA, (1993)
12. Newman, D. J., Hettich, S., Blake, C. L., and Merz, C.J.: UCI Repository of machine learning databases. [http://www.ics.uci.edu/ mlearn/ MLRepository.html]. Irvine, CA: University of California, Department of Information and Computer Science (1998)

Semi-structured Decision Rules in Object-Oriented Rough Set Models for Kansei Engineering

Yasuo Kudo[1] and Tetsuya Murai[2]

[1] Dept. of Computer Science and Systems Eng., Muroran Institute of Technology
Mizumoto 27-1, Muroran 050-8585, Japan
kudo@csse.muroran-it.ac.jp
[2] Graduate School of Information Science and Technology, Hokkaido University
Kita 14, Nishi 9, Kita-ku, Sapporo 060-0814, Japan
murahiko@main.ist.hokudai.ac.jp

Abstract. Decision rule generation from Kansei data using rough set theory is one of the most hot topics in Kansei engineering. Usually, Kansei data have various types of scheme, however, Pawlak's "traditional" rough set theory treats structured data mainly, that is, decision tables with fixed attributes and no hierarchy among data. On the other hand, Kudo and Murai have proposed the object-oriented rough set model which treats structural hierarchies among objects. In this paper, we propose semi-structured decision rules in the object-oriented rough set model to represent structural characteristics among objects, which enable us to consider characteristics of hierarchical data by rough sets.

Keywords: semi-structured decision rules, object-oriented rough sets, hierarchical data.

1 Introduction

Prof. Pawlak's rough set theory [5,6] provides useful framework of approximation and reasoning about data. In the research field of Kansei engineering, various methods of decision rule generation from Kansei data using rough set theory are widely studied [4]. Usually, Kansei data have various types of scheme, however, "traditional" rough set theory treats structured data mainly, that is, decision tables with fixed attributes and no hierarchy among data. On the other hand, Kudo and Murai have proposed the object-oriented rough set model (for short, OORS) [2], and also proposed decision rule generation in OORS [3]. The object-oriented rough set model introduces object-oriented paradigm (cf. [1]) to the "traditional" rough set theory, which illustrates hierarchical structures between classes, names and objects based on is-a and has-a relationships. Moreover, decision rules in OORS illustrates characteristic combination of objects as parts of some objects. However, in the previous paper [3], hierarchical characteristics among objects are not represented in decision rules sufficiently.

J.T. Yao et al. (Eds.): RSKT 2007, LNAI 4481, pp. 219–227, 2007.

In this paper, we propose semi-structured decision rules in the object-oriented rough set model to represent hierarchical characteristics among objects, which enable us to consider characteristics of hierarchical data by rough sets.

2   The Object-Oriented Rough Set Model

We briefly review the object-oriented rough set model. Note that the contents of this section are entirely based on the authors' previous papers [2,3].

2.1   Class, Name, Object

OORS consists of the following three triples: a *class structure* $\mathcal{C}$, a *name structure* $\mathcal{N}$ and an *object structure* $\mathcal{O}$, respectively:

$$\mathcal{C} = (C, \ni_C, \sqsupseteq_C), \quad \mathcal{N} = (N, \ni_N, \sqsupseteq_N), \quad \mathcal{O} = (O, \ni_O, \sqsupseteq_O),$$

where C, N and O are finite and disjoint non-empty sets such that $|C| \leq |N|$ ($|X|$ is the cardinality of X). Each element $c \in C$ is called a *class*. Similarly, each $n \in N$ is called a *name*, and each $o \in O$ is called an *object*. The relation $\ni_X$ ($X \in \{C, N, O\}$) is an acyclic binary relation on X, and the relation $\sqsupseteq_X$ is a reflexive, transitive, and asymmetric binary relation on X. Moreover, $\ni_X$ and $\sqsupseteq_X$ satisfy the following property:

$$\forall x_i, x_j, x_k \in X, \ x_i \sqsupseteq_X x_j, \ x_j \ni_X x_k \ \Rightarrow \ x_i \ni_X x_k.$$

These three structures have the following characteristics, respectively:

- The class structure illustrates abstract data forms and those hierarchical structures based on part / whole relationship (has-a relation) and specialized / generalized relationship (is-a relation).
- The name structure introduces numerical constraint of objects and those identification, which provide concrete design of objects.
- The object structure illustrates actual combination of objects.

Two relations $\ni_X$ and $\sqsupseteq_X$ on $X \in \{C, N, O\}$ illustrate hierarchical structures among elements in X. The relation $\ni_X$ is called a *has-a relation*, and $x_i \ni_X x_j$ means "x_i has-a x_j", or "x_j is a part of x_i". For example, $c_i \ni_C c_j$ means that "the class c_i has a class c_j", or "c_j is a part of c_i". On the other hand, the relation $\sqsupseteq_X$ is called an *is-a relation*, and $x_i \sqsupseteq_X x_j$ means that "x_i is-a x_j". For example, $\sqsupseteq_C$ illustrates relationship between superclasses and subclasses, and $c_i \sqsupseteq_C c_j$ means that "c_i is a superclass of c_j", or "c_j is a subclass of c_i".

2.2   Well-Defined Structures

Each object $o \in O$ is defined as an instance of some class $c \in C$, and the class of o is identified by the *class identifier* function. The class identifier id_C is a *p-morphism* between $\mathcal{O}$ and $\mathcal{C}$ (cf. [7], p.142), that is, the function $id_C : O \longrightarrow C$ satisfies the following conditions:

1. $\forall o_i, o_j \in O, o_i \ni_O o_j \Rightarrow id_C(o_i) \ni_C id_C(o_j)$.
2. $\forall o_i \in O, \forall c_j \in C, id_C(o_i) \ni_C c_j \Rightarrow \exists o_j \in O$ s.t. $o_i \ni_O o_j$ and $id_C(o_j) = c_j$,

and the same conditions are also satisfied for $\sqsupseteq_O$ and $\sqsupseteq_C$. $id_C(o) = c$ means that the object o is an instance of the class c.

The object structure $\mathcal{O}$ and the class structure $\mathcal{C}$ are also connected through the name structure $\mathcal{N}$ by the *naming function* $nf : N \longrightarrow C$ and the *name assignment* $na : O \longrightarrow N$. The naming function provides names to each class, which enable us to use plural instances of the same class simultaneously. On the other hand, the name assignment provides names to every objects, thus we can treat objects by using their names.

Formally, the naming function $nf : N \longrightarrow C$ is a surjective p-morphism between $\mathcal{N}$ and $\mathcal{C}$, and satisfies the following *name preservation constraint*:

- For any $n_i, n_j \in N$, if $nf(n_i) = nf(n_j)$, then $H_N(c|n_i) = H_N(c|n_j)$ is satisfied for all $c \in C$,

where $H_N(c|n) = \{n_j \in N \mid n \ni_N n_j, f(n_j) = c\}$ is the set of names of c that n has. These characteristics of the naming function nf imply that (1) there is at least one name for each class, (2) the name structure reflects all structural characteristics of the class structure, and (3) all names of the parts of any class are uniquely determined.

On the other hand, the name assignment $na : O \longrightarrow N$ is a p-morphism between $\mathcal{O}$ and $\mathcal{N}$, and satisfies the following *uniqueness condition*:

- For any $x \in O$, if $H_O(x) \neq \emptyset$, the restriction of na into $H_O(x)$:
  $na|_{H_O(x)} : H_O(x) \longrightarrow N$ is injective,

where $H_O(x) = \{y \in O \mid x \ni_O y\}$ is the set of objects that x has. $na(x) = n$ means that the name of the object x is n. The uniqueness condition requires that all distinct parts $y \in H_O(x)$ have different names.

We say that $\mathcal{C}, \mathcal{N}$ and $\mathcal{O}$ are *well-defined* if and only if there exist a naming function $nf : N \longrightarrow C$ and a name assignment $na : O \longrightarrow N$ such that $id_C = nf \circ na$, that is, $id_C(x) = nf(na(x))$ for all $x \in O$.

In well-defined structures, if a class c_i has m objects of a class c_j, then any instance o_i of the class c_i has exactly m instances $o_{j1}, \cdots, o_{jm}$ of the class c_j [2]. This good property enables us the following description for clear representation of objects. Suppose we have $o_1, o_2 \in O$, $n_1, n_2 \in N$, and $c_1, c_2 \in C$ such that $o_1 \ni_O o_2$, and $na(o_i) = n_i$, $nf(n_i) = c_i$ for $i \in \{1, 2\}$. We denote $o_1.n_2$ instead of o_2 by means of "the instance of c_2 named n_2 as a part of o_1".

2.3   Indiscernibility Relations in OORS

We consider indiscernibility relations in OORS based on the concept of *equivalence as instances*. In [2], to evaluate equivalence of instances, an indiscernibility relation $\sim$ on O are recursively defined as follows:

$$x \sim y \iff \begin{array}{l} x \text{ and } y \text{ satisfy the following two conditions:} \\ 1. \ id_C(x) = id_C(y), \text{ and,} \\ 2. \ \begin{cases} x.n \sim y.n, \ \forall n \in H_N(na(x)) & \text{if } H_N(na(x)) \neq \emptyset, \\ Val(x) = Val(y) & \text{otherwise,} \end{cases} \end{array} \qquad (1)$$

where $H_N(na(x))$ is the set of names that $na(x)$ has. $Val(x)$ is the "value" of the "value object" x. Because C is a finite non-empty set and $\ni_C$ is acyclic, there is at least one class a such that a has no other class c. We call such class a an *attribute*, and for any instance x of the attribute a, we call x a *value object* of a. The value object x of a represents a "value" of the attribute a. Moreover, we assume that we can compare "values" of value objects of the same attribute.

$x \sim y$ means that the object x is equivalent to the object y as an instance of the class $id_C(x)$. Using the indiscernibility relation $\sim$, an indiscernibility relation $\sim_B$ with respect to a given subset $B \subseteq N$ of names is defined as follows:

$$x \sim_B y \iff \begin{array}{l} x \text{ and } y \text{ satisfy the following two conditions:} \\ 1 \ B \cap H_N(na(x)) = B \cap H_N(na(y)), \text{ and,} \\ 2. \ \forall n[n \in B \cap H_N(na(x)) \Rightarrow x.n \sim y.n]. \end{array} \qquad (2)$$

$x \sim_B y$ means that x and y are equivalent as instances of the class $id_C(x)$ in the sense that, for all $n \in B \cap H_N(na(x))$, x and y have equivalent instances of the class $id_C(x.n)$. Equivalence classes $[x]_{\sim_B}$ by $\sim_B$ are usually defined.

2.4   Decision Rules in OORS

Decision rules in OORS illustrate characteristic combination of objects as parts of some objects. Decision rules in OORS are defined as follows [3]:

$$c \wedge c.n_1 \sim x.n_1 \wedge \cdots \wedge c.n_i \sim x.n_i \Rightarrow c.m_1 \sim x.m_1 \wedge \cdots \wedge c.m_j \sim x.m_j, \quad (3)$$

where $c \in C$, $x \in O$ such that $id_C(x) = c$, $n_1, \cdots, n_i \in N_{CON} \cap H_N(na(x))$ $(i \geq 0)$ and $m_1, \cdots, m_j \in N_{DEC} \cap H_N(na(x))$ $(j \geq 1)$. N_{CON} is the set of names that may appear in antecedents of decision rules (called condition names), and N_{DEC} is the set of names that may appear in conclusions of decision rules (called decision names). Note that $N = N_{CON} \cup N_{DEC}$ and $N_{CON} \cap N_{DEC} = \emptyset$. We call this rule a *decision rule of the class c by the object x*, and denote $DR(c; x)$.

The decision rule $DR(c; x)$ means that, for any object $y \in O$, if y is an instance of c and each part $y.n_k$ is equivalent to $x.n_k$ $(k \leq i)$, then all parts $y.m_l$ are also equivalent to $x.m_l$ $(l \leq j)$, respectively. Thus, $DR(c; x)$ describes a certain property about combination of objects as an instance of the class c.

Example 1. We need to check three desk-top personal computers (PCs) and three lap-top PCs about the price (p: low or high), maker (m: X or Y) and clock (c: low or high) of CPUs, and size (s: A4 notebook size or B5 notebook size, only lap-top PCs), and find some rules about prices of these PCs. Here, we consider the following class structure $\mathcal{C} = (C, \ni_C, \sqsupseteq_C)$ with $C = \{$PC, LPC, Price, Size, CPU, Maker, Clock$\}$, where Price, Size, Maker and Clock are attributes. Next, the

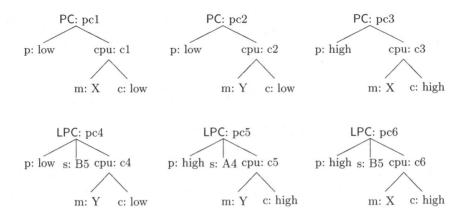

Fig. 1. Has-a relationship in Example 1

name structure is $\mathcal{N} = (N, \ni_N, \sqsupseteq_N)$ with $N = \{$pc, lpc, price, size, cpu, maker, clock$\}$. Finally, the object structure is $\mathcal{O} = (O, \ni_O, \sqsupseteq_O)$ such that all objects appear in Fig. 1. Suppose that these structures are well-defined.

Figure 1 illustrates has-a relationship between objects, and assignment of names to objects. For example, the object "pc1" is an instance of the PC class, and has an instance "low"(=pc1.price) of the attribute Price, and an instance "c1"(=pc1.cpu) of the CPU class. Moreover, the object "c1" has an instance "X"(=pc1.cpu.maker) of the attribute Maker, and an instance "low"(=pc1.cpu. clock) of the attribute Clock.

Equivalence classes about the set $N_{DEC} = \{$price$\}$ of names defined by (2) represents the following two decision classes about the price of PCs:

$$[pc1]_{N_{DEC}} = \{pc1, pc2, pc4\}, \ [pc3]_{N_{DEC}} = \{pc3, pc5, pc6\},$$

where $[pc1]_{N_{DEC}}$ is the set of low price PCs, and $[pc3]_{N_{DEC}}$ is the set of high price PCs, respectively. Similarly, equivalence classes by $B = \{$cpu$\}$ are as follows:

$$[pc1]_B = \{pc1\}, \ [pc2]_B = \{pc2, pc4\}, [pc3]_B = \{pc3, pc6\}, [pc5]_B = \{pc5\}.$$

Thus, using these equivalence classes, we get the following six decision rules:

- $DR(PC, pc1)$: PC$\wedge$ PC.cpu $\sim$ c1 $\Rightarrow$ PC.price $\sim$ low,
- $DR(PC, pc2)$: PC$\wedge$ PC.cpu $\sim$ c2 $\Rightarrow$ PC.price $\sim$ low,
- $DR(PC, pc3)$: PC$\wedge$ PC.cpu $\sim$ c3 $\Rightarrow$ PC.price $\sim$ high,
- $DR(LPC, pc4)$: LPC$\wedge$ LPC.cpu $\sim$ c4 $\Rightarrow$ LPC.price $\sim$ low,
- $DR(LPC, pc5)$: LPC$\wedge$ LPC.cpu $\sim$ c5 $\Rightarrow$ LPC.price $\sim$ high,
- $DR(LPC, pc6)$: LPC$\wedge$ LPC.cpu $\sim$ c6 $\Rightarrow$ LPC.price $\sim$ high.

In Fig. 1, it is obvious that price of PCs and clock of CPUs have clear relationship, that is, if the clock is low (high), then price is low (high). However, these decision rules can not represent this relationship sufficiently. We consider that this is because decision rules in OORS (and indiscernibility relations defined by (2)) can only treat *equivalence of parts which consist in directly*. Thus, when we consider the equivalence of PCs, we can only treat value objects of the Price attribute and instances of the CPU class, and can not treat value objects of the Clock attribute. Therefore, to represent structural characteristics by decision rules, we need to treat hierarchical structures among objects recursively.

3   Semi-structured Decision Rules in OORS

In this section, we propose *semi-structured decision rules* in OORS to represent hierarchical characteristics among objects by decision rules. First, we introduce a concept of *consistent sequences of names* which illustrate hierarchical structures in the given name structure correctly, and define an indiscernibility relation based on such sequences. Next, using the indiscernibility relation, we propose semi-structured decision rules in OORS.

3.1   Indiscernibility Relations by Consistent Sequences of Names

Definition 1. *Let* C, $\mathcal{N}$ *and* $\mathcal{O}$ *be well-defined class, name and object structures, respectively. A sequence of names* $n_1. \cdots .n_k$ *with length* k $(k \geq 1)$ *such that* $n_i \in N$ $(1 \leq i \leq k)$ *is called a consistent sequence of names if and only if either* (1) $k = 1$, *or* (2) $k \geq 2$ *and* $n_{j+1} \in H_N(n_j)$ *for each name* n_j $(1 \leq j \leq k-1)$. *We denote the set of all consistent sequences of names in* $\mathcal{N}$ *by* N^+.

Hereafter, we concentrate consistent sequences of names. Note that all names $n \in N$ are consistent sequences with length 1, thus we have $N \subset N^+$.

Consistent sequences of names have the following good property, which guarantees that consistent sequences describe hierarchical structures among objects correctly.

Proposition 1. *For any consistent sequence of names* $n_1. \cdots .n_k \in N^+$ $(k \geq 1)$ *and any object* $x \in O$, *if* $n_1 \in H_N(na(x))$, *then there exist* $x_i \in O$ $(1 \leq i \leq k)$ *such that* $na(x_i) = n_i$ *and* $x_j \ni_O x_{j+1}$ $(0 \leq j \leq k-1)$, *where* $x = x_0$.

By this property, we can easily check that, for any object x and any consistent sequence $n_1. \cdots .n_k$, whether the sequence $n_1. \cdots .n_k$ "connects" to the object x, and find the object $y(= x.n_1. \cdots .n_k)$ by tracing the has-a relation $\ni_O$ such that $x \ni_O \cdots \ni_O y$.

Using consistent sequences of names, we define indiscernibility relations which directly treat "nested" parts of objects as follows.

Definition 2. *Let* $S \subseteq N^+$ *be a non-empty set of consistent sequences of names. We define a binary relation* $\approx_S$ *on* O *as follows:*

$$x \approx_S y \iff \begin{array}{l} x \text{ and } y \text{ satisfy the following two conditions:} \\ 1.\ \forall n_1. \cdots .n_k \left[\begin{array}{l} n_1. \cdots .n_k \in S \Rightarrow \\ \{n_1 \in H_N(na(x)) \Leftrightarrow n_1 \in H_N(na(y))\} \end{array} \right], \\ 2.\ \forall n_1. \cdots .n_k \left[\begin{array}{l} n_1. \cdots .n_k \in S \text{ and } n_1 \in H_N(na(x)) \\ \Rightarrow x.n_1. \cdots .n_k \sim y.n_1. \cdots .n_k \end{array} \right], \end{array} \quad (4)$$

where $\sim$ is the indiscernibility relation defined by (1).

It is easy to check that $\approx_S$ is an equivalence relation.

Theorem 1. *The relation $\approx_S$ defined by (4) is an equivalence relation on O.*

The condition 1 in (4) requires that the object x and y concern the same sequences in S, which means that x and y have the same structure partially illustrated by such sequences. The condition 2 requires that, for all sequences $n_1 \cdots .n_k \in S$ that connect both x and y, the object $x.n_1 \cdots .n_k$ as a nested part of x is equivalent to the object $y.n_1 \cdots .n_k$ as a nested part of y.

We intend that the relation $\approx_S$ is an extension of the indiscernibility relation $\sim_B$ by using sequences of names. Actually, Proposition 2 below shows that $\sim_B$ is a special case of $\approx_B$.

Proposition 2. *Suppose $B \subseteq N$ is a non-empty set of names, that is, a set of consistent sequences of names with length 1, $\sim_B$ is the indiscernibility relation defined by (2), and $\approx_B$ is the indiscernibility relation defined by (4). Then, $x \approx_B y$ if and only if $x \sim_B y$ for any objects $x, y \in O$.*

3.2 Semi-structured Decision Rules

Using consistent sequences of names and indiscernibility relations defined by (4), we propose semi-structured decision rules in OORS. Similar to the case of decision rules in OORS, we need to divide the set of consistent sequences of names N^+ into the set N^+_{CON} of consistent sequences which may appear in antecedents and the set N^+_{DEC} of consistent sequences which may appear in conclusions. Note that $N^+ = N^+_{CON} \cup N^+_{DEC}$ and $N^+_{CON} \cap N^+_{DEC} = \emptyset$. The set N^+_{DEC} provides decision classes as equivalence classes based on the equivalence relation $\approx_{N^+_{DEC}}$. We define semi-structured decision rules in OORS as follows.

Definition 3. *A semi-structured decision rule in the object-oriented rough set model has the following form:*

$$c \wedge c.n_{11} \cdots .n_{1k_1} \sim x.n_{11} \cdots .n_{1k_1} \wedge \cdots \wedge c.n_{s1} \cdots .n_{sk_s} \sim x.n_{s1} \cdots .n_{sk_s}$$
$$\Rightarrow c.m_{11} \cdots .m_{1l_1} \sim x.m_{11} \cdots .m_{1l_1} \wedge \cdots \wedge c.m_{t1} \cdots .m_{tl_t} \sim x.m_{t1} \cdots .m_{tl_t}$$
$$(5)$$

where $c \in C$, $x \in O$ such that $id_C(x) = c$, $n_{i1} \cdots .n_{ik_i} \in N^+_{CON}$ $(1 \le i \le s, k_i \ge 1)$, $m_{j1} \cdots .m_{jl_j} \in N^+_{DEC}$ $(1 \le j \le t, l_j \ge 1)$, and all sequences which appear in (5) have to connect to the object x. We call this rule a semi-structured decision rule of the class c by the object x, and denote $SSDR(c;x)$.

The semi-structured decision rule $SSDR(c;x)$ means that, for any object $y \in O$, if y is an instance of c and each part $y.n_{i1} \cdots .n_{ik_i}$ $(1 \le i \le s)$ is equivalent to $x.n_{i1} \cdots .n_{ik_i}$, then all parts $y.m_{j1} \cdots .m_{jl_j}$ $(1 \le j \le t)$ are also equivalent to $x.m_{j1} \cdots .m_{jl_j}$, respectively. Thus, $SSDR(c;x)$ describes certain characteristics of combination of parts of x even though the parts do not consist in x directly, which enable us to represent hierarchical characteristics of parts by decision rules.

Example 2. We consider to generate semi-structured decision rules from well-defined structures illustrated in Example 1. The set N_{CON}^+ is constructed as follows:

$$N_{CON}^+ = (N \setminus \{\text{price}\}) \cup \left\{ \begin{array}{l} \text{pc.price, pc.cpu, pc.cpu.maker, pc.cpu.clock,} \\ \text{lpc.price, lpc.size, lpc.cpu, lpc.cpu.maker,} \\ \text{lpc.cpu.clock, cpu.maker, cpu.clock} \end{array} \right\}.$$

Equivalence classes based on the set $N_{DEC}^+ = \{\text{price}\}$ are identical to the case of the equivalence classes by (2) in Example 1. On the other hand, equivalence classes about the set $S = \{\text{cpu.clock}\}$ are as follows:

$$[\text{pc1}]_S = \{\text{pc1, pc2, pc4}\}, \quad [\text{pc3}]_S = \{\text{pc3, pc5, pc6}\}.$$

Therefore, using these equivalence classes, we get the following four semi-structured decision rules:

- $SSDR(\text{PC, pc1})$: PC$\wedge$ PC.cpu.clock $\sim$ low $\Rightarrow$ PC.price $\sim$ low,
- $SSDR(\text{PC, pc3})$: PC$\wedge$ PC.cpu.clock $\sim$ high $\Rightarrow$ PC.price $\sim$ high,
- $SSDR(\text{LPC, pc4})$: LPC$\wedge$ LPC.cpu.clock $\sim$ low $\Rightarrow$ LPC.price $\sim$ low,
- $SSDR(\text{LPC, pc5})$: LPC$\wedge$ LPC.cpu.clock $\sim$ high $\Rightarrow$ LPC.price $\sim$ high.

Consequently, from these rules, we can interpret the characteristic "if the clock of CPU is low (high), then the price of PC is low (high)."

4   Conclusion

In this paper, we have proposed semi-structured decision rules in OORS based on consistent sequences of names. Semi-structured decision rules represent certain characteristics of combination of parts even though the parts do not consist in directly, and consistent sequences of names capture hierarchical characteristics of objects correctly. Development of algorithms to find consistent sequences of names which appear in antecedents of semi-structured decision rules, and application of rule generation methods of semi-structured decision rules to various Kansei data are important future issues.

References

1. Budd, T. A.: An Introduction of Object-Oriented Programming, 2nd Edition. Addison Wesley Longman (1997)
2. Kudo, Y. and Murai, T.: A Theoretical Formulation of Object-Oriented Rough Set Models. Journal of Advanced Computational Intelligence and Intelligent Informatics, **10**(5) (2006) 612–620
3. Kudo, Y. and Murai, T.: A Method of Generating Decision Rules in Object-Oriented Rough Set Models. Rough Sets and Current Trends in Computing, LNAI **4259**, Springer (2006) 338–347

4. Mori, N., Tanaka, H. and Inoue, K.: Rough Sets and Kansei —Knowledge Acquisition and Reasoning from Kansei Data—. Kaibundo (2004) (in Japanese)
5. Pawlak, Z.: Rough Sets. International Journal of Computer and Information Science, **11** (1982) 341–356
6. Pawlak, Z.: Rough Sets: Theoretical Aspects of Reasoning about Data. Kluwer Academic Publisher (1991)
7. Popkorn, S.: First Steps in Modal Logic. Cambridge University Press (1994)

Functional Data Analysis and Its Application

Masahiro Mizuta[1] and June Kato[2]

[1] Hokkaido University, Japan
mizuta@iic.hokudai.ac.jp
[2] NTT, Japan
kato.jun@lab.ntt.co.jp

Abstract. In this paper, we deal with functional data analysis including functional clustering and an application of functional data analysis. Functional data analysis is proposed by Ramsay *et al.* In functional data analysis, observed objects are represented by functions. We give an overview of functional data analysis and describe an actual analysis of *Music Broadcast Data* with functional clustering.

Keywords: Functional Clustering, Music Broadcast Data, Power Law.

1 Introduction

Most methods for data analysis assume that the data are sets of numbers with structure. For example, typical multivariate data are identified as a set of n vectors of p real numbers and dissimilarity data on pairs of n objects are as $n \times n$ matrix. However, requests for analysis of data with new models become higher, as the kind and quantity of the data is increased. In concert with the requests, Ramsay *et al.* proposed Functional Data Analysis (FDA) in the 1990's, which treats data as functions. Many researchers proposed various methods for functional data, including functional regression analysis, functional principal components analysis, functional clustering, functional multi dimensional scaling *etc.* We can study these methods from http://www.psych.mcgill.ca/misc/fda/, [16] and [17].

Data mining has been a field of active research and is defined as the process of extracting useful and previously unknown information out of large complex data sets. Cluster analysis is a powerful tool for data mining. The purpose of cluster analysis is to find relatively homogeneous clusters of objects based on measured characteristics. There are two main approaches of cluster analysis: hierarchical clustering methods and nonhierarchical clustering methods. Single Linkage is a kind of hierarchical clustering and is a fundamental method. k-means is a typical nonhierarchical clustering method. We have studied clustering methods for functional data: functional single linkage [10], functional k-means [11] etc.

In this paper, we deal with functional data analysis, especially functional clustering methods with an actual example.

J.T. Yao et al. (Eds.): RSKT 2007, LNAI 4481, pp. 228–235, 2007.

2   Functional Data and Functional Data Analysis

Functions are dealt with as data in FDA and have domain and range. We discuss about *functions* from the viewpoint of domain and range. It is possible to understand it from considering the argument of functional data to be the time axis, in the case that the domain is one dimensional real number space. When the range is one dimensional real space, the functional data represent the value that changes timewise. When the range is two or three dimensional Euclidean space, the functional data can be taken as the motion on the plane or the space respectively. Most studies on FDA use these kinds of functional data. When the dimension of the domain is more than one, i.e. multidimensional arguments, there are many issues to deal with functional data. In the section 22.3 (Challenges for the future) of [17], they show the topics on it.

There are many techniques in FDA including functional regression analysis [3] [4] [18], functional principal components analysis, functional discriminant analysis, functional canonical correlation and functional clustering [1][19][20]. We can get excellent lists of bibliography on FDA from [17] and http://www.psych.mcgill.ca/misc/fda/bibliography.html. We have proposed several methods for functional data: functional multidimensional scaling [8], extended functional regression analysis [18] etc. Functional data can be considered as infinity-dimensional data. Most methods in the book (Ramsay and Silverman [17]) for functional data are based on an approximation with finite expansions of the functions with basis functions. Once the functional data can be thought as finite linear combinations of the basis functions, functional data analysis methods for the functional data are almost the same as those of conventional data analysis methods. But, there is some possibility of using different approaches.

3   Clustering and Functional Clustering

Sometimes, we divide methods of cluster analysis into two groups: hierarchical clustering methods and nonhierarchical clustering methods.

Hierarchical clustering refers to the formation of a recursive clustering of the objects data: a partition into two clusters, each of which is itself hierarchically clustered. We usually use data set of dissimilarities or distances; $S = \{s_{ij}; i, j = 1, 2, \ldots, n\}$, where s_{ij} are dissimilarity between object i and object j, and n is the size of the objects. Single linkage is a typical hierarchical clustering method. We start with each object as a single cluster and in each step merge two of them together. In each step, the two clusters that have minimum distance are merged. At the first step, the distances between clusters are defined by the distance between the objects. However, after this, we must determine the distances between new clusters. In single linkage, the distance between two clusters is determined by the distance of the two closest objects in the different clusters. The results of hierarchical clustering are usually represented by dendrogram. The height of dendrogram represents the distances between two clusters. It is well known that the result of Single Linkage and the minimal spanning tree (MST) are equivalent from the computational point of view[5].

Nonhierarchical clustering partitions the data based on a specific criterion. Most nonhierarchical clustering methods do not deal with a set of dissimilarities directly. They use a set of p-dimensional : $X = \{\boldsymbol{x}_i; i = 1, 2, \ldots, n\}$. k-means method is a kind of nonhierarchical clustering and is to choose initial seeds points and assign the cases to them using a minimum of the trace of within variances.

We will describe clustering methods for functional data. Typically, there are two types of functional data to be applied to clustering methods. The first type of functional data is p-dimensional functions. The other type is a set of dissimilarity functions. We denote the functions for n objects depending on a variable t as $X(t) = \{\boldsymbol{x}_i(t)\}(i = 1, 2, \ldots, n)$ and denote dissimilarity data as $S(t) = \{s_{ij}(t); i, j = 1, 2, \ldots, n\}$, where $s_{ij}(t)$ are dissimilarity functions between object i and object j. We also use the notations $X = \{\boldsymbol{x}_i\}$ and $S = \{s_{ij}\}$ for ordinary data.

3.1   Using Conventional Clustering Methods

Simple idea for clustering methods for functional data is to transform functional data to ordinary data and we apply conventional clustering methods to these data.

There are several methods to derive ordinary data $S = \{s_{ij}\}$ or $X = \{\boldsymbol{x}_i\}$ from functional dissimilarity data $S(t)$ or p-dimensional functional data $X(t)$. The most natural method may be to use integration in the domain or max operator:

$$s_{ij} = \int \| \boldsymbol{x}_i(t) - \boldsymbol{x}_j(t) \|^2 \, dt,$$

$$s_{ij} = \int s_{ij}(t) dt,$$

$$s_{ij} = max_t s_{ij}(t),$$

$$\boldsymbol{x}_i = \int \boldsymbol{x}_i(t) dt.$$

Then we can use ordinary clustering method to set of dissimilarity data $S = \{s_{ij}\}$ or p-dimensional data $X = \{\boldsymbol{x}_i\}$.

3.2   Functional k-Means Method

Functional k-means method is proposed in this section. We assume that we have p-dimensional functional data $X(t) = \{\boldsymbol{x}_i(t)\}(i = 1, 2, \ldots, n)$. It is realistic that values $X(t_j), j = 1, 2, \ldots, m$ are given. We restrict ourselves to two dimensional functional data and the one dimensional domain. The number of clusters k is prespecified as a user parameter.

The idea of functional k-means method is repetitive procedure of the conventional k-means method. At first, we apply conventional k-means method to $X(t_j)$ for each t_j. Then, we adjust labels of clusters. Even if we fix clustering method, there is freedom of labeling. We denote $C_i(t)$ as the cluster label of the i-th object at t for fixed K. We discuss about adjusting the cluster label-ing of $C_i(t_2)$ with fixed $C_i(t_1)$. $C_i^*(t_2)$ are new cluster labels of the object that $\Sigma_i \sharp\{C_i(t_1) = C_i^*(t_2)\}$ takes the maximum value, where $\sharp$ indicates the size of the set. A simple method for adjusting the cluster labels is to use the cluster centers of the previous clustering for initial guesses for the cluster centers.

We must pay attention to the fact that even if two objects belong to the same cluster at t_j, it is possible that the two objects belong to different clusters at t'_j.

The results of the proposed functional k-means method can be represented graphically. We use the three dimensional space for the representation. Two dimensions are used for the given functional data, one is for t, and the clusters are shown with colors. If we may use dynamic graphics, the results can be analyzed more effectively with interactive operations: rotations, slicing, zooming etc. We can analyze the result deeply.

3.3   Functional Single Linkage Method

We introduce the algorithm of single linkage for functional dissimilarity data $S(t) = \{s_{ij}(t); i, j = 1, 2, \ldots, n\}$. The basic idea of this method is that we apply conventional single linkage to $S(t)$ and get functional MST, say MST(t). Then we calculate functional configuration and adjust labels of objects. We mention the detail of these two steps in the following.

We must construct a functional configuration of vertices of MST(t); functional multidimensional scaling [8] is useful to get them. The MST(t) is represented on the 2 dimensional space. It is possible to represent the results of functional single linkage for the given number of clusters K. Especially when we can use dynamic graphical display, the representation is much interactive.

But, even if we fix hierarchical clustering method, there is freedom of labeling of clusters. We adjust the cluster labels used the same method as functional k-means.

3.4   Moving Functional Clustering

In the previous sections, we deal with clustering method for functional data. When we must analyze one dimensional functional data $x_i(t)(i = 1, \cdots, n)$, functional k-means method or functional single linkage method can be applicable formally. But, these methods do not use multidimensional information of the data. We propose a method for clustering for one dimensional functional data.

In order to actively use the information from the functional data, we apply *windows* of the domain of the functions. We define the dissimilarities functions of the two functions:

$$s_{ij}(t)_d = \int_{t-d}^{t+d} (x_i(u) - x_j(u))^2 du.$$

These $s_{ij}(t)_d$ are called moving dissimilarities functions with windows range d. $s_{ij}(t)_d$ represent the degree of the closeness between functions $x_i(t)$ and $x_j(t)$ in the interval $[t - d, t + d]$. We can apply functional single linkage clustering, in the subsection 3.3, to these functional dissimilarities $s_{ij}(t)_d$. The parameter d affects the $s_{ij}(t)_d$ and the clusters as results. But, in general, cluster analysis is an exploratory method. So the freedom of d is important for finding interesting results. For example, when we analysis stocks data, we can see interesting structures using $d = 3$ (days).

We assumed the dimension of the domain of the functions is one. It is easy to extend this method to more than one dimensional domain. The integration domain in the definition of $s_{ij}(t)_d$ is changed to multidimensional neighborhood; sphere or hyper sphere.

4   Actual Data Analysis: Music Broadcast Data

We will show an example of functional clustering with actual data. *Music Broadcast Data* are collected from Japanese seven FM stations in Tokyo during 2003. There are about 350000 records on broadcasting contained playlist ID, music ID and title, start time, end time etc. Frequency of broadcast of musics for each day is changing under some pattern. The frequency is usually increasing until the day of release and takes peak some later date, then decreasing (Figure 1). Figure 2 shows the double logarithmic plot of ranking versus rate of broadcast and reveals Power law.

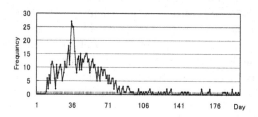

Fig. 1. Typical pattern of frequency of broadcast

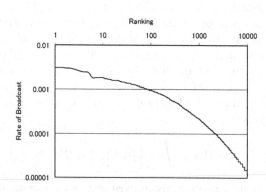

Fig. 2. Plot of ranking versus rate of broadcast

The top 100 musics (Figure 3) are applied Functional clustering described in subsection 4.1. The number of cluster is 4 and we adopted hierarchical clustering. Figure 4 shows four clusters.

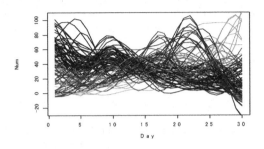

Fig. 3. Top 100 musics (Dec. 2003

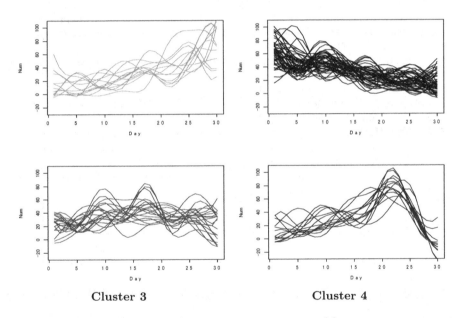

Fig. 4. Result of functional clustering [7]

5   Concluding Remarks

We mainly discussed about clustering methods for functional data. We would summarize the method of functional clustering.

If you would like to use conventional clustering methods, the functional data are transformed to the ordinal data set with integration on t or maximum operation. When the functional dissimilarities data $S(t)$ are given, functional single linkage method is available. The results of functional clustering can be represented by functional minimum spanning tree (MST). When p-dimensional functions $X(t)$ are given, functional k-means is a candidate for the method. Specially when $p = 1$, moving functional clustering is helpful in analysis.

There is a strong possibility of extending the concept of functional data analysis. For example, *mapping* is a generalization of function. Formally, we can treat the contingency table as discrete functional data. These concepts may also help to improve original functional data analysis.

References

1. Abraham, C., Cornillon, P.A., Matzner-Lober, E., Molinari, N.: Unsupervised curve clustering using B-splines. Scand. J. Statist, **30** (2003) 581–595
2. Bosq, D.: Linear processes in functional spaces: theory and applications. Lecture Notes in Statistics **149**, New York: Springer-Verlag (2000)
3. Cardot, H., Ferraty, F., Sarda, P.: Spline estimators for the functional linear model. Statistica Sinica, **13**, (2003) 571–591
4. Ferraty, F., Vieu, P.: Nonparametric models for functional data, with application in regression, time-series prediction and curve discrimination. Nonparametric Statistics, **16(1-2)** (2004) 111–125
5. Gower, J.C., Ross,G.J.S.: Minimum spanning trees and single linkage cluster analysis. Appl. Stat., **18** (1969) 54–64
6. Hiro, S., Komiya, Y., Minami, H., Mizuta, M.: An application of relative projection pursuit for functional data to human growth. Proceedings in Computational Statistics 2006, Physica-Verlag, A Springer Company (2006) 1113–1120
7. Hoshika, H., Kato, J., Minami, H., Mizuta, M.: Analysis of Music Broadcast Data with functional clustering (in Japanese). Proceedings in Japanese Computational Statistics (2006) 43–46
8. Mizuta, M.: Functional multidimensional scaling. Proceedings of the Tenth Japan and Korea Joint Conference of Statistics (2000) 77–82
9. Mizuta, M.: Cluster analysis for functional data. Proceedings of the 4th Conference of the Asian Regional Section of the International Association for Statistical Computing (2002) 219–221
10. Mizuta, M.: Hierarchical clustering for functional dissimilarity data. Proceedings of the 7th World Multiconference on Systemics, Cybernetics and Informatics, Volume **V** (2003) 223–227
11. Mizuta, M.: K-means method for functional data. Bulletin of the International Statistical Institute, 54th Session, Book **2** (2003) 69–71
12. Mizuta, M.: Clustering methods for functional data. Proceedings in Computational Statistics 2004, Physica-Verlag, A Springer Company (2004) 1503–1510
13. Mizuta, M.: Multidimensional scaling for dissimilarity functions with several arguments. Bulletin of the International Statistical Institute, 55th Session (2005) p.244
14. Mizuta, M.: Discrete functional data analysis. Proceedings in Computational Statistics 2006, Physica-Verlag, A Springer Company (2006) 361–369
15. Nason, G.P.: Functional Projection Pursuit. Computing Science and Statistics, **23** (1997) 579–582,http://www.stats.bris.ac.uk/~guy/Research/PP/PP.html
16. Ramsay, J.O., Silverman, B.W.: Applied Functional Data Analysis – Methods and Case Studies –. New York: Springer-Verlag (2002)

17. Ramsay, J.O., Silverman, B.W.: Functional Data Analysis. 2nd Edition. New York: Springer-Verlag (2005)
18. Shimokawa, M., Mizuta, M., Sato, Y.: An expansion of functional regression analysis (in Japanese). Japanese Journal of Applied Statistics **29-1** (2000) 27–39
19. Tarpey, T., Kinateder, K.K.J.: Clustering functional data. J. of Classification, **20** (2003) 93–114
20. Tokushige,S., Inada,K., Yadohisa,H.: Dissimilarity and related methods for functional data. Proceedings of the International Conference on New Trends in Computational Statistics with Biomedical Applications (2001) 295–302
21. Yamanishi, Y., Tanaka, Y.: Geographically weighted functional multiple regression analysis: A numerical investigation. Proceedings of the International Conference on New Trends in Computational Statistics with Biomedical Applications (2001) 287–294

Evaluation of Pictogram Using Rough Sets

Kei Masaki and Hisao Shiizuka

Graduate School of Informatics, Kogakuin University
Nishishinjuku 1-24-2, Shinjuku-ku, Tokyo 163-8677, Japan
shiizuka@cc.kogakuin.ac.jp

Abstract. In this paper, we consider pictograms from a point of view of Kansei engineering and rough sets. We propose an evaluation method to the understanding level of and the pictogram to good quality further. The feature of this text is to use rough sets for the method. In this text that treats information obtained from a lot of pictograms, using rough-set-based-rule extraction. The targeted pictogram used is chart signs for a standard guide "Locker room."

Keywords: Kansei engineering, Pictogram, Rough sets.

1 Introduction

In all situations of our life, to give neither inconvenience nor the unpleasantness, and to compose an easy, safe, comfortable commodity and space, the universal design attracts attention. The center concern might be "A lot of people availably design the product, the building, and the space." As for pictogram considered here, it is one of the information transmission means when seeing from the aspect of such a universal design, and the utility will be expected more and more in the contemporary society in the future.

The pictogram is a means to be translated into "Emoticon" and "Picture word", etc., and to visually convey information by figure. As for the pictogram, when the pictogram was adopted for the first time as a visual information that showed the athletic event in Tokyo Olympics in 1964, the pictogram etc. that expressed the rest room were made from the advantage "It was possible to impart information frankly regardless of the age or the learning environment, etc." in the Osaka exposition in 1970 in Japan. Various the one was made, and it was used by various scenes when came to be used widely in Japan after 1980. As for the pictogram that had been made in this age, the one not understood easily because standardization had not been done in one side was not few though there were a lot of good quality one for the sight person to understand the expression object easily.

The Ministry of Land, Infrastructure and Transport and the traffic ecology mobility foundation standardized the pictogram to solve such a problem in 2001, and 125 chart-sign items for a standard guide[1] were enacted. The pictogram from which there was no standard up to now by this standardization and it was judged to be difficult to obtain information became a good quality information transmission means with high

---

[1] See http://www.ecomo.or.jp/symbols/, for the chart sign guideline of a standard guide.

J.T. Yao et al. (Eds.): RSKT 2007, LNAI 4481, pp. 236–243, 2007.

understanding level at which sight person's experience and knowledge were put to use. However, the one with a difficult expression of transmission information in these standardized pictograms is not few. Moreover, it is proven that a lot of pictograms with low understanding level exist from understanding level investigation of the chart-sign for a standard guide.

This paper proposes the pictogram generally used based on such a current state now and it proposes the evaluation method to the understanding level of and the pictogram to good quality further. The feature of this text is to use the rough set for the evaluation method. In this text that treats information obtained from a lot of pictograms, using the rough set that is the rule extraction means that a lot of in formation can be analyzed becomes a major plus. Moreover, proposing a new expression standard of the pictogram from which it is evaluated that the rule in the if-then form is extracted concerning the validity of straightening and the standardized pictogram, and has been thought the expression to be difficult before becomes possible.

Moreover, the method by annexation was first done by using the rough set in this text, and it analyzed it by the combination table based on the result. It is a purpose to improve the reliability of the evaluation by using two or more analysis methods. The targeted pictogram used chart sign for a standard guide "Locker room".

First, in Section 2, the decision and making the pictogram targeted by this paper are shown. Next, in Section 3, the method of the understanding level investigation that uses the pictogram is shown. In Section 4, then, the process of extracting the decision rule of the rough set is shown. Then, in Section 5, the decision rule obtained as a result is shown. From these, in Section 6, the evaluation and consideration are actually done. As a result, it is shown that "Locker room" fills three of four extracted attribute values. In addition, the attribute value of the remainder showed the numerical value with the highest column score in the expression of figure. When this result images "Locker room", it is shown to become a very big factor. It is shown that it is a pictogram with low understanding level in "Locker room" because it lacks the attribute value from this.

Next, the guess by the rough set that is another possibility is considered in seven. The pictogram that filled all of the four extracted attribute values was made figure, and it compared, and it investigated, saying that chart sign for a standard guide "Locker ro om". As a result, it has been understood that the understanding level of the pictogram derived by the guess is higher. In a word, it was clarified that the pictogram guessed from the rough set was a good quality information transmission means with high understanding level.

2 Decision and Making of Pictogram

In this Section, the evaluated pictogram is decided. The targeted pictogram is assumed to be the one to meet the following requirement.

    (*a*) The one that understanding level is low in understanding level investigation.
    (*b*) The one that expression object is not adopted in figure as it is.
    (*c*) The one that doesn't depend on knowledge and experience alone.
    (*d*) The one that figure can be operated.
    (*e*) The one frequently used in public domain.

As for the pictogram that does not correspond to this condition, because the factor that depends on knowledge and the experience increases when transmission information is understood, the act "The expression object is guessed from the feature of figure" is not extracted easily. Moreover, figure in the pictogram should be the one replaced by something related to transmission information. Therefore, the pictogram that expresses "Locker room" standardized by chart-sign for a standard guide is targeted as a pictogram that meets the requirement of (a)-(e) in described the above.

"Locker room" is made to be imaged indirectly by pictogram. "Locker room" is not drawing the expression object in figure, and drawing the suspender and the person. Moreover, it is easy to be simple figure, and to change to various figures. It is thought that the understanding level is a pictogram not comparatively understood easily for the side estimated to be 50% low in investigation concerning the understanding level[2]. It is necessary to make two or more pictograms used by the understanding level investigation to use the rough set in this text. When making it, figure used in the pictogram was assumed to be five "Locker", "Suspender", "Towel", "Person", and "Chair" based on "Image of the locker room" that had been obtained from the image research into "Locker room". The pictogram of 16 kinds of the combination was made and these figures were used for the understanding level investigation. Moreover, it referred to chart sign guideline for a standard guide when the pictogram was made.

3 Understanding Level Investigation

The understanding level investigation used 16 pictograms shown in Section 2. Each pictogram was presented, and the answer form was assumed to be *"(1) It was easy to understand"* and *"(2) It is not easy to understand"* by the one to ask whether of each being able to image "Locker room". It was thought it was improper in a case this time when the understanding level was asked and omitted it though it thought the selection item "There was not both" was added to give answered continuousness when the rough set was treated.

The understanding level investigation was done to 20 people (The boy: 16 people and girl: 4 people) in the total of the man and woman in his/her twenties.

4 Rough Sets

4.1 Decision of Attributes and Attribute Values

The index of the attribute was a feature of figure described by three and information on the easiness of figure to see. Moreover, "Side availability" was installed as an evaluation to the visual check level so that the pictogram had to make the other party intuitively understand transmission information. The attribute and the attribute value are shown in Table 1.

Table 1. Attribute and attribute value list

Attribute	Attribute value	Sign
Locker	There is a display.	R1
	There is no display.	R2
Suspender	There is a display.	H1
	There is no display.	H2
Towel	There is a display.	T1
	There is no display.	T2
Chair	Display under the left	C1
	Display under the right	C2
	There is no display.	C3
Man	Left display	P1
	Right display	P2
	There is no display.	P3
Number of figures	One	G1
	Two	G2
	Three	G3
Side availability	20% or less	U1
	20%~35%	U2
	35%~40%	U3
	40%~45%	U4
	45% or more	U5

4.2 Decision Rule

It applies it to the attribute value in which each 16 pictograms used to investigate is requested by 4.1. Afterwards, the result of the survey is assumed to be a decision attribute and the decision table is made. Here, the decision attribute is assumed to be two kinds of "Y=1: It borrows the solution easily" and "Y=2: It is incomprehensible".

The individual decision rule conditional part is extracted based on the obtained decision table for 20 to decision attribute Y=1 and Y=2.

5 Analysis of Data and Its Result

We analyze it based on the decision rule conditional part obtained in the previous Section. In this paper, annexation, combination table, and two analysis methods are used. Because a different analysis method is used, the output result is different though both are the techniques that the large number of people decision rule extracts from an individual decision rule. The reliability of the evaluation is improved by using these two kinds of analysis methods.

5.1 Analysis Result by Annexation

The individual decision rule conditional part of 20 people who obtained it in the previous Section is annexed to large number of people's decision rule. Annexation brings the output individual rule together, and shows how much each decision rule conditional part contributes to others.

Table 2 shows the rule of the output annexation rule and the C.I. value (high rank).

Table 2. Condition part of annexation rule and C.I. value

Y=1	
Annexation rule	S.C.I value
C3P1T2U2	0.777777
G2H1R1U5	0.611111
C3P1R1U2	0.611111
H1P1T2U2	0.444444
G2H1R1U1	0.444444
G3P1R1U3	0.388888
P1R1T2U3	0.388888
C2P1R1U3	0.388888
G2H1P3T2U5	0.388888
C3G3P3R1T2	0.388888
C3G2P3T2U5	0.388888
C3G3P3T2U2	0.388888
C3P3R1T2U5	0.388888
C3P3R1T2U4	0.388888
C3H1P3R1U5	0.388888

Table 3. Combination table: result

Attribute	Attribute value	Y=1
Locker	It displays and it exists: R1.	4.986508
	There is no display: R2.	2.555952
Towel	It displays and it exists: T1.	0.833333
	There is no display: T2.	5.853283
Suspender	It displays and it exists: H1.	4.010318
	There is no display: H2.	1.267532
Person	Left display: P1	1.528571
	Right display: P2	2.685714
	There is no display: P3.	1.616667
Chair	Right display: C1	0
	Left display: C2	0.4
	There is no display: C3.	3.861833
Number of figures	One: G1	0.5
	2: G2	2.075902
	3: G3	1.761905
Side availability	20% or less: U1.	1.028571
	20%~35%:U2	0
	35%~40%:U3	0.4
	40%~45%U4	1.495238
	45% or more: U5.	0
Combination rate		44.4444
Column score threshold		2.787041
Distribution score threshold		0.628817

5.2 Analysis Result by Combination Table

The analysis method by the combination table requests large number of people's decision rule from individual information as well as annexation.

First of all, it calculates the column score based on the CI value to the decision rule. In addition, it is a technique for analyzing each attribute value based on them based on the value for the threshold of the column score and the threshold of the distribution score. Table 3 shows the analysis result of decision attribute Y=1 (It borrows the solution easily) obtained according to the decision rule conditional part and the CI value.

Table 4. Combination pattern

Attribute value	Combination pattern
R1	R1H1P2C3
T2	T2R2H2P2C3G2
H1	R1H1
C3	C3R1T2

6 Evaluation and Consideration

In the analysis result in 5.1, it annexes it among extracted the annexation rules C.I . The value high and 15 titles are adopted and evaluated. As for the reason, the following two are thought: (a) Because the number of annexation rules increases extremely when the rule after that is evaluated; (b) Because reliability is scarce because the annexed CI value is low. The condition of requesting it is evaluated from the pictogram of the chart sign for the standard guide shown in Figure 1 based on these of information.

Fig. 1. Pictogram that corresponds to annexation

The obtained annexation rule is examined as follows. We examined the annexation rule that the value of S.C.I. is high. As a result, it has been understood that the annexation rule (C3P1T2U2) for S.C.I.=077 corresponds to the pictogram shown in figure 1. This means the answer that 70 percent investigated is comprehensible. Moreover, it has been understood that there are three criteria to make "Locker room" imaged from other annexation rules, as follow.

(1) The expression of the chair is excluded from figure.
(2) The number of figures must be two or more.
(3) Put the suspender in figure.

The pictogram used for the chart sign for a standard guide satisfies these three conditions, as follows from Figure 1.

The attribute value whose column score calculated from Table 3 is higher than the threshold 3.5 is four "R1", "T2", "H1", and "C3". In this evaluation table where the combination table was used, the one to meet four requirements is a pictogram with high understanding level. The attribute value of "R1" is not satisfied though the pictogram of Figure 1 satisfies three attribute values. Moreover, the corresponding pattern was not output from the combination pattern in three attribute values that corresponded to Figure 1. Chart sign for a standard guide "Locker room" is evaluated in consideration of these two results. This pictogram has raised the visual check level by making figure simple. It makes "Locker room" it imaged by using comprehensible figure for person, which consists of hanger and person. This adopts the suspender in figure, and transmits information with a difficult expression of point well raising the visual check level. However, it is necessary to expand information in figure more than the improvements of the visual check level to make "Locker room" imaged like understanding resultants of 5.2. Therefore, it is judged, the cause with low understanding level of this pictogram is from "Information is a little in the chart".

7   Proposal of Pictogram by Guess

It is not easy to pack transmission information in figure small as understood from the evaluation result. However, this problem can be solved by using the guess that is the feature of the rough set. The guess extracts, and makes the pictogram that agrees with the condition by the annexation rule and the combination table obtained by six. The pictogram is made based on the condition when not corresponding. Figure 2 is given as a pictogram that meets the above-mentioned requirement.

Fig. 2. Example of guessed pictogram

These pictograms were the one that "5.1: an annexation rule and 3 standards necessary to make it image" and "5.2: the combination pattern to 4 attribute values and 4 attribute values in the combination table" is filled, and the pictogram at the right of Figure 2 succeeded when these understanding levels were investigated in obtaining an understanding level that was higher than the pictogram of Figure 1. In a word, it is thought that the condition of a good quality pictogram can be derived by using the rough set for the feature extraction of the pictogram.

8 Summary

In this paper, it proposed to use the rough set for the evaluation method of the pictogram. When figure used is expressed because the pictogram is an information transmission means to depend only on the sight, it is necessary to consider Canonical View. It becomes easy to image information from 2 to 3 dimensions by correcting this respect, and the improvement of a further understanding level can be expected.

The device will be done repeatedly to the feature extraction method in the future. Not a single pictogram but two or more pictograms are evaluated from various angles. It has in a different expression though the same chart is adopted when the same chart is used in the pictogram with a different expression object and it has arrived to cite instances. Moreover, it is necessary to propose a good quality pictogram with high understanding level about other pictograms with a difficult expression.

References

1. Mori,N., TanakamH., and Inoue,K.: Rough Sets and Kansei, Kaibundo (2004) (in Japanese)
2. Chart sign of symbol sign for standard guide that understands at one view, Traffic Ecology Mobility Foundation
3. Kaiho,H.: Psychology Technique of Expression That Understands from Document, Diagram, and Illustration glance, Kyoritsushuppan (1992) (in Japanese)

A Logical Representation of Images by Means of Multi-rough Sets for Kansei Image Retrieval

Tetsuya Murai[1], Sadaaki Miyamoto[2], and Yasuo Kudo[3]

[1] Division of Computer Science
Graduate School of Information Science and Technology,
Hokkaido University, Sapporo 060-0814, Japan
murahiko@main.ist.hokudai.ac.jp
[2] Department of Risk Engineering
Faculty of Systems and Information Engineering
University of Tsukuba, Ibaraki 305-8573, Japan
miyamoto@risk.tsukuba.ac.jp
[3] Department of Computer Science and Systems Engineering
Muroran Institute of Technology, Muroran 050-8585, Japan
kudo@csse.muroran-it.ac.jp

Abstract. In this paper, we examine a logical representation of images by means of multi-rough sets. We introduce a level of granularization into a color space and then define approximations of images. Further we extend the idea to conditionals in images, which is expected to give a basis of image indexing and retrieval by images themselves.

Keywords: Multi-rough sets, Representation of images, Approximation of images, Conditionals between images, Kansei retrieval.

1 Introduction

Rough set theory proposed by Pawlak[5] is now regarded as providing one of the most theoretically elegant and powerful tool of dealing with granularity in various areas of science and technology. Recently Miyamoto[3] has studied multisets and rough sets. Further Miyamoto et al.[4] has described multi-roughness in a clear way in information systems as well as tables in relational databases[6]. In this paper, we examine a logical representation of images themselves in the framework as an application of Miyamoto's concepts of multi-rough sets.

2 Multi-rough Sets

We give a brief review on multi-rough sets according to Miyamoto[3,4].

2.1 Information Tables

Let $\mathcal{A}$ be a fixed set of m attributes, say, $\mathcal{A} = \{a_1, \cdots, a_m\}$. Assume that each attribute a_i has its corresponding domain D_i. Then, any subset T in the

J.T. Yao et al. (Eds.): RSKT 2007, LNAI 4481, pp. 244–251, 2007.

Cartesian product $D_1 \times \cdots \times D_n$ is called an *information table* with respect to $\mathcal{A}$. An element in an information table is called a *tuple* of the table. For any tuple $t \in \mathcal{T}$, denote its i-th component by $t(a_i)$, then t can be written as $t = (t(a_1), \cdots, t(a_m))$.

2.2 Projection

For any non-empty subset $A = \{a_{i_1}, \cdots, a_{i_r}\} \subseteq \mathcal{A}$, we can define the projection of a tuple t with respect to A by

$$t(A) \stackrel{\text{def}}{=} (t(a_{i_1}), \cdots, t(a_{i_r})).$$

Further it can be extended for a set of tuples $T \subseteq \mathcal{T}$ as

$$T(A) \stackrel{\text{def}}{=} \{t(A) \mid t \in T\}.$$

Then, we have the following proposition.

Proposition 1. ([4]) $t \in T \Longrightarrow t(A) \in T(A)$.

Note that the converse does not necessarily hold[1].

2.3 Multisets

As well known in database theory[6], the operation of projection usually induces *multisets*. In general, a multiset $\widetilde{T}$ in a universe U is characterized by the count function

$$Ct_{\widetilde{T}}(x) : U \to \mathbb{N},$$

where $\mathbb{N}$ is the set of natural numbers including 0. We also use the following notation of multisets as

$$\widetilde{T} = \{\frac{Ct_{\widetilde{T}}(x)}{x} \mid x \in U\}.$$

By $\widetilde{T} \widetilde{\subseteq} U$, we mean $\widetilde{T}$ is a multiset in U.

For any non-empty subset $A \subseteq \mathcal{A}$ of attributes, we can define an equivalence relation R_A by

$$tR_At' \stackrel{\text{def}}{\Longleftrightarrow} t(A) = t'(A)$$

for any t, $t' \in \mathcal{T}$. Then we have the quotient set $\mathcal{T}/R_A$ defined by

$$\mathcal{T}/R_A = \{[t]_{R_A} \mid t \in \mathcal{T}\},$$

where $[t]_{R_A}$ is the equivalence class of t with respect to R_A:

$$[t]_{R_A} = \{t' \in \mathcal{T} \mid tR_At'\} = \{t' \in \mathcal{T} \mid t(A) = t'(A)\}.$$

Then we define a multiset $\widetilde{T}(A)$ in T characterized by

$$Ct_{\widetilde{T}(A)}(t(A)) = |\,[t]_{R_A}|,$$

where $|\cdot|$ is a number of elements. If every $[t]_{R_A}$ is a singleton, then $T(A) = \widetilde{T}(A)$.

---

[1] See Example 1 in Section 3.

2.4 Approximation

For arbitrary subset of tuples $T \subseteq \mathcal{T}$, its lower and upper approximations can be defined by

$$[A]T \overset{\text{def}}{=} \{t \in \mathcal{T} \mid [t]_{R_A} \subseteq T\},$$

$$\langle A \rangle T \overset{\text{def}}{=} \{t \in \mathcal{T} \mid [t]_{R_A} \cap T \neq \emptyset\}.$$

Then we have the following characterizations of the apporximations:

Proposition 2. ([4])

$$[A]T = \{t \in \mathcal{T} \mid t(A) \in T(A) - T^C(A)\},$$

$$\langle A \rangle T = \{t \in \mathcal{T} \mid t(A) \in T(A)\},$$

$$\langle A \rangle T - [A]T = \{t \in \mathcal{T} \mid t(A) \in T(A) \cap T^C(A)\},$$

where $T^C(A) = \{t(A) \mid t \in T^C\}$.

Assume we have already had a multiset $\widetilde{T}$ in $\mathcal{T}$ by another equivalence relation $R_{A'}$ with respect to a set of attributes $A' \subseteq \mathcal{A}$. Miyamoto[3] proposed the following definitions of approximations:

$$[A]\widetilde{T} \overset{\sim}{\subseteq} \mathcal{T} : Ct_{[A]\widetilde{T}}(t) \overset{\text{def}}{=} \min\{ \mid [t']_{R_A} \mid \mid t' \in [t]_{R_{A'}} \},$$

$$\langle A \rangle \widetilde{T} \overset{\sim}{\subseteq} \mathcal{T} : Ct_{\langle A \rangle \widetilde{T}}(t) \overset{\text{def}}{=} \max\{ \mid [t']_{R_A} \mid \mid t' \in [t]_{R_{A'}} \}.$$

In many application areas, in general, the crisp definition of approximations is too strict and, to solve the problem, Ziarko[7] proposed *variable precision rough set models*, in which graded approximations are defined by

$$[A]_\alpha T \overset{\text{def}}{=} \{t \in \mathcal{T} \mid \text{Inc}([t]_{R_A}, T) \geq \alpha\},$$

$$\langle A \rangle_\alpha T \overset{\text{def}}{=} \{t \in \mathcal{T} \mid \text{Inc}([t]_{R_A}, T^C) < \alpha\},$$

where $\text{Inc}(X, Y)$ is a inclusion measure such as, say, $\text{Inc}(X, Y) = |X \cap Y|/|X|$. We have $[A]_1 T = [A]T$ and $\langle A \rangle_1 T = \langle A \rangle T$.

2.5 Conditionals

Following Chellas's conditional logics[1], for two subsets $T, T'(\subseteq \mathcal{T})$ of tuples, we can define a *conditional* by

$$T' \Rightarrow T \overset{\text{def}}{=} \{t \in \mathcal{T} \mid f(T', t) \subseteq T\},$$

where f is a selection function

$$f : 2^{\mathcal{T}} \times \mathcal{T} \to 2^{\mathcal{T}}.$$

The simplest example of a selection function is

$$f(T, t) = T.$$

Table 1. Images as information tables

(a)

X	Y	R	G	B
0	0	255	255	255
0	1	255	253	0
......				
$w-1$	$h-2$	0	0	255
$w-1$	$h-1$	0	0	255

(b)

ID	R	G	B
0	255	255	255
1	255	253	0
......			
$wh-2$	0	0	255
$wh-1$	0	0	255

Similarly we can introduce variable precision model:

$$T' \Rightarrow_\alpha T \stackrel{\text{def}}{=} \{t \in \mathcal{T} \mid \text{Inc}(f(T', t), T) \geq \alpha\}.$$

By the definition, obviously we have

$$(T' \Rightarrow_1 T) = (T' \Rightarrow T),$$
$$\alpha \leq \beta \ (\alpha, \beta \in (0, 1]) \Longrightarrow (T' \Rightarrow_\beta T) \subseteq (T' \Rightarrow_\alpha T).$$

3   Logical Representation of Images

3.1   Images as Information Tables

A color image data of RGB 24 bit/pixel can be regarded as an information table with the attributes of positions (X, Y) and colors (R, G, B). If the size of a given image is $w \times h$, where w is the width of the image and h is a height, then

$$D_X = \{x \in \mathbb{N} \mid 0 \leq x < w - 1\},$$
$$D_Y = \{y \in \mathbb{N} \mid 0 \leq y < h - 1\},$$

where $\mathbb{N}$ is the set of natural numbers including 0. And, for colors, say

$$D_R = D_G = D_B = \{c \in \mathbb{Z} \mid 0 \leq c < 2^8 = 256\}.$$

An example of a color image data is shown in Table 1(a) where two attributes X and Y are the candidate key of the table in Table 1(a). By formula $ID = w * X + Y$, we can replace each position (X, Y) by its one to one corresponding ID number whose domain is $D_{ID} = \mathbb{N}$. It is also the candidate key[2] in Table 1(b). Such an example is shown in Table 1(b). We can use the ID number as an index number for each tuple like $t_0 = (0, 255, 255, 255)$.

Definition 1. *Let us define the* atomic color space $\mathbb{C}$ *by*

$$\mathbb{C} \stackrel{\text{def}}{=} D_R \times D_G \times D_B.$$

---

[2] In the example in Table 1(b), the domain of ID numbers is actually $\{i \in \mathbb{N} \mid 0 \leq i < wh\}$.

Then a color image img *is represented as an information table in* $\mathbb{N} \times \mathbb{C}$ *with respect to* $\mathcal{A} = \{ID, R, G, B\}$:

$$\text{img} \subseteq \mathbb{N} \times \mathbb{C}.$$

3.2   Projection and Approximation of Images

By the operation of projection[3], we have a subset or multiset of img in the atomic color space $\mathbb{C}$:

$$\text{img}(RGB) = \{t(RGB) \mid t \in \text{img}\} \subseteq \mathbb{C},$$
$$\widetilde{\text{img}}(RGB) \widetilde{\subseteq} \mathbb{C} : Ct_{\widetilde{\text{img}}(RGB)}(t(RGB)) = |\,[t]_{R_{RGB}}|.$$

Further, by confining ourselves to a subset $A \subseteq RGB$, we have the following lower approximations:

$$[A]\text{img} \stackrel{\text{def}}{=} \{t \in \text{img} \mid [t]_{R_A} \subseteq \text{img}\},$$
$$\langle A\rangle\text{img} \stackrel{\text{def}}{=} \{t \in \text{img} \mid [t]_{R_A} \cap \text{img} \neq \emptyset\}.$$

Example 1. Consider the following image as an information table:

ID	R	G	B
0	255	0	0
1	255	255	0
2	255	255	0
3	255	0	0
4	255	0	255

Then for a subset img$' = \{t_0, t_1, t_2\}(\subseteq$ img$)$, *we have*

$$\text{img}'(RGB) = \{(255,0,0), (255,255,0), (255,0,255)\} \subseteq \text{img}(RGB) \subseteq \mathbb{C}.$$

We can easily see that $t_3 \notin$ img$'$, *but* $t_3(RGB) = (255,0,0) \in$ img$'(RGB)$. *Next, we define an equivalence relation* R_{RGB}, *then*

$$\text{img}/R_{RGB} = \{\,\{t_0, t_3\},\ \{t_1, t_2\},\ \{t_4\}\,\},$$

and thus

$$\widetilde{\text{img}}'(RGB) = \{\frac{2}{(255,0,0)},\ \frac{2}{(255,255,0)},\ \frac{1}{(255,0,255)}\} \subseteq \widetilde{\text{img}}(RGB) \widetilde{\subseteq} \mathbb{C}.$$

Further, for $A = GB$, *we have*

$$[GB]\,\text{img}' = \{t_1, t_2\},$$
$$\langle GB\rangle\text{img}' = \{t_0, t_1, t_2, t_3\}.$$

---

[3] We use the usual abbreviations for subsets in database theory, for instance, RG for $\{R, G\}$.

Original   Lower apporximations

$\alpha = 0.01$ $\alpha = 0.011$ $\alpha = 0.017$ $\alpha = 0.0245$

Original   Lower approximations

$\alpha = 0.01$ $\alpha = 0.011$ $\alpha = 0.017$ $\alpha = 0.0245$

Fig. 1. Examples of images and their lower apporximations (black part) with level 2 of base granularity and several precision value $\alpha = 0.01$, 0.011, 0.017, and 0.0245

3.3   Granularization of the Atomic Color Space

We can introduce another kind of granularity of colors into the atomic color space $\mathbb{C}$. Let us consider a division of each color component R, G, amd B into equally n parts[4]. That is, we divide $\mathbb{C}$ equally into $n \times n \times n$ parts and let R_n be the equivalence relation on $\mathcal{T}$ that corresponds to the partition. And define

$$\mathbb{C}_n \stackrel{\text{def}}{=} \mathbb{C}/R_n,$$

and let $\mathbb{C}_n = \{C_1, \cdots, C_{n \times n \times n}\}$. We call n a *level* of *base* granularity.

We can define two kinds of approximations[5] of a given image img :

$$[\text{img}]_n \stackrel{\text{def}}{=} \{C \in \mathbb{C}_n \mid C \subseteq \text{img}(RGB)\},$$

$$\langle \text{img} \rangle_n \stackrel{\text{def}}{=} \{C \in \mathbb{C}_n \mid C \cap \text{img}(RGB) \neq \emptyset\}.$$

Example 2. We show several examples of lower approximation for images in Figure 1.

---

[4] Of course, we can give different level of granularity, say, (n_r, n_g, n_b) to red, green, and blue, respectively.

[5] To be more precise, they should be denoted as $[R_n]\text{img}(RGB)$ and $\langle R_n \rangle \text{img}(RGB)$, respectively.

3.4 Conditionals Between Images

Then for given two images img and img′, we can formulate the following four kinds of conditionals:

$$[\text{img}']_n \Rightarrow [\text{img}]_n, \quad [\text{img}']_n \Rightarrow \langle\text{img}\rangle_n, \quad \langle\text{img}'\rangle_n \Rightarrow [\text{img}]_n, \quad \langle\text{img}'\rangle_n \Rightarrow \langle\text{img}\rangle_n$$

under a given level n of the base granularity.

In this paper, we take the first type of conditional as a first step, and we define the following two indices between of two images img and img′:

$$\text{COND}_n(\text{img}, \text{img}') \overset{\text{def}}{=} \sup_{\alpha \in (0,1]}\{\alpha \mid [\text{img}']_n \Rightarrow_\alpha [\text{img}]_n\},$$

$$\text{REL}_n(\text{img}, \text{img}') \overset{\text{def}}{=} \min(\text{COND}_n(\text{img}, \text{img}'), \text{COND}_n(\text{img}, \text{img}')).$$

These indices are useful to make some ranking among images.

Example 3. Given four images and their lower appoximations in Fig. 2, we can calculate indices of COND_n and REL_n between the images in Table 2.

Four original images

Lower approximations of the four original images

Fig. 2. Examples of four images and their lower approximation

Table 2. Degrees of conditionals and relevance between images

(a) COND(img, img').

	1	2	3	4
1	1.0000	0.3547	0.3786	0.4589
2	0.3449	1.0000	0.3671	0.3315
3	0.3614	0.3604	1.0000	0.2837
4	0.3801	0.2824	0.2462	1.0000

(b) REL(img, img').

	1	2	3	4
1	1.0000	0.3449	0.3614	0.3801
2		1.0000	0.3604	0.2824
3			1.0000	0.2462
4				1.0000

4   Concluding Remarks

In this paper, we presented a basic framework of dealing with image data by means of multi-rough sets. We are now planning a Kansei-based component of images retrieval systems in the framework. In the component user-system interaction is carried out directly by using images relevant to users' Kansei, and without any word like index terms. In fact, we have many tasks for the plan and among them:

– Indexing and retrieval of images *by images themselves.*
– Meaning and applications of rules extracted from images as information tables.

Finally it should be noted that manipulation of several images as information tables at the same time is also related with the concept of multi-agent rough sets proposed by Inuiguchi[2].

References

1. Chellas, B.F.: Modal Logic: An Introduction. Cambridge University Press (1980).
2. Inuiguchi, M:, Rule Induction Based on Multi-Agent Rough Sets. Proceedings of Computational Intelligence and Industrial Applications, Jinan University Press (2006) 320-327.
3. Miyamoto, S.: Generalizations of Multisets and Rough Approximations, *International Journal of Intelligent Systems*, **19** (2004) 639-652.
4. Miyamoto, S., Murai, T., Kudo, Y.: A Family of Polymodal Systems and Its Application to Generalized Possibility Measures and Multi-Rough Sets, *Journal of Advanced Computational Intelligence and Intelligent Informatics*, **10** (2006) 625-632.
5. Pawlak, Z.: Rough Sets: Theoretical Aspects of Reasoning about Data, Kluwer Academic Publishers (1991).
6. Ullman, J.D.: Principles of Database and Knowledge-base Systems, Volume 1: Classical Database Systems, Computer Science Press (1988).
7. Ziarko, W.: Variable Precision Rough Set Model, *Journal of Computer and System Sciences*, **46**(1993), 39–59.

A Batch Rival Penalized EM Algorithm for Gaussian Mixture Clustering with Automatic Model Selection

Dan Zhang[1] and Yiu-ming Cheung[2]

[1] Faculty of Mathematics and Computer Science
HuBei University, WuHan, China
mathzhang52@yahoo.com.cn
[2] Department of Computer Science
Hong Kong Baptist University, Hong Kong SAR, China
ymc@comp.hkbu.edu.hk

Abstract. Cheung [2] has recently proposed a general learning framework, namely Maximum Weighted Likelihood (MWL), in which an adaptive Rival Penalized EM (RPEM) algorithm has been successfully developed for density mixture clustering with automatic model selection. Nevertheless, its convergence speed relies on the value of learning rate. In general, selection of an appropriate learning rate is a nontrivial task. To circumvent such a selection, this paper further studies the MWL learning framework, and develops a batch RPEM algorithm accordingly provided that all observations are available before the learning process. Compared to the adaptive RPEM algorithm, this batch RPEM need not assign the learning rate analogous to the EM, but still preserve the capability of automatic model selection. Further, the convergence speed of this batch RPEM is faster than the EM and the adaptive RPEM. The experiments show the efficacy of the proposed algorithm.

Keywords: Maximum Weighted Likelihood, Rival Penalized Expectation-Maximization Algorithm, Learning Rate.

1 Introduction

As a typical statistical technique, clustering analysis has been widely applied to a variety of scientific areas such as data mining, vector quantization, image processing, and so forth. In general, one kind of clustering analysis can be formulated as a density mixture clustering problem, in which each mixture component represents the probability density distribution of a corresponding data cluster. Subsequently, the task of clustering analysis is to identify the dense regions of the input (also called *observation* interchangeably) densities in a mixture.

In general, the Expectation-Maximum (EM) algorithm [3] provides an efficient way to estimate the parameters in a density mixture model. Nevertheless, it needs to pre-assign a correct number of clusters. Otherwise, it will almost always lead to a poor estimate result. Unfortunately, from the practical viewpoint,

J.T. Yao et al. (Eds.): RSKT 2007, LNAI 4481, pp. 252–259, 2007.

it is hard or even impossible to know the exact cluster number in advance. In the literature, one promising way is to develop a clustering algorithm that is able to perform a correct clustering without pre-assigning the exact number of clusters. Such algorithms include the RPCL algorithm [4] and its improved version, namely RPCCL[1]. More recently, Cheung [2] has proposed a general learning framework, namely Maximum Weighted Likelihood (MWL), through which an adaptive Rival Penalized EM (RPEM) algorithm has been proposed for density mixture clustering. The RPEM learns the density parameters by making mixture component compete each other at each time step. Not only are the associated parameters of the winning density component updated to adapt to an input, but also all rivals' parameters are penalized with the strength proportional to the corresponding posterior density probabilities. Subsequently, this intrinsic rival penalization mechanism enables the RPEM to automatically select an appropriate number of densities by fading out the redundant densities from a density mixture. The numerical results have shown its outstanding performance on both of synthetic and real-life data. Furthermore, a simplified version of RPEM has included RPCL and RPCCL as its special cases with some new extensions.

In the papers [2], the RPEM algorithm learns the parameters via a stochastic gradient ascending method, i.e., we update the parameters immediately and adaptively once the current observation is available. In general, the adaptiveness of the RPEM makes it more applicable to the environment changed over time. Nevertheless, the convergence speed of the RPEM relies on the value of learning rate. Often, by a rule of thumb, we arbitrarily set the learning rate at a small positive constant. If the value of learning rate is assigned too small, the algorithm will converge at a very slow speed. On the contrary, if it is too large, the algorithm may even diverge. In general, it is a nontrivial task to assign an appropriate value to the learning rate, although we can pay extra efforts to make the learning rate dynamically changed over time, e.g. see [5].

In this paper, we further study the MWL learning framework, and develop a batch RPEM algorithm accordingly provided that all observations are available before the learning process. Compared to the adaptive RPEM, this batch one need not assign the learning rate analogous to the EM, but still preserve the capability of automatic model selection. Further, the convergence speed of this batch RPEM is faster than the EM and the adaptive RPEM. The experiments have shown the efficacy of the proposed algorithm.

2   Overview of Maximum Weighted Likelihood (MWL) Learning Framework

Suppose an input $\mathbf{x}$ comes from the following density mixture model:

$$P(\mathbf{x}|\boldsymbol{\Theta}) = \sum_{j=1}^{k} \alpha_j p(\mathbf{x}|\boldsymbol{\theta}_j), \quad \sum_{j=1}^{k} \alpha_j = 1, \alpha_j > 0, \quad \forall 1 \leq j \leq k, \tag{1}$$

where $\boldsymbol{\Theta}$ is the parameter set of $\{\alpha_j, \boldsymbol{\theta}_j\}_{j=1}^{k}$. Furthermore, k is the number of components, α_j is the mixture proportion of the jth component, and $p(\mathbf{x}|\boldsymbol{\theta}_j)$ is a multivariate probability density function (pdf) of $\mathbf{x}$ parameterized by $\boldsymbol{\theta}_j$. In the MWL learning framework, the parameter set $\boldsymbol{\Theta}$ is learned via maximizing the following Weighted Likelihood (WL) cost function:

$$l(\boldsymbol{\theta}) = \int \sum_{j=1}^{k} g(j|\mathbf{x},\boldsymbol{\Theta}) \ln[\alpha_j p(\mathbf{x}|\boldsymbol{\theta}_j)] dF(\mathbf{x}) - \int \sum_{j=1}^{k} g(j|\mathbf{x},\boldsymbol{\Theta}) \ln h(j|\mathbf{x},\boldsymbol{\Theta}) dF(\mathbf{x}),$$

(2)

with

$$h(j|\mathbf{x},\boldsymbol{\Theta}) = \frac{\alpha_j p(\mathbf{x}|\boldsymbol{\theta}_j)}{P(\mathbf{x}|\boldsymbol{\Theta})}$$

(3)

to be the posterior probability of $\mathbf{x}$ coming from the jth density as given $\mathbf{x}$, where $g(j|\mathbf{x},\boldsymbol{\Theta})$s are the designable weights satisfying the two conditions:

(Condition 1) $\sum_{j=1}^{k} g(j|\mathbf{x},\boldsymbol{\Theta}) = 1$, and
(Condition 2) $\forall j, g(j|\mathbf{x},\boldsymbol{\Theta}) = 0$ if $h(j|\mathbf{x},\boldsymbol{\Theta}) = 0$.

Suppose a set of N observations, denoted as $\chi = \{\mathbf{x}_1, \mathbf{x}_2, \ldots, \mathbf{x}_N\}$, comes from the density mixture model in Eq.(1), the empirical WL function of Eq.(2), written as $\Upsilon(\boldsymbol{\Theta}; \chi)$, can be further expanded as:

$$\Upsilon(\boldsymbol{\Theta};\chi) = \frac{1}{N} \sum_{t=1}^{N} \sum_{j=1}^{k} g(j|\mathbf{x}_t,\boldsymbol{\Theta}) \ln[\alpha_j p(\mathbf{x}_t|\boldsymbol{\theta}_j)] - \frac{1}{N} \sum_{t=1}^{N} \sum_{j=1}^{k} g(j|\mathbf{x}_t,\boldsymbol{\Theta}) \ln h(j|\mathbf{x}_t,\boldsymbol{\Theta}).$$

(4)

In [2], the weights $g(j|\mathbf{x}_t,\boldsymbol{\Theta})$ have been generally designed as:

$$g(j|\mathbf{x}_t,\boldsymbol{\Theta}) = (1 + \varepsilon_t) I(j|x_t,\boldsymbol{\Theta}) - \varepsilon_t h(j|\mathbf{x}_t,\boldsymbol{\Theta})$$

(5)

where ε_t is a coefficient varying with the time step t. Please note that $g(j|\mathbf{x}_t,\boldsymbol{\Theta})$ in Eq.(5) can be negative as well as positive. For simplicity, we hereinafter set ε_t as a constant, denoted as ε. Furthermore, $I(j|x_t,\boldsymbol{\Theta})$ is an indicator function with

$$I(j|\mathbf{x}_t,\boldsymbol{\Theta}) = \begin{cases} 1, & \text{if } j = c = \arg\max_{1 \le i \le k} h(j|\mathbf{x}_t,\boldsymbol{\Theta}); \\ 0, & \text{otherwise.} \end{cases}$$

(6)

Subsequently, the earlier work [2] has presented the adaptive RPEM to learn $\boldsymbol{\Theta}$ via maximizing the WL function of Eq.(4) using a stochastic gradient ascent method. Interested readers may refer to the paper [2] for more details. In the following, we just summarize the main steps of the adaptive RPEM as follows:

Step 1.   Given the current input $\mathbf{x}_t$ and the parameter estimate, written as $\boldsymbol{\Theta}^{(n)}$, we compute $h(j|\mathbf{x}_t,\boldsymbol{\Theta}^{(n)})$ and $g(j|\mathbf{x}_t,\boldsymbol{\Theta}^{(n)})$ via Eq.(3) and Eq.(5), respectively.

Step 2.   Given $h(j|\mathbf{x}_t,\boldsymbol{\Theta}^{(n)})$ and $g(j|\mathbf{x}_t,\boldsymbol{\Theta}^{(n)})$, we update $\boldsymbol{\Theta}$ by

$$\boldsymbol{\Theta}^{(n+1)} = \boldsymbol{\Theta}^{(n)} + \eta \frac{q_t(\boldsymbol{\Theta}; \mathbf{x}_t)}{\boldsymbol{\Theta}} |_{\boldsymbol{\Theta}^{(n)}},$$

(7)

with

$$q_t(\boldsymbol{\Theta}; \mathbf{x}_t) = \sum_{j=1}^{k} g(j|\mathbf{x}_t, \boldsymbol{\Theta}) \ln[\alpha_j p(\mathbf{x}_t|\boldsymbol{\theta}_j)], \qquad (8)$$

where η is a small positive learning rate.

Step 3.   Let $n = n + 1$, and go to Step 1 for the next iteration until $\boldsymbol{\Theta}$ is converged.

The experiments have shown the superior performance of the adaptive RPEM, in particular the capability of automatic model selection. Nevertheless, the convergence speed of this adaptive algorithm relies on the value of learning rate. Under the circumstances, we will present a batch version without the learning rate in the next section.

3   Batch RPEM Algorithm

3.1   Algorithm

As shown in Step 2 of Section 2, we actually update the parameters via maximizing the first term of Eq.(4), whereas the second term is just a conditional entropy of the densities and can be regarded as the constant when updating the parameters. In the following, we denote the first term of Eq.(4) as:

$$\zeta(\boldsymbol{\Theta}; \chi) = \frac{1}{N} \sum_{t=1}^{N} \sum_{j=1}^{k} g(j|\mathbf{x}_t, \boldsymbol{\Theta}) \ln[\alpha_j p(\mathbf{x}_t|\boldsymbol{\theta}_j)]. \qquad (9)$$

Hence, we need to solve the following nonlinear optimization problem:

$$\widetilde{\boldsymbol{\Theta}} = \arg\max_{\boldsymbol{\Theta}}\{\zeta(\boldsymbol{\Theta}; \chi)\} \qquad (10)$$

subject to the constraints as shown in Eq.(1). To solve this optimal problem with equality constraint, we adopt Lagrange method analogous to the EM by introducing a Lagrange multiplier λ into the Lagrange function. Subsequently, we have:

$$F(\boldsymbol{\Theta}, \lambda) = \zeta(\boldsymbol{\Theta}; \chi) + \lambda(\sum_{j=1}^{k} \alpha_j - 1) \qquad (11)$$

with $\boldsymbol{\Theta} = (\alpha_j, \boldsymbol{\theta}_j)_{j=1}^{k}$.

In this paper, we concentrate on the Gaussian mixture model only, i.e., each component $p(j|\mathbf{x}|\boldsymbol{\theta}_j)$ in Eq.(1) is a Gaussian density. We then have

$$p(j|\mathbf{x}_t, \boldsymbol{\theta}_j) = \frac{1}{(2\pi)^{d/2}|\boldsymbol{\Sigma}|^{1/2}} exp[-\frac{1}{2}(\mathbf{x}_t - \mathbf{m}_j)^T \boldsymbol{\Sigma}_j^{-1}(\mathbf{x}_t - \mathbf{m}_j)], \qquad (12)$$

where $\boldsymbol{\theta}_j = (\mathbf{m}_j, \boldsymbol{\Sigma}_j)$, $\mathbf{m}_j$ and $\boldsymbol{\Sigma}_j$ are the mean (also called *seed points* interchangeably) and the covariance of jth density, respectively.

Through optimizing Eq.(11), we then finally obtain the batch RPEM algorithm as follows:

Step 1.   Given $\Theta^{(n)}$, we compute $h(j|\mathbf{x}_t, \Theta^{(n)})$s and $g(j|\mathbf{x}_t, \Theta^{(n)})$s for all $\mathbf{x}_t$s via Eq.(3) and Eq.(5), respectively.

Step 2.   Given $h(j|\mathbf{x}_t, \Theta^{(n)})$s and $g(j|\mathbf{x}_t, \Theta^{(n)})$s computed in Step 1, we update Θ by

$$\alpha_j^{(n+1)} = \aleph_j^{(n)} / \sum_{j=1}^k \aleph_j^{(n)}, \quad \mathbf{m}_j^{(n+1)} = \frac{1}{\aleph_j^{(n)}} \sum_{t=1}^N \mathbf{x}_t g(j|\mathbf{x}_t, \boldsymbol{\theta}^{(n)})$$

$$\boldsymbol{\Sigma}_j^{(n+1)} = \frac{1}{\aleph_j^{(n)}} \sum_{t=1}^N g(j|\mathbf{x}_t, \boldsymbol{\theta}^{(n)})(\mathbf{x}_t - \mathbf{m}_j^{(n)})(\mathbf{x}_t - \mathbf{m}_j^{(n)})^T, \quad (13)$$

where $\aleph_j^{(n)} = \sum_{t=1}^N g(j|\mathbf{x}_t, \boldsymbol{\theta}^{(n)})$. If the covariance matrix $\boldsymbol{\Sigma}_j^{(n+1)}$ is singular, it indicates that the corresponding j^{th} density component is degenerated and can be simply discarded without being learned any more in the subsequent iterations. In this case, we have to normalize those remaining $\alpha_j^{(n+1)}$s so that their sum is always kept to be 1.

Step 3.   Let $n = n + 1$, and go to Step 1 for the next iteration until Θ is converged.

In the above batch RPEM, we need to assign a value to ε in the weight design as shown in Eq.(5). A new question is how to assign an appropriate value of ε? The next sub-section will answer this question.

3.2   How to Assign Parameter ε?

We rewrite Eq.(5) as the following form:

$$g(j|\mathbf{x}_t, \Theta) = \begin{cases} h(j|\mathbf{x}_t, \Theta) + (1 + \varepsilon)(1 - h(j|\mathbf{x}_t, \Theta)), & \text{if } j = c \\ h(j|\mathbf{x}_t, \Theta) - (1 + \varepsilon)h(j|\mathbf{x}_t, \Theta), & \text{otherwise,} \end{cases} \quad (14)$$

where the term $(1 + \varepsilon)(1 - h(j|\mathbf{x}_t, \Theta))$ is the award of the winning density component (i.e. the cth density with $I(c|\mathbf{x}_t, \Theta) = 1$), and meanwhile the term $-(1 + \varepsilon)h(j|\mathbf{x}_t, \Theta)$ is the penalty of the rival components (i.e., those densities with $I(j|\mathbf{x}_t, \Theta) = 0$). Intuitively, it is expected that the award is positive and the penalty is negative, i.e., ε should be greater than -1. Otherwise, as $\varepsilon < -1$, we will meet an awkward situation: the amount of award is negative and the penalty one becomes positive. This implies that we will penalize the winner and award the rivals, which evidently violates our expectations. Furthermore, as $\varepsilon = -1$, both of the award and penalty amount becomes zero. In this special case, the batch RPEM is actually degenerated into the EM without the property of automatic model selection.

In addition, it is noticed that the larger the ε, the stronger the award and penalty are. This property makes the algorithm converge faster with a larger value of ε, but the algorithm is more prone to a sub-optimal solution. Our empirical studies have shown that the covariance matrix of a rival density is prone to singular if ε is too large. Hence, an appropriate selection of ε in the batch

RPEM would be a negative value. Further, our empirical studies have found that the algorithm has a poor capability of automatic model selection if ε is close to zero. As we can see, the smaller the $|\varepsilon|$, the smaller difference among the rival densities is. For example, if we set $|\varepsilon| = 0$, we get $g(j|\mathbf{x}_t, \boldsymbol{\Theta}) = I(j|\mathbf{x}_t, \boldsymbol{\Theta})$. Subsequently, the batch RPEM degenerates to the hard-cut EM without the capability of automatic model selection. Hence, by a rule of thumb, an appropriate selection of ε should be within the range of $(-1, -0.4)$. In the next section, we will arbitrarily set ε at -0.5.

4   Experimental Results

Because of the space limitation, we will conduct two experiments only to demonstrate the performance of the batch RPEM. In these two experiments, we used the same $1,000$ observations that were generated from a mixture of three bivariate Gaussian distributions, whose true parameters were:

$$\alpha_1^* = 0.3, \quad \alpha_2^* = 0.4, \quad \alpha_3^* = 0.3$$
$$\mathbf{m}_1^* = [1.0, 1.0]^T, \quad \mathbf{m}_2^* = [1.0, 2.5]^T, \quad \mathbf{m}_3^* = [2.5, 2.5]^T$$
$$\boldsymbol{\Sigma}_1^* = \begin{pmatrix} 0.20, 0.05 \\ 0.05, 0.30 \end{pmatrix}, \quad \boldsymbol{\Sigma}_2^* = \begin{pmatrix} 0.2 & 0.0 \\ 0.0 & 0.20 \end{pmatrix}, \quad \boldsymbol{\Sigma}_3^* = \begin{pmatrix} 0.2 & -0.1 \\ -0.1 & 0.2 \end{pmatrix}. \quad (15)$$

4.1   Experiment 1

This experiment was to investigate the convergence speed of the batch RPEM algorithm. We set $k = 3$, and the three seed points were randomly allocated in the observation space. Furthermore, all α_js and $\boldsymbol{\Sigma}_j$s were initialized at $\frac{1}{k}$ and the identity matrix, respectively. For comparison, we also implemented the EM under the same experimental environment. After all parameters were converged, both of the batch RPEM and EM gave the correct parameter estimates. Nevertheless, as shown in Fig. 1(a) and (b), the batch RPEM converges at 25 epochs while the EM needs 60 epochs. That is, the convergence speed of the former is significantly faster than the latter. This indicates that the intrinsic rival-penalization scheme of the batch RPEM, analogous to the RPCL [4], RPCCL [1] and the adaptive RPEM [2], is able to drive away the rival seed points so that they can be more quickly towards the other cluster centers. As a result, the batch RPEM converges much faster than the EM. Furthermore, we also compared it with the adaptive RPEM, in which we set the learning rate $\eta = 0.001$. Fig. 1(c) shows that the adaptive RPEM converges at 40 epochs, slower than the proposed batch version.

4.2   Experiment 2

This experiment was to investigate the capability of the batch RPEM on model selection. We set $k = 10$, and randomly allocated the 10 seed points, $\mathbf{m}_1, \mathbf{m}_2, \ldots,$ $\mathbf{m}_{10}$ into the observation space as shown in Fig. 2(a). During the learning process, we discarded those densities whose covariance matrices $\boldsymbol{\Sigma}_j$s were singular. After

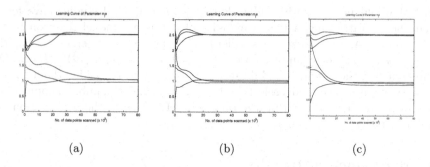

(a)   (b)   (c)

Fig. 1. Learning curves of m_js by (a) EM, (b) Batch RPEM, and (c) Adaptive RPEM, respectively

90 epochs, we found that 5 out of 10 density components have been discarded. The mixture proportions of the remaining components were converged to $\alpha_1 = 0.0061$, $\alpha_2 = 0.3508$, $\alpha_3 = 0.3222$, $\alpha_4 = 0.3161$, $\alpha_5 = 0.0048$. Furthermore, the corresponding m_js and Σ_js were:

$$m_1 = [3.1292, 1.3460]^T, m_2 = [0.9462, 2.5030]^T, m_3 = [1.0190, 0.9788]^T,$$
$$m_4 = [2.5089, 2.5286]^T, m_5 = [1.8122, 1.8955]^T$$

$$\Sigma_1 = \begin{pmatrix} 0.1920, 0.0310 \\ 0.0310, 0.0088 \end{pmatrix}, \Sigma_2 = \begin{pmatrix} 0.1708, 0.0170 \\ 0.0170, 0.1489 \end{pmatrix}, \Sigma_3 = \begin{pmatrix} 0.1894, 0.0461 \\ 0.0461, 0.2892 \end{pmatrix},$$

$$\Sigma_4 = \begin{pmatrix} 0.2155, -0.1101 \\ -0.1101, \ \ 0.2116 \end{pmatrix}, \Sigma_5 = \begin{pmatrix} 0.0027, -0.0053 \\ -0.0053, \ \ 0.0213 \end{pmatrix}. \tag{16}$$

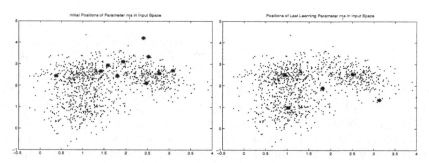

(a) The initial seed positions marked by '*', where seed points are those data points to be learned towards the cluster centers in the observation space.

(b) The converged seed positions learned by the batch RPEM algorithm

Fig. 2. The Results of the Batch RPEM Algorithm

It can be seen that the three seed points $\mathbf{m}_2, \mathbf{m}_3, \mathbf{m}_4$ together with the corresponding proportions $\alpha_2, \alpha_3, \alpha_4$ have provided a good estimate of the true parameters as shown in Fig. 2(b), where $\mathbf{m}_2$, $\mathbf{m}_3$, $\mathbf{m}_4$ are stabled at the center of the clusters. In contrast, $\mathbf{m}_1$ and $\mathbf{m}_5$ have been driven away and located at the boundary of the clusters with a very small mixture proportions: $\alpha_1 = 0.0061$ and $\alpha_5 = 0.0048$. Actually, the 1st and 5th density components have been gradually faded out from the mixture.

5 Conclusion

In this paper, we have developed a batch RPEM algorithm based on MWL learning framework for Gaussian mixture clustering. Compared to the adaptive RPEM, this new one need not select the value of learning rate. As a result, it can learn faster in general and still preserve the capability of automatic model selection analogous to the adaptive one. The numerical results have shown the efficacy of the proposed algorithm.

Acknowledgment

This work was fully supported by the grants from the Research Grant Council of the Hong Kong SAR with the Project Codes: HKBU 2156/04E and HKBU 210306.

References

1. Cheung Y.M.: Rival Penalized Controlled Competitive Learning for Data Clustering with Unknown Cluster Number. Proceedings of Ninth International Conference on Neural Information Processing (ICONIP'02), Paper ID: 1983 in CD-ROM Proceeding (2002)
2. Cheung Y.M.: Maximum Weighted Likelihood via Rival Penalized EM for Density Mixture Clustering with Automatic Model Selection. IEEE Transactions on Knowledge and Data Engineering, Vol. 17, No. 6 (2005) 750–761
3. Dempster A. Laird N. and Rubin D.: Maximum Likelihood from Incomplete Data via the EM Algorithm. Journal of Royal Statistical Society, Vol. 39, No. 1 (1977) 1–38
4. Xu L., Krzyżak A. and Oja E.: Rival Penalized Competive Learning for Clustering Analysis, RBF Net, and Curve Detection. IEEE Transactions on Neural Networks, Vol. 4, (1993) 636–648
5. Zhao X.M., Cheung Y.M., Chen L. and Aihara K.: A New Technique for Adjusting the Learning Rate of RPEM Algorithm Automatically. to appear in the Proceedings of The Twelfth International Symposium on Artificial Life and Robotics, Japan (2007)

A Memetic-Clustering-Based Evolution Strategy for Traveling Salesman Problems

Yuping Wang[1] and Jinhua Qin[2]

[1] School of Computer Science and Technology,
Xidian University, Xi'an, China
ywang@xidian.edu.cn
[2] Department of Mathematics Science,
Xidian University, Xi'an, China
abcxyz-999@163.com

Abstract. A new evolution strategy based on clustering and local search scheme is proposed for some kind of large-scale travelling salesman problems in this paper. First, the problem is divided into several subproblems with smaller sizes by clustering, then the optimal or the approximate optimal tour for each subproblem is searched by a local search technique. Moreover, these tours obtained for the subproblems are properly connected to form a feasible tour based on a specifically-designed connection scheme. Furthermore, a new mutation operator is designed and used to improve each connected feasible tour further. The global convergence of the proposed algorithm is proved. At last, the simulations are made for several problems and the results indicate the proposed algorithm is effective.

Keywords: TSP, evolutionary algorithm, clustering, memetic algorithm.

1   Introduction

The traveling salesman problem (TSP) is one of the most famous combinatorial optimization problems. Given n cities and a distance matrix $D = [d_{ij}]$, where d_{ij} is the distance between city i and city j, TSP requires finding a tour (i.e., a permutation of cities) through all of the cities, visiting each exactly once, and returning to the originating city such that the total distance traveled is minimized.

In this paper we consider the two-dimensional (2-D) Euclidean TSP, in which the cities lie in R^2 and the distance between two cities is calculated by Euclidean distance. The 2-D Euclidean TSP is known to be NP-hard ([1],[2],[3]). It was proven that there is no polynomial-time approximation scheme for TSP unless $P = NP$ ([4]). However, the TSP and its variants have a diverse practical applications. More than 1700 related papers have been published during the past five years (see the INSPEC database: http://www.inspec.org). They are one of the most actively studied topics in the evolutionary computation community, too. Many papers have published in this field (e.g., [5]~ [10]). One of the most successful evolutionary algorithms (EAs) for TSP is the hybrid EAs (or memetic

J.T. Yao et al. (Eds.): RSKT 2007, LNAI 4481, pp. 260–266, 2007.

EAs)that incorporate local search scheme into EAs (e.g., [7],[8],[9]). To the best of our knowledge, Freisleben and Merz's hybrid EA ([8],[9]) with powerful Lin-Kernighhan (LK) local search algorithm ([10]) is in practice one of the best EAs for TSP among the published algorithms. However, LK local search algorithm needs a lot of computation and can not be applied to general large scale problems. For some special kind of TSP problems, it is possible to design effective evolutionary algorithms.

In this paper, we focus our attention on some special TSP problems, i.e., the problems in which the cities can be classified into several groups, and in each group the cities crowd together, and try to design an effective evolutionary algorithm. To do so, we first divided cities into several groups by using clustering technique. Second, we look for the optimal or approximate optimal tour for each group by a local search technique. Third, we connect the tours found for these groups to form a feasible tour based on a specifically-designed connection scheme. Fourth, we design a new mutation operator and use it to improve the feasible tour. Based on these, a novel evolution strategy based on clustering and local search is proposed for this kind of TSP problems. At last, the simulations are made and the results demonstarte the effectiveness of the proposed algorithm.

2   Classification of All Cities into Several Groups by Clustering

For traveling salesman problems in which the cities can be classified into several groups, and in each group the cities crowd together, it seems that the salesman should go through all cities in best way in one group, then move to some other group and also go through all cities in this group in optimal way. Repeat this process until he goes through all groups and returns the starting city. Based on this idea, we have to classify all cities into several groups. In this paper we use the following clustering algorithm: K-mean clustering ([11]).

Algorithm 1

1. Randomly choose k cities as the initial centers of k clusters, where k is a parameter. Let $t = 0$.
2. Calculate the distance between each city and each center of the clusters. Classify the city into a cluster in which the distance between the city and the center of the cluster is the shortest.
3. Re-compute the center for each cluster. Let $t = t + 1$.
4. If the center for every cluster is the same as the center for this cluster in previous iteration $t - 1$, then stop; otherwise, go to step 2.

3   Local Search

For each cluster, a local search algorithm, $2-opt$ algorithm, is used to search high quality tour. To explain the idea of $2 - opt$ algorithm clearly, we first introduce the following definitions.

Definition 1. *For a given tour P, if λ links (edges) of tour P are replaced by other λ links and the resulted graph is also a feasible tour, then the operation of exchanging λ links is called a λ-exchange, and the resulted tour is called a λ-exchange tour.*

Definition 2. *A tour is said to be λ-optimal (or simply $\lambda-opt$) if it is impossible to obtain a shorter tour by replacing any λ of its links (edges) by any other set of λ links.*

The following is a local search algorithm for each cluster of cities.

Algorithm 2 (Local search heuristics based on λ-exchange)

1. Set $L_{opt} = M$, where M is a large enough positive number and $P_{opt} = \emptyset$. Let $t1 = 0$.
2. Arbitrarily take a feasible tour P.
3. Check whether there is a λ-exchange tour $P_{\lambda-ex}$ which is better than current tour P. If there is such a tour, let $P = P_{\lambda-ex}$. Go to step 3; otherwise, go to step 4.
4. Tour P is the λ-optimal tour for the initial tour, let $P_1 = P$. Let $L1$ is the length of tour P_1.
5. If $t1 < 2$, go to step 6; otherwise, P_{opt} is an approximate optimal tour.
6. If $L1 < L_{opt}$, let $L_{opt} = L1$, $P_{opt} = P_1$, go to step 2; otherwise, let $t1 = t1+1$, go to step 2.

In our simulation we take $\lambda = 2$. Note that for a given tour there are total $\frac{n(n-3)}{2}$ λ-exchange tours.

4   Scheme for Connecting Clusters

In the previous two sections, all cities of a special kind of TSP problems are divided into several clusters, and for the cities contained in each cluster an optimal or approximate optimal tour is found. To get a high quality tour for all cities it is necessary to design an effective connection scheme to connect all these tours. Our motivation is that the connection is made in such a way that the connected tour is as short as possible. The detail is as follows.

Algorithm 3 (Connection Scheme)

1. Let k denotes the number of clusters and let $\mathrm{L} = 2$.
2. Choose two clusters whose centers are the nearest. Their corresponding tours are denoted as $T1$ and $T2$, respectively.
3. Choose an edge $A_1B_1 \in T1$ and an edge $A_2B_2 \in T2$ satisfying

$$|A_1A_2|+|B_1B_2|-|A_1B_1|-|A_2B_2| = \min\{|C_1C_2|+|D_1D_2|-|C_1D_1|-|C_2D_2|\},$$

where edge $C_1D_1 \in T1$ and edge $C_2D_2 \in T2$, respectively, and $|A_1A_2|$ denotes the length of edge A_1A_2.

4. Connect A_1 and A_2 (i.e., add edge A_1A_2), and B_1 and B_2. Remove edges A_1B_1 and A_2B_2, respectively. Then a new tour T through all cities of two chosen clusters is formed by connecting $T1$ and $T2$, and a new cluster is formed by the union of these two chosen clusters.
5. If $L < k$, then let $L = L + 1$, go to step 2; otherwise tour T is a feasible tour through all cities. Stop.

5   Mutation Operator

To further improve the quality of the tours obtained by section 4, we design a mutation operator and use it to each tour obtained. The detail is as follows.

Algorithm 4 (Mutation Operator)

1. For each tour $i_1i_2\cdots i_n$ obtained by section 4, where $i_1i_2\cdots i_n$ is a permutation of $1, 2, \cdots, n$.
2. For $q = 1, 2, \cdots, n$, randomly change i_q into any element, denoted by j_q, in $\{1, 2, \cdots, n\}/\{j_1, j_2, \cdots, j_{q-1}\}$ with equal probability, where

$$\{1, 2, \cdots, n\}/\{j_1, j_2, \cdots, j_{q-1}\}$$

represents a set whose elements belong to set $\{1, 2, \cdots, n\}$ but do not to set $\{j_1, j_2, \cdots, j_{q-1}\}$.
3. Tour $j_1j_2\cdots j_n$ is the offspring of tour $i_1i_2\cdots i_n$.

It can be seen from this mutation operator that, for any feasible tours $i_1i_2\cdots i_n$ and $j_1j_2\cdots j_n$, the probability of generating $j_1j_2\cdots j_n$ via $i_1i_2\cdots i_n$ by mutation operator is $\frac{1}{n} > 0$.

6   The Proposed Algorithm

Based on algorithms in the previous four sections, the proposed evolution strategy can be described as follows:

Algorithm 5 (A Memetic-Clustering-Based Evolution Strategy)

1. (Initialization) Given population size N, maximum generations g_{max}, and N positive integers $k_1, k_2, \cdots, k_N$. For each k_i for $i = 1, 2, \cdots, N$, do the following:
   - Generate k_i clusters by algorithm 1, and generate a tour for each cluster by using Algorithm 2.
   - Connect all these tours for clusters into one feasible tour for all cities by using Algorithm 3.

   All these N feasible tours constitute the initial population $P(0)$, Let $t = 0$.

2. (Mutation)For each individual

$$T_r = i_1 i_2 \cdots i_n \in P(t),$$

generate an offspring

$$O_r = j_1 j_2 \cdots j_n,$$

$r = 1, 2, \cdots, N$. The set of all offspring is denoted as $O(t)$.

3. (Selection) Select best N individuals from $P(t) \cup O(t)$ as the next generation population $P(t+1)$.

4. (Stop Criterions) If $t > g_{max}$ and the best tour obtained can not improved in successive 10 generations, then stop; otherwise, let $t = t+1$, go to step 2.

7   Global Convergence

First, we introduce the concept of the global convergence as follows.

Definition 3. *Let $a^* \in A$ denote the chromosome which corresponds to an optimal tour. If*

$$prob\{\lim_{t \to \infty} a^* \in P(t)\} = 1,$$

then the proposed genetic algorithm is called to converge to the global optimal solution with probability one, where $prob\{\}$ represents the probability of random event $\{\}$.

For any feasible tours $i_1 i_2 \cdots i_n$ and $j_1 j_2 \cdots j_n$, note that

$$prob\{M(i_1 i_2 \cdots i_n) = j_1 j_2 \cdots j_n\} = \frac{1}{n} > 0,$$

and the best tour found so far will be kept by the selection process, where $M(i_1 i_2 \cdots i_n)$ represents the offspring of $i_1 i_2 \cdots i_n$ by mutation. It can be proved by making use of the conclusions in [12] that the proposed algorithm has the following property of the global convergence.

Theorem 1. *The proposed evolution strategy (Algorithm 5) converges to the global optimal solution with probability one.*

8   Simulations

8.1   Test Problems

In the simulations, We execute the proposed algorithm to solve six standard benchmark problems: $nrw1379$, a 1379-city problem, $rL1889$, a 1889-city problem $u2319$, a 2319-city problem, $pr2392$, a 2392-city problem, $pcb3038$, a 3038-city problem, and $rL5915$, a 5915-city problem. These problems are available from TSPLIB at http://www.iwr.uni-heidelberg.de/iwr/comopt/soft/TSPLIB95/TSPLIB.html

8.2   Parameter Values

We adopt the following parameter values: $N = 20$, $k_1 = \cdots = k_N = s$, and $s = 5, 8, 10$ respectively for problems with fewer than 3000 cities, and $s = 5, 8, 10$ and 15 respectively for problems with more than 3000 cities. $g_{max} = 300$ for problems with fewer than 3000 cities and 500 for problems with more than 3000 cities.

8.3   Results

The simulations are carried out on a AthlonXP1800+512M PC and we program the proposed algorithm in Matlab language. For each benchmark problem, we perform 20 independent executions. We record the following data:

- The best, the average and the worst lengths over 20 runs, denote them by Best, Mean and Worst, respectively.
- Average CPU time over 20 runs, denoted by CPU.
- Percentage of the amount the best tour found surpasses the optimal tour over the optimal tour, denoted by %.

Table 1. Results obtained by proposed algorithm, where $Opt - tour$ represents the length of the optimal tour, and K is the number of clusters used

TSP	Opt-tour	K	Best	Mean	Worst	CPU	%
		5	59770	60236	61069	65.3	5.5
nrw1379	56638	8	59922	60177	60353	57.6	5.8
		10	59981	60389	60873	65.6	5.9
		5	342447	347407	353226	147.7	8.2
rL1889	316536	8	344107	350777	358637	135.4	8.7
		10	344513	349777	355378	130.5	8.8
		5	243379	243997	244465	227.7	3.9
u2319	234256	8	243167	243843	245016	216.1	3.8
		10	242677	243243	243829	209.8	3.6
		5	389288	394309	396554	246.8	2.9
pr2392	378032	8	394863	398895	404705	232.1	4.4
		10	398937	402243	405692	225.6	5.5
		5	148031	149461	150881	433.0	7.5
pcb3038	137694	8	149209	150126	151309	410.4	8.3
		10	148867	149681	150905	408.7	8.1
		15	148326	149853	151286	398.1	7.7
		5	625686	629688	638488	2229.8	10.6
rL5915	565530	8	613469	625109	635527	2040.0	8.5
		10	617823	627439	637816	1828.7	9.2
		15	624123	633052	655479	1965.5	10.3

The results are given in Table 1. It can be seen from Table 1 that the percentage of the amount the best tour found surpasses the optimal tour over the optimal tour is relatively small although the proposed algorithm has not found the optimal tours for these problems. This indicates the proposed algorithm is effective.

9   Conclusions

In this paper we deal with some special TSP problems, i.e., the problems in which the cities can be classified into several groups, and within each group the cities crowd together. We designed an effective and globally convergent evolutionary algorithm for this kind of problems. The proposed algorithm has the ability of finding close-to-optimal solutions with high speed and a relatively small amount of computation. The simulation results demonstrate the effectiveness of the proposed algorithm.

Acknowledgements. The authors wold like to thank the anonymous reviewers for their insightful comments and helpful suggestions that have improved the paper greatly. This work was supported by National Natural Science Foundation of China (No. 60374063).

References

1. S.Jung, B.R.Moon.: Toward minimal restriction of renetic coding and crossovers for the 2-D eunclidean TSP. *IEEE Trans. on Evolutionary Compuation* 6(6) (2002) 557-565.
2. R.Baraglia, J.I.Hidalgo, R.Perego.: A hybrid Hheuristic for the traveling salesman problem. *IEEE Trans.on Evolutionary Compuation* 5(6) (2001) 613-622.
3. M.R.Garey, R.L.Graham, D.S.Johnson.: Some NP-complete geometric problems. In *Proc. 8th Annu. ACM Symp. Theory of Computing*, New York, NY, USA, ACM Press (1976) 10-22
4. S.Arora, C.Lund, R.Motwani, et al.: Proof verification and intractability of approximation problems. In *Proc. 33th IEEE Symp. Foundations of Computer Science*, Pittsburgh, Pennsylvania, USA, IEEE Press (1992) 3-22.
5. K.Katayama, H.Narihisa.: Iterated local search approach using genetic trasformation to the traveling salesman problem. In *Proc. Genetic and Evolutionary compuation Conf.* New York, USA, Morgan Kaufmann (1999) 321-328.
6. Y.Nagata, S.Kobayashi.: Edge assembly crossover: A high-power genetic algorithm for the travrling salesman problem. In *Proc. 7th Int. Conf. Genetic Algorithm*, San Mateo, CA: USA, Morgan Kaufmann (1997) 450-457.
7. D.S.Johnson.: Local optimization and the traveling salesman problem. In *Proc. 17th Automata, Languages, and Programming*, Warwick University, England, Springer-Verlag (1990) 446-461.
8. B.Freisleben, P.Merz.: New genetic local search operators for the traveling salesman problem. In *Proc. 4th Conf. Parallel Problem solving fron Nature*, Berlin, Germany, Springer-Verlag (1996) 616-621.
9. P.Merz, B.Freisleben.: Genetic local search for the TSP:New results. In *Proc. of the 1997 International Conferenece on Evolutionary Computation*, Piscataway,NJ:IEEE Press (1997) 159-163.
10. S.Lin, B.W.Kernighan.: An effective heuristic algorithm for the traveling salesman problem. *Operations Research* 21(2) (1973) 495-516
11. A.K. Jain, M.N. Murty, P.J. Flyn.: Data clustering: A review. *ACM Computing Survey* 31(3) (1999) 264-323.
12. T. Bäck.: *Evolutionay algorithms in theory and practice*. New York: Oxford Univ. Press (1996).

An Efficient Probabilistic Approach to Network Community Mining

Bo Yang[1] and Jiming Liu[2]

[1] College of Computer Science and Technology, Jilin University, Changchun, P.R.China
[2] School of Computer Science, University of Windsor, Windsor, Ontario, Canada
ybo@jlu.edu.cn, jiming@uwindsor.ca

Abstract. A network community refers to a group of vertices within which the links are dense but between which they are sparse. A network community mining problem (NCMP) is the problem to find all such communities from a given network. A wide variety of applications can be formalized as NCMPs such as complex network analysis, Web pages clustering as well as image segmentation. How to solve a NCMP efficiently and accurately remains an open challenge. Distinct from other works, the paper addresses the problem from a probabilistic perspective and presents an efficient algorithm that can linearly scale to the size of networks based on a proposed Markov random walk model. The proposed algorithm is strictly tested against several benchmark networks including a semantic social network. The experimental results show its good performance with respect to both speed and accuracy.

Keywords: Social networks, Community, Markov chain, Semantic Web.

1 Introduction

A network community refers to a group of vertices within which the links are dense but between which they are sparse, as illustrated in Fig.1. A network community mining problem (NCMP) is the problem to find all such communities from a given network. A wide variety of problems can be formalized as NCMPs such as social network analysis[1], linked-based Web clustering[2], biological network analysis[3] as well as image segmentation[4]. Network communities in different contexts have different meanings and serve different functions. For instances, communities in a social network are interest-related social groups within which people have more contacts with each other; communities in the World Wide Web are groups of topic-related Web pages; communities in a protein regulatory network indicate groups of function-related proteins. Therefore, the ability to discover such a kind of hidden pattern from those networks can help us understand their structures and utilize them more effectively.

So far, many methods have been proposed to address the NCMP. They can be generally divided into three main categories: (1) bisection methods, mainly including spectral methods [4,5] and the Kernighan-Lin algorithm[6]; (2) hierarchical methods, including agglomerative methods based on some structural measures[7] and divisive

J.T. Yao et al. (Eds.): RSKT 2007, LNAI 4481, pp. 267–275, 2007.
© Springer-Verlag Berlin Heidelberg 2007

methods such as the GN algorithm[8], the Tyler algorithm[9] and the Radicchi algorithm[10]; (3) linked-based methods for mining Web communities mainly including the MFC algorithm[2] and the HITS algorithm[11].

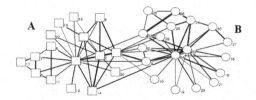

Fig. 1. A schematic representation of network communities. This simple network describes the social interactions among the members of a karate club at an American university, constructed by Wayne Zachary in the 1970s[1]. Different width of edges denotes different interaction strength. The club eventually split into two communities. One (community A) was led by its administrator (vertex 1) whose members are represented by squares, and another (community B) was led by its teacher (vertex 33) whose member are represented by circles. As we can observe, the social interactions within communities are much more than those between them.

The above mentioned methods have their advantages and limitations. The authors hold the opinion that a method for solving the NCMP is said to be efficient if it integrates three features at the same time, that is, be fast, be accurate, and be insensitive to parameters. In the paper, we will present a novel method to address the NCMP from a perspective of Markov random walk process, which demonstrates the three features as mentioned. The remainder of the paper is organized as follow: Section 2 describes our method. Section 3 gives main experimental results. Section 4 concludes the paper by highlighting the major advantages of our method.

2   The Probabilistic Approach

First of all, we observe that if we can extract the community containing a specified vertex from a complete network, we will be able to find all communities hidden in the network by means of recursively extracting communities one by one. However, in order to do so, we need to solve three key problems as follows: (1) How to specify a vertex? (2) How to correctly extract the community containing the specified vertex from a complete network? (3) When to stop the recursive extracting process? In what follows, we will answer these questions with the aid of a Markov random walk model. Based on this model, we propose our method for solving the NCMP.

2.1 Markov Random Walk Model

Consider a random walk process in a given network N, in which an agent wanders freely from one vertex to another along the links between them. After the agent arrives at one vertex, it will select one neighbor at random and move there. Let $N = (V, E)$

and $V = \{v_1, \cdots, v_n\}$. Let $X = \{X_l, l \geq 0\}$ denote a random series, and $P\{X_l = v_{i_l}, 1 \leq i_l \leq n\}$ denote the probability that the agent arrives at the vertex v_{i_l} after l steps. We have $P(X_l = v_{i_l} \mid X_0 = v_{i_0}, X_1 = v_{i_1}, \cdots, X_{l-1} = v_{i_{l-1}}) = P(X_l = v_{i_l} \mid X_{l-1} = v_{i_{l-1}})$. Thus, the random walk process is a finite and discrete *Markov chain*, denoted as $\xi(N)$, and its state space is V. In a network, the probability of the agent walking from vertex i to vertex j is $p_{ij} = (a_{ij}/d_i) \cdot I_{\{<i,j>\in E\}} + 0 \cdot I_{\{<i,j>\notin E\}}$ where I denotes the condition of different case, a_{ij} is the weight on the edge $<i, j>$ and $d_i = \sum_j a_{ij}$ is the degree of vertex i. Furthermore, we have $P(X_l = v_j \mid X_{l-1} = v_i) = p_{ij}$, so $\xi(N)$ is a *homogeneous Markov chain*. Let $A = (a_{ij})_{n\times n}$ be the adjacency matrix of network N and $D = diag(d_1, \cdots d_n)$. Let P be the *one-step transfer matrix* of $\xi(N)$, we have $P = (p_{ij})_{n\times n} = D^{-1}A$. Let $p_{ij}^{(l)}$ be the probability that the agent starting from vertex i can arrive at vertex j after exact l steps. We have $p_{ij}^{(l)} = P^l(i, j)$ where P^l is the l-step transfer matrix of $\xi(N)$. We have proven the following: Suppose that network $N = (V, E)$ is a connected network and contains at least a loop with an odd path length, we have $\lim_{l\to\infty} p_{ij}^{(l)} = d_j / \sum_k d_k$.

A network is called *ergodic* if the Markov random walk process in the network is *ergodic*. Most real networks, especially social networks and the Web, are *ergodic* because they usually contain a lot of triangles due to their high *clustering coefficient*. The fact tells us that a random-walk agent wondering in an *ergodic* network will reach a specified destination vertex (called *sink*) with a fixed limit probability independent of the starting vertex. The community containing the sink is called a *sink community*, denoted as C_{sink}. Note that the density of edges within communities is much denser than that between communities, thus it is more likely for the random walk to reach the sink if it sets out from the vertex within the sink community, as it has more redundant paths to choose in order to reach its destination. On the contrary, if the random-walk agent starts from a vertex outside the sink community, it is very difficult for the agent to reach the sink because it has to take much more efforts to escape from other communities and get into the sink community through only a few "bottleneck" edges connecting them. In that case, the random walk will arrive at the sink with a quite low probability. Therefore, we have the following property for the above-mentioned Markov random walk model: Given a sink v_j, there exists an integer L, when $l > L$ we have $p_{ij}^{(l)} > p_{kj}^{(l)}$ for $v_i \in C_j$ and $v_k \notin C_j$.

Based on this property, the procedure for extracting the community containing a specified vertex can be described as follows: (1) specifying a vertex v_j as the sink; (2) calculating the L-step transfer probability $p_{ij}^{(L)}$ for each vertex I; (3) ranking all vertices into a non-decreasing permutation according to these probabilities. After the above three steps, all members belonging to the sink community are accumulated together in the bottom of the ranked permutation. Then, the members of the sink community can be separated from others using a predefined cut strategy.

2.2 Selecting a Sink and Computing the L-Step Transfer Probabilities

There are no special requirements about the selection of a sink, and we can randomly select one. While in a practical application, in addition to mining all communities, we also hope to identify their respective centers. For this reason, we will manage to select the center of a community as a sink in each bipartition. From the perspective of the random walk, a vertex with a maximum limit probability is likely the center of a network because all walks randomly wandering in the network will eventually come together at such a vertex with the largest probability wherever they sets out. Therefore, we can choose a sink according to $\text{sink} = \arg\max\{\pi_i\} = \arg\max\{d_i / \sum_k d_k\} = \arg\max\{d_i\}$.

All $p_{ij}^{(L)}$ can be found by calculating the L-step transfer matrix P^L using at least $O(n^3)$ time. Here, we present an efficient algorithm for computing all L-step transfer probabilities to a sink with $O(n+Lm)$ time, where m is the number of edges. Let $P^{(l)}(i) = p_{i,sink}^{(l)}$, we have

$$\begin{cases} P^{(0)}(i) = 0 \cdot I_{\{i \neq sink\}} + 1 \cdot I_{\{i = sink\}} \\ P^{(l)}(i) = (\sum_{<v_i,v_j> \in E} a_{ij} \cdot P^{(l-1)}(j))/d_i \end{cases}$$

Now the problem can be stated as follows: Given a sink v_j, how to decide a reasonable value of L. Instead of calculating an exact value, we can estimate a large enough value using a simple error-control method, in which L is the minimum value of l satisfying $error^{(l)} < \varepsilon$ where $error^{(l)} = \sum_i | p_{i,sink}^{(l)} - \pi_{sink} |$ and ε is a predefined constant such as 10^{-3}. Briefly speaking, based on the Dobrushin convergent theorem we can obtain the following:

$$L = \min\{l \mid error^{(l)} < \varepsilon\} = \lceil k \cdot (log_{10}\varepsilon - log_{10}\alpha)/log_{10}(C(P^k)) \rceil$$

where α, $C(P) = (\sup_{i,k} \sum | p_{ij} - p_{kj} |)/2$, and $k = \arg\min\{C(P^k) < 1\}$ are three Dobrushin constants which are insensitive to the dimension of the network.

So for a given network and a predefined error, the value of L can be considered as a constant and the time of computing the L-step transfer probabilities will be $O(n + Lm) = O(n + m)$. In practice, L can be quite small for most real-world

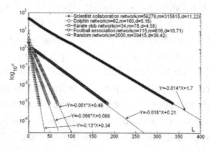

Fig. 2. The linear relationship between steps L and the logarithm of control error ε in different networks with different dimensions

networks. We can verify this through the experiments as presented in Fig.2. In the legend, n, m, and d correspond, respectively, to the number of vertices, the number of edges, and the average degree of networks. Using a linear regression approach, we can respectively figure out the linear function of $log_{10}(\varepsilon)=aL+b$, where $a = log_{10}(C(P^k))/k$ and $b = log_{10}\alpha$, that best fits the experimental data of each network. For all cases, the error will be below $\varepsilon = 10^{-3}$ after at most 350 steps.

2.3 Probability Cut and Stopping Criterion

Now our next problem is, based on the sorted $P^{(L)}$, how to find a cut position that can perfectly distinguish the sink community from others. In this section, we present a new cut criterion, *probability cut*, to efficiently complete this task.

Let (V_1,V_2) be a bipartition of N, where $V_1 \cup V_2 = V$ and $V_1 \cap V_2 = \phi$. The *escaping probability*, $EP(V_1,V_2)$, is defined as the probability of a random walk transitioning from set V_1 to set V_2 in one step, and can be calculated as $EP(V_1,V_2) = \sum_{i \in V_1, j \in V_2} p_{ij} / |V_1|$.

Specifically, $EP(V_1,V_1)$ is called a *trapping probability*, which is the probability of the random walk being trapped in set V_1 in one step. We have $EP(V_1,V_1)+EP(V_1,V_2)=1$. From the viewpoint of community structure, if (V_1, V_2) is a "good" bipartition of the entire network, the *escaping probability* should be very small and at the same time the *trapping probability* should be very large because it is very hard for a random-walk agent to escape from the trap of one community in which it currently stays and get into another via only a few "bridge links" connecting them. Based on this idea, we define the probability cut, *Pcut*, as $Pcut(V_1,V_2) = EP(V_1,V_2) + EP(V_2,V_1)$.

Given a cut position $pos(1 \leq pos < n)$, the bipartition of sorted $P^{(L)}$ divided by pos is denoted as $(V_1, V_2) = ([1, pos], [pos+1, n])$. Now, the problem of finding a good cut position of a sorted $P^{(L)}$ is essentially that of finding its bipartition that minimizes the *Pcut* value, that is, $cut(P^{(L)}) = \arg \min_{1 \leq pos < n} \{Pcut([1, pos],[pos+1,n])\}$.

Such a cut position can be efficiently computed by at most thrice scanning the matrix with $O(n+m)$ time in term of the adjacency list.

If a given network N can be nicely divided into two parts, the escaping probability will be much less than the trapping probability. Otherwise, no further division is required because there is no obvious community structure in the network such that it is easy for a random walk to escape from one community and get into another. Based on this fact, a reasonable criterion to stop the recursive process of extracting sink communities is $EP(V_1,V_2) \geq EP(V_1,V_1)$ *and* $EP(V_2,V_1) \geq EP(V_2,V_2)$. Furthermore, we have $EP(V_1,V_2) \geq 0.5$ *and* $EP(V_2,V_1) \geq 0.5$, where (V_1, V_2) is the optimal probability cut of network N as discussed above. We have proven that the worst time complexity of the above algorithm is $O((2K-1)(n\log n + m))$, where K is the number of communities in the network. Usually, K is much smaller than n. Also note that the value of $\log n$ is extremely small and can be ignored as compared with the value of n. Thus, the worst time complexity of the NCMA algorithm is approximately $O(n+m)$, or $O(n)$ in a sparse network.

2.4 Main Algorithm and Time Complexity

Algorithm. NCMA: Algorithm for Mining Network Communities

Input: A, the adjacency matrix of the network to be mined
Output: B, the adjacency matrix of the clustered network
1. select the vertex with the maximum degree as the sink;
2. compute the transfer probability vector $P^{(L)}$ with respect to a given error;
3. sort $P^{(L)}$ into a non-decreasing permutation $SP^{(L)}$;
4. transform adjacency matrix A into matrix B based on $SP^{(L)}$;
5. compute the optimal cut position and two escaping probabilities based on B;
6. return matrix B if the stopping criterion is satisfied; otherwise, divide matrix B into sub-matrices B_1 and B_2 based on the cut position;
7. NCMA (B_1);
8. NCMA (B_2).

---

3 Validation of the Probabilistic Algorithm for Solving NCMA

We have validated our proposed probabilistic algorithm for solving the NCMA by using various benchmark datasets, including real and artificial networks. As limited by the space, here we will discuss only two of them.

3.1 The Karate Club Network

Fig.3 presents the community structure of the karate club network extracted by the probabilistic NCMA algorithm. The number on the left of each row denotes the vertex identifier and the binary tree denotes the hierarchy of communities, in which three communities are detected. We can see that the two largest groups are exactly identical with the actual division as shown in Fig.1. In addition, the NCMP also predicts a potential division of the group led by the administrator in the future.

Fig. 3. The output of the probabilistic NCMA algorithm for the karate club network of Fig.1

3.2 Applying the Probabilistic NCMA Algorithm to a Semantic Social Network

Flink is a social network with semantic information. It describes the scientific collaborations among 681 Semantic Web researchers[12]. It is constructed based on Semantic Web technologies and all related semantic information is extracted automatically from

"Web-accessible information sources", such as "Web pages, FOAF profiles, email lists and publication archives. The directed arc between vertices, as shown in Fig. 4(a), denotes the "know relationship" indicating scientific activities happened between researchers. The weight on each arc quantitatively measures the strength of such a "know relationship". We select Flink as the experimental testbed for our algorithm mainly because it contains rich semantic information and statistical information of the network which allow us to analyze the practical meaning of each community as detected by our algorithm. Additionally, so far Flink has not provided useable information related to its community structure, and we hope our work can supplement this important information to it.

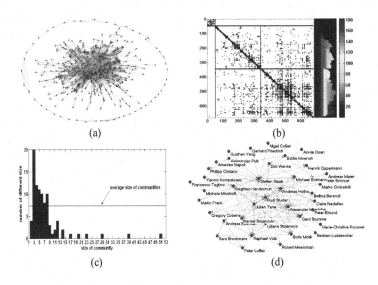

Fig. 4. The results of the probabilistic NCMA algorithm for the Flink network

Fig.4(a) shows the network structure of the Flink, and Fig.4(b) presents the output of the NCMA algorithm, in which each black dot denotes an arc ignoring its weight and the grey bar at the right denotes the hierarchical community structure with different grey level. Three biggest groups are detected which contains 51, 294, and 336 vertices, respectively. The first group is almost completely separated from the entire network, and its 49 member constitute the outer cycle of Fig.4(a). In total, 93 communities are detected and the average size of community is 7.32, as shown in Fig.4(c). Approximately, we can observe a power-law emerging: Most communities have a small size, while a small number of communities contain quite a large number of members. As an example, we take a look at the second largest communities in which Steffen Staab, with the biggest centrality, is the center, as shown in Fig.4(d). After carefully checking their profiles one by one, we can discover an interesting fact: the community is highly location-related. In it, 22 of 39 come from Germany, 21 come from the same city, Karlsruhe, and 12 out of such 21 come from the same university, the University of Karlsruhe where Steffen Staab works. Also we can note that the community is research topic related. Currently, Flink does not cover the research interest information of all people, we have to estimate their respective research interests according to their published papers. In this community, most members are interested in the topic related to learning.

4 Conclusions

In this paper, we have presented a probabilistic NCMP algorithm to mine network communities based on a Markov random walk model. We have validated it using various types of networks to show confirm its performance. As demonstrated in our experiments, the algorithm is very efficient with a linear time, with respect to the size of networks. In addition, it is accurate with a good clustering capability. Finally, our algorithm can automatically detect a hierarchical community structure without any prior knowledge, such as the number and the size of communities.

Acknowledgements

The authors would like to acknowledge the following support for their research on network community mining: The Center for e-Transformation Research at Hong Kong Baptist University; Major State Basic Research Development Program of China (973 Program) (2003CB317001); National Natural Science Foundation of China grant (G60503016).

References

[1] Zachary, W.W.: An Information Flow Model for Conflict and Fission in Small Groups. J. Anthropological Research 33 (1977) 452–473

[2] Flake, G.W., Lawrence, S., Giles, C.L., Coetzee, F.M.: Self-Organization and Identification of Web Communities. IEEE Computer 35 (2002) 66-71

[3] Wilkinson, D.M., Huberman, B.A.: A method for finding communities of related genes. Proc. Nat'l Academy of Science 101 (2004) 5241-5248

[4] Shi, J., Malik, J.: Normalized Cuts and Image Segmentation. IEEE Tans. On Pattern analysis and machine Intelligent 22 (2000) 888-904

[5] Fiedler, M.: Algebraic Connectivity of Graphs. Czechoslovakian Math. J. 23 (1973) 298-305

[6] Kernighan, B.W., Lin, S.: An Efficient Heuristic Procedure for Partitioning Graphs. Bell System Technical 49 (1970) 291-307

[7] Scott, J.: Social Network Analysis: A Handbook. 2nd edn. Sage Publications, London, (2000)

[8] Girvan, M., Newman, M.E.J.: Community Structure in Social and Biological Networks. Proc. Nat'l Academy of Science 9 (2002) 7821-7826

[9] Tyler, J.R., Wilkinson, D.M., Huberman, B.A.: Email as Spectroscopy: Automated Discovery of Community Structure within Organizations. Proc. 1st Int'l Conf. Communities and Technologies, Kluwer (2003)

[10] Radicchi, F., Castellano, C., Cecconi, F., Loreto, V., Parisi, D.: Defining and Identifying Communities in Networks. Proc. Nat'l Academy of Science, 101 (2004) 2658-2663

[11] Kleinberg, J.M.: Authoritative Sources in a Hyperlinked Environment. Proc. 9th Ann. ACM-SIAM Symp. Discrete Algorithms (1998) 668-677

[12] http://flink.semanticweb.org/

A New Approach to Underdetermined Blind Source Separation Using Sparse Representation

Hai-Lin Liu and Jia-Xun Hou

Faculty of Applied Mathematics,
Guangdong University of Technology,
Guangzhou, China
lhl@scnu.edu.cn, jiaxun_hou@yahoo.com.cn

Abstract. This paper presents a new approach to blind separation of sources using sparse representation in an underdetermined mixture. Firstly, we transform the observations into the new ones within the generalized spherical coordinates, through which the estimation of the mixing matrix is formulated as the estimation of the cluster centers. Secondly, we identify the cluster centers by a new classification algorithm, whereby the mixing matrix is estimated. The simulation results have shown the efficacy of the proposed algorithm.

Keywords: Sparse representation, blind source separation, underdetermined mixture.

1 Introduction

Blind separation of sources using sparse representation in an underdetermined mixture has recently received wide attention in the literature, e.g. see [1]-[7], because of its attractive applications in wireless communication, time series analysis, bioinformatics, to name a few. In this paper, we consider the following underdetermined blind source separation (BSS) model only:

$$x(t) = As(t), \tag{1}$$

where $A \in \Re^{m \times n}$ is an unknown mixing matrix, $s(t) = [s_1(t), s_2(t), \ldots, s_n(t)]^T$ is an n-dimensional hidden source vector at time step t, and $x(t) = [x_1(t), x_2(t), \ldots, x_m(t)]^T$ is the corresponding m-dimensional observation vector with $m < n$. The objective of the problem is to estimate the wave-form of the source signals from the q observations, denoted as $x(1), x(2), \ldots, x(q)$ without knowing A.

Typically, such a problem can be performed with two stages: (1) the estimation of the mixing matrix A, and (2) identify the source signals based on the estimate of A. If the source signals are strictly sparse, i.e., there is one and only one component of $s(t)$ for each t, $t = 1, 2, \ldots, q$, to be non-zero, a two-stage algorithm can successfully identify the source signals. Unfortunately, from a practical viewpoint, the sources may not be sparse enough. Under the circumstances, it is nontrivial to estimate the source signals. In the literature, some works have been

J.T. Yao et al. (Eds.): RSKT 2007, LNAI 4481, pp. 276–283, 2007.

done towards solving this problem. For example, Bofill and Zibulevsky [1] have proposed a contrast function to estimate the mixing matrix A and then identified the source signals using a short-path separation criterion. The paper [1] has demonstrated the success of this method in two-dimensional observation space, i.e. $m = 2$. However, it is not applicable to the situations as $m \geq 3$. Further, this method introduces a lot of parameters, whose values are hard to be determined in advance. In the work of [2], the time-frequency masks are created to partition one of the mixtures into the original sources. The experimental results on speech mixtures have verified this technique, which, however, is somewhat sensitive to the sparseness of the sources. Subsequently, the paper [3] has further developed a generalized version of the algorithm in [2]. Nevertheless, this generalized algorithm involves a number of parameters and the computation is laborious. The basic procedure in literature [6] and [7] is to parametrize the mixing matrix in spherical coordinates, to estimate the projections of the maxima of the multidimensional PDF that describes the mixing angles through the marginals, and to reconstruct the maxima in the multidimensional space from the projections.

In this paper, we will present a new two-stage algorithm for the underdetermined BSS problem as described in Eq.(1). In this algorithm, we first transform the observations into the new ones within the generalized spherical coordinates, whereby the estimation of the mixing matrix can be formulated as the estimation of the cluster centers. Then, we identify the cluster centers by using a new classification algorithm, through which the mixing matrix is therefore estimated. Compared to the existing methods, the advantages of the proposed algorithm are as follows:

1. Its performance is somewhat insensitive to the source sparseness in comparison with the k-means based BSS algorithms and those in [2,5];
2. The number of parameters involving in the proposed algorithm is less than the one in [3], whereby saving the computation costs.

The simulation results have shown the efficacy of the proposed algorithm.

2   Transformation of Observations into Generalized Spherical Coordinates

Following the model of Eq.(1), we can re-write $x(t)$, $t = 1, 2, \ldots, q$, by:

$$x(t) = s_1(t)a_1 + s_2(t)a_2 + \ldots + s_n(t)a_n, \tag{2}$$

where a_j is the jth column of the mixing matrix A, i.e., $A = (a_1, a_2, \ldots, a_n)$. When the sources are sparse enough, there is one and only one to be non-zero among $s_1(t)$, $s_2(t)$, $\ldots$, $s_n(t)$. Under the circumstances, a clustering based method can provide an efficient way to estimate the value of A. Unfortunately, the sources are often not sparse. To make the sources become sparse, one possible

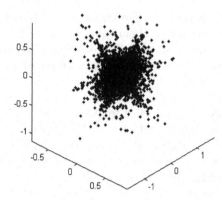

Fig. 1. Three observed signals applied a discrete wavelet transformation

way is to transform the sources into the spectrum ones via a Fourier transformation or a wavelet packet transformation. However, such a simple transformation cannot make the sources sparse enough as shown in Figure 1, where the sources are four music signals and are transformed into time-frequency by applying a discrete wavelet packets transformation to three observed signals. In the following, we will map the observations into the generalized spherical coordinates. For simplicity, we assume that the first component $x_1(t)$ of an $\boldsymbol{x}_t$ is non-negative, i.e., all observations $\boldsymbol{x}(t)$s are in half the space of $\Re^m$. Otherwise, we may let $\boldsymbol{x}(t) = -\boldsymbol{x}(t)$ as $x_1(t) < 0$. Subsequently, we make the following transformation:

1. As $m = 2$, we make the polar-coordinate transformation:

$$\begin{cases} x_1(t) = \rho(t)\cos\theta_1(t) \\ x_2(t) = \rho(t)\sin\theta_1(t), \end{cases} \tag{3}$$

where $0 \leq \theta_1(t) \leq \pi$.
2. As $m \geq 3$, we make the generalized-spherical-coordinate transformation:

$$\begin{cases} x_1(t) &= \rho(t)\cos\theta_1(t) \\ x_2(t) &= \rho(t)\sin\theta_1(t)\cos\theta_2(t) \\ \quad\vdots &\quad\vdots \\ x_{m-1}(t) &= \rho(t)\sin\theta_1(t)\sin\theta_2(t)\ldots\sin\theta_{m-2}(t)\cos\theta_{m-1}(t) \\ x_m(t) &= \rho(t)\sin\theta_1(t)\sin\theta_2(t)\ldots\sin\theta_{m-2}(t)\sin\theta_{m-1}(t), \end{cases} \tag{4}$$

where $0 \leq \theta_1(t) \leq \frac{\pi}{2}$, $0 \leq \theta_2(t),\ldots,\theta_{m-1}(t) \leq 2\pi$. Subsequently, we can compute the values of $\rho(t)$ and $\theta_j(t)$s with $t = 1, 2, \ldots, q$, and $j = 1, 2, \ldots, m-1$.

Among all $\mathbf{s}(t)$s, there must exist many of source column vectors that are sparse enough, i.e. there is only one nonzero entry in each column vector. Suppose there are q_k source column vectors with only the jth component being nonzero, we re-write them as $s(i_1^k), s(i_2^k), \ldots, s(i_{q_k}^k)$ $(k = 1, 2, \ldots, n)$. According to Eq.(2), we then have:

$$\begin{aligned}
\mathbf{x}(t) &= s_1(t)\mathbf{a}_1, \ t = i_1^1, i_2^1, \ldots, i_{q_1}^1; \\
\mathbf{x}(t) &= s_2(t)\mathbf{a}_2, \ t = i_1^2, i_2^2, \ldots, i_{q_2}^2; \\
&\vdots \\
\mathbf{x}(t) &= s_n(t)\mathbf{a}_n, \ t = i_1^n, i_2^n, \ldots, i_{q_n}^n.
\end{aligned} \tag{5}$$

When $t = i_1^k, i_2^k, \ldots, i_{q_k}^k$, the vectors $(\rho(t), \theta_1(t), \theta_2(t), \ldots, \theta_{m-1}(t))^T$s transformed by $\mathbf{x}(t)$s via generalized spherical-coordinate have the same values in all the components except for the first one $\rho(t)$s. We let them be a vector without the first one:

$$\theta^k = (\theta_1^k, \theta_2^k, \ldots, \theta_{m-1}^k)^T, (k = 1, 2, \ldots, n). \tag{6}$$

Therefore, the problem of estimating the n column vectors of mixing matrix $\mathbf{A}$ has been changed into the problem of finding out the n points under the coordinate of $\theta_1, \theta_2, \ldots, \theta_{m-1}$ by the generalized-spherical-coordinate transformation.

3   Algorithm for Estimating the Mixing Matrix

As shown in Fig.1, the sources near origin are not so sparse. In order to remove these sources which are not sparse in the time-frequency domain, we set an appropriate parameter r_0, and remove those $\mathbf{x}(t)$s that satisfy the condition $\rho(t) < r_0$. Subsequently, the number of the corresponding points $(\theta_1(t), \theta_2(t), \ldots, \theta_{m-1}(t))^T$ $(t = 1, 2, \ldots, q)$ under the coordinate constructed by $\theta_1, \theta_2, \ldots, \theta_{m-1}$ is reduced. It is well known that many points stay around the points $(\theta^1, \theta^2, \ldots, \theta^n)$ in Eq.(5); while the others are dispersal. This feature coincides the insufficient sparseness of the source signals. Hence, we propose the following algorithm to estimate the mixing matrix based on above property.

Firstly, we consider the entries of the first axis and remove the disturbing ones. Let the total number of the points in the coordinate $\theta_1, \theta_2, \ldots, \theta_{m-1}$ be q^1. Also, let m^1, M^1 be the minimum and maximum of the first component corresponding to these points, respectively. Then, we divide the interval $[m^1, M^1]$ into L slide windows with equal size and count the number of points in each window. Suppose the average number in each slide window is p^1 with $p^1 = \frac{q^1}{L}$. We remove those windows in which the number of points is less than rp^1. Here r is a parameter greater than 1. Subsequently, we combine the adjacent subintervals. As a result, we obtain l_1 new slide windows, denoted as $I_1^1, I_2^1, \ldots, I_{l_1}^1$. Secondly, we remove the disturbing entries of the second axis. Similar to the process of the first axis, we let q_j^2 be the total number of the points, and m_j^2, M_j^2 be the minimum and maximum of the second component corresponding to these points in each slide window I_j^1, respectively. Again, we divide each window $[m_j^2, M_j^2]$ into L slide window with equal size. Then, we remove the windows with the number of points less than rp_j^2, $(p_j^2 = \frac{q_j^2}{L})$, and combine the adjacent subintervals to obtain new slide windows.

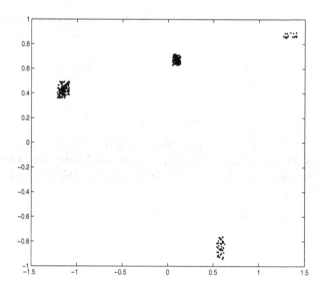

Fig. 2. Three observed signals deleted disturbing points

In this way, we can remove the disturbing entries from the 3rd to nth axis, and we obtain many slide windows. The each window is processed once again by the above way, and finally obtain the results as shown in Fig.2. After removing all of disturbing entries, we denote the final slide windows as $I_1^n, I_2^n, \ldots, I_{l_n}^n$, where each window I_j^n contains Q_j points with:

$$
(\theta_1(i_1), \theta_2(i_1), \ldots, \theta_{m-1}(i_1))^T,
$$
$$
(\theta_1(i_2), \theta_2(i_2), \ldots, \theta_{m-1}(i_2))^T,
$$
$$
\vdots
$$
$$
(\theta_1(i_{Q_j}), \theta_2(i_{Q_j}), \ldots, \theta_{m-1}(i_{Q_j}))^T,
$$
$$
(j = 1, 2, \ldots, l_n).
$$

Since the source signals are not sparse enough, we calculate

$$
mean(\theta_1^j) = \frac{1}{Q_j} \sum_{k=1}^{Q_j} \theta_1(i_k),
$$
$$
mean(\theta_2^j) = \frac{1}{Q_j} \sum_{k=1}^{Q_j} \theta_2(i_k),
$$
$$
\vdots
$$
$$
mean(\theta_{m-1}^j) = \frac{1}{Q_j} \sum_{k=1}^{Q_j} \theta_{m-1}(i_k),
$$
$$
(j = 1, 2, \ldots, l_n).
$$

Let $\rho(t) = 1, \theta_1(t) = mean(\theta_1^j), \theta_2(t) = mean(\theta_2^j),\ldots,\theta_{m-1}(t) = mean(\theta_{m-1}^j)$, we can get the estimated vectors of a_j from Eq.(3) or Eq.(4) $(j = 1, 2, \ldots, l_n)$.

Algorithm :

Step 1: Use the Fourier or discrete wavelet packets transformation to transform the observed signals to the frequency domain or time-frequency domain, and make them sparse. In the following, we assume $\mathbf{x}(t)$ is the data vector on frequency domain or time-frequency domain $(t = 1, 2, \ldots, q)$.

Step 2: Transform the points into the sphere coordinates, and define the vector as $(\rho(t), \theta_1(t), \theta_2(t), \ldots, \theta_{m-1}(t))^T$ with $t = 1, 2, \ldots, q$.

Step 3: Given the parameter r_0, we remove the column vectors $(\rho(t), \theta_1(t), \theta_2(t), \ldots, \theta_{m-1}(t))^T$, $(t = 1, 2, \ldots, q)$ in the data vectors, which satisfy $\rho(t) < r_0$. Also, let the remaining vectors be $(\rho(t), \theta_1(t), \theta_2(t), \ldots, \theta_{m-1}(t))^T$, (t = 1, 2, ..., q_1)

Step 4: We set $k = 1$.

Step 5: Set the parameter L and r. We calculate $m^k = \min_{1 \leq t \leq q} \theta_k(t)$ and $M^k = \max_{1 \leq t \leq q} \theta_k(t)$. Further, we divide the interval $[m^k, M^k]$ into L slide windows with equal-size, and let them be $I_i^k = [m^k + (j-1)\dfrac{M^k - m^k}{L}, m^k + j\dfrac{M^k - m^k}{L})$ $(j = 1, 2, \ldots, L-1)$, $I_L^k = [m^k + (L-1)\dfrac{M^k - m^k}{L}, m^k + L\dfrac{M^k - m^k}{L}]$. We remove the windows I_j^k if $q_j^k < r\frac{q_1}{L}$ (j=1,2,...,L), where q_j^k is the number of vectors the number of vectors of $\theta_k(t)$ belong to the window I_j^k.

Step 6: Combine the adjacent slide windows, and re-write them as $I_j^k, j = 1, 2, \ldots, L_k$.

Step 7: If $k < m - 1$, let $k = k + 1$, and goto **Step 5**.

Step 8: If $k = m - 1$, after removing the disturbing entries, the vectors $(\theta_1(t), \theta_2(t), \ldots, \theta_{m-1}(t))^T$ belong to L_{m-1} hyperrectangle regions. For the points of each hyperrectangle region, process **Step 4** to **Step 7** once more.

Step 9: Calculate the average value of vectors $(\rho(t), \theta_1(t), \theta_2(t), \ldots, \theta_{m-1}(t))^T$ $(t = 1, 2, \ldots, q_1)$ from the second axes to the last axes, which belong to the slide window I_j^{m-1} $(j = 1, 2, \ldots, L_{m-1})$, and let $\rho(t) = 1, \theta_1(t) = mean(\theta_1), \theta_2(t) = mean(\theta_2), \ldots, \theta_{m-1} = mean(\theta_{m-1})$. Then we get the estimated vectors of $a_j, j = 1, 2, \ldots, L_{m-1}$ from Eq.(3) or Eq.(4). (Note that $L_{m-1} = n$)

4   Simulation Examples

To demonstrate the efficacy of our algorithm, we compared the estimated mixing matrix with the original mixing matrix in the examples. In most cases, the sources are not sparse enough. To obtain a more sparse representation, we should apply the Fourier transformation or a wavelet packet transformation to them. Note that the sparseness of them influences the precision of our algorithm. The following examples come from Bofill's web site, *http://www.ac.upc.es/homes/pau/*.

Example 1. In this example, the observations are three signals mixed by four music signals via the mixing matrix.

$$A = \begin{bmatrix} -0.7797 & -0.3703 & 0.1650 & 0.5584 \\ -0.0753 & 0.8315 & 0.6263 & 0.3753 \\ -0.6215 & -0.4139 & 0.7619 & -0.7397 \end{bmatrix}.$$

We apply the discrete wavelet packet transformation to make it sparse by transforming these signals from the time domain to the time-frequency domain. Subsequently, the estimated matrix obtained by our algorithm of is:

$$\tilde{A} = \begin{bmatrix} -0.7797 & -0.3773 & 0.1652 & 0.5435 \\ -0.0737 & 0.8284 & 0.6271 & 0.3788 \\ -0.6239 & -0.4137 & 0.7611 & -0.7489 \end{bmatrix}.$$

Comparing the columns of $\tilde{A}$ with those of A, we can see that the algorithm works well.

Example 2. In this example, the observations are three signals mixed by six flutes signals with

$$A = \begin{bmatrix} -0.1670 & -0.3558 & 0.5703 & -0.7712 & -0.5104 & -0.1764 \\ 0.6337 & -0.9265 & 0.2944 & -0.5643 & 0.2496 & -0.5603 \\ -0.7552 & -0.1221 & -0.7668 & -0.2943 & 0.8228 & -0.8092 \end{bmatrix}.$$

We apply the Fourier transformation to make it sparse by transforming these signals from the time domain to the frequency domain. The estimated matrix obtained by our algorithm is

$$\tilde{A} = \begin{bmatrix} 0.1695 & 0.3588 & 0.5659 & 0.7713 & 0.5100 & 0.1776 \\ -0.6358 & 0.9288 & 0.3003 & 0.5680 & -0.2480 & 0.5580 \\ 0.7529 & 0.1245 & -0.7677 & 0.2869 & -0.8236 & 0.8105 \end{bmatrix}.$$

Comparing the columns of $\tilde{A}$ with those of $\mathbf{A}$, we can see once again that the algorithm works well.

5   Conclusion

This paper has presented a new algorithm to blindly separate the sources using sparse representation in an underdetermined mixture. In our method, we transform the observations into the new ones within the generalized spherical coordinates, through which the estimation of the mixing matrix is formulated as the estimation of the cluster centers. Subsequently, we identify the cluster centers by a new classification algorithm, whereby the mixing matrix is estimated. The numerical simulations have demonstrated the efficiacy of our algorithm.

References

1. P. Bofill and M. Zibulevsky.: Underdetermined Blind Source Separation Using Sparse Representation, Signal Processing. Vol. **81**, No. **11** (2001) 2353–2362
2. O. Yilmaz and S. Rickard.: Blind Separation of Speech Mixtures via Time-Frequency Masking, IEEE Transactions on Signal Processing. Vol. **52**, No. **7** (2004) 1830–1847
3. Y. Li, S. Amari, A. Cichocki, Daniel W.C. Ho and S. Xie.: Underdetermined Blind Source Separation Based on Sparse Representation, IEEE Transactions on Signal Processing. Vol. **54**, No. **22** (2006) 423–437
4. M. Zibulevsky and B.A. Pearlmutter.: Blind Source Separation by Sparse Decomposition, Neural Computation. Vol. **13**, No. **4** (2001) 863–882
5. F. Abrard and Y. Deville.: A Time-Frequency Blind Signal Separation Method Applicable to Underdetermined Mixtures of Dependent Sources, Signal Processing. Vol. **85**, No. **77**(2005) 1389–1403
6. L. Vielva, Y. Pereiro, D. Erdogmus, J. Pereda, and J. C. Principe.: Inversion techniques for underdetermined BSS in an arbitrary number of dimensions. In: Proc. 4th International Conference on Independent Component Analysis and Blind Source Separation (ICA'03), Nara, Japan. Apr (2003) 131-136
7. L. Vielva, I. Santamaria, D. Erdogmus, and J. C. Principe.: On the estimation of the mixing matrix for underdetermined blind source separation in an arbitrary number of dimensions. In: Proc. 5th International Conference on Independent Component Analysis and Blind Source Separation (ICA'04), Granada, Spain. Sep (2004) 185-192

Evolutionary Fuzzy Biclustering of Gene Expression Data

Sushmita Mitra[1], Haider Banka[2], and Jiaul Hoque Paik[1]

[1] Machine Intelligence Unit, Indian Statistical Institute, Kolkata 700 108, India
{sushmita,jia_t}@isical.ac.in
[2] Centre for Soft Computing Research, Indian Statistical Institute,
Kolkata 700 108, India
hbanka_r@isical.ac.in

Abstract. Biclustering or simultaneous clustering attempts to find maximal subgroups of genes and subgroups of conditions where the genes exhibit highly correlated activities over a range of conditions. The possibilistic approach extracts one bicluster at a time, by assigning to it a membership for each gene-condition pair. In this study, a novel evolutionary framework is introduced for generating optimal fuzzy possibilistic biclusters from microarray gene expression data. The different parameters controlling the size of the biclusters are tuned. The experimental results on benchmark datasets demonstrate better performance as compared to existing algorithms available in literature.

Keywords: Microarray, Genetic algorithms, Possibilistic clustering, Optimization.

1 Introduction

It is often observed in microarray data that a subset of genes are coregulated and coexpressed under a subset of conditions, but behave almost independently under other conditions. Here the term "conditions" can imply environmental conditions as well as time points corresponding to one or more such environmental conditions. Biclustering attempts to discover such local structure inherent in the gene expression matrix. It refers to the simultaneous clustering of both genes and conditions in the process of knowledge discovery about local patterns from microarray data [1]. This also allows the detection of overlapped groupings among the biclusters, thereby providing a better representation of the biological reality involving genes with multiple functions or those regulated by many factors. For example, a single gene may participate in multiple pathways that may or may not be co-active under all conditions[2].

A good survey on biclustering is available in literature [3], such as: i)Iterative row and column clustering combination [4], ii) Divide and conquer [5], iii) Greedy iterative search [1,6], iv) Exhaustive biclustering enumeration [7], v) Distribution parameter identification [8].

J.T. Yao et al. (Eds.): RSKT 2007, LNAI 4481, pp. 284–291, 2007.

The pioneering work by Cheng and Church [1] employs a set of heuristic algorithms to find one or more biclusters, based on a uniformity criteria. One bicluster is identified at a time, iteratively. There are iterations of masking null values and discovered biclusters (replacing relevant cells with random numbers), coarse and fine node deletion, node addition, and the inclusion of inverted data. The computational complexity for discovering k biclusters is of the order of $O(mn \times (m + n) \times k)$, where m and n are the number of genes and conditions respectively. Here similarity is computed as a measure of the coherence of the genes and conditions in the bicluster. Although the greedy local search methods are by themselves fast, they often yield suboptimal solutions.

Genetic algorithms, with local search, have been developed for identifying biclusters in gene expression data [9]. Sequential evolutionary biclustering (SEBI) [10] detects high quality overlapped biclusters by introducing the concept of penalty into the fitness functions, in addition to some constrained weightage among the parameters.

Recently a fuzzy possibilistic biclustering has been developed [11] to include the concept of membership for realistically representing each gene-condition pair. We present here a novel approach to evolutionary tuning of the different parameters that control the size of the generated biclusters.

The rest of the paper is organized as follows. Section 2 introduces the concept of biclustering and the fuzzy possibilistic approach. The proposed evolutionary modeling is described in Section 3. The experimental results, along with comparative study, are provided in Section 4. Finally, Section 5 concludes the article.

2   Fuzzy Possibilistic Biclustering

A bicluster is defined as a pair (g, c), where $g \subseteq \{1, \ldots, p\}$ is a subset of genes and $c \subseteq \{1, \ldots, q\}$ is a subset of conditions/time points. The optimization task [1] is to find the largest bicluster that does not exceed a certain homogeneity constraint. The size (or volume) $f(g, c)$ of a bicluster is defined as the number of cells in the gene expression matrix E (with values e_{ij}) that are covered by it. The homogeneity $\mathcal{G}(g, c)$ is expressed as a mean squared residue score; it represents the variance of the selected genes and conditions.

A user-defined threshold δ represents the maximum allowable dissimilarity within the bicluster. For a good bicluster, we have $\mathcal{G}(g, c) < \delta$ for some $\delta \geq 0$.

Often centralized clustering algorithms impose a *probabilistic constraint*, according to which the sum of the membership values of a point in all the clusters must be equal to one. Although this competitive constraint allows the unsupervised learning algorithms to find the barycenter (center of mass) of fuzzy clusters, the obtained evaluations of membership to clusters are not interpretable as a *degree of typicality*. Moreover isolated outliers can sometimes hold high membership values to some clusters, thereby distorting the position of the centroids. The possibilistic approach to clustering [12] assumes that the membership function of a data point in a *fuzzy* set (or cluster) is absolute, *i.e.*, it is an evaluation of a degree of typicality not depending on the membership values of the same point in other clusters.

Here we generalize the concept of biclustering in the possibilistic framework [11]. For each bicluster we assign two vectors of membership, one for the rows and one other for the columns, denoting them respectively by **a** and **b**. In a crisp set framework row i and column j can either belong to the bicluster ($a_i = 1$ and $b_j = 1$) or not ($a_i = 0$ or $b_j = 0$). An element x_{ij} of input data matrix X belongs to the bicluster if both $a_i = 1$ and $b_j = 1$, *i.e.*, its membership u_{ij} to the bicluster may be defined as

$$u_{ij} = \text{and}(a_i, b_j), \tag{1}$$

with the cardinality of the bicluster expressed as

$$n = \sum_i \sum_j u_{ij} . \tag{2}$$

A fuzzy formulation of the problem can help to better model the bicluster and also to improve the optimization process. Now we allow membership u_{ij}, a_i and b_j to lie in the interval $[0, 1]$. The membership u_{ij} of a point to the bicluster is obtained by the average of row and column membership as

$$u_{ij} = \frac{a_i + b_j}{2}. \tag{3}$$

The fuzzy cardinality of the bicluster is again defined as the sum of the memberships u_{ij} for all i and j from eqn. (2). In the gene expression framework we have

$$\mathcal{G} = \sum_i \sum_j u_{ij} d_{ij}^2, \tag{4}$$

where

$$d_{ij}^2 = \frac{(e_{ij} + e_{gc} - e_{ic} - e_{gj})^2}{n}, \tag{5}$$

$$e_{ic} = \frac{\sum_j u_{ij} e_{ij}}{\sum_j u_{ij}}, \tag{6}$$

$$e_{gj} = \frac{\sum_i u_{ij} e_{ij}}{\sum_i u_{ij}}, \tag{7}$$

$$e_{gc} = \frac{\sum_i \sum_j u_{ij} e_{ij}}{\sum_i \sum_j u_{ij}}. \tag{8}$$

The objective is to maximize the bicluster cardinality n while minimizing the residual $\mathcal{G}$ in the fuzzy possibilistic paradigm. Towards this aim we treat one bicluster at a time, with the fuzzy memberships a_i and b_j being interpreted as typicality degrees of gene i and condition j with respect to the bicluster. These requirements are fulfilled by minimizing the functional J_B as

$$J_B = \sum_i \sum_j \left(\frac{a_i + b_j}{2}\right) d_{ij}^2 + \lambda \sum_i (a_i \ln(a_i) - a_i) + \mu \sum_j (b_j \ln(b_j) - b_j). \quad (9)$$

The parameters λ and μ control the size of the bicluster by penalizing small membership values.

Setting the derivatives of J_B with respect to the memberships a_i and b_j to zero, we have the necessary conditions

$$a_i = \exp\left(-\frac{\sum_j d_{ij}^2}{2\lambda}\right), \quad (10)$$

$$b_j = \exp\left(-\frac{\sum_i d_{ij}^2}{2\mu}\right). \quad (11)$$

for the minimization of J_B.

PBC-Algorithm (Possibilistic Biclustering)

1. Initialize the memberships **a** and **b**.
2. Compute $d_{ij}^2 \ \forall i, j$ using eqn. (5).
3. Update $a_i \ \forall i$ using eqn. (10).
4. Update $b_j \ \forall j$ using eqn. (11).
5. **If** $\|\mathbf{a}^{t+1} - \mathbf{a}^t\| < \varepsilon$ and $\|\mathbf{b}^{t+1} - \mathbf{b}^t\| < \varepsilon$ **then Stop.**
   else jump to **Step 2**.

Here the parameter ε is a threshold controlling the convergence of the algorithm, $\mathbf{a}^t$ and $\mathbf{b}^t$ are the vectors of row and column memberships respectively after the t-th iteration. The memberships initialization can be made randomly or by using some a priori information about relevant genes and conditions. After convergence the memberships **a** and **b** can be defuzzified with respect to a threshold (say, 0.5) for subsequent comparative analysis.

3   Evolutionary Possibilistic Biclustering

The size of the biclusters generated by the PBC algorithm is very sensitive to the choice of the values of parameters λ and μ. It may so happen that for some values of parameters either of the following cases occur:

1. No bicluster can be found.
2. A bicluster with a very large mean square residue can be found.

We find that the conditions of eqns. (10)-(11) are necessary for the functional J_B of eqn. (9) to be minimum. Now if the values of λ and μ are increased gradually, the presence of second and third terms in the r.h.s. cause the functional value to

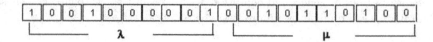

Fig. 1. An equivalent encoded chromosome representing $\lambda = 0.56348$ and $\mu = 180$

go on decreasing. In other words, for unrestricted values of parameters the functional J_B fails to attain a minimum value over a real space. Therefore a proper choice of λ and μ plays a crucial role during the extraction of a large bicluster (within a pre-specified tolerable mean square residue). The proposed evolutionary method determines the optimal values for these parameters by minimizing a fitness function.

Each parameter is represented using 10 bits in a chromosome. We assume that λ takes real values between 0 and 1 and μ takes integer values. A chromosome corresponds to the candidate solution (values of parameters) in the extraction of the largest bicluster. The initial population of encoded parameters is generated randomly. But a more informed initialization scheme (based on some prior knowledge about the problem domain) may be helpful in generating a good quality bicluster faster. Fig. 1 depicts such an encoding scheme for a chromosome.

Let $g_{\lambda,\mu}$, $c_{\lambda,\mu}$ be the selected genes and selected conditions, with cardinality respectively, $|g|_{\lambda,\mu}$, $|c|_{\lambda,\mu}$ and $\mathcal{G}_{\lambda,\mu}$ be the mean square residue, in the bicluster retrieved by the *PBC* algorithm for a particular values of λ and μ. The bicluster size is now expressed as $n_{\lambda,\mu} = |g|_{\lambda,\mu} \times |c|_{\lambda,\mu}$. Threshold δ is used to incorporate a measure of homogeneity of the extracted bicluster. The aim is to minimize the fitness function [9]

$$F(g_{\lambda,\mu}, c_{\lambda,\mu}) = \begin{cases} \frac{1}{|g|_{\lambda,\mu} \times |c|_{\lambda,\mu}} & \text{if } \mathcal{G}_{\lambda,\mu} \leq \delta, \\ \frac{\mathcal{G}_{\lambda,\mu}}{\delta} & \text{otherwise,} \end{cases} \tag{12}$$

such that the size of the bicluster is simultaneously maximized.

Evolutionary PBC Algorithm

GA is employed over an initial population of P chromosomes with single point crossover. The steps of the algorithm are as follows.

1. Generate a random population of P chromosomes encoding parameters λ and μ.
2. Run **PBC** algorithm to compute residual $\mathcal{G}_{\lambda,\mu}$ by eqn. (4).
3. Compute fitness F by eqn. 12.
4. Generate offspring population of size P through selection, crossover and mutation.
5. **Repeat Steps 2 to 4** to tune the parameters λ and μ, **until** the GA converges to a minimum value for the fitness function.

Table 1. Best biclusters for *Yeast* data based on fitness function F

λ	μ	No.of genes	No.of conditions	Bicluster size	Mean squared residue	Fitness value ($\times 10^{-6}$)
0.3251	102	902	11	9922	279.21	100.79
0.2000	300	662	17	11254	298.30	88.86
0.2988	116	871	13	11323	278.81	88.32
0.2880	130	905	13	11765	284.94	85.00
0.2304	190	739	16	11824	299.17	84.57
0.2958	132	977	13	12701	294.19	78.73
0.3020	130	1003	13	13039	298.25	76.69
0.3025	130	1006	13	13078	298.89	76.46
0.3300	**109**	**1020**	**13**	**13260**	**299.35**	**75.41**

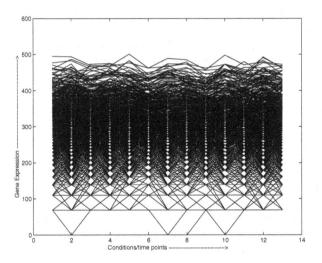

Fig. 2. Largest bicluster profile of size 13260 on *Yeast* data

4 Experimental Results

The two soft computing based biclustering algorithms described in Sections 2 and 3 were implemented on the benchmark gene expression *Yeast cell cycle* data[1] is a collection of 2884 genes (attributes) under 17 conditions (time points), having 34 null entries with -1 indicating the missing values. All entries are integers lying in the range of 0 to 600. The missing values are replaced by random number between 0 to 800, as in [1]. We selected δ=300 for fair comparison with the existing literature.

The crossover and mutation probability were selected as .8 and .02 after several experiments with random seeds. The GA was run for around 150 generations,

---

[1] http://arep.med.harvard.edu/biclustering

Table 2. Comparative study on *Yeast* data

Method	Average residue	Average bicluster size	Average no. of genes	Average no. of conditions	Largest bicluster size
Cheng-Church [1]	204.29	1576.98	167	12	4485
GA [13]	52.87	570.86	191.12	5.13	1408
SEBI [10]	205.18	209.92	13.61	15.25	–
PBC [11]	-	-	-	-	12,857
EPBC	292.35	12018	898.33	13.55	13,260

with the population size varying between 50 and 150 chromosomes. Since GA involves randomness, we have shown the average results. Table 1 summarizes some of the best biclusters using fitness function F, with $P = 50$ after 100 generations. The largest bicluster is found at $\lambda = 0.3300$ and $\mu = 109$, with a minimum fitness value of 75.41×10^{-6} for a size of 13,260 with 1020 genes and 13 conditions. Figure 2 depicts the gene expression profile for the largest bicluster. We observe that the genes in this bicluster demonstrate high coregulation, with expression values in the range 220–400 for the selected set of conditions of the bicluster.

Table 2 provides a comparative study with related algorithms. We observe that the proposed evolutionary PBC gives the best results both in terms of mean square residue (homogeneity) as well as the bicluster size.

5   Conclusions and Discussion

A gene expression data set contains thousands of genes. However, biologists often have different requirements on cluster granularity for different subsets of genes. For some purpose, biologists may be particularly interested in some specific subsets of genes and prefer small and tight clusters. While for other genes, people may only need a coarse overview of the data structure. However, most of the existing clustering algorithms only provide a crisp set of clusters and may not be flexible to different requirements for cluster granularity on a single data set. It is here that the overlapping membership concept of fuzzy biclusters becomes useful.

In this study we have presented a novel evolutionary approach for selecting the different parameters that control the size of the generated fuzzy possibilistic biclusters. We next plan to work on Bezdek's [14] simultaneous clustering of rows and columns of a data matrix.

References

1. Cheng, Y., Church, G.M.: Biclustering of gene expression data. In: Proceedings of ISMB 2000. (2000) 93–103
2. : Special Issue on Bioinformatics. Pattern Recognition **39** (2006)

3. Madeira, S.C., Oliveira, A.L.: Biclustering algorithms for biological data analysis: A survey. IEEE Transactions on Computational Biology and Bioinformatics **1** (2004) 24–45
4. Getz, G., Gal, H., Kela, I., Notterman, D.A., Domany, E.: Coupled two-way clustering analysis of breast cancer and colon cancer gene expression data. Bioinformatics **19** (2003) 1079–1089
5. Hartigan, J.A.: Direct clustering of a data matrix. Journal of American Statistical Association **67(337)** (1972) 123–129
6. Yang, J., Wang, H., Wang, W., Yu, P.: Enhanced biclustering on expression data. In: Proceedings of the Third IEEE Symposium on BioInformatics and Bioengineering (BIBE'03). (2003) 1–7
7. Tanay, A., Sharan, R., Shamir, R.: Discovering statistically significant biclusters in gene expression data. Bioinformatics **18** (2002) S136–S144
8. Lazzeroni, L., Owen, A.: Plaid models for gene expression data. Statistica Sinica **12** (2002) 61–86
9. Bleuler, S., Prelić, A., Zitzler, E.: An EA framework for biclustering of gene expression data. In: Proceedings of Congress on Evolutionary Computation. (2004) 166–173
10. Divina, F., Aguilar-Ruiz, J.S.: Biclustering of expression data with evolutionary computation. IEEE Transactions on Knowledge and Data Engineering **18** (2006) 590–602
11. Filippone, M., Masulli, F., Rovetta, S., Mitra, S., Banka, H.: Possibilistic approach to biclustering: An application to oligonucleotide microarray data analysis. In: Computational Methods in Systems Biology. Volume 4210 of LNCS. Springer Verlag, Heidelberg (2006) 312–322
12. Krishnapuram, R., Keller, J.M.: A possibilistic approach to clustering. Fuzzy Systems, IEEE Transactions on **1** (1993) 98–110
13. Mitra, S., Banka, H.: Multi-objective evolutionary biclustering of gene expression data. Pattern Recognition **39** (2006) 2464–2477
14. Bezdek, J.C.: Pattern Recognition with Fuzzy Objective Function Algorithms. Plenum Press, New York (1981)

Rough Clustering and Regression Analysis

Georg Peters

Munich University of Applied Sciences
Faculty of Computer Science and Mathematics
Lothstrasse34, 80335 Munich, Germany
georg.peters@muas.de

Abstract. Since Pawlak introduced rough set theory in 1982 [1] it has gained increasing attention. Recently several rough clustering algorithms have been suggested and successfully applied to real data. Switching regression is closely related to clustering. The main difference is that the distance of the data objects to regression functions has to be minimized in contrast to the minimization of the distance of the data objects to cluster representatives in k-means and k-medoids. Therefore we will introduce rough switching regression algorithms which utilizes the concepts of rough clustering algorithms as introduced by Lingras at al. [2] and Peters [3].

Keywords: Rough sets, switching regression analysis, clustering.

1 Introduction

The main objective of cluster analysis is to group similar objects together into one cluster while dissimilar objects should be separated by putting them in different clusters.

Besides many classic approaches [4,5] cluster algorithms that utilize soft computing concepts have been suggested, e.g. Bezdek's fuzzy k-means [6] or Krishnapuram and Keller's possibilistic approach [7]. Recently also rough cluster algorithms have gained increasing attention and have been successfully applied to real life data [2,8,9,10,11,12,3,13,14].

Switching regression models [15,16] are closely related to cluster algorithms. However, while cluster algorithms, like the k-means, minimize the cumulated distance between the means and the associated data objects the objective of switching regression analysis is to minimize the cumulated distance between the K regression functions Y_k $(k = 1, ..., K)$ and their associated data objects (Figure 1).

The objective of the paper is to transfer the concepts of rough clustering algorithms to switching regression models and introduce rough versions. We also briefly specify possible areas of applications.

The paper is structured as follows. In the following Section we give a short overview on switching regression models and rough cluster algorithms. In Section 3 we introduce rough switching regression models. In the last Section we give a brief discussion and conclusion.

J.T. Yao et al. (Eds.): RSKT 2007, LNAI 4481, pp. 292–299, 2007.

2   Fundamentals: Switching Regression and Rough Clustering

2.1   Switching Regression Models

Switching regression models were introduced in the fifties of the last century [15]. In the meantime these classic, probabilistic based models have been accompanied by switching regression models that utilize soft computing concepts like fuzzy set theory.

Classic Switching Regression Models. Let us consider a simple data constellation as depicted in Figure 1. Obviously two linear regression functions (Y_1 and Y_2) should adequately represent these data:

$$Y_1(x) = a_{10} + a_{11}x \quad \text{and} \quad Y_2(x) = a_{20} + a_{21}x \tag{1}$$

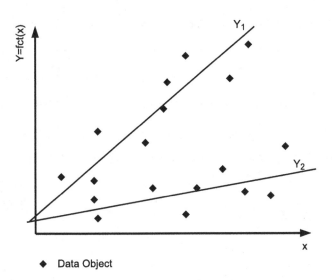

Fig. 1. Switching Regression Analysis

The challenge is to determine which of the two regression functions should represent a certain observation y_i:

$$y_i = \widehat{Y_1}(x_i) = a_{10} + a_{11}x_i + \mu_{1i} \quad \text{or} \quad y_i = \widehat{Y_2}(x_i) = a_{20} + a_{21}x_i + \mu_{2i} \tag{2}$$
$$\text{with} \quad \mu_{1i} \quad \text{and} \quad \mu_{2i} \quad \text{error terms.}$$

To solve this problem in switching regression analysis - the estimation of the parameters a - one can apply Goldfeld and Quandt's D-method [17].

Fuzzy Switching Regression Models. Besides classic switching regression models Hathaway and Bezdek [18] suggested a fuzzy switching regression model which is closely related to Bezdek's fuzzy k-means [6]. Jajuga [19] also proposed a linear switching regression model that consists of a two step process: (1) the data are clustered with the fuzzy k-means, (2) the obtained membership degrees are used as weights in weighted regression analysis.

2.2   Rough Clustering Algorithms

Lingras' Rough k-Means. Lingras et al. rough clustering algorithm belongs to the branch of rough set theory with a reduced set of properties [20]:

1. A data object belongs to no or one lower approximation.
2. If a data object is no member of any lower approximation it is member of two or more upper approximations.
3. A lower approximation is a subset of its underlying upper approximation.

The part of an upper approximation that is not covered by a lower approximation is called boundary area. The means are computed as weighted sums of the data objects $\boldsymbol{X_n}(n = 1, ..., N)$:

$$m_k = \begin{cases} w_L \sum\limits_{\boldsymbol{X_n} \in \underline{C_k}} \frac{\boldsymbol{X_n}}{|\underline{C_k}|} + w_B \sum\limits_{\boldsymbol{X_n} \in C_k^B} \frac{\boldsymbol{X_n}}{|C_k^B|} & \text{for } C_k^B \neq \emptyset \\ w_L \sum\limits_{\boldsymbol{X_n} \in \underline{C_k}} \frac{\boldsymbol{X_n}}{|\underline{C_k}|} & \text{otherwise} \end{cases} \qquad (3)$$

where $|\underline{C_k}|$ is the number of objects in lower approximation and $|C_k^B| = |\overline{C_k} - \underline{C_k}|$ ($\overline{C_k}$ the upper approximation) in the boundary area of cluster k ($k = 1, ..., K$). Then rough cluster algorithm goes as follows:

1. Define the initial parameters: the weights w_L and w_B, the number of clusters K and a threshold ϵ.
2. Randomly assign the data objects to one lower approximation (and per definitionem to the corresponding upper approximation).
3. Calculate the means according to Eq (3).
4. For each data object, determine its closest mean. If other means are not reasonably farer away as the closest mean (defined by ϵ) assign the data object to the upper approximations of these close clusters. Otherwise assign the data object to the lower and the corresponding upper approximation of the cluster of its closest mean (see Figure 2).
5. Check convergence. If converged: STOP otherwise continue with STEP 3.

Extensions and Variations of the Rough k-Means. Lingras rough k-means was refined and extended by an evolutionary component. Peters [12,3] presented a refined version of the rough k-means which improves its performance in the presence of outliers, its compliance to the classic k-means, its numerical stability besides others. To initialize the rough k-means one has to select the weights

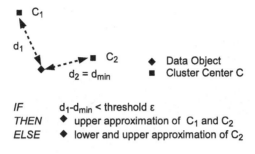

Fig. 2. Lingras' Rough k-Means

of the lower approximation and the boundary area as well as the number of clusters. Mitra [11] argued that a good initial setting of these parameters is one of the main challenges in rough set clustering. Therefore she suggested an evolutionary version of Lingras rough k-means which automates the selection of the initial parameters. And, recently Mitra et al. [10] introduced a collaborative rough-fuzzy k-means.

3   Rough Switching Regression Models

The new rough switching regression models utilize the concepts of rough clustering as suggested by Lingras [8] and Peters [3]. First let us define some terms and abbreviations:

- Data set: $S_n = (y_n, x_n) = (y_n, x_{0n}, ..., x_{Mn})$ for the nth observation and $S = (S_1, ..., S_N)^T$ with $n = 1, ..., N$. The variable y_n is endogenous while $x_n = (x_{0n}, ..., x_{Mn})$ with $m = 0, ..., M$ (features) and $x_{0n} := 1$ represent the exogenous variables.
- Y_k the kth function: $\hat{y}_{kn} = Y_k(x_n) = \sum_{m=0}^{M} a_{km} x_{mn}$ for $k = 1, ..., K$.
- Approximations: $\underline{Y_k}$ is the lower approximation corresponding to the regression function Y_k, $\overline{Y_k}$ the upper approximation and $Y_k^B = \overline{Y_k} - \underline{Y_k}$ the boundary area. This implies $\underline{Y_k} \subseteq \overline{Y_k}$.
- The distance in y between the data object S_n and the regression function Y_k is given by $d(S_n, Y_k) = |y_n - \hat{y}_{kn}|$.

3.1   A First Rough Switching Regression Algorithm Based on Lingras' k-Means

First we present a rough switching regression model based on Lingras' k-Means.

- **Step 0: Initialization**
  (i) Determine the number K of regression functions.
  (ii) Define the weights for the lower approximations and the boundary areas: w_L and w_B with $w_L + w_B = 1$.

(iii) Randomly assign each data object S_n to one lower approximation $\underline{Y_k}$ of the corresponding regression function Y_k.

– **Step 1: Calculation of the New Regression Coefficients**

The new regression coefficients a_{km} are calculated using weighted regression analysis with weights defined as follows:

$$w_{kn} = \begin{cases} w_B & \text{for } S_n \in Y_k^B \\ w_L & \text{for } S_n \in \underline{Y_k} \\ 0 & \text{else} \end{cases} \tag{4}$$

– **Step 2: Assignment of the Data Objects to the Approximations**

(i) For an object S_n determine its closest regression function Y_h (Figure 3):

$$y_{hn}^{min} = d(S_n, Y_h) = \min_k d(S_n, Y_k). \tag{5}$$

Assign S_n to the upper approximation of the function Y_h: $S_n \in \overline{Y_h}$.

(ii) Determine the regression functions Y_t that are also close to S_n. They are not farther away from S_n than $d(S_n, Y_h) + \epsilon$ with ϵ a given threshold:

$$T = \{t : d(S_n, Y_k) - d(S_n, Y_h) \leq \epsilon \wedge h \neq k\}. \tag{6}$$

- If $T \neq \emptyset$ (S_n is also close to at least one other regression function Y_t besides Y_h)
  Then $S_n \in \overline{Y_t}$, $\forall t \in T$.
- **Else $S_n \in \underline{Y_h}$.**

– **Step 3: Checking the Convergence**

The algorithms has converged when the assignments of all data objects to the approximations remain unchanged in the latest iteration i in comparison to iteration $i - 1$.

- **If** the algorithm has not converged **Then** continue with Step 1.
- **Else** STOP.

However, the algorithm has similar weaknesses as Lingras' k-means (see Peters [3] for a detailed discussion). E.g., please note that the algorithm does not enforce that each regression function has two or more data objects in its lower approximation.

3.2   A Rough Switching Regression Algorithm Based on Peters Rough k-Means

– **Step 0: Initialization**

(i) Determine the number K of the regression functions which is limited by: $2 \leq K \leq \frac{N}{2}$ since each regression function should be defined by at least two data points.

(ii) Randomly assign each data object S_n to one and only one lower approximation $\underline{Y_k}$ of the corresponding regression function Y_k so that each regression function has at least two data objects in its lower approximation.

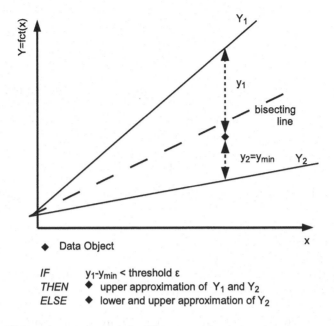

IF y_1-y_{min} < threshold ε
THEN   ◆ upper approximation of Y_1 and Y_2
ELSE   ◆ lower and upper approximation of Y_2

Fig. 3. Assignment of the Objects to the Approximations

- **Step 1: Calculation of the New Regression Coefficients**
  The new regression coefficients a_{km} are calculated using weighted regression analysis (see Eq 4). The weights are defined as follows:

  (i) A data object S_n in lower approximations of a regression functions k is weighted by 1: $w_L = 1$.
  (ii) A data object S_n that is member of b boundary areas is weighted by $w_B = \frac{1}{b}$.

  Alternatively the weights of the lower approximation w_L and the boundary area w_B can be determined by the user.

- **Step 2: Assignment of the Data Objects to the Approximations**

  (i) Assign the data object that best represents a regression function to its lower and upper approximation.
    (a) Find the minimal distance between all regression functions Y_k and all data objects S_n and assign this data object S_l to lower and upper approximation of the regression function Y_h:

$$d(S_l, Y_h) = \min_{n,k} d(S_n, Y_k) \Rightarrow S_l \in \underline{Y_k} \wedge S_l \in \overline{Y_k}. \qquad (7)$$

    (b) Exclude S_l. If this is the second data object that has been assigned to the regression function Y_h exclude Y_h also. If regression functions are left - so far, in Step (a) no data object has been assigned to them - go back to Step (a). Otherwise continue with Step (ii).

(ii) For each remaining data points $S'_{n'}$ ($n' = 1, ..., N'$, with $N' = N - 2K$) determine its closest regression function Y_h:

$$y^{min}_{hn'} = d(S'_{n'}, Y_h) = \min_k d(S'_{n'}, Y_k). \qquad (8)$$

Assign $S'_{n'}$ to the upper approximation of the function h: $S'_{n'} \in \overline{Y_h}$.

(iii) Determine the regression functions Y_t that are also close to $S'_{n'}$. Take the relative distance as defined above where ζ is a given relative threshold:

$$T' = \left\{ t : \frac{d(S'_{n'}, Y_k)}{d(S'_{n'}, Y_h)} \leq \zeta \wedge h \neq k \right\}. \qquad (9)$$

- **If** $T' \neq \emptyset$ ($S'_{n'}$ is also close to at least one other regression function Y_t besides Y_h)
  Then $S'_{n'} \in \overline{Y_t}, \forall t \in T'$.
- **Else** $S'_{n'} \in \underline{Y_h}$.

- **Step 3: Checking the Convergence**
  The algorithms has converged when the assignments of all data objects to the approximations remain unchanged in the latest iteration i in comparison to iteration $i - 1$.
  - **If** the algorithm has not converged **Then** continue with Step 1.
  - **Else** STOP.

4   Discussion and Conclusion

In the paper we proposed rough switching regression models which are based on rough clustering. While classic switching regression models have been extensively applied in economics (e.g. [21,22]) applications to bioinformatics can hardly be found. However Qin et al. [23] suggested the related CORM method (Clustering of Regression Models method) and applied it to gene expression data.

Therefore future work can go in different directions. First, the rough switching regression model should be applied to real life data and compared to classic models, especially in the field of economics. Second, the potential of switching regression (classic, fuzzy, rough) for bioinformatics could be further evaluated.

References

1. Pawlak, Z.: Rough sets. International Journal of Information and Computer Sciences **11** (1982) 145–172
2. Lingras, P., West, C.: Interval set clustering of web users with rough k-means. Technical Report 2002-002, Department of Mathematics and Computer Science, St. Mary's University, Halifax, Canada (2002)
3. Peters, G.: Some refinements of rough k-means clustering. Pattern Recognition **39**(8) (2006) 1481–1491
4. Hartigan, J.: Clustering Algorithms. John Wiley & Sons, Inc., New York, New York (1975)

5. Mirkin, B.: Mathematical Classification and Clustering. Kluwer Academic Publishers, Boston (1996)
6. Bezdek, J.: Pattern Recognition with Fuzzy Objective Algorithms. Plenum Press, New York (1981)
7. Krishnapuram R., K.J.: A possibilistic approach to clustering. IEEE Transactions on Fuzzy Systems **1** (1993) 98–110
8. Lingras, P., West, C.: Interval set clustering of web users with rough k-means. Journal of Intelligent Information Systems **23** (2004) 5–16
9. Lingras, P., Yan, R., West, C.: Comparison of conventional and rough k-means clustering. In: International Conference on Rough Sets, Fuzzy Sets, Data Mining and Granular Computing. Volume 2639 of LNAI., Berlin, Springer (2003) 130–137
10. Mitra, S., Banka, H., Pedrycz, W.: Rough-fuzzy collaboration clustering. IEEE Transactions on Systems, Man and Cybernetics - Part B: Cybernetics **36**(4) (2006) 795–805
11. Mitra, S.: An evolutionary rough partitive clustering. Pattern Recognition Letters **25** (2004) 1439–1449.
12. Peters, G.: Outliers in rough k-means clustering. In: Proceed. First International Conference on Pattern Recognition and Machine Intelligence. Volume 3776 of LNCS., Kolkata, Springer Verlag (2005) 702–707
13. Voges, K., Pope, N., Brown, M.: Cluster Analysis of Marketing Data Examining On-Line Shopping Orientation: A Comparision of k-Means and Rough Clustering Approaches. In: Heuristics and Optimization for Knowledge Discovery. Idea Group Publishing, Hershey PA (2002) 207–224
14. Voges, K., Pope, N., Brown, M.: A rough cluster analysis of shopping orientation data. In: Proceedings Australian and New Zealand Marketing Academy Conference, Adelaide (2003) 1625–1631
15. Page, E.S.: A test for a change in a parameter occurring at an unknown point. Biometrika **42** (1955) 523–527
16. Quandt, R.: The estimation of the parameters of a linear regression system obeying two separate regimes. Journal of the American Statistical Association **53** (1958) 873–880
17. Goldfeld, S., Quandt, R.: Nonlinear Methods in Econometrics. North-Holland, Amsterdam (1972)
18. Hathaway, R., Bezdek, J.: Switching regression models and fuzzy clustering. IEEE Transactions on Fuzzy Systems **1**(3) (1993) 195–204
19. Jajuga, K.: Linear fuzzy regression. Fuzzy Sets and Systems **20** (1986) 343–353
20. Yao, Y., Li, X., Lin, T., Liu, Q.: Representation and classification of rough set models. In: Proceedings Third International Workshop on Rough Sets and Soft Computing, San Jose, CA (1994) 630–637
21. Fuglie, K., Bosch, D.: Economic and environmental implications of soil nitrogen testing: A switching-regression analysis. American Journal of Agricultural Economics **77**(4) (1995) 891–90
22. Ohtani, K., Kakimoto, S., Abe, K.: A gradual switching regression model with a flexible transition path. Economics Letters **32**(1) (1990) 43–48
23. Qin, L., Self, S.: The clustering of regression model method with applications in gene expression data. Biometrics **62**(2) (2006) 526–533

Rule Induction for Prediction of MHC II-Binding Peptides

An Zeng[1], Dan Pan[2], Yong-quan Yu[1], and Bi Zeng[1]

[1] Faculty of Computer, Guangdong University of Technology, Guangzhou,
510006 Guangdong, China
csanzeng@gmail.com, yyq@gdut.edu.cn, z9215@163.com
[2] China Mobile Group Guangdong Co., Ltd., Guangzhou, 510100 Guangdong, China
pandan@gd.chinamobile.com

Abstract. Prediction of MHC (Major Histocompatibility Complex) binding peptides is prerequisite for understanding the specificity of T-cell mediated immunity. Most prediction methods hardly acquire understandable knowledge. However, comprehensibility is one of the important requirements of reliable prediction systems of MHC binding peptides. Thereupon, SRIA (Sequential Rule Induction Algorithm) based on rough set was proposed to acquire understandable rules. SRIA comprises CARIE (Complete Information-Entropy-based Attribute Reduction algorithm) and ROAVRA (Renovated Orderly Attribute Value Reduction algorithm). In an application example, SRIA, CRIA (Conventional Rule Induction Algorithm) and BPNN (Back Propagation Neural Networks) were applied to predict the peptides that bind to HLA-DR4(B1*0401). The results show the rules generated with SRIA are better than those with CRIA in prediction performance. Meanwhile, SRIA, which is comparable with BPNN in prediction accuracy, is superior to BPNN in understandability.

1 Introduction

T lymphocytes play a key role in the induction and regulation of immune responses and in the execution of immunological effector functions [1]. Binding of peptides to MHC (Major Histocompatibility Complex) molecules conveys critical information about the cellular milieu to immune system T cells. Different MHC molecules bind distinct sets of peptides, and only one in 100 to 200 potential binders actually binds to a certain MHC molecules. And it is difficult to obtain sufficient experimental binding data for each human MHC molecule. Therefore, computational modeling of predicting which peptides can bind to a specific MHC molecule is necessary for understanding the specificity of T-cell mediated immunity and identifying candidates for the design of vaccines.

Recently, many methods have been introduced to predict MHC binding peptides. They could be classified as 4 categories: 1) Prediction method based on motif [2]; 2) Prediction method based on quantitative matrices [3]; 3) Prediction method based on structure [4]; 4) Prediction method based on machine learning [5]. Because the methods in category 4 consider the interactive effect among amino acids in all positions of

J.T. Yao et al. (Eds.): RSKT 2007, LNAI 4481, pp. 300–307, 2007.
© Springer-Verlag Berlin Heidelberg 2007

the peptide, their prediction performance has been improved a lot. The involved machine learning approaches are mainly from ANNs (artificial neural networks) and HMMs (Hidden Markov Models). Brusic et al. proposed PERUN method, which combines the expert knowledge of primary anchor positions with an EA (evolutionary algorithm) and ANNs, for prediction of peptides binding to HLA-DRB1*0401 [5].

Category 4 has better prediction performance than other categories when much structure information cannot be obtained since category 4 owns the strongest non-linearity processing capability and generalization ability and self-organization specialty among the four categories. However, category 4 has been mainly focused on the application of ANNs so far. Meanwhile, it is very hard to understand the weights in ANNs and it is very difficult to provide the rules for the experts to review and modify so as to aid them to understand the reasoning processes in another way.

Rough set theory (RS), which was advocated by Pawlak Z. [6] in 1982, gives an approach to automatic rule acquisition, i.e., one might use RS to find the rules describing dependencies of attributes in database-like information systems, such as a decision table. The basic idea of RS used for rule acquisition is to derive the corresponding decision or classification rules through data reduction (attribute reduction and attribution value reduction) in a decision table under the condition of keeping the discernibility unchanged.

The rest of the paper is organized as follows: Section 2 proposes the methodology for prediction of MHC II-binding peptides, which consists of two subparts: peptide pre-processing and the SRIA (Sequential Rule Induction Algorithm) algorithm based on rough set theory. Section 3 describes and discusses the comparable experiment results of various algorithms. Section 4 summarizes the paper.

2 Methodology

The process of prediction of MHC II-binding peptides is composed of two phases: 1) an immunological question is converted into a computational problem with peptide pre-processing, 2) SRIA, which consists of Complete Information-Entropy-based Attribute Reduction sub-algorithm (CARIE) and Renovated Orderly Attribute Value Reduction sub-algorithm (ROAVRA), is advocated to acquire sequential rules from pre-processed peptides.

2.1 Peptide Pre-processing

MHC class II molecules bind peptides with a core region of 13 amino acids containing a primary anchor residue. Analysis of binding motifs suggests that only a core of nine amino acids within a peptide is essential for peptide/MHC binding [7]. It was found that certain peptide residues in anchor positions are highly conserved, and contributed significantly to the binding by their optimal fit to residues in the MHC binding groove [8]. Moreover, evidence further shows that MHC class II-binding peptides contain a single primary anchor, which is necessary for binding, and several secondary anchors that affect binding [5,7]. Thereupon, all peptides with the variable lengths could be reduced to putative binding nonamer cores (core sequences of nine amino acids) or non-binding nonamers.

In terms of domain knowledge about primary anchor positions in reported binding motifs [7], position one (1) in each nonamer corresponds to the primary anchor. Each non-binder is resolved into as many putative non-binder nonamers as its first position is occupied by primary anchor residue. And for binders, after the position one (1) as primary anchor residue is fixed, each binder yields many putative nonamer subsequences. Among these subsequences, the highest scoring nonamer subsequence scored by the optimized alignment matrix is regarded as pre-processed result of the corresponding binding peptide. Here, just like the description in the paper [5], an EA is utilized to obtain the optimized alignment matrix. In this way, the problem of predicting MHC class II-binding peptides is converted into the classification problem. The detailed description of peptide pre-processing is shown in paper [5].

With the pre-processed peptides (nonamers), we can form a decision table where every object represents a nonamer and the numbers of decision attributes and condition attributes respectively are one and 180 (nine positions by 20 possible amino acids at each position, i.e., amino acids are represented as binary strings of length 20, of 19 zeros and a unique position set to one). The values of condition attributes are $\{0, 1\}$ and the values of the decision attribute are $\{0, 1\}$, which corresponds to peptide classes (0: non-binders; 1: binders).

2.2 Sequential Rule Induction Algorithm

Complete Information-Entropy-Based Attribute Reduction Sub-algorithm
Here, we proposed CAR^{IE}. It can acquire an attribute reduct only comprising the essential condition attributes with higher importance measured by the information entropy, while the algorithm in [9] could obtain an attribute subset with redundancy.

Given a decision table $T=(U, A, C, D)$ and a partition of U with classes X_i, $1 \le i \le n$. Here, $C, D \subset A$ be two subsets of attributes A, called condition and decision attributes respectively. We define the entropy of attributes B as in formula (1) [9]:

$$H(B) = -\sum_{i=1}^{n} p(X_i) \log(p(X_i))$$ (1)

where $p(X_i) = |X_i| / |U|$. Here, $|X|$ means the cardinality of set X.

The conditional entropy of D ($U/\text{Ind}(D)=\{Y_1, Y_2, \cdots, Y_m\}$) with reference to another attribute set $B \subseteq C$ ($U/\text{Ind}(B)=\{X_1, X_2, \cdots, X_n\}$) is defined as in formula (2):

$$H(D|B) = -\sum_{i=1}^{n} p(X_i) \sum_{j=1}^{m} p(Y_j | X_i) \log(p(Y_j | X_i))$$ (2)

where $p(Y_j | X_i) = |Y_j \cap X_i| / |X_i|$, $1 \le i \le n$, $1 \le j \le m$.

The relative importance degree of an attribute a for a condition attribute set B ($B \subseteq C$) is defined as in formula (3):

$$Sgf(a, B, D) = H(D|B) - H(D|B \cup \{a\})$$ (3)

For a given decision table $T=(U, A, C, D)$, the detailed steps of the CAR^{IE} sub-algorithm are as follows:

(1) Calculate the conditional entropy $H(D|C)$.

(2) Compute the RDCT, which was detailed in paper [10]. Here, assume the RDCT be DD, and let R be an attribute reduct of the decision table T.

(3) All of the columns of DD are summed transversely and the result is CC.

(4) Among the rows of DD, find out the rows of $\{r_1, r_2, \cdots, r_k\}$ corresponding to the rows of the locations of the minimal elements among CC.

(5) If the minimal element among CC is one, find out the columns $\{c_{r_1}, c_{r_2}, \cdots, c_{r_n}\}$ where the element 1s in the rows $\{r_1, r_2, \cdots, r_k\}$ are located and initialize R be the attribute set responding to the columns $\{c_{r_1}, c_{r_2}, \cdots, c_{r_n}\}$. Otherwise, initialize R be empty.

(6) Compute the conditional entropy $H(D|R)$. Here, if R is empty, $H(D|R) = H(D)$.

(7) Judge if $H(D|R)$ is equal to $H(D|C)$; if not, repeat the steps i~iii till $H(D|R) = H(D|C)$.

   i. $E = C - R$;

   ii. For every attribute a $(a \in E)$, compute the relative importance degree of a for R $SGF(a, R, D)$ according to formula (3).

   iii. Find out maximum $SGF(a, R, D)$, let $R = R \cup \{a\}$.

(8) Call the RJ algorithm in paper [10], and decide whether R is a reduct or not.

(9) If so, the attribute reduct is R. **STOP**.

(10) If not, obtain the possibly redundant attributes with RJ algorithm, and delete the condition attributes with the least $SGF(a, R, D)$ among them one by one till R is an attribute reduct. **STOP**.

Here, RJ algorithm is the complete algorithm for judgment of attribute reduct in paper [10], which can be used to completely and correctly judge whether an attribute subset is an attribute reduct or not. Paper [10] gives the detailed steps and proof about RJ algorithm.

Renovated Orderly Attribute Value Reduction Sub-algorithm

In order to more efficiently acquire the rule set with the stronger generalization capability, we advocate ROAVRA, which combines OAVRA [10] with domain knowledge of the primary anchor positions.

In ROAVRA, firstly, one object in a decision table is taken out in a certain sequence one at a time, and the current object's attribute value is classified. Secondly, a rule is generated according to the classification result. Finally the objects that are consistent with the current rule in the decision table are deleted. The above steps repeat until the decision table is empty. Thus, compared with OAVRA, ROAVRA need not classify the attribute values of the objects consistent with obtained rules so that it can reduce the scanning costs to a great extent.

The description of ROAVRA is as follows:

(1)   Select an object in the decision table in a certain sequence one at a time. Here, we adopted a random sequence.

(2)   For the selected object, classify the attribute value as three classes.

(3)   According to the classification results for the selected object, judge whether the first-class attribute values are enough to constitute a correct rule. If not, one at a time precedently choose the second-class attribute value correspond-ing to an amino acid on the primary anchor position. If all attribute values of the second class have used to compose a rule and a correct rule can't be formed yet, one at a time prece-dently choose the attribute value corresponding to an amino acid on the primary an-chor position among the third class until a correct rule can be generated. Save the obtained rule in the rule set.

(4)   Delete all the objects that are consistent with the current rule in the decision table. The rest of the decision table is saved as the decision table.

(5)   Repeat step (1) – (4) until the decision table is empty. STOP.

The obtained rules are sequential and have the priority order, i.e. the rule generated earlier has the higher priority order. When the rule set is used to make a decision for an unseen instance, the rules must be used in the same sequence as they were pro-duced. If a rule is qualified to making a decision, the others with the lower priority order than its need not to be used.

3   Experiment Results and Discussions

The data set is composed of 650 peptides to bind or not bind to *HLA-DR4 (B1*0401)*, which is provided by Dr. Vladimir Brusic. The lengths of peptides are variable from 9 to 27 amino acids. With the help of SYFPEITHI software [3], the primary anchor of peptides binding to *HLA-DR(B1*0401)* can be obtained. The alignment matrix [5] is used to score each nonamer within the initial peptide after fixing the first position into any one among F, Y, W, I, L, V or M. The highest scoring nonamer sequence is seen as pre-processed results of the corresponding peptide.

Here, 915 pre-processed nonamers are obtained. There are some nonamers with unknown affinity and some inconsistence nonamers (i.e. the same nonamers have different binding affinity) among the 915 nonamers. After removing the inconsistent and unknown nonamers from 915 pre-processed peptides, we have 853 nonamers remained to analyze. The decision table is composed of 853 nonamers (553 non-binders, 300 binders). The numbers of condition attributes and decision attributes are 180 and one respectively.

In the experiment, the decision table is divided into two parts by a 4-fold stratified cross-validation sample method. The following experimentation consists of eight 4-fold stratified cross-validations.

CAR^{IE} sub-algorithm is called to compute an attribute reduct. According to the re-sulting attribution reduct, ROAVRA is used to acquire sequential rules. The rules have been examined and the results are shown in Table 1.

For comparison purposes, two different algorithms are utilized to process the same decision table. The first is CRIA consisting of attribute reduction sub-algorithm [11] and attribute value reduction sub-algorithm [12]. The second is BPNN.

Table 1. Test Results with SRIA

No. of Test	sensitivity (%)	specificity (%)	precision (%)	accuracy (%)
1	81.333	88.788	81.879	86.166
2	81.000	90.054	83.219	86.870
3	81.000	88.969	82.373	86.166
4	80.667	89.693	81.757	86.518
5	79.667	87.884	81.293	84.994
6	78.667	88.427	81.661	84.994
7	78.667	90.235	82.517	86.166
8	78.667	90.958	83.688	86.635
Average (%)	79.959	89.376	82.298	86.064

Table 2. Test Results with CRIA

No. of Test	sensitivity (%)	specificity (%)	precision (%)	accuracy (%)
1	62.333	76.130	83.111	71.278
2	63.667	76.492	80.591	71.981
3	64.000	79.566	82.759	74.091
4	64.667	76.673	79.835	72.450
5	65.333	76.673	86.344	72.685
6	65.667	74.503	82.083	71.395
7	65.667	75.226	82.427	71.864
8	65.667	74.684	78.800	71.512
Average (%)	64.625	76.243	81.994	72.157

Table 3. Test Results with BPNN

No. of Test	sensitivity (%)	specificity (%)	precision (%)	accuracy (%)
1	79.667	91.682	83.860	87.456
2	84.667	91.501	84.385	89.097
3	83.000	91.139	83.557	88.277
4	78.333	90.958	82.456	86.518
5	77.000	92.224	84.307	86.870
6	78.667	89.512	80.272	85.698
7	80.333	92.405	85.159	88.159
8	80.333	92.405	85.159	88.159
Average (%)	80.250	91.478	83.644	87.530

With the help of CRIA, the rules have been acquired with the training part and examined with the test part. The results are shown in Table 2.

The structure of ANNs is 180-4-1 style, i.e., the input layer and hidden layer consist of 180 nodes and 4 nodes respectively, and output layer with a single node. The learning procedure is error back-propagation, with a sigmoid activation function.

Values for learning rate and momentum are 0.2 and 0.9 respectively. The prediction performance of ANNs is shown in Table 3.

From comparisons of the test results listed in table 1, 2 and 3, we can see that the sensitivity, specificity, precision and accuracy with SRIA are much higher than those with CRIA, and very close to those with BPNN. This suggests that SRIA is much better than CRIA in the generalization capability of induced rules though the both algorithms can obtain the plain and understandable rules. In addition, compared with BPNN, SRIA can provide the comprehensible rules that can help experts to understand the basis of immunity.

Table 4 shows a part of rules generated from SRIA in the experimentation.

Table 4. A part of rules generated with SRIA

Rule No.	Antecedents	Consequent
1	1L(1)&2A(0)&2L(0)&2R(0)&2T(0)&3Q(1)&4M(0)&4Q(0)&4V(0)&5A (0)&5L(0)&6A(0)&6S(0)&6T(0)&7L(0)&7P(0)&8L(0)&8R(0)&8S(0)& 9A(0)&9G(0)&9S(0)&9V(0)&9W(0)	0
2	1F(1)&2A(0)&2R(0)&4M(0)&4Q(0)&4V(0)&5A(0)&5L(0)&6A(0)&6T (0)&7L(1)&8R(0)&9A(0)&9G(0)&9S(0)&9V(0)	0
3	2R(0)&6A(1)&8R(0)&9V(1)	1

Here, we can write the third rule in table 4 as "2R(0)&6A(1)&8R(0)&9V(1) 1", i.e., if amino acid code "R" does not appear in the second position of a nonamer and "A" appears in the sixth position and "R" does not appear in the eighth position and "V" appears in the ninth position, the nonamer is classified into "binders".

4 Conclusions

In order to minimize the number of peptides required to be synthesized and assayed and to advance the understanding for the immune response, people have presented many computational models mainly based on ANNs to predict which peptides can bind to a specific MHC molecule. Although the models work well in prediction performance, knowledge existing in the models is very hard to understand because of the inherent "black-box" nature of ANNs and the difficulty of extraction for the symbolic rules from trained ANNs. In fact, comprehensibility is one of the very important requirements of reliable prediction systems of MHC binding peptides.

Thus, SRIA based on RS theory is proposed to acquire the plain and understandable rules. The CAR^{IE} algorithm, which is adopted as a sub-algorithm of SRIA, could compute an attribute reduct only comprising essential and relatively important condition attributes in a decision table composed of 180 condition attributes. The ROAVRA in SRIA is used to extract sequential rules from the reduced decision table based on the attribute reduct. Experimental results suggest SRIA is comparable to the conventional computational model based on BPNN and is obviously superior to CRIA in prediction performance. Moreover, the SRIA algorithm can extract plain rules that help experts to understand the basis of immunity while BPNN cannot.

Acknowledgements

The authors gratefully acknowledge Dr. Vladimir Brusic for providing the data set used in this paper and Prof. Dr. Hans-Georg Rammensee for providing his paper. This work is supported in part by the NSF of Guangdong under Grant 6300252, and the Doctor Foundation of GDUT under Grant 063001.

References

1. Peter, W.: T-cell Epitope Determination. Current Opinion in Immunology. 8 (1996) 68–74
2. Falk, K., Rötzschke, O., Stevanovié, S., Jung, G., Rammensee, H.G.: Allele-specific Motifs Revealed by Sequencing of Self-peptides Eluted from MHC Molecules. Nature. 351 (1991) 290–296
3. Rammensee, H.G., Bachmann, J., Emmerich, N.P., Bachor, O.A., Stevanovic, S.: SYFPEITHI: Database for MHC Ligands and Peptide Motifs. Immunogenetics. 50 (1999) 213–219
4. Jun, Z., Herbert, R.T., George, B.R.: Prediction Sequences and Structures of MHC-binding Peptides: A Computational Combinatorial Approach. Journal of Computer-Aided Molecular Design. 15 (2001) 573–586
5. Brusic, V., George, R., Margo, H., Jürgen, H., Leonard, H.: Prediction of MHC Class II-binding Peptides Using An Evolutionary Algorithm and Artificial Neural Network. Bioinformatics. 14 (1998) 121–130
6. Pawlak, Z: Rough Sets. International Journal of Information and Computer Sciences. 11 (1982) 341–356
7. Rammensee, H.G., Friede, T., Stevanovic, S.: MHC Ligands and Peptide Motifs: First Listing. Immunogenetics. 41 (1995) 178–228
8. Madden, D.R.: The Three-dimensional Structure of Peptide MHC Complexes. Annu. Rev. Immunol.. 13 (1995) 587–622
9. Guo-Yin, W.: Rough Reduction in Algebra View and Information View. International Journal of Intelligent Systems. 18 (2003) 679–688
10. Dan, P., Qi-Lun, Z., An., Z, Jing-Song, H.: A Novel Self-optimizing Approach for Knowledge Acquisition. IEEE Transactions on Systems, Man, And Cybernetics- Part A: Systems And Humans. 32 (2002) 505–514
11. Fu-Bao, W., Qi, L., Wen-Zhong, S.: Inductive Learning Approach to Knowledge Representation System Based on Rough Set Theory" (in Chinese). Control & Decision. 14 (1999) 206–211
12. Pawlak, Z., Slowinski, R.: Rough Set Approach to Multi-attribute Decision Analysis. European Journal of Operational Reaserch. 72 (1994) 443–459

Efficient Local Protein Structure Prediction

Szymon Nowakowski[1] and Michał Drabikowski[2]

[1] Infobright Inc., ul. Krzywickiego 34 pok. 219
02-078 Warszawa, Poland
s.nowakowski@mimuw.edu.pl
[2] Institute of Informatics
Warsaw University, Poland
m.drabikowski@mimuw.edu.pl

Abstract. The methodology which was previously used with success in genomic sequences to predict new binding sites of transcription factors is applied in this paper for protein structure prediction. We predict local structure of proteins based on alignments of sequences of structurally similar local protein neighborhoods. We use Secondary Verification Assessment (SVA) method to select alignments with most reliable models. We show that using Secondary Verification (SV) method to assess the statistical significance of predictions we can reliably predict local protein structure, better than with the use of other methods (log-odds or p-value). The tests are conducted with the use of the test set consisting of the CASP 7 targets.

Keywords: statistical significance, SV method, SVA method, assessing predictions, model assessment, protein structure prediction, CASP 7.

1   Introduction

Predicting protein structure has been for many years in the very center of attention of many researchers. In this paper we focus on a little easier problem of predicting *local* protein structure. However, it is clear, that having done sufficiently many local predictions it should be possible to select the self-consistent subset of local structure predictions (in other words, the subset of local structure predictions which structurally agree with one another) and construct the global structure. In this paper we show that it is possible, indeed. We present an example of global protein structure prediction on one of the hardest CASP 7 targets. The main focus of this paper remains, however, on the local protein structure prediction.

For doing the predictions we use the SVA (*Secondary Verification Assessment*) and SV (*Secondary Verification*) methods. They were originally developed in [1] for the use in genomic sequences. In this paper we show that they can be adapted to protein sequences. Both methods are shown to be more successful than other methods used routinely nowadays.

We use a database of protein motifs [2]. A motif constitutes of many fragments, not adjacent on the sequence, but neighboring in 3D structure. In this way it is

J.T. Yao et al. (Eds.): RSKT 2007, LNAI 4481, pp. 308–315, 2007.

possible to use many local predictions to predict global structure: finding only a few local structures which agree with one another gives us the prediction of majority of amino acids in the protein [3]. It is known that the number of fragments grows exponentially with motif length for contiguous fragment databases [4]. But thanks to building our database out of fragments adjacent in 3D space, not on the sequence, we can have long motifs (on average, our motifs constituted of 28 amino acids), while keeping the size of representative motif set reasonably small (less than 3000 motifs are needed to cover evenly and densely the space of all possible protein motifs). Currently used methods which use contiguous protein fragments [5] as motifs are limited to short fragments, and contiguous and short fragments which agree with one another build a contiguous and short prediction.

2 Materials and Methods

2.1 Data

Descriptor Database Preparation. *Local descriptors* are the structural building blocks that we use to construct our database of motifs. They are units encompassing 3 or more short contiguous backbone *segments* consisting of at least 5 residues that are close in 3D space but not necessarily along the protein sequence. For each residue, from each considered protein with a known structure, a descriptor is constructed as a set of segments which are located close to the selected residue in a selected protein. A detailed description of the descriptor construction may be found in [2,3]. In short, descriptors are created for all amino acids in all proteins in a representative database.

Creating a Motif and Its Sequential Profile. In this step we base on the assumption that structurally similar local descriptors have also similar sequence. To measure the structural similarity of any two descriptors we calculate the root mean square deviation ($RMSD$) in the best alignment of these descriptors. If this alignment contains at least 75% of residues belonging to each descriptor and the $RMSD$ score is not greater than 2.5Å then we consider descriptors to be *structurally similar*.

Based on the structural similarity, for each constructed descriptor we define a *group* of similar descriptors. The structure of the descriptor for which a given group g was calculated will be further considered representative for g and called a *motif* representing g. Since within each group descriptors are aligned, a sequence profile can be computed [6]. This profile is in the form of a PSSM (*Position Specific Score Matrix*) and defines the probability of seeing each amino acid in each column of a group, assuming that columns are independent. The PSSM representation will further be considered a *sequence model* associated with a motif.

Preparing Training and Test Sets
Training Set. To construct a database of descriptors we used the ASTRAL 1.63 database [7] restricted to domains having no more than 40% sequence identity

to one another. Based on these descriptors the set of 127078 motifs containing at least 10 descriptors and 3 segments was constructed.

Additional Training Set. Domains from ASTRAL 1.65 which do not appear in ASTRAL 1.63 form the additional training set. In this set we identified *positive* and *negative* sequences for each motif. Sequences corresponding to local neighborhoods with the structure similar to the motif are positive examples, while sequences corresponding to local neighborhoods with the structure not similar to the motif are negative examples (of that motif).

For each positive and negative sequence we calculate its log-odds score (see equation (1) in Section 2.2) using the model M associated with the fixed motif and using the background distribution B. Next we estimate the *pdf* (probability density function) of the normal distribution of scores in positive sequences (this *pdf* is denoted f_+) and in negative sequences (this *pdf* is denoted f_-). Then we estimate P^+ and P^-, the *a priori* probabilities that a sequence is positive or negative, respectively, before one even sees the sequence. To do that we simply divide the number of positive (or, respectively, negative) sequences in the additional training set by the number of all sequences (positive and negative).

Test Set. As a test set we used all proteins from CASP 7 experiment except those classified to the easiest FM (*Free Modeling*) category. This set of 71 proteins is independent from our training sets (ASTRAL 1.63 and 1.65 databases were published before the CASP 7 experiment) and covers a wide range of known structures.

2.2   Making Predictions

In this Section we describe how we make local structure predictions. It is detailed how to find neighborhoods in the query sequence with high affinity to the fixed motif. The predictions can be repeated for all motifs one wants to consider.

Selecting Sequences - Candidates to Become Predictions. Suppose that the motif that we use for local structure prediction consists of one segment only. For a given protein sequence we use a simple sliding window and all subsequences of the same length as the motif are considered as the candidates.

Since we use representative descriptors containing several segments, the number of possible candidates is exponential with respect to the number of segments. We showed that the problem of finding the best assignment is NP-complete [8], even with the assumption that all segments can be assigned independently. To deal with this problem we proposed the dynamic programming algorithm which allows limiting the number of candidates efficiently [8].

Making Predictions and Assessing Their Statistical Significance. In this Section we focus on making predictions, i.e. selecting only the sequences with the highest scores (according to the scoring method) out of all candidates selected in previous Section.

Let us fix the motif and its PSSM model M with N columns. The PSSM model of i-th column of M will be denoted M_i. Let us also fix the background model B (as B we use average amino acid frequencies in ASTRAL 1.65).

Three methods of making predictions are compared in our paper. The most straightforward method is the *log-odds score* of a sequence.

Log-odds Score. The log-odds score of the amino acid sequence $S = s_1 s_2 \ldots s_N$ of length N is (thanks to the column independence assumption):

$$\text{log-odds}(S) = \log \frac{P(S|M)}{P(S|B)} = \sum_{i=1}^{N} \log \frac{P(s_i|M_i)}{P(s_i|B)}. \tag{1}$$

We use notation $P(S|X)$ when we want to make explicit dependence of the probability measure on a model X. X can be a PSSM model or the background distribution.

Since the above score is a sum of many similar values, we make a simplifying assumption (based on the Central Limit Theorem) that it is normally distributed [1].

Other two methods assess statistical significance of log-odds score and require the existence of the additional training set. They start by identifying positive and negative populations for the fixed motif and then estimate f_+, f_-, P^+ and P^- for that motif as in Section 2.1.

P-value Method. The p-value of a log-odds score s is the probability that the score this high (or higher) could be obtained by chance. Thus the p-value is calculated as a 'right tail' of the *cdf* (cumulative distribution function) of negative distribution estimated as in Section 2.1, p-value$(s) = 1 - F_-(s)$. The sequences with the lowest p-values of the log-odds score (i.e. the most significant) are selected to become predictions. Since we want to score better predictions higher, in our paper we use

$$\text{PV}(s) = -\log(\text{p-value}(s)) = -\log(1 - F_-(s)). \tag{2}$$

under the name *p-value scoring method*.

Secondary Verification Method. In [1] it is pointed out that in cases of similar positive and negative distributions of log-odds scores it may be not informative to simply have a look at a negative distribution: a score which is statistically significant may not be likely to be positive, either. One would require to consult both distributions to tell if the score s is significant. To this end Secondary Verification (SV) score was proposed, which assesses the log-odds score (1) by the probability that the sequence with this score is positive. The SV score of a log-odds score s is thus (we make the use of Bayes Theorem):

$$\text{SV}(s) = P(+|s) = \frac{f_+(s)P^+}{f_+(s)P^+ + f_-(s)P^-}. \tag{3}$$

2.3   Assessing Model Quality

Every local structural motif from our database of structural motifs was associated with a PSSM model, as it is described in Section 2.1. Some of those models are obviously of very poor quality and if they were used to predict local structure of proteins from the test set, the proportion of false positives would increase. In order to minimize this effect, the PSSM models are scored with the use of Secondary Verification Assessment (SVA) method [1] and only models with the highest SVA scores are used for predictions.

To calculate the SVA score for a given motif we use the sets of positive and negative examples in the additional training set (see Section 2.1). We start by estimating f_+, f_-, P^+ and P^- for the fixed motif as in Section 2.1.

The SVA score of a model M (as it was in detail explained in [1]) is then given by

$$\mathrm{SVA}(M) = E^+(\mathrm{SV}) = \int_{-\infty}^{\infty} \mathrm{SV}(s) f_+(s)\, ds = \int_{-\infty}^{\infty} \frac{f_+^2(s) P^+\, ds}{f_+(s) P^+ + f_-(s) P^-}. \quad (4)$$

In other words, the SVA score is the expected Secondary Verification score over the population of positive sequences. Observe that this measure is very appealing intuitively: the best models M are the ones which can distinguish positive sequences faultlessly and this very feature is captured in the SVA score.

Some of the motifs may be rare and other motifs may be frequent in proteins. To predict the local protein structure most reliably we want not only the motif models to be of the highest quality, but we also want to use only the most frequent motifs (having fixed a constraint that the selected motifs span the possible protein motif space). In other words, if we could choose the very rare motif with a high quality model, it would be better to replace it with a much more frequent motif having a model of slightly worse quality. To this end, when scoring the model M, we used a modified SVA score:

$$\mathrm{mSVA}(M) = \#(\text{positive sequences}) \cdot \mathrm{SVA}(M), \quad (5)$$

where #(positive sequences) is the number of positive motif occurances in the additional training set.

2.4   Selecting Best Motifs Spanning the Structural Space

In order to reduce the number of considered motifs we need to identify a subset of good and structurally different motifs representing all local structures of proteins. We introduce two criteria.

Structural Criterion. The main idea behind this approach is to promote big groups (with many descriptors) representing frequently occurring substructures. Using algorithm presented in [3] we selected 2569 motifs which cover the space of all possible motifs. This set of structural motifs is referred to as the set picked up with "the structural criterion".

Sequential Criterion. Based on the modified SVA score (5) we selected structurally different motifs which have the most reliable models. Using algorithm presented in [3] we selected 2771 motifs which cover the space of all posible motifs. This set of motifs is referred to as the set picked up with "the sequential criterion".

3   Experiments

3.1   Visualizing Results

The results of the experiments are presented in Figure 1. It presents the number of true positives (correct predictions) and the corresponding number of false positives (wrong predictions) for all proteins in the test set as a function of a changing threshold value (when the score of a chosen method is higher than the threshold, we consider it a local structure prediction). For maximal threshold obviously no predictions are made (no score value can be higher). As the threshold gets smaller the number of predictions increases - and the number of correct predictions increases. Simultaneously the number of incorrect predictions increases. We want the number of true positives to be as large as possible while we want the number of false positives to be as small as possible. The obtained curve is similar to the ROC (Receiver Operating Characteristic) curve known from signal processing field: in short, the steeper the curve close to the $(0,0)$ point, the better the prediction quality.

3.2   Structural vs. Sequential Criterion for Selecting Motifs

In this Section we compare two sets of motifs: motifs and their models picked up with the structural criterion and motifs and their models picked up with the sequential criterion (see Section 2.4). Both sets are used to make predictions of local structure of proteins from the test set. We used the SV method and the log-odds method to make predictions.

We can see in Figure 1 (Left and Middle) that the sequential criterion by far outperforms the structural criterion for both methods.

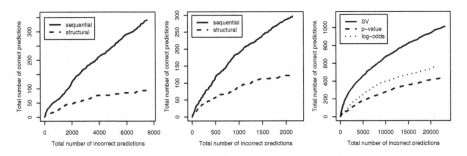

Fig. 1. Comparison of structural and sequential criteria for selecting the motifs to do predictions with the log-odds method (Left) and the Secondary Verification method (Middle). Comparison of three methods of predicting local structure for sequential criterion (Right).

3.3   Predicting Local Protein Structure

In this Section we use the sequential criterion (see Section 2.4) for picking up the motifs and their models. We make predictions using log-odds, p-value and SV method. We predict local structure of proteins from the test set.

We can see in Figure 1 (Right) that the SV method by far outperforms two other methods. In fact, for a fixed number of false positives, the number of true positives (i.e. correct predictions) is almost twice as large for the SV method as for the log-odds method, the better one of the remaining two.

3.4   Future Work: Global Protein Structure Predictions

After identifying the the most reliable self-consistent local predictions the global structure can be predicted. This process allows eliminating false positive predictions and building the global prediction only from structurally correct motifs. An example of global prediction obtained in such a way is presented in Figure 2. We predicted the structure of CASP 7 domain T0347_2 (qualified as FM, i.e. with no structural template). We predicted 47 amino acids out of 71 with $RMSD$ equal to 4.4Å. This global prediction is the result of finding a consensus over 9 self-consistent local predictions.

Fig. 2. The prediction of CASP 7 T0347_2 domain. Our prediction is presented in black, while the original structure is presented in grey.

4   Conclusions

In this paper we show that the SV and SVA methods, which were used before in the case of genomic sequences, can be used also in the case of protein sequences with great success. We show it predicting local protein structure which may be the first step in predicting global protein structure. Predicting global protein structure is currently one of the most important problems in bioinformatics. As we have seen on the example of the real protein (and note that it was shown on one of the hardest CASP 7 targets), it is possible to efficiently use the approach presented in this paper to predict protein 3D structure.

Two main results of this paper include showing that using the SVA score to select only motifs with the best model quality is a great improvement over selecting motifs only based on their structural properties. It is connected with the fact, which was many times pointed out by different authors in the past [1,9], that the models of alignments are frequently of quality too bad to make it possible to differentiate them from the background.

The second result is showing the superiority of the SV score (which incorporates positive sequence population) over routinely used log-odds scoring and p-value scoring (which does statistical significance assessment of the results, but is based solely on the negative sequence population).

References

1. Nowakowski, S., Tiuryn, J.: A new approach to the assessment of the quality of predictions of transcription factor binding sites (in press). Journal of Biomedical Informatics (2006) doi:10.1016/j.jbi.2006.07.001.
2. Hvidsten, T.R., Kryshtafovych, A., Komorowski, J., Fidelis, K.: A novel approach to fold recognition using sequence-derived properties from sets of structurally similar local fragments of proteins. Bioinformatics 19 **Suppl 2** (Oct 2003) II81–II91
3. Drabikowski, M., Nowakowski, S., Tiuryn, J.: Library of local descriptors models the core of proteins accurately (in press). Proteins: Structure, Function, and Bioinformatics (2007)
4. Unger, R., Harel, D., Wherland, S., Sussman, J.L.: A 3D building blocks approach to analyzing and predicting structure of proteins. Proteins: Struct. Funct. Genet. **5** (1989) 355–373
5. Yang, A., Wang, L.: Local structure prediction with local structure-based sequence profiles. Bioinformatics 19 (2003) 1267–1274
6. Sjölander, K., Karplus, K., Brown, M., Hughey, R., Krogh, A., Mian, I.S., Haussler, D.: Dirichlet mixtures: a method for improved detection of weak but significant protein sequence homology. CABIOS **12** (1996) 327–345
7. Brenner, S.E., Koehl, P., Levitt, M.: The ASTRAL compendium for protein structure and sequence analysis. Nucl. Acids Res. **28** (2000) 254–256
8. Drabikowski, M.: Analysis of groups and signals used to build protein structures from local descriptors (written in Polish). PhD thesis, Institute of Informatics, Warsaw University, Poland (2006)
9. Rahmann, S., Müller, T., Vingron, M.: On the power of profiles for transcription factor binding sites detection. Statistical Applications in Genetics and Molecular Biology **2**(1) (2003)

Roughfication of Numeric Decision Tables: The Case Study of Gene Expression Data

Dominik Ślęzak[1] and Jakub Wróblewski[2,1]

[1] Infobright Inc.
218 Adelaide St. W, Toronto, ON, M5H 1W8 Canada
[2] Polish-Japanese Institute of Information Technology
Koszykowa 86, 02-008 Warsaw, Poland

Abstract. We extend the standard rough set-based approach to be able to deal with huge amounts of numeric attributes versus small amount of available objects. We transform the training data using a novel way of non-parametric discretization, called *roughfication* (in contrast to *fuzzification* known from fuzzy logic). Given *roughfied* data, we apply standard rough set attribute reduction and then classify the testing data by voting among the obtained decision rules. Roughfication enables to search for reducts and rules in the tables with the original number of attributes and far larger number of objects. It does not require expert knowledge or any kind of parameter tuning or learning. We illustrate it by the analysis of the gene expression data, where the number of genes (attributes) is enormously large with respect to the number of experiments (objects).

Keywords: Rough Sets, Discretization, Reducts, Gene Expression Data.

1 Introduction

DNA microarrays provide a huge quantity of information about genetically conditioned susceptibility to diseases [1,2]. However, a typical gene expression data set, represented as an information system $\mathbb{A} = (U, A)$ [9,14], has just a few objects-experiments $u \in U$, while the number of attributes-genes $a \in A$ is counted in thousands. Moreover, preciseness of measuring gene expressions, i.e. the values $a(x) \in \mathbb{R}$, is still to be improved. Both these issues yield a problem for methods assuming data to be representative enough.

We solve the above problems using *roughfication*,[1] already applied to gene clustering [6] and classification [13]. Foe a given $\mathbb{A} = (U, A)$, we produce a new system $\mathbb{A}^* = (U^*, A^*)$, where U^* corresponds to $U \times U$ and A^* – to the original A. Every $a^* \in A^*$ labels a given $(x, y) \in U^*$ with symbolic value "$\geq a(x)$" iff $a(y) \geq a(x)$, and "$< a(x)$" otherwise. This simple trick provides us with a larger universe, where the number of attributes remains unchanged. This way, it is very different from other *discretization* techniques, which keep U unchanged while exploding the amount of possible attributes in A (cf. [5,8]).

---

[1] In [6,13] we also used the terms *rough discretization* and *rank-based approach*.

J.T. Yao et al. (Eds.): RSKT 2007, LNAI 4481, pp. 316–323, 2007.
© Springer-Verlag Berlin Heidelberg 2007

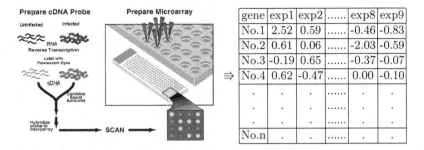

gene	exp1	exp2	......	exp8	exp9
No.1	2.52	0.59	......	-0.46	-0.83
No.2	0.61	0.06	......	-2.03	-0.59
No.3	-0.19	0.65	......	-0.37	-0.07
No.4	0.62	-0.47	......	0.00	-0.10
.	.	.	......	.	.
.	.	.	......	.	.
.	.	.	......	.	.
No.n	.	.	......	.	.

Fig. 1. Microarrays provide the gene expression data. A sample of 9 experiments from *Synovial Sarcoma* data is illustrated. We have $n = 5520$ genes in this data set (cf. [6]).

Given *roughfied*, symbolic data, we apply the rough set methods for finding optimal (approximate) decision reducts and rules [11,18]. We extend our research reported in [13] by thorough experimental analysis, involving different reduction and voting options. We show that the proposed simple mechanism provides results comparable to far more complex methods (cf. [15]).

The paper is organized as follows: Section 2 contains basics of gene expression data. Section 3 – basics of rough sets and attribute reduction. Section 4 – *roughfication* and its usage in classification. Section 5 – results of 8-fold cross-validation analysis of the breast cancer-related data with 24 biopsies-objects, 12,625 genes-attributes, and binary decision. Section 6 concludes the paper.

2   Gene Expression Data

The DNA microarray technology [1] enables simultaneous analysis of characteristics of thousands of genes in the biological samples of interest. It is automated, much quicker, and less complicated than the previous methods of molecular biology, allowing scientists to study no more than a few genes at a time.

Microarrays rely on DNA sequences fabricated on glass slides, silicon chips, or nylon membranes. Each slide (DNA chip) contains samples of many genes in fixed spots. It may represent cDNA (most popular, used also in this paper), DNA or oligonucleotide. Microarray production starts with preparing two samples of mRNA. The sample of interest is paired with a healthy control sample. Fluorescent labels are applied to the control (green) and the actual (red) samples. Then the slide is washed and the color intensities of gene-spots are scanned, indicating to what extent particular genes are expressed. Figure 1 illustrates the process.

The analysis of such prepared data can lead to discoveries of important dependencies in gene sequences, structures, and expressions. The cDNA microarray data sets are often analyzed to track down the changes of the gene activations for different types of tumors. This information could be then applied to identifying tumor-specific and tumor-associated genes. However, a large number of gathered numerical data makes this analysis particularly hard.

3   Rough Sets, Reducts, Classification

In the rough set theory [9], we analyze information system $\mathbb{A} = (U, A)$, where attributes $a \in A$ correspond to functions $a : U \to V_a$ from universe U into value sets V_a. In this paper, A corresponds to the set of genes, U – to the set of experiments, and functions $a : U \to \mathbb{R}$, reflect gene expressions of humans or other organisms, measured in certain conditions. For classification purposes, we distinguish decision $d \notin A$ to be determined using A. Then, we talk about decision systems $\mathbb{A} = (U, A \cup \{d\})$. In the case of gene expression data d is usually symbolic, reflecting some classes of diseases or behaviors. For instance, in Section 5, d labels two types of behaviors related to breast cancer.

Rough set-based classifiers are a good illustration of tradeoff between accuracy and complexity. In this paper, we focus on decision reducts – minimal subsets $B \subseteq A$ that (almost) determine d. Smaller reducts induce shorter and more general rules. Often, it is even better to remove attributes to get shorter rules at a cost of slight loss of decision determination [11], which can be expressed in many ways. We refer to the original rough set positive region [9] and to its *probabilistic* counterpart [12]. Both measures base on the indiscernibility classes $[u]_B \subseteq U$ defined, for every $u \in U$ and $B \subseteq A$, as $[u]_B = \{x \in U : \forall_{a \in B} \, a(x) = a(u)\}$. The positive region is defined as $POS(B) = \{u \in U : [u]_B \subseteq [u]_{\{d\}}\}$. Each $u \in POS(B)$ induces an "if-then" rule saying that if a new object is equal to u on all attributes $a \in B$, then it can be classified to the u's decision class. One can use also *inexact* rules, especially if the exact ones require too many attributes. Even given no $[u]_B \subseteq [u]_{\{d\}}$ we can still refer to *rough memberships* $\mu_B^v(u) = |\{x \in [u]_B : d(x) = v\}|/|[u]_B|$,[2] where $v \in V_d$ [10]. Whenever a new object is equal to u on all $a \in B$, our belief that its value on d equals to v relates to $\mu_B^v(u)$. In [12] it is shown that such a classification strategy relates to *probabilistic* positive region $POS_\mu(B) = \{(u, \mu_B^{d(u)}(u)) : u \in U\}$ and, more precisely, to its *cardinality* represented as in the theory of fuzzy sets [19]:[3]

$$|POS_\mu(B)| = \sum_{u \in U} \mu_B^{d(u)}(u) = \sum_{u \in U} \frac{|[u]_{B \cup \{d\}}|}{|[u]_B|} \tag{1}$$

By a decision reduct we mean $B \subseteq A$ such that $POS(B) = POS(A)$ and there is no $C \subsetneq B$ such that $POS(C) = POS(A)$ [9]. In this paper,[4] by an *ε-approximate decision* and *ε-approximate μ-decision reducts* we mean subsets $B \subseteq A$ such that, respectively, $|POS(B)| \geq (1-\varepsilon)|POS(A)|$ and $|POS_\mu(B)| \geq (1-\varepsilon)|POS_\mu(A)|$, and there are no $C \subsetneq B$ holding analogous inequalities. In both cases, threshold $\varepsilon \in [0, 1)$ expresses willingness to reduce more attributes (and simplify rules) on the cost of losing the (probabilistic) positive region's strength.[5] In the special case of $\varepsilon = 0$, we are interested in complete preserving of the regions, i.e. in equalities $POS(B) = POS(A)$ and $POS_\mu(B) = POS_\mu(A)$, respectively.

---

[2] By $|X|$ we denote cardinality of the set X.

[3] Formula (1), after additional division by $|U|$, was denoted in [12] by $E(B)$.

[4] There are many formulations of approximate reducts in the literature [12,18].

[5] $|POS|, |POS_\mu| : 2^A \to [0, |U|]$ are monotonic with respect to inclusion [9,12].

$|POS(B)| \geq (1-\varepsilon)|POS(A)|$	$|POS(B)| \geq (1-\varepsilon)|POS(A)|$
Only exact decision rules used	Only exact decision rules used
$\lvert[u]_B\rvert$ added for $d(u)$	1 added for $d(u)$
$|POS_\mu(B)| \geq (1-\varepsilon)|POS_\mu(A)|$	$|POS_\mu(B)| \geq (1-\varepsilon)|POS_\mu(A)|$
Also inexact decision rules used	Also inexact decision rules used
$\lvert[u]_B\rvert \cdot \mu_B^v(u)$ added for each v	$\mu_B^v(u)$ added for each v

Fig. 2. Four techniques of classification, each described by three lines. Line 1 indicates constraints for subsets $B \subseteq A$ that can be contained in the classifier. (E.g., in Section 5 we always choose 5 best found subsets.) Line 2: the types of rules. (Note that they are related to the types of reducts.) Line 3: the weights being added to decision classes, whenever a rule generated by B is applicable to u. The decision for u is finally chosen as $v \in V_d$ with the highest overall weight, summed up over all applicable rules.

The rough set classifier is based on a collection of *optimal* (approximate) decision reducts and the resulting rules [14,18]. Relaying on a larger set of reducts improves both applicability and accuracy of rough set classifiers. A chance that a new object is recognized by at least one out of many rules is getting higher when operating with diversified subsets of attributes. Appropriate synthesis of information based on different attributes also helps in predicting the right decisions (cf. [3]). In this paper, we compare two techniques of voting – by rules' *supports* and by rules' counting, referring to both of the above-considered types of approximate reducts. We obtain four variants described in Figure 2. We apply them in Section 5 to classification of real-life data.

4 Roughfication

Standard rule-based methods are hardly applicable to real-valued data systems unless we use discretization [7,8] or switch to more advanced techniques, requiring more parameters and/or expert knowledge (cf. [16,19]). Machine learning methods have serious problems while dealing with disproportions between attributes and objects. Gene expression technology is still quite imprecise, which causes additional problems with data representativeness. As a result, many approaches, including those in Section 3, cannot be applied straightforwardly.

We suggest a new way of data preparation, called *roughfication*. Given $\mathbb{A} = (U, A \cup \{d\})$, we create a new system $\mathbb{A}^* = (U^*, A^* \cup \{d^*\})$, where $U^* \equiv U \times U$ and $A^* \cup \{d^*\} \equiv A \cup \{d\}$. It is illustrated by Figure 3. For every $a \in A$, $V_a = \mathbb{R}$, a new attribute $a^* \in A^*$ has the value set $V_a^* = \bigcup_{u \in U}\{``\geq a(x)", ``< a(x)"\}$. For every $x, y \in U$, we put $a^*(x, y) = ``\geq a(x)"$ iff $a(y) \geq a(x)$, and $a^*(x, y) = ``< a(x)"$ otherwise. For symbolic attributes, in particular for $d \notin A$, we put $d^*(x, y) = d(y)$, i.e. $V_d^* = V_d$. It is important that V_a^* can be treated as a symbolic-valued domain during further calculations, e.g., those described in Section 3. When the classifier is ready, the values in V_a^* begin to be interpreted in a non-symbolic way again. Continuing with analogy to fuzzy sets [19], we may say that the data is

	a	b	d
u1	3	7	0
u2	2	1	1

	a	b	d
u3	4	0	1
u4	0	5	2

====>

	a^*	b^*	d^*
$(u1,u1)$	≥ 3	≥ 7	0
$(u1,u2)$	< 3	< 7	1
$(u1,u3)$	≥ 3	< 7	1
$(u1,u4)$	< 3	< 7	2
$(u2,u1)$	≥ 2	≥ 1	0
$(u2,u2)$	≥ 2	≥ 1	1
$(u2,u3)$	≥ 2	< 1	1
$(u2,u4)$	< 2	≥ 1	2

	a^*	b^*	d^*
$(u3,u1)$	< 4	≥ 0	0
$(u3,u2)$	< 4	≥ 0	1
$(u3,u3)$	≥ 4	≥ 0	1
$(u3,u4)$	< 4	≥ 0	2
$(u4,u1)$	≥ 0	≥ 5	0
$(u4,u2)$	≥ 0	< 5	1
$(u4,u3)$	≥ 0	< 5	1
$(u4,u4)$	≥ 0	≥ 5	2

IF $a \geq 3$ AND $b \geq 7$ THEN $d = 0$
IF $a \geq 3$ AND $b < 7$ THEN $d = 1$
IF $a \geq 2$ AND $b < 1$ THEN $d = 1$
IF $a < 2$ AND $b \geq 1$ THEN $d = 2$
IF $a \geq 4$ AND $b \geq 0$ THEN $d = 1$
IF $a \geq 0$ AND $b < 5$ THEN $d = 1$

Fig. 3. Top left: Original $\mathbb{A} = (U, A \cup \{d\})$, $U = \{u1, u2, u3, u4\}$, $A = \{a, b\}$, $V_d = \{0, 1, 2\}$. Right: *Roughfied* $\mathbb{A}^* = (U^*, A^* \cup \{d^*\})$. Bottom left: Exact rules based on A^*.

first *roughfied*, then the learning process is performed, and finally the resulting classifier is *deroughfied* to deal with new objects.

Figure 3 illustrates a decision system obtained using roughfication. Region $POS(a^*, b^*) = \{(u1, u1), (u1, u3), (u2, u3), (u2, u4), (u3, u3), (u4, u2), (u4, u3)\}$ is covered by six exact rules. Although formally we should write, e.g., IF $a^* = "\geq 3"$ AND $b^* = "\geq 7"$ THEN $d^* = 0$, we use a *deroughfied* notation, as used while classifying new objects. Note that one set of attributes can induce more than one rule. For a new object $u \notin U$, the rules may point at different decisions. For instance, if $a(u) = 4$ and $b(u) = 7$, then the first rule in Figure 3 will yield $d = 0$ while the fifth one will give us $d = 1$. Also, some rules may be more applicable than the others. For example, whenever conditions of the third rule are satisfied, the sixth rule is applicable too. We may say that such rules are able to represent more and less typical patterns for decision classes. It indicates that roughfication should be followed by well-designed voting mechanisms.

Continuing with Figure 3, let us consider $POS(\{a^*\}) = \{(u2, u4), (u3, u3)\}$ and $POS(\{b^*\}) = \{(u1, u1), (u2, u3), (u4, u2), (u4, u3)\}$. Given that $|POS(A^*)|$ equals 7, the set $\{a^*\}$ begins to be an ε-approximate reduct for $\varepsilon \geq 3/7$ and $\{b^*\}$ – for $\varepsilon \geq 5/7$. Note that $POS(\{a^*\})$ and $POS(\{b^*\})$ are disjoint, covering together 6 our of 7 elements of $POS(A^*)$. Hence, $\{a^*\}$ and $\{b^*\}$ might work well together, as a reduct collection. The only missing element is $(u1, u3)$ and, accordingly, the second rule in Figure 3 cannot be shortened.

Finally, we briefly illustrate $POS_\mu(A^*)$ in Figure 4. The rules not occurring in Figure 3 are, e.g.: IF $a < 3$ AND $b < 7$ THEN $d = 0$ with weight 0.5 AND $d = 2$ with weight 0.5; or, e.g.: IF $a \geq 2$ AND $b \geq 1$ THEN $d = 0$ with weight 0.5 AND $d = 1$ with weight 0.5. As illustrated in Figure 2, such decision weights participate in voting in the case of inexact rule-based reduction/classification.

5   Classification Results

We study the breast cancer data downloaded from Gene Expression Omnibus GEO, http://www.ncbi.nlm.nih.gov/projects/geo/gds/gds_browse.cgi?gds=360, analyzed in [15]. It contains 24 core biopsies taken from patients, who are

$u1$	a^*	b^*	d^*	μ	$u2$	a^*	b^*	d^*	μ	$u3$	a^*	b^*	d^*	μ	$u4$	a^*	b^*	d^*	μ
$u1$	≥ 3	≥ 7	0	1	$u1$	≥ 2	≥ 1	0	$1/2$	$u1$	< 4	≥ 0	0	$1/3$	$u1$	≥ 0	≥ 5	0	$1/2$
$u2$	< 3	< 7	1	$1/2$	$u2$	≥ 2	≥ 1	1	$1/2$	$u2$	< 4	≥ 0	1	$1/3$	$u2$	≥ 0	< 5	1	1
$u3$	≥ 3	< 7	1	1	$u3$	≥ 2	< 1	1	1	$u3$	≥ 4	≥ 0	1	1	$u3$	≥ 0	< 5	1	1
$u4$	< 3	< 7	2	$1/2$	$u4$	< 2	≥ 1	2	1	$u4$	< 4	≥ 0	2	$1/3$	$u4$	≥ 0	≥ 5	2	$1/2$

Fig. 4. The $POS_\mu(A^*)$ coefficients displayed in the column "μ". The objects in U^* are now split onto subtables with respect to the first elements in pairs (ui, uj), $i, j = 1, ..., 4$.

resistant (14 objects) or sensitive (10 objects) to the docetaxel treatment. There are 12,625 genes-attributes. Figure 5 shows the results obtained using the 8-fold cross validation (CV-8). For each split onto 21 training and 3 testing objects, we repeated the following steps: 1) *Roughfy* the training data; 2) Calculate 5 best reducts of a given type; 3) Classify the testing data using voting of a given type. The reported numbers are averaged over the 40 independent CV-8 data splits, with standard deviations low enough to provide the results' credibility.

Given the number of genes-attributes, we applied very simple heuristics to optimize ε-approximate (μ-)decision reducts: 1) Start with $B^* = \emptyset$ and keep randomly adding attributes until we get $|POS(B^*)| \geq (1 - \varepsilon)|POS(A^*)|$ (or $|POS_\mu(B^*)| \geq (1 - \varepsilon)|POS_\mu(A^*)|$); 2) Also randomly, keep removing attributes from B – if a given attribute cannot be removed without losing the above inequality, keep it in B and try with the next one; 3) Out of 15 such randomly generated reducts,[6] select 5 with minimal cardinality. Obviously, more advanced techniques can be applied to get better collections for each $\varepsilon \in [0, 1)$ [17,18].

The obtained results are fully comparable with those obtained using far more complicated methodologies, which usually remain difficult to understand for the domain experts (cf. [4,15]). A huge advantage of the proposed method is that it simply bases on collections of genes' subsets and "if-then" rules operating on inequalities. Besides a need of tuning the level of $\varepsilon \in [0, 1)$, which is actually related to the *simplicity versus accuracy* tradeoff specific for particular users, there are no parameters requiring additional tuning or expert knowledge, including the ones related to discretization [7,8]. Even the most basic variant (b) in Figure 5, the one using only exact rules and very simplified voting mechanism, provides good results. Operating with $|POS_\mu|$ instead of $|POS|$ enables to reduce attributes faster along the ε-axis, though we need to be more careful with accuracy. The two considered voting mechanisms give quite similar outcomes, though (b) and (d) seem to slightly better than (a) and (c), respectively. Explanation may lay in a fact that the collections of rules derived from roughfied data are able to express more and less typical areas of particular decision classes by themselves (see Section 4), hence any additional weights assigned to the rules are not so necessary. Last but not least, we do not report percentage of *unrecognized* testing objects because it is practically equal to 0. Also, we do not consider any additional filtration of decision rules [14,18] because we experimentally measured

---

[6] Previously mentioned monotonicity of functions $|POS|$ and $|POS_\mu|$, now defined over 2^{A^*}, guarantees that the obtained sets are reducts of appropriate type in $\mathbb{A}^*$.

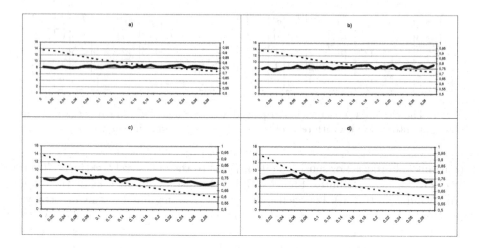

Fig. 5. Accuracy of classification within the CV-8 scheme (solid line) and average the reduct length (dotted line). The following attribute reduction and rule voting settings correspond to those illustrated in Figure 2: a) reduction measure $|POS|$ and voting by the rules' supports; b) reduction measure $|POS|$ and voting by numbers of rules; c) reduction measure $|POS_\mu|$ and voting by the rules' supports times rough memberships to decision classes; d) reduction measure $|POS_\mu|$ and voting by rough memberships. In each case, the horizontal axis shows the values of ε, for which the results were obtained.

its insignificance to the results. As a summary, the proposed method is very easy to design and implement, comparing to other approaches.

6   Conclusions and Further Research

We introduced a new method for dealing with real-valued decision tables, especially useful in the cases when the sets of attributes are significantly larger than the sets of objects, with very limited applicability of standard classification methods based on both parameterized and non-parameterized models. We showed how to combine it with a classical framework for construction of rough set-based classifiers. We tested the obtained approach against the real-life gene expression data set with 12,625 genes-attributes and 24 measurements-objects. The results turn out to be comparable with those of other classification techniques, while our method remains far simpler in interpretation for the domain experts. Further research is required to fully prove its practical usefulness over a wider range of benchmark data. We also consider adapting the proposed framework for real-valued decisions, as well as dominance/preference-related attributes.

Acknowledgements. Research of the second co-author was supported by the Research Center at the Polish-Japanese Inst. of Information Technology and by grant 3T11C00226 from Polish Ministry of Sci. Research and Higher Education.

References

1. Baldi P., Hatfield W.G.: DNA Microarrays and Gene Expression: From Experiments to Data Analysis and Modelling. Cambridge University Press (2002)
2. Chang J.C. et al.: Gene expression profiling for the prediction of therapeutic response to docetaxel in patients with breast cancer. The Lancet, **362** (2003)
3. Dietterich T.: Machine learning research: four current directions. AI Magazine, **18/4** (1997)
4. Draghici, S.: Data Analysis Tools for DNA Microarray. Chapman & Hall (2003)
5. Ganter B., Wille R.: Formal Concept Analysis: Mathematical Foundations. Springer (1997)
6. Gruźdź A., Ihnatowicz A., Ślęzak D.: Interactive gene clustering: A case study of breast cancer microarray data. Information Systems Frontiers, **8** (2006)
7. Fang J., Grzymala-Busse J.W.: Leukemia Prediction from Gene Expression Data – A Rough Set Approach. In: Proc. of ICAISC'06 (2006)
8. Nguyen H.S.: Approximate Boolean Reasoning: Foundations and Applications in Data Mining. Transactions on Rough Sets V, Springer, LNCS, **4100** (2006)
9. Pawlak Z.: Rough sets: Theoretical aspects of reasoning about data. Kluwer (1991)
10. Pawlak Z., Skowron A.: Rough membership functions. In: R.R. Yaeger, M. Fedrizzi, and J. Kacprzyk (eds.), Advances in the Dempster Shafer Theory of Evidence. Wiley (1994)
11. Ślęzak D.: Approximate reducts in decision tables. In: Proc. of IPMU'96, vol. **3** (1996)
12. Ślęzak D.: Various approaches to reasoning with frequency-based decision reducts: a survey. In: L. Polkowski, S. Tsumoto, T.Y. Lin (eds.), Rough Set Methods and Applications. Physica-Verlag (2000)
13. Ślęzak D., Wróblewski J.: Rough Discretization of Gene Expression Data. In: Proc. of ICHIT'06, vol. **2** (2006)
14. Słowiński R., Greco S., Matarazzo B.: Rough Set Based Decision Support. Introductory Tutorials on Optimization, Search and Decision Support Methodologies, Springer (2005)
15. Valdes J.J., Barton A.J.: Relevant Attribute Discovery in High Dimensional Data: Application to Breast Cancer Gene Expressions. In: Proc. of RSKT'06 (2006)
16. Wojna A.: Analogy-Based Reasoning in Classifier Construction. Transactions on Rough Sets IV, Springer, LNCS (2005)
17. Wróblewski J.: Theoretical Foundations of Order-Based Genetic Algorithms. Fundamenta Informaticae, **28/3-4** (1996)
18. Wróblewski J.: Ensembles of classifiers based on approximate reducts. Fundamenta Informaticae, **47/3-4** (2001)
19. Zadeh L.A.: Fuzzy Sets. Information and Control, **8** (1965)

Ubiquitous Customer Relationship Management (uCRM)

Sang-Chan Park[1], Kwang Hyuk Im[1], Jong Hwan Suh[1], Chul Young Kim[1], and Jae Won Kim[2]

[1] Korea Advanced Institute of Science and Technology (KAIST)
373-1 Gu-Seong-dong, Yu-Seong-gu, Daejeon, Korea 305-701
{sangchanpark,gunni,suhjonghwan,fezero}@kaist.ac.kr
[2] National Information Society Agency (NIA)
77 Moogyo-dong, Joong-gu, Seoul, Korea
kimjw@nia.or.kr

Abstract. The U-commerce service is "context-aware," and it focuses more on actively sensing different customer's roles through both time and location specificity [1] [2]. In U-commerce environment, we can make decisions proactively and intelligently by automatically detecting users' contextual data such as time, identity, location, entity. Context-aware technology can provide personalization services that reference the user's context and preferences. Proactive service and high personalization will enable a great number of improvements in the current CRM processes and open a new area of customer satisfaction. uCRM must pay due regard to 'context-aware' characteristics of U-commerce. In this paper, we define the term "context" and "context-aware computing." In addition, we suggest a practical framework of uCRM as equipped with context data warehouse correspondingly.

Keywords: u-commerce, CRM, context aware, context data warehouse.

1 Context and Context-Aware Computing

Humans are quite successful at conveying ideas to each other and reacting appropriately. This is due to many factors: the richness of the language they share, the common understanding of how the world works, and an implicit understanding of everyday situations. When humans talk with humans, they are able to use implicit situational information, or context, to increase the conversational bandwidth. Unfortunately, this ability to convey ideas does not transfer well to humans interacting with computers. In traditional interactive computing, users have an impoverished mechanism for providing input to computers. Consequently, computers are not currently enabled to take full advantage of the context of the human-computer dialogue. By improving the computer's access to context, we increase the richness of communication in human-computer interaction and make it possible to produce more useful computational services.

Context can be specifically classified as TILE: time, identity, location and entity [3]. **First**, time context is usable when the user enters a service zone and if the service is available at that time, then the service is working for the user. **Second**, identity

J.T. Yao et al. (Eds.): RSKT 2007, LNAI 4481, pp. 324–330, 2007.

context indicates with whom the user is communicating. **Third**, location context means a location of an identity in which the user is interested. **Last**, entity context is the information of any things that the user may be currently using. The context can be co-opted to create an inferable compound context. For example, the user's activity can be inferred from the user's current location and entity context [4].

Context-aware computing is becoming more crucial in mobile distributed computing systems. These systems aim to provide people with context-aware access to information, communication and computation. For example, CAMP (context-aware mobile portal) is a modular mobile internet portal enhanced with context awareness features, such as user preferences, location, and temperature [5]. Location-aware applications are other examples of context-awareness which take advantage of location-awaking sensors. Some excellent work in this area includes PARCTab at Xerox PARC [6], the InfoPad project at Berkeley [7], the Olivetti Active Badge system [8] and the Personal Shopping Assistant [9] [10].

2 uCRM

Mobile or ubiquitous computing services contain the following additional features in comparison with conventional electronic commerce, which runs on fixed networks. Successful mobile or ubiquitous commerce depends on the realization of their unique characteristics such as mobility, proactiveness, embededness, context-awareness, as well as the characteristics-business model fit. Characteristics increase the timeliness and promptness of the legacy services, and hence create new service areas.

Ubiquity can be simply defined as the state of being everywhere at once with any networks and devices. To do so, the service should be mobile: the users can access to the service via wireless network through portable client devices or smart objects. Moreover, the service system may be embedded in the client device or in smart objects, which can communicate and coordinate with each other. Decision-orientedness means to what extent a service embeds decision technologies and automates decision making process on behalf of the user. The watermarked area in the upper-right side of the figure indicates the mobile or ubiquitous decision support oriented service: mSCM, mCRM, mobile recommender, context-aware comparative shopping, and uCRM. Current mobile or ubiquitous computing services still stress information-based services such as location based services (LBS). uCRM are in their early age of adopting ubiquitous computing technologies, such as context-aware computing technology and wearable computing technology [4].

3 ContextDW : Integrated Customer Information

Customer information is usually stored in multiple customer databases. For example, customer's purchase history, such as demographic, production category, and payment method, is stored in a purchase database. In a Web log file, customer's navigation behavior is logged. When customers want to refund their purchase or revoke complaints, their records are saved into a customer complaint database. The greater the number of data sources for the customer information, the richer the information available to the business. Thus the integrated customer information is regarded as a CRM critical success factor [11][12].

In U-commerce environment, we name the integrated customer information the contextDW. The contextDW consists of basic information, sales information and behavioral patterns. As shown in Figure 1, basic information contains customer data, product data, statistical sales data and accounting data. Sales information contains dynamic sales data, event & campaign data and complains & claim data. Behavioral patterns information contains logical patterns and physical patterns. Logical patterns data is extracted from web access log and physical patterns data is extracted from user's contextual data such as current location, time, physiological state, personal profile.

E-commerce can use not only basic and sales information but also logical behavioral patterns data such as web access log. eCRM segments customers into several groups and provide customers with personalized service such as recommendation service using web access log. We can push personalized information such as "Customers who bought this product also bought" service using eCRM functions [13] [14] [15]. Because eCRM can use only logical pattern data such as web access log, purchase patterns, it is difficult that we connect on-line patterns with off-line behavioral patterns.

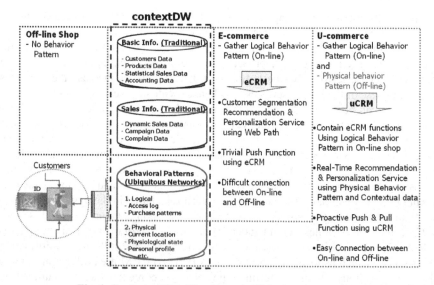

Fig. 1. Contextual DataWarehouse in Ubiquitous computing

On the other hand, U-commerce can use not only basic and sales information but also logical behavioral patterns data and physical contextual data such as **TILE: T**ime, **I**dentity, **L**ocation, **E**ntity. Because uCRM can use both logical patterns and physical patterns, it can provide customers with real-time recommendation service and high personalized service such as Personal Shopping Assistant (PSA). In U-commerce, companies can push personalized service and customers can pull appropriate service using uCRM functions. Also, it is easy that we connect on-line patterns with off-line behavioral patterns.

To utilize the context information for the personalization services, we need the context entity attribute, the current situation context, the situation context history and the situation context database. First, the database of the context entity attribute information is composed of the customer's profile information. The current situation context is the context entity, the information of the customer's current activity, and it is used for the context recognition service. The context entity history information is used for the recommendation of the objects through analyzing the customer's preferences. The environment context information in U-commerce is provided by the host of the shopping mall to get the information of the context entity which is the customer's location or the selected items.

4 Practical Framework of uCRM

Figure 2 shows data flow and process in ubiquitous environment. The publishers, such as government, companies, schools, hospitals and so on, attach RFID tag to belongings of customers. RFID tag can be attached to almost everything such as credit cards, industrial products, agricultural and marine products and so on.

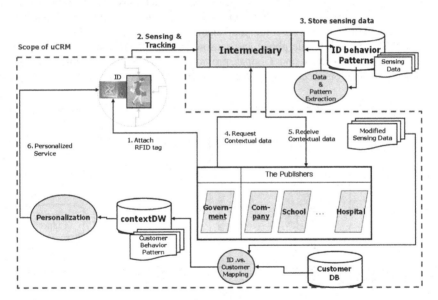

Fig. 2. Scope of uCRM in Ubiquitous environment

We name RFID-attached thing a smart object. The publishers have the RFID tag ID, the smart object and the customer information. While the customer moves with the smart object, intermediary are sensing and tracking RFID tab ID continuously. Intermediary is an organization that has an infrastructure such as sensing receiver, wireless local area network, mobile communication network and so on. That is, government, mobile communication companies, schools, department stores, hospitals can be intermediary. Intermediary stores sensing data that has contextual information such as time, identity location, and entity. Intermediary has not smart object information and customer information but only RFID tag ID and contextual data of ID. The publisher requests sensing contextual data to Intermediary in real time or at regular intervals. After Intermediary extracts sensing data of requested ID and modifies the data, it provides the publisher with modified sensing data. The modified sensing data contains contextual data of requested ID. The publisher executes mapping ID in modified sensing data and customer information in customer database. After mapping process is completed, customers' contextual data is stored in contextDW. Therefore, the publisher can know customer's behavior patterns. Using customer's behavior patterns, the publisher can provide the customer with high personalized service in real time or at regular intervals. Actually, OMRON Corporation, in a joint effort with PIA CORPORATION and TO-KYU CORPORATION (Tokyu Railways), has launched an information service experiment called Goopas that utilizes automated turnstiles at the stations of Tokyu Railways. Train passengers who use seasonal passes to go through the turnstiles get time- and place-dependent information (event notices, shop ads and coupon tickets) in the form of emails addressed to their mobile phones. For example, a passenger who has just entered a station will receive current time-sensitive information concerning his destination. And passengers leaving a station will get similar information about the areas around the terminal. The road towards services of this nature has been paved by the rapid spread of mobile phones with push technology and automated turnstiles that are connected to ubiquitous networks. In this case, the publishers are PIA CORPORATION and TOKYU CORPORATION and mobile phone and season pass are smart objects. Intermediary is TOKYU CORPORATION and Goopas provide the customer with time- and place-dependent information concerning customer's current state as personalized service [16].

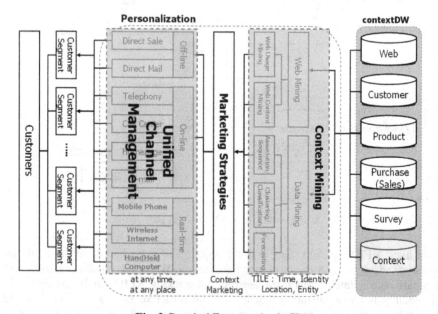

Fig. 3. Practical Framework of uCRM

The greater part of the contextual data can be analyzed with traditional data mining techniques but the new mining techniques can be needed to analyze the rest. So, we name the mining techniques containing data mining, web mining and the new mining for the rest **context mining** [13][14][17]. Context mining contains web mining techniques such as web usage mining and web content mining and data mining techniques such as association/sequence, clustering/classification, and forecasting. Context mining techniques must handle nominal or symbolic data expertly because contextual data has nominal values.

With analysis results, marketing strategies, for example, context marketing, benchmark marketing, and collaborative marketing, will be developed and delivered to proper customer segment. From the viewpoint of marketing, the realm of customer information handled by enterprises will expand dramatically under ubiquitous computing. In addition to the records they currently maintain on customer attributes and purchase histories, companies will gather real-time data that give clues to the circumstances surrounding each customer (the hour, place, activities, and other details). This is information that can have drastically stronger marketing effects if skillfully utilized. Moreover, interactive contacts with customers will become closer and shift to a real-time basis. Companies can show merchandise in the planning stage to potential buyers through an online customer community and carry forward with improvements based on the returned assessments. This will enable a company to produce goods whose design includes the "participation" of customers.

References

1. Lyytinen, K., Yoo, Y.: Issues and Challenges in Ubiquitous Computing, Communications of ACM 45(12) (2002) 63-65.
2. Fano, A., Gershman, A.: The future of business services in the age of ubiquitous computing, Communications of ACM 45(12) (2002) 63-87.
3. Abowd, G.D. Software engineering issues for ubiquitous computing, Proceedings of the 21st international conference on software engineering (1999) 75-84.
4. Kwon, O.B.: The potential roles of context-aware computing technology in optimization-based intelligent decision-making, Expert Systems with Applications 31(3) (2005) 629-642.
5. Mandato, D., Kovacs, E., Hohl, F., Amir-Alikhani, H.: CAMP: A context-aware mobile portal, IEEE Communications Magazine 40(1) (2002) 90-97.
6. Want, R., Hopper, A., Falcao, V., Gibbons, J.: The active badge location system, ACM Transactions on Information Systems 10(1) (1992) 91-102.
7. Long, S., Aust, D., Abowd, G.D., Atkeson, C.G.: Rapid prototyping of mobile context-aware applications: The cyberguide case study, Proceedings of the 1995 conference on human factors in computing systems (CHI'96), (1996) 293-294.
8. Schilit, W.N.: System architecture for context-aware mobile computing, PhD thesis, Columbia University (1995).
9. Asthana, A., Cravatts, M., Krzyzanouski, P.: An indoor wireless system for personalized shopping assistance, Workshop on mobile computing systems and applications (1994) 69-74.
10. Kwon, O.B., Sadeh, N.: Applying case-based reasoning and multi-agent intelligent system to context-aware comparative shopping, Decision Support Systems 37(2) (2004) 199-213.

11. Bae, S.M.: Collaborative CRM: Channel Management in Information-Based EC. Doctoral Thesis, KAIST (2003).
12. Kim, T.H., Park, S.C.: Metadata and Information Asset for Infomediary Business Model on Primary Product Market, Lecture Notes in Computer Science 3828 (2005) 926 – 935.
13. Bae, S.M., Ha, S.H., Park, S.C.: Fuzzy Web Ad Selector Based on Web Usage Mining, IEEE Expert - Intelligent Systems and Their Applications 18(6) (2003) 62 – 69.
14. Lee, J.H., Park, S.C.: Agent and data mining based decision support system and its adaptation to a new customer-centric electronic commerce, Expert Systems with Applications: An International Journal 25(4) (2003) 619 – 635.
15. Park, J.H., Park, S.C.: Agent-based merchandise management in Business-to-Business Electronic Commerce, Decision Support Systems 35(3) (2003) 311 – 333.
16. Nakajima, H.: Marketing Strategy in the Era of Ubiquitous Networks, NRI Papers 44. March 1 (2002).
17. Park, S.C., Piramuthu, P., Shaw, M.J., Dynamic Rule Refinement in Knowledge-Based Data Mining Systems, Decision Support Systems 31(2) (2001) 205 – 222.

Towards the Optimal Design of an RFID-Based Positioning System for the Ubiquitous Computing Environment

Joon Heo[1], Mu Wook Pyeon[2], Jung Whan Kim[1], and Hong-Gyoo Sohn[1]

[1] Yonsei University, School of Civil and Environmental Eng., Seoul, Korea
{jheo,sohn1}@yonsei.ac.kr, roak7@hanmail.net
[2] Konkuk University, Department of Civil Eng. Seoul, Korea
neptune@konkuk.ac.kr

Abstract. Positioning technology is a core component of the ubiquitous computing environment. Radio Frequency IDentificaiton (RFID) is recognized as a powerful means of replacing not only barcodes for product identification but also conventional positioning systems in unfavorable situations such as indoor and obstacle-laden sites. The objective of this study was to determine the design parameters of an RFID-based positioning system and to conduct a series of simulations in order to evaluate the impact of individual parameters on positional accuracy according to RMSE. The conclusions drawn from the experimentation are that (1) geometric distribution (triangular or square) of RFID tags is not a determining factor; (2) multiple-level ranging of reference tags can significantly improve positional accuracy; (3) the detection rate is the most significant factor in positional accuracy; (4) a smaller-percent standard deviation accompanies a larger fluctuation of positional accuracy; and (5) the detection range over the neighboring nodes is sufficient to achieve a reasonable accuracy.

Keywords: Positioing, RFID, Simulation, Accuracy.

1 Introduction

Radio Frequency IDentificaiton (RFID) is a means of storing and retrieving data through electromagnetic transmission to compatible reading devices. RFID originated back in World War II, but has only recently become popular in many applications such as automated inspection/identification of products, real-time inventory management, distribution tracking, access control, and location-sensing systems. RFID is recognized as a powerful means not only of replacing barcodes for product identification but also of assisting in conventional positioning systems in unfavorable situations. The advent of wireless technologies and mobile computing has required the knowledge of the physical location of objects for location-aware services. This is mainly because contextual information obtained from a sensor network is meaningful only when the physical location of the source of that information is determined [1]. The Global Positioning System (GPS) is the best known, best established, and most successful positioning solution. However, as GPS is satellite-dependent, it has the intrinsic problem of positioning objects in signal-blocked environments such as the inside of buildings and obstacle-laden

J.T. Yao et al. (Eds.): RSKT 2007, LNAI 4481, pp. 331–338, 2007.
© Springer-Verlag Berlin Heidelberg 2007

sites. RFID is one of the solutions for overcoming such a limitation, and it has been studied by many research groups [2, 3, 4, 5]. Most of the research has focused on feasibility studies of RFID-based positioning systems. However, very little progress has been made in determining the performance factors necessary to achieve the best system performance. To that end, the objective of this study was to present the design parameters of an effective RFID-based positioning system, to conduct a series of simulations in order to evaluate the impact of individual parameters on the expected positional-accuracy performance measures, and finally to produce more meaningful design guidelines for an RFID-based positioning system.

2 RFID-Based Positioning System

2.1 Overview of Positioning Technology

Position is typically represented in terms of a set of coordinates within a semantic reference system. Many positioning systems have been designed to determine the

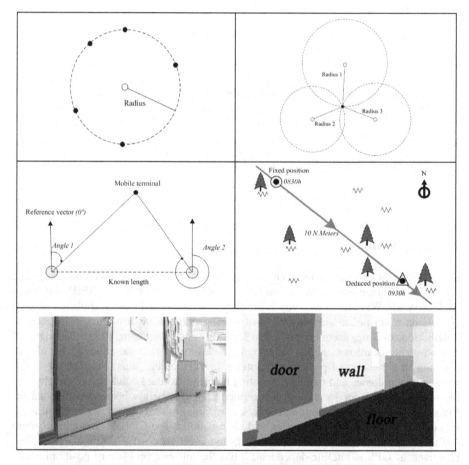

Fig. 1. Conceptual views of positioning algorithms

position of objects. They are categorized into three major groups: global location systems such as GPS, wide-area location systems, and indoor location systems [6]. Numerous technologies have been deployed for those positioning systems, such as infrared, ultrasonic, video surveillance, and wireless local area network (WLAN). Regardless of the choice of technology, positioning is based on one of four algorithms: (1) proximity sensing; (2) triangulation; (3) pattern matching; and (4) dead reckoning. Fig. 1 offers a conceptual view of each algorithm. Proximity sensing is based on a single base station's coverage area. Triangulation combines lateration and angulation. Lateration computes the position of an object by measuring its distance from multiple reference positions. Complementarily, angulation computes the position of that object by measuring the angles with which it sees those multiple reference points. For lateration, the governing equation in 2D is the distance between I and J: $IJ = \sqrt{(x_i - x_j)^2 + (y_i - y_j)^2}$, which is linearized for solving unknown coordinates using least squares estimation. On the other hand, when no explicit distance between objects is available, the k-neighboring algorithm can be used. Thereby, the coordinates of the target point can be obtained by

$$(x, y) = \sum_{i=1}^{k} w_i(x_i, y_i) \Big/ \sum_{i=1}^{k} w_i \qquad (1)$$

where x_j, y_i are the coordinates of the ith reference point and w_i is the weighting factor. The choice of weighting factor is inverse-proportional to the distance between the reference point and the target point, in order to ensure minimal error in coordinate estimation.

2.2 RFID Technology for Positioning

RFID is composed of tags, a reader with an antenna, and software. The main function of the RFID reader is to retrieve the information stored on the tags. If the tags store the coordinates of the locations where they are installed, the location of the reader that communicates with the tags can be estimated. There are two methods of communicating between readers and tags: inductive coupling and electromagnetic waves. Inductive coupling is also called passive, in which the antenna coil of the reader induces a magnetic field in the antenna of the tag. The induced energy is used to transmit the data of the tag to the reader. The reader is very light, inexpensive, and battery-free. However, it has a very limited detection range, and consequently, it is not appropriate to positioning systems requiring a longer range. Instead, electromagnetic waves or active tags that can communicate with the reader over a long range are ideally suited for the positioning system. For example, Mantis tag solution [6] has a maximum read range of 1,000 feet for communication between tags and readers. With the capability of reading multiple tags and signal strength, it is feasible to obtain the location of the RFID reader through triangulation and the k-neighboring algorithm as well as conventional proximity sensing. A recent development makes it feasible to measure the distance between the RFID tag and the reader using Time Difference of Arrival (TDOA) [7]. For all of these reasons, an active RFID system was assumed in the simulation. For performance measures of positioning accuracy, RMSE in 2D which is defined as the closeness to the true position, was used.

$$RMSE = \sqrt{\frac{\sum_{i=1}^{n}[(\hat{x}_i - x_i)^2 + (\hat{y}_i - y_i)^2]}{n-1}} \tag{2}$$

$(\hat{x}_i, \hat{y}_i)$ is the estimated coordinates of a point in the true coordinates of (x_i, y_i).

3 Simulation

3.1 Simulation Setup

In the simulation of the proposed RFID-based positioning system, it was assumed that reference tags were installed in the space of interest and that they stored the coordinates of the locations. The moving object (target object) was equipped with a reader for communication with the reference tags and a computing machine for the estimation of the coordinates. Using lateration, proximity sensing, or the k-neighboring algorithm, the location of the reader could be estimated. The opposite setup – positioning moving-tags with installed reference readers – is rather popular for the purposes of inventory monitoring and distribution tracking, where all of the other processing is generally conducted at a centralized server. In contrast, the proposed setup is more suited for navigation systems, where additional processing is conducted at the moving object. The current RFID technology can measure the distance between reader and tag, but it requires a sufficiently long range for reasonable accuracy. In other words, it is not a generic solution for all environments – indoor and outdoor, close range and long range – as yet. For that reason, the k-neighboring algorithm was selected for positioning computation in the simulation.

3.2 Simulation Design

The objective of the simulation was to uncover the governing factors in RFID-based positioning systems and to present a design guide that could maximize the performance measure of positional accuracy according to RMSE. The authors determined to investigate the following five factors: (1) geometric placement of reference RFID tags; (2) multiple-level ranging; (3) tag detection rate; (4) percent standard deviation of detection range; and (5) detection range. First of all, triangle-based and square-based regular assignments of the reference tags were simulated and compared. For a fair comparison, interval of points was determined to make Voronoi diagrams of the points have the same size, in other words, affecting region of a reference tag have the same size. The ratio between the square and the triangle was 1:1.07457. The second consideration was the detection range, which is directly associated with the number of neighboring reference tags. With respect to the given configurations of triangles and squares, detection ranges from 8m to 27m were tested. The third design factor was the ranging capability of the RFID reader. Signal power analysis can be used for multiple leveling of range, and is common in WLAN-based positioning systems [6]. For instance, if three levels – close, medium, and far – of range can be measured, the weight factors will be 1, 1/2, and 1/3, respectively, and the result will be expected to be more accurate than that for a single level or two levels. The fourth factor to be considered was detection rate. A 100% detection rate is ideal but hardly achievable in reality for many reasons. Power shortage and mechanical/electromagnetic breakdown, signal interference, signal collision, multi-path, obstacles, and other harsh conditions are examples. The last factor is the variation of the behavior of tags. It was assumed that all of the tags would have the same detection range and signal strength, but they vary in reality [5]. In order to reflect that effect in the simulation, the detection range was treated as a random variable following a Gaussian distribution, which employs the mean of the given detection range and the standard deviation of 10% of the detection range in most simulations. For example, when the detection range of 15m is given, the detection range will be randomly generated under the condition of $d\sim N$ *(15m, 1.5m)*.

It was simulated that a target object equipped with an RFID reader was randomly moving in the space where the reference tags were installed, as shown in Fig. 2. The detection range of each reference tag was generated from the Gaussian distribution. Any tags that included the target object within its detection range were considered for the *k*-neighboring algorithm. For each parameter setup, the simulation was repeated *ten thousands times* to report positional accuracy according to RMSE.

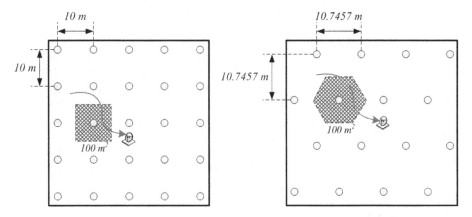

Fig. 2. Geometric placement of reference tags

3.3 Simulation Results and Analysis

(1) Geometric placement of reference RFID tags

For the evaluation of the geometric placement of reference RFID tags – either triangular or square, of single-level ranging capability and with 100% of the detection rate fixed – the average detection range as well as a standard deviation of detection range of 10%, from 8m to 27m were applied to the simulation. Fig. 3(A) illustrates the simulation results. The average RMSE over a 1 m detection range were 1.93m for triangles and 1.98m for squares. This shows that triangular placement could yield a slightly better accuracy, but that geometric placement is not an important factor in differentiating the positional accuracy.

(2) Multiple-level ranging

For this simulation, it was assumed that the RFID reader could simultaneously analyze the signal power and multiple-level discrete ranges. With respect to the different number of range levels, the changes of RMSE were monitored through the simulation of triangular placement reference tags. Again, 100% of the detection rate was fixed, and the average detection range as well as a standard deviation of detection range of 10%, from 8 m to 27 m, was applied to the simulation. Fig. 3(B) illustrates the simulation results. Certainly, it is an important design factor that can differentiate the positional accuracy. It is also noted that the marginal benefit of the number of discrete levels decreases.

(3) Detection rate

With respect to five different detection rates – 100%, 95%, 90%, 85%, and 80% – the proposed RFID-based positioning system was simulated. Triangular placement, single-level ranging, and the average detection range as well as a standard deviation of detection range of 10%, from 13m to27 m were applied. The impact of the changes in detection rate was very discernible. As shown in Fig. 3(C), the positional accuracy almost linearly increases as detection rate is improved. In other words, improvement of detection rate can produce a clear upgrade of positional accuracy.

(4) Percent standard deviation of detection range

The next parameter to be tested was percent standard deviation, which is associated with RFID system reliability. Up to this point, 10% standard deviation of detection range under Gaussian assumption was used for the simulation. Triangular placement, single-level ranging, 100% detection rates, and a detection range from 10m to 27m were assumed. Fig. 3(D) summarizes the results of the simulation. It presents the interesting result that a smaller standard deviation accompanies a larger fluctuation of positional accuracy. It was also an important finding that the RMSE cycle should be considered in the design, particularly when installed RFID tags are expected to perform stably and consistently in radar signal emission.

(5) Detection range

Detection range, before the simulations, was supposed to be the most important factor. However, surprisingly it is not a dominating factor in the positional accuracy of RFID-based positioning systems. All four graphs in Fig. 3, in a variety of combinations of parameter setup, substantiate the insignificance of detection range. Provided that the range reaches the neighboring node, any addition of detected reference tags basically is redundant. RMSE is highly stable with detection ranges over 14m, regardless of a change of parameters. The insignificant impact on RMSE is dramatically illustrated in Fig. 3(D). For instance, the detection range of 13 m with no uncertainty (0% standard deviation) produced a 0.85m RMSE, whereas 16 m generated a 1.67m RMSE. That is, the increase of detection range doubled the RMSE.

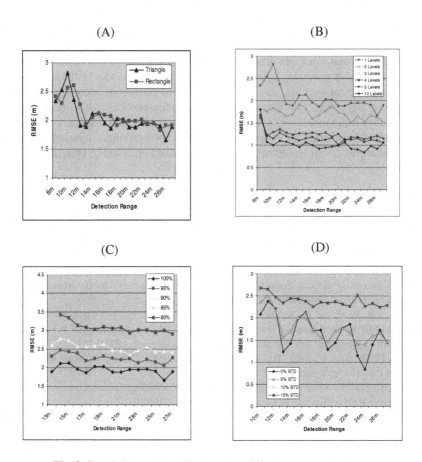

Fig. 3. Simulation results with respect to different parameter setups

4 Conclusions and Future Study

Determination of the physical location of objects is one of the key issues in development of context-aware applications in ubiquitous computing. In this study, the conceptual framework of an RFID-based positioning system based was simulated and tested for find a guideline for the optimal design. The moving object (target object) was equipped with an RFID reader for communication with the reference RFID tags and with a computing machine for the estimation of the coordinates. The k-neighboring algorithm was used for coordinate estimation of the moving target. RMSE was chosen for comparative measurement of the performances of the positioning system. In the simulation, five parameters were tested–geometric distribution of reference tags, multiple-level ranging, detection rate, percent standard deviation of detection range, and detection range. The results of the simulation revealed that (1) the geometric distribution (triangular or square) of the RFID tags is not a determining factor; (2) multiple-level ranging can significantly improve the positional accuracy; (3) the detection rate is the most significant factor in positional accuracy; (4) a smaller-percent standard deviation accompanies a larger fluctuation of positional accuracy; and (5) the detection range is sufficient to reach the neighboring nodes and any increase of that does not produce a significant gain of positional accuracy. Another finding worth reporting was that about 20% of the intervals of the reference tags in the given regular spacing was the upper limit of positional accuracy under the k-neighboring algorithm. For future research, an integration of the k-mean algorithm and trajectory estimation techniques such as Kalman filter in 3D space will be considered in order to construct an enhanced RFID-based positioning system and to produce more meaningful design guidelines for an RFID-based positioning system.

References

1. Fukuju, Y., Minami, M., Morikawa, H., and Aoyama, T.: DOLPHIN: An Autonomous Indoor Positioning System in Ubiquitous Computing Environment. WSTFES2003, Hakodate, Japan (2003) 53-56
2. Kubitz, O., Berger, M., Perlick, M., and Dumoulin, R.: Application of Radio Frequency Identification Devices to Support Navigation of Autonomous Mobile Robots. IEEE 47th Vehicular Technology Conference (1997) 126-130
3. Hightower, J., Vakili, C., Borriello, G., and Want, R.: Design and Calculation of the SpotON ad-hoc location sensing system. UW CSE 00-02-02, University of Washington, Department of Computer Science and Engineering, Seattle, WA. (2001)
4. Chon, H. D., Jun, S., Jung, H., and An, S. W.: Using RFID for Accurate Positioning. J. Global Positioning Systems, Vol. 3. (2004) 32-39
5. Ni, L.M., Liu, Y., Lau, Y.C., and Patil, A. P.: LANDMARC: Indoor Location Sensing Using Active RFID. Wireless Network, Vol. 10. (2004) 701-710
6. Xiang, Z., Song, S., Chen, J., Wang, H., Huang, J., and Gao, X.: A wireless LAN-based indoor positioning technology. IBM J. Res. And Dev. Vol. 48. (2004) 617-626.
7. RF Code Inc.: 433 MHz Mantis Tags. http://www.rfcode.com/433mantis_tags.asp (2006)
8. AeroScout Inc.: AeroScout Visibility System Overview. http://www.aeroscout.com/content.asp?page=SystemOverview (2006)

Wave Dissemination for Wireless Sensor Networks

Jin-Nyun Kim[1], Ho-Seung Lee[2], Nam-Koo Ha[2],
Young-Tak Kim[1], and Ki-Jun Han[2]

[1] Department of Information and Communication Engineering, Yeungnam University,
214-1, Dae-Dong, Kyungsan-Si, Kyungbook, 712-749, Korea
[2] Department of Computer Engineering, Kyungpook National University, 1370 Sankyuk-Dong,
Book-Gu, Daegu, 702-710, Korea
duritz1979@yahoo.co.kr, hslee@netopia.knu.ac.kr,
adama2@netopia.knu.ac.kr, ytkim@yu.ac. kr,kuhan@bh.knu.ac.kr

Abstract. Wireless sensor networks consist of a large number of the distributed nodes organized in a multihop structure. Energy consumption in sensor nodes that are generally battery-operated is very important. This paper proposes an energy efficient routing/MAC integrated scheme, called Wave Dissemination, to prolong the network lifetime by having all nodes uniformly consume their energies without increasing delay and degrading throughput. Wave dissemination scheme generates the virtual wave lines that induce the wave-to-wave networking. Wave dissemination consists of three network phases: network initialization, query wave and data wave. Simulation study assures that wave dissemination scheme outperforms the traditional protocols; it achieves a higher energy saving and a lower message delay.

Keywords: Wireless sensor network, energy efficiency, routing protocol, media access control (MAC).

1 Introduction

Wireless sensor networks consist of large number of distributed nodes that organize themselves into a multihop wireless network. Sensor nodes are generally battery-operated, and deployed in a variety of terrains. With many nodes placed in their target application, changing batteries becomes difficult or impossible. So, energy consumption is very important.

Energy conservation in wireless sensor networks involves two dominant approaches. The first is at the medium access control (MAC) and networking layers. In MAC layer, one of the main mechanisms to save energy is the use of sleep modes, by powering off the radio since transmitting, receiving, and listening to idle channel are functions which require a comparable amount of power. In network layer, routing protocol must be lightweight, and designed energy-constrainedly. The second is data aggregation (fusion) through in-network processing, whereby correlations in data are exploited to reduce the size of data, and different data packets are mixed into one to reduce the number of data transmissions.

J.T. Yao et al. (Eds.): RSKT 2007, LNAI 4481, pp. 339–346, 2007.
© Springer-Verlag Berlin Heidelberg 2007

This paper proposes a new routing scheme, called Wave Dissemination, to prolong the network lifetime by reducing the number of exchanged messages, the relayed messages, the collision and the overhearing. Wave dissemination can integrate MAC and routing into a single layer, like a GeRaF [1,2] and [3]. All messages in wave dissemination are transmitted in multi-hop manner by using intermediate sensors as relay nodes. Wave dissemination constitutes the virtual wave lines that induce the wave-to-wave networking. Wave dissemination is based on four elementary functions: wave transmission, aggressive data aggregation, proactive data path reservation, and predictable active-sleep.

2 Wave Dissemination

A network in wave dissemination is viewed like the waves on the windy ocean as illustrated in Fig. 1. The traffic direction in traditional wired and mobile ad hoc network is wayward; they cannot expect who sends message and receives message. On the other hand, the message direction in wireless sensor networks is relatively deterministic. There are a sink and sensor nodes. The direction of message communication is sink-to-nodes for query message and nodes-to-sink for data message. We can constitute the virtual wave lines in network to energy-efficiently transmit the messages. All networking processes and message transmissions are performed in wave-to-wave.

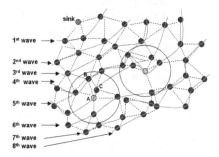

Fig. 1. Virtual wave lines in wireless sensor network

Wave dissemination consists of three network phases: network initialization, query wave and data wave phases. A network initialization process is performed only once when the network is deployed. Then, query wave and data wave phases are repeated. In wave dissemination, query wave phase may be omitted, so only data wave phase may occur periodically.

2.1 Network Initialization

When all sensor nodes in network turn on their powers, networking information is collected for network initialization. For this, dummy query messages are flooded into the whole network, and dummy data messages are followed in order to obtain the networking information as illustrated in Fig. 2. At this time, all nodes compute the minimum hop count from the sink to themselves in order to constitute virtual wave lines. Hop count is based on radio radius. All nodes share the minimum hop count information. Also, nodes share their residual battery information. Nodes with the same minimum hop count are connected by virtual wave line.

In general, sharing of whole topology information imposes a heavy burden on battery-powered sensor nodes. Thus, in our scheme, nodes maintain the information of only nodes at one level upper or down wave. For this, each node relays its own information to nodes at one level upper or down wave. This is enough for message routing in our scheme.

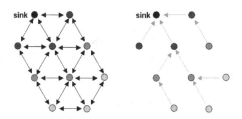

Fig. 2. Network initialization: (a) Dummy query messages. (b) Dummy data message.

Nodes that are located in the same wave level are called as associate. There can be multiple paths from a node to the sink, so a node can have different multiple hop counts to the sink. But, there is no possibility of any muddle due to the wave line built based on the minimum hop count. As illustrated in Fig. 1, node A has two paths through node B and node C to reach the sink. But, A uses B located at inbound wave, to deliver message to the sink. So, the minimum hop count from A to the sink is 5. Node A and C are in the same wave level (5-th wave level). All messages except dummy query messages are exchanged with only nodes at one level upper or one level down wave, not with associate. Node-to-node round trip time (RTT) is computed in the network initialization phase. RTT can be computed simply by using time-stamp in dummy query and dummy data message. In order to obtain an accurate RTT value, dummy query and dummy data messages must be sized as the same as the actual query and data message. This RTT value is used for predicting the time of receiving the data message. In our scheme, the data path of the next round is reserved in a proactive manner by using the dummy data message as illustrated in Fig. 2 (b). Each node knows the residual battery information of nodes on one level upper wave; it is acquired from dummy query message. In order to reserve the data path for the next round, nodes choose the most powerful node on its one level upper wave. Each node maintains the path information of just nodes at one level upper wave, do not the whole path information. Data path for the next round is reserved in every round. This beforehand-reserved data path makes the aggressive data aggregation and predictable active-sleep.

2.2 Query Wave

In wave dissemination, the number of query messages transmitted and relayed is reduced through the wave transmitting. When sensing task is demanded by application, the sink sends the query messages to sensor nodes. In wave dissemination, it is assumed that the query is transmitted to all nodes in network and all nodes receiving the query sense the event and have the sensed data to send to the sink. While the query message including the amount of residual battery is transmitted, each node shares the battery information with nodes at one level down wave. So, each node can know the most powerful node of the one level upper wave. Query message is relayed only to nodes at one level down wave and is not relayed to its associates in the same level as shown in Fig. 3 (a). In our scheme, nodes may turn the radio off just after relaying the query message in order to reduce the energy waste by overhearing and collision. Associates on the same wave line send the message and turn the radio off almost simultaneously. All other networking processes are performed in wave-to-wave fashion. Each node transmits the amount of its own residual battery. It does not relay the information to nodes at one level upper wave. For data path, it is enough that node knows the information on its neighbor among nodes at one level upper wave. The roof problem of routing is naturally excluded because query messages are transmitted to nodes at one level down wave, and data messages are transmitted to nodes at one level upper wave. In wave-to-wave transmission, broadcast storm problem can be relieved. Query messages are relayed to neighbor among nodes at one level down wave. At this time, the query messages transmitted from its associate are not relayed. After relaying the query message, it turns off the radio and sets the TTL (Time To Live) field to the wave level of the destination (e.g., TTL value for the 3rd wave nodes is 3).

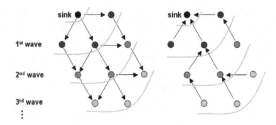

Fig. 3. Network phases: (a) Query messages in query wave. (b) Data messages in data wave.

2.3 Data Wave

Data messages wave to the sink through beforehand-reserved data paths after sensing task ends as shown in Fig. 3 (b). Data paths for the next round in wave dissemination are reserved beforehand. The sink-to-node and node-to-sink traffic feature of wireless sensor networks permits the data path to be reserved. This is proactive manner. For this, each node maintains the information on its neighbor at one level upper or down wave. End-to-end path or whole topology information is not required in our scheme. In our scheme, the most powerful node at one level upper wave is reserved for the next round. Wave dissemination assumes that all nodes always have the data to send. However, when only a few nodes have the data to send, the reserved data path can be changed every round to keep the network energy-balanced.

Data messages are relayed to the reserved node among nodes at one level upper wave. Considering that transmission of data message is generally unicast, the amount of data transmissions are uniformly balanced over all nodes in the network by the aggressive data aggregation. In wave dissemination, each sensor node transmits its data message only one time per round through the aggressive data aggregation. Each node can know who will send the data message to itself by using the reserved data paths. Each node is waiting until receiving all data messages from nodes at one level down wave, aggregates those data and its own data, and then transmits it to the upper wave. This aggressive data aggregation can increase the latency, but in energy-constrained sensor networks, the number of message transmissions can be reduced and be evenly balanced over all nodes. The radio may be turned off after data message is relayed to mitigate the overhearing and collision from nodes at down wave.

2.4 Predictable Active-Sleep Period

One important factor for energy saving is idle listening, i.e., listening to receive possible traffic that is not sent. During an active period, node is always in the idle listen state. Idle listening can be reduced by shortening active period. In our scheme, the active period is shortened by predicting the receive time of the data message. Wave dissemination uses the predictable dynamic active-sleep period, while PS mode of 802.11 uses a fixed duty cycle or uses an adaptive duty cycle that chooses uniformly active (idle) period (e.g., 1 second ~ 8 second). In wireless sensor networks, the message communication time and traffic pattern are relatively deterministic, and a message size is relatively constant comparing with mobile ad hoc and wired IP networks. These features make the message latency more constant. Wave dissemination predicts the time when each node receives the data message by using wave-to-wave round trip time (RTT) to prolong the sleep period for energy saving.

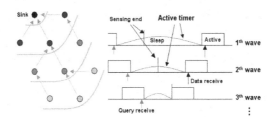

Fig. 4. Predictable active-sleep period

The process of predictable active-sleep technique is as follows and also as shown Fig. 4. Each sensor node receives the query message from the sink. The node that is closer to the sink more quickly receives the query message. As soon as the node relays this query message to nodes at one level down wave, it turns the radio off and transits into the sleep mode. Then, the active timer in the node starts. The active timer is used for determining the time when it should be active (idle) again. When the active timer expires, it wakes up and transits into the active mode to receive the data message from down wave. Timeout value of the active timer is given by

$$T_{timeout} = T_{query} + T_{RTT} + T_{sensing} \qquad (1)$$

where T_{query} is the time when the node receives the query message, T_{RTT} is RTT value from itself to the farthest wave which is computed in network initialization phase, and

$T_{sensing}$ means the sensing period spent on sensing task. For more energy saving, the predictable active-sleep may be used together with 802.11 PS mode. For example, each node can be usually in PS mode and then it can perform the predictable active-sleep only when it receives the query message. Predictable active-sleep may be more efficient in the application that needs the long sensing time like tracking the mobile target for quite a time. The predictable active-sleep can reduce the energy consumption by shortening the active period and reduce the message latency.

3   Performance Evaluation

To evaluate the performance of wave dissemination scheme, we have implemented an ns-2 simulator. This includes four functional modules; wave transmission, aggressive data aggregation, proactive data path reservation and prediction of active-sleep period.

3.1   Simulation Model

We used a random topology which consists of initially 20-node in a 600×600m area. Each sensor node has a radio range of 250m. Each node has its own ID. In our simulation, 0-node is the sink and others are sensor nodes. We used 802.11 DCF (Distributed Coordination Function) as MAC protocol. The radio power characteristics are given in Table 1. These values are taken from the specifications for the TR1000 radio from RF Monolithics [5]. The sink generates query messages in CBR (Constant Bit Rate). The destination address of query message is the broadcast address. The sink generates one query message for every sensor node. So, 19 query messages are generated at the sink for 19 sensor nodes. These query messages are transmitted and relayed to sensor nodes in a form of wave. Nodes that receive it send out the data message to the sink after sensing task ends. Data messages are aggregated at nodes. Query and data messages have a length of 36 bytes and 64 bytes, respectively.

Table 1. Radio Power Characteristics

Radio mode	Power consumption (mW)
Transmit (Tx)	14.88
Receive (Rx)	12.50
Idle	12.36
Off	0.016

3.2   Experimental Results

Fig. 5 shows the average dissipated energy which means the dissipated energy per node per second (dissipated total energy in network / number of nodes / simulation time). The wave dissemination is compared with other traditional protocols such as flooding of RREQ (Route Request) used in AODV (Ad hoc On Demand Distance Vector) and 802.11 PS mode. We can see that the wave dissemination consumes the less energy than the flooding scheme in Fig. 5 (a). This is because in the wave dissemination, each node receives the query messages from only nodes upper by one level while the flooding protocol uses the broadcast ID for reducing the rebroadcast overhead. For example, the node at the second wave level receives the query messages from only nodes at the first wave level. All messages from its associate and other wave level are dropped. Also, the wave dissemination uses the wave level of the destination

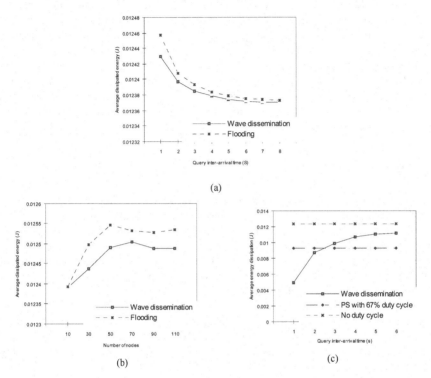

Fig. 5. Average energy dissipation: (a) Impact of the query loads. (b) Impact of the number of nodes. (c) Impact of the query loads.

for the TTL field for reducing the rebroadcast overhead. For example, TTL value for third wave nodes is 3. In this 20-node network, four levels of waves are generated. So, our scheme reduces the broadcast storm. This advantage can be observed more apparently as the traffic becomes heavier. As the traffic is heavier, the wave dissemination consumes much less energy than the flooding scheme. Fig. 5 (b) shows the average dissipated energy as the number of nodes in network increase. Query message is generated every second. This figure indicates that the wave dissemination outperforms the flooding scheme because wave dissemination re-broadcasts the relatively less messages through wave transmitting. We can observe that the average dissipated energy becomes saturated as the traffic load increases. Fig. 5 (b) shows a saturation point where there are around 50 nodes in the network. This results from the characteristics of CSMA/CA. In the saturation situation, messages are collided, are dropped and are not relayed. So, no more energy is consumed. It should be noted that the wave dissemination consumes less energy than the flooding scheme regardless of saturation because less messages are relayed than the flooding scheme.

Fig. 5 (c) shows that our scheme consumes the less energy than the PS mode with the duty cycle of 67% at high traffic intensities. But, we can see a contrary result at low traffic intensities. In the simulation, sensing period is assumed 0.8 second. In the PS mode, it is assumed that the sleep and idle periods are 2 and 8 seconds, respectively. And, idle period is uniformly selected. In Fig. 5 (c), "No duty cycle" represents that it does not have any radio on-off scheme; radio is always turned on. CSMA/CA protocol always consumes the energy because it is always working in the idle listening state except transmit and receive states. Idle listening state consumes almost the same energy as the receive state. So, the number of the received messages rarely affects the energy consumption in sensor node. Only the number of transmitted messages impacts the amount of energy dissipation in sensor node. In idle listening state, the traffic density makes only a little effect on the total energy dissipation since 802.11 DCF always consumes the energy.

4   Conclusion

In this paper, we proposed an energy-efficient MAC-routing integrated protocol, called wave dissemination, for wireless sensor networks. Wave dissemination induces the wave-to-wave networking by virtual wave lines. Our scheme is carried out by four basic functions; wave transmission, aggressive data aggregation, proactive data path reservation and prediction of active-sleep period. Experimental results using the ns-2 simulator show that wave dissemination achieves a higher energy savings and lower message latency than the traditional protocols.

Acknowledgments. "This research was supported by the MIC (Ministry of Information and Communication), Korea, under the ITRC (Information Technology Research Center) support program supervised by the IITA (Institute of Information Technology Advancement)" (IITA-2007-(C1090-0603-0002)).

References

1. M. Zorzi, R.R. Rao.: Geographic Random Forwarding (GeRaF) for Ad Hoc and Sensor Networks: Energy and Latency Performance. IEEE Trans. On Mobile Computing, vol. 2. (2003)
2. M. Zorzi, R.R. Rao.: Geographic Random Forwarding (GeRaF) for Ad Hoc and Sensor Networks: Multihop Performance. IEEE Trans. On Mobile Computing, vol. 2, (2003)
3. R. Rugin, G. Mazzini.: A Simple and Efficient MAC-Routing Integrated Algorithm for Sensor Network. IEEE Comm. Society (2004)
4. Network Simulator (ns), http://www.isi.edu/nsnam/ns
5. Ash Transceiver's Designers Guide, http://www.rfm.com (2004)

Two Types of a Zone-Based Clustering Method for Wireless Sensor Networks

Kyungmi Kim[1], Hyunsook Kim[2], and Kijun Han[2]

[1] Global Leadership School, Handong Global University
Bukgu, Pohang, 791-708, Korea
kmkim@handong.ac.kr
[2] Department of Computer Engineering, Kyungpook National University,
1370, Sangyuk-dong, Buk-gu, Daegu, 702-701, Korea
hskim@netopia.knu.ac.kr, kjhan@knu.ac.kr

Abstract. In this paper, we propose a zone-based clustering method to balance the amount of energy consumptions over all sensor nodes in the wireless sensor network field. In our method, a network field is divided into several zones with two types of shape: arc and square. And, the number of clusterheads selected in each zone is determined depending on its area. These contribute to distributing clusterheads evenly over the network field. Simulation results show that our zone method can outperform LEACH and PEGASIS in terms of network lifetime and connectivity.

Keywords: wireless sensor network, clustering, clusterhead selection, multiple hop transmission, zone, routing.

1 Introduction

Persistent advances in hardware and wireless network technologies have led us to another era where small wireless gadgets will allow access to prompt information anytime, anywhere. So, wireless sensor networks have gathered a great research interest in recent years mainly due to their possible wide applicability. Especially, they have advantages in inaccessible environments, such as difficult terrains, or on a spaceship [1]. In addition potential applications for such large-scale wireless sensor networks exist in a variety of fields, including environmental monitoring [2, 3], surveillance, home security, military operations, and object tracking [4]. Generally a sensor node consists of sensing elements, microprocessor, limited memory, battery, and low power radio transmitter and receiver. Sensor nodes are usually unattended, resource-constrained, and unrechargeable in wireless sensor networks. And thus distributing power consumption to all nodes is a major design factor because the system lifetime of wireless sensor networks is limited [5]. And also, locating sensor nodes over network fields efficiently is one of the most important topics in wireless sensor networks.

To solve this problem, several clustering approaches in wireless sensor networks have been proposed. Clustering approaches can reduce the energy used to communicate data

J.T. Yao et al. (Eds.): RSKT 2007, LNAI 4481, pp. 347–354, 2007.

from sensor nodes to the sink [6]. The essential operation in clustering approaches is to select a set of clusterheads from the set of sensor nodes in the network, and then group the remaining sensor nodes with these clusterheads. Clusterheads are responsible for coordination among sensor nodes within their clusters and aggregation of their data, and communication with each other with external observers on behalf of their clusters. A good clustering scheme should preserve its structure of cluster as much as possible [7].

In this paper, we propose a clustering method to evenly distribute clusterheads over the network field to reduce the energy consumption and the computational overhead. To distribute the clusterheads almost evenly, we propose two different methods for dividing the network field. One method uses an arc-shaped zones and the other uses a square-shaped zones. The key idea proposed here is that the network field is divided into several zones and the number of clusterheads to be included in each zone is determined in proportion to its area.

This paper is organized as follows. We discuss some related works in section 2 and present an overview and discussion of our method in section 3. In section 4, we compare our method with the existing protocols and show the results. Finally, we conclude the paper in section 5.

2 Related Works

Low-Energy Adaptive Clustering Hierarchy (LEACH) is an efficient routing of data in wireless sensor networks. In LEACH, sensor nodes elect themselves as clusterheads with some probability and broadcast their decisions. Each sensor node determines to which cluster it wants to belong by choosing the clusterhead that requires the minimum communication energy. The algorithm is run periodically, and the probability of becoming a clusterhead for each period is chosen to ensure that every sensor node becomes a clusterhead at least once within 1/p rounds, where p is 5 percent of the number of all nodes [8]. The positive aspect of LEACH is that sensor nodes will randomly consume their power supply, and they should randomly die throughout the network. On the other hand, the randomized clusterheads will make it very difficult to achieve optimal results. Moreover, direct transmission from clusterheads to the sink leads to deplete a plenty of transmission energy.

Power-Efficient GAthering in Sensor Information Systems (PEGASIS) [9] is near optimal for this data gathering application. The key idea in PEGASIS is to form a chain among sensor nodes so that each node will receive from and transmit to a close neighbor node. Gathered data moves from a sensor node to another node, get fused, and eventually a designated node transmits to the sink. Nodes take turns transmitting to the sink so that the average energy spent by each node per round is reduced. The PEGASIS shows improvement compared to the LEACH protocol.

HEED [10] uses an iterative cluster formation algorithm, where sensor nodes assign themselves a "clusterhead probability" that is a function of their residual energy and a "communication cost" that is a function of neighbor nearness. The advantages of HEED are that sensor nodes only require neighborhood information to form the clusters, the algorithm terminates in $O(1)$ iterations, the algorithm guarantees that every sensor nodes is part of just one cluster, and the clusterheads are distributed evenly.

3 Our Proposed Method

We propose a clustering method for wireless sensor networks to balance energy consumptions over all sensor nodes in the network field. We call it "zone-based clustering method" in which a network field is divided into several zones with two types of shape: arc and square. The network field is first divided into several zones, and the number of clusterheads is determined in proportion to its area. These lead to distributing clusterheads evenly over the network field. Each sensor node transmits data to the nearest neighbor node or clusterhead in its zone. Each clusterhead aggregates data and sends it to the nearest clusterhead in the previous zone towards the sink as illustrated in Figure 1. The sink accumulates all packets from clusterheads. The sink is an essential component with complex computational abilities. On the other hand, sensor nodes are considered to perform very simple and cost effective functions.

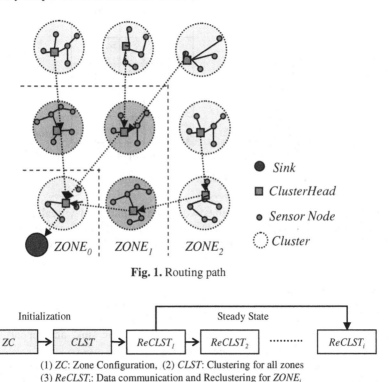

Fig. 1. Routing path

(1) *ZC*: Zone Configuration, (2) *CLST*: Clustering for all zones
(3) *ReCLST$_i$*: Data communication and Reclustering for *ZONE$_i$*

Fig. 2. Three stages for the proposed method

We make some assumptions for our method. First, all sensor nodes in the network are uniformly distributed, static, homogeneous and energy constrained. Second, all sensor nodes know their location. Third, the sink node is immobile and controls to select clusterheads. Fourth, all data sent by the previous node are aggregated with a constant bit size. Our schemes are implemented by the following three stages. (a) Zone configuration stage, (b) Clustering stage, (c) Data communication and reclustering stage as shown in Figure 2.

3.1 Zone Configuration Stage

The essential task in this stage is to divide the network field into several zones. It is divided into several zones based on the zone range(r) which is determined by considering the network size, transmission range, and distribution density of the nodes. In the arc-shaped zone scheme, the first zone is created starting from the sink. The first zone includes sensor nodes which are located within zone range(r) from the sink. The next zone, $ZONE_1$, contains sensor nodes whose distances to the sink are greater than r but less than $2r$. The i-th zone, $ZONE_i$, includes sensor nodes whose distances to the sink are greater than $i \times r$ but less than $(i+1) \times r$. Finally, the last zone covers all remaining sensor nodes which have not yet been included in the previously established zones (e.g. $ZONE_5$ in Figure 3a).

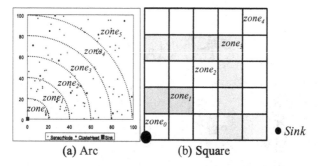

(a) Arc   (b) Square

Fig. 3. Zone configuration

In the square-shaped zone scheme, we create a number of virtual grids by simply making squares with widths (or depths) of zone range(r) as illustrated in Figure 3b. After zone configuration, the sink broadcasts the zone information to allow each node to know which zone it will be assigned to.

3.2 Clustering Stage

This stage consists of the clusterhead selection, the cluster setup, and the formation of routing paths. The clusterhead selection is done by the sink, and repeats until the desired number of clusterheads in each zone is attained.

In arc-shaped zone scheme, selection of clusterheads is based on two parameters: the density of neighbor nodes within a transmission range, and the distance from previously selected clusterhead in a specific zone. The former implies that a highly dense node which has many neighbor nodes around it would be selected as a clusterhead. There may be two or more clusterheads in a zone depending on the zone size. The number of clusterheads selected in a zone is proportional to its size relative to size of the zone to which the sink belongs. By not only selecting clusterheads in this manner, but by also considering the distance from previously selected clusterhead in a zone, the allocation of clusterheads is almost evenly distributed.

In the square-shaped zone scheme, however, selection of clusterheads in a square is done in a round robin fashion. By selecting clusterheads in this way, clusterheads can be almost evenly distributed over the entire network field.

The number of clusterheads in an arc-shaped zone is obtained by

$$N\_CH_i = \frac{Area(ZONE_i)}{Area(ZONE_1)}, \qquad i \geq 1 \tag{1}$$

where N_CH_i is the number of clusterheads selected in $ZONE_i$. In the square-shaped zone scheme, the number of clusterheads in each zone is obtained by

$$N\_CH_i = \frac{Area(ZONE_i)}{Area(ZONE_0)}, \qquad i \geq 0 \tag{2}$$

Cluster setup operation in this stage means that each sensor node has to make a logical connection to the clusterhead in the same square. Once the clusters and the clusterheads have been identified, the sink determines the routing path for any two adjacent clusterheads. Transmission can be done via a multihop path between cluster-heads in adjacent zones. It leads to diminish communication energy as well as setting optimal routing path from sensor nodes to the sink.

3.3 Data Communication and Reclustering Stage

The main action in this stage is data communication whose task includes data gather-ing, data fusion and data forwarding, and reclustering for a single zone. Each sensor node transmits the sensed information to its clusterhead on a multiple-hop path. Once a clusterhead receives data from any node, it performs data fusion on the collected data to reduce the amount of raw data that needs to be sent to the sink.

A sensor node transmits its data to the nearest neighbor node within the cluster that it belongs to. The neighbor node aggregates the data with its own data, and transmits it to the next node until reaching to clusterhead. Similarly, the clusterhead sends its aggregated data to the nearest clusterhead in the next zone until arriving at the sink. There is an exception for $ZONE_0$ where data is transmitted to the sink directly. We assume that the sink request or query is generated every round.

At the reclustering part of this stage, the clusterheads are reselected in a single zone for every round. After this, each node joins the closest clusterhead in the same zone. In other words, reclustering is performed for only a single zone but not in all zones every round. This contributes to reducing the computation overheads. Once the clus-ters and the clusterheads have been reclustered, the sink determines the routing path for any two nearby clusterheads. After as many rounds as the number of zones, all clusterheads will be replaced once for all zones.

4   Simulation Results

Throughout the simulation, we consider a 100 x 100 network configuration with 200 nodes and 300 nodes. Some important simulation parameters are listed in Table 1. In

Table 1. Simulation Parameters

Parameter	Value	Parameter	Value
Network size	100 x 100	Transmission energy	50 *nJ/bit*
Number of nodes	200 / 300	Data Aggregation energy	5 *nJ/bit/message*
Packet size	2000 *bits*	Transmit amplifier energy	100 *pJ/bit/m²*
Initial energy	1 *J*	Zone range(*r*)	20(arc),14(square)

our simulations, all sensor nodes are assumed to carry out sensing operation at a fixed rate and always have data to send. It is also assumed that all data sent by the previous nodes are aggregated into a data segment with a constant size of 2000 bits.

Figure 4 shows the number of rounds when a sensor node becomes dead for the first time and all sensor nodes are dead. Our two schemes outperform LEACH and PEGASIS in terms of the system lifetime.

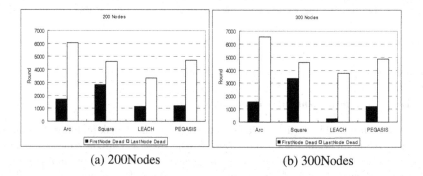

(a) 200Nodes   (b) 300Nodes

Fig. 4. The number of rounds until the first or the last sensor node dies

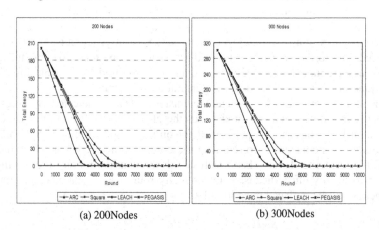

(a) 200Nodes   (b) 300Nodes

Fig. 5. The amount of residual energy as time goes on

Figure 5 shows the amount of used energy at all sensor nodes during network lifetime. This plot shows that our two schemes have more desirable energy expenditure than those of LEACH and PEGASIS since a multi-hop routing path is allowed for data transmission, and thus the distance required for data transmission is closer than that of LEACH. Consequently, the short transmission distance in our two schemes allows less energy consumption at all sensor nodes.

Figure 6 shows the connection failure ratio. The x-axis represents the time steps in rounds, and the y-axis represents the connection failure ratio which is obtained by the number of failure counts divided by the number of live nodes. Our proposed method shows a low connection failure ratio comparable with PEGASIS until 30% nodes are alive. As the number of dead nodes is increased, the PEGASIS protocol shows a lower connection failure ratio as compared with the zone-based method because all nodes can be connected in a chain with a minimum distance from neighbor nodes in PEGASIS. Although PEGASIS appears to perform better when more than 70% of nodes are dead, our proposed methods still have a longer lifetime.

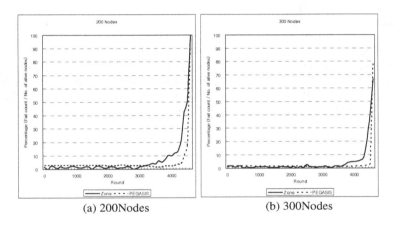

(a) 200Nodes   (b) 300Nodes

Fig. 6. The connection failure ratio

5 Conclusions

In this paper, we propose a clustering method based on zone with a shape of arc or square for wireless sensor networks aiming at balancing energy consumption over all sensor nodes in the network field. To estimate the performance of proposed methods, we compared their performance with other cluster-based protocols, LEACH and PEGASIS. Simulation results show that our arc-based zone method and square-based zone method outperform LEACH and PEGASIS in terms of the system lifetime, the energy expenditure but compete with PEGASIS in terms of the connectivity.

References

1. Ibriq, J., Mahgoub, I.: Cluster-Based Routing in Wireless Sensor Networks: Issues and Challenges, Proceedings of the 2004 Symposium on Performance Evaluation of Computer Telecommunication Systems, March (2004) 759-766.
2. 3GPP Web Site. http://www.gpp.org, (2000).

3. Agarwal, P., Procopiuc, C.: Exact and Approximation Algorithms for Clustering, Proceedings of the Ninth Annual ACM_SIAM Symposium on Discrete Algorithms, January (1999) 658-667.
4. Perillo, M., Heinzelman, W.: Wireless Sensor Network Protocols, Fundamental Algorithms and Protocols for Wireless and Mobile Networks, CRC Hall, (2005).
5. Beyens, P., Nowe, A., Steenhaut, K.: High-Density Wireless Sensor Network: a New Clustering Approach for Prediction-based Monitoring, Proceedings of the Second European Workshop on Wireless Sensor Network, February (2005) 188-196.
6. Muruganathan, S. D., Daniel, C.F., Ma, N. R. I., Fapojuwo, A. O.: A Centralized Energy-Efficient Routing Protocol for Wireless Sensor Networks, IEEE Communications Magazine, 43 March (2005) 8-13.
7. Chatterjee, M., Sajal, K. D., Turgut, D.: WCA: A Weighted Clustering Algorithm for Mobile Ad Hoc Networks, Journal of Cluster Computing (Special Issue on Mobile Ad hoc Networks), 5(2) April (2002) 193-204.
8. Heinzelman W.: Application-Specific Protocol Architectures for Wireless Networks, Ph.D. thesis, Massachusetts Institute of Technology, (2000).
9. Lindsey, S., Cauligi, R. S.: PEGASIS: Power-Efficient Gathering in Sensor Information Systems, Proceedings of the IEEE Aerospace Conference, 3 March (2002) 1125-1130.
10. Younis, O., Fahmy, S.: Distributed clustering in ad-hoc sensor networks: A hybrid, energy efficient approach, Proceedings of the Twenty-Third Annual Joint Conference of the IEEE Computer and Communications Societies (INFOCOM), 3(5) December (2004) 366-379.

Set Approximations in Multi-level Conceptual Data

Ming-Wen Shao[1,2], Hong-Zhi Yang[3], and Shi-Qing Fan[3]

[1] School of Information Technology, Jiangxi University of Finance & Economics,
Nanchang, Jiangxi 330013, P. R. China
shaomingwen1837@163.com
[2] Department of Automation, Tsinghua University, Beijing, 100084, P. R. China
[3] Faculty of Science, Xi'an Jiaotong University, Xi'an, Shaan'xi 710049, P. R. China
yzxz@163.com, fsqsd147@yahoo.com.cn

Abstract. In this paper, two kinds of multi-level formal concepts are discussed. Based on the proposed Multi-level formal concepts, we present two pairs of rough set approximations within fuzzy formal contexts. By the proposed rough set approximations, we not only approximate a crisp set, but also approximate a fuzzy set with the multi-level concepts. We discuss the properties of the proposed two pairs of approximation operators in details.

Keywords: Rough sets, formal concept analysis, concept lattice, lower approximation, upper approximation.

1 Introduction

The theory of formal concept analysis (FCA) proposed by Wille [1,2] has been studied intensively, and obtained results have played an important role in conceptual data analysis and knowledge processing. Wille's definition of a concept be a (objects, attributes) pair, the set of objects is referred to as the extension and the set of attributes as the intension of formal concept. They uniquely determine each other [1,2].

FCA is analyzed based on a formal context, which is a binary relation between a set of objects and a set of attributes with the value 0 and 1. However, in many practical applications, the binary relation is a fuzzy set represented by a membership degrees, instead of a single value in $\{0,1\}$. For this fuzzy binary relation, several generalizations to formal concept can be found in the existent literatures [3,4,5,6,7,8,9,10]. In [5], Elloumi defined a Lukasiewicz based fuzzy Galois connection. Belohlavek [6,7,8] proposed fuzzy conceps in fuzzy formal context based on residuated lattice. Moreover, Georgescu and Popescu [9,10] discussed a general approach to fuzzy FCA.

The theory of Rough sets, proposed by Pawlak [11], as a method of set approximation, it has continued to flourish as a tool for data mining and data analysis. The basic operators in rough set theory are approximations. Using the concepts of lower and upper approximations, knowledge hidden in information

J.T. Yao et al. (Eds.): RSKT 2007, LNAI 4481, pp. 355–363, 2007.
© Springer-Verlag Berlin Heidelberg 2007

tables may be unraveled and expressed in the form of decision rules. Many authors have generalized the rough set model to rough fuzzy sets and fuzzy rough sets models (see [12,13,14]).

FCA and Rough set are both analyzed based on binary information tables. In recent years, many efforts have been made to compare and combine the two theories [15,16,17,18,19]. The combination of formal concept analysis and rough set theory can provide related and complementary approaches for data analysis.

In this paper,based on the proposed Multi-level formal concepts, we present two pairs of rough set approximations within fuzzy formal contexts. By the proposed rough set approximations, we not only approximate a crisp set, but also approximate a fuzzy set with the multi-level concepts. Furthermore, for a fixed set, we have different lower and upper approximation according to different precision level. We discuss the properties of the proposed two pairs of approximation operators in details.

2   Two Kinds of Multi-level Formal Concepts

The notion of residuated lattice provides a very general truth structure for fuzzy logic and fuzzy set theory. In the following, we list its definition and basic properties.

Definition 1. *[7] A residuated lattice is a structure $(L, \vee, \wedge, \otimes, \rightarrow, 0, 1)$ such that*

(1) $(L, \vee, \wedge, 0, 1)$ *is a lattice with the least element 0 and the greatest element 1;*

(2) $(L, \otimes, 1)$ *is a commutative monoid;*

(3) *for all $a, b, c \in L, a \leq b \rightarrow c$ iff $a \otimes b \leq c$.*

Residuated lattice L is called complete if $(L, \vee, \wedge)$ is a complete lattice.

Lemma 1. *[7,9] In any complete residuated lattice $(L, \vee, \wedge, \otimes, \rightarrow, 0, 1)$, $\rightarrow$ is antitone in the first and isotone in the second argument.*

Let L be a residuated lattice. An $L - set$ A on a universe set U is any map A: $U \rightarrow L, A(x)$ being interpreted as the truth degree of the fact " x belongs to A". By L^U denote the set of all $L - set$ in U. For any $X_1, X_2 \in L^U$, $X_1 \subseteq X_2$ if and only if $X_1(x) \leq X_2(x)$ $(\forall x \in U)$. Operations $\vee$ and $\wedge$ on L^X are defined by:

$$(X_1 \vee X_2)(x) = X_1(x) \vee X_2(x), \quad (X_1 \wedge X_2)(x) = X_1(x) \wedge X_2(x), \quad \forall X_1, X_2 \in L^U.$$

Example 1. Let $U = \{x_1, x_2, \ldots, x_n\}$ be a set of n elements, An $L - set$ A on U is denoted by $A = \{x_1/A(x_1), x_2/A(x_2), \ldots, x_n/A(x_n)\}$, where $A(x)$ represents the degree to which an element $x \in U$ is an element of A.

A fuzzy formal context is defined as a triple (U, M, R), where U and M are the object and attribute sets, $R \in L^{U \times M}$ is a fuzzy relation between U and M.

Table 1. A fuzzy formal context (U, M, R)

R	a	b	c	d
x_1	0.4	0.4	0.9	0.6
x_2	0.8	0.2	0.7	0.8
x_3	0.5	0.4	0.7	0.9
x_4	0.8	0.2	0.7	0.7

Example 2. Table 1 represents a fuzzy formal context (U, M, R) with $U = \{x_1, x_2, x_3, x_4\}$ and $M = \{a, b, c, d\}$, the fuzzy relation R defined as in Table 1.

Let (U, M, R) be a fuzzy formal context and we denote the power set of U by $\mathcal{P}(U)$. For any $\delta \in (0, 1]$, the operators: $* : \mathcal{P}(U) \longrightarrow L^M$, $\star : L^M \longrightarrow \mathcal{P}(U)$, defined for $X \in \mathcal{P}(U), B \in L^M$ and $a \in M$ by [20]:

$$X^*(a) = \delta \to \bigwedge_{x \in X} R(x, a),$$
$$B^\star = \{x \in U \mid \bigwedge_{a \in M}(B(a) \to R(x, a)) \geq \delta\}.$$

The following property list the basic properties of the adjoint pair of operators.

Theorem 1. *Let* (U, M, R) *be a fuzzy formal context,* $X, X_1, X_2, X_i \in \mathcal{P}(U), B,$ $B_1, B_2, B_i \in L^M$, *then*

(i) $X_1 \subseteq X_2 \Rightarrow X_2^* \subseteq X_1^*$, $B_1 \subseteq B_2 \Rightarrow B_2^\star \subseteq B_1^\star$;

(ii) $X \subseteq X^{**}$, $B \subseteq B^{\star\star}$;

(iii) $X^* = X^{***}$, $B^\star = B^{\star\star\star}$;

(iv) $(\bigcup_{i \in I} X_i)^* = \bigwedge_{i \in I} X_i^*$, $(\bigvee_{i \in I} B_i)^\star = \bigcap_{i \in I} B_i^\star$.

Proof. It is omitted.

A crisp-fuzzy concept [20] of (U, M, R) is a pair of $(X, B) \in \mathcal{P}(U) \times L^M$, such that $X^* = B$ and $B^\star = X$. For a set of objects $X \subseteq \mathcal{P}(U)$ and a fuzzy set of attributes $B \subseteq L^M$, from Theorem 1 (iii) we have that (X^{**}, X^*) and $(B^\star, B^{\star\star})$ are crisp-fuzzy concepts. We have different level crisp-fuzzy concepts with different precision level δ. For two crisp-fuzzy concepts (X_1, B_1) and (X_2, B_2), $(X_1, B_1) \leq (X_2, B_2)$, if and only if $X_1 \subseteq X_2$ (or equivalently, $B_2 \subseteq B_1$). All the crisp-fuzzy concepts of (U, M, R) forms a complete lattice in which infimum and supremum are defined by:

$$(X_1, B_1) \vee (X_2, B_2) = ((X_1 \cup X_2)^{**}, B_1 \cap B_2)$$
$$= ((B_1 \cap B_2)^\star, B_1 \cap B_2);$$
$$(X_1, B_1) \wedge (X_2, B_2) = ((X_1 \cap X_2), (B_1 \cup B_2)^{\star\star})$$
$$= ((X_1 \cap X_2), (X_1 \cap X_2)^*).$$

Example 3. In *Example 2*, let $\to$ be the Lukasiewicz implication, ie., for $x, y \in [0, 1]$,

$$x \to y = \begin{cases} 1, & x \leq y, \\ 1 - x + y & x > y; \end{cases}$$
$$x \otimes y = (x + y - 1) \vee 0.$$

Table 2. Crisp-fuzzy concepts for $\delta = 1$

Label	Objects × properties
FC_0	$\emptyset \times \{a/1.0, b/1.0, c/1.0, d/1.0\}$
FC_1	$\{x_1\} \times \{a/0.4, b/0.4, c/0.9, d/0.6\}$
FC_2	$\{x_2\} \times \{a/0.8, b/0.2, c/0.7, d/0.8\}$
FC_3	$\{x_3\} \times \{a/0.5, b/0.4, c/0.7, d/0.9\}$
FC_4	$\{x_1, x_3\} \times \{a/0.4, b/0.4, c/0.7, d/0.6\}$
FC_5	$\{x_2, x_3\} \times \{a/0.5, b/0.2, c/0.7, d/0.8\}$
FC_6	$\{x_2, x_4\} \times \{a/0.8, b/0.2, c/0.7, d/0.7\}$
FC_7	$\{x_2, x_3, x_4\} \times \{a/0.5, b/0.2, c/0.7, d/0.7\}$
FC_8	$\{x_1, x_2, x_3, x_4\} \times \{a/0.4, b/0.2, c/0.7, d/0.6\}$

When $\delta = 1$, by computation we obtain the crisp-fuzzy concepts presented in Table 2.

Let (U, M, R) be a fuzzy formal context, for any $\delta \in (0, 1]$, the operators: $^{\#}$: $L^U \longrightarrow \mathcal{P}(M)$, $^{\triangle} : \mathcal{P}(M) \longrightarrow L^U$, defined for $X \in L^U$, $B \in \mathcal{P}(M)$ and $x \in U$ by [20]:

$$X^{\#} = \{a \in M | \bigwedge_{x \in U}(X(x) \to R(a, x)) \geq \delta\},$$
$$B^{\triangle}(x) = \delta \to \bigwedge_{a \in B} R(x, a).$$

Theorem 2. *Let (U, M, R) be a fuzzy formal context, $X, X_1, X_2, X_i \in L^U, B, B_1, B_2, B_i \in \mathcal{P}(M)$, then*

    (i) $X_1 \subseteq X_2 \Rightarrow X_2^{\#} \subseteq X_1^{\#}$, $B_1 \subseteq B_2 \Rightarrow B_2^{\triangle} \subseteq B_1^{\triangle}$;

    (ii) $X \subseteq X^{\#\triangle}$, $B \subseteq B^{\triangle\#}$;

    (iii) $X^{\#} = X^{\#\triangle\#}$, $B^{\triangle} = B^{\triangle\#\triangle}$;

    (iv) $(\bigvee_{i \in I} X_i)^{\#} = \bigcap_{i \in I} X_i^{\#}$, $(\bigcup_{i \in I} B_i)^{\triangle} = \bigwedge_{i \in I} B_i^{\triangle}$.

Proof. It is similar to the proof of Theorem 1.

A fuzzy-crisp concept [20] of (U, M, R) is a pair of $(X, B) \in L^U \times \mathcal{P}(M)$, such that $X^{\#} = B$ and $B^{\triangle} = X$. For a fuzzy set of objects $X \subseteq L^U$ and a set of attributes $B \subseteq \mathcal{P}(M)$, from Theorem 2 (iii) we have that $(X^{\#\triangle}, X^{\#})$ and $(B^{\triangle}, B^{\triangle\#})$ are fuzzy-crisp concepts. We have different level fuzzy-crisp concepts with different precision level δ. For two fuzzy-crisp concepts (X_1, B_1) and (X_2, B_2), $(X_1, B_1) \leq (X_2, B_2)$, if and only if $X_1 \subseteq X_2$ (or equivalently, $B_2 \subseteq B_1$). All the fuzzy-crisp concepts of (U, M, R) forms a complete lattice in which infimum and supremum are defined by:

$$(X_1, B_1) \vee (X_2, B_2) = ((X_1 \cup X_2)^{\#\triangle}, B_1 \cap B_2)$$
$$= ((B_1 \cap B_2)^{\triangle}, B_1 \cap B_2);$$
$$(X_1, B_1) \wedge (X_2, B_2) = ((X_1 \cap X_2), (B_1 \cup B_2)^{\triangle\#})$$
$$= ((X_1 \cap X_2), (X_1 \cap X_2)^{\#}).$$

Example 4. Continuing from *Example 3*, when $\delta = 1$, by calculation we obtain the fuzzy-crisp concepts presented in Table 3.

Table 3. Fuzzy-crisp concepts for $\delta = 1$

Label	Objects × properties
FC_0	$\{x_1/1.0, x_2/1.0, x_3/1.0, x_4/1.0\} \times \emptyset$
FC_1	$\{x_1/0.4, x_2/0.8, x_3/0.5, x_4/0.8\} \times \{a\}$
FC_2	$\{x_1/0.9, x_2/0.7, x_3/0.7, x_4/0.7\} \times \{c\}$
FC_3	$\{x_1/0.6, x_2/0.8, x_3/0.9, x_4/0.7\} \times \{d\}$
FC_4	$\{x_1/0.4, x_2/0.8, x_3/0.5, x_4/0.7\} \times \{a, d\}$
FC_5	$\{x_1/0.6, x_2/0.7, x_3/0.7, x_4/0.7\} \times \{c, d\}$
FC_6	$\{x_1/0.4, x_2/0.7, x_3/0.5, x_4/0.7\} \times \{a, c, d\}$
FC_7	$\{x_1/0.4, x_2/0.2, x_3/0.4, x_4/0.2\} \times \{a, b, c, d\}$

3   The Approximation Operators Based on Multi-level Formal Concepts

In this section, based on above discussed two pair of adjoint operators we introduced two pair of lower and upper approximation operators in fuzzy formal contexts.

Definition 2. *Let (U, M, R) be a fuzzy formal context. For any set $X \subseteq U$, a pair of upper and lower approximations, $\overline{apr}(X)$, $\underline{apr}(X)$, is defined by*

$$\overline{apr}(X) = X^{**}, \quad \underline{apr}(X) = \sim \overline{apr}(\sim X).$$

Operators, $\underline{apr}, \overline{apr} \colon \mathcal{P}(U) \longrightarrow \mathcal{P}(U)$, are referred to as lower and upper approximation operators for object sets, and the pair $(\underline{apr}(X), \overline{apr}(X))$ is referred to as a generalized rough object set.

Where $\sim$ denotes the complement of a set.

Theorem 3. *Let (U, M, R) be a fuzzy formal context. The generalized lower and upper approximation satisfy the following properties: for any $X, Y \subseteq U$,*

$$
\begin{aligned}
(L_1) \quad & \underline{apr}(X) = \sim (\overline{apr}(\sim X)), \\
(U_1) \quad & \overline{apr}(X) = \sim (\underline{apr}(\sim X)); \\
(L_2) \quad & \underline{apr}(\emptyset) = \overline{apr}(\emptyset) = \emptyset, \\
(U_2) \quad & \overline{apr}(U) = \underline{apr}(U) = U; \\
(L_3) \quad & \underline{apr}(X \cap Y) \subseteq \underline{apr}(X) \cap \underline{apr}(Y), \\
(U_3) \quad & \overline{apr}(X \cup Y) \supseteq \overline{apr}(X) \cup \overline{apr}(Y); \\
(L_4) \quad & X \subseteq Y \Longrightarrow \underline{apr}(X) \subseteq \underline{apr}(Y), \\
(U_4) \quad & X \subseteq Y \Longrightarrow \overline{apr}(X) \subseteq \overline{apr}(Y); \\
(L_5) \quad & \underline{apr}(X \cup Y) \supseteq \underline{apr}(X) \cup \underline{apr}(Y), \\
(U_5) \quad & \overline{apr}(X \cap Y) \subseteq \overline{apr}(X) \cap \overline{apr}(Y); \\
(L_6) \quad & \underline{apr}(X) \subseteq X, \\
(U_6) \quad & X \subseteq \overline{apr}(X); \\
(L_7) \quad & \underline{apr}(\underline{apr}(X)) = \underline{apr}(X), \\
(U_7) \quad & \overline{apr}(\overline{apr}(X)) = \overline{apr}(X).
\end{aligned}
$$

Proof. Properties (L_1) and (U_1) show that approximation operators $\underline{apr}$ and $\overline{apr}$ are dual to each other. Properties with the same number may be regarded as dual properties. Thus, we only need to prove one of them.

(L_1) and (L_2) follows immediately from the definition of lower approximation. From the definition of upper approximation, we have

$$\overline{apr}(X \cup Y) = (X \cup Y)^{**} = \{x \in U | \forall\, a \in M, \delta \to R(x,a) \geq \delta \to \bigwedge_{y \in X \cup Y} R(y,a)\}.$$

Since

$$\bigwedge_{y \in X} R(y,a) \geq \bigwedge_{y \in X \cup Y} R(y,a), \quad \bigwedge_{y \in Y} R(y,a) \geq \bigwedge_{y \in X \cup Y} R(y,a)$$

from Lemma 1, we have

$$\delta \to \bigwedge_{y \in X} R(y,a) \geq \delta \to \bigwedge_{y \in X \cup Y} R(y,a), \quad \delta \to \bigwedge_{y \in Y} R(y,a) \geq \delta \to \bigwedge_{y \in X \cup Y} R(y,a)$$

which implies $(X \cup Y)^{**} \supseteq X^{**}$, $(X \cup Y)^{**} \supseteq Y^{**}$. Then, $(X \cup Y)^{**} \supseteq X^{**} \cup Y^{**}$. Thus, (U_3) holds.

From Theorem 1 (i), we have

$$X \subseteq Y \Rightarrow Y^* \subseteq X^* \Rightarrow X^{**} \subseteq Y^{**}.$$

Property (U_4) holds.

Similar to the proof of Property (U_3), we have

$$\overline{apr}(X \cap Y) = (X \cap Y)^{**} = \{x \in U | \forall\, a \in M, \delta \to R(x,a) \geq \delta \to \bigwedge_{y \in X \cap Y} R(y,a)\}.$$

Since

$$\bigwedge_{y \in X \cap Y} R(y,a) \geq \bigwedge_{y \in X} R(y,a), \quad \bigwedge_{y \in X \cap Y} R(y,a) \geq \bigwedge_{y \in Y} R(y,a),$$

then from Lemma 1 we have

$$\delta \to \bigwedge_{y \in X \cap Y} R(y,a) \geq \delta \to \bigwedge_{y \in X} R(y,a), \quad \delta \to \bigwedge_{y \in X \cap Y} R(y,a) \geq \delta \to \bigwedge_{y \in Y} R(y,a),$$

which implies $(X \cap Y)^{**} \subseteq X^{**}$, $(X \cap Y)^{**} \subseteq Y^{**}$. Then, $(X \cap Y)^{**} \supseteq X^{**} \cap Y^{**}$. Thus, (U_5) holds.

Property (U_6) follows directly from Theorem 1 (ii).

Since $\overline{apr}(\overline{apr}(X)) = (X^{**})^{**} = X^{****}$, from Theorem 1 (iii) we conclude that (U_7) holds.

Definition 3. *Let (U, M, R) be a fuzzy formal context. For any set $X \subseteq L^U$, another pair of upper and lower approximations, $\overline{Apr}(X)$, $\underline{Apr}(X)$, is defined by*

$$\overline{Apr}(X) = X^{\#\triangle}, \quad \underline{Apr}(X) = \sim \overline{Apr}(\sim X).$$

Operators, $\underline{Apr}$, $\overline{Apr} \colon L^U \longrightarrow L^U$, are referred to as lower and upper approximation operators for fuzzy object sets, and the pair $(\underline{Apr}(X), \overline{Apr}(X))$ is referred to as a generalized rough fuzzy object set.

Where $\sim$ denotes the complement of a fuzzy set. For example, in *Example 1*, $\sim A = \{x_1/1 - A(x_1), x_2/1 - A(x_2), \ldots, x_n/1 - A(x_n)\}$.

Theorem 4. *Let (U, M, R) be a fuzzy formal context. The generalized lower and upper approximation of fuzzy object sets satisfy the following properties: for any $X, Y \subseteq L^U$,*

$$
\begin{array}{ll}
(FL_1) & \underline{Apr}(X) = \sim (\overline{Apr}(\sim X)), \\
(FU_1) & \overline{Apr}(X) = \sim (\underline{Apr}(\sim X)); \\
(FL_2) & \underline{Apr}(\emptyset) = \overline{Apr}(\emptyset) = \emptyset, \\
(FU_2) & \overline{Apr}(U) = \underline{Apr}(U) = U; \\
(FL_3) & \underline{Apr}(X \cap Y) \subseteq \underline{Apr}(X) \cap \underline{Apr}(Y), \\
(FU_3) & \overline{Apr}(X \cup Y) \supseteq \overline{Apr}(X) \cup \overline{Apr}(Y); \\
(FL_4) & X \subseteq Y \Longrightarrow \underline{Apr}(X) \subseteq \underline{Apr}(Y), \\
(FU_4) & X \subseteq Y \Longrightarrow \overline{Apr}(X) \subseteq \overline{Apr}(Y); \\
(FL_5) & \underline{Apr}(X \cup Y) \supseteq \underline{Apr}(X) \cup \underline{Apr}(Y), \\
(FU_5) & \overline{Apr}(X \cap Y) \subseteq \overline{Apr}(X) \cap \overline{Apr}(Y); \\
(FL_6) & \underline{Apr}(X) \subseteq X, \\
(FU_6) & X \subseteq \overline{Apr}(X); \\
(FL_7) & \underline{Apr}(\underline{Apr}(X)) = \underline{Apr}(X), \\
(FU_7) & \overline{Apr}(\overline{Apr}(X)) = \overline{Apr}(X).
\end{array}
$$

Proof. Properties with the same number may be regarded as dual properties, we only need to prove one of them.

Property (FL_1) and Property (FL_2) are evident by the definition of lower approximation.

For any $x \in U$, we have

$$
\begin{aligned}
\overline{Apr}(X \cup Y)(x) &= (X \cup Y)^{\#\triangle}(x) \\
&= \delta \rightarrow \bigwedge_{a \in (X \cup Y)^{\#}} R(x, a) \\
&= \bigwedge_{a \in X^{\#} \cap Y^{\#}} (\delta \rightarrow R(x, a)).
\end{aligned}
$$

Since $X^{\#} \cap Y^{\#} \leq X^{\#}$, $X^{\#} \cap Y^{\#} \leq X^{\#}$, from Lemma 1 we have

$$
\begin{aligned}
\bigwedge_{a \in X^{\#} \cap Y^{\#}} (\delta \rightarrow R(x, a)) &\geq \bigwedge_{a \in X^{\#}} (\delta \rightarrow R(x, a)) = X^{\#\triangle}(x), \\
\bigwedge_{a \in X^{\#} \cap Y^{\#}} (\delta \rightarrow R(x, a)) &\geq \bigwedge_{a \in Y^{\#}} (\delta \rightarrow R(x, a)) = Y^{\#\triangle}(x),
\end{aligned}
$$

then $\overline{Apr}(X \cup Y) \supseteq \overline{Apr}(X) \cup \overline{Apr}(Y)$. Thus, (FU_3) holds.

Properties (FU_4) follows directly from Theorem 2 (i).

For any $x \in U$, we have

$$
\begin{aligned}
\overline{Apr}(X \cap Y)(x) &= (X \cap Y)^{\#\triangle}(x) \\
&= \delta \rightarrow \bigwedge_{a \in (X \cap Y)^{\#}} R(x, a) \\
&= \bigwedge_{a \in (X \cap Y)^{\#}} (\delta \rightarrow R(x, a)).
\end{aligned}
$$

From Theorem 2 (i) we have $X^{\#} \leq (X \cap Y)^{\#}$, $Y^{\#} \leq (X \cap Y)^{\#}$, which implies

$$
\begin{aligned}
\bigwedge_{a \in (X \cap Y)^{\#}} (\delta \rightarrow R(x, a)) &\leq \bigwedge_{a \in X^{\#}} (\delta \rightarrow R(x, a)) = X^{\#\triangle}(x), \\
\bigwedge_{a \in (X \cap Y)^{\#}} (\delta \rightarrow R(x, a)) &\leq \bigwedge_{a \in Y^{\#}} (\delta \rightarrow R(x, a)) = Y^{\#\triangle}(x),
\end{aligned}
$$

then $\overline{Apr}(X \cap Y) \subseteq \overline{Apr}(X) \cap \overline{Apr}(Y)$, and (FU_5) holds.

Property (FU_6) follows directly from Theorem 2 (ii).

Since $\overline{Apr}(\overline{Apr}(X)) = (X^{\#\triangle})^{\#\triangle} = X^{\#\triangle\#\triangle}$, from Theorem 2 (iii) we conclude that (U_7) holds.

By the definition of $\overline{apr}(X)$ and $\overline{Apr}(X)$, $\overline{apr}(X)$ is the extent of the crisp-fuzzy concept derived from X, and $\overline{Apr}(X)$ is the extent of the fuzzy-crisp concept derived from X. For any crisp subset of the universe U we can approximate it by the extent of the crisp-fuzzy concept, and for any fuzzy subset of the universe U we can approximate it by the extent of the fuzzy-crisp concept. Furthermore, for a fixed object set, we have different lower and upper approximation according to different precision level δ.

Theorem 5. *Let* (U, M, R) *be a fuzzy formal context,* $X \in \mathcal{P}(U)$, $Y \in L^U$, *then*

(1) $\overline{apr}(X) = X$ *iff* X *is the extent of a crisp-fuzzy concept;*

(2) $\overline{Apr}(Y) = Y$ *iff* Y *is the extent of a fuzzy-crisp concept.*

Proof. Straightforward.

Example 5. In *Example 3*, let $X = \{x_1, x_2\}$ and $Y = \{x_1/0.5, x_2/0.6, x_3/0.8, x_4/0.4\}$. When $\delta = 1$, by calculation we obtain

$$\overline{apr}(X) = X^{**} = \{x_1, x_2, x_3, x_4\}, \quad apr(X) = \sim \overline{apr}(\sim X) = \{x_1\},$$
$$\overline{Apr}(Y) = Y^{\#\triangle} = \{x_1/0.6, x_2/0.8, x_3/0.9, x_4/0.7\},$$
$$\underline{Apr}(Y) = \sim \overline{Apr}(\sim Y) = \{x_1/0.4, x_2/0.3, x_3/0.3, x_4/0.3\}.$$

4   Conclusions

Multi-level formal concept analysis is an important tool that can be applied to deal with uncertainty contained in conceptual fuzzy data analysis and knowledge processing. In the paper, based on the discussed multi-level formal concept lattice, we present two pairs of rough fuzzy set approximations within fuzzy formal contexts. By the proposed rough set approximations, we not only approximate a crisp set, but also approximate a fuzzy set with the multi-level concepts. Furthermore, for a fixed set, we have different lower and upper approximation according to different precision level. The proposed rough set approximations are different from former set approximations which are based on a equivalence relation or a fuzzy similarity relation. The equivalence relation or the fuzzy similarity relation seems too strict so that it can't be satisfied in most situations and this constrains its applications. The applications of the proposed lower and upper approximation operators are our next research.

Acknowledgments

This paper was supported by the National 973 Program of China (No.2002CB 312200) and Technology Program of Jiangxi Education Office (No. [2006]321).

References

1. Wille, R.: Restructuring lattice theory: an approach based on hierarchies of concepts. In: *Ordered Sets*, Rival, I.(Ed.), Reidel, Dordrecht-Boston (1982) 445–470.
2. Gediga, G., Wille, R.: Formal Concept Analysis. Mathematic Foundations. Springer, Berlin (1999).
3. Burusco, A., Fuentes-Gonzalez, R.: Construction of the L-Fuzzy concept lattice, Fuzzy Sets and systems, 97(1998) 109–114.
4. Burusco, A., Fuentes-Gonzalez, R.: Concept lattices defined from implication operators, Fuzzy Sets and systems, 114(3)(2000) 431–436.
5. Elloumi, S., Jaam, J., Hasnah, A., Jaoua, A., Nafkha, I.: A multi-level conceptual data reduction approach based on the Lukasiewicz implication, Information Sciences, 163 (2004) 253–262.
6. Belohlavek, R.: Fuzzy closure operators, I.J.Math.Anal.Appl 262 (2001) 473–489..
7. Belohlavek, R.: Concept lattice and order in fuzzy logic, Annals of pure and Apll.Logic, 128(1-3) (2004), 277–298.
8. Belohlavek, R.: Logic precision in concept lattices, Journal of Logic and Computatiion 12 (2002) 137–148.
9. Popescu, A.: A general approach to fuzzy concept, Math.Logic Quaterly 50(3)(2001) 1–17.
10. Georgescu, G., Popescu, A.: Non-dual fuzzy connections, Archive for Mathematic Logic, 43(8) (2004) 1009–1039.
11. Pawlak, Z.: Rough sets. International Journal of Computer and Information Science 11 (1982) 341–356.
12. Radzikowska, A. M., Kerre, E. E.: A comparative study of fuzzy rough sets, Fuzzy sets and Systems 126(2002) 137–155.
13. Wu, W. Z., Mi, J. S., Zhang, W. X., Generalized fuzzy rough sets, Information Sciences 151(2003) 263–282.
14. Yao, Y.Y.: A comparative study of fuzzy sets and rough sets, Information Sciences 109(1998) 227–242.
15. Hu, K., Sui, Y., Lu, Y., Wang, J., Shi, C.: Concept approximation in concept lattice. Knowledge Discovery and Data Mining, Proceedings of the 5th Pacific-Asia Conference, PAKDD 2001. Lecture Notes in Computer Science 2035 (2001) 167–173.
16. Saquer, J., Deogun, J.S.: Formal rough concept analysis. New directions in Rough Sets, Data Mining, and Granular-Soft Computer Science 1711. Springer, Berlin (1999) 91–99.
17. Yao, Y.Y.: Concept lattices in rough set theory. In: Proceedings of 2004 Annual Meeting of the North American Fuzzy Information Processing Society, (2004) 796–801.
18. Wolff, K.E.: A conceptual view of knowledge bases in rough set theory. Rough Sets and Current Trends in Computing, Second International Conference, RSCTC 2000. Lecture Notes in Computer Science 2005. Springer, Berlin (2001) 220–228.
19. Yao, Y.Y. and Chen, Y.: Rough set approximations in formal concept analysis. In: Proceedings of 2004 Annual Meeting of the North American Fuzzy Information Processing Society, (2004) 73–78.
20. Fan, S.Q.: Fuzzy Concept Lattice and Variable Precision Concept Lattice. Ph. M. Thesis, Faculty of Science, Xi'an Jiaotong University (2006).

Knowledge Reduction in Generalized Consistent Decision Formal Contexts

Hong Wang[1] and Wei-Zhi Wu[2]

[1] College of Mathematics and Computer Science,
Shanxi Normal University, Linfen, Shanxi 041004, P. R. China
wangh@sxnu.edu.cn
[2] School of Mathematics, Physics and Information Science,
Zhejiang Ocean University, Zhoushan, Zhenjiang 316004, P. R. China
wuwz@zjou.edu.cn

Abstract. This paper deals with knowledge reduction in generalized consistent decision formal contexts. The concept of a generalized consistent decision formal context is first introduced. Its equivalent definitions are also examined. The judgement theorem and discernibility matrix which are helpful for computing reducts are then established, from which we can obtain an approach to knowledge reduction in generalized consistent decision formal contexts.

Keywords: Concept lattices, Formal contexts, Knowledge reduction, Rough sets.

1 Introduction

The theory of formal concept analysis, proposed by Wille [8], has very important meaning in mathematical thinking for conceptual data analysis and knowledge processing. A concept lattice generated from a formal context reflects relationship of generalization and specialization among concepts, it thereby is more intuitional and more effective to research on reducing and discovering knowledge. As a kind of very effective methods for data analysis, formal concept analysis has been widely applied in machine learning, artificial intelligence and knowledge discovery.

In the rough set theory [5], knowledge reduction has been studied extensively in recent years from various point of perspectives and each of them aims at some basic requirement [4,7,9,11].

There are strong connections between formal concept lattice theory and rough set theory. Proposals have been made to combine the two theories in a common framework. The notions of rough set approximations have been introduced into formal concept analysis [1,3,6], and the notions of formal concept and concept lattice have also been introduced into rough set theory by considering different types of formal concepts [10].

Comparing with the studies on knowledge reduction in rough set theory, there is less effort investigated on the issue in concept lattice theory. In [12], concepts of

J.T. Yao et al. (Eds.): RSKT 2007, LNAI 4481, pp. 364–371, 2007.

strong consistent decision formal context and consistent decision formal context were defined, and approaches to knowledge reduction with two requirements in these decision formal contexts were proposed. We try to investigate in this paper knowledge reduction in generalized consistent decision formal contexts.

2   Preliminaries

A formal context is a triplet (U, A, I), where $U = \{x_1, x_2, \ldots, x_n\}$ is a non-empty, finite set of objects called the universe of discourse, $A = \{a_1, a_2, \ldots, a_m\}$ is a non-empty, finite set of attributes, and $I \subseteq U \times A$ is a binary relation between U and A. In a formal context (U, A, I), for each $(x, a) \in U \times A$, if $(x, a) \in I$, we write xIa, we say that x has attribute a, or the attribute a is possessed by object x.

For $X \subseteq U$ and $B \subseteq A$, we denote

$$X^* = \{a \in A : \forall x \in X, (x, a) \in I\},$$
$$B^* = \{x \in U : \forall a \in B, (x, a) \in I\}.$$

X^* is the maximal set of attributes shared by all objects in X. Similarly, B^* is the maximal set of objects that have all attributes in B. For $x \in U$ and $a \in A$, we denote $x^* = \{x\}^*, a^* = \{a\}^*$. Thus x^* is the set of attributes possessed by x, and a^* is the set of objects having attribute a.

Definition 1. ([2]) *A pair $(X, B), X \subseteq U, B \subseteq A$, is called a formal concept of the context (U, A, I) if $X^* = B$ and $B^* = X$. X and B are respectively referred to as the extent and the intent of the concept (X, B).*

Theorem 1. ([2]) *Let (U, A, I) be a formal context, $X, X_1, X_2 \subseteq U, B, B_1, B_2 \subseteq A$. Then*
(1) $X_1 \subseteq X_2 \Longrightarrow X_2^* \subseteq X_1^*$;
(2) $B_1 \subseteq B_2 \Longrightarrow B_2^* \subseteq B_1^*$;
(3) $X \subseteq X^{**}$, $B \subseteq B^{**}$;
(4) $X^* = X^{***}$, $B^* = B^{***}$;
(5) $(X_1 \cup X_2)^* = X_1^* \cap X_2^*$, $(B_1 \cup B_2)^* = B_1^* \cap B_2^*$.

Let $L(U, A, I)$ denote all concepts of formal context (U, A, I), denote

$$(X_1, B_1) \leq (X_2, B_2) \Longleftrightarrow X_1 \subseteq X_2 \Longleftrightarrow B_1 \supseteq B_2,$$

then $\leq$ is an order relation on $L(U, A, I)$.

Theorem 2. ([2]) *Let (U, A, I) be a formal context, $(X_1, B_1), (X_2, B_2) \in L(U, A, I)$, then*

$$(X_1, B_1) \wedge (X_2, B_2) = (X_1 \cap X_2, (B_1 \cup B_2)^{**}),$$

$$(X_1, B_1) \vee (X_2, B_2) = ((X_1 \cup X_2)^{**}, B_1 \cap B_2).$$

are concepts. Thus $L(U, A, I)$ is a complete lattice called the concept lattice of the context (U, A, I).

3   Generalized Consistent Decision Formal Contexts

In this section, we first introduce a new partial ordered relation on the set of all concept lattices with the same universe, and we then propose the notion of a generalized consistent decision formal context.

Theorem 3. *Let U be the universe of discourse and $P(U)$ the power set of U, denote*

$$\mathcal{H} = \{H \subseteq P(U) : \emptyset \in H, X, Y \in H \Longrightarrow X \cap Y \in H\}.$$

For any $H_1, H_2 \in \mathcal{H}$, we define

$$R = \{(X_i, X_j) \in H_1^2 : X_i \subseteq Y \Longleftrightarrow X_j \subseteq Y \ (Y \in H_2)\}.$$

Then R is an equivalence relation on H_1, it partitions H_1 into equivalence classes

$$H_1/R = \{L(Y) \neq \emptyset : Y \in H_2\},$$

where

$$L(Y) = \{X \in H_1 : X \subseteq Y, \forall Y' \in H_2, Y' \subset Y, X \nsubseteq Y'\}.$$

Proof. For any $X \in H_1$, denote

$$L_0(X) = \{Y' \in H_2 : X \subseteq Y'\}, \quad Y = \cap L_0(X).$$

Then, for each $X \in H_1$, there exists $Y \in H_2$ such that $X \in L(Y)$, i.e., $\bigcup_{Y \in H_2} L(Y) = H_1$.

For $Y_1, Y_2 \in H_2$, if $Y_1 \neq Y_2$, we have $L(Y_1) \cap L(Y_2) \neq \emptyset$, then there exists $X \subseteq U$ such that $X \in L(Y_1)$ and $X \in L(Y_2)$, hence $X \subseteq Y_1 \cap Y_2$. Since $Y_1 \neq Y_2$, by the definition of $L(Y)$ we obtain $X \notin L(Y_1)$ or $X \notin L(Y_2)$, This is a contradiction. Hence $\{L(Y) \neq \emptyset : Y \in H_2\}$ forms a partition of H_1. Evidently, $X_1, X_2 \in L(Y)$ if and only if $(X_1, X_2) \in R$.

Definition 2. *Let $H_1, H_2 \in \mathcal{H}$. If for any $Y \in H_2, L(Y) \neq \emptyset$, then H_1 is said to be finer than H_2 and is denoted as $H_1 \leq^* H_2$.*

Let U be a finite nonempty set, then $(\mathcal{H}, \leq^*)$ is an partial ordered set.

Definition 3. *$S = (U, A, I, C, J)$ is said to be a decision formal context, where $A \cap C = \emptyset$, $I \subseteq U \times A$ and $J \subseteq U \times C$, A and C are called the conditional attribute set and decision attribute set respectively.*

Let (U, A, I, C, J) be a decision formal context, for any $D \subseteq A$, denote $I_D = I \cap (U \times D)$, then (U, D, I_D) is a formal context. For $X \subseteq U$, we denote

$$I_A = I, \quad X^{*A} = X^*, \quad X^{*D} = X^{*A} \cap D = X^* \cap D,$$

$$X^{*C} = \{a \in C : \forall x \in X, (x, a) \in J\}.$$

Let $K = (U, A, I)$ be a formal context. Then $L_u(U, A, I) = \{X \subseteq U : X^{**} = X\} \in \mathcal{H}$.

Definition 4. *Let $L(U, A_1, I_1)$ and $L(U, A_2, I_2)$ be two concept lattices, if*

$$L_u(U, A_1, I_1) \leq^* L_u(U, A_2, I_2),$$

then $L(U, A_1, I_1)$ is said to be finer than $L(U, A_2, I_2)$ and is denoted as $L(U, A_1, I_1) \leq^* L(U, A_2, I_2)$.

Definition 5. *Let $S = (U, A, I, C, J)$ be a decision formal context, if*

$$L(U, A, I) \leq^* L(U, C, J),$$

we say that S is a generalized consistent decision formal context. For any $D \subseteq A$, if

$$L(U, D, I_D) \leq^* L(U, C, J),$$

we say that D is a consistent set of S. If D is consistent set and no proper subset of D is a consistent set of S, then D is referred to as a reduct of S.

4   Knowledge Reduction in Generalized Consistent Decision Formal Contexts

This section provides an approach to knowledge reduction in a generalized consistent decision formal context.

Theorem 4. *Let $S = (U, A, I, C, J)$ be a decision formal context and $D \subseteq A$. Denote*

$$H_D = L_u(U, D, I_D), H = L_u(U, C, J),$$
$$R_D = \{(X_i, X_j) \in H_D^2 : X_i \subseteq Y \Longleftrightarrow X_j \subseteq Y (Y \in H)\}.$$

Then:

(1) R_D is an equivalence relation,

(2) R_D partitions H_D into a family of disjoint subsets H_D/R_D called a quotient set of H_D:

$$H_D/R_D = \{L_D(Y) \neq \emptyset : Y \in H\},$$

where $L_D(Y)$ is the equivalence class determined by Y with respect to (w.r.t.) D, i.e.,

$$L_D(Y) = \{X \in H_D : X \subseteq Y, \forall Y' \in H, Y' \subset Y, X \nsubseteq Y'\}.$$

Proof. It follows immediately from Theorem 3.

Theorem 5. *Let $S = (U, A, I, C, J)$ be a formal context, $D \subseteq A$. Then:*

(1) S is a generalized consistent decision formal context iff $L_A(Y) \neq \emptyset$ for all $Y \in H$,

(2) D is a consistent set iff $L_D(Y) \neq \emptyset$ for all $Y \in H$.

Proof. It follows immediately from Theorem 3 and Definition 5.

Definition 6. *Let* $S = (U, A, I, C, J)$ *be a generalized consistent decision formal context, denote*

$$R_A = \{(X_i, X_j) \in H_A^2 : X_i \subseteq Y \Longleftrightarrow X_j \subseteq Y(Y \in H)\},$$

$$H_A/R_A = \{L_A(Y_1), L_A(Y_2), \ldots, L_A(Y_k)\},$$

where $Y_l \in H, l = 1, 2, \ldots, k$. *For* $(X_i, B_i), (X_j, B_j) \in L(U, A, I)$, *we define*

$$D((X_i, B_i), (X_j, B_j)) = \begin{cases} B_i \, \triangle \, B_j, & X_i \in L_A(Y_l), X_j \notin L_A(Y_l), \\ \emptyset, & \text{otherwise.} \end{cases} \quad (l = 1, 2, \ldots, k)$$

where $B_i \triangle B_j = B_i \cup B_j - B_i \cap B_j$, *then* $D((X_i, B_i), (X_j, B_j))$ *is referred to as the discernibility attribute set of the two concepts* (X_i, B_i) *and* (X_j, B_j). *And*

$$\mathcal{D} = \{D((X_i, B_i), (X_j, B_j)) : (X_i, B_i), (X_j, B_j) \in L(U, A, I)\}$$

is referred to as the discernibility matrix of the formal context.
   Denote

$$\mathcal{D}_0 = \{D((X_i, B_i), (X_j, B_j)) : D((X_i, B_i), (X_j, B_j)) \neq \emptyset\}$$

Theorem 6. *Let* $S = (U, A, I, C, J)$ *be a generalized consistent decision formal context,* $D \subseteq A, D \neq \emptyset$. *Then the following conditions are equivalent:*
   (1) D *is a consistent set,*
   (2) $D \cap D((X_i, B_i), (X_j, B_j)) \neq \emptyset$ *for all* $D((X_i, B_i), (X_j, B_j)) \in \mathcal{D}_0$,
   (3) *For any* $B \subseteq A$, *if* $B \cap D = \emptyset$, *then* $B \notin \mathcal{D}_0$.

Proof. "(1) $\Rightarrow$ (2)" If D is a consistent set, then for any $Y_i, Y_j \in H$ we have

$$L_D(Y_i) = \{X_i \in H_D : X_i \subseteq Y_i, \forall Y_i' \in H, Y_i' \subset Y_i, X_i \nsubseteq Y_i'\},$$

$$L_D(Y_j) = \{X_j \in H_D : X_j \subseteq Y_j, \forall Y_j' \in H, Y_j' \subset Y_j, X_j \nsubseteq Y_j'\}.$$

Since $Y_i \neq Y_j$, $L_D(Y_i) \cap L_D(Y_j) = \emptyset$, it follows that $X_i \neq X_j$. Then there exist $C_i, C_j \subseteq D$ such that $(X_i, C_i), (X_j, C_j) \in L(U, D, I_D)$ and $(X_i, C_i) \neq (X_j, C_j)$, hence $C_i \neq C_j$, but

$$C_i = X_i^{*D} = X_i^* \cap D = B_i \cap D, C_j = X_j^{*D} = X_j^* \cap D = B_j \cap D,$$

so $B_i \cap D \neq B_j \cap D$.
   If $H_A = H_D$, then for any $X_i \in L_A(Y_l)$ and $X_j \notin L_A(Y_l), l = 1, 2, \ldots, k$, it is clear that $B_i \cap D \neq B_j \cap D$.
   If $H_A \neq H_D$, that is, $H_D \subset H_A$, then it is easy to verify that $L_D(Y) \subset L_A(Y)$ for all $Y \in H$, hence we have $X_i \in L_A(Y)$ for all $X_i \in L_D(Y)$. It should be noted that $X_i^* \cap D = X^* \cap D$ for all $X \in L_A(Y)$. Consequently, for any $X_i \in L_A(Y_l)$, $X_j \notin L_A(Y_l), l = 1, 2, \ldots, k$, we have $B_i \cap D \neq B_j \cap D$. Therefore

$$B_i \cap D - B_j \cap D = D \cap B_i \cap \overline{B_j} \neq \emptyset,$$

(where $\overline{B_j}$ is the complement of B_j in A) or

$$B_j \cap D - B_i \cap D = D \cap B_j \cap \overline{B_i} \neq \emptyset.$$

It follows that

$$
\begin{aligned}
D \cap D((X_i, B_i), (X_j, B_j)) &= D \cap (B_i \cup B_j - B_i \cap B_j) \\
&= D \cap (B_i \cup B_j) \cap (\overline{B_i} \cup \overline{B_j}) \\
&= (D \cap B_j \cap \overline{B_i}) \cup (D \cap B_i \cap \overline{B_j}) \neq \emptyset.
\end{aligned}
$$

"(2)$\Rightarrow$(1)" To prove that D is a consistent set, we only need to prove that

$$L_D(Y) = \{X' \in H_D : X' \subseteq Y, \forall Y' \in H, Y' \subset Y, X' \not\subseteq Y'\} \neq \emptyset, \forall Y \in H.$$

Since S is a consistent decision formal context, we have

$$L_A(Y) = \{X \in H_A : X \subseteq Y, \quad \forall Y' \in H, Y' \subset Y, X \not\subseteq Y'\} \neq \emptyset, \quad \forall Y \in H.$$

Hence, for any $Y \in H$, we can find $X \in H_A$ such that $X \subseteq Y$, i.e., $(X, B) \in L(U, A, I)$.

Now we are to prove that $(X, B \cap D) \in L(U, D, I_D)$. Obviously, $X^{*D} = X^* \cap D = B \cap D$. If $(B \cap D)^* \neq X$, notice that $((B \cap D)^*, (B \cap D)^{**}) \in L(U, A, I)$, we have $(X, B) \neq ((B \cap D)^*, (B \cap D)^{**})$, and in turn, $B \neq (B \cap D)^{**}$. Since $B_i \cap D \neq B_j \cap D$, we have $B \cap D \neq (B \cap D)^{**} \cap D$, and

$$
\begin{aligned}
B \cap D \subseteq B &\Longrightarrow (B \cap D)^* \supseteq B^* \\
&\Longrightarrow (B \cap D)^{**} \subseteq X^* = B \\
&\Longrightarrow (B \cap D)^{**} \cap D \subseteq B \cap D.
\end{aligned}
$$

On the other hand,

$$B \cap D \subseteq (B \cap D)^{**} \Longrightarrow B \cap D = B \cap D \cap D \subseteq (B \cap D)^{**} \cap D.$$

Therefore $B \cap D = (B \cap D)^{**} \cap D$, which contradicts the fact that $B \cap D \neq (B \cap D)^{**} \cap D$. Consequently, $(B \cap D)^* = X$. Hence for any $Y \in H$ we have

$$L_D(Y) = \{X' \in H_D : X' \subseteq Y, \forall Y' \in H, Y' \subset Y, X' \not\subseteq Y'\} \neq \emptyset.$$

Thus D is a consistent set.

"(2)$\Leftrightarrow$(3)" is obvious.

Theorem 6 provides an approach to knowledge reduction in a generalized consistent decision formal context. Now we present an example to illustrate this approach.

Example 1. Table 1 depicts an example of a decision formal context $S = (U, A, I, C, J)$, where $U = \{1, 2, 3, 4\}$, $A = \{a, b, c, d, e\}$ is the conditional attribute set, and $C = \{f, g\}$ is the decision attribute set. The conditional concept lattice and the decision concept lattice are depicted as Figure 1 and Figure 2 respectively.

Table 1. A decision formal context S

U	a	b	c	d	e	f	g
1	1	1	0	1	1	0	1
2	1	1	1	0	0	1	0
3	0	0	0	1	0	1	0
4	1	1	1	0	0	1	0

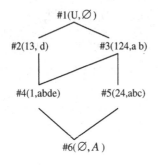

Fig. 1. The concept lattice $L(U, A, I)$

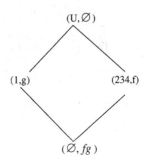

Fig. 2. The concept lattice $L(U, C, J)$

Following the standard notions in formal concept analysis, set notions are separator-free in the sequel to follow, e.g., 24 stands for $\{2, 4\}$.

It is easy to observe

$$H_A = L_u(U, A, I) = \{U, 13, 124, 1, 24, \emptyset\},$$
$$H = L_u(U, C, J) = \{U, 1, 234, \emptyset\}.$$

Also,

$$L_A(U) = \{U, 13, 124\}, \qquad L_A(234) = \{24\},$$
$$L_A(1) = \{1\}, \qquad L_A(\emptyset) = \{\emptyset\}.$$

Obviously, for any $Y \in H, L_A(Y) \neq \emptyset$, i.e., $L(U, A, I) \leq^* L(U, C, J)$. Thus S is consistent. Now we give the discernibility matrix of S as Table 2.

By Theorem 6 we conclude that $D_1 = \{a, c, d\}, D_2 = \{b, c, d\}, D_3 = \{c, e\}$ are the reducts of S.

Table 2. The discernibility matrix

	X_1	X_2	X_3	X_4	X_5	X_6
X_1	$\emptyset$	$\emptyset$	$\emptyset$	$abde$	abc	$abcde$
X_2	$\emptyset$	$\emptyset$	$\emptyset$	abe	$abcd$	$abce$
X_3	$\emptyset$	$\emptyset$	$\emptyset$	de	c	cde
X_4	$abde$	abe	de	$\emptyset$	cde	c
X_5	abc	$abcd$	c	cde	$\emptyset$	de
X_6	$abcde$	$abce$	cde	c	de	$\emptyset$

5   Conclusion

Knowledge reduction is one of the main issues for the study of data mining and knowledge discovery in databases. Comparing with the studies on knowledge reduction in rough set theory, there is less results on this issue in concept lattice theory. In this paper, we have introduced a new partial ordered relation on the set of concept lattices with the same universe. We have also proposed a new method to knowledge reduction in generalized consistent decision formal contexts, from which relationship between concepts in conditional part and decision part of contexts can be analyzed and knowledge in the sense of decision rules about the concepts hidden in decision formal contexts can be discovered.

Acknowledgements

This work was supported by a grant from the Major State Basic Research Development Program of China (973 Program No. 2002CB312200) and a grant from the National Natural Science Foundation of China (No. 60673096).

References

1. Duntsch, I., Gediga, G.: Approximation operators in qualitative data analysis. Lecture Notes in Computer Science 2929, Springer, New York, 2003, pp.214–230
2. Ganter, B., Wille, R.: Formal Concept Analysis, Mathematical Foundations. Springer, Belin, 1999
3. Kent, R.E.: Rough concept analysis: a synthesis of rough sets and formal concept analysis. Fundamenta Informaticae 27(1996) 169–181
4. Kryszkiewicz, M.: Comparative study of alternative types of knowledge reduction in inconsistent systems. International Journal of Interligent Systems 16(2001) 105–120
5. Pawlak, Z.: Rough Sets: Theoretical Aspects of Reasoning about Data. Kluwer Academic Publishers, Boston, 1991
6. Saquer, J., Deogun, J.S.: Formal rough concept analysis. Lecture Notes in Computer Science 1711, Springer, Berlin, 1999, pp.91–99
7. Slezak, D.: Approximate reducts in decision tables. In: Proceedings of IPMU'96, Granada, Spain, vol.3, 1996, pp.1159–1164
8. Wille, R.: Restructuring lattice theory: an approach based on hierarchies of concepts. In: Rival, I.(ed.): Ordered Sets, Reidel, Dordrecht-Boston, 1982, pp.445–470
9. Wu, W.-Z., Zhang, M., Li, H.-Z., Mi, J.-S.: Knowledge reduction in random information systems via Dempster-Shafer theory of evidence. Information Sciences 174(2005) 143–164
10. Yao, Y.Y.: Concept lattices in rough set theory. In: Dick, S., et al.(eds.): Proceedings of 2004 Annual Meeting of the North American Fuzzy Information Processing Society (NAFIPS 2004), June 27-30, 2004, pp.796–801
11. Zhang, W.-X., Mi, J.-S., Wu, W.-Z.: Approaches to knowledge reductions in inconsistent systems. International Journal of Intelligent Systems 21(2003) 989–1000
12. Zhang, W.-X., Wei, L., Qi, J.-J.: Attribute reduction theory and approach to concept lattice. Science in China E 48(2005) 713–726

Graphical Representation of Information on the Set of Reducts

Mikhail Moshkov[1] and Marcin Piliszczuk[2]

[1] Institute of Computer Science, University of Silesia
Będzińska 39, 41-200 Sosnowiec, Poland
moshkov@us.edu.pl
[2] ING Bank Śląski S.A.
Sokolska 34, 40-086 Katowice, Poland
marcin.piliszczuk@ingbank.pl

Abstract. In this paper we study properties of the graph $G(T)$, associated with an arbitrary decision table T. The set of vertices of $G(T)$ coincides with the set of conditional attributes of T belonging to at least one decision reduct for T, and the set of edges coincides with the set of pairs of attributes which do not belong to any decision reduct for T.

Keywords: rough sets, decision tables, decision reducts.

1 Introduction

The set of decision reducts [5] for a decision table can give us various information on relationships among conditional attributes. Unfortunately, there is no polynomial algorithm which for a given decision table T constructs the set of decision reducts for T. However, there exist polynomial algorithms for obtaining indirect but useful information on the set of reducts, and representation of this information in simple graphical form.

For example, in [6] it is shown that there exists a polynomial algorithm which for a given decision table T constructs so-called pairwise core graph. The set of vertices of this graph coincides with the set of conditional attributes of T, and the set of edges coincides with the set of pairs of attributes such that each attribute from the pair does not belong to the core of T (the intersection of all reducts for T), but each reduct contains at least one attribute from the pair.

In [3] it is shown that that there exists a polynomial algorithm which for a given decision table T constructs a graph $G(T)$. The set of vertices of this graph coincides with the set of attributes belonging to at least one reduct for T, and the set of edges coincides with the set of pairs of attributes which do not belong to any reduct for T.

This paper is devoted to consideration of properties of the graph $G(T)$. We show that any undirected graph can be represented as the graph $G(T)$ for an appropriate decision table T. We study experimentally dependencies of parameters of the graph $G(T)$ on parameters of table T, possibilities of the use of degree of an attribute in the graph $G(T)$ as a characteristic of the attribute importance,

J.T. Yao et al. (Eds.): RSKT 2007, LNAI 4481, pp. 372–378, 2007.

and possibilities of the use of changes of the graph $G(T)$ after adding of a new object to the table T for evaluation of influence of this new object on T.

2   Graph $G(T)$

A *decision table* T is a finite table in which each column is labeled by a *conditional attribute* (name of conditional attribute). Rows of the table T are interpreted as tuples of values of conditional attributes on some objects. Each row is labeled by a *decision* which is interpreted as the value of the *decision attribute*.

Let A be the set of conditional attributes (the set of names of conditional attributes) of T. We will say that a conditional attribute $a \in A$ *separates* two rows if these rows have different values at the intersection with the column labeled by a. We will say that two rows are *different* if at least one attribute $a \in A$ separates these rows. Denote by $P(T)$ the set of unordered pairs of different rows from T which are labeled by different decisions.

A subset R of the set A is called a *test* for T if for each pair of rows from $P(T)$ there exists an attribute from R which separates this pair. A test R for T is called a *reduct* for T if each proper subset of R is not a test for T. Really, we deal with decision reducts but we will omit the word "decision".

Let us describe a graph $G(T)$. The set of vertices of this graph is equal to the set of attributes $a \in A$ for each of which there exists a reduct R for T such that $a \in R$. Two different vertices a_1 and a_2 of $G(T)$ are linked by an edge if and only if there is no reduct R for T such that $a_1, a_2 \in R$.

Example 1. Let us denote by T_Z the decision table "Zoo" [4] with 16 conditional attributes $a_1, \ldots, a_{16}$ (we ignore first attribute "animal name") and 101 rows. Only attributes $a_1, a_3, a_4, a_6, \ldots, a_{14}, a_{16}$ are vertices of the graph $G(T_Z)$. The set of reducts for T_Z is represented in Table 1. The graph $G(T_Z)$ is represented in Fig. 1.

Note that there exists close analogy between the graph $G(T)$ and so-called cooccurance graph [1] for positive Boolean function f. The set of vertices of this

Table 1. The set of reducts for the decision table T_Z ("Zoo")

$\{a_3, a_4, a_6, a_8, a_{13}\}$	$\{a_3, a_6, a_8, a_9, a_{12}, a_{13}\}$	$\{a_3, a_6, a_8, a_{13}, a_{16}\}$
$\{a_3, a_4, a_6, a_9, a_{13}\}$	$\{a_1, a_3, a_6, a_7, a_{10}, a_{12}, a_{13}\}$	$\{a_3, a_6, a_9, a_{13}, a_{16}\}$
$\{a_3, a_6, a_8, a_{10}, a_{13}\}$	$\{a_3, a_4, a_6, a_7, a_{10}, a_{12}, a_{13}\}$	$\{a_4, a_6, a_8, a_{11}, a_{13}, a_{16}\}$
$\{a_1, a_3, a_6, a_9, a_{10}, a_{13}\}$	$\{a_1, a_6, a_8, a_{10}, a_{12}, a_{13}\}$	$\{a_4, a_6, a_9, a_{11}, a_{13}, a_{16}\}$
$\{a_1, a_3, a_6, a_8, a_{11}, a_{13}\}$	$\{a_1, a_6, a_9, a_{10}, a_{12}, a_{13}\}$	$\{a_3, a_6, a_7, a_{10}, a_{11}, a_{13}, a_{16}\}$
$\{a_1, a_3, a_6, a_9, a_{11}, a_{13}\}$	$\{a_1, a_3, a_6, a_{10}, a_{13}, a_{14}\}$	$\{a_1, a_6, a_8, a_{10}, a_{11}, a_{13}, a_{16}\}$
$\{a_3, a_6, a_8, a_9, a_{11}, a_{13}\}$	$\{a_3, a_4, a_6, a_{10}, a_{13}, a_{14}\}$	$\{a_1, a_6, a_9, a_{10}, a_{11}, a_{13}, a_{16}\}$
$\{a_1, a_3, a_6, a_8, a_{12}, a_{13}\}$	$\{a_3, a_6, a_8, a_{11}, a_{13}, a_{14}\}$	$\{a_3, a_6, a_7, a_{10}, a_{12}, a_{13}, a_{16}\}$
$\{a_4, a_6, a_8, a_{12}, a_{13}\}$	$\{a_3, a_6, a_8, a_{12}, a_{13}, a_{14}\}$	$\{a_3, a_6, a_{10}, a_{13}, a_{14}, a_{16}\}$
$\{a_1, a_3, a_6, a_9, a_{12}, a_{13}\}$	$\{a_1, a_6, a_{10}, a_{12}, a_{13}, a_{14}\}$	$\{a_1, a_6, a_{10}, a_{11}, a_{13}, a_{14}, a_{16}\}$
$\{a_4, a_6, a_9, a_{12}, a_{13}\}$	$\{a_4, a_6, a_{10}, a_{12}, a_{13}, a_{14}\}$	$\{a_4, a_6, a_{10}, a_{11}, a_{13}, a_{14}, a_{16}\}$

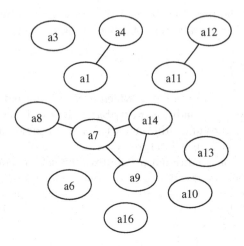

Fig. 1. Graph $G(T_Z)$ for the decision table T_Z ("Zoo")

graph is equal to the set of variables of f. Two different variables are linked by an edge if and only if f has a prime implicant containing these variables.

3   Representation of Undirected Graphs by Graphs $G(T)$

Let us show that any undirected graph G can be represented as the graph $G(T)$ for an appropriate decision table T. It means that the graph $G(T)$ can give us reach information about the set of reducts for a decision table T.

Proposition 1. *Let A be a finite set of names of conditional attributes, and $G = (V, E)$ be an undirected graph, where $V \subseteq A$ is the set of vertices of G and E is the set of edges of G such that each edge of G is a two-element subset of V. Then there exists a decision table T with the set of names of conditional attributes A such that $G(T) = G$.*

Proof. Let $\mathcal{R}$ be a set of subsets of A. This set is called independent if there are no two subsets $B_1, B_2 \in \mathcal{R}$ such that $B_1 \subset B_2$. It is known (see, for example, [2]) that $\mathcal{R}$ is independent if and only if a decision table T exists such that the set of names of conditional attributes of T coincides with A, and the set of reducts for T coincides with $\mathcal{R}$.

Let $V = \emptyset$ and T be a decision table with the set of names of conditional attributes A such that all rows of T are labeled by the same decision. Then, evidently, $G(T) = G$.

Let now $V \neq \emptyset$, E' be the set of two-element subsets of V each of which does not belong to E, and $a_1, \ldots, a_r$ be all vertices in the graph $G' = (V, E')$ without incident edges. Let us denote

$$\mathcal{R} = \{\{a_1\}, \ldots, \{a_r\}\} \cup E'.$$

It is clear that the set $\mathcal{R}$ is independent. Then there exists a decision table T with the set of names of conditional attributes A such that the set of reducts for T coincides with $\mathcal{R}$. One can show that $G(T) = G$.   □

4   On Parameters of Graphs $G(T)$

In this section we consider results of experiments with randomly generated decision tables filled by numbers from the set $\{0, 1\}$. The probability of appearance of 1 is a number p from the set $\{0.05, 0.1, 0.2, 0.3, 0.4\}$. Values of the decision attribute belong to the set $\{0, 1\}$. The value of the decision attribute is equal to 1 with the probability 0.5. The number of conditional attributes is equal to 20. The number n of rows belongs to the set $\{20, 50, 100\}$.

We generate randomly 10 decision tables T for each pair of parameters (n, p), where $n \in \{20, 50, 100\}$ and $p \in \{0.05, 0.1, 0.2, 0.3, 0.4\}$, and find average number of vertices in the graph $G(T)$ (see left hand subtable of Table 2) and average number of edges in $G(T)$ (see right hand subtable of Table 2).

Table 2. Average number of vertices and edges in graphs $G(T)$

Number	Probability p						Number	Probability p				
of rows n	0.05	0.10	0.20	0.30	0.40		of rows n	0.05	0.10	0.20	0.30	0.40
20	10.4	14.1	20.0	20.0	20.0		20	142.1	102.6	6.8	0.6	0.2
50	15.7	17.0	20.0	20.0	20.0		50	73.7	54.8	1.1	0	0
100	19.1	19.4	19.6	20.0	20.0		100	17.1	11.5	8.9	0	0

Results of experiments show that in randomly generated decision tables T with fixed number of conditional attributes as the rule the average number of vertices in graphs $G(T)$ increases with the growth of the number of rows, and the average number of edges in graphs $G(T)$ decreases with the growth of the number of rows. Note also that the alignment of probabilities of appearance of 0 and 1 in tables leads to increasing of the average number of vertices and to decreasing of average number of edges.

The obtained results show also that sometimes it is appropriate to represent graphically not the graph $G(T)$ but the graph $G'(T)$ which is the complement of $G(T)$. The graph $G'(T)$ has the same set of vertices as $G(T)$, and two different vertices in $G'(T)$ are adjacent (linked by an edge) if and only if these vertices in $G(T)$ are not adjacent.

5   On Degree of Attributes in $G(T)$

In this section we compare the degree of an attribute in the graph $G(T)$ (the number of edges incident to this attribute) and the number of reducts for T containing this attribute (last parameter is considered often as attribute importance). We study three decision tables from [4]: T_Z "Zoo" (see Example 1), T_L

"Lymphography" with 18 conditional attributes $a_1, \ldots, a_{18}$ and 148 rows (each of the considered attributes is a vertex of the graph $G(T_L)$), and T_S "Soybean-small" with 35 conditional attributes $a_1, \ldots, a_{35}$ and 47 rows (only attributes $a_1, \ldots, a_{10}, a_{12}, a_{20}, \ldots, a_{28}, a_{35}$ are vertices of the graph $G(T_S)$).

In Table 3 for each attribute a_i, which is a vertex of $G(T_Z)$ $(G(T_L), G(T_S))$, the degree of a_i and the number of reducts for T_Z (T_L, T_S) containing a_i are considered. In Table 4 for each appropriate value of degree the average number of reducts for T_Z (T_L, T_S) containing fixed attribute of this degree is considered.

Table 3. Comparison of degree of attribute (third row of each subtable) and number of reducts containing this attribute (fourth row of each subtable)

"Zoo"												
a_1	a_3	a_4	a_6	a_7	a_8	a_9	a_{10}	a_{11}	a_{12}	a_{13}	a_{14}	a_{16}
1	0	1	0	3	1	2	0	1	1	0	2	0
13	21	10	33	4	13	11	17	11	13	33	9	11

"Lymphography"																	
a_1	a_2	a_3	a_4	a_5	a_6	a_7	a_8	a_9	a_{10}	a_{11}	a_{12}	a_{13}	a_{14}	a_{15}	a_{16}	a_{17}	a_{18}
0	1	1	8	1	0	1	1	1	1	0	1	0	0	0	0	0	0
283	232	183	5	176	252	37	216	32	201	190	203	270	356	238	219	220	265

"Soybean–small"																			
a_1	a_2	a_3	a_4	a_5	a_6	a_7	a_8	a_9	a_{10}	a_{12}	a_{20}	a_{21}	a_{22}	a_{23}	a_{24}	a_{25}	a_{26}	a_{27}	a_{28}	a_{35}
0	0	3	0	0	0	1	0	0	0	0	0	5	4	1	1	0	4	4	1	0
227	178	156	163	210	304	165	232	209	349	305	232	47	39	167	167	193	121	121	142	289

Table 4. Comparison of degree of attribute (second row of each subtable) and average number of reducts containing fixed attribute with this degree (third row of each subtable)

"Zoo"			
0	1	2	3
23	12	10	4

"Lymphography"		
0	1	8
254.77	160	5

"Soybean–small"				
0	1	3	4	5
240.91	160.25	156	93.66	47

The obtained results show that there exists a correlation between the degree of an attribute and the number of reducts containing this attribute.

6   Changes of $G(T)$

In this section we study possibilities to use changes of the graph $G(T)$ after adding of a new object to the decision table T as an indicator of changes of the set of reducts for T. The presence of changes of the set of reducts can be considered as a noticeable influence of the new object to the table T.

First experiment is connected with the decision table T_Z "Zoo". Let us denote $T_Z^{(0)}$ empty table, and for $i = 1, \ldots, 100$ denote $T_Z^{(i)}$ the decision table containing first i rows of T_Z. In Table 5 for $i = 1, \ldots, 100$ the column with number i contains the sign "+" at the intersection with the row "graph" if $G(T_Z^{(i-1)}) \neq G(T_Z^{(i)})$, and contains the sign "+" at the intersection with the row "reducts" if the set of reducts for $T_Z^{(i-1)}$ is not equal to the set of reducts for $T_Z^{(i)}$. One can see that in 78% of cases changes in the set of reducts entail changes in the considered graph.

Table 5. Changes in the decision table T_Z "Zoo"

i	1	2	3	4	5	6	7	8	9	10	11	12	13	14	15	16	17	18	19	20	21	22	23	24	25
graph			+		+					+		+		+	+		+		+	+		+		+	+
reducts			+		+					+		+		+	+		+		+	+		+		+	+

i	26	27	28	29	30	31	32	33	34	35	36	37	38	39	40	41	42	43	44	45	46	47	48	49	50
graph	+		+					+				+			+		+	+							+
reducts	+					+		+	+			+			+	+	+	+	+					+	+

i	51	52	53	54	55	56	57	58	59	60	61	62	63	64	65	66	67	68	69	70	71	72	73	74	75
graph		+							+				+										+		
reducts		+							+				+	+									+		

i	76	77	78	79	80	81	82	83	84	85	86	87	88	89	90	91	92	93	94	95	96	97	98	99	100
graph						+	+					+				+	+								
reducts		+	+			+	+					+				+	+								

Second experiment is similar to the first one and is connected with the decision table T_L "Lymphography". There are 133 changes in the set of reducts and 32 changes in the considered graph. So in 24% of cases changes in the set of reducts entail changes in the graph.

The obtained results show that there exists a correlation between changes of the set of reducts for T and changes of the graph $G(T)$ after adding of a new object to T.

7 Conclusions

In the paper some properties of the graph $G(T)$ are studied. It is proved that any undirected graph can be represented as the graph $G(T)$ for an appropriate decision table T. Experimental results show that there exist correlations between the degree of attribute in the graph $G(T)$ and the number of reducts for T containing this attribute (last parameter is considered often as attribute importance), and between changes of $G(T)$ and changes of the set of reducts for T after adding of new object to T (the presence of changes of the set of reducts can be considered as a noticeable influence of new object to the decision table).

References

1. Boros, E, Gurvich, V., Hammer, P.L.: Dual subimplicants of positive Boolean functions. Optimization Methods and Software **10** (1998) 147–156
2. Moshkov, M.: Elements of mathematical theory of tests (methodical instructions). Gorky State University, 1986 (in Russian)
3. Moshkov, M., Skowron, A., Suraj, Z.: On covering attribute sets by reducts. International Conference Rough Sets and Emerging Intelligent Systems Paradigms. In Memoriam Zdzislaw Pawlak, Warsaw, Poland, June 28-30, 2007 (submitted)
4. Newman, D.J., Hettich, S., Blake, C.L., Merz, C.J.: UCI Repository of machine learning databases http://www.ics.uci.edu/~mlearn/MLRepository.html. University of California, Irvine, Department of Information and Computer Sciences (1998)
5. Pawlak, Z.: Rough Sets – Theoretical Aspects of Reasoning about Data. Kluwer Academic Publishers, Dordrecht, Boston, London, 1991
6. Wróblewski, J.: Pairwise cores in information systems. Proceedings of the 10th International Conference Rough Sets, Fuzzy Sets, Data Mining, and Granular Computing, Part 1. Regina, Canada. Lecture Notes in Artificial Intelligence **3641**, Springer-Verlag, Heidelberg (2005) 166–175

Minimal Attribute Space Bias for Attribute Reduction

Fan Min, Xianghui Du, Hang Qiu, and Qihe Liu

School of Computer Science and Engineering,
University of Electronic Science and Technology of China, Chengdu 610054, China
{minfan,xianghd,qiuhang,qiheliu}@uestc.edu.cn

Abstract. Attribute reduction is an important inductive learning issue addressed by the Rough Sets society. Most existing works on this issue use the minimal attribute bias, i.e., searching for reducts with the minimal number of attributes. But this bias does not work well for datasets where different attributes have different sizes of domains. In this paper, we propose a more reasonable strategy called the *minimal attribute space bias*, i.e., searching for reducts with the minimal attribute domain sizes product. In most cases, this bias can help to obtain reduced decision tables with the best space coverage, thus helpful for obtaining small rule sets with good predicting performance. Empirical study on some datasets validates our analysis.

Keywords: Attribute reduction, bias, space coverage, rule set.

1 Introduction

Practical machine learning algorithms are known to degrade in performance (prediction accuracy) when faced with many attributes that are not necessary for rule discovery [1]. It is therefore not surprising that much research has been carried out on attribute reduction [2], particularly by people in the Rough Sets society. A reduct is a subset of attributes that is jointly sufficient and individually necessary for preserving the same information (in terms of positive region [3], class distribution [4] among others) under consideration as provided by the entire set of attributes [5].

A commonly used reduct selection/construction strategy, called the minimal attribute bias [1][3][6], is to searching for a reduct with the minimal number of attributes. In some cases, especially when different attribute have approximately the same size of domain, this bias may be helpful for obtaining small rule sets with good performance. However, for data in reality where attribute domain sizes vary, this strategy is *unfair* since attributes with larger domains tend to have better discernibility or other significance measures [7], and it has severe implications when applied blindly without regarding for the resulting induced concept [1].

To cope with these problems, in this paper we propose a new bias called the *minimal attribute space bias* which is intended to minimize the attribute space.

J.T. Yao et al. (Eds.): RSKT 2007, LNAI 4481, pp. 379–386, 2007.

We argue that this bias is more reasonable, thus more helpful for obtaining small rule set, than the minimal attribute bias. Empirical study on some datasets in the UCI library [8] validates our analysis.

2 Preliminaries

In this section we enumerate basic concepts introduced by Pawlak [3] through an example.

Formally, a *decision table* is a triple $S = (U, C, \{d\})$ where $d \notin C$ is the decision attribute and elements of C are called *conditional attributes* or simply *conditions*. Table 1 lists a decision table where $U = \{t1, \ldots, t8\}$, $C = \{$Shape, Material, Weight, Color$\}$ and $d = $ Size.

Table 1. An exemplary decision table

Toy	Shape	Material	Weight	Color	Size
t1	round	wood	light	red	small
t2	round	plastic	heavy	black	small
t3	round	wood	heavy	white	large
t4	round	wood	light	white	small
t5	triangle	wood	light	green	small
t6	triangle	plastic	heavy	blue	large
t7	triangle	plastic	light	pink	large
t8	triangle	plastic	heavy	yellow	large

Any $\emptyset \neq B \subseteq C \cup \{d\}$ determines a binary relation $I(B)$ on U, which will be called an *indiscernibility relation*, and is defined as follows:

$$I(B) = \{(x_i, x_j) \in U \times U | \forall a \in B, a(x_i) = a(x_j)\}, \tag{1}$$

where $a(x)$ denotes the value of attribute a for element x.

A partition determined by B is denoted by $U/I(B)$, or simply by U/B. Let $\underline{B}X$ denotes $B-lower\ approximation$ of X, the positive region of $\{d\}$ with respect to $B \subseteq C$ is defined as $POS_B(\{d\}) = \bigcup_{X \in U/\{d\}} \underline{B}(X)$.

A reduct is the minimal subset of attributes that enables the same classification of elements of the universe as the whole set of attributes. This can be formally defined as follows:

Definition 1. *Any $B \subseteq C$ is called a* decision relative reduct *of $S = (U, C, \{d\})$ iff:*

1. $POS_B(\{d\}) = POS_C(\{d\})$, *and*
2. $\forall a \in B, POS_{B-\{a\}}(\{d\}) \subset POS_C(\{d\})$.

A decision relative reduct can be simply called a relative reduct, or a reduct for briefness if the decision attribute is obvious from the context.

According to Definition 1, the exemplary decision table has two reducts: $R_1 = \{$Shape, Material, Weight$\}$ and $R_2 = \{$Weight, Color$\}$.

3   The Minimal Attribute Bias

This bias is described as follows: A reduct R is *optimal* iff $|R|$ is minimal, where $|\cdot|$ denotes the cardinality of a set.

According to this bias, $R_2 =\{$Weight, Color$\}$ is an optimal reduct of Table 1 because $|R_1| = 3$ and $|R_2| = 2$. For the sake of clarity, we use the term *minimal reduct* instead of *optimal reduct* in the following context.

The main advantage of this bias is simple and tend to give short rules. For datasets where different attributes have approximately the same size of domains, it may be also helpful for obtaining small rule sets with good predicting performance.

However, it also has the following drawbacks:

1. Unfair for different attributes. For example, in Table 1, attribute Color is the most important attribute from the viewpoint of discernibility. But this is due to its relatively large domain (7 values versus 2 of others) rather than its intrinsic importance.
2. Too many optimal reducts. For example, the Mushroom dataset [8] (further discussed in Subsection 5.1) has 14 minimal reducts. Some of them perform well in terms of further rule set generation and/or decision tree construction, but others do not. The bias does not indicate a more detailed strategy for choosing among those reducts.

4   The Minimal Attribute Space Bias

We propose the minimal attribute space bias as follows: A reduct R is *optimal* iff $\Pi_{a\in R}|V_a|$ is minimal, where V_a is the domain of attribute a.

According to this bias, $R_1 = \{$Shape, Material, Weight$\}$ is an optimal reduct of Table 1 because $\Pi_{a\in R_1}|V_a| = 8$ and $\Pi_{a\in R_2}|V_a| = 14$. We also use the term *minimal space reduct* instead of *optimal reduct*.

Remark 1. If $V_{a_i} = V_{a_j}$ for any $a_i, a_j \in C$, then the minimal attribute space bias coincides with the minimal attribute bias.

Now we explain why this bias is more reasonable than the minimal attribute bias using the exemplary decision table. Each object in the table corresponds with a decision rule. For example, t1 corresponds with

Shape=round $\wedge$ Material=wood $\wedge$ Weight $=$ light $\wedge$ Color $=$ red $\Rightarrow$ Size $=$ small.

This type of rules will be called *original rules* since no inductive learning algorithm has been introduced. Because no object pairs are indiscernible, there are a total of 8 original rules. On the other hand, the attribute space of the decision table is $|$Shape$| \times |$Material$| \times |$Weight$| \times |$Color$| = 2 \times 2 \times 2 \times 7 = 56$. Therefore objects in the decision table only cover $8/56 = 1/7 = 0.143$ of the attribute space, and the original rule set may have poor performance in terms of coverage.

As an inductive approach, attribute reduction can reduce the number of attribute; or more importantly, it can reduce the attribute space. The attribute space of $S(R_1) = \{\{t1, \ldots, t8\}, R_1, \{Size\}\}$ is $|Shape| \times |Material| \times |Weight| = 2 \times 2 \times 2 = 8$, while it has two indiscernible object pairs: t1 and t4 as well as t6 and t8, hence only $8 - 2 = 6$ original rules could be obtained, incurring $6/8 = 0.75$ of space coverage. The attribute space of $S(R_2) = \{\{t1, \ldots, t8\}, R_2, \{Size\}\}$ is $|Weight| \times |Color| = 2 \times 7 = 14$, and no indiscernible object pairs exist, hence 8 original rules could be obtained, incurring $8/14 = 0.571$ of space coverage. From this viewpoint, both reducts have notable generalization ability, and R_1 performs better (with space coverage 0.75 versus 0.571 of R_2).

Table 2. Rule sets generated from $S(R_1)$ and $S(R_2)$

rule No.	rule	support
r1	Material = wood ∧ Weight = light ⇒ Size = small	3
r2	Shape = triangle ∧ Material = plastic ⇒ Size = large	3
r3	Shape = round ∧ Weight = light ⇒ Size = small	2
r4	Shape = triangle ∧ Weight = heavy ⇒ Size = large	2
r5	Shape = round ∧ Material = plastic ⇒ Size = small	1
r6	Material = wood ∧ Weight = heavy ⇒ Size = large	1
r7	Shape = triangle ∧ Material = wood ⇒ Size = small	1
r8	material = plastic ∧ weight = light ⇒ Size = large	1
r9	Color = red ⇒ Size = small	1
r10	Color = black ⇒ Size = small	1
r11	Weight = heavy ∧ Color = white ⇒ Size = large	1
r12	Weight = light ∧ Color = white ⇒ Size = small	1
r13	Color = green ⇒ Size = small	1
r14	Color = blue ⇒ Size = large	1
r15	Color = pink ⇒ Size = large	1
r16	Color = yellow ⇒ Size = large	1

It should be noted that better generalization ability does not ensure smaller rule sets. In fact, using the exhaustive algorithm [9] we obtained 8 rules for either reduced decision tables, as listed in Table 2, where the former 8 rules corresponds with $S(R_1)$. But it should be noted further that each rule generated from $S(R_2)$ is supported by only 1 object, while in all rules generated from $S(R_1)$, 2 rules (r1 and r2) are supported by 3 objects, and another 2 rules (r3 and r4) are supported by 2 objects. In other words, R_1 is more helpful for generating *strong* rules.

One can observe that in this case both rule sets cover the whole attribute space. While for larger datasets, reducts with smaller attribute spaces often result in smaller rule sets with better space coverage.

Formally, the space coverage of $S = (U, C, \{d\})$ is defined as

$$SC(S) = \frac{|U/C|}{\Pi_{a \in C} |V_a|}, \tag{2}$$

which serves as an important factor for further rule generation / decision tree construction.

From the viewpoint of space coverage, the goal of attribute reduction should be searching for a reduct R such that $SC((U, R, \{d\}))$ is minimal. One approach is to maximize $|U/R|$, but $|U/R|$ has an upper bound $|U|$, and $|U/R|$ does not vary too much for different reducts. Thus this approach does not make sense. The other approach is to minimize $\Pi_{a \in R}|V_a|$, which coincides with the minimal attribute space bias.

In most cases, minimal attribute space results in maximal space coverage. One might construct a counterexample as follows: A decision table $S = (U, C, \{d\})$ and two reducts R_1, R_2 satisfying $|U/R_1| > |U/R_2|$, $\Pi_{a \in R_1}|V_a| > \Pi_{a \in R_2}|V_a|$ and $SC((U, R_1, \{d\})) > SC((U, R_2, \{d\}))$; but this situation is quite unlikely to happen in real data.

5   Experiments and Comparisons

There are very few datasets (e.g., Nursery) with complete coverage of the attribute space. For many datasets, attribute reduction is a very important approach to improving the space coverage.

We tested these two reduct selection biases on some datasets of the UCI library [8] using RSES [9] and a software developed by us called RDK (Rough Developer's Kit). For some datasets the set of all minimal space reducts coincides with the set of all minimal reducts. These datasets can be classified as follows:

1. One reduct datasets, e.g., Zoo, solar-flare and Monks;
2. Datasets whose all conditional attributes have the same size of domain, e.g., Letter Recognition and Tic-Tac-Toe, and
3. Other Datasets, e.g., bridges.

In what follows experiments on two datasets will be discussed in more detail.

5.1   Experiments on Mushroom

The Mushroom dataset [8] contains 8416 objects and 22 conditional attributes. The domain sizes of attributes vary from 1 (veil-type, the attribute value UNIVERSAL announced in agaricus-lepiota.names never appeared in the dataset) to 12 (gill-color). It has 292 reducts, and all minimal space reducts are also minimal reducts. For each minimal reduct, LEM2 was employed (with the cover parameter set to 1.0) to generate rule sets on reduced decision tables. Furthermore, we used CV-5 (the rule generation algorithm was also LEM2) to test the performance of those reducts. For all reducts tested, both the coverage and the accuracy of respective rule sets were 1.0. Other results are listed in Table 3.

For this dataset, the number of minimal attribute reducts is much less than the number of minimal reducts. Also, minimal attribute reducts are more helpful for obtaining smaller rule sets.

Table 3. Experimental results of Mushroom

	minimal attribute bias	minimal attribute space bias
optimal reducts	14	2
minimal rule set size	19	19
maximal rule set size	30	26
average rule set size	26.5	22.5

5.2   Experiments on Soybean

The Large Soybean Database [8] contains two parts: the training set with 307 instances, and the testing set with 376 instances. There are 35 nominal conditional attributes, with domain sizes varying from 2 to 8, and some of them have missing (unknown) values. The domain size (i.e., 18) of the decision attribute is rather large.

Due to the relatively large number of attributes, when we tried to use the exhaustive algorithm [9] to obtain the set of all reducts, an error "out of memory" was reported. So we used the genetic algorithm on the training set to obtain reducts, with the speed set to low and the number of reducts set to 100. In this way, 100 reducts were obtained, 35 of which were minimal reducts, and 9 of which were minimal space reducts. 8 out of 9 minimal space reducts contain 9 attributes, hence they were also minimal reducts. Rule sets were generated through employing the exhaustive algorithm on reduced decision tables, then they were tested on the testing set. Some results are listed in Table 4.

Table 4. Experimental results of Soybean (Bolded Values Indicate the Best Results)

	minimal attribute reducts			minimal attribute space reducts		
	minimal	maximal	average	minimal	maximal	average
rule set size	**1201**	2563	1902	1324	**1806**	**1577**
coverage	0.818	**0.960**	**0.930**	**0.912**	0.949	0.926
accuracy	0.544	**0.766**	0.668	**0.626**	**0.766**	**0.692**
F-measure	0.522	**0.740**	0.644	**0.602**	**0.740**	**0.668**

Since the minimal attribute reduct set is much larger than the minimal attribute space reduct set, it contains both the "best" and the "worst" reducts. In general, for this dataset the minimal attribute space bias outperforms the minimal attribute bias in terms of average rule set size (325 less), average accuracy (0.024 more) and averge F-measure (0.024 more). It is quite interesting that the latter bias outperforms (0.004 more than) the former in terms of average coverage. Two reducts drew our special attention:

1. The best reduct. It helped obtaining a rule set with a predication accuracy of 0.766 and F-measure of 0.740, and it was included in both reduct sets; and
2. The minimal attribute space reduct with 10 attributes. Although not a minimal reduct, it helped obtain a rule set containing 1789 rules, with the predication coverage 0.949, accuracy 0.658 and F-measure 0.641. The results are fairly good compared with that of minimal reducts.

6   Discussion

In this section we discuss these two biases from a broader viewpoint.

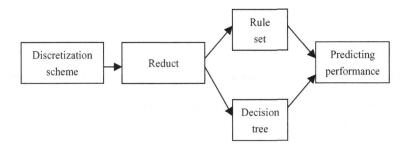

Fig. 1. A typical inductive learning scenario

As depicted in Fig. 1, the ultimate goal of inductive learning is to obtain good predicting performance, defined by the coverage, the accuracy, or the combination of both (e.g., F-measure) on the data.

But the predicting performance can be obtained only after rule set was generated or decision tree [10] was constructed (for the sake of simplicity, other approaches such as neural network or kNN are not included in Fig. 1). According to Occam's Razor, smaller rule sets, or simpler decision trees (with least nodes) are preferred.

In order to obtain a small rule set or a simple decision tree, also according to Occam's Razor, the simplest reduct is desired. But the key issue is: What is the metric of evaluating the simplicity of a reduct? Aforementioned biases are two metrics, among which the new bias seems to be closer to the essence. Then why the minimal attribute bias worked well for so many applications? In fact, many reduct construction algorithms use the following strategy: "[i]f two attributes have the same performance with respect to the criterion described above then the one having less values is selected" [1], and it is quite possible that a minimal space reduct be constructed while a minimal reduct is required. Moreover, even if the minimal reduct obtained is not a minimal space reduct, its attribute spaces is not too large compared with that of a minimal space reduct. In other words, the minimal attribute bias is often a good approximation of the minimal attribute space bias.

7   Conclusions and Further Works

Compared with the minimal attribute bias, the minimal attribute space bias is closer to goal of constructing simple reducts from viewpoints of attribute space and attribute space coverage. Also, it does not incur the fairness problem. Experiments on two datasets showed that the new bias can help to narrow the scope of optimal reducts, and more importantly, it can help to obtain better rule sets in terms of accuracy and F-measure.

Since the definition of a bias is a quite fundamental issue, many research works, e.g., reduct constructing algorithms, on the new bias are expected in the near future.

Acknowledgement

Fan Min was supported by an information distribution project under grant No. 9140A06060106DZ223 and the Youth Foundation of UESTC.

The authors would like to thank Zichun Zhong and Yue Liu for their help in experiments and paper proofing.

References

1. Zhong, N., Dong, J.: Using rough sets with heuristics for feature selection. Journal of Intelligent Information Systems, **16** (2001) 199–214
2. Pawlak, Z.: Rough sets. International Journal of Computer and Information Sciences **11** (1982) 341–356
3. Pawlak, Z.: Some issues on rough sets. In Peters, J.F., Skowron, A., Grzymała-Busse, J.W., Kostek, B., Świniarski, R.W., Szczuka, M.S., eds.: Transactions on Rough Sets I. LNCS 3100. Springer-Verlag, Berlin Heidelberg (2004) 1–58
4. Zhang, W., Mi, J., Wu, W.: Knowledge reductions in inconsistent information systems. Chinese Journal of Computers **26**(1) (2003) 12–18
5. Yao, Y., Yan, Z., Wang, J.: On reduct construction algorithms. In Wang, G., Peters, J.F., Skowron, A., Yao, Y., eds.: RSKT 2006. LNCS 4062, Berlin Heidelberg, Springer-Verlag (2006) 297–304
6. Skowron, A., Rauszer, C.: The discernibility matrices and functions in information systems. In Słowiński, R., ed.: Intelligent Decision Support–Handbook of Applications and Advances of the Rough Sets Theory. Kluwer, Academic Publishers, Dordrecht (1992) 331–362
7. Xu, C., Min, F.: Weighted reduction for decision tables. In Wang, L., Jiao, L., Shi, G., Li, X., Liu, J., eds.: FSKD 2006. LNCS 4223, Berlin Heidelberg, Springer-Verlag (2006) 246–255
8. Blake, C.L., Merz, C.J.: UCI repository of machine learning databases, http://www.ics.uci.edu/~mlearn/mlrepository.html (1998)
9. Bazan, J., Szczuka, M.: The RSES homepage, http://alfa.mimuw.edu.pl/~rses (1994–2005)
10. Quinlan, J.R.: Induction of decision trees. Machine Learning **1** (1986) 81–106

Two-Phase β-Certain Reducts Generation

Pai-Chou Wang[1] and Hsien-Kuang Chiou[2]

[1] Department of International Business
[2] Department of Information Management,
Southern Taiwan University of Technology,
No.1, Nantai St., Yung-Kang City, Tainan County, Taiwan 710 ROC
pwang@mail.stut.edu.tw, hchiou@mail.stut.edu.tw

Abstract. Reducts are applied to represent the knowledge without superfluous attributes in rough set. In this paper, a two-phase β-certain reducts generation is developed to preserve the original classification of each decision class in the table under the majority inclusion relation with a user defined admissible error β. The first phase finds the initial solutions of β-certain reducts. Initial solutions are passed to the second phase and β-certain reducts are found by generating certain reducts of the second pseudo decision when all sub categories based on these certain reducts in the non-β-positive region are totally rejected under the β criterion. No verification is needed when β-certain reducts are found.

Keywords: β-certain reducts, Variable precision rough set model, Rough set theory, Information reduction.

1   Introduction

Rough set theory was developed by Zdzislaw Pawlak [8] and it uses a strict mathematical formulism to represent, analyze and manipulate knowledge in a decision table. Rough set theory is a powerful tool to analyze the imprecise and ambiguous data. Variable precision rough set model (*VPRS* model) [13] is an extension of the rough set theory which takes partially incorrect classification into account. *VPRS* adapts the notation of the majority inclusion relation, which means an indiscernible category will be taken in *VPRS* if the classification error of assigning all instances to the class of the largest proportion in this category is less than or equal to the admissible error, β. β is defined by the user and it is within the range [0.0, 0.5). Based on the *VPRS*, a β reduct is defined using the β degree of dependency [13]. Beynon [2] further investigated β reduct selection and considered the whole domain of the β value and the different levels of quality of classification. In [3], Beynon also introduced the notion of (l, u)-graphs to classify groups of objects in condition classes to certain decision classes for a choice of l and u values.

In this paper, we adapt similar definition from Kryszkiewicz [6] and Inuiguchi [4]. Kryszkiewicz defined a $\underline{\beta}$ reduct for *VPRS* model then she applied the discernibility function [10] to compute $\underline{\beta}$ reducts incrementally. Inuiguchi [4] defined similar reducts using a discernibility function and reducts are generated by searching prime implicants in the function. The way of generating reducts using the discernbility

J.T. Yao et al. (Eds.): RSKT 2007, LNAI 4481, pp. 387–394, 2007.

matrix may not be feasible when table becomes large (e.g. with over 10,000 instances) because it contains too many candidates [7]. β-certain reduct is defined in this paper which preserves the original classification of each decision class in the table under the user defined admissible error, β. We call these reducts as β-certain because they preserve the original classification of the table like the certain reducts [1, 5, 9]. A two-phase β-certain reducts generation is developed in this paper.

The first phase creates a pseudo decision table and certain reducts of the first pseudo decision table are taken as the initial solutions of β-certain reducts. The second phase takes initial solutions from the first phase and β-certain reducts are certain reducts of the second pseudo decision if all sub categories based on these certain reducts in the non-β-positive region are totally rejected under the β condition. Pseudo decision table was proposed by Wang [11] and certain reducts of the table are generated by merging candidate attributes into an attribute set recursively. Related propositions are shown in this paper and the generation of any certain reduct of consistent or inconsistent decision table takes $O(k^2 n \log n)$ [11] where k is the total number of condition attributes and n is the total number of instances of the table. The generation of a certain reduct can be further reduced to $O(k^2 n)$ if the hash data structure proposed by Wang [12] is applied.

The rest of the paper is organized as follows: Section 2 reviews the basic preliminaries of rough set and the β-certain reduct is defined. Section 3 and 4 presents the first and the second phase of the β-certain reducts generation.

Table 1. A simple decision table with 11 instances and 4 condistions

U	c_1	c_2	c_3	c_4	D
x_1	1	1	1	2	1
x_2	1	1	2	4	3
x_3	1	1	3	2	1
x_4	1	1	3	2	1
x_5	1	1	3	2	2
x_6	1	1	3	2	2
x_7	2	1	2	2	2
x_8	3	1	1	2	1
x_9	1	2	1	3	1
x_{10}	1	2	1	3	2
x_{11}	1	2	1	3	2

2 Rough Set Preliminaries and β-Certain Reduct

In this section, basic notations related to information systems, decision systems and rough set theory are introduced. Let $T = (U, A)$ be an information table where $U = \{x_1,..., x_n\}$ is a nonempty finite set of instances called the universe and $A = \{a_1,..., a_k\}$ is a nonempty finite set called attributes. $V = \bigcup_{a \in A} V_a$ is a set of attribute values and V_a is the value set of attribute a. An information function $f: U \times A \rightarrow V$ returns values of attributes in instances. With any subset $B \subseteq A$, an indiscernibility relation, denoted by $IND(B)$, called the B-indiscernibility relation can be defined as $IND(B) = \{(x, y) \in U^2 :$ for every $a \in B$, $f(x, a) = f(y, a)\}$. Instances x, y satisfying relation $IND(B)$ are indiscernible from the subset of attributes B. In rough set, indiscernible relations are used to represent knowledge in an information table. Instances indiscernible with regard to the attribute set B of the table are denoted as $I_B(x)$ where $I_B(x) = \{y \in U: (x, y) \in IND(B)\}$. In regarding of classification using knowledge from these indiscernible relations, approximation [9] is introduced. When $B \subseteq A$ and $X \subseteq U$, the B-lower and B-upper approximation of X in table T are denoted as $\underline{B}X = \bigcup\{I_B(x): I_B(x) \subseteq X\}$ and $\overline{B}X = \bigcup\{I_B(x): I_B(x) \cap X \neq \varnothing\}$. B-lower approximation of X, $\underline{B}X$, contains the indiscernible relations definitely define X and B-upper approximation of X, $\overline{B}X$, contains the indiscernible relations possibly define X.

A decision table T is denoted as $T = (U, C, D)$ where $U = \{x_1, ..., x_n\}$ is a nonempty finite set of instances, $C = \{c_1, ..., c_k\}$ is a nonempty finite sets of attributes called condition and D is the decision. The decision D is considered as a singleton set and $D \cap C = \varnothing$. In each decision class i, the β-lower approximation of $X_i = \{x \in U: f(x, D) = i\}$ for $P \subseteq C$ is defined as

$$\underline{P}_\beta X_i = \bigcup \{I_P(x) \mid card(I_P(x) \cap \overline{X}_i) / card(I_P(x)) \leq \beta\}, \ \forall \ x \in X_i \text{ and } \overline{X}_i = U \setminus X_i$$

Function $f(x, D)$ returns the decision class of instance x. $\underline{P}_\beta X_i$ contains all indiscernible categories satisfying the β condition in decision class i and $\overline{X}_i$ contains the complement instances of X_i in the table. The union of all β-lower approximations in all decision classes is called the β-positive region of the table which contains original classification of the table and can be denoted as $POS(T, C, \underline{\beta}) = \bigcup_{d_i \in D} \underline{C}_\beta X_i$. We can take Table 1 as an example. If $\beta = 1/3$, $\underline{C}_\beta X_2$ contains $\{x_7, x_9, x_{10}, x_{11}\}$ and the β-positive region of Table 1, $POS(T, C, 1/3)$, contains $\{x_1, x_2, x_7, x_8, x_9, x_{10}, x_{11}\}$.

A β-certain reduct preserves the classification of each decision class in the original table under the admissible error β and is defined as

Definition 1. Let $T = (U, C, D)$ be a decision table, β is a user defined admissible error, $P \subseteq C$, and $P \neq \varnothing$. P is a β-certain reduct of table T such that

$$\forall \ D_i \in D, \underline{P}_\beta D_i = \underline{C}_\beta D_i \text{ and } \forall Q \subset P, Q \neq \varnothing, \ \underline{Q}_\beta D_i \neq \underline{C}_\beta D_i$$

Definition 1 shows β-certain reducts preserve the classification of each decision class in the β-positive region and this indicates all indiscernible categories in the β-positive region must retain the original classification and none of the indiscernible categories based on these reducts in the non-β-positive region can be accepted under β condition. If $\beta = 1/3$, original classification of each decision class in the β-positive region are $\underline{C}_\beta D_1 = \{x_1, x_8\}$, $\underline{C}_\beta D_2 = \{x_7, x_9, x_{10}, x_{11}\}$, and $\underline{C}_\beta D_3 = \{x_2\}$. The attribute set $\{c_1, c_2, c_3\}$ in Table 1 preserve the original classification of each decision class and no subsets of attribute from $\{c_1, c_2, c_3\}$ can preserve the original classification of each decision class in the β-positive region of Table 1. As a result, $\{c_1, c_2, c_3\}$ is a β-certain reduct of Table 1. Attribute set $\{c_3, c_4\}$ is another β-certain reduct of Table 1. In this paper, β criterion is used to represent the majority inclusion relation with the admissible error β for an indiscernible category and a two-phase β-certain reducts generation is developed. In the next section, the first phase of the β-certain reducts generation is presented to find the initial solutions of the β-certain reducts.

3 First Phase of the β-Certain Reducts Generation

The first phase of the β-certain reducts generation finds initial solutions for the β-certain reducts and these solutions can be derived from searching the certain reducts in a consistent pseudo decision table. Pseudo decision table is proposed by Wang [11] to replace the original table in generating certain reducts. Wang [11] applied a consistent pseudo decision table to replace the original decision table by converting the decision classes of all inconsistent categories in the original table to a new decision class. Certain reducts of the original table are generated by merging candidate attributes into an attribute set recursively until they reach all consistencies of the pseudo decision table with minimal number of condition attributes. The search of consistencies of the table is based on the sort and the generation of any certain reduct of consistent or inconsistent decision table takes $O(k^2 n \log n)$ [11]. The generation of certain reducts can be further improved if the hash data structure proposed by Wang [12] is applied. Two hash tables can be created from the hash-based pseudo decision table and the search of consistencies for any attribute set of the table takes $O(kn)$. The generation of any certain reduct based the hash data structure proposed by Wang [12] is reduced to $O(k^2 n)$.

Before we present the certain reducts generation in the consistent pseudo decision table, proposition 1 is proposed to show any subset of attributes can not be a β-certain reduct if indiscernible categories containing instances from the β-positive and non-β-positive region exist using the given subset of attributes.

Proposition 1. Let $T = (U, C, D)$ be a decision table, β is a user defined admissible error, $x \in U$, $P \subseteq C$, and $P \neq \emptyset$. Any subset of P can not be a β-certain reduct if $\exists I_P(x)$, $I_P(x) = \{X \cup Y \mid X \subseteq POS(T, C, \underline{\beta}), Y \subseteq \{U \setminus POS(T, C, \underline{\beta})\}, X \neq \emptyset, Y \neq \emptyset\}$

Proposition shows initial solutions of the β-certain reducts contain no indiscernible category having instances from both the β-positive and the non-β-positive region. The proof of proposition 1 is trivial. If an indiscernible category has instances from both the β-positive and the non-β-positive region, this category is either accepted or rejected under the β criterion and this will move instances from β-positive region to non-β-positive region or from non-β-positive region to β-positive region. This violates the definition of the β-certain reduct.

β-positive region of Table 1 contains $\{x_1, x_2, x_7, x_8, x_9, x_{10}, x_{11}\}$ when $\beta = 1/3$. The attribute set $\{c_1, c_2, c_4\}$ can not be an initial solution of the β-certain reducts because an indiscernible category, $\{x_1, x_3, x_4, x_5, x_6\}$, containing instances from both the β-positive and the non-β-positive region. Category, $\{x_1, x_3, x_4, x_5, x_6\}$, does not satisfy the β criterion and it moves instance x_1 from the β-positive to the non-β-positive region. Any subsets of $\{c_1, c_2, c_4\}$ can not be an initial solution of the β-certain reducts. To generate initial solutions from proposition 1, a consistent pseudo decision table is constructed from the original table by assigning all instances in the β-positive region to one decision class and assigning all other instances in the non-β-positive region to another decision class. Proposition 2 shows initial solutions of β-certain reducts of the original table are the certain reducts of the consistent pseudo decision table.

Proposition 2. Let $T = (U, C, D)$ be a decision table and $T_1 = (U, C, D')$ is its consistent pseudo decision table by assigning all instances in $POS(T, C, \beta)$ to one class and assigning all other instances in $\{U \setminus POS(T, C, \beta)\}$ to another class. Any certain reduct of T_1 is the initial solution of β-certain reducts of table T.

The correctness of the proposition 2 can be shown by proving any proper subset of the certain reduct of T_1 satisfy the proposition 1. We know this is true because any proper subset of certain reducts of T_1 causes inconsistencies and this shows indiscernible categories containing instances from both the β-positive and the non-β-positive region of table T exist. This proves proposition 2.

In generating certain reducts of T_1, we use the concept of the current rules size, **CRS**, proposed by Wang [11] and it can be applied to generate certain reducts of consistent and inconsistent tables. **CRS** excludes false candidates from generating certain reducts. Next, the definition of **CRS** and related propositions are introduced.

Definition 2. Let $T = (U, C, D)$ be a decision table and $P \subseteq C$. The current rules size of P, **CRS**(P), is derived by adding the size of the positive region, $|POS_P(D)|$, and the number of inconsistent categories using P where $POS_P(D)$ contains instances which are certainly classified using P. [11]

Proposition 3. Let $T = (U, C, D)$ be a consistent pseudo decision table and $P \subseteq C$. [11]

> (a) An attribute $a \in \{C\text{-}P\}$ can not be a candidate of the certain reducts starting with P if **CRS**$(P \cup \{a\}) = $ **CRS**(P).
>
> (b) If S is a set of candidate attributes derived from proposition 3.a for certain reducts starting with P and we merge an attribute $a \in S$ into P. New candidate attributes for certain reducts starting with $P \cup \{a\}$ can be found in S-$\{a\}$.
>
> (c) If attribute $a \in \{C\text{-}P\}$ satisfies proposition 3.a for certain reducts starting with P, we say a is not a candidate attribute of P if $\exists q \in P$ such that **CRS**$(\{P\text{-}\{q\}\} \cup \{a\})=$**CRS**$(P \cup \{a\})$.

Proposition 3 finds candidate attributes for a subset of attributes P in generating certain reducts and certain reducts generation using **CRS** finds reducts by merging candidate attributes into an attribute set until the **CRS** value is equal to the table size. Example of applying proposition 3 to generate certain reducts will be shown in section 4 and we will skip the algorithm in this paper. User can refer Wang's paper [11] for more details. Having generated initial solutions of the β-certain reducts, we will present the second phase of the β-certain reducts generation.

4 Second Phase of the β-Certain Reducts Generation

The second phase of β-certain reducts generation takes the initial solutions from the first phase and the second pseudo decision table is constructed using categories in the original β-positive region. After we create the second pseudo decision table, certain reducts of the second pseudo decision table are generated using proposition 3 by merging attributes into the initial solutions until all categories in the second pseudo decision table become consistent. Certain reducts of the second pseudo decision table are β-certain reducts if all sub categories based on these certain reducts in the non-β-positive region of the original table are totally rejected under the β criterion. Next, proposition is proposed to show the condition of certain reducts of the second pseudo decision table to become β-certain reducts of the original table.

Proposition 4. Let $T = (U, C, D)$ be a decision table. $T_2 = (U', C, D'')$ contains all instances in $POS(T, C, \underline{\beta})$ and T_2 is a consistent pseudo decision table by assigning all instances in $\underline{C}_\beta D_i$ to class i for each decision class. Certain reducts of T_2 are β-certain reducts of T if all sub categories based on these certain reducts in the non-β-positive region of T are totally rejected under β criterion.

The proof of proposition 4 is trivial. Certain reducts of T_2 contain minimal number of attributes to ensure the original classification of each decision class. Any subset of certain reducts of T_2 generates inconsistent categories which contain instances from different decision classes of the original classification of T. Under the majority inclusion relation in $VPRS$, some inconsistent categories will be moved to the non-β-positive region if they violate the β criterion or instances in some categories will change their original classification if categories satisfy the β criterion. From both cases, they violate the definition of the β-certain reduct. In order to say certain reducts of the second decision table are β-certain reducts, we have to make sure none of sub categories based on these certain reducts in the non-β-positive region of T are accepted under the β criterion. This proves proposition 4. To apply proposition 4 to generate all β-certain reducts, we have to show all β-certain reducts can be generated using proposition 4 and this is shown in proposition 5.

Proposition 5. Let $T = (U, C, D)$ be a decision table. T_2 is the consistent pseudo decision table of T generated from proposition 4. All β-certain reducts of T are subset of all certain reducts of T_2.

All β-certain reducts of T preserve the original classification of each decision class in the β-positive region of the original table. Proposition 5 is proved.

Table 2. The second pseudo decision table of Table 1

U	c_1	c_2	c_3	c_4	D
x_1	1	1	1	2	1
x_8	3	1	1	2	1
x_7	2	1	2	2	2
x_9	1	2	1	3	2
x_{10}	1	2	1	3	2
x_{11}	1	2	1	3	2
x_2	1	1	2	4	3

Table 3. Non-β-positive region of Table 1

U	c_1	c_2	c_3	c_4	D
x_3	1	1	3	2	1
x_4	1	1	3	2	1
x_5	1	1	3	2	2
x_6	1	1	3	2	2

An example of two-phase β-certain reducts generation is shown using Table 1. If $\beta=1/3$, attribute c_3 is the only initial solution from the first phase. The second consistent pseudo decision table of Table 1 is shown in Table 2 by assigning all instances in the category, $\{x_9, x_{10}, x_{11}\}$, in the β-positive to decision class 2. Initial solution $\{c_3\}$ is not a β-certain reduct of Table 1 because inconsistent categories, $\{x_1, x_8, x_9, x_{10}, x_{11}\}$ and $\{x_2, x_7\}$, exist in Table 2. From the definition 2, $\textbf{\textit{CRS}}(c_3)$ of Table 2 is 2. When proposition 3.a is applied to the initial solution $\{c_3\}$, three candidate attributes $\{c_1, c_2, c_4\}$ are found and the attribute set $\{c_3, c_4\}$ is a certain reduct of Table 2 because the value of $\textbf{\textit{CRS}}(\{c_3, c_4\})$ is equal to the size of Table 2 and attribute c_3 is indispensable. Next, attribute c_1 is merged into attribute c_3 and attribute c_2 is the only candidate for $\{c_3, c_1\}$ based on proposition 3.a for $\textbf{\textit{CRS}}(\{c_3, c_1\})=4$ and $\textbf{\textit{CRS}}(\{c_3, c_1, c_2\})=7$. Attribute set $\{c_3, c_1, c_2\}$ is another certain reduct of Table 2 because $\textbf{\textit{CRS}}(\{c_3, c_1, c_2\})$ is equal to the size of Table 2 and no attribute in $\{c_1\}$ satisfies proposition 3.c. Table 2 contains two certain reducts, $\{c_1, c_2, c_3\}$ and $\{c_3, c_4\}$. Certain reducts of Table 2 are applied to Table 3 which contains instances in the non-β-positive region of Table 1 and all sub categories based on the certain reducts of Table 2 are totally rejected under the β criterion. Two β-certain reducts, $\{c_1, c_2, c_3\}$ and $\{c_3, c_4\}$, are found for Table 1 with $\beta = 1/3$.

Next, the algorithm of the two-phase β-certain reducts generation is presented.

Algorithm. Two-phases β-certain reducts generation

'$T = (U, C, D)$ is a decision table

Construct the first and the second pseudo decision tables T_1 and T_2;

FOR each certain reduct, R, in T_1 DO

    IF R is not a β-certain reduct THEN

        FOR each certain reduct, S, of T_2 generated by merging attributes from $\{C\text{-}R\}$ into R using proposition 3 in T_2 DO

            IF all sub categories based on S in the non-β-positive region of T are total rejected under the β criterion THEN

                S is a β-certain reduct;

            ELSE

                S is not a β-certain reduct;

            END

        END

    END

END

---

5 Conclusion

In this paper, a two-phase β-certain reducts generation algorithm is presented. In the first phase, we find certain reducts of the first pseudo decision table and set them as initial solutions of β-certain reducts. The second phase takes initial solutions from the first phase and β-certain reducts are found by generating certain reducts of the second pseudo decision when all sub categories based on these certain reducts in the non-β-positive region are totally rejected under the β criterion. No verification is needed when β-certain reducts are found.

References

1. Bazan, J.G., Nguyen, H.S., Nguyen, S.H., Synak, P., Wroblewski, J.: Rough set algorithms in classification problems. In: Polkowski, L., Tsumoto, S., Lin, T.Y. (eds.): Rough Set Methods and Applications: New Developments in Knowledge Discovery in Information Systems, Physical-Verlag, Heidelberg (2000) 49-88.
2. Beynon, M.: Reducts with the variable precision rough set model: A further investigation. European Journal of Operating Research.134 (2001) 592-605.
3. Beynon, M.: The introduction and utilization of (l, u)-graphs in the extended variable precision rough set model. International Journal of Intelligent System. 18 (2003) 1035-1055.
4. Inuiguchi, M.: Attribute Reduction in Variable Precision Rough Set Model. International Journal of Uncertainty, Fuzziness and Knowledge-Based Systems. 14 (4) (2006) 461-479.
5. Komorowski, J., Pawlak, Z., Polkowski, L., Skowron, A.: Rough sets: a tutorial. In: Pal S. K., Skowron, A. (eds.): Rough Fuzzy Hybridization: A New Trend in Decision-Making. Springer-Verlag, Telos (1999) 3-98.
6. Kryszkiewicz, M.: Maintenance of reducts in the variable precision rough set model. ICS Research Report. Warsaw University of Technology (June 1994) 31/94.
7. Nguyen, S.H., Nguyen, H.S.: Some efficient algorithms for rough set methods. In: Proc. of Information Processing and Management of Uncertainty in Knowledge-Based Systems. (1996) 1451-1456.
8. Pawlak, Z.: Rough sets. International Journal of Computer and Information Sciences. 11 (1982) 341-356.
9. Pawlak, Z.: Rough Sets: Theoretical Aspects of Reasoning about Data. Kluwer, Netherlands (1991).
10. Skowron, A., Rauszer, C.: The discernibility matrices and functions in information Systems. Fundamenta Informaticae 15 (2) (1991) 331-362.
11. Wang, P.C.: Highly scalable rough set reducts generation. Journal of Information Science and Engineering 23 (4) (2007) (to appear).
12. Wang, P.C.: Monotonic reducts generation using fbHash in Rough Set. (submitted to Data and Knowledge Engineering).
13. Ziarko, W.: Variable precision rough set model. Journal of Computer and System Sciences 46 (1) (1993) 39-59.

Formal Concept Analysis and Set-Valued Information Systems

Xiao-Xue Song[1,2] and Wen-Xiu Zhang[1]

[1] Institute for Information and System Sciences, Faculty of Science,
Xi'an Jiaotong University, Xi'an, Shaan'xi, P.R. China
[2] Department of Computer, Xianyang Normal College, Xianyang,
Shaan'xi, P.R. China
sxx1669@163.com
wxzhang@mail.xjtu.edu.cn

Abstract. This paper discusses the relationship between formal contexts and set-valued information systems, and indicates that each formal context can be transformed into a set-valued information system. The extensions of the object concepts in a formal context are equivalent to the dominance classes in the set-valued information system induced by the formal context. It is proved that the granular reduct in a formal context is equivalent to the attribute reduct in the set-valued information system. And then a comparative study of the two kinds of reduct in a formal context is presented. Finally the characteristics of three types of granular attributes in a formal context are analyzed.

Keywords: Formal contexts, set-valued information systems, granular reduct, attribute reduct, attribute characteristics.

1 Introduction

Rough set theory (RST) is proposed by Pawlak Z. in 1982 [1]. It provides a new mathematical approach to deal with inexact, uncertain or vague knowledge. The rough set theory that based on the conventional indiscernibility relation is not useful for analyzing incomplete information. By an incomplete information system (IIS), we mean a system with unknown data or partly-known data. In the real world, many information systems (IS) are incomplete. In this situation, some attributes values may be subsets of attributes domain. This kind of system can be regarded as a set-valued information system (a set-valued IS).

Formal concept analysis(FCA) is proposed by Wille R. in the same year [2]. The basis of FCA are formal concepts and concept lattices. A concept lattice is an ordered hierarchical structure of formal concepts which are induced by a binary relation between a set of objects and a set of attributes.

Although RST and FCA are different theories, they have much in common, in terms of both goals and methodologies. Many efforts have been made to compare and combine the two theories [3-8]. Düntsch and Gediga study various forms of set approximations via the modal-style operators [3]. Saquer and Deogun present

J.T. Yao et al. (Eds.): RSKT 2007, LNAI 4481, pp. 395–402, 2007.
© Springer-Verlag Berlin Heidelberg 2007

a novel approach for approximating concepts in the framework of formal concept analysis [4]. Yao introduces the notion of rough set approximations into formal concept analysis, and presents a comparative study of rough set theory and formal concept analysis [5,6]. Wolff studies the differences between the "partition oriented" rough set theory and the "order oriented" formal concept analysis [7].

In this paper, we discuss the relationship between formal contexts and set-valued information systems, and show that each formal context can be transformed into a set-valued IS. The relationship between the extensions of the object concepts in a formal context and the dominance classes in the set-valued IS induced by the formal context is investigated, and then proved that the granular reduct in a formal context is equivalent to the attribute reduct in the set-valued IS induced by the formal context. The relationship between the two kinds of reduct in a formal context is studied. The characteristics of three types of granular attributes in a formal context are analyzed.

2   Preliminaries

First, to make this paper self-contained, the involved notions both in FCA and RST are introduced (see[8-10]).

2.1   Basic Definitions in Formal Concept Analysis

A *formal context* is a triplet (U, A, I), where $U = \{x_1, x_2, \cdots, x_n\}$ is a nonempty finite set of objects and $A = \{a_1, a_2, \cdots, a_m\}$ is a nonempty finite set of attributes, and I is a relation between U and A, which is a subset of the Cartesian Product $U \times A$, $(x, a) \in I$ means that object x has attribute a. In this paper, $(x, a) \in I$ is denoted by 1, and $(x, a) \notin I$ is denoted by 0. This, a formal context can be represented by a table only with 0 and 1.

For $X \subseteq U$, $B \subseteq A$, we define a pair of dual operators (see [2]):

$$X^* = \{a \in A \mid \forall x \in X, \ (x, a) \in I\},$$

$$B^* = \{x \in U \mid \forall a \in B, \ (x, a) \in I\}.$$

A pair (X, B), $X \subseteq U, B \subseteq A$, is called *formal concept* of the context (U, A, I) if $X^* = B, B^* = X$. X is called the *extension* and B is called the *intension* of the formal concept (X, B). For an object $x \in U$, (x^{**}, x^*) is called the *object concept*. Correspondingly, for an attribute $a \in A$, (a^*, a^{**}) is called the *attribute concept*.

The set of all formal concepts forms a complete lattice called *a concept lattice* and is denoted by $L(U, A, I)$.

For $x \in U$ and $a \in A$, we denote $\{x\}^* = x^*$ and $\{a\}^* = a^*$.

The basic extension set $\{x^{**} \mid x \in U\}$ is called *the granules* of the concept lattice $L(U, A, I)$. Obviously, $\{x^{**} \mid x \in U\}$ forms a cover of the set of objects U.

Definition 1. Let (U, A, I) be a formal context. For any $C \subseteq A$, we can obtain a formal context (U, C, I_C) which is called a *sub-context* of (U, A, I), where $I_C = I \cap (U \times C)$.

For $X \subseteq U$, $B \subseteq C$,

$$X^{*C} = \{a \in C | \ \forall x \in X, \ (x, a) \in I\},$$

$$B^{*C} = \{x \in U | \ \forall a \in B, \ (x, a) \in I\}.$$

Clearly, $X^{*C} = X^* \cap C$ and $X^{*A} = X^*$.(see [10])

Definition 2. Let (U, A, I) be a formal context. An attribute subset $B \subseteq A$ is referred to as a *granular consistent set* of (U, A, I) if $x^{*B*B} = x^{*A*A}$ for all $x \in U$. And if for any $E \subset B$, E is not a granular consistent set of (U, A, I), B is referred to as a *granular reduct* in (U, A, I). (see [10])

Granular reduct in a formal context is the smallest attribute subset that preserves the granules of the concept lattice.

Let (U, A, I) be a formal context, $x \in U$, and $a \in A$. We denote

$$x_a^* = \begin{cases} \{a\}, & (x, a) \in I, \\ \emptyset, & otherwise. \end{cases}$$

And for $B \subseteq A$, note

$$R_B^* = \{(x, y) \in U \times U | x_a^* \subseteq y_a^*, \forall a \in B\}.$$

Let (U, A, I) be a formal context and $(x, y) \in U \times U$. Define

$$M^*(x, y) = \{a \in A | (x, y) \notin R_a^*\}.$$

$M^*(x, y)$ is referred to as *the granular discernibility attribute set* of x and y in (U, A, I) and $\mathcal{M}^* = \{M^*(x, y) | (x, y) \in U \times U\}$ is called *the granular discernibility matrix* of (U, A, I). (see [10])

2.2 Set-Valued Information Systems

A set-valued information system (see [11]) is a triple (U, A, F), where $U = \{x_1, x_2, \cdots, x_n\}$ is a nonempty finite set of objects called universe and $A = \{a_1, a_2, \cdots, a_m\}$ is a nonempty finite set of attributes,

$$F = \{f_a | a \in A\},$$

where $f_a : U \longrightarrow \mathcal{P}_0(V_a)(a \in A)$ is an attribute value function, V_a is a domain of attribute a, $\mathcal{P}_0(V_a)$ is the whole nonempty subsets of V_a.

Let (U, A, F) is a set-valued IS, $\forall B \subseteq A$, define a relation as follow:

$$R_B^{\subseteq} = \{(x_i, x_j) \in U \times U | f_a(x_i) \subseteq f_a(x_j) \ (\forall a \in B)\}.$$

The relation is called a *dominance relation* ,and note

$$[x_i]_B^{\subseteq} = \{x_j \in U | (x_i, x_j) \in R_B^{\subseteq}\} = \{x_j \in U | f_a(x_i) \subseteq f_a(x_j) \ (\forall a \in B)\}$$

called *dominance class*.

Definition 3. Let (U, A, F) be a set-valued IS. A set $B \subseteq A$ is an *attribute consistent set* of (U, A, F) if $R_{\overline{B}}^{\subseteq} = R_{\overline{A}}^{\subseteq}$. If B is an attribute consistent set, and no proper subset of B is consistent, then B is referred to as an *attribute reduct* in (U, A, F).

Denote

$$D(x_i, x_j) = \{a \in A | f_a(x_i) \not\subseteq f_a(x_j)\} \ (x_i, x_j \in U).$$

$D(x_i, x_j)$ is called *the discernibility attribute set* with respect to the relation $R_{\overline{A}}^{\subseteq}$, $\mathcal{D} = \{D(x_i, x_j) | (x_i, x_j) \in U \times U\}$ is called *the attribute discernibility matrix* with respect to $R_{\overline{A}}^{\subseteq}$. (see [8])

3   The Set-Valued Information System Induced by a Formal Context

Let (U, A, I) be a formal context, define a map as follow:

$$f_a(x) = \begin{cases} \{0, 1\}, & (x, a) \in I, \\ \{0\}, & (x, a) \notin I. \end{cases}$$

Then a set-valued IS (U, A, F) can be obtained from the formal context (U, A, I). (U, A, F) is called *the set-valued information system induced by* (U, A, I). Obviously, R_B^* is equivalent to $R_{\overline{B}}^{\subseteq}$.

Example 1. A formal context (U, A, I) is shown as Table 1, the set-valued IS (U, A, F) induced by (U, A, I) is shown as Table 2.

Table 1. A Formal Context (U, A, I)

	a	b	c	d	e
x_1	1	0	1	1	1
x_2	1	0	1	0	0
x_3	0	1	0	0	1
x_4	0	1	0	0	1
x_5	1	0	0	0	0
x_6	1	1	0	0	1

Proposition 1. *Let* (U, A, I) *be a formal context. Then*

$$y \in x^{*A*A} \text{ iff } y^{*A} \supseteq x^{*A} \ (x, y \in U).$$

Proposition 2. *Let* (U, A, I) *be a formal context,* (U, A, F) *is the set-valued information system induced by* (U, A, I). *Then* $x^{*A*A} = [x]_{\overline{A}}^{\subseteq} (x \in U)$.

Table 2. The Set-Valued IS Induced By (U, A, I)

	a	b	c	d	e
x_1	$\{0,1\}$	$\{0\}$	$\{0,1\}$	$\{0,1\}$	$\{0,1\}$
x_2	$\{0,1\}$	$\{0\}$	$\{0,1\}$	$\{0\}$	$\{0\}$
x_3	$\{0\}$	$\{0,1\}$	$\{0\}$	$\{0\}$	$\{0,1\}$
x_4	$\{0\}$	$\{0,1\}$	$\{0\}$	$\{0\}$	$\{0,1\}$
x_5	$\{0,1\}$	$\{0\}$	$\{0\}$	$\{0\}$	$\{0\}$
x_6	$\{0,1\}$	$\{0,1\}$	$\{0\}$	$\{0\}$	$\{0,1\}$

Remark. From Proposition 2 we can immediately conclude that Definition 3 and Definition 2 are equivalent, i.e. the granular consistent set of a formal context is equivalent to the attribute consistent set of the set-valued IS induced by the formal context.

Corollary 1. *Let* (U, A, I) *be a formal context,* $B \subseteq A$. *If* $R_B^* = R_A^*$, *then* $x^{*B*B} = x^{*A*A}$.

Corollary 2. *Let* (U, A, I) *be a formal context, then*
  (1) *if* $B_1 \subseteq B_2 \subseteq A$, $R_{B_1}^* \supseteq R_{B_2}^* \supseteq R_A^*$.
  (2) *if* $B_1 \subseteq B_2 \subseteq A$, $x^{*B_1*B_1} \supseteq x^{*B_2*B_2} \supseteq R^{*A*A}$.
  (3) $\mathcal{I} = \{x^{*B*B} | x \in U\}$ *constitute a covering of* U $(B \subseteq A)$.

4   A Comparative Study of Two Kinds of Reduct in a Formal Context

This section gives the relationship between the granular reduct and the attribute reduct in a formal context. We first introduce the definition of the attribute reduct in a formal context (see [9]).

Let (U, A_1, I_1) and (U, A_2, I_2) be two formal context. We say (U, A_2, I_2) is *extension coarser* than (U, A_1, I_1) if for any $(X, B) \in L(U, A_2, I_2)$, there exists $(X, B') \in L(U, A_1, I_1)$ and denoted by

$$L(U, A_1, I_1) \leq L(U, A_2, I_2).$$

If $L(U, A_1, I_1) \leq L(U, A_2, I_2)$ and $L(U, A_2, I_2) \leq L(U, A_1, I_1)$, we say that $L(U, A_1, I_1)$ and $L(U, A_2, I_2)$ are *isomorphic*, and denoted by

$$L(U, A_1, I_1) \cong L(U, A_2, I_2).$$

Definition 4. Let (U, A, I) be a formal context. A set $B \subseteq A$ is an *attribute consistent set* of (U, A, I) if $L(U, B, I_B) \cong L(U, A, I)$. If B is an attribute consistent set of (U, A, I), and no proper subset of B is consistent, then B is referred to as an *attribute reduct* in (U, A, I).

An attribute reduct in a formal context is a minimal attribute subset preserving immovable of the structure of the concept lattice. That is the same lattice structure can be obtained through a reduced attribute.

Let (U, A, I) be a formal context, $(X_i, B_i), (X_j, B_j) \in L(U, A, I)$. The *discernibility attributes set* between $(X_i, B_i), (X_j, B_j)$ are defined by

$$DIS((X_i, B_i), (X_j, B_j)) = B_i \cup B_j - B_i \cap B_j.$$

and

$$\mathcal{D} = (DIS((X_i, B_i), (X_j, B_j)), (X_i, B_i), (X_j, B_j) \in L(U, A, I))$$

is called the *discernibility matrix* of the context.

Definition 5. Let (U, A, I) be a formal context. We define (U, A, I^c) as follow: for any $x \in U$ and $a \in A$, if $(x, a) \notin I$, then $(x, a) \in I^c$. (U, A, I^c) is called the *complement formal context* of (U, A, I).

Proposition 3. *Let (U, A, I) be a formal context, (U, A, I^c) be the complement formal context of (U, A, I). Then the granular reduct in (U, A, I) is equivalent to the granular reduct in (U, A, I^c).*

Proposition 4. *Let (U, A, I) be a formal context, (U, A, I^c) be the complement formal context of (U, A, I). Then the attribute reducts in (U, A, I) and (U, A, I^c) are all the granular reducts in (U, A, I).*

Example 2. Table 3 shows the complement formal context (U, A, I^c) of the formal context (U, A, I) described by Table 1, Table 4 gives the discernibility matrix of the set-valued IS induced by the formal context (U, A, I).

Table 3. The complement formal context (U, A, I^c) of the Table 1

	a	b	c	d	e
x_1	0	1	0	0	0
x_2	0	1	0	1	1
x_3	1	0	1	1	0
x_4	1	0	1	1	0
x_5	0	1	1	1	1
x_6	0	0	1	1	0

From Table 4, we can obtain $B_1 = \{a, b, c, d\}$ and $B_2 = \{a, b, c, e\}$ are reducts in the set-valued IS presented by Table 2. So they are also the granular reducts in the formal context (U, A, I) presented by Table 1. The conclusion is consistent with the conclusion obtained in [10].

Analogously, we can obtain that $B_1 = \{a, b, c, d\}$ and $B_2 = \{a, b, c, e\}$ are also the granular reducts in the complement formal context (U, A, I^c).

We can also obtain that the attribute reduct in formal context (U, A, I) presented by Table 1 is $B_1 = \{a, b, c, d\}$, and the attribute reduct in the complement formal context (U, A, I^c) presented by Table 3 is $B_2 = \{a, b, c, e\}$.

Table 4. The discernibility matrix of set-valued IS induced by (U, A, I)

$x \backslash y$	x_1	x_2	x_3	x_4	x_5	x_6
x_1		de	acd	acd	cde	cd
x_2			ac	ac	c	c
x_3	b	be			be	
x_4	b	be			be	
x_5			a	a		
x_6	b	be	a	a	be	

5 The Granular Attributes Characteristics in a Formal Context

Let $\mathcal{B} = \{B_k | k \leq l\}$ is the set of all granular reducts in a formal context (U, A, I), denote

$$C = \bigcap_{k \leq l} B_k, K = \bigcup_{k \leq l} B_k - C, J = U - (K \cup C),$$

C is called the set of *granular core attribute* in (U, A, I), K is called the set of *granular relative indispensable attribute* in (U, A, I), J is called the set of *granular dispensable attribute* in (U, A, I).

Proposition 5. *Let (U, A, I) is a formal context, then the follow conclusions are equivalent*:

    (1) a is granular core attribute;

    (2) $\exists x, y \in U, M^*(x, y) = \{a\}$;

    (3) $R^*_{A-\{a\}} \not\subseteq R^*_A$;

    (4) $x^{*(A-\{a\})*(A-\{a\})} \not\subseteq x^{*A*A}$.

Proposition 6. *Let (U, A, I) is a formal context, then the follow conclusions are equivalent*:

    (1) a is a granular dispensable attribute;

    (2) $R^*(a) \subseteq R^*_a$, where $R^*(a) = \cup \{R^*_{B-\{a\}} | R^*_B \subseteq R^*_A, B \subseteq A\}$;

    (3) $\cup \{x^{*(B-\{a\})*(B-\{a\})} | \ x^{*B*B} \subseteq x^{*A*A}, B \subseteq A\} \subseteq x^{*a*a}$.

Proposition 7. *Let (U, A, I) is a formal context, then*

    (1) a is granular core attribute iff $R^*_{A-\{a\}} \not\subseteq R^*_A$;

    (2) a is granular relative indispensable attribute iff $R^*_{A-\{a\}} \subseteq R^*_A$, and $R^*(a) \not\subseteq R^*_a$;

    (3) a is granular dispensable attribute iff $R^*(a) \subseteq R^*_a$.

6 Conclusion

Formal concept analysis (FCA) and Rough set theory (RST) are two different methods for knowledge representation and knowledge discovery. They can be studied in a common framework. In this paper, the relationship of a formal context and a set-valued IS is investigated. It is showed that each formal context can be transformed into a set-valued IS, and then proved that the granular reduct in a formal context is equivalent to the attribute reduct in the set-valued IS induced by the formal context. A comparative study of two kinds of reduct in a formal context is discussed.

Acknowledgment. The paper was supported by the National 973 Program of China(no.2003CB312206).

References

1. Pawlak, Z.: Rough Sets: Theoretical Aspects of Reasoning about Data. Boston: Kluwer Academic Publishers(1991)
2. Ganter,B., Wille,R.: Formal concept analysis, Mathematical Foundations, Springer, Belin (1999)
3. Düntsch, I., Gediga, G.: Modal-style operators in qualitative data analysis. Proceedings of the 2002 IEEE International Conference on Data Mining, TARSKI,(2002)155-162
4. Saquer, J., Deogun, J.S.: Formal rough concept analysis. In: N. Zhong, A. Skowron, S. Ohsuga (Eds.): RSFDGrC99, LNAI 1711, Springer, Berlin(1999)91-99
5. Yao, Y.Y.: A comparative study of formal concept analysis and rough set theory in Data analysis. In: Tsumoto, S. et al. (Eds.): RSCTC 2004, Lecture Notes in Computer Science 3066. Springer, Berlin (2004) 59-68
6. Yao, Y.Y., Chen, Y.H.: Rough set approximation in formal concept analysis. In: Dick, S., Kurgan, L., Pedrycz, W., Reformat, M. (eds.): Proceedings of 2004 Annual Meeting of the North American Fuzzy Information Processing Society. IEEE (2004)73-78
7. Wolff, K.E.: A comceptual view of knowledge bases in rough set theory. In: W. Ziarko and Y. Yao (Eds.): RSCTC 2000, LNAI 2005, Springer, Berlin(2001) 220-228
8. Zhang, W.X., Yao, Y.Y., Leung, Y.(Eds.): Rough Sets and Concept Lattices. Xi'an Jiaotong University Press, Xi'an (2006)(in chinese)
9. Zhang, W.X., Wei, L., Qi, J.J.: Attribute Reduction Theory and Approach to Concept Lattice. Science in China Ser.F Information Sciences (2005) 713-726
10. Wu, W.Z.: knowledge Reduction in Formal Contexts.(Submitted to Approximation Reasoning)
11. Zhang, W.X. Leung, Y., Wu, W.Z.(Eds.): Information Systems and Knowledge Discovery. Science Press, Beijing (2003)(in chinese)

Descriptors and Templates in Relational Information Systems

Piotr Synak[1,2] and Dominik Ślęzak[2]

[1] Polish-Japanese Institute of Information Technology
Koszykowa 86, 02-008 Warsaw, Poland
[2] Infobright Inc.
218 Adelaide St. W, Toronto, ON, M5H 1W8 Canada
synak@pjwstk.edu.pl, slezak@infobright.com

Abstract. We discuss descriptors and templates within relational information systems. We provide examples how to use attribute-specific similarity measures to equip standard information systems with relational structures. We introduce new types of descriptors, which take an advantage of available relations, forming cliques or stars of interrelated attribute values. We also show how to build relational templates from collections of such relational descriptors. We illustrate the proposed framework with possible applications.

Keywords: Descriptors, Templates, Relational Information Systems.

1 Introduction

The notion of a template occurs frequently in the data mining literature, to describe multi-feature regularities that are highly supported across the whole data or specific for particular data groups [4,5,7]. Within the framework of rough sets and information systems [8,14], templates are understood as conjunctions of descriptors built over particular attributes [11,12]. It should be then obvious that specification of a descriptor should depend on the attribute's data type, possibly with different types of descriptors involved into the same template. By now, however, there is no significant research on this topic.

One of possibilities to handle different data types within a unified framework is to equip attributes with relations over their domains, specific for real values, multi-sets, attributes with missing values etc. [1,6,18,19,20]. It leads to relational information systems [3] and to the need of extending basic concepts of rough sets and data mining to deal with different kinds of relations.

In [21], we extended the notion of a template towards dealing with tolerance relations in data. We basically noticed that, although tolerances (similarities) have been already widely studied for information systems [9,17,24], nobody considered them in the context of efficient representation and search of templates. Tolerances, however, are not sufficient to model many real-life data challenges, e.g., those related to temporal or preference-based attributes [19]. In this paper, we provide a more general framework for defining descriptors and templates within systems with different types of relations.

J.T. Yao et al. (Eds.): RSKT 2007, LNAI 4481, pp. 403–410, 2007.

2   Information Systems and Templates

In rough set theory [13,14,15], data is represented by *information system* $\mathbb{A} = (U, A)$, with the universe U of *objects* and the *attribute set* A. Each $a \in A$ is a function $a : U \rightarrow V_a$. V_a is the *value set* of a. Each $B \subseteq A$ induces *indiscernibility relation* $IND(B)$. We say that $x, y \in U$ are *indiscernible* with respect to B, if and only if $a(x) = a(y)$ for each $a \in B$. For each $x \in U$, its attribute values form an *elementary pattern* (*information signature*) $Inf_B(x) = \{(a, a(x)) : a \in B\}$. Hence, $(x, y) \in IND(B) \Leftrightarrow Inf_B(x) = Inf_B(y)$. By $INF(A) = \{Inf_B(x) : x \in U, B \subseteq A\}$ we denote the set of all signatures occurring in $\mathbb{A}$.

One of the tasks of data mining is searching for patterns in data [4,5,7]. We consider one kind of such patterns – *templates*, which are propositional formulas $T = \bigwedge(a_i = v_i)$, where $a_i \in A$, $a_i \neq a_j$ for $i \neq j$, and $v_i \in V_{a_i}$ [11,12]. Assuming $A = \{a_1, \ldots, a_m\}$, $card(A) = m$, one can represent any template

$$T = (a_{i_1} = v_{i_1}) \wedge \ldots \wedge (a_{i_k} = v_{i_k}) \tag{1}$$

by m-dimensional vector, where each p-th position is either v_p, if $p = i_1 \ldots i_k$, or "*" (don't care symbol) otherwise. We say that $x \in U$ *satisfies* the descriptor $(a = v)$ if $a(x) = v$. Further, x satisfies (*matches*) the template T if it satisfies all the descriptors of T. By $\|T\|_{\mathbb{A}}$ we denote the T's semantics, i.e., the set of all $x \in U$ that satisfy T. By $length(T)$ we denote the number of different descriptors $(a = v)$ occurring in T. By $supp_{\mathbb{A}}(T) = card(\|T\|_{\mathbb{A}})$, we denote T's *support*. The set of all templates having non-empty support in $\mathbb{A}$ is in one-to-one correspondence with the set of signatures $INF(A)$.

The above ideas can be extended onto *generalized templates* $T = \bigwedge(a_i \in V_i)$, where $V_i \subseteq V_{a_i}$. As before, we can write:

$$T = (a_{i_1} \in V_{i_1}) \wedge \ldots \wedge (a_{i_k} \in V_{i_k}). \tag{2}$$

The difference here is that we consider many-valued descriptors. We say that $x \in U$ satisfies a generalized descriptor $(a \in V)$ if $a(x) \in V$. As before, x satisfies the generalized template T if it satisfies its descriptors.

Length and support of generalized templates are defined in the same way. The quality, however, requires more study. In the case of simple templates T, the quality is a function of support and length. For example, if we search for strong patterns in the whole information system, it could be the product of $supp_{\mathbb{A}}(T)$ and $length(T)$. In the case of generalized templates, the quality function should additionally take into account *precision* of particular descriptors. Intuitively, by a precise descriptor $(a \in V)$ we mean the one with relatively low cardinality of V. Formally, precision $s_{\mathbb{A}}$ of descriptor $(a \in V)$ in $\mathbb{A}$ is defined as follows:

$$s_{\mathbb{A}}((a \in V)) = \begin{cases} \frac{card(V_a) - card(V)}{card(V_a) - 1} & card(V_a) \neq 1 \\ 1 & card(V_a) = 1 \end{cases} \tag{3}$$

Consequently, precision of generalized template T is defined as

$$S_{\mathbb{A}}(T) = \sum_{(a \in V) \in T} s_{\mathbb{A}}((a \in V)) \tag{4}$$

If all descriptors in T are fully precise, then $S_{\mathbb{A}}(T)$ equals to its length. Generally, we can consider quality of T as a function of its support and precision. Often, after identifying the objects satisfying T, it turns out that some of T's descriptors $(a \in V)$ can be *precisiated*, i.e. replaced by $(a \in V')$, $V' \subsetneq V$, without decreasing T's support. Generalized templates with non-empty support, fully precisiated in the above sense, are in one-to-one correspondence with *generalized signatures* – the elements of $INF^*(A) = \{Inf_B(X) : X \subseteq U, B \subseteq A\}$, where $Inf_B(X) = \{(a, a(X)) : a \in B\}$ and $a(X) = \{a(x) : x \in X\}$.

3   Similarities and Relational Information Systems

If the only information about objects is their signature, we naturally use indiscernibility relation to deal with. Moreover, as indiscernibility is an equivalence relation, it enables to easily operate with disjoint classes of indiscernible objects. However, in many cases this approach can be insufficient to deal with data, for example, when we have real-valued attributes and objects tend to potentially differ from the others. The indiscernibility classes could be then too specific, few-element sets, disabling efficient data-based knowledge representation.

One of solutions is based on *similarity measures* $\delta_a : V_a \times V_a \to [0,1]$, $a \in A$. The ranges could obviously differ – we normalize them to $[0,1]$ for simplicity. Different attributes can be equipped with different measures. We can consider, e.g., distances between values. We can define similarities for fuzzy sets, multi-sets, ordered domains, and other complex attributes [1,2,3,19].

Once similarity measures are defined for particular attributes, there are various possibilities to extend indiscernibility. We list a few examples (cf. [12]).

Example 1. For $B = \{a_1, ..., a_k\}$ and functions δ_i, $i = 1, ..., k$, we put:

1. $(x, y) \in r_B^{(1)}(w_0) \Leftrightarrow \forall_{i=1,...,k} \delta_i(a_i(x), a_i(y)) \geq w_0$
2. $(x, y) \in r_B^{(2)}(w_1, \ldots, w_k) \Leftrightarrow \forall_{i=1,...,k} \delta_i(a_i(x), a_i(y)) \geq w_i$
3. $(x, y) \in r_B^{(3)}(w_0) \Leftrightarrow \prod_{i=1}^{k} \delta_i(a_i(x), a_i(y)) \geq w_0$
4. $(x, y) \in r_B^{(4)}(w_0, w_1, \ldots, w_k) \Leftrightarrow \prod_{i=1}^{k} \delta_i(a_i(x), a_i(y))^{w_i} \geq w_0$

For $w_0, w_1, ..., w_k \in [0,1)$ (again we normalize for simplicity [12]), if we assume that $\delta_i(v, v) = 1$ and $\delta_i(v_1, v_2) = \delta_i(v_2, v_1)$, then all relations in Example 1 are *tolerances* over the universe U, i.e., they are reflexive, symmetric, but not necessarily transitive. The rough set approach can be, obviously, extended onto any type of binary relation [9,18,23,24], not necessarily tolerance. (For example, similarity does not need to be symmetric, i.e., $\delta_i(v_1, v_2) \neq \delta_i(v_2, v_1)$.) On the other hand, tolerances are certainly most popular relations applied to generalize indiscernibility [10,12,17,19].

Relations can be defined for any $B \subseteq A$ at the level of objects (like we did in examples above), their signatures, or even whole sets of (generalized) signatures. In any case, we would like such relations to satisfy some properties, e.g., the following *monotonicity*, here formulated for relations defined over objects:

$$\forall_{B,C \subseteq A} \forall_{x,y \in U} B \subseteq C \wedge (x, y) \in r_C \Rightarrow (x, y) \in r_B. \tag{5}$$

Such property is important because we usually assume that if objects are indiscernible with respect to a larger set of attributes, then they remain indiscernible also after removing some attributes. Let us note that all relations in Example 1 satisfy (5) for normalized similarities and weights. However, we do not even need to assume symmetry of similarities here.

Instead of constructing multi-attribute relations from local, single-attribute similarity measures, we can also define single-attribute relations first and then induce the multi-attribute ones automatically as their conjunctions, i.e.:

$$(x, y) \in r_B \Leftrightarrow \forall_{a \in B}(a(x), a(y)) \in r_a. \tag{6}$$

Certainly, such an approach has less expressive power in defining relations.

Example 2. The first two relations in Example 1 can be constructed as follows:

1. $(x, y) \in r_B^{(1)}(w_0) \Leftrightarrow \forall_{i=1,\ldots,k}(x, y) \in r_i(w_0)$
2. $(x, y) \in r_B^{(2)}(w_1, \ldots, w_k) \Leftrightarrow \forall_{i=1,\ldots,k}(x, y) \in r_i(w_i)$

where, for any $a_i \in B$, we put $(x, y) \in r_i(w) \Leftrightarrow \delta_i(a_i(x), a_i(y)) \geq w$. However, the last two relations in Example 1 cannot be built in this way.

On the other hand, formula (6) enables us to control better particular attributes and descriptors, and remains far more straightforward extension of the notion of indiscernibility. Any family of relations r_B, $B \subseteq A$ defined by (6) satisfies monotonicity property (5). In particular, if relations r_a are tolerances over V_a, then r_B defined by (6) is always a tolerance over U.

Consequently, from now on, we consider *relational information systems* $\mathbb{A} = (U, A, R)$, where $R = \{r_a : a \in A, r_a \subseteq V_a \times V_a\}$ is the set of binary relations corresponding to attributes $a \in A$. Let us note, that we do not require any $r_a \in R$ to meet additional criteria, even reflexivity. Nor, r_a is required to be related with any other relation – it is related only with particular attribute and its domain. In the simplest case, each r_a can be an indiscernibility relation. In general, however, each $a \in A$ can be equipped with r_a of different properties. For example, in the context of missing value analysis there are considered binary relations which are reflexive only, transitive only, or reflexive and transitive [3,6,20]. Relation r_a can be also a preference order relation [19].

4   Templates in Relational Information Systems

Let us now present, how to extend the notion of (generalized) template to the relational case. Additional relational structures enable to form a wide range of types of patterns to be extracted from data, reflecting the nature of particular attributes. Let $\mathbb{A} = (U, A, R)$ and $B \subseteq A$, $card(B) = k$, be given. We define a *relational template* over $\mathbb{A}$ as a conjunction of *relational descriptors*, i.e.:

$$T = (a_{i_1}, V_{i_1})^{(\#)} \wedge \ldots \wedge (a_{i_k}, V_{i_k})^{(\#)}, \tag{7}$$

where $\#$ determines the type of descriptor and it may be actually different for different attributes in B. Let us list some examples of interpretations of $\#$:

- $(a, V)^{(0)}$ denotes a simple descriptor $(a = v)$, i.e., $V = \{v\}$
- $(a, V)^{(1)}$ denotes a generalized descriptor $(a \in V)$
- $(a, V)^{(2)}$ denotes a *strong clique*, i.e., $\forall_{v, v' \in V}(v, v') \in r_a \land (v', v) \in r_a$
- $(a, V)^{(2')}$ denotes a *weak clique*, i.e., $\forall_{v, v' \in V}(v, v') \in r_a \lor (v', v) \in r_a$
- $(a, V)^{(3)}$ denotes a *strong star*, i.e., $\exists_{v \in V}\forall_{v' \in V}(v, v') \in r_a \land (v', v) \in r_a$
- $(a, V)^{(3')}$ denotes a *weak star*, i.e., $\exists_{v \in V}\forall_{v' \in V}(v, v') \in r_a \lor (v', v) \in r_a$
- $(a, V)^{(3'')}$ denotes a *directed star*, i.e., $\exists_{v \in V}\forall_{v' \in V}(v', v) \in r_a$

One can imagine other types too. Descriptor's type determines the type of pattern to be extracted from data. For example, when choosing clique-like descriptors, we search for patterns of values fully related to each other with respect to the given relation. The notion of descriptor's satisfiability can also vary depending on the type of descriptor and requirements. On one hand, we can define satisfiability very strictly, e.g., saying that $x \in U$ satisfies $(a, V)^{(2)}$, iff $a(x)$ is strongly related with all elements of a in terms of r_a. On the other hand, the support of $(a, V)^{(\#)}$ may be measured also in many other ways, depending on the context. For example, we may say that $x \in U$ satisfies descriptor:

- $(a, \{v\})^{(0)}$ iff $a(x) = v$, but also, e.g., iff $(a(x), v) \in r_a \lor (v, a(x)) \in r_a$
- $(a, V)^{(1)}$ iff $a(x) \in V$, but also, e.g., iff $\exists_{v \in V}(a(x), v) \in r_a \land (v, a(x)) \in r_a$
- $(a, V)^{(3'')}$ iff $\exists_{v \in V}\forall_{v' \in V \cup \{a(x)\}}(v', v) \in r_a$, but also iff $\exists_{v \in V}(a(x), v) \in r_a$

When we visualize different concepts of satisfiability for particular types of descriptors, we obtain different shapes. For example, in the first case above, we can obtain a totally homogeneous class $\{x : a(x) = v\}$, but also a larger star of objects with values strongly related to v. In the second case, we can obtain a generalized class $\{x : a(x) \in V\}$, but also a collection of stars around particular elements of V. In the third above case, we can obtain a star of objects with values related to the central $v \in V$, but also a kind of *snowflake* with elements related either directly to the center or to one of its neighbors in V.

Let $T = \bigwedge\{(a, V)^{(\#)}\}$ be a relational template. As before, we say that $x \in U$ satisfies T if it satisfies its all descriptors. The support of T, $supp_\mathbb{A}(T)$, is defined as number of objects that satisfy T and length of T is simply the number of its descriptors. Analogously to generalized templates, we should take into account precision of particular descriptors. Formula for precision should depend on the type of descriptor. For $\# = 0, 1$ we may use standard definitions from Section 2. For *structured* types of relational descriptors, i.e., $\# = 2, 2', 3, 3', 3''$, we suggest the following formula, where $V \subseteq V_a$ is said to be $\#$-*valid* with respect to r_a, if $(a, V)^{(\#)}$ actually satisfies the requirements of a given type of descriptor:

$$s^{(\#)})_\mathbb{A}((a, V) = \begin{cases} \frac{card(V)}{card(V_a)} & V \text{ is } \#\text{-valid w.r.t. } r_a \\ 0 & \text{otherwise} \end{cases} \tag{8}$$

When feasible, we can also use V_{max} – a maximal strong/weak clique/star with respect to r_a such that $V \subseteq V_{max}$ – instead of V_a in (8). Unlike in case of generalized descriptors $(a \in V)$, penalized for $card(V)$, now we would like to operate with possibly large sets V, of course until they remain $\#$-valid. From this perspective, the above descriptors are more like single-valued than generalized ones. We could also consider generalized relational descriptors as the sets of cliques or stars. However, such examples are beyond the scope of this paper.

5   Algorithms for Relational Templates Generation

Given precision of relational descriptors, we can generalize the formula (4) onto the case of relational templates. As noted before, the quality of template T, here denoted as $Q(T)$, may become a function of T's support, $supp_{\mathbb{A}}(T)$, and precision, $S_{\mathbb{A}}(T)$. For example, we may consider $Q(T) = supp_{\mathbb{A}}(T) \cdot S_{\mathbb{A}}(T)$. To generate T we can use a greedy algorithm that tries to add descriptors that maximize $Q(T)$. We assume that for each attribute there is chosen type (#) of required descriptor and definition of its support.

We start with an empty T and iteratively find either a new descriptor to be added to T or a value that would extend an existing descriptor. We repeat those steps till some stopping criteria are satisfied, e.g., quality of template is significantly decreasing in consecutive steps or it reaches some required level.

Algorithm 1 Relational template generation

Input: relational information system $\mathbb{A} = (U, A, R)$, *quality threshold th*
Output: Relational template T
   1. $T = T_{next} = \emptyset$
   2. **while** $Q(T) < th$ **begin**
   3.   **for each** $a \in A$ **begin**
   4.   *Choose optimal value* $v \in V_a$
   5.   **if** *there is no descriptor in* T *corresponding to* a **then**
   6. $T_{temp} = T \cup \{(a, \{v\})^{\#}\}$
   7.   **else**
   8. $T_{temp} = T \setminus (a, V)^{(\#)} \cup (a, V \cup \{v\})^{(\#)}$
   9.   **if** $Q(T_{temp}) > Q(T_{next})$ **then**
   10. $T_{next} = T_{temp}$
   11.   **end for**
   12. $T = T_{next}$
   13. **end while**

In the first few iterations of the algorithm, T is usually constructed from one-value descriptors. Then the value sets are being extended. Let us note that the choice of optimal value for a given attribute (Step 4) strongly depends on type (#) of the corresponding descriptor. For example, for $\# = 0$, once some descriptor is added to T the corresponding attribute is not considered any longer. If descriptor is of type 1 (generalized descriptor) then next iterations of the algorithm can possibly decrease its precision. For $\# = 2$, value v must form a clique with all values already chosen in previous steps. Analogously, for star-like descriptors we choose v that is in the corresponding relation with value that was chosen at first. To choose optimal value in Step 4 several known strategies can be adopted (see [10]) but they are specific to the type of descriptor.

The ideas presented above can be utilized by other algorithms [12,10]. One of the most natural extensions is to make the choice of optimal descriptor non-deterministic. Basing on qualities of k best descriptors, for some integer k, we can define their probabilities that can be used in the selection process.

One can also consider the case where T initially contains some one-value descriptors. Those descriptors are generated from some randomly chosen base object $x \in U$. Then, in each iteration we can either add or remove descriptors. There can be defined several strategies of adding and/or removing descriptors which are well known in the literature, e.g., sequential forward search (SFS), sequential backward search (SBS), remove l – add r [5].

Yet another modification can be based on following a permutation of A while choosing the consecutive attributes, for which optimal descriptors are being constructed and added to a template. We refer the readers to [11,22], where the algorithms for searching for optimal attribute permutations are presented.

6 Conclusions

We provided a framework for representing and extracting descriptors and templates within relational information systems. We discussed examples of how attribute-specific relations can be created and then used for descriptors' and templates' formation. Both the lists of such examples and of algorithms extracting relational templates from data remain open. Relational information systems and their special cases (e.g., relational decision tables) may be also considered with respect to other types of regularities and knowledge/decision models, not necessarily based on templates. We believe that the proposed framework can be applied also to other models, not only related to rough set methodology.

Acknowledgements. The research of the first co-author was supported by the Research Center at the Polish-Japanese Inst. of Information Technology and by grant 3T11C00226 from Polish Ministry of Sci. Research and Higher Education.

References

1. A. Bargiela and W. Pedrycz. *Granular Computing. An Introduction.* The International Series in Engineering and Computer Science, Vol. 717. Kluwer Academic Publishers, Boston, MA, USA, 2002.

2. J. G. Bazan, A. Skowron, D. Ślęzak, and J. Wróblewski. Searching for the complex decision reducts: The case study of the survival analysis. In Z. W. Raś, N. Zhong, S. Tsumoto, and E. Suzuku, editors, *International Symposium on Methodologies for Intelligent Systems ISMIS, LNAI,* **2871**, pp. 160–168, 2003. Springer-Verlag.

3. S. Demri and E. Orlowska. *Incomplete Information: Structure, Inference, Complexity.* Monographs in Theoretical Computer Science. An EATCS Series. Springer-Verlag, Berlin, Germany, 2002.

4. U. M. Fayyad, G. Piatetsky-Shapiro, P. Smyth, and R. Uthurusamy, editors. *Advances in Knowledge Discovery and Data Mining.* The AAAI Press, 1996.

5. J. H. Friedman, T. Hastie, and R. Tibshirani. *The Elements of Statistical Learning: Data Mining, Inference, and Prediction.* Springer-Verlag, Germany, 2001.
6. J. W. Grzymała-Busse. Incomplete data and generalization of indiscernibility relation, definability, and approximations. In D. Slezak, G. Wang, M. Szczuka, I. Duntsch, and Y. Yao, editors, *Rough Sets, Fuzzy Sets, Data Mining and Granular Computing RSFDGrC, LNAI,* **3641**, pp. 244–253, 2005. Springer-Verlag.
7. W. Kloesgen and J. Żytkow, editors. *Handbook of Knowledge Discovery and Data Mining.* Oxford University Press, Oxford, UK, 2002.
8. J. Komorowski, L. Polkowski, and A. Skowron. Rough sets: A tutorial. In S. K. Pal and A. Skowron, editors, *Rough Fuzzy Hybridization: A New Trend in Decision-Making,* pages 3–98. Springer-Verlag, Singapore, 1999.
9. K. Krawiec, R. Słowiński, and D. Vanderpooten. Learning decision rules from similarity based rough approximations. In Polkowski and Skowron [16], pp. 37–54.
10. S. H. Nguyen. *Regularity Analysis and Its Applications in Data Mining.* PhD thesis, Warsaw University, Warsaw, Poland, 2000.
11. S. H. Nguyen, L. Polkowski, A. Skowron, P. Synak, and J. Wróblewski. Searching for approximate description of decision classes. In *International Workshop on Rough Sets, Fuzzy Sets and Machine Discovery RSFD,* pp. 153–161, 1996.
12. S. H. Nguyen, A. Skowron, and P. Synak. Discovery of data patterns with applications to decomposition and classification problems. In L. Polkowski and A. Skowron [16], pages 55–97.
13. Z. Pawlak. Information systems - theoretical foundations. *Information Systems,* 6:205–218, 1981.
14. Z. Pawlak. Rough sets. *International Journal of Computer and Information Sciences,* 11:341–356, 1982.
15. Z. Pawlak. *Rough Sets: Theoretical Aspects of Reasoning about Data,* volume 9 of *D: System Theory, Knowledge Engineering and Problem Solving.* Kluwer Academic Publishers, Dordrecht, The Netherlands, 1991.
16. L. Polkowski and A. Skowron, editors. *Rough Sets in Knowledge Discovery 2: Applications, Case Studies and Software Systems,* volume 19 of *Studies in Fuzziness and Soft Computing.* Physica-Verlag, Heidelberg, Germany, 1998.
17. A. Skowron and J. Stepaniuk. Tolerance approximation spaces. *Fundamenta Informaticae,* 27(2-3):245–253, 1996.
18. D. Ślęzak and P. Wasilewski. Granular sets: Foundations and case study of tolerance spaces. 2007. In preparation.
19. R. Słowiński, S. Greco, and B. Matarazzo. Rough set analysis of preference-ordered data. In J. J. Alpigini, J. F. Peters, A. Skowron, and N. Zhong, editors, *International Conference on Rough Sets and Current Trends in Computing RSCTC, LNAI,* **2475**, pp. 44–59, 2002. Springer-Verlag.
20. J. Stefanowski and A. Tsoukias. Incomplete information tables and rough classification. *Computational Intelligence,* 17(3):545–566, 2001.
21. P. Synak and D. Ślęzak. Tolerance based templates for information systems. In G. Lee, D. Ślęzak, T. hoon Kim, and P. Sloot, editors, *International Conference on Hybrid Information Technology ICHIT,* 2006. SERSC.
22. J. Wróblewski. Genetic algorithms in decomposition and classification problem. In Polkowski and Skowron [16], chapter 24, pages 471–487.
23. Y. Y. Yao. Information granulation and rough set approximation. *International Journal od Intelligent Systems,* 16(1):87–104, 2001.
24. Y. Y. Yao, S. M. Wong, and T. Y. Lin. A review of rough set models. In T. Y. Lin and N. Cercone, editors, *Rough Sets and Data Mining. Analysis of Imprecise Data,* pages 47–75. Kluwer Academic Publishers, Boston, MA, USA, 1997.

ROSA: An Algebra for
Rough Spatial Objects in Databases

Markus Schneider* and Alejandro Pauly

University of Florida
Department of Computer and Information Science and Engineering
Gainesville, FL 32611, USA
{mschneid,apauly}@cise.ufl.edu

Abstract. A fundamental data modeling problem in geographical information systems and spatial database systems refers to an appropriate treatment of the *vagueness* or *indeterminacy* features of spatial objects. Geographical applications often have to deal with spatial objects that cannot be adequately described by the determinate, crisp concepts exclusively available in these systems since these objects have an intrinsically indeterminate and vague nature. The goal of this paper is to show that rough set theory can be leveraged in an elegant manner to seamlessly model this kind of spatial data. Our approach introduces novel *rough spatial data types* for *rough points*, *rough lines*, and *rough regions* that can be employed as attribute types in database schemas. These data types are part of a data model called *ROSA (ROugh Spatial Algebra)*. Their formal framework is based on already existing, general, exact models of crisp spatial data types, which simplifies the definition of the rough spatial model. In addition, we obtain *executable specifications* for the operations on rough spatial objects; these can be immediately used as implementations. This paper gives a formal definition of the three rough spatial data types as well as some basic operations.

Keywords: Spatial database, spatial vagueness, rough spatial data type.

1   Introduction

Geographical information systems (GIS) and spatial database systems (SDBS) are currently confronted with two main data modeling problems. Beside an appropriate integration of the temporal aspect, the feature of *spatial vagueness* or *spatial indeterminacy* is inherent to many geometric and geographic data [1]. The current mapping of spatial phenomena of the real world to exclusively crisp, i.e., precisely determined, spatial objects has turned out to be an insufficient abstraction process for many geographic applications. So far, applications based on indeterminate spatial data cannot be supported by current GIS and SDBS.

These systems assume, often contrary to reality, that the positions of points, the locations and routes of lines, and the extent and hence the boundary of

---

* This work was partially supported by the National Science Foundation under grant number NSF-CAREER-IIS-0347574.

J.T. Yao et al. (Eds.): RSKT 2007, LNAI 4481, pp. 411–418, 2007.

regions are precisely determined and universally recognized. Examples are especially man-made spatial objects (e.g., monuments, highways, buildings) and immaterial spatial objects (e.g., countries, districts, land parcels with their political, administrative, and cadastral boundaries). We denote this kind of entities as *crisp* or *determinate spatial objects*.

However, to an increasing degree, there are many geometric applications in which positions of points are not exactly known, the locations and routes of lines are unclear, and regions do not have sharp boundaries, or their boundaries cannot be precisely determined. Examples are social or natural phenomena (e.g., terrorists' refuges and escape routes, population density, unemployment rate, soil quality, vegetation, oceans, oil fields, biotopes, deserts). We denote this kind of entities as *rough* or *indeterminate spatial objects*.

In GIS and SDBS[1], *spatial data types* (see [2] for a survey) like *point, line,* or *region* provide fundamental abstractions for modeling the structure of geometric entities, their relationships, properties, and operations. This paper presents an object model for defining *rough spatial data types* for *rough points, rough lines,* and *rough regions.* We use the term *rough* for the characterization of these types since we leverage concepts of *rough set theory* [3] as a formal framework. The types are part of a novel data model called *ROSA* (*Rough Spatial Algebra*). The model rests on "traditional" (i.e., exact) modeling techniques and extends, rather than replaces, the current theory of SDBS and GIS. Further, moving from an exact to a rough domain does not necessarily invalidate conventional (computational) geometry for executing spatial operations; it is merely an extension. Hence, current exact object models can be considered as special cases of our rough spatial object model. We show in this paper that all rough spatial data types and some main rough spatial operations can be defined generically, i.e., without type-specific definitions. Since our rough spatial data types and operations are based on their crisp counterparts and can be expressed by them, we obtain *executable specifications* that can be directly used as an implementation. In this paper, we do not aim at developing a type system with a "complete" set of operations and predicates. The goal is more to demonstrate the power, simplicity, and expressiveness of our rough spatial data model.

Section 2 discusses related work. Section 3 informally introduces the concept of rough spatial objects and motivates it by giving some application examples. Section 4 gives a generic definition of rough spatial data types and rough spatial operations. Finally, Section 5 draws some conclusions and addresses future work.

2   Related Work

The problem of spatial vagueness, indeterminacy, uncertainty and imprecision with their many different nuances and diversities has been a research topic in GIS (but not SDBS) for a long time. As a trend, GIS clearly advocate *fuzzy set theory* [4] as an appropriate formal framework for solving this problem. Despite

---

[1] It is important to understand that SDBS deal exclusively with vector data in contrast to image databases that exclusively handle raster data.

many concepts and ideas, unfortunately, no overall and satisfying solution has been found so far. In particular, the implementation of fuzzy concepts turns out to be rather difficult in the spatial domain.

We regard *rough set theory* [3] as an appropriate, alternative option and compromise for representing spatial vagueness. This approach is based on representing a set X by a pair of determinate lower and upper approximations. From the lower approximation, we know that its elements belong definitely to X. The upper approximation is the set of elements that possibly belong to X. In this paper, we demonstrate how this concept can be seamlessly transferred to a model of rough spatial objects. The use of rough sets for representing vague spatial data has been proposed before in a few publications [5,6,7]. These approaches have in common that they are based on image data but not on vector data.

A benefit of the fact that the lower and upper approximations of a set are crisp implies for the transferal to the spatial domain that we can leverage the existing definitions, techniques, data structures, and algorithms of exact spatial object models. As an example, in the past, we have developed the *ROSE* (*Robust Spatial Extension*) *Algebra* [2,8], which provides crisp spatial data types like *point*, *line*, and *region* together with a comprehensive collection of spatial operations and predicates. Such an algebra (type system) can be taken to define our Rough Spatial Algebra ROSA and to obtain an executable specification of rough spatial operations.

3   What Are Rough Spatial Objects?

As indicated before, our concept of rough spatial objects necessitates a general, underlying crisp spatial object model which incorporates the determinate spatial data types *point*, *line*, and *region*. These data types must be defined in a way so that they are closed under (appropriately defined) *geometric union*, *geometric intersection*, *geometric difference*, and *geometric complement* operations. Such crisp type systems have, e.g., been formally defined in [2,8,9,10], and we will leverage them in this paper. Informally, these models represent a *point* object as a finite set of individual points, a *line* object as a finite set of disjoint *blocks* where each block represents a finite set of curves, and a *region* object as a finite set of disjoint, connected areal components called *faces* possibly with disjoint holes (see Figure 1). Examples of *point* objects are collections of lighthouses, collections of junctions, and collections of landmarks. Examples of *line* objects are streets, railways, and waterways. Examples of *regions* are districts, land parcels, and parks.

The central idea of *rough spatial objects* is to represent them by a lower approximation, which specifies those object parts that definitely belong to the rough spatial object, and an upper approximation, which, in addition, specifies those object parts that possibly partially or completely belong to the rough spatial object. As an illustrating example, we consider a homeland security scenario to introduce our concept for dealing with spatial vagueness and to demonstrate its usability. Secret services (should) have knowledge of the whereabouts of

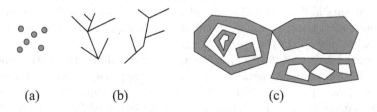

Fig. 1. Examples of a crisp point object (a), a crisp line object (b), and a crisp region object (c). Each collection of components forms a single crisp spatial object.

terrorists. For each terrorist, some of their refuges are precisely known, some are not and only conjectures. We can model these locations as a *rough point* object where the precisely known locations are called the *lower point approximation* and all possible locations are called the *upper point approximation*. Secret services are also interested in the routes a terrorist takes to move from one refuge to another. These routes can be modeled as *rough line* objects. Some paths have definitely been identified as terrorist routes. They form the *lower line approximation* of the rough line object. All possible paths that terrorists can have taken yield its *upper line approximation*. Knowledge about areas of terroristic activities is also important for secret services. From some areas it is well known that a terrorist operates in them; we call them the *lower region approximation*. From other areas we can only assume that they are the target of terroristic activity. We summarize all possible areas in the *upper region approximation*. Figure 2 gives some examples. Grey shaded areas, straight lines, and grey points indicate lower approximations; areas with white interiors, dashed lines, and white points refer to vague parts and form the upper approximations together with the determinate parts.

Based on this scenario and taking into account spatial vagueness, we are able to pose interesting queries. We can ask for the locations where any two terrorists have taken the same refuge. We can determine those terrorists that operated in the same area. We can compute the locations where routes taken by different terrorists crossed each other. Many further queries are possible. Vague concepts offer a greater flexibility for modeling properties of spatial phenomena in the real

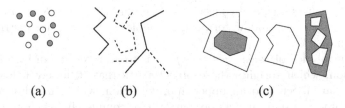

Fig. 2. Examples of a rough point object (a), a rough line object (b), and a rough region object (c). Each collection of components forms a single rough object.

world than determinate concepts do. Still, vague concepts comprise the modeling power of determinate concepts as a special case.

In this sense, many scenarios can be found that could make meaningful use of the concept of rough spatial objects. They all have in common that a rough spatial object (e.g., a rough line) is described by a pair of two crisp spatial objects (e.g., two crisp lines) where the first object is topologically contained in or inside of the second object. The first crisp spatial object, called the *lower object approximation*, describes the determinate parts of the rough spatial object, i.e., the parts that definitely and always belongs to the rough object. The second crisp spatial object, called the *upper object approximation*, describes the possible parts of the rough spatial object, i.e., the parts that definitely or perhaps belong to the rough object.

4 A Generic Definition of Rough Spatial Data Types and Rough Spatial Set Operations

Based on the motivation in the previous section, in Section 4.1, we first introduce some needed concepts from crisp (i.e., determinate) spatial type systems. Afterwards, we give a formal definition of rough spatial data types and rough spatial set operations (Section 4.2). An interesting observation is that these definitions can be given in a generic manner, i.e., type-specific considerations are unnecessary.

4.1 Crisp Spatial Data Types and Crisp Spatial Set Operations

Since rough sets are based on the concept of sets, the first issue is how crisp spatial objects can be modeled as sets. In this paper, we are exclusively dealing with two-dimensional spatial objects in the plane. Hence, crisp spatial objects are formally defined as point sets and subsets of the Euclidean space $\mathbb{R}^2$ that have to satisfy certain topological constraints in order to be well defined for geographic applications. The definition of the crisp spatial data types *point*, *line*, and *region* [10] is based on point set theory and point set topology [11]. We obtain:

(i) $point = \{P \subset \mathbb{R}^2 \mid P \text{ is finite}\}$

(ii) $line = \{L \subset \mathbb{R}^2 \mid$ (a) $L = \bigcup_{i=1}^{n} f_i([0,1])$ with $n \in \mathbb{N}_0$

                       (b) $\forall 1 \leq i \leq n : f_i : [0,1] \to \mathbb{R}^2$ is a continuous mapping

                       (c) $\forall 1 \leq i \leq n : |f_i([0,1])| > 1\}$

(iii) $region = \{R \subset \mathbb{R}^2 \mid$ (a) R is regular closed

                       (b) R is bounded

                       (c) The number of connected sets of R is finite$\}$

A set is *regular closed* if it is equal to the closure of its interior [11]. The above definition gives an *unstructured* view of crisp spatial objects. Another equivalent definition leads to a *structured* view [10], which we only describe informally here. The unstructured and structured views for the data type *point* coincide. A *line* object is assembled from a finite number of connected components called *blocks*; each block contains a finite number of curves. A *region* object consists of a finite number of disjoint *faces*; each face possibly contains a finite number of disjoint holes. An illustration of these informal descriptions is shown in Figure 1.

For $\alpha \in \{point, line, region\}$, each type α is closed under the *geometric set operations union* ($\oplus : \alpha \times \alpha \rightarrow \alpha$), *intersection* ($\otimes : \alpha \times \alpha \rightarrow \alpha$), and *difference* ($\ominus : \alpha \times \alpha \rightarrow \alpha$) [10]. The partial order $(\alpha, \subseteq)$ is a distributive lattice $(\alpha, \oplus, \otimes)$. But it is not a complemented lattice, and consequently not a Boolean algebra, since the spatial data types are not closed under the operation *complement* ($\sim$). That is, $\forall v \in \alpha : \sim v \notin \alpha$. The identity of $\otimes$ is denoted by $\mathbf{1}$, which corresponds to $\mathbb{R}^2$. The identity of $\oplus$ is presented by $\mathbf{0}$, which corresponds to the empty spatial object (empty point set). The geometric set operations of the type *point* are equal to the standard set operations. The operation $\oplus$ is equal to the set operation $\cup$ for the types *line* and *region*. The geometric operations $\otimes$ and $\ominus$ require a *regularization* step so that they cannot produce geometric anomalies.

4.2   Rough Spatial Data Types and Rough Spatial Set Operations

Syntactically, the extension of a crisp spatial data type to a corresponding rough spatial type is given by a type constructor ρ as follows:

$$\rho(\alpha) = \alpha \times \alpha \qquad \forall \alpha \in \{point, line, region\}$$

That is, each rough spatial data type is represented as a pair of corresponding crisp spatial data types. For example, for $\alpha = point$ we obtain $\rho(point) = point \times point$, which we also name *rpoint*. Accordingly, the data types *rline* and *rregion* are defined. For a rough spatial object $R = (\underline{R}, \overline{R}) \in \rho(\alpha)$, we call $\underline{R} \in \alpha$ the *lower object approximation* of R, and $\overline{R} \in \alpha$ denotes the *upper object approximation* of R.

Semantically, the lower (minimal, guaranteed) object approximation represents the determinate, crisp part of R, i.e., the area which *definitely* belongs to R. The upper (maximally possible, speculative) object approximation describes the *potential* spatial extent of R. Hence, we know that $\mathbf{1} - \overline{R} = \mathbb{R}^2 - \overline{R}$ does *definitely not* belong to R. Therefore, the area of *spatial determinacy* is $\underline{R} \cup (\mathbb{R}^2 - \overline{R})$. The *area of spatial vagueness* extends over $\overline{R} - \underline{R}$. It represents the indeterminate, vague part of R, i.e., the area for which we cannot say with any certainty whether it or parts of it belong to R or not. *Maybe* it or parts of it belong to R, *maybe* this is not the case. We could also say that this is *unknown* or *unclear*. Note that, in general, $\overline{R} - \underline{R} \notin \alpha$ and $\overline{R} \ominus \underline{R} \in \alpha$ hold. To enable the intended semantics described above, we require:

$$\forall \alpha \in \{point, line, region\} \ \forall R = (\underline{R}, \overline{R}) \in \rho(\alpha) : \underline{R} \subseteq \overline{R}$$

In the spatial domain, subset relationships are expressed by so-called *topological relationships* [10], which characterize the relative position of spatial objects to each other. Examples of such relationships are *disjoint*, *meet*, and *overlap*. As [10] shows, the subset relationship depends on the combination $\alpha \times \alpha$ of spatial data types considered and corresponds to several topological relationships. In an abbreviated form, on the basis of so-called *clustered* topological relationships (indicated by a subscript '*c*'), the above requirement can be rewritten as:

$$\forall\,\alpha \in \{point, line, region\}\ \forall\,R = (\underline{R}, \overline{R}) \in \rho(\alpha) :$$
$$equal_c(\underline{R}, \overline{R})\ \vee\ inside_c(\underline{R}, \overline{R})\ \vee\ coveredBy_c(\underline{R}, \overline{R})$$

For $\alpha = point$, we obtain three (unclustered) topological relationships, for $\alpha = line$ ten relationships, and for $\alpha = region$ five relationships.

Let $points : \rho(\alpha) \to \mathbb{R}^2$ be an auxiliary function that yields the (unknown) point set of a rough spatial object $R = (\underline{R}, \overline{R}) \in \rho(\alpha)$. We can conclude that

$$\underline{R} \subseteq points(R) \subseteq \overline{R}$$

If $\underline{R} = \overline{R}$, R is either the empty rough spatial object $\mathbf{0} = (\varnothing, \varnothing)$ or corresponds to the crisp spatial object $\underline{R}$. The 1-element is $\mathbf{1} = (\mathbb{R}^2, \mathbb{R}^2)$ and corresponds to the Euclidean plane. Even if we do not know the exact point set of R, we assume and require that $points(R)$ is not arbitrary but compatible to α, i.e.,

$$points(R) \in \alpha \quad \text{and} \quad points(R) \ominus \underline{R} \in \alpha$$

Using the characteristic function χ deciding about the existence or non-existence of an element in a set, we obtain $\chi(p) = 1$ for all $p \in \underline{R}$, $\chi(p) = 0$ for all $p \in \mathbb{R}^2 - \overline{R}$, $\chi(p) = 1 \vee \chi(p) = 0$ for all $p \in \overline{R} - \underline{R}$, and $\chi(p) = 1$ for all $p \in points(R) \in \alpha$. Note the deliberate use of set-theoretic operations. In particular, possible common boundary points of $\underline{R}$ and $\overline{R} \ominus \underline{R}$ are mapped to 1.

Let $R = (\underline{R}, \overline{R}), S = (\underline{S}, \overline{S}) \in \rho(\alpha)$. We then define the geometric set operations **union**, **intersection**, and **difference** on rough spatial objects as follows:

$$R \text{ union } S = (\underline{R} \oplus \underline{S}, \overline{R} \oplus \overline{S})$$
$$R \text{ intersection } S = (\underline{R} \otimes \underline{S}, \overline{R} \otimes \overline{S})$$
$$R \text{ difference } S = (\underline{R} \ominus \overline{S}, \overline{R} \ominus \underline{S})$$

The equalities and subset relationships used in the following are derived from [3]. For the operation **union**, we define the lower object approximation in a more pessimistic and stricter way since $\underline{R} \oplus \underline{S} \subseteq \underline{R \oplus S}$. For the upper object approximation, there is no difference since $\overline{R} \oplus \overline{S} = \overline{R \oplus S}$. For the operation **intersection**, we are rather optimistic for the upper object approximation since $\overline{R} \otimes \overline{S} \supseteq \overline{R \otimes S}$. For the lower object approximation, $\underline{R} \otimes \underline{S} = \underline{R \otimes S}$ holds. For the operation **difference**, for the lower object approximation, we must subtract both the intersecting determinate and indeterminate parts of S from $\underline{R}$. Only the result of this computation can be definitely part of the difference of R and S. For computing the upper object approximation, we only have to subtract the definite parts of S from $\overline{R}$ since they can definitely not belong to R. The particular reason why all operations make use of $\underline{R}$, $\overline{R}$, $\underline{S}$, and $\overline{S}$ is that these approximations are known and can be used in an implementation.

5   Conclusions

In this paper, we have made a first step towards a simple but expressive data model of points, lines, and regions that is capable of describing many different aspects of spatial vagueness. It is based on rough set theory and a canonical extension of determinate spatial data models. This facilitates the treatment of rough and exact objects in one single model. Since our approach is based on exact spatial modeling concepts, it allows us to build upon existing work and simplifies many definitions. In particular, we can leverage already existing implementations of crisp spatial type systems (like the ROSE Algebra) to realize rough spatial objects with only minimal effort by executable specifications.

The data model presented in this paper is part of ROSA, our Rough Spatial Algebra. So far, ROSA is incomplete. We plan to supplement it by further spatial operations, topological predicates, directional predicates, numerical operations, and more. A later step refers to the implementation of ROSA on the basis of our ROSE Algebra and to the embedding of ROSA into a database query language.

References

1. Burrough, P.A., Frank, A.U., eds.: Geographic Objects with Indeterminate Boundaries. GISDATA Series, vol. 2. Taylor & Francis (1996)
2. Schneider, M.: Spatial Data Types for Database Systems - Finite Resolution Geometry for Geographic Information Systems. Volume LNCS 1288. Springer-Verlag (1997)
3. Pawlak, Z.: Rough Sets. Int. Journal of Computer and Information Sciences **11** (1982) 341–356
4. Zadeh, L.A.: Fuzzy Sets. Information and Control **8** (1965) 338–353
5. Ahlqvist, O., Keukelaar, J., Oukbir, K.: Rough Classification and Accuracy Assessment. Int. Journal of Geographical Information Sciences **14** (2004) 475–496
6. Beaubouef, T., Ladner, R., Petry, F.: Rough Set Spatial Data Modeling for Data Mining. Int. Journal of Intelligent Systems **19** (2004) 567–584
7. Worboys, M.: Computation with Imprecise Geospatial Data. Computational, Environmental and Urban Systems **22** (1998) 85–106
8. Güting, R.H., Schneider, M.: Realm-Based Spatial Data Types: The ROSE Algebra. VLDB Journal **4** (1995) 100–143
9. Clementini, E., Di Felice, P.: A Model for Representing Topological Relationships between Complex Geometric Features in Spatial Databases. Information Systems **90** (1996) 121–136
10. Schneider, M., Behr, T.: Topological Relationships between Complex Spatial Objects. ACM Trans. on Database Systems **31** (2006) 39–81
11. Gaal, S.: Point Set Topology. Academic Press (1964)

Learning Models Based on Formal Concept

Guo-Fang Qiu

School of Management, Xi'an University of Architecture and Technology,
Shaanxi 710055, P.R. China
qiugf_home@163.com

Abstract. From the classic view in which a concept consists of the set of extents and the set of intents, a concept learning system extended from a formal context is introduced and two concepts such as an under concept and an over concept are defined. Any pair of subsets from extents and intents in this concept learning system can be changed to an under or an over concept. Further it can be changed to a concept by learning from the set of extents or from the set of intents. It is proved that the concept learned in this framework is an optimal concept. This process of learning a concept describes the recognizing ability from unclear to clear.

Keywords: Concept learning system, Formal context, Formal concept analysis, Learning models.

1 Introduction

Learning theory is original used to the empirical study from the behaviorist paradigm in psychology. Later, computational learning theory developed and applied by many scholars coming from other fields such as computer science, mathematics etc. Their studies focused on learning-theoretic epistemology include logical reliability [1,2] and means-ends epistemology [7].

Concept learning was primarily a function of contiguity and stimulus-response generalization. Merrill described a model that focuses on attributes and examples [3]. One of the major goals of this model was to reduce three typical errors in concept formation: undergeneralization, overgeneralization and misconception. Tennyson suggest a model for concept teaching that has three stages [11], this model acknowledges the declarative and procedural aspects of cognition.

There are many theoretical views of concepts, concept formation and learning [5,9,10,12]. The classical view treats concepts as a pair of sets of necessary and sufficient conditions [13,16], in which every concept is described as two parts, the intent and the extent. The intent of a concept consists of all attributes that are valid for all those objects to which the concept applies. The extent of a concept is the set of objects which are concrete instances of the concept.

In my view, the essences of concept learning are investigative strategies of concept formations. Different strategies will deduce different combinations of objects and attributes. These combinations may be approximated recognizing to a concept [4,6,8,14]. I try to introduce a concept learning system based on

J.T. Yao et al. (Eds.): RSKT 2007, LNAI 4481, pp. 419–426, 2007.

thoughts of Yao [15], and define some special combinations which can be approximated to a concept. For any pair of subsets from objects and attributes in this learning system, an under concept, an over concept and a concept can be obtained by learning from models established in this paper. This process of learning a concept describes the recognizing ability from unclear to clear.

The paper is organized as follows. Section 2 recalls basic definitions of the formal concept. Section 3 introduces a concept learning system extended from a formal context and gives definitions of under and over concepts. Section 4 establishes models which can be learned to form concepts. Section 5 concludes the paper.

2   Formal Concept

Definition 1. *Let U be a finite set of objects, A be a finite set of attributes, and $I \subseteq U \times A$ be the set of relations between objects and attributes, then the triple (U, A, I) is called a formal context. Further, if $\forall x \in U$, $\exists a_1, a_2 \in A$, such that $(x, a_1) \in I$, $(x, a_2) \notin I$, and $\forall a \in A$, $\exists x_1, x_2 \in U$, such that $(x_1, a) \in I$, $(x_2, a) \notin I$, then the formal context is called regular.*

A regular formal context describes an object has at least one attribute and reversely an attribute is possessed by at least one object. For simplicity, we suppose the formal context mentioned in this paper is regular.

Let (U, A, I) be a formal context, if $X \subseteq U$, $B \subseteq A$, we define algorithms on the set of objects and the set of attributes as follows:

$$X^* = \{a \in A : \forall x \in X, (x, a) \in I\}. \tag{1}$$
$$B^* = \{x \in U : \forall a \in B, (x, a) \in I\}. \tag{2}$$

X^* represents the set of objects X having common attributes from A and B^* represents the set of attributes B is possessed by the objects from U.

Obviously, if $X, X_1, X_2 \subseteq U$, $B, B_1, B_2 \subseteq A$ and $\emptyset_U, \emptyset_A$ denote empty set on U, A respectively, then these two operators have the following properties:

(1) $U^* = \emptyset_A$, $\emptyset_A^* = U$, $\emptyset_U^* = A$, $A^* = \emptyset_U$.
(2) $(X_1 \cup X_2)^* = X_1^* \cap X_2^*$, $(B_1 \cup B_2)^* = B_1^* \cap B_2^*$.
(3) $X^{**} \supseteq X$, $B^{**} \supseteq B$.

Definition 2. *Let (U, A, I) be a formal context, $\forall X \subseteq U$, $B \subseteq A$, if $X^* = B$, $B^* = X$, then (X, B) is called a formal concept.*

A formal concept is a pair of subsets from objects and attributes, in which objects and attributes are reflected each other uniquely. Generally, the set of objects is called the extent and the set of attributes is called the intent.

3   Concept Learning System

Definition 3. *Let U be a finite set of objects, A be a finite set of attributes, $L: 2^U \to 2^A$ is called an extent-intent operator if L satisfies:*

$$L(\emptyset_U) = A, \ L(U) = \emptyset_A. \tag{3}$$
$$L(X_1 \cup X_2) = L(X_1) \cap L(X_2) \ (X_1, X_2 \subseteq U). \tag{4}$$

and $H: 2^A \to 2^U$ is called an intent-extent operator if H satisfies:

$$H(\emptyset_A) = U, \ H(A) = \emptyset_U. \tag{5}$$
$$H(B_1 \cup B_2) = H(B_1) \cap H(B_2) \ (B_1, B_2 \subseteq A). \tag{6}$$

if the two operators further satisfy:

$$H(L(X)) \supseteq X, \ L(H(B)) \supseteq B. \tag{7}$$

Then the quadruplet (U, A, L, H) is called a concept learning system.

A concept learning system has following properties:

(1) $X_1 \subseteq X_2 \Rightarrow L(X_2) \subseteq L(X_1)$.
(2) $B_1 \subseteq B_2 \Rightarrow H(B_2) \subseteq H(B_1)$.
(3) $L(H(L(X))) = L(X), \ H(L(H(B))) = H(B)$.

Theorem 1. *Let (U, A, I) be a formal context, if $L(X) = X^*$, $H(B) = B^*$, then (U, A, L, H) forms a concept learning system. Conversely, if (U, A, L, H) is a concept learning system, there exists a formal context (U, A, I), $\forall X \subseteq U$, $B \subseteq A$, such that $X^* = L(X)$, $B^* = H(B)$.*

Proof. A formal context changed to a concept learning system is obviously. We prove the converse side. Denote $I = \{(x, a) : x \in H(\{a\})\}$. For $B \subseteq A$, we have

$$\begin{aligned} B^* &= \{x \in U : \forall a \in B, \ (x, a) \in I\} \\ &= \{x \in U : \forall a \in B, \ x \in H(\{a\})\} \\ &= \{x \in U : x \in \bigcap_{a \in B} H(\{a\}) = H(B)\} \\ &= H(B). \end{aligned}$$

If $x \in H(\{a\})$, then $L(\{x\}) \supseteq L(H(\{a\}))$, therefore $L(\{x\}) \supseteq \{a\}$ by formula (7), i.e. $a \in L(\{x\})$ and vice versa. Then $I = \{(x, a) : a \in L(\{x\})\}$. Similarly for $X \subseteq U$, we have $X^* = L(X)$.

From Theorem 1, a concept learning system is corresponded to a formal context, i.e. it is an extension from a formal context.

Denote all concepts in (U, A, L, H) by $\mathcal{C} = \{(X, B) : L(X) = B, \ H(B) = X\}$. Let $(X_1, B_1), (X_2, B_2) \in \mathcal{C}$ and $(X_1, B_1) \leq (X_2, B_2)$ represents $X_1 \subseteq X_2$, then

$(\mathcal{C}, \leq)$ is a partial order relation and forms a complete lattice. The meet and join of the two concepts are defined as:

$$(X_1, B_1) \wedge (X_2, B_2) = (X_1 \cap X_2, L(H(B_1 \cup B_2))). \tag{8}$$

$$(X_1, B_1) \vee (X_2, B_2) = (H(L(X_1 \cup X_2)), B_1 \cap B_2). \tag{9}$$

We call (X_1, B_1) as a sub-concept of (X_2, B_2), conversely call (X_2, B_2) as a sup-concept of (X_1, B_1).

Definition 4. *Let (U, A, L, H) be a concept learning system, $\forall X \subseteq U$, $B \subseteq A$, (X, B) is called an under concept if $L(X) \subseteq B$, $H(B) \subseteq X$, and (X, B) is called an over concept if $L(X) \supseteq B$, $H(B) \supseteq X$.*

Denote all under and over concepts as:

$$\mathcal{C}_{\mathcal{UN}} = \{(X, B): L(X) \subseteq B, \ H(B) \subseteq X\}. \tag{10}$$

$$\mathcal{C}_{\mathcal{OV}} = \{(X, B): L(X) \supseteq B, \ H(B) \supseteq X\}. \tag{11}$$

$(\mathcal{C}_{\mathcal{UN}}, \leq)$ or $(\mathcal{C}_{\mathcal{OV}}, \leq)$ doesn't form a partial order relation, but the meet operator (8) and the join operator (9) also hold.

Theorem 2. *Let (U, A, L, H) be a concept learning system, $\forall X \subseteq U$, $B \subseteq A$, under concepts are obtained by:*

$$(H(L(X)), L(X) \cup B) \in \mathcal{C}_{\mathcal{UN}}. \tag{12}$$

$$(X \cup H(B), L(H(B))) \in \mathcal{C}_{\mathcal{UN}}. \tag{13}$$

Proof. Because $L(H(L(X))) = L(X) \subseteq L(X) \cup B$ and $H(L(X) \cup B) = H(L(X)) \cap H(B) \subseteq H(L(X))$, then $(H(L(X)), L(X) \cup B) \in \mathcal{C}_{\mathcal{UN}}$ by formula (10). Similarly, $(X \cup H(B), L(H(B))) \in \mathcal{C}_{\mathcal{UN}}$.

Theorem 3. *Let (U, A, L, H) be a concept learning system, $\forall X \subseteq U$, $B \subseteq A$, over concepts are obtained by:*

$$(X \cap H(B), B \cup L(X)) \in \mathcal{C}_{\mathcal{OV}}. \tag{14}$$

$$(X \cup H(B), B \cap L(X)) \in \mathcal{C}_{\mathcal{OV}}. \tag{15}$$

Proof. Because $X \cap H(B) \subseteq X$ and $X \cap H(B) \subseteq H(B)$, then $L(X \cap H(B)) \supseteq L(X) \cup L(H(B)) \supseteq L(X) \cap B$. And $H(B \cup L(X)) = H(B) \cap H(L(X)) \supseteq H(B) \cap X$. Thus $(X \cap H(B), B \cup L(X)) \in \mathcal{C}_{\mathcal{OV}}$ by formula (11). Similarly, $(X \cup H(B), B \cap L(X)) \in \mathcal{C}_{\mathcal{OV}}$.

Example 1. A formal context (U, A, I), where $U = \{1, 2, 3, 4\}$, $A = \{a, b, c, d\}$, $I \subseteq U \times A$.

From Theorem 1, there exists a concept learning system (U, A, L, H), in which $L(X) = X^*$, $H(B) = B^*$, corresponding to the formal context (U, A, I). All concepts $\mathcal{C}$, under concepts $\mathcal{C}_{UN}$ and over concepts $\mathcal{C}_{OV}$ are illustrated in Fig.1, Fig.2 and Fig.3.

Table 1. A formal context

U	a	b	c	d
1	1	0	1	1
2	1	1	0	0
3	0	0	1	0
4	1	1	0	0

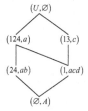

Fig. 1. All concepts $\mathcal{C}$

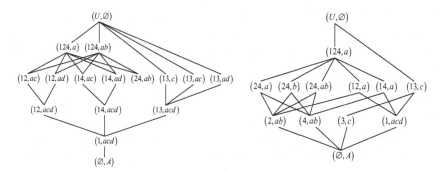

Fig. 2. All under concepts $\mathcal{C}_{UN}$　　　　　**Fig. 3.** All over concepts $\mathcal{C}_{OV}$

4   Concept Learning Models

For a concept learning system (U, A, L, H), $\forall X \subseteq U$, $B \subseteq A$, a pair (X, B) can be changed to an under or over concept by Theorem 2 or 3, but it is just a partial acquaintance with a concept. This section, we further study how it can be changed to a concept.

Theorem 4. *Let (U, A, L, H) be a concept learning system, $\forall X_1 \subseteq U$, $B_1 \subseteq A$, $H(B_1) \neq \emptyset$, a pair (X_1, B_1) can be changed to a concept by learning from the set of extents:*

$$X_n = X_{n-1} \cup H(B_{n-1}) \quad (n \geq 2) \tag{16}$$
$$B_n = L(X_n) \quad (n \geq 2) \tag{17}$$

then there must exist m, such that $(X_m, B_m) \in \mathcal{C}$.

Proof. Suppose $(X_1, B_1) \notin C$. Due to $B_n = L(X_n)$ $(n \geq 2)$, we have $H(B_n) \supseteq X_n$. X_n is monotonic non-decreased, U is a finite set, so there must exist m, such that $X_{m+1} = X_m$. Then $X_{m+1} = X_m = X_m \cup H(B_m)$, i.e. $H(B_m) \subseteq X_m$. Therefore, $H(B_m) = X_m$, i.e. $(X_m, B_m) \in C$.

Theorem 5. *Let* (U, A, L, H) *be a concept learning system,* $\forall X_1 \subseteq U$, $B_1 \subseteq A$, $L(X_1) \neq \emptyset$, *a pair* (X_1, B_1) *can be changed to a concept by learning from the set of intents:*

$$B_n = B_{n-1} \cup L(X_{n-1}) \quad (n \geq 2) \tag{18}$$
$$X_n = H(B_n) \quad (n \geq 2) \tag{19}$$

then there must exist m, *such that* $(X_m, B_m) \in C$.

Proof. It can be proved similarly as Theorem 4.

Theorem 6. *Let* (U, A, L, H) *be a concept learning system,* $\forall X_1 \subseteq U$, $B_1 \subseteq A$, $L(X_1) \neq \emptyset$, $H(B_1) \neq \emptyset$, *denote:*

$$C_{EX}(X_1, B_1) = \{(X, B) : X \supseteq X_1 \cup H(B_1)\}. \tag{20}$$
$$C_{IN}(X_1, B_1) = \{(X, B) : B \supseteq B_1 \cup L(X_1)\}. \tag{21}$$

Then $\forall (X, B) \in C_{EX}(X_1, B_1)$, $(X_m, B_m) \leq (X, B)$ *holds, where* (X_m, B_m) *from Theorem 4. And similarly,* $\forall (X, B) \in C_{IN}(X_1, B_1)$, $(X_m, B_m) \geq (X, B)$ *holds, where* (X_m, B_m) *from Theorem 5.*

Proof. Suppose $(X, B) \in C_{EX}$, because $X \supseteq X_1 \cup H(B_1)$, then $X \supseteq X_1$, $X \supseteq B_1$, therefore, $X \supseteq X_2$. Due to $B_2 = L(X_2) \supseteq L(X) = B$, consequently, $H(B_2) \subseteq H(B) = X$, so $X \supseteq X_2 \cup H(B_2)$. For $n \geq 1$, we have $X \supseteq X_n \cup H(B_n)$. Thus $X \supseteq X_m$, i.e. $(X_m, B_m) \leq (X, B)$. Similarly, $\forall (X, B) \in C_{IN}(X_1, B_1)$, $(X_m, B_m) \geq (X, B)$ holds.

Example 2. For the concept learning system (U, A, L, H) illustrated in Example 1, if $X_1 = \{12\}$, $B_1 = \{ab\}$, due to $L(X_1) = L(12) = \{a\}$, $H(B_1) = H(ab) = \{24\}$, a pair (X_1, B_1) isn't an under or over concept. Using the above learning process, a concept $(124, a)$ and a concept $(24, ab)$ are obtained by Theorem 4 and Theorem 5. And we know that $(124, a)$ and $(24, ab)$ are minimal sup-concept and maximal sub-concept respectively in Fig.1.

If $\forall X_1 \subseteq U$, $B_1 \subseteq A$, $(X_1, B_1) \in C_{UN}$, then a concept can also be learned by following models.

Theorem 7. *Let* (U, A, L, H) *be a concept learning system,* $\forall X_1 \subseteq U$, $B_1 \subseteq A$, $H(B_1) \neq \emptyset$, *a concept can be obtained by learning from the set of extents:*

$$X_n = X_{n-1} \cap H(B_{n-1}) \quad (n \geq 2) \tag{22}$$
$$B_n = L(X_n) \quad (n \geq 2) \tag{23}$$

then there must exist m, *such that* $(X_m, B_m) \in C$.

Proof. It can be proved similarly as Theorem 4.

Theorem 8. *Let (U, A, L, H) be a concept learning system, $\forall X_1 \subseteq U$, $B_1 \subseteq A$, $L(X_1) \neq \emptyset$, a concept can be obtained by learning from the set of intents:*

$$B_n = B_{n-1} \cap L(X_{n-1}) \quad (n \geq 2) \tag{24}$$

$$X_n = H(B_n) \quad (n \geq 2) \tag{25}$$

then there must exist m, such that $(X_m, B_m) \in \mathcal{C}$.

Proof. It can be proved similarly as Theorem 4.

Theorem 9. *Let (U, A, L, H) be a concept learning system, $\forall X_1 \subseteq U$, $B_1 \subseteq A$, $L(X_1) \neq \emptyset$, $H(B_1) \neq \emptyset$, denote:*

$$\mathcal{C}_{EX}(X_1, B_1) = \{(X, B) : X \subseteq X_1 \cap H(B_1)\}. \tag{26}$$

$$\mathcal{C}_{IN}(X_1, B_1) = \{(X, B) : B \subseteq B_1 \cap L(X_1)\}. \tag{27}$$

Then $\forall(X, B) \in \mathcal{C}_{EX}(X_1, B_1)$, $(X_m, B_m) \geq (X, B)$ holds, where (X_m, B_m) from Theorem 7. Then $\forall(X, B) \in \mathcal{C}_{IN}(X_1, B_1)$, $(X_m, B_m) \leq (X, B)$ holds, where (X_m, B_m) from Theorem 8.

Proof. It can be proved similarly as Theorem 6.

Example 3. For the concept learning system (U, A, L, H) illustrated in Example 1, we choice $X_1 = \{12\}$, $B_1 = \{ad\}$ in Fig.2. Then $(12, ad)$ is an under concept. Using the above learning process, a concept $(1, acd)$ and a concept $(124, a)$ are obtained by Theorem 7 and Theorem 8. And we know that $(1, acd)$ and $(124, a)$ are maximal sub-concept and minimal sup-concept respectively in Fig.1.

5   Conculsion

The problem of concept learning originated from cognitive science has been paid attention to many researchers in many other fields. We study the problem of concept formations based on the theory of formal concept analysis. A concept learning system is extended from a formal context and two kinds of concepts are defined in this system. Any pair (X, B) in this system not only can be changed to the two concepts and also can be changed to a concept. This process simulates different investigative strategies of concept formations.

Acknowledgement

This work is supported by the National Natural Science Foundation of China under grant No. 60673096 and the author gratefully acknowledges the comments from reviewers.

page 426 G.-F. Qiu

References

1. Glymour, C.: The hierarchies of knowledge and the mathematics of discovery. Minds and Machines. **1** (1991) 75–95
2. Kelly, K.: The Logic of Reliable Inquiry. Oxford: Oxford University Press. (1996)
3. Merrill, M., Tennyson, R.: Concept Teaching: An Instructional Design Guide. Englewood Cliffs, NJ: Educational Technology. (1977)
4. Pawlak, Z.: Rough sets. International Journal of Computer and Information Sciences. **11** (1982) 341–356
5. Peikoff, L.: Objectivism: The Philosophy of Ayn Rand. New York: Dutton. (1991)
6. Qiu, G. F., Li, H. Z., et al.: A knowledge processing method for intelligent systems based on inclusion degree. Expert Systems. **20(4)** (2003) 187–195
7. Schulte, O.: Means-ends epistemology. The British Journal for the Philosophy of Science. **50** (1999) 1–31
8. Skowron, A.: The rough set theory and evidence theory. Fundamenta Informatica. XIII (1990) 245–162
9. Smith, E.: Concepts and induction. In: Foundations of Cognitive Science. Posner, M.(Ed.). Massachusetts: The MIT Press. (1989) 501–526
10. Sowa, J.: Conceptual Structures, Information Processing in Mind and Machine. Massachusetts: Addison-Wesley. (1984)
11. Tennyson, R., Cocchiarella, M.: An empirically based instructional design theory for teaching concepts. Review of Educational Research. **56(1)** (1986) 40–71
12. Van Mechelen, I., Hampton J. et al.(Eds.): Categories and Concepts, Theoretical Views and Inductive Data Analysis, New York: Academic Press. (1993)
13. Wille, R.: Restructuring lattice theory: an approach based on hierarchies of concepts. In: Ordered Sets, Rival, I. (Ed.). Dordrecht-Boston: Reidel. (1982)
14. Yao, Y. Y.: Concept Lattices in Rough Set Theory. Proceedings of 2004 Annual Meeting of the North American Fuzzy Information Processing Society (NAFIPS 2004). Dick, S., Kurgan, L. et al.(Eds.). IEEE Catalog Number: 04TH8736, June 27-30, (2004) 796–801
15. Yao, Y. Y.: Constructive and algerbraic methods of the theory of rough sets. Information Sciences. **109** (1998) 21–47
16. Zhang, W. X., Wei, L. et al.: Attribute reduction in concept lattice based on discernibility matrix. In: Slezak D. et al. (Eds.): Proceeding of RSFDGrC 2005, LNAI 3642, (2005) 157–165

Granulation Based Approximate Ontologies Capture

Taorong Qiu[1,2], Xiaoqing Chen[1,2], Qing Liu[2], and Houkuan Huang[1]

[1] School of Computer and Information Technology, Beijing Jiaotong University,
Beijing 100044, China
taorongqiu@163.com
[2] Department of Computer, Nanchang University, Nanchang, Jiangxi 330031, China

Abstract. Ontologies are of vital importance to the successful realization of semantic Web. Currently, the existing concepts in ontologies are not approximate but clear. However, in real application domains many concepts are difficult to define explicitly. In order to fulfill semantic Web, it's not only necessary but also important to study approximate concepts and approximate ontologies generated from the approximate concepts. In this paper, based on the principle of granular computing, a granulation model for representing approximate ontologies was constructed. Then algorithms for capturing approximate concepts and generating approximate ontologies were proposed and illustrated with a real example.

Keywords: Concept approximation, granular computing, ontologies.

1 Introduction

Generally, an ontology is an explicit, agreed specification about a shared conceptualization. It gathers a set of concepts that are considered relevant to a given domain. However, in real application domains many concepts are difficult to define explicitly. Recently, some researchers have studied how to transfer the representation of ontologies from the distinguished logistic categorization into fuzzy and approximate structure. Stoilos and Straccia used fuzzy theory to extend fuzzy concepts and fuzzy axioms, and fulfilled the reasoning about imprecise concepts and roles [1,2,3]. Doherty [4] also constructed a formal framework for defining approximate concepts, approximate ontologies and approximate operations based on generalizing rough set theory. Stepaniuk and Skowron [8] studied granulated information systems and granular approximate space, and discussed the granular framework of approximation and dependency relationship between concepts.

Though ontologies are now widely used in the information technology community. Several barriers must be overcome before ontologies become practical and useful tools. A critical issue is the ontology construction, i.e., the task of identifying, defining, and entering the concept definitions. In case of large and complex application domains this task can be lengthy, costly, and controversial. Therefore, Many techniques of machine learning have been applied to constructing ontologies from data sources with different types like structural, semi-structural or non-structural data sources [15].

J.T. Yao et al. (Eds.): RSKT 2007, LNAI 4481, pp. 427–434, 2007.

A comprehensive methodology for developing ontologies includes ontology capture, ontology coding and integrating existing ontologies. Ontology capture mainly involves the identification of the key concepts and relationships in the domain of interest; The main objectives of this paper are to build a granulation model for representing approximate concepts and approximate ontologies, and present algorithms for automatically generating approximate ontologies based on granular computing from information tables.

The rest of the paper is organized as follows. In the second section, based on granular computing, a granulation model for representing approximate ontologies is introduced. In the third section, algorithms for capturing approximate concepts and generating approximate ontologies are described. A real world example that proves the algorithms useful is illustrated in the fourth section. Finally the paper gives conclusion and directs the future work.

2 Granulation Model for Approximate Ontologies

Generally, an ontology O (called lightweight ontology) can be defined as a 3-tuple: $O = (C, I, R)$, where C is a set of classes, which define the concepts used to the real object description; I is a set of instances, which represents the instance of the concept defined in the set of classes; R is a set in relations on the set of classes. So, capturing concepts and their relations from the given data source is a main procedure in constructing the ontology of the domain of interest. In the paper, we discuss the generating of approximate ontologies from information tables or structural data sources. We firstly create a granulation model of approximate ontologies.

Definition 1. (Information Table) Let IT = (U, A, V, f) be an information table, where U is the universe of discourse objects, $A = \{a_1, ..., a_{|A|}\}$ is a set of attributes, $V = \{V_{a_1}, ..., V_{a_{|A|}}\}$ is a set of attribute values, and f: $U \times A \to V$ is an information function.

Definition 2. (Tolerance Function) Let $\tau: U \times U \to [0,1]$ be a tolerance function on universal set U, such that $\forall x, y \in U$, $\tau(x, x)=1$ and $\tau(x, y)=\tau(y, x)$.

Definition 3. (Parameterized Tolerance Relation) Given a tolerance function, the parameterized tolerance relation is defined as: $\tau_p=\{(x, y)| \tau(x, y) \geq p \}$, where $p \in [0,1]$ is a real number called threshold value.

Definition 4. (Tolerance Space) A tolerance space is defined as a 3-tuple $TS=<U, \tau, p>$.

Definition 5. (Tolerance Class) Let $Y \subseteq U$ be a tolerance class satisfying $x \tau_p y | \forall x, y \in Y$.

Definition 6. (Neighborhood Function) Neighborhood Function with respect to τp is defined as: $n^{\tau p} (u \in U) \overset{def}{=} \{u' \in U | \tau_p (u, u')\}$

Considering a tolerance relation between sets, we introduce the following definitions.

Definition 7. (Extension Inclusion Function) Let 3-tuple $TS=<U, \tau, p>$ be a tolerance space and $U_1, U_2 \subseteq U$, then an extension inclusion function on the tolerance space TS is defined as : $V_{TS}(U_1, U_2) \overset{def}{=} \{\begin{matrix} \frac{|\{u_1 \in U_1 | \exists u_2 [u_1 \in n^{\tau_p}(u_2)]\}|}{|U_1|} & if \quad U_1 \neq \phi \\ 1 & otherwise \end{matrix}$

Definition 8. (Power Tolerance Space) Let 3-tuple $TS=<2^U, \tau, p>$ be a power tolerance space, such that $(1) U_1, U_2 \in 2^U, \quad \tau(U_1, U_2) \overset{def}{=} \min\{V_{TS}(U_1, U_2), V_{TS}(U_2, U_1)\}$ $(2) p \in [0, 1]$

Definition 9. (Extension Neighborhood Function) An extension neighborhood function is defined as $n^{\tau_p}(u \in 2^U) \overset{def}{=} \bigcup u' \in 2^U | \tau_p(u, u')$

Information granulation is a grouping of elements based on their indistinguishability, similarity, proximity or functionality [13]. This definition does not reflect the fact that information granules should be considered as being semantically distinct from the granulated entities. However, on practical point of view, information granulation should be considered as a semantically meaningful grouping of elements. So, in our research, we define a concept granule as follows:

Definition 10. (Concept Granule) Let 3-tuple CG=(EG,IG,RG) be a concept granule, where EG is the extension of the concept granule, IG is the intension of the concept granule, RG is the abstract representation or the semantic representation of the concept granule, standing for some classifying (or fuzzy classifying) characters.

The EG is related to a tolerance class or a neighborhood. IG and RG are specified according to the context of application domains. For example, let U be a set of students, and EG is a subset of U, if IG is a condition that satisfies $x \in EG | students' grade \leq 6$, then RG may be denoted by "pupil" to reflect the abstract meaning of EG. In this paper, IG denoted by a vector stands for the center point of the concept granule. RG is described by a 2-tuple and stands for the meaning of the concept granule. In the given real world example of this paper, IG and RG are generated by function operations. We call concept granules generated from the tolerance space as approximate concepts.

Definition 11. (Granularity of Concept Granule) Let $G_1 = (eg_1, ig_1, rg_1)$ and $G_2 = (eg_2, ig_2, rg_2)$ are two concept granules, if $eg_1 \subseteq eg_2$, namely G_1 included in G_2, then we call that the granularity of the G_1 is finer than that of the G_2 in the intent of the concept granule, in other words, the intent of the G_1 is more concrete than that of the G_2 or the granularity of the G_2 is coarser than that of the G_1, denoted by $G_1 \prec G_2$.

The granularity of concept granules not only reflect the inclusion relation between concept granules, but also show that there is a hierarchy among concept granules.

Definition 12. (Concept super-granule and sub-granule) Let $G_1 = (eg_1, ig_1, rg_1)$ and $G_2 = (eg_2, ig_2, rg_2)$ are two concept granules. If $G_1 \prec G_2$, then the G_2 is defined as a concept super-granule of G_1, and G_1 is a concept sub-granule of G_2. If there does not exist a concept granule $G_3 = (eg_3, ig_3, rg_3)$, such that $G_1 \prec G_3 \prec G_2, G_1 \neq G_3$ and $G_2 \neq G_3$, then G_2 is called as the direct concept super-granule of G_1 (parent-granule, for short), and G_1 is the direct concept sub-granule of G_2 (child-granule, for short).

Definition 13. (Approximate Ontology) Let $TS = (U, \tau, p)$ be a tolerance space, then 3-tuple $O_{TS} = (U, CG, \tau_p)$ is defined as an approximate ontology on the tolerance space, where CG is a set of concept granules (or approximate concepts).

3 Algorithms for Approximate Ontologies Building

After a granulation model of approximate ontologies is introduced, in this section, capturing approximate concepts and generating approximate ontologies are discussed. Firstly, according to the given tolerance relation and the given tolerance threshold value, information granules, called the initial concept granules, are formed. Secondly, we try to obtain new concept granules on the basis of existing concept granules by computing each extension neighborhood function until no new concept granule is formed. Finally, we can build approximate ontologies by generating the tree structure of concept granules.

Let IS= (U, A, V, f) be an information table, where the number of individuals is |U|=N, the number of attributes is |A|=M. Supposing every attribute is a quantitative attribute and the attribute values of the uth individual is denoted by $[v_{u1}, ..., v_{uM}]$, then the intension of the concept granule is described by IG[1,...,M].

3.1 Generating Intension and Abstract Meaning of Concept Granules

Function F_IG(u):
Input: the extension of concept granules;
Output: the intension of concept granules;
F_IG (EG) {IG [1... M] = [0... 0]   ; //initiate the intension of a concept granule
   For all u∈ EG do { IG[1... M] =IG[1... M] +$[v_{u1}, ..., v_{uM}]$;}
   IG[1,...,M]=IG[1,...,M]/|EG|; // |EG| is the cardinality of set EG
   Return IG[1,...,M];}
Function F_RG(u):
Input: the extension of concept granules
Output: the abstract meaning of concept granules
F_RG(EG){Min=Max=0;
   For all u∈ EG do{If Min>($v_{u1} + v_{u2} + ... + v_{uM}$) then Min=$v_{u1} + v_{u2} + ... + v_{uM}$
     Else if Max<($v_{u1} + v_{u2} + ... + v_{uM}$) then Max=$v_{u1} + v_{u2} + ... + v_{uM}$ }
   return {Min, Max};}

3.2   Capturing Approximate Concepts and Generating Approximate Ontologies

Algorithm 1. Capturing the initial concept granules from the given information table

Input: information table (U, A, V, f), (let |A|=M,|U|=N), tolerance functionτand tolerance threshold value p.

Output: the set CG0 of the initial concept granules

Algorithm description:

```
{  CG0={ };i=1;
     For all u1∈ U do{
        EG={u} ; //initiate the extension of a concept granule
        while (all u2∈ U and u1<>u2) do{
          if (τ(u1,u2)>p) then {
             EG=EG ∪ {u2};
          }}}
     IG[1,...,M]=F_IG(EG); //generate the intension of the concept granule
     RG=F_RG(EG); //generate the abstract meaning of the concept granule
     G=(EG,IG,RG); //generate the concept granule
     CG0=CG0 ∪ G;
}
```

Algorithm 2. Generating new concept granules that belong to the domain ontology from the initial concept granules

Input: the initial concept granules

Output: all concept granules

Algorithm description:

```
{ CG=the set that composes of the extension of concept granules in the initial concept
granules CG0
    C1={ };
    Do{ For all u∈ CG do{
          New_EG= n^{τp} (u) ;//extension neighborhood function
          If New_EG∉ CG then { //if it is a new concept granule
             C1=C1 ∪ New_EG; CG=CG ∪ New_EG}
          else  New_EG={ };
      }
    }while New_EG!={ }
    for all EG∈ C1 do{ //compute the intension and abstract meaning of new concept
granule
       New_IG=F_IG(EG);
     New_RG=F_RG(EG);
     G=(New_EG,New_IG,New_RG); CG0=CG0 ∪ G;}
  CG0=CG0 ∪ U;
}
```

Algorithm 3. Constructing tree structure of concept granules

Input: the set CG0 of concept granules

Output: every parent concept granule of concept granules

Note: something like X_EG stands for the extension of the concept granule X, and f(u∈ CG0) means the parent concept granule of the concept granule u.

Algorithm description:
{ f(root)={ } ; //the root node has no parent concept granule
  Temp=CG0-{root}; Temp1=CG0;
  For all u∈ Temp do{
  f(u)=root;//initiate the parent concept granule of the concept granule u.
    Temp1=Temp1-{u};
   While temp1!={ } do {
      u1∈ Temp1; temp1=temp1-{u1}
      if (u_EG⊂u1_EG and u1_EG⊂ f(u)_EG) then f(u)=u1
}}}

From the above analysis, it is clear that the worst time complexity of the algorithms is $O(N^2)$, where N is the number of individuals in a given information table.

4 A Real World Example

The information table about students' assessments is showed as follows. Having been preprocessed, every domain of attribute values is classified into five grades (that is 1, 2, 3, 4, 5). The bigger the number is, the higher assessment the students would get. Let IS=(U, A, V, f) stand for the information table, where U={1,2,3,4,5, 6,7,8,9,10,11,12}, A={moral, discipline, practice}, and the values of attributes V={1, 2, 3, 4, 5}.

Table 1. Information table

No.	Moral	Discipline	Practice
1	2	4	1
2	2	4	1
3	2	2	1
4	2	4	1
5	4	2	4
6	2	2	4
7	4	4	4
8	4	4	1
9	2	2	1
10	4	4	4
11	4	2	4
12	2	4	4

In order to compute conveniently, according to the characters of the real world example, this paper defines the tolerance relation as follows:
$\tau(u_1,u_2) = 1 - \left| \sum_{i=1}^{M} x_i - \sum_{i=1}^{M} y_i \right| / N$, where $u_1, u_2 \in U$, x_i is the ith attribute value of individual u_1, and y_i is the ith attribute value of individual u_2. Let the tolerance parameter p=0.8.

According to the tolerance relation and the neighborhood function, we can obtain tolerance classes as follows:

n(1)=n(2)=n(4)={1,2,3,4,6,8,9};n(3)=n(9)={1,2,3,4,9};

n(5)=n(11)=n(12)={5,6,7,8,10,11,12};n(6)=n(8)={1,2,4,5,6,8,11,12};n(7)=n(10)= {5,7,10,11,12}

According to algorithm 1, the set of the initial concept granules is:

CG0={(({1,2,3,4,6,8,9}, (2.29,3.13,1.43),{5,9}), ({1,2,3,4,9}, (2,3.2,1), {5,7}), ({5,6,7,8,10,11,12}, (3.43,3.14,3.57),{8,12}), ({1,2,4,5,6,8,11,12}, (2.75,3.25,2.5), {7,10}), ({5,7,10,11,12},(3.6,3.2,4), {10,12}))}

From algorithm 2, the set of new concept granules is:

C1={(({1,2,4,5,6,7,8,10,11,12}, (3,3.4,3),{7,12}))}, And the set of concept granules is:

CG0 = {(({1,2,3,4,6,8,9}, (2.29,3.13,1.43), {5,9}), ({1,2,3,4,9}, (2,3.2,1), {5,7}), ({5,6,7,8,10,11,12}, (3.43,3.14,3.57), {8,12}), ({1,2,4,5,6,8,11,12}, (2.75,3.25,2.5), {7,10}), ({5,7,10,11,12},(3.6,3.2,4), {10,12}), ({1,2,4,5,6,7,8,10,11,12}, (3,3.4,3), {7,12}), ({1,2,3,4,5,6,7,8,9,10,11,12}, (2.83,3.17,2.5),{5,12}))}

By using algorithm 3, the tree structure of concept granules is constructed, as shown in figure 1.

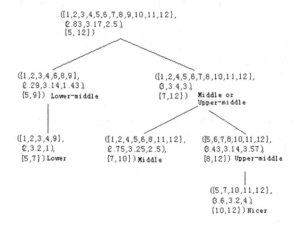

Fig. 1. Approximate ontology obtained from the given information table

From figure 1, it's obvious that the concepts and their relations are represented. And it is convenient to capture the student's general assessments in the context of the information table, namely, the concepts of the real world example can be classified to six concepts which are called lower, lower-middle, middle, upper-middle, middle/upper-middle and good, respectively.

5 Conclusion and Future Work

Approximate ontologies are the generalization of standard ontologies, and are also considered essential to the success of the semantic Web. In this paper, a granulation model for the representation of approximate ontologies is built. Based the principle of granular computing, algorithms for capturing approximate concepts and generating approximate ontologies are proposed and illustrated with a real example. It shows that the proposed algorithms are useful and effective.

Further work involves some aspects, such as perfecting the algorithms, studying the general framework of approximate ontologies based on granular computing and doing experiments, etc.

Acknowledgement. This study is supported by National Natural Science Foundation of P.R. China (NSFC- 60173054).

References

1. Klinov, P., Mazlack, L.J.: Granulating Semantic Web Ontologies. In: The Proceedings of International Conference on Granular Computing, IEEE, Atlanta, Georgia, USA (2006) 225-229
2. Stoilos, G. Stamou, G., Tzouvaras, V., Pan, J., Horrocks, I.: Fuzzy OWL: Uncertainty And The Semantics Web. In: In Proc. Of the International Workshop on OWL: Experience and Directions (2005)
3. Straccia, U.:Towards A Fuzzy Description Logic For The semantic Web. In: In Proc. 2nd European Semantic Web Conference(ESWC-05), Lecture Notes in Computer Science, 3532, A. Gomez-Perez, J. Euzenat(Eds.), Springer Verlag, Crete (2005)167-181
4. Doherty, P., Grabowski, M., Lukaszewicz, W., Szalas, A.: Towards A Framework For Approximate Ontologies. Fundamenta Informaticae, 57 (2003) 147-165
5. Liu, Q.: Rough Sets and Rough Reasoning. Science Press. Beijing, China (2003)
6. Liu, Q.: Granules and Applications of Granular Computing in Logical Reasoning. Journal of Computer Research and Development (In Chinese) 41 (2004) 546-551
7. Zadeh, L.A.: Fuzzy Sets and Information Granularity. In Advances in Fuzzy Sets Theory and Applications, edited by M.Gupta, R. Ragade and R. Yager, North Holland, Amsterdam, (1979) 3-18
8. Stepaniuk, J., Skowron, A.: Ontological Framework for Approximation. In: RSFDGrC, LNAI 3641 (2005) 718-727
9. Yao, Y.Y.: Perspectives of Granular Computing. In: Proceedings of IEEE International Conference on granular computing, Beijing, China (2005) 85-90
10. Studer R., Benjamins, V. R., Fensel, D.: Knowledge Engineering Principles and Methods. Data and Knowledge Engineering 25 (1998) 161-197
11. Poli, P.: Ontological Methodology. Int. J. Human-Computer Studies 56 (2002) 639-664
12. Deng, Z. H., Tang, S. W., Zhang, M.: Overview of Ontology. Journal of Peking University (natural Science, In Chinese) 38 (2002) 730-737
13. Yao, Y.Y.: Granular Computing. Computer Science (in Chinese) 31(2004) 1-5
14. Lin, T.Y.: Granular Computing II: Infrastructures for AI-Engineering. In: The Proceedings of International Conference on Granular Computing, IEEE, Atlanta, USA (2006) 2-7
15. Du, X. Y., Li, M., Wang, S.: A Survey on Ontology Learning Research. Journal of Software (in Chinese) 17 (2006) 1837-1847

Fuzzy-Valued Transitive Inclusion Measure, Similarity Measure and Application to Approximate Reasoning

Hong-Ying Zhang and Wen-Xiu Zhang

Institute for Information and System Sciences, Faculty of Science,
Xi'an Jiaotong University, Xi'an, Shaan'xi 710049, P.R. China
{zhyemily,wxzhang}@mail.xjtu.edu.cn

Abstract. In fuzzy set theory, inclusion measure indicates the degree to which a given fuzzy set is contained in another fuzzy set. Many inclusion measures taking values in [0,1] have been made in the literature. This paper proposes a series of fuzzy-valued inclusion measures which, by a relation view, are reflexive, antisymmetric and $\mathscr{T}$-transitive where $\mathscr{T}$ is a left-continuous triangular norm; In addition, they possess most of the axiomatic properties which are postulated by Sinha and Dougherty for an inclusion measure. Fuzzy-valued similarity measures are also defined by the fuzzy-valued inclusion measures; They have $\mathscr{T}$-transitivity and properties introduced by Liu for a similarity measure. Lastly two methods for inference in approximate reasoning based on the fuzzy-valued inclusion measure and the fuzzy-valued similarity measure are studied.

Keywords: Fuzzy-valued inclusion measure, Fuzzy-valued similarity measure, Fuzzy inference.

1 Introduction

Inclusion measure is an important concept in the area of fuzzy sets. It is a generalization of the existed approximate reasoning, such as probability reasoning, fuzzy inference, evidential reasoning and so on [1]. It has also been introduced successfully into rough set theory in [2-7] and fuzzy concept lattice theory in [8]. It surfaces in knowledge discovery, tuning rules and determining the coincidence measure of rules in fuzzy logic. A related concept is a measure of similarity between two fuzzy sets. The similarity measure is a relation which can be seen as a fuzzification of a crisp equivalence relation.

The study of inclusion measure is developed by at least two ways, namely the constructive and axiomatic approaches. In the constructive approach, the inclusion measures were constructed by fuzzy implication operators and conditional probability [1-4,9-13]. The constructive approach is suitable for practical applications of inclusion measure [2-4,9-10].

On the other hand, the axiomatic approach is appropriate for studying the structures of inclusion measure. In this approach, a set of axioms [9,10,13,14] are used to characterize inclusion operators that are the same as the ones produced

J.T. Yao et al. (Eds.): RSKT 2007, LNAI 4481, pp. 435–442, 2007.

by using the constructive approach . Sinha and Dougherty [13] first listed nine axiomatic properties that a reasonable inclusion measure should have.

Although the inclusion measures mentioned in these papers have been used successfully for various applications, they have shortcomings. First, they lack transitivity. In [15], B.De Baets et al. mentioned that it was surprising that the fuzzy inclusion measure of Kosko[16] and Fan et al. [10] were even not Z- transitive. In [17], A.Kehagias et al. proposed that it was reasonable to assume that when fuzzy set A is included in fuzzy set B to a degree x and B is included in fuzzy set C to a degree y, then A is included in C to a degree equal to or greater than min(x,y). As for crisp set, we know the inclusion relation has transitivity, namely, $A \subseteq B, B \subseteq C \Rightarrow A \subseteq C$. We think it's natural that fuzzy inclusion measure should have transitivity at some degree. In addition, scalar inclusion measures may be quite sensitive to small changes in the membership of elements (sometimes even of a single element) to the fuzzy sets considered [17]. So it is an interesting question whether an inclusion measure can be introduced which is transitive and not overly sensitive to the membership of single elements. A.Kehagias et al. introduced a L-fuzzy valued transitive inclusion measure $I(A,B)$ which took values in a partial order set of vectors composed of 0 and 1. This paper will show L-fuzzy valued inclusion measure in [17] is too special to show difference between fuzzy sets.

In section 2, the basic notions are introduced. In section 3, the proposed generalized fuzzy-valued inclusion measures prove to be $\mathscr{T}$-transitive and satisfy most of the axioms postulated by Sinha and Dougherty. Example illustrates fuzzy-valued inclusion measure defined by us can show the difference between fuzzy sets. The related fuzzy-valued similarity measures are defined by the $\mathscr{T}$-transitive fuzzy-valued inclusion measure and their properties are discussed. In section 4, we propose a method for inference in approximate reasoning based on the fuzzy-valued inclusion measure and the fuzzy-valued similarity measure. Section 5, the conclusion.

2   Preliminaries

By an implicator we mean a function $\mathscr{I} : I^2 \to I$ satisfying $\mathscr{I}(1,0) = 0$ and $\mathscr{I}(1,1) = \mathscr{I}(0,1) = \mathscr{I}(0,0) = 1$.

Some axioms have been postulated by Smets and Magrez [18] as axiomatically appropriate for an implicator. Such as Hybrid Monotonicity: $\forall (x,y) \in [0,1]^2$, $\mathscr{I}(.,y)$ is decreasing, yet $\mathscr{I}(x,.)$ is increasing; Confinement Principle(CP principle): $\forall (x,y) \in [0,1]^2, x \leq y \Longleftrightarrow \mathscr{I}(x,y) = 1$ and Border Principle: $\forall x \in [0,1], \mathscr{I}(1,x) = x$.

An R-implicator (residual implicator) based on a left-continuous t-norm $\mathscr{T}$ if for every x, y $\in [0,1]$, $\mathscr{I}(x,y) = \sup\{\gamma \in [0,1], \mathscr{T}(x,\gamma) \leq y\}$.

Proposition 1. ([19]) *Every R-implicator is Hybrid monotonic, Border and CP.*

In the following sections, $(\mathscr{F}(X), \leq)$ is a partial order set where $A \leq B \Leftrightarrow A(x_i) \leq B(x_i), \forall x_i \in X = \{x_1, x_2, ..., x_n\}, [a] = (a, ..., a); A \cap B$ and $A \cup B$ are denoted by $(A \cap B)(x) = \min(A(x), B(x))$ and $(A \cup B)(x) = \max(A(x), B(x))$.

Definition 1. *A fuzzy relation R is a $\mathscr{T}$-transitive fuzzy relation iff $R(A,C) \geq \sup_{B \in \mathscr{F}(X)} \mathscr{T}(R(A,B), R(B,C))$, for all $A, B, C \in \mathscr{F}(X)$; A fuzzy relation R is called a fuzzy $\mathscr{T}$ similarity relation if it is reflexive, symmetric and $\mathscr{T}$-transitive; R is a $\mathscr{T}$-fuzzy order relation if it is reflexive, antisymmetric, and $\mathscr{T}$-transitive.*

3 $\mathscr{T}$-Transitive Fuzzy-Valued Inclusion Measure, Similarity Measure

3.1   Fuzzy-Valued Inclusion Measure

A. Kehagias et al [17] gave some examples to show that the locally determined scalar inclusion measure enjoyed many properties at the cost of great sensitivity to isolated membership values of two fuzzy sets. They defined a global fuzzy inclusion measure in the sense that the influence of the membership of one element is averaged over all elements as follows:

Definition 2. [17] *For all $A, B \in \mathscr{F}(X)$, the measure of inclusion of A in B is a fuzzy relation, denoted by $\mathrm{I}(A,B)$. The value of $\mathrm{I}(A,B)$ is defined for each $x \in X$ by $\mathrm{I}_x(A,B) = 1$ iff $A(x) \leq B(x)$ and $\mathrm{I}_x(A,B) = 0$ else.*

Example 1. Three fuzzy sets A, B, C, $A = 0.1/x_1 + 0.8/x_2 + 0.9/x_3$, $B = 0.1/x_1 + 0.1/x_2 + 0.1/x_3$ and $C = 0.1/x_1 + 0.79/x_2 + 0.89/x_3$. Then we have $I(A,B) = (1,0,0) = I(A,C) = (1,0,0)$.

It is obvious that the inclusion measure ignored the great difference between fuzzy sets B and C.

From the above analysis, we propose a series of fuzzy-valued inclusion measures which keep the transitivity pointed out in [17]. In addition, these inclusion measures are more precise to show the difference between fuzzy sets and possess most of axioms made for inclusion measure in [13].

Definition 3. *For all $A, B \in \mathscr{F}(X)$, a fuzzy-valued inclusion measure is denoted by*

$$I(A,B) : \mathscr{F}(X) \times \mathscr{F}(X) \to \mathscr{F}(X) \tag{1}$$

where $I(A,B)(x) = I(A(x), B(x)) = \mathscr{I}(A(x), B(x))$ for each $x \in X$ and $\mathscr{I}$ a R-implicator.

Example 2. Continuing with the fuzzy sets in example 3.1, we take Łukasiewicz implicator $\mathscr{I}_L(x,y) = \min(1, 1 - x + y)$. We have $I(A,B) = (1, 0.3, 0.2) \leq I(A,C) = (1, 0.99, 0.99)$ which displays the difference between B and C.

Lemma 1. *Let a, b, $c \in [0,1]$, $\mathscr{T}$ is a left-continuous t-norm , then we have $\sup_{b \in [0,1]} \mathscr{T}(\mathscr{I}(a,b), \mathscr{I}(b,c)) \leq \mathscr{I}(a,c)$ where $\mathscr{I}$ is the R-implicator based on $\mathscr{T}$.*

Theorem 1. *For all $A, B, C, D \in \mathscr{F}(X)$, $\mathcal{I}$ is a finite index set, the inclusion measure $I(A,B)$ defined by formula (1) has the following properties:*

(I1)$I(A, B) = [1] \Leftrightarrow A \leq B$;*(I2)*$A \leq B \Rightarrow I(B, C) \leq I(A, C)$;*(I3)*$A \leq B \Rightarrow I$
$(C, A) \leq I(C, B)$;*(I4)*$I(A, \bigcap_{i \in I} B_i) = \bigcup_{i \in I} I(A, B_i)$;*(I5)* $I(\bigcup_{i \in I} A_i, B) = \bigcap_{i \in I}$
$I(A_i, B)$;*(I6)* if $[1/2] \leq A$, then $I(A, A^c) = [0] \Leftrightarrow A = X$;*(I7)* $I(A, B) \leq I(A \cap$
$C, B \cap C)$;*(I8)*$I(A, B) \leq I(A \cup C, B \cup C)$;*(I9)* $I(A, B) \leq I(\mathscr{I}(A, C), \mathscr{I}(B, C))$
(I10)$\bigcup_{i \in I} I(A, B_i) \leq I(A, \bigcup_{i \in I} B_i)$;*(I11)*$\bigcup_{i \in I} I(A_i, B) \leq I(\bigcap_{i \in I} A_i, B)$;*(I12)*$I$
$(A, B) \cap I(C, D) \leq I(A \cap C, B \cap D)$;*(I13)*$I(A, B) \cap I(C, D) \leq I(A \cup C, B \cup D)$;
(I14)$\bigcap_{i \in I} I(A_i, B_i) \leq I(\bigcap_{i \in I} A_i, \bigcap_{i \in I} B_i)$;*(I15)*$\bigcap_{i \in I} I(A_i, B_i) \leq I(\bigcup_{i \in I} A_i, \bigcup_{i \in I} B_i)$.

Proof. (I1)For all $A, B \in \mathscr{F}(X)$, $A \leq B \Leftrightarrow A(x) \leq B(x)$, $\forall x \in X$, then
$I_x(A(x), B(x)) = \mathscr{I}(A(x), B(x)) = 1$, so $I(A, B) = [1]$. We can easily get I2-I5,
I7-I15 by the properties of R-implicator and t-norm. We just show the proof of
I6 as follows:

(I6) When A=X, we have $\mathscr{I}(X(x), \emptyset(x)) = 0$, then: $I(X, \emptyset)(x) = \mathscr{I}(X(x),$
$\emptyset(x)) = \mathscr{I}(1, 0) = 0$. Namely $I(X, \emptyset) = [0]$. On the contrary, for $[1/2] \leq A$
and $I(A, A^c) = [0]$, which implies $I(A, 1 - A)(x) = 0, \forall x \in X$. Suppose that
$A \neq X$, then there exists some $x \in X$ satisfies then we have $1/2 \leq A(x) < 1$ and
$0 < 1 - A(x) \leq 1/2$. By the border property and hybrid monotonic property of
$\mathscr{I}$, we have $I(A, 1-A)(x) = \mathscr{I}(A(x), 1-A(x)) \geq \mathscr{I}(1, 1-A(x)) = 1-A(x) \neq 0$.
It is a contradiction to $I(A, A^c) = [0]$; So the suppose is not true, namely, $A = X$.
By properties of t-norm and R-implicator, it's not difficult to prove the rest.

Remark 1. $I(A, B)$ defined by formula (1) has most of the axioms postulated in
[13] by Theorem 1.

Theorem 2. $I(A, B)$ *defined by formula (1) is a $\mathscr{T}$-fuzzy order relation on*
$\mathscr{F}(X)$.

Proof. Take any $A, B, C \in \mathscr{F}(X)$. We have the following.

1. Reflexivity: For $\forall x \in X$, we have $A(x) \leq A(x)$, then $I(A, A)(x) = \mathscr{I}(A(x),$
$A(x)) = 1 \Rightarrow I(A, A) = [1]$.

2. Antisymmetry: For all $B, A \in \mathscr{F}(X)$, $I(A, B) = I(B, A)$,namely $I(A, B)(x)$
$= I(B, A)(x)$. For any $x \in X$, $A(x)$ and $B(x)$ must satisfy one of the two conditions: $A(x) \leq B(x)$ or $A(x) \geq B(x)$. So we suppose $A(x) \leq B(x)$ is true, then
we have that $\mathscr{I}(B(x), A(x)) = I(B, A)(x) = I(A, B)(x) = \mathscr{I}(B(x), A(x)) = 1$,
then we get $B(x) \leq A(x)$ by the CP principle. Combined with the suppose that
$A(x) \geq B(x)$, then we have $A(x) = B(x)$, for all $x \in X$. Namely, we have $A = B$.

3. $\mathscr{T}$-transitivity: By Lemma 3.1, for all $A, C \in \mathscr{F}(X)$ and $\forall x \in X$, we
have $\sup_{B \in \mathscr{F}(X)} \mathscr{T}(\mathscr{I}(A(x), B(x)), \mathscr{I}(B(x), C(x))) \leq \mathscr{I}(A(x), C(x))$. Namely,
for $\forall A, C \in \mathscr{F}(X)$, $\sup_{B \in \mathscr{F}(X)} \mathscr{T}(I(A, B)(x), I(B, C)(x)) \leq I(A, C)(x)$. That
mean $\sup_{B \in \mathscr{F}(X)} \mathscr{T}(I(A, B), I(B, C)) \leq I(A, C)$.

Since some kind of transitivity of fuzzy-valued inclusion measure is the requirement of the rational generalization of crisp inclusion relation, we have proved that
the fuzzy-valued inclusion measure has $\mathscr{T}$-transitivity and many good properties.

3.2   Fuzzy-Valued Similarity Measure

Model recognition and rule matching are common to expert system, fuzzy control
and fuzzy neural network. The similarity measure is employed to determine

whether a model or a rule should be matched with the expectation. Many authors [20-25]have paid much attention to the theoretic study and application of fuzzy similarity measure.In this section, we first state definition of Liu [23].

Definition 4. ([23]) *A fuzzy-valued function* $S : \mathscr{F}(X) \times \mathscr{F}(X) \rightarrow \mathscr{F}(X)$ *is called a fuzzy-valued similarity measure, if S has the following properties:*
   (SP1) $S(A, A) = [1], \forall A, B \in \mathscr{F}(X)$;
   (SP2) $S(A, B) = S(B, A), \forall A, B \in \mathscr{F}(X)$;
   (SP3) $S(D, D^c) = [0], \forall D \in \mathscr{P}(X)$;
   (SP4) $\forall A, B, C \in \mathscr{F}(X)$, *if* $A \leq B \leq C$, *then* $S(A, C) \leq S(A, B) \cap S(B, C)$.

Definition 5. *For all* $A, B \in \mathscr{F}(X)$, *we define* $S(A, B) = \mathscr{T}(I(A, B), I(B, A))$ *which is determined by*

$$S(A, B)(x) = \mathscr{T}(I(A, B)(x), I(B, A)(x)) \tag{2}$$

for each $x \in X$, *where* I *is a fuzzy-valued inclusion measure defined by formula (1) and* $\mathscr{T}$ *is a left continuous t-norm.*

Theorem 3. *For all* $A, B \in \mathscr{F}(X)$, *S(A,B) defined by formula (2) is a fuzzy-valued similarity measure.*

Proof. (SP1) and (SP2) are obviously true.
   (SP3) For $D \in \mathscr{P}(X)$ and $\forall x \in X$, we have $I(D, D^c)(x) = 0(x \in D)$ or $I(D^c, D)(x) = 0(x \in D^c)$. So $S(A, B)(x) = \mathscr{T}(I(D, D^c)(x), I(D^c, D)(x)) = 0$, then: $S(D, D^c) = [0], \forall D \in \mathscr{P}(X)$;
   (SP4) $\forall A, B, C \in \mathscr{F}(X)$, if $A \leq B \leq C$, then $S(A, C) = \mathscr{T}(I(A, C), I(C, A))$ $= I(C, A)$. By the same way, we have $S(A, B) \cap S(B, C) = I(B, A) \cap I(C, B)$. Since $I(C, A) \leq I(B, A), I(C, A) \leq I(C, B)$, we get $S(A, C) \leq S(A, B) \cap S(B, C)$.

Theorem 4. *For all* $A, B \in \mathscr{F}(X)$, $I(A, B)$ *is a fuzzy-valued inclusion measure defined by formula (1), then* $S(A, B) = I(A, B) \cap I(B, A)$ *is a fuzzy* $\mathscr{T}$-*similarity relation on* $\mathscr{F}(X)$ *where* $\mathscr{T}$ *is a left-continuous t-norm and* $\mathscr{I}$ *is the R-implicator based on* $\mathscr{T}$.

Proof. For all $A, B, C \in \mathscr{F}(X)$, it's obvious that $S(A, B) = I(A, B) \cap I(B, A)$ is reflexive and symmetric. Now we show S is $\mathscr{T}$-transitive, $\forall x \in X$, $\mathscr{T}(S(A, B)(x)$, $S(B, C)(x)) = \mathscr{T}(I(A, B)(x) \wedge I(B, A)(x), I(B, C)(x) \wedge I(C, B)(x)) \leq \mathscr{T}(I(A, B)(x), I(B, C)(x)) \leq I(A(x), C(x))$. By the same way, we get $\mathscr{T}(S(A, B)(x), S(B, C)(x)) \leq \mathscr{T}(I(B(x), A(x)), I(C(x), B(x))) \leq I(C(x), A(x))$. So we have $\sup_{B \in \mathscr{F}(X)} \mathscr{T}(S(A, B), S(B, C)) \leq S(A, C)$.

Corollary 1. *The similarity measure denoted by* $S(A, B) = I(A, B) \cap I(B, A)$ *satisfies that* $S(A, C) = I(A, B) \cap I(B, C)$ *for* $\forall A \leq B \leq C$.

Proof. Let $\mathscr{T} = \wedge$, we have $S(A, C) = I(A, B) \cap I(B, C)$ by Theorem 3 and Theorem 4.

Theorem 5. *For all* $A, B, C, D \in \mathscr{F}(X)$, $S(A, B) = I(A, B) \wedge I(B, A))$ *where* $\mathcal{I}$ *is the index set,* $\mathscr{I}$ *is a R-implicator. we have:*

(S1)$S(X, \emptyset) = S(\emptyset, X) = [0], S(\emptyset, \emptyset) = S(X, X) = [1]$.

(S2)$S(A, B) = [1] \Leftrightarrow A = B$.

(S3)$S(A, B) \cap S(C, D) \leq S(A \cap C, B \cap D)$.

(S4)$S(A, B) \cap S(C, D) \leq S(A \cup C, B \cup D)$.

(S5)$\bigcap_{i \in \mathcal{I}} S(A_i, B_i) \leq S(\bigcap_{i \in \mathcal{I}} A_i, \bigcap_{i \in \mathcal{I}} B_i)$

(S6)$\bigcap_{i \in \mathcal{I}} S(A_i, B_i) \leq S(\bigcup_{i \in \mathcal{I}} A_i, \bigcup_{i \in \mathcal{I}} B_i)$

It is not hard to prove Theorem 5 by the properties of implicator and fuzzy-valued inclusion measure.

4 Method for Inference in Approximate Reasoning Based on Fuzzy-Valued Inclusion Measure and Fuzzy-Valued Similarity Measure

Approximate reasoning is the process or processes by which a possible imprecise conclusion is deduced from a collection of imprecise premises. In this subsection we present an algorithm for obtaining the conclusion of the GMP(generalized modus ponens):

> If x is A then y is B
> If x is A'
> _____
> y is B'

Definition 6. *Let* $I(A_1, A_2)$ *and* $S(A_1, A_2)$ *be fuzzy-valued inclusion measure and similarity measure defined by formulas (1) and (2) respectively; For fuzzy sets* $A, A' \in \mathscr{F}(X)$ *and* $B \in \mathscr{F}(Y)$, *the conclusion is generated in the following ways:*

$$B'(y) = \bigwedge_{x_i \in X} \mathscr{T}(B(y), I(A', A)(x_i)), \tag{3}$$

$$B''(y) = \bigwedge_{x_i \in X} \mathscr{T}(B(y), S(A', A)(x_i)). \tag{4}$$

By the properties of fuzzy-valued inclusion measure, similarity measure and T-norm, such as the $\mathscr{T}$-transitivity, $\mathscr{T}(x, \bigvee_{i \in I} y_i) = \bigvee_{i \in I} \mathscr{T}(x, y_i)$ and so on, it is not difficult to obtain the following properties.

Proposition 2. *The fuzzy inference rule defined by formula (3) has the following characteristics:*

(P1) If $A' = A$, then $B' = B$; (P2) If $A' < A$, then $B' = B$;

(P3) For all A', A, then $B' \leq B$; (P4) If $A'_1 \leq A'_2$, then $B'_1 = B'_2$;

(P5) If $A_1 \leq A_2$, then $B'_1 \leq B'_2$; (P6) If $B'(y) = \bigwedge_{x_i \in X} \mathscr{T}(B(y), I(A', \bigcup_{j=1}^n A_j)(x_i))$, $B'_j(y) = \bigwedge_{x_i \in X} \mathscr{T}(B(y), I(A', A_j)(x_i))$, then $\bigcup_{j=1}^n B'_j \leq B'$;

*(P7) If $B'(y) = \bigwedge_{x_i \in X} \mathscr{T}(B(y), I(A', \bigcap_{j=1}^{n} A_j)(x_i))$, $B'_j(y) = \bigwedge_{x_i \in X} \mathscr{T}(B(y),$
$I(A', A_j)(x_i))$, then $\bigcup_{j=1}^{n} B'_j \leq B'$; (P8) If $B'(y) = \bigwedge_{x_i \in X} \mathscr{T}(B(y), I(\bigcup_{j=1}^{n} A'_j, A)$
$(x_i))$, $B'_j(y) = \bigwedge_{x_i \in X} \mathscr{T}(B(y), I(A'_j, A)(x_i))$, then $B' \leq \bigcap_{j=1}^{n} B'_j$;*

*(P9) If $B'(y) = \bigwedge_{x_i \in X} \mathscr{T}(B(y), I(\bigcap_{j=1}^{n} A'_j, A)(x_i))$, $B'_j(y) = \bigwedge_{x_i \in X} \mathscr{T}(B(y),$
$I(A'_j, A)(x_i))$, then $\bigcup_{j=1}^{n} B'_j \leq B'$;*

(P10) If $B'_C(y) = \bigwedge_{x_i \in X} \mathscr{T}(B(y), I(A' \cup C, A \cup C)(x_i))$, then $B' \leq B'_C$;

*(P11) If $B'_{\mathscr{T}}(y) = \bigwedge_{x_i \in X} \mathscr{T}(B(y), I(\mathscr{T}(A', C), \mathscr{T}(A, C))(x_i))$, then $B' \leq$
$B'_{\mathscr{T}}$;*

*(P12) If $B'(y) = \bigwedge_{x_i \in X} \mathscr{T}(B(y), I(A', A)(x_i))$, $B''(y) = \bigwedge_{x_i \in X} \mathscr{T}(B_1(y), I(A,$
$A_1)(x_i))$, $B'''(y) = \bigwedge_{x_i \in X} \mathscr{T}(B_1(y), I(A', A_1)(x_i))$, then $\bigwedge_{y \in Y} \mathscr{T}(B_1, \mathscr{T}(B', B''))$
$\leq B'''$.*

Proposition 3. *The fuzzy inference rule defined by formula (4) has the following characteristics:*

(L1) If $A' = A$, then $B'' = B$; (L2) If $A' < A$, then $B'' < B$;

*(L3) For all A', A, then $B'' \leq B$; (L4) If $A = D \in \mathscr{P}(X), A' = D^c$, then
$B'' = \emptyset$;*

*(L5) If $B''(y) = \bigwedge_{x_i \in X} \mathscr{T}(B(y), S(A', \bigcap_{j=1}^{n} A_j)(x_i))$ and $B''_j(y) = \bigwedge_{x_i \in X} \mathscr{T}$
$(B(y), S(A', A_j)(x_i))$, then $B'' \leq \bigcup_{j=1}^{n} B''_j$;*

*(L6) If $B''(y) = \bigwedge_{x_i \in X} \mathscr{T}(B(y), S(\bigcup_{j=1}^{n} A'_j, A)(x_i))$ and $B''_j(y) = \bigwedge_{x_i \in X} \mathscr{T}$
$(B(y), S(A'_j, A)(x_i))$, then $B'' \leq \bigcup_{j=1}^{n} B''_j$.*

5   Conclusion

We have introduced a group of fuzzy-valued transitive inclusion measures which are less sensitive to the small change of the membership value of single point and have proved that they are $\mathscr{T}$-fuzzy order relation. They possess most of the axioms postulated by Sinha and Dougherty. Furthermore, fuzzy-valued similarity measure defined by fuzzy-valued inclusion measure has proved to be the rational generalization of relation of equality. Finally, two methods for fuzzy inference based on the fuzzy-valued inclusion measure and the fuzzy-valued similarity measure have been introduced and studied in detail.

Acknowledgement

This work was supported by the National 973 Program of China(No.2002 CB312200).

References

1. Zhang, W.-X., Leung, Y.: The Uncertainty Reasoning Principles, Xi'an Jiaotong University Press, Xi'an (1996a)
2. Ma, Z., Zhang W., Ma W.: Assessment of data redundancy in fuzzy relational databases based on semantic inclusion degree, Information Processing Letters, 72 (1999) 25-29

3. Xu, Z.-B., Liang, J.-Y., Chen, D.-D. and Chin, K., Inclusion degree: a perspective on measures for rough set data analysis, Information Sciences, 141 (2002) 227-236
4. Qiu, G.,-F, Li, H.,-Z, Xu, L.,-D., Zhang, W.-X.: A knowledge processing method for intelligent systems based on inclusion degree, 20 (2003) 187-195
5. S. Bodjanova: Approximation of fuzzy concepts in decision making, Fuzzy Sets and Systems 85 (1997) 23-29
6. Zhang, W.-X., Qiu, G.,-F.: Uncertain Decision Making Based On Rough Sets, Tsinghua University Press, Beijing, 2005
7. Zhang, W.-X., Leung, Y., Wu, W.,-Z.: Information Systems and Knowledge Discovery, Science Press,Beijing, 2003
8. Fan, S.-Q., Zhang, W.-X., Xu, W.: Fuzzy inference based on fuzzy concept lattice, Fuzzy Sets and Systems 157(2006)3177-318
9. V.R. Young: Fuzzy subsethood, Fuzzy Sets and Systems 77 (1996) 371-384
10. Fan, J.-L , Xie, W., Pei, J.: Subsethood measure: new definitions, Fuzzy Sets and Systems 106 (1999) 201-209
11. R. Willmott: On the transitivity of containment and equivalence in fuzzy power set theory, J. Math. Anal. Appl. 120 (1986) 384-396
12. Zhang, W.,-X., Leung, Y.: Theory of including degrees and its applications to uncertainty inferences, Soft Computing in Intelligent Systems and Information Processing, New York: IEEE,(1996b) 496-501
13. D. Sinha, E.R. Dougherty: Fuzzication of set inclusion: theory and applications,Fuzzy Sets and Systems 55 (1993) 15-42
14. Chris Cornelis, Carol Van der Donck, Etienne kerre: Sinha-dougherty approach to the fuzzication of set incluaion revisited, Fuzzy Sets and Systems 134 (2003) 283-295
15. B.De Baets, H. De Meyer, H.naessens: On rational cardinality-based inclusion measure, Fuzzy Sets and Systems 128 (2002) 169-183
16. B.Kosko: Neural Networks and Fuzzy Systems, Prentice-Hall, Englewood Cliffs, NJ, 1992
17. A.Kehagias, M.Konstantinidou: L-fuzzy valued inclusion measure, L-fuzzy similarity and L-fuzzy distance, Fuzzy Sets and Systems, 136 (2003) 313-332
18. Smets P, Magrez P.: Implication in fuzzy logic, Int.J.Approximate reasoning, 1 (1987) 327-347
19. A.M. Radzikowska, E.E. Kerre: A comparative study of fuzzy rough sets, Fuzzy Sets and Systems 126 (2002) 137-155
20. C.P. Pappis, N.I. Karacapilidis: Application of a similarity measure of fuzzy sets to fuzzy relational equations,Fuzzy Sets and Systems 75 (1992) 135-142
21. C.P.Pappis, N.I. Karacapilidis: A comparative assessment of measures of similarity of fuzzy values,Fuzzy Sets and Systems 56 (1993) 171-174
22. Liu, X.-C.: Entropy, distance measure and similarity measure of fuzzy sets and their relations, Fuzzy Sets and Systems 52 (1992) 305-318
23. Fan, J.-L , Xie, W.: Some notes on similarity measure and proximity measure,Fuzzy Sets and Systems 101 (1999) 403-412
24. Wang, B.de Baets, E. Kerre: A comparative study of similarity measures,Fuzzy Sets and Systems 73 (1995) 259-268
25. Wang,W.-J.: New similarity measures on fuzzy sets and elements, Fuzzy Sets and Systems 85 (1997) 305-309

Model Composition in Multi-dimensional Data Spaces

Haihong Yu[1,2], Jigui Sun[1,2], Xia Wu[1,2], and Zehai Li[1,2]

[1] College of Computer Science and Technology, Jilin University,
130012 Changchun, China
[2] Key Laboratory of Symbolic Computation and Knowledge Engineering
of Ministry of Education,
130012 Changchun, China
{yuhh,jgsun,wuxia,zhli}@jlu.edu.cn

Abstract. Model composition is an important problem in model management. In this paper, we propose a new method to support model composition in multi-dimensional data spaces. We define a model as a 6-tuple with input interface and output interface. An algorithm for model composition and execution is given. Moreover, the method has been applied into a practical project. The running statistics showed that there had been 105 instances of model composition, and 89 decision problems had been effectively solved.

Keywords: Decision Support System, Model Management, Model Composition, Multi-dimension Data.

1 Introduction

Model-base is at the heart of Decision Support System (DSS), which consists of dialogue component, database and model-base. Model is viewed as computer executable program that typically requires data inputs[1]. For certain decision-making situations, a sequence of models is executed in order to solve a decision problem. Typically, inputs to a model in the sequence are obtained from outputs of other models upstream in the sequence and from database retrievals. The problem of composing a sequence of models is known as the model composition problem[2].

Data Warehouse (DW) and On-Line Analytical Processing (OLAP) are supporting technology for analytical processing. DSS integrated with DW and OLAP is known as DW-based DSS[3]. There are historical mass data organized in the form of multi-dimension in DW. The model execution based on DW improves the analytic ability of DSS. Multi-dimensional data became the main data source for models. Therefore, it is very important for integrating DW with DSS to research model composition in multi-dimension spaces. At present, the data source of OLAP comes directly from DW or database. But some useful data can not be obtained from DW, which has to be gotten by model computation. Model computation embedded in OLAP is helpful to improve analytical ability.

J.T. Yao et al. (Eds.): RSKT 2007, LNAI 4481, pp. 443–450, 2007.

Few papers on model composition are available in the literature[2,4]. There are mainly four types of logical representation of composite models: relational, graphical, knowledge-based, or script-based. In the relational approach[5], models are treated as virtual relations whose tuples are generated on demand. Two models are linked together by joining their corresponding virtual relations, thereby facilitating model composition. The relational approach does not provide any mechanisms to differentiate between two models that map the same inputs to outputs, even when they differ on underlying assumptions and mapping functions. In the graphical approaches[6,7], the interconnection of models is represented by the graphs so that to form the composite models. Models are viewed as nodes or edges of a graph. Graphical representations with the notable exception of Basu and Blanning[7] are typically weak in representing domain knowledge such as model preconditions or assumptions. In knowledge-based approaches[8], models are represented as components of a knowledge base and various reasoning mechanisms are used to construct composite models. Knowledge-based approaches are capable of representing model preconditions, model assumptions, and specifications pertaining to model input and output. In script-based approaches[9], model composition is achieved via predefined scripts. These scripts specify the sequence of model executions and data-formatting requirements.

In [10], Chari presented a new theoretical construct called filter spaces as the basis to support the capabilities of model composition. The filter spaces provide a means to represent constraints on data, and support for an inferential mechanism to determine if a collection of database and model resources satisfy stated constraints on data. But, the filter spaces must work only in relational data environment. We extended the approach to support multi-dimensional data.

This paper is organized as follows. Section 2 gives the representations of the model. Model composition in multi-dimension data environment are presented in section 3. Moreover, an algorithm for model composition and execution is given. Section 4 shows the results the method applied into a practical project. Finally, a conclusion is drawn.

2   Model Representations

In [11], we proposed a novel dimension model to support the modeling of irregular dimensions by defining a partial mapping from a child level to a parent level. We defined the operations of the isomorphism, project, select, join, union and intersection of multi-dimensional data set, as well as the dimension aggregation, roll-up and drill-down operations. The related concepts and notions in this paper can refer to [11].

Domain knowledge about models and data is used in selecting the "right" models and data during model composition and execution. This knowledge can be represented in DSS using a meta language. In [10], Chari presented an approach to support model composition based on the relational data. He defined the meta language as a first-order logic language. The related concepts, such as the filter clause, filter list, etc, were given. A filter clause is a binary predicate $P(x_1, x_2)$ when $P \in \{LT(<), LE(\leq), EQ(=), GT(>), GE(\geq)\}$ and x_1, x_2 are non-predicate terms. Or a filter clause is a disjunctive form of the binary predicates above. Usually, it needs more than one filter clauses to represent model assumptions and data scope in the practice problem. The filter clauses of the same problem are conjunctive relation. A set of these filter clauses is called filter list, which is in the conjunctive normal form. Furthermore, the filter space is used to describe data scope. It is in fact the value set of variables satisfied the filter list.

Chari[10] presented the approach of model composition in the relational data environment, but not in multi-dimensional data spaces. So we extend his approach to support multi-dimensional data. The relational data is viewed as a simple situation of multi-dimensional data[11]. Namely our approach can support both relational data and multi-dimensional data.

Definition 1. *The multi-dimensional data interface is a 4-tuple $MDI = (\Omega, L, \Lambda, \Sigma)$ where,*

(1)$\Omega = \{d_1, d_2, \cdots, d_n\}$ is a set of dimensions, where d_i is a dimension;

(2)$L = \{(d_i, l_{d_i}) | d_i = (D_i, \prec_i) \in \Omega, l_{d_i} \in D_i, 1 \leq i \leq n\}$ is a set, and d_i is on the level l_{d_i};

(3)$\Lambda = \{m_1, m_2, \cdots, m_k\}$ is a set of the facts, and $m_j = (M_j, agr_j), (1 \leq j \leq k)$;

(4)$\Sigma = \{P_1, P_2, \cdots, P_k\}$ is a filter list including dimension variables

The definition 1 shows the multi-dimensional data set $C = (\Omega, \Lambda, f)$ has multiple MDIs, and every MDI provides data for different queries. By rolling-up or drilling-down, the dimensions are associated with their levels specified by L. Then, the data set specified by the MDI can be gotten by the operations, such as project, select etc, on C.

Definition 2. *A model M is a six-tuple, $M = (N, I, O, A, F, S)$, where N is a string to identify the model, I is the MDI set representing model input data, O is the MDI set representing model output data, A is a filter list representing model assumptions, which are not related to model inputs, F is the filter list representing constraints among model inputs, S is a string encoding the physical location and access method of the model executable.*

A model includes the model identifier, followed by model input, output, assumptions, and model access information. For the set of model input MDIs, let $I = (MDI_1^i, MDI_2^i, \cdots, MDI_s^i), MDI_j = (\Omega_j^i, L_j^i, \Lambda_j^i, \Sigma_j^i), j = 1, 2, ..., s$, s is the number of model input MDIs. Every Σ_j^i can be satisfied and has the maximum dimensional property. Furthermore, two model inputs cannot have different filter clauses of Type 1b[10], namely the filter clauses with same predicate symbol and variable. This requirement maintaining the consistency among model inputs imposes restrictions on the type of models being represented. For the set of model output MDIs, let $O = (MDI_1^o, MDI_2^o, \cdots, MDI_t^o), MDI_j^o = (\Omega_j^o, L_j^o, \Lambda_j^o, \Sigma_j^o)$, $j = 1, 2, ..., t$, t is the number of model output MDIs. Every Σ_j^o can be satisfied and has the maximum dimensional property. And any Σ_j^o has additional *binding clauses*[10] from $\bigcup_{k=1}^{s} \Sigma_k^i$ and F such that $\Sigma_j^o \supseteq \bigcup_{k=1}^{s} \Sigma_k^i \cup F$. F can only contain clauses of Type 1a[10]. F is satisfied when $|F| \neq 0$. Model assumptions not related to model inputs are specified in A. The clauses in A are represented as well-formed formulas, thus, automated reasoning is possible during model selection. The location as well as the access method of a model is encoded as a string represented by S.

Definition 3. *Given two MDIs , $MDI_1 = (\Omega_1, L_1, \Lambda_1, \Sigma_1)$ and $MDI_2 = (\Omega_2, L_2, \Lambda_2, \Sigma_2)$, MDI_1 matches with MDI_2 ,denoted by $MDI_1 \subseteq MDI_2$,if and only if:*

(1) $\Omega_1 = \Omega_2$, MDI_1 has the same dimensions with MDI_2 ;

(2) $L_1 = L_2$, the same dimensions of MDI_1 and MDI_2 are on the same levels;

(3)$\Lambda_1 \subseteq \Lambda_2$, the fact set of MDI_1 is contained in the fact set of MDI_2 ;

(4) Σ_1 is covered in Σ_2

$MDI_1 \subseteq MDI_2$ shows that MDI_2 can provide data for MDI_1. If MDI_1 is a multi-dimensional data query or a input data interface of a model, MDI_2 satisfies MDI_1 .

Theorem 1. *Given a MDI of a query q, $MDI^q = (\Omega^q, L^q, \Lambda^q, \Sigma^q)$ and Σ^q is satisfiable. $C = (\Omega, \Lambda, f)$ is a multi-dimensional data set. If $\Omega^q \subseteq \Omega$, $\Lambda^q \subseteq \Lambda$, C provides data for MDI^q.*

Proof. For the multi-dimensional data set $C = (\Omega, \Lambda, f)$, because $\Omega^q \subseteq \Omega$, we suppose that there are t dimensions $d_{i_1}, d_{i_2}, \cdots, d_{i_t} \in \Omega$, but $d_{i_1}, d_{i_2}, \cdots, d_{i_t} \notin \Omega^q$. For each d_{i_j}, we can delete it from Ω by dimension aggregation $Agg(C, d_{i_j}, h, all)$,where h is any hierarchy attribute of d_{i_j}. Then we get the new multi-dimensional data set $C_a = (\Omega_a, \Lambda, f_a)$. Since $\Lambda^q \subseteq \Lambda$, we can project C_a on Λ^q, suppose $Project(C_a, \Lambda^q) = C_p = (\Omega_a, \Lambda_q, f_p)$. By properly rolling-up or drilling-down, the dimensions of C_p can locate on their levels specified by L^q. Since Σ^q is satisfiable, it is deduced the multi-dimensional data set matching with MDI^q by $Select(C_p, \Sigma^q)$.

3   Model Composition and Execution Algorithm

Model composition and execution entails searching the domain knowledge for data sources in response to the query. The algorithm matches the query to multi-dimensional data sets in DW. If the multi-dimensional data set can't provide data for the query, it will examine whether there are combined multi-dimensional data sets to satisfy the query. Otherwise, it will match the query to the model output MDIs. If a relevant model is found, it takes one of the model input MDIs as another query and continues the above process. Once all the model input data is obtained, the model will execute and output the result data. The algorithm will not stop until the original query is satisfied.

Given a multi-dimensional data query q. Let $MDI^q = (\Omega^q, L^q, \Lambda^q, \Sigma^q)$ be the MDI of q, and Σ^q is satisfiable. Suppose MStack is a stack to save the models. Fstack is a stack to store the constraints among the model inputs. InputDataStack is used to store MDIs of the model inputs. Let MDI_x be temporary variable of the multi-dimensional data set interface, MDResultSequ be a sequence to hold the result multi-dimensional data. The initial values are all NULL. Let $MDI_x = MDI^q$

Step 1: Search for the multi-dimensional data matching with MDI_x in the domain knowledge. For a data set $C = (\Omega, \Lambda, f)$:

1.1 If $\Omega^q \subseteq \Omega$,$\Lambda^q \subseteq \Lambda$

(1) For $\forall d_i \in \Omega$ and $d_i \notin \Omega^q$, run the aggregation operation on a dimensional hierarchy of d_i :$Agg(C, d_i, h, all) = (\Omega_a, \Lambda, f_a)$, and ignore the dimension of d_i. Let $\Omega_a = \Omega^q$.

(2) For $\forall(d_i, l_{d_i}) \in L^q$, do rolling-up or drilling-down to make d_i locate on the level l_{d_i} .

(3) Do the project operation: $Project((\Omega_a, \Lambda, f_a), \Lambda^q) = C'$.

(4) Do the operation $Select(C', \Sigma^q) = C^q$.

(5) Bring data from C^q to data buffer area, go to step 4.

1.2 If $\Omega^q \subseteq \Omega$,$\Lambda^q \cap \Lambda = \pi \neq \emptyset$,

(1) For $\forall d_i \in \Omega$ and $d_i \notin \Omega^q$, run the aggregation operation on a dimensional hierarchy of d_i :$Agg(C, d_i, h, all) = (\Omega_a, \Lambda, f_a)$, and ignore the dimension of d_i. Let $\Omega_a = \Omega^q$.

(2) For $\forall(d_i, l_{d_i}) \in L^q$, make d_i to locate on the level l_{d_i} by rolling-up or drilling-down.

(3) Do the project operation: $Project((\Omega_a, \Lambda, f_a), \Lambda^q) = C'$.

(4) Do the operation: $Select(C', \Sigma^q) = C^q$.

(5) Take data from C^q to data buffer area, let $\Lambda' = \Lambda^q - \pi$, $MDI_x = (\Omega^q, L^q, \Lambda', \Sigma^q)$, go to step 1, continue to search for the data set matching with MDI_x.

Step 2: Search for the model which output interface matches with MDI_x . If successful, the model M is found, customize the model to obtain the MDI set of all model inputs as well as an updated set of interinput model constraints FSet. Push M into the MStack, push all model input MDIs into InputDataStack, Push FSet into FStack. Otherwise, go to step 6.

Step 3: Pop a MDI from InputDataStack to MDI_x, go to step 1.

Step 4: If MStack is not empty, pop a model from MStack to M, else go to step 6.

Step 5: Examine whether the data in the buffer area satisfy the inputs of the model M. If successful, take the data in the buffer area to the inputs of model M and execute model M, save the outputs data into the buffer area. Go to step 4. If failed, push M into MStack, and go to step 3.

Step 6: Deal with the data in the buffer area, output the results and exit.

4   Implementation

Grain Management Information Intelligent Decision Support System (GMI-IDSS) is a DSS based on DW for grain trade. The DW-based subsystem of OLAP is one of its subsystems. We implement our model composition approach in its subsystem. The domain knowledge and model composition algorithm are implemented using Visual Prolog, whereas models are implemented in Visual C++. The database and data warehouse used in the subsystem are respectively Sybase ASE12.5 and Sybase IQ. We implement the decision model base by model dictionary and model files. Model dictionary is saved in data-base. Model files are stored in the file system. OLAP functions are implemented in Business Objects 5.0.

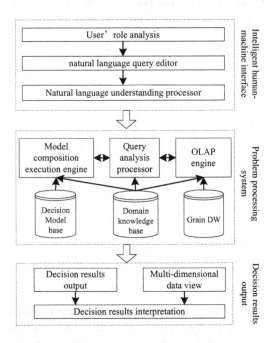

Fig. 1. Architecture of the sub-system

The subsystem architecture is showed in Fig. 1. There are 3 parts: Intelligent Human-Machine Interface(IHMI), Problem Processing System(PPS) and decision results output. The task of the intelligent human-machine interface is to analyze the information of users' characteristic and the information about what users expect. The query analysis processor in PPS processes the interior form of the inputs by using the knowledges in the domain knowledge base. If the grain data warehouse provides the data for the query (user's inputs), the OLAP engine processes the inputs and extracts data from the grain DW. The results are showed in multi-dimensional data view. If a model or model composition is needed to solve the user's query, model composition and execution engine will give the answer. The PPS sends the result data to the decision result output part. This part shows the results data and the explaining text to help users understand the results.

GMI-IDSS has run for more than one year. The running statistics of GMI-IDSS showed that there had been 105 instances of model composition, and 89 decision problems had been effectively solved by GMI-IDSS. The model composition approach can solve the practical problem effectively. It has important practicability and application value. Table 1 shows some of composite models GMI-IDSS generated.

Table 1. Some composite models that GMI-IDSS generated

Problem	Model number	Component model	Runtime	Applied
P_1	1	M_1	00:01:05	yes
P_2	2	M_2,M_3	00:00:17	yes
P_3	5	M_4,M_5,M_6,M_7,M_8	00:10:45	yes
P_4	3	M_9,M_{10},M_{11}	00:01:24	yes
P_5	3	M_4,M_5,M_{12}	00:07:29	yes
P_6	4	M_4,M_5,M_6,M_{12}	00:09:08	yes
P_7	2	M_{13},M_{14}	00:02:31	yes

In Table 1, P_1 is the problem of forecast for grain production of Jilin province of China,P_2 is the problem of making decision of grain inventory updating by-turns,P_3 is the problem of forecast for grain inventory,P_4 is the problem of allotting and dispatching grain, P_5 is the problem of making the plan of purchase grain,P_6 is the problem of making the plan of grain sales, P_7 is the problem of analysis of every variety of grain cost in every grain depot in every year. And M_1 denotes artificial neural network forecast model, M_2 denotes data classification model, M_3 denotes rule inference model, M_4 denotes forecast model for grain production,M_5 denotes forecast model for grain price, M_6 denotes forecast model of grain demand, M_7 denotes model of grain inventory updating by-turns, M_8 denotes EOQ model, M_9 denotes grain allot model, M_{10} denotes graph search model, M_{11} denotes integer programming model, M_{12} denotes dynamic programming model, M_{13} denotes mathematical statistics model, M_{14} denotes grain cost analysis model.

5 Conclusion

Model composition is the core problem of model management in DSS. In this paper, we introduce a new approach of model composition in multi-dimensional data spaces. A 6-tuple with input interface and output interface, which are represented by multi-dimensional data interface, is defined. An algorithm for the model composition and execution is presented. The algorithm can find relevant data set, and select appropriate models, and automate model composition and execution by searching the domain knowledge. The method has been applied into a practical project named "GMI-IDSS". The running statistics of GMI-IDSS showed that the model composition approach can solve many problems effectively so that we believe it will have important application prospect.

The DW schema is the relational schema when all multi-dimensional data sets in the DW are relational schema[11]. Our approach can not only work in multi-dimensional data spaces, but also work in relational data environment. So, Chari's approach[10] is a special case of our method.

Acknowledgments. This paper was supported by NSFC Major Research Program 60496321: Basic Theory and Core Techniques of Non Canonical Knowledge (Grant No.60273080 and 60473003), the Science and Technology Development Program of Jilin Province of China (Grant No.20040526), Specialized Research Fund for the Doctoral Program of Higher Education (Grant No.20050183065).

References

1. Sprague R.H., Jr.E.D.Carlson. Building Effective Decision Support Systems. Upper Saddle River New Jersey: Prentice Hall (1982)
2. Krishnan R. Model management: Survey, future research directions and a bibliography. ORSA CSTS Newsletter. 14(1993) 8-16
3. Daniel R. Dolk. Integrated model management in the data warehouse era. European Journal of Operational Research. 122(2000)199-218
4. Bala Iyer, G. Shankaranarayanan, Melanie L. Lenard. Model management decision environment: a Web service prototype for spreadsheet models. Decision Support Systems. 40(2005)283-304
5. Blanning R. H. A relational framework for join implementation in model management systems. Decision Support Systems. 1(1985)69-81
6. Basu A., R.W.Blanning. Model integration using metagraphs. Information Systems Research. 5(1994)195-218.
7. Basu A., R.W.Blanning. The analysis of assumptions in model bases using metagraphs. Management Science.44(1998)982-995.
8. Roger Alan Pick, Gary Klein. Model management as a component of a knowledge management system: capturing modeling knowledge in the enterprise. In proc. of 8th Americas Conference on Information Systems.Dallas(2002)
9. Bhargava H.K., R.Krishnan, R.Muller. Decision support on demand: Emerging electronic markets for decision technologies. Decision Support Systems. 19(1997) 193-214
10. Kaushal Chari. Model Composition Using Filter Spaces. Information Systems Research.13(2002)15-35
11. LiZehai, SunJigui, ZhaoJun, YuHaihong. Modeling Irregular Dimensions in OLAP. Journal of Computer Research and Development (in Chinese). 43(2006)301-306.

An Incremental Approach for Attribute Reduction in Concept Lattice

Bin Yang, Baowen Xu, and Yajun Li

School of Computer Science and Engineering, Southeast University,
210096 Nanjing, P.R. China
ybstudent@163.com
{bwxu,lyj}@seu.edu.cn

Abstract. One of the key problems of knowledge discovery is knowledge reduction. Concept lattice is an effective tool for data analysis and knowledge processing. The existing works on attribute reduction in concept lattice have mainly been focused on static database. This paper presents an incremental approach to identify reductions from dynamic database. The properties of attributes are discussed within the framework of equivalence classes and the determinant theorem of attribute reduction is presented. Based on the theorem, the reductions can be easily derived. The experimental results validate the effectiveness of the approach.

Keywords: concept lattice, knowledge reduction, equivalence classes.

1 Introduction

Formal concept analysis (FCA) [1] is a field of applied mathematics based on a lattice-theoretic formalization of the notions of concept and conceptual hierarchy. In FCA, a pair of sets of objects and sets of attributes common to these objects is called a concept. The family of all concepts is structured in the form of a lattice called concept lattice. The hierarchy of concepts in a lattice can be interpreted as the possibility to generalize or specialize a concept. As an effective tool for data analysis and knowledge processing, concept lattice has been widely used in many fields [8~14,19], such as machine learning, information retrieval, software engineering and knowledge processing.

Concept lattice is formulated based on formal context, which is a binary relation between a set of objects and a set of attributes. The abundant relationship between attributes makes some attributes have no effect on the discovery of knowledge. These attributes should be reduced without modifying the lattice structure. The notion of attribute reduction in concept lattice is to find minimal attributes set that can determine all concepts and their hierarchy structure [2].

Since first introduced by Wille R in 1980s, concept lattice has been extensively researched over the past two decades [3-6,8-10,23]. Ganter [2] presents reducible attributes and objects from the point of view of row and column of formal context.

J.T. Yao et al. (Eds.): RSKT 2007, LNAI 4481, pp. 451–459, 2007.

Wille [1] provides an approach that employs "up-down" arrow to purify formal context. Li et al. [7] discuss the properties of the irreducible objects on extent closure system and present an approach to classify and reduce attribute. Zhang et al [15,18] study the theory of attribute reduction in concept lattice. They take the idea similar to Skoworn and Rauszer's discernibility matrix [20]. To build Zhang's discernibility matrix for a given context, the user needs to compute the set of all concepts.

These works have mainly been focused on static database. In many practical cases, the databases are dynamic and may be modified by adding new dataset. In this paper we present an incremental approach to compute the attribute reduction in concept lattice. The properties of attributes are discussed and the determinant theorem of reduction in concept lattice is presented within the framework of equivalence classes. Based on the determinant theorem, we can easily derive the set of all reductions.

This paper is organized as follows: The next section describes the basic notions of formal concept analysis. Section 3 presents the attribute reduction in concept lattice within the framework of equivalence classes. In Section 4, an incremental approach for attribute reduction in concept lattice is proposed. Section 5 provides the experimental results and the final section contributes to the conclusions.

2 Formal Concept Analysis

Formal concept analysis [1,2] is a mathematical technique used for identifying meaningful groupings of objects that have common attributes.

Concept analysis starts with a formal context $K=(O,A,I)$, where O and A are finite sets of objects and sets of attributes, respectively, and $I \subseteq O \times A$ is a binary relation between O and A.

Let $X \subseteq O$, the set of common attributes of the objects contained in X is defined by

$$X' = \{a \in A \mid \forall\, o \in X, (o,a) \in I \} \tag{1}$$

Similarly, let $Y \subseteq A$, the set of common objects of the attributes contained in Y is:

$$Y^* := \{o \in O \mid \forall\, a \in Y, (o,a) \in I\} \tag{2}$$

Table 1. Example of formal context (O,A,I)

		a	b	c	d	e
(O_1,A,I_1)	1	×	×	×		×
	2	×				
	3		×	×	×	
(O_2,A,I_2)	4		×	×	×	
	5	×	×	×		×
	6	×				

A pair of sets of objects and sets of attributes (X,Y) is called a concept if $X'=Y$ and $Y^*=X$. That is, a concept is a maximal collection of objects sharing common attributes. X is called the extent and Y is called the intent of (X,Y).

The set of all concepts of a given formal context forms a partial order via

$$(X_1,Y_1) \leq (X_2,Y_2) \Leftrightarrow X_1 \subseteq X_2 \Leftrightarrow Y_1 \supseteq Y_2 \tag{3}$$

The set of all concepts and the partial order '$\leq$' form a concept lattice $L(K)$.

Table 1 gives an example of a formal context $K=(O,A,I)$, where $O=\{1,2,3,4,5,6\}$ and $A=\{a,b,c,d,e\}$. There're six concepts: $(123456,\emptyset)$, $(1256,a)$, $(1345,bc)$, $(15,abce)$, $(34,bcd)$ and $(\emptyset,abcde)$. The lattice is shown in Fig.1.

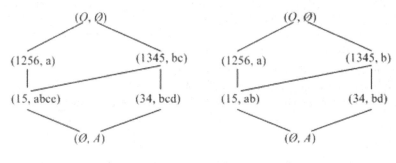

Fig. 1. $L(O,A,I)$ **Fig. 2.** $L(O,D_1,I_{D_1})$

3 Attribute Reduction in Concept Lattice

It has been proved that the reduction in concept lattice can be transformed into the reduction of formal context while the hierarchy of concepts is not changed [1]. In this section, we will discuss the reduction from the point of view of formal context.

Let $K=(O,A,I)$ be a formal context. For $\forall D \subseteq A$ and $I_D=I\cap(O\times D)$, $K_D=(O,D,I_D)$ is also a formal context. It's easy to prove that for any concept $(U,Z)\in L(K_D)$, there exists $(X,Y)\in L(K)$ such that $X=U$.

Definition 1. Let $K=(O,A,I)$ be a formal context and $D\subseteq A$. For $K_D=(O,D,I_D)$ and $\forall (X,Y)\in L(K)$, If there exists a concept $(U,Z)\in L(K_D)$ such that $X=U$, then D is called a consistent set in $L(K)$. And further, if $D-\{d\}$ is not a consistent set for $\forall d \in D$, then D is called a reduct of $L(K)$ [18].

The reduct certainly exists for any concept lattice but unnecessary exclusively.

Definition 2. Let $Red=\{D_i|\ D_i$ is a reduct, $i\in J\}$(J is an index set) be the set of all reducts in $L(K)$, $C=\bigcap_{i\in J}D_i$, $R=\bigcup_{i\in J}D_i-\bigcap_{i\in J}D_i$, $U=A-\bigcup_{i\in J}D_i$. 1) If $a\in C$, then a is a core attribute; 2) If $a\in R$, then a is a dispensable attribute; 3) If $a\in U$, then a is a unnecessary attribute [7,18].

The attribute that is not a core attribute is called a reducible attribute, which is either a dispensable attribute or an unnecessary attribute.

Theorem 1. Let $K=(O,A,I)$ be a formal context. For $D\subseteq A$, D is a consistent set in $L(K)$ iff $(a^{*'}\cap D)^*=a^*$ for $\forall a\in A-D$. And further, D is a reduct iff $(d^{*'}\cap(D-\{d\}))^*\neq d^*$ for $\forall d \in D$. The proof is referred to [15,18].

It directly follows that $\forall a\in A$, a is a core attribute iff $(a^{*'}\cap(A-\{a\}))^*\neq a^*$.

Definition 3. Let $K=(O,A,I)$ be a formal context. For $\forall\, a,b\in A$, a is equivalent to b iff $a^*=b^*$. The equivalence class of a in K is denoted as $[a]=\{b\in A \mid b^*=a^*\}$.

The attributes set A can be divided into several disjoint subsets with equivalence relation. Each subset is an equivalence class of an attribute.

By definition 3, if a is a reducible attribute for $\forall\, a\in A$, then the attributes contained in $[a]$ are also reducible attributes.

For a given formal context K, if a is a dispensable attribute in K, then $[a]\supset\{a\}$. However, the contrary assertion is not necessarily true.

Theorem 2. Let $K=(O,A,I)$ be a formal context. For $\forall\, a\in A$,
  1) a is an unnecessary attribute $\Leftrightarrow (a^{*'}\cap(A-[a]))^*=a^*$
  2) a is a dispensable attribute $\Leftrightarrow [a]\supset\{a\}$ and $(a^{*'}\cap(A-[a]))^*\neq a^*$

Proof. 1) ($\Rightarrow$) Obviously, for any reduct D, $D\subseteq A-[a]$. Thus, $a^*=(a^{*'}\cap D)^* \supseteq (a^{*'}\cap(A-[a]))^* \supseteq a^*$, i.e., $(a^{*'}\cap(A-[a]))^*=a^*$.

($\Leftarrow$) Suppose that a is not an unnecessary attribute, there certainly exists a reduct D such that $a\in D$. Let $M=a^{*'}\cap(A-[a])$. Because $M=M\cap(A-\{a\})=(M\cap (D-\{a\}))\cup(M\cap(A-D))$, thus $M^*=(M\cap(D-\{a\}))^*\cap(M\cap(A-D))^*$. Since $M\subseteq a^{*'}$, we have that $(M\cap(D-\{a\}))^*\supseteq(a^{*'}\cap(D-\{a\}))^*$. For $\forall\, e\in M\cap(A-D)$, we have that $e\in M$, $a^*\subset e^*$ and $a^{*'}\supset e^{*'}$. Hence, $a\notin e^{*'}$ and $(e^{*'}\cap D)^*= (e^{*'}\cap(D-\{a\}))^*\supseteq(a^{*'}\cap (D-\{a\}))^*$. By definition, $(M\cap(A-D))^* = \bigcap_{e\in M\cap(A-D)} e^*$, thus $(M\cap(A-D))^*\supseteq(a^{*'}\cap (D-\{a\}))^*$. Therefore, $M^*\supseteq(a^{*'}\cap(D-\{a\}))^*$. By Theorem 1, it follows that $(a^{*'}\cap (D-\{a\}))^*\supset a^*$ since D is a reduct. That is, $M^* =(a^{*'}\cap(A-[a]))^*\supset a^*$. But it is in contradiction with the given condition. Thus, a is an unnecessary attribute.

  2) Follows immediately from 1). ∎

The set of all dispensable attributes R for a given context $K=(O,A,I)$ can be transformed into the union of disjoint equivalence classes of attributes, $R=\bigcup_{i=1}^{n}R_i$, where $\exists\, a\in R$, $R_i=[a]$, and $R_i\cap R_j=\varnothing$ $1\leq i,j\leq n$. Let $F=R_1\times R_2\times...\times R_n=\{\{f_1,f_2,...,f_n\} \mid f_i\in R_i, 1\leq i\leq n\}\}$, we will prove that Lemma 1 and Theorem 3 hold.

Lemma 1. Let $K=(O,A,I)$ be a formal context. For $\forall\, a\in A$, $F_i\in F$, $(a^{*'}\cap R)^*=(a^{*'}\cap F_i)^*$.

Proof. Suppose that $F_i=\{\{f_{i1},...,f_{in}\} \mid f_{ik}\in R_i, 1\leq k\leq n\}$ for $\forall\, F_i\in F$. Then, for $\forall\, a\in A$, we have $(a^{*'}\cap R)^*=\bigcap_{i=1}^{n}(a^{*'}\cap R_i)^* =\bigcap_{i=1}^{n}(\{f_{ik}\})^*=(a^{*'}\cap F_i)^*$. ∎

Theorem 3. Let $K=(O,A,I)$ be a formal context, and C be the set of all core attributes of $L(K)$. Let Red be the set of all reducts, then $Red=\{F_i\cup C\mid \forall\, F_i\in F\}$.

Proof. Let U be the set of all unnecessary attributes. Since $A-F_i\cup C=(R-F_i)\cup U$, we have $a^*=(a^{*'}\cap(R\cup C))^*=(a^{*'}\cap R)^*\cap(a^{*'}\cap C)^*=(a^{*'}\cap F_i)^*\cap(a^{*'}\cap C)^*$ for $\forall\, a\in U$. That is, $a^*=(a^{*'}\cap(F_i\cup C))^*$; for $\forall\, a\in R-F_i$, we have $a^*=(a^{*'}\cap R)^*=(a^{*'}\cap F_i)^*$. Hence, $F_i\cup C$ is a consistent set in $L(K)$.

In addition, for any $a \in F_i \cup C$. If $a \in C$, then $(a^{*'} \cap (F_i \cup C - \{a\}))^* \neq a^*$; If $a \in F_i$ and $(a^{*'} \cap (F_i \cup C - \{a\}))^* = a^*$, then a must be an unnecessary attribute. It contradicts with the definition of R. Thus, $F_i \cup C$ is a reduct. It's easy to prove that each reduct can be denoted by the form of $F_i \cup C$.   ∎

4   An Incremental Approach for Attribute Reduction in Concept Lattice

In many practical cases, the database will be modified by appending some new data. Rather than computing the reducts from scratch we present an incremental approach to obtain the reducts in concept lattice. The basic notion of our incremental approach is to perform independent computation on the new part of the dataset, and then join the results to obtain the new reducts.

Definition 4. Let $K_1 = (O_1, A, I_1)$ and $K_2 = (O_2, A, I_2)$ are two contexts with common attributes. If $O = O_1 \cup O_2$ and $I = I_1 \cup I_2$, then $K = (O, A, I)$ is also a formal context. K is called the vertical merger of K_1 and K_2.

Two cases exist for the join of O_1 and O_2: $O_1 \cap O_2 = \emptyset$ and $O_1 \cap O_2 \neq \emptyset$. In this paper, we assume that O_1 and O_2 are two disjoint sets.

The formal context formed by adding new dataset can be regarded as the vertical merger of the original context and the new part.

Definition 5. Let $K = (O, A, I)$ be a formal context. For $a \in A$, the super set of a is defined by E_a: $E_a = \{ b \in A \mid b^* \supset a^* \}$.

Let $K = (O, A, I)$ is the vertical merger of K_1 and K_2. For the sake of simplicity, the mapping * on objects set and ' on attributes set in K_1, K_2, K are denoted by $*_1, *_2, *$ and $'_1, '_2, '$, respectively. For any attribute a, the equivalence class and superset of a in K_1, K_2, K are denoted by $[a]_1, [a]_2, [a]$ and E_1, E_2, E, respectively. Then, we have that $a^* = a^*_1 \cup a^*_2$.

Theorem 4. Let K is the vertical merger of two contexts K_1 and K_2. For any attribute a, if a is a core attribute in K_1 or K_2, then a is also a core attribute in K.

Proof. Assume that the set of core attributes in K_1 and K_2 are denoted by C_1 and C_2, respectively. For $\forall a \in C_1 \cup C_2$, if $a \in C_1$, then $(a^* \cap (A - \{a\}))^*_1 = (a^*_1 \cap a^*_2 \cap (A - \{a\}))^*_1 \supseteq (a^*_1 \cap (A - \{a\}))^*_1 \supset a^*_1$. Therefore, $(a^* \cap (A - \{a\}))^* = (a^* \cap (A - \{a\}))^*_1 \cup (a^* \cap (A - \{a\}))^*_2 \supset a$; The proof is similarly for $a \in C_2$.   ∎

Theorem 5. Let K is the vertical merger of K_1 and K_2. For any attribute a, a is an unnecessary attribute in K if one of the following conditions holds.
   1) $[a]_1 \cap E_2 \neq \emptyset$ and $(([a]_1 \cap E_2) \cup (E_1 \cap E_2))^*_2 = a^*_2$
   2) $[a]_2 \cap E_1 \neq \emptyset$ and $(([a]_2 \cap E_1) \cup (E_1 \cap E_2))^*_1 = a^*_1$
   3) $([a]_1 \cap E_2) \cup ([a]_2 \cap E_1) = \emptyset$ and $(E_1 \cap E_2)^*_1 = a^*_1$ and $(E_1 \cap E_2)^*_2 = a^*_2$

Proof. If a is an unnecessary attribute in K, then $a^* = (a^* \cap (A - [a]))^*$. Since $a^* = a^*_1 \cup a^*_2$, we need to prove that $a^*_1 = (a^* \cap (A - [a]))^*_1 = (a^*_1 \cap a^*_2 \cap (A - [a]))^*_1$ and $*_2 = (a^* \cap (A - [a]))^*_2 = (a^*_1 \cap a^*_2 \cap (A - [a]))^*_2$, which follow directly from the above three conditions. Further, if a is an unnecessary attribute, then a is also a reducible attribute in K_1 and K_2. Therefore, at least one of the three conditions holds. ∎

The following corollaries are simple consequence of the above Theorem. If a is not an unnecessary attribute in K,

Corollary 1. If $[a]_1 \cap [a]_2 \supset \{a\}$, then a is a dispensable attributes in K.

Corollary 2. If $[a]_1 \cap [a]_2 = \{a\}$, then a is a core attribute in K.

The set of core attributes C and the set of dispensable attributes R in K are:

$$C = C_1 \cup C_2 \cup \{ a \in A - C_1 \cup C_2 \mid [a]_1 \cap [a]_2 = \{a\}, (a^{*'} \cap (A - \{a\}))^* \neq a^* \}.$$
$$R = \{ a \in A - C \mid [a]_1 \cap [a]_2 \supset \{a\}, (a^{*'} \cap (A - [a]))^* \neq a^* \}$$

Thus, by Theorem 3, we can easily obtain the set of all reducts in $L(K)$.

Example. The formal context K in Table 1 is formed by the merger of $K_1 = (O_1, A, I_1)$ and $K_2 = (O_2, A, I_2)$. The attributes a, d are core attributes in K_1 and K_2. The attributes b, c are dispensable attributes in K since $[b] = \{bc\}$ and $E_1 = E_2 = \emptyset$. Since $[e] = \{e\}$ and $E_1 = E_2 = \{abc\}$, thus e is an unnecessary attribute. Therefore, the reducts in $L(K)$ are $D_1 = \{abd\}$ and $D_2 = \{acd\}$. Figure 2 shows the lattice $L(O, D_1, I_{D1})$.

5   Experimental Results

In this section, we will evaluate the incremental algorithm in comparison with the discernibility matrix-based algorithm in [18]. The two algorithms were implemented in Java and the experiments were performed on a table PC with $P_4 1.6MHz$, 512MB memory and Windows 2000 installed. The same data structure was used for each algorithm in order minimize its effect on the performance.

Fig. 3 illustrates the behavior of the two algorithms. The CPU time measured for the incremental algorithm is the total time necessary to compute the reducts by adding the objects one by one using the incremental process. The experiments were done with $\|A\| = 20, 40$ and $\|O\|$ varies from 20 to 120. The number of attributes possessed by each object has an upper limit of 5.

In Fig.3, the discernibility matrix-based algorithm outperforms the incremental algorithm although the difference is relatively small while the values of $\|O\|$ is enough smaller. There is always a crossover point, beyond which the incremental algorithm becomes the best. This crossover point depends on the density of the relation. When the number of attributes is increased, the crossover point becomes bigger.

Table 2. The six datasets collected from UCI library

Datasets	Objects	Attributes	k
Vote	435	32	16
hypothyroid	106	228	57
audiology	226	161	69
soybean	687	133	35
flag	194	300	27
hepatitis	155	364	19

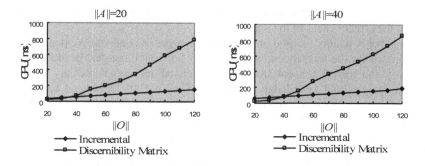

Fig. 3. CPU time for simulation with uniform distribution

Table 2 shows the six dataset collected from UCI library, where k is the average number of attributes possessed by objects. Figure 4 shows the results of the two algorithms. It's clear that the incremental algorithm outperforms the discernibility matrix-based algorithm. Because the discernibility matrix is constructed on concepts, the measured CPU time is bigger than 5s except the datasets *vote* and *hypothyroid*. However, the incremental algorithm has a more stable behavior than the discernibility matrix-based algorithm, which is an important consideration for applications.

The above results show that the incremental approach is a more attractive algorithm if good performance and stable behavior are an application requirement. We can take advantage of the incremental approach when the dataset is dynamic.

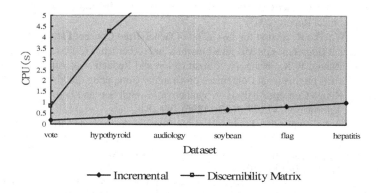

Fig. 4. CPU time for the six UCI datasets

6 Conclusions

Concept lattice is an effective tool for knowledge discovery. The attribute reduction in concept lattice makes the discovery of knowledge easier and the representation simpler. This paper discusses the properties of attributes based on the notion of equivalence classes and presents an incremental approach. The experimental results validate the effectiveness of the approach.

This paper is mainly focused on the vertical merger of contexts. We plan to explore the approach for the formal contexts formed by horizontal merger. In addition, rough set theory and formal concept analysis offer related and complementary approaches for data analysis and knowledge processing. They are all formulated based on the notion of context. We plan to compare the similarities and differences in the two theories from the point of view of reduction in the future.

Acknowledgments. This work was supported in part by National Natural Science Foundation of China (60373066, 90412003), Young Scientist's Foundation of NSFC (60425206) and Natural Science Foundation of Jiangsu Province (BK2005060).

References

1. Wille, R.: Restructuring Lattice theory: An Approach Based on Hierarchies of Concepts. In: Rival I, ed. Ordered Sets. Dordrecht-Boston: Reidel (1982) 445-470
2. Ganter, B., Wille, R.: Formal Concept Analysis: Mathematical Foundations. Springer (1999)
3. Godin, R., Missaoui, R., Alaoui, H.: Incremental Concept Formation Algorithms Based on Galois Lattices. Computational Intelligence, Vol.2, No.11 (1995) 246-267
4. Chen, Y.H., Yao, Y.Y.: Formal Concept Analysis and Hierarchical Classes Analysis. Annual Meeting of the North American Fuzzy Information Processing Society (2005) 276-281
5. Burmeister, P.: Formal Concept Analysis with ConImp: Introduction to The Basic Features. Technical Report (2003)
6. Shao, M.W.: The Reduction for Two Kind of Generalized Concept Lattice. Proceedings of International Conference on Machine Learning and Cybernetics (2005) 2217-2222
7. Li, H.R., Zhang, W.X., Wang, H.: Classification and Reduction of Attributes in Concept Lattices. IEEE International Conference on Granular Computing (2006) 142-147
8. Kent, R.: Rough Concept Analysis: A synthesis of Rough Sets and Formal Concept Analysis. Fund. Information, Vol.27, No.2, (1996) 169-181
9. Yao, Y.Y.: A Comparative Study of Formal Concept Analysis and Rough Set Theory in Data Analysis. Proceedings of Rough Sets and Current Trends in Computing (2004) 59-68
10. Yao, Y.Y.: Concept Lattices in Rough Set Theory. In: NAFIPS2004, IEEE Catalog Number: 04TH8736 (2004) 796-801
11. Jin, J.Y., Qin, K., Pei, Z.: Reduction-Based Approaches Towards Constructing Galois (Concept) Lattice. G.Wang et al. (eds.). RSKT 2006, LANI 4062 (2006) 107-113
12. Eisenbarth, T., Koschke, R.: Locating Features in Source Code. IEEE Transactions on Software Engineering Vol.29, Issue.3 (2003) 195-209
13. Snelting, G., Tip, F.: Reengineering Class Hierarchies Using Concept Analysis. Proc Sixth Int'l Symp. Foundations of Software Eng (1998)
14. Siff, M., Reps, T.: Identifying Modules via Concept Analysis. IEEE Transactions on Software Engineering, Volume 25 Issue 6 (1999)
15. Zhang, W.X., Liu, W., Qi, J.J.: The Theory and Approach of Attribute Reduction in Concept Lattice. China Science, Informational Science, Volume 35 Issue 6 (2005) 628-639

16. Stumme, G., Taouil, R., Bastide, Y., Pasquier, N., Lakhal, L.: Intelligent Structuring and Reducing of Association Rules with Formal Concept Analysis. Proc KI'2001. LNAI 2174. Springer-Verlag, Berlin Heidelberg (2001) 335-350
17. Pawlak, Z.: Rough Sets: Theoretical Aspects of Reasoning about Data. Kluwer Academic Publishers (1991)
18. Zhang, W.X., Liu, W., Qi, J.J.: Attribute Reduction in Concept Lattice Based on Discernibility Matrix. Rough Sets,Fuzzy Sets, Data Mining and Granular Computing: 10th International Conference, RSFDGrC (2005) 157-165
19. Godin, R., Missaoui, R., April, A.: Experimental Comparison of Navigation in Galois Lattice with Conventional Information Retrieval Methods. International Journal of Man-Machine Studies Vol.38 (1993)
20. Skowron, A., Rauszwer, C.: The Discernibility Matrices and Functions in Information Systems. Intelligent Decision Support: Handbook of Applications and Advances of the Rough Set Theory, Dordrecht: Kluwer Academic Publishers (1992) 331-362
21. Pagliani, P.: From Concept Lattices to Approximation Spaces: Algebraic Structures of Some Spaces of Partial Objects. Fundamental Information, Vol.18 (1993) 1-25
22. Hu, K., Sui, Y., Lu, Y., Wang, J., Shi, C.: Concept Approximation in Concept Lattice. Knowledge Discovery and Data Mining PAKDD (2001) 167-173
23. Deogun, J.S., Saquer, J.: Concept Approximations for Formal Concept Analysis. In: Stumme, G. (ed.), Working with Conceptual Structures. ICCS (2000) 73-83

Topological Space for Attributes Set of a Formal Context

Zheng Pei[1] and Keyun Qin[2]

[1] School of Mathematics & Computer Engineering, Xihua University,
Chengdu, Sichuan, 610039, China
[2] Department of Mathematics, Southwest Jiaotong University,
Chengdu, Sichuan 610031, China
pqyz@263.net
keyunqin@263.net

Abstract. Generating concept lattice is investigated for long years. Whether formal concepts and concept lattice can be generated in topological space for objects set G or attributes set M of a formal context $F = (G, M, I)$ is an interesting problem. In this paper, our discussions are concentrated on constructing a topology for M of F, the topology is generated by an approximation space on M. Then there exists one-to-one mapping between the approximation space on M and the topology for M, and it is obtained that attributes of each formal concept is open set in the topological space.

Keywords: Concept Lattice, Approximation Space, Topological Space.

1 Introduction

Formal concept analysis (FCA) presented by Wille [1] is a discipline that studies the hierarchical structures induced by a binary relation between a pair of sets. As the core of a mathematical theory of FCA, formal concepts or concept lattices can be used to represent relationship between objects and attributes or conceptual hierarchies which are inherent in data. FCA is now considered as the mathematical backbone of conceptual knowledge processing (CKP), a theory located in computer science, having as task to provide methods and tools for human oriented, concept-based knowledge processing [13]. FCA starts with a formal context T, also called (conceptual) information systems [2]-[4]. Nowadays, FCA is widely adopted by data analysis (see for instance, [5]-[7]), information retrieval (see for instance, [8]-[12]), knowledge discovery (see for instance, [13]-[15]), ontology engineering (see for instance [13], [16]-[19], [23]).

A formal context is expressed by $F = (G, M, I)$, in which, G and M be nonempty finite sets and $I : G \times M \longrightarrow \{0, 1\}$ a binary relation. $\forall g \in G$ is called object, $\forall m \in M$ is attribute. $I(g, m) = 1$ means that g has m. Let

$$\uparrow : 2^G \longrightarrow 2^M, \ X^\uparrow = \{m \in M| \ \forall g \in X, I(g, m) = 1\}, \tag{1}$$

$$\downarrow : 2^M \longrightarrow 2^G, \ Y^\downarrow = \{g \in G| \ \forall m \in Y, I(g, m) = 1\}. \tag{2}$$

J.T. Yao et al. (Eds.): RSKT 2007, LNAI 4481, pp. 460–467, 2007.
© Springer-Verlag Berlin Heidelberg 2007

In $F = (G, M, I)$, a formal concept is a pair $(A, B) \in 2^G \times 2^M$ such that $A^\uparrow = B$ and $B^\downarrow = A$. The set A is called its extent, the set B its intent. Denote that L_F is the set of all formal concepts, then $(L_F, \vee, \wedge)$ is called the formal concept lattice of $F = (G, M, I)$, it is a complete lattice. Generating concept lattices is investigated for long years (see for instance, [23]-[27]). In generating formal concepts or concept lattice, the follows closure properties are important [20]-[23]: Given $F = (G, M, I)$ and $(A, B) \in (L_F, \vee, \wedge)$, then $A^\uparrow = ((A^\uparrow)^\downarrow)^\uparrow$ and $B^\downarrow = ((B^\downarrow)^\uparrow)^\downarrow$. $\downarrow\uparrow$: $M \longrightarrow M$ is closure operator on M, the set $H_{\downarrow\uparrow} = \{B \subseteq M | \downarrow\uparrow (B) = B\}$ is a closure system on M and $(L_F, \vee, \wedge) = \{(B^\downarrow, B^{\downarrow\uparrow}) | B \subseteq M\}$. The properties make us to generate concept lattice by attributes set (or objects set) of $F = (G, M, I)$. Whether formal concepts and concept lattice can be generated in topological space for objects set G or attributes set M of $F = (G, M, I)$ is an interesting problem. In this paper, our discussions are concentrated on constructing a topology for attributes set M.

2   Approximation Space on M of $T = (G, M, I)$

The follows relation on M is used to obtain an approximation space on M.

Definition 1. Let $F = (G, M, I)$ be a formal context. $\forall m \in M$, denotes

$$M_m = \{m' \in M | \forall g \in \{g \in G | I(g, m) = 1\}, I(g, m') = 1\}, \tag{3}$$

then a binary relation on M induced by $M_m (m \in M)$ is as follows, $\forall m_1, m_2 \in M$,

$$R_M(m_1, m_2) = \begin{cases} 1, & if \quad m_2 \in M_{m_1}, \\ 0, & if \quad m_2 \notin M_{m_1}. \end{cases} \tag{4}$$

Example 1. [21] $F = (G, M, I)$ is a formal context, where $G = \{1, 2, 3, 4, 5, 6, 7, 8\}$, $M = \{a, b, c, d, e, f, g, h, i\}$ and I be described by Table 1.

Table 1. The binary relation of $F = (G, M, I)$

	a	b	c	d	e	f	g	h	i
1	×	×					×		
2	×	×					×	×	
3	×	×	×				×	×	
4	×		×				×	×	×
5	×	×		×		×			
6	×	×	×	×		×			
7	×		×	×	×				
8	×		×	×		×			

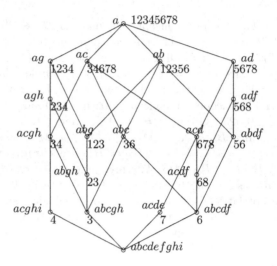

Fig. 1. Concept (or Galois) lattice corresponding to $F = (G, M, I)$

Its concept lattice is showed in Fig. 1. According to (3) and (4), then $M_a = \{a\}, M_b = \{a, b\}, M_c = \{a, c\}, M_d = \{a, d\}, M_e = \{a, c, d, e\}, M_f = \{a, d, f\}, M_g = \{a, g\}, M_h = \{a, g, h\}, M_i = \{a, c, g, h, i\}$, and the binary relation R_M on M induced by $M_m (m \in M)$ is as follows

R_M	a	b	c	d	e	f	g	h	i
a	1	0	0	0	0	0	0	0	0
b	1	1	0	0	0	0	0	0	0
c	1	0	1	0	0	0	0	0	0
d	1	0	0	1	0	0	0	0	0
e	1	0	1	1	1	0	0	0	0
f	1	0	0	1	0	1	0	0	0
g	1	0	0	0	0	0	1	0	0
h	1	0	0	0	0	0	1	1	0
i	1	0	1	0	0	0	1	1	1

Property 1. Let $F = (G, M, I)$ be a formal context. $\forall m \in M$, M_m is defined by (3) and $m_2 \in M_{m_1}$, then $M_{m_2} \subseteq M_{m_1}$.

Proof. According to (3), $M_{m_2} = \{m' \in M | \forall g \in \{g \in G | I(g, m_2) = 1\}, I(g, m') = 1\}$. By $m_2 \in M_{m_1}$, then $\forall g \in \{g \in G | I(g, m_1) = 1\}$ and $I(g, m_2) = 1$ hold, this means $\{g \in G | I(g, m_1) = 1\} \subseteq \{g \in G | I(g, m_2) = 1\}$. Hence, $\forall m' \in M_{m_2}$, m' is such that $\forall g \in \{g \in G | I(g, m_1) = 1\}$, $I(g, m') = 1 \implies m' \in M_{m_1}$, i.e., $M_{m_2} \subseteq M_{m_1}$.

Property 2. Let $F = (G, M, I)$ be a formal context. R_M is defined by (4), then R_M is a reflexive and transitive relation on M.

Proof. (1) $\forall m \in M$, according to (3), $m \in \{m' \in M | \forall g \in \{g \in G | I(g, m) = 1\}, I(g, m') = 1\} = M_m$, i.e., $R_M(m, m) = 1$, R_M is reflexive.

(2) $\forall m_1, m_2, m_3 \in M$, let $R_M(m_1, m_2) = 1$ and $R_M(m_2, m_3) = 1$, then $m_2 \in M_{m_1}$ and $m_3 \in M_{m_2}$. According to Proposition 1, $m_3 \in M_{m_2} \subseteq M_{m_1}$, this means $R_M(m_1, m_3) = 1$, i.e., R_M is transitive.

From Example 1, generally, R_M is not symmetrical relation on M.

Definition 2. *Let $F = (G, M, I)$ be a formal context. (M, R_M) is called an approximation space of $F = (G, M, I)$, and $\forall B \subseteq M$, $\overline{R_M}(B)$ and $\underline{R_M}(B)$, which are called upper approximation and lower approximation of B about (M, R_M), respectively, are defined as following, in which, M_m is defined by (3).*

$$\overline{R_M}(B) = \{m \in M | B \cap M_m \neq \emptyset\}, \quad \underline{R_M}(B) = \{m \in M | M_m \subseteq B\}, \qquad (5)$$

If R_M is an equivalence relation, then M_m is a equivalent class, and $\forall B \subseteq M$, $\overline{R_M}(B) = \{m \in M | [m]_{R_M} \cap B \neq \emptyset\}, \underline{R_M}(B) = \{m \in M | [m]_{R_M} \subseteq B\}$, which are Pawlak's upper approximation and lower approximation. The follows lemmas (Lemma 1-3) have been discussed in [29]-[31].

Lemma 1. *For any $R \subseteq M \times M$, $B, B_1, B_2 \subseteq M$, (1) $\underline{R}(M) = M$, $\overline{R}(\emptyset) = \emptyset$; (2) If $B_1 \subseteq B_2$, then $\overline{R}(B_1) \subseteq \overline{R}(B_2), \underline{R}(B_1) \subseteq \underline{R}(B_2)$; (3) $\underline{R}(B_1 \cap B_2) = \underline{R}(B_1) \cap \underline{R}(B_2)$, $\overline{R}(B_1 \cup B_2) = \overline{R}(B_1) \cup \overline{R}(B_2)$; (4) The pair $(\overline{R}, \underline{R})$ is dual, i.e., $\overline{R}(B) = \sim \underline{R}(\sim B)$, where $\sim B = M - B$, $\underline{R}$ and $\overline{R}$ are defined by (5).*

Lemma 2. *Let R be an arbitrary relation on M. (1) R is reflexive $\iff \forall B \subseteq M$, (a) $\underline{R}(B) \subseteq B$; (b) $B \subseteq \overline{R}(B)$. (2) R is transitive $\iff \forall B \subseteq M$, (a) $\underline{R}(\underline{R}(B)) \supseteq \underline{R}(B)$; (b) $\overline{R}(\overline{R}(B)) \subseteq \overline{R}(B)$. (3) If R is reflexive and transitive, then (a) $\underline{R}(\underline{R}(B)) = \underline{R}(B)$; (b) $\overline{R}(\overline{R}(B)) = \overline{R}(B)$ hold.*

Lemma 3. *Let M be a nonempty finite set and $R \subseteq M \times M$ a reflexive and transitive relation on M. $\forall B_j \subseteq M$, $j \in I$ (I is an index set),*

$$\underline{R}(\bigcup_{j \in I} \underline{R}(B_j)) = \bigcup_{j \in I} \underline{R}(B_j). \qquad (6)$$

If $R \subseteq M \times M$ is a reflexive and transitive relation on M, then $\forall m \in M$, define

$$M_m = \{m' \in M | R(m, m') = 1\} \qquad (7)$$

Property 3. Let M be a nonempty finite set, R a reflexive and transitive relation on M. The relation R_M on M is decided by (4) according to (7), then $R_M = R$.

Proof. $\forall m_1, m_2 \in M$, $R_M(m_1, m_2) = 1 \iff m_2 \in M_{m_1} \iff R(m_1, m_2) = 1$.

3 Topology for M Induced by Approximation Space (M, R_M)

In [28], fuzzy topologies which are generated by approximation space based on reflexive and transitive fuzzy relation are discussed. Here, classical topology is generated by (M, R_M).

Theorem 1. *Let $F = (G, M, I)$ be a formal context. $T_{R_M} = \{\underline{R_M}(B)|B \subseteq M\}$ is a topology for M, and (M, T_{R_M}) a topological space. T_{R_M} is called the topology generated by approximation space (M, R_M).*

Proof. (1) By (1) of Lemma 1, $\underline{R_M}(M) = M \in T_{R_M}$ holds. (2) If $\underline{R_M}(B_1)$, $\underline{R_M}(B_2) \in T_{R_M}$, by (3) of Lemma 1, $\underline{R_M}(B_1) \cap \underline{R_M}(B_2) = \underline{R_M}(B_1 \cap B_2) \in T_{R_M}$ holds. (3) If $\underline{R_M}(B_j) \in T_{R_{S_M}}$, $j \in I$ (I is an index set), by Proposition 2 and Lemma 3, $\bigcup_{j \in I} \underline{R_M}(B_j) = \underline{R_M}(\bigcup_{j \in I} \underline{R_M}(B_j)) \in T_{R_M}$ holds. According to (1)-(3), T_{R_M} is a topology for M, and (M, T_{R_M}) a topological space.

Theorem 2. *Let $F = (G, M, I)$ be a formal context. $T_{R_M} = \{\underline{R_M}(B)|B \subseteq M\}$ is a topology for M. $\forall B \subseteq M$, $\underline{R_M}(B) = i(B) = \cup\{\underline{R_M}(B')|\underline{R_M}(B') \subseteq B\}$, $\overline{R_M}(B) = c(B) = \cap\{\sim \underline{R_M}(B')| \sim \underline{R_M}(B') \supseteq B\}$, where i and c are interior operator and closure operator of (M, T_{R_M}).*

Proof. (1) By $\cup\{\underline{R_M}(B')|\underline{R_M}(B') \subseteq B\} \subseteq B$, consequently, $\underline{R_M}(\cup\{\underline{R_M}(B')| \underline{R_M}(B') \subseteq B\}) \subseteq \underline{R_M}(B)$. By Lemma 3, $\underline{R_M}(\cup\{\underline{R_M}(B')|\underline{R_M}(B') \subseteq B\}) = \cup\{\underline{R_M}(B')| \underline{R_M}(B') \subseteq B\}$, so, $\cup\{\underline{R_M}(B')|\underline{R_M}(B') \subseteq B\} \subseteq \underline{R_M}(B)$. By Lemma 2, $\underline{R_M}(B) \subseteq B$, $\underline{R_M}(B) \subseteq \cup\{\underline{R_M}(B')|\underline{R_M}(B') \subseteq B\}$ holds, hence, $\cup\{\underline{R_M}(B')| \underline{R_M}(B') \subseteq B\} = \underline{R_M}(B)$.

(2) By (1) and the duality of $\underline{R_M}$ and $\overline{R_M}$, $\overline{R_M}(B) = \sim \underline{R_M}(\sim B) = \sim (\cup\{\underline{R_M}(B')|\underline{R_M}(B') \subseteq \sim B\}) = \cap\{\sim \underline{R_M}(B')| \sim \underline{R_M}(B') \supseteq B\} = c(B)$.

Theorem 3. *Let $T_{R_M} = \{\underline{R_M}(B)|B \subseteq M\}$ be a topology. $\forall m_1, m_2 \in M$, $R_M(m_1, m_2) = 1 \Longleftrightarrow m_1 \in c(\{m_2\})$.*

Proof. By Definition 1 and Theorem 2, $R_M(m_1, m_2) = 1 \Longleftrightarrow m_2 \in M_{m_1} \Longleftrightarrow \{m_2\} \cap M_{m_1} \neq \emptyset \Longleftrightarrow m_1 \in \overline{R_M}(\{m_2\}) \Longleftrightarrow m_1 \in c(\{m_2\})$.

Example 2. Continues Example 1. According to (5) and Theorem 2, for attributes g and h, $c(\{g\}) = \overline{R_M}(\{g\}) = \{m \in M|\{g\} \cap M_m \neq \emptyset\} = \{m \in M|g \in M_m\} = \{g, h, i\}$, $c(\{h\}) = \overline{R_M}(\{h\}) = \{m \in M|\{h\} \cap M_m \neq \emptyset\} = \{m \in M|h \in M_m\} = \{h, i\}$, and $R_M(g, h) = 0 \Longleftrightarrow g \notin c(\{h\})$, $R_M(h, g) = 1 \Longleftrightarrow h \in c(\{g\})$.

Corollary 1. *Let R be a reflexive and transitive relation on M, $T_R = \{\underline{R}(B)| B \subseteq M\}$ the topology. $\forall m_1, m_2 \in M$, define a relation $R_{T_R}(m_1, m_2) = 1$ if and only if $m_1 \in c(\{m_2\})$, then $R = R_{T_R}$.*

4   Approximation Spaces Generated by Topology for M

In this section, discussions are focused on how to generate approximation space (M, R) by a topology T for M. Let i and c are interior operator and closure operator of topological space (M, T), respectively. Defining a binary relation R_T on M as follows: $\forall m_1, m_2 \in M$,

$$R_T(m_1, m_2) = 1 \Longleftrightarrow m_1 \in c(\{m_2\}). \tag{8}$$

Theorem 4. *The relation R_T on M decided by (8) is reflexive and transitive.*

Proof. (1) $\forall m \in M$, due to $m \in c(\{m\})$, $R_T(m, m) = 1$ holds, i.e., R_T is reflexive. (2)$\forall m_1, m_2, m_3 \in M$, let $R_T(m_1, m_2) = 1$ and $R_T(m_2, m_3) = 1$, then

$m_1 \in c(\{m_2\}) = \bigcap_{\sim B \in T}\{B|B \supseteq \{m_2\}\}$ and $m_2 \in c(\{m_3\}) = \bigcap_{\sim B' \in T}\{B'|B' \supseteq \{m_3\}\}$, hence, $m_1 \in \bigcap_{\sim B \in T}\{B|B \supseteq \{m_2\}\} = (\bigcap_{\sim B \in T}^{m_3 \notin B}\{B|B \supseteq \{m_2\}\}) \cap (\bigcap_{\sim B \in T}^{m_3 \in B}\{B|B \supseteq \{m_2\}\})$. Suppose that there exists B such that $\sim B \in T, m_3 \in B$ and $m_2 \notin B$, then $m_2 \notin \bigcap_{\sim B' \in T}\{B'|B' \supseteq \{m_3\}\} \implies m_2 \notin c(\{m_3\})$, a contradiction. Hence, $\bigcap_{\sim B \in T}^{m_3 \in B}\{B|B \supseteq \{m_2\}\}) = \bigcap_{\sim B' \in T}\{B'|B' \supseteq \{m_3\}\} = c(\{m_3\})$, this means that $m_1 \in c(\{m_3\})$ holds, i.e, $R_T(m_1, m_3) = 1$, R_T is transitive.

Based on the theorem, (M, R_T) is an approximation space, called the approximation space generated by T. $\forall m \in M$, denotes $M_m = \{m'|R_T(m, m') = 1\}$.

Theorem 5. *Let T be a topology for finite set M and the relation R_T decided by (8). $\forall B \subseteq M$, 1. $\overline{R_T}(B) = c(B)$; 2. $\underline{R_T}(B) = i(B)$. In which, $\underline{R_T}$ and $\overline{R_T}$ are decided by (5).*

Proof. Firstly, $\forall m' \in M$, according to (5), $\overline{R_T}(\{m'\}) = \{m \in M|\{m'\} \cap M_m \neq \emptyset\} = \{m \in M|m' \in M_m\} = \{m \in M|R_T(m, m') = 1\} = c(\{m'\})$.

(1) $\forall B \subseteq M$, $B = \bigcup_{m \in B}\{m\}$. According to Lemma 1 and finiteness of M, $\overline{R_T}(B) = \overline{R_T}(\bigcup_{m \in B}\{m\}) = \bigcup_{m \in B}\overline{R_T}(\{m\}) = \bigcup_{m \in B}c(\{m\}) = c(\bigcup_{m \in B}\{m\}) = c(B)$.

(2) By the duality of $\overline{R_T}$ and $\underline{R_T}$, i and c, and (1), it is obtained that $\underline{R_T}(B) =\sim \overline{R_T}(\sim B) =\sim c(\sim B) = i(B)$.

Corollary 2. *Let T be a topology on M, R_T the relation decided by (8), then $T_{R_T} = \{\underline{R_T}(B)|B \subseteq M\} = T$.*

Proof. $\forall B \subseteq M$, $B \in T \iff B = i(B) \iff B = \underline{R_T}(B) \iff B \in T_{R_T}$.

Corollary 3. *There exists one-to-one mapping between $\Gamma = \{R|R$ is a reflexive and transitive relation on $M\}$ and $\Sigma = \{T|T$ is a topology for $M\}$.*

Proof. Define $F : \Gamma \longrightarrow \Sigma$ such that $F(R) = T_R$. Firstly, if $R_1 \neq R_2$, then $T_{R_1} \neq T_{R_2}$. Otherwise, $R_1 = R_{T_{R_1}} = R_{T_{R_2}} = R_2$ due to Corollary (1), a contradiction, this means F is injective. On the other hand, $\forall T \in \Sigma$, R_T is a reflexive and transitive relation on M according to Theorem 4, i.e., $R_T \in \Gamma$ and $F(R_T) = T_{R_T} = T$ due to Corollary 2, consequently, F is surjective.

Theorem 6. *let $F = (G, M, I)$ be a formal context. $\forall B \subseteq M$, $\underline{R_M}(B^{\downarrow\uparrow}) = B^{\downarrow\uparrow}$.*

Proof. Firstly, $\underline{R_M}(B^{\downarrow\uparrow}) \subseteq B^{\downarrow\uparrow}$ is trivial. On the other hand, $\forall m \in B^{\downarrow\uparrow}$, according to (1), (2) and (3), $\forall g \in B^{\downarrow}$, $I(g, m) = 1$, and $M_m = \{m' \in M|\forall g \in \{g \in G|I(g, m) = 1\}, I(g, m') = 1\}$. Hence, $\forall m' \in M_m$ and $\forall g \in B^{\downarrow}$, $I(g, m') = 1$, this means $m' \in B^{\downarrow\uparrow}$, i.e., $M_m \subseteq B^{\downarrow\uparrow} \implies m \in \underline{R_M}(B^{\downarrow\uparrow}) \implies B^{\downarrow\uparrow} \subseteq \underline{R_M}(B^{\downarrow\uparrow})$, and $\underline{R_M}(B^{\downarrow\uparrow}) = B^{\downarrow\uparrow}$ holds.

Theorem 6 means that for every formal concept $(B^{\downarrow}, B^{\downarrow\uparrow})$ of a formal context $F = (G, M, I)$, its attribute set $B^{\downarrow\uparrow}$ must be an open set in topological space (M, T_{R_M}), i.e., $B^{\downarrow\uparrow} \in T_{R_M}$. This property makes us to generate formal concepts in topology for M.

Example 3. Continues Example 1. For formal concept $(6, \{a, c, d, e, f\})$, $i(\{a, c, d, e, f\}) = \underline{R_M}(\{a, c, d, e, f\}) = \{a, c, d, e, f\}$.

5 Conclusion

In this paper, a topology for M of $F = (G, M, I)$ is generated. An interesting conclusion is that attributes of each formal concept is open set in the topological space. Next, generating formal concepts and concept lattice in the topological space will be considered.

Acknowledgments

This work is supported by the excellent younger foundation of sichuan province (grant no. 06ZQ026-037) and a Project Supported by Scientific Reserch Fund of SiChuan Provincial Education Department(grant no. 2005A121, 2006A084).

References

1. Wille, R.: Restructuring the lattice theory: an approach based on hierarchies of concepts. Rival, I.(Ed.): Ordered Sets, Reidel, Dordrecht, Boston. (1982) 445-470
2. Vogt, F., Wille,R.: TOSCANA-a graphical tool for analyzing and exploring data. Lecture Notes in Computer Science, 894, Springer, Heidelberg. (1995) 226-233
3. Guan, J. W., Bell, D. A.: Rough computational methods for information systems. Artificial Intelligence. 105 (1998) 77-103
4. Pei, Z., Resconi, G., Van Der Wal, A. J., Qin, K. Y., Xu, Y.: Interpreting and extracting fuzzy decision rules from fuzzy information system and their inference. Information Sciences. 176 (2006) 1869-1897
5. Carpineto, C., G. Romano: Concept Data Analysis: Theory and Applications. New York: Wiley. (2004)
6. Elloumi, S., Jaam, J., Hasnah, A., Jaoua, A., Nafkha, I.: A multi-level conceptual data reduction approach based on the Lukasiewicz implication. Information Sciences. 163 (2004) 253-262
7. Prediger, S.: Formal Concept Analysis for general objects. Discrete Applied Mathematics. 127 (2003) 337–355
8. Peng, D., Wang, X., Zhou, A.: A concept-based approach for retrieving alternate web services. International Journal of Business Process Integration and Management. 1(3) (2006) 219-230
9. Kim, M., Compton, P.: Evolutionary document management and retrieval for specialized domains on the web. Int. J. Human-Computer Studies. 60 (2004) 201-241
10. Patrick, D. B., Derek, B.: Collaborative Recommending using Formal Concept Analysis. Knowledge-Based Systems. 19 (2006) 309-315
11. Tho, Q. T., Hui, S. C., Fong, A. C. M.: A citation-based document retrieval system for finding research expertise. Information Processing and Management. 43 (2007) 248-264
12. Beydoun, G., Kultchitsky, R., Manasseh, G.: Evolving semantic web with social navigation. Expert Systems with Applications. 32 (2007) 265-276
13. Stumme, G.: Off to newshores: conceptual knowledge discovery and processing. Int. J. Human-Computer Studies. 59 (2003) 287-325
14. Priss, U.: Formal concept analysis in information science. Annual Review of Information Science and Technology. 40 (2006) 521-543

15. Cole, R., Eklund, P., Stumme, G.: Document retrieval for e-mail search and discovery using formal concept analysis. Applied Artificial Intelligence. 17(3) (2003) 257-280

16. Stumme, G., Mdche, A.: FCA Merge: bottom-up merging of ontologies. Proceedings 17th International Conference on Artificial Intelligence (IJCAI01), Seattle, WA, USA. (2001) 225-230

17. Nanda, J.,Simpson, T. W., Kumara, S. R. T., Shooter, S. B.: A methodology for product family ontology developmentusing formal concept analysis and web ontology language. Journal of Computing and Information Science in Engineering. 6(2) (2006) 103-113

18. Quan, T. T., Hui, S. C., Fong, A. C. M., Cao, T. H.: Automatic fuzzy ontology generation for semantic web. IEEE Transactions on knowledge and data engineering. 18(6) (2006) 842-865

19. Formica, A.: Ontology-based concept similarity in Formal Concept Analysis. Information Sciences. 176 (2006) 2624-2641

20. Jin, J., Qin, K. Y., Pei, Z.: Reduction-Based Approaches Towards Constructing Galois (Concept) Lattices. Lecture Notes in Artificial Intelligence, 4062, Springer, Berlin. (2006) 107-113

21. Ganter, B., Wille, R. (eds.): Formal Concept Analysis: Mathematical Foundations, Springer, Berlin, 1999.

22. Berry, A., SanJuan, E., Sigayret, A.: Generalized domination in closure systems. Discrete Applied Mathematics. 154 (2006) 1064-1084

23. Stumme, G., Taouil, R., Bastide, Y., Pasquier, N., Lakhal, L.: Computing iceberg concept lattices with TITANIC. Data & Knowledge Engineering. 42 (2002) 189-222

24. Valtchev, P., Missaoui, R., Lebrun, P.: A partition-based approach towards constructing Galois (concept) lattices. Discrete Mathematics. 256 (2002) 801-829

25. Berry, A., Sigayret, A.: Representing a concept lattice by a graph. Discrete Applied Mathematics. 144 (2004) 27-42

26. Kuznetsov, S. O., Obiedkov, S. A.: Comparing performance of algorithms for generating concept lattices. Journal of Experimental and Theoretical Artificial Intelligence. 14(2-3) (2002) 189-216

27. Kuznetsov, S. O.: Complexity of learning in concept lattices from positive and negative examples. Discrete Applied Mathematics. 142 (2004) 111-125

28. Qin, K. Y., Pei, Z.: On the topological properties of fuzzy rough sets. Fuzzy Sets and Systems. 151 (2005) 601-613

29. Mi, J. S., Wu, W. Z., Zhang, W. X.: Constructive and axiomatic approches for the study of the theory of rough sets. Pattern Recognition and Artificial Intelligence. 15 (2002) 280-284

30. Thiele, H.: On axiomatic characterization of fuzzy approximation operators I, the fuzzy rough set based case. Proc. RSCTC 2000, Banff Park Lodge, Bariff, Canada. (2000) 239-247

31. Wu, W. Z.,Mi, J. S., Zhang, W. X.: Generalized fuzzy rough sets. Information Sciences. 151 (2003) 263-282

32. Kelley, J. L.: General topology. Springer-Verlag. (1955)

Flow Graphs as a Tool for Mining Prediction Rules of Changes of Components in Temporal Information Systems

Zbigniew Suraj[1,2] and Krzysztof Pancerz[3,4]

[1] Chair of Computer Science, University of Rzeszów
Rejtana Str. 16A, 35-310 Rzeszów, Poland
zsuraj@univ.rzeszow.pl
[2] Institute of Computer Science, State School of Higher Education
Czarnieckiego Str. 16, 37-500 Jarosław, Poland
[3] Chair of Computer Science Foundations
University of Information Technology and Management
Sucharskiego Str. 2, 35-225 Rzeszów, Poland
kpancerz@wsiz.rzeszow.pl
[4] Chair of Computer Science and Knowledge Engineering
College of Management and Public Administration
Akademicka Str. 4, 22-400 Zamość, Poland

Abstract. Flow Graphs proposed by Z. Pawlak are a very useful tool for reasoning from data. In this paper, an application of flow graphs is presented. We use them to mine rules describing component changes in the consecutive time windows which a given temporal information system is split into. Such rules can be helpful for predicting future changes of components in the system and what follows, predicting the future behavior of the analyzed system.

1 Introduction

One of the important aspects of data mining is the analysis of data changing in time (temporal data). Many of the systems change their properties as time goes. Then, some models of systems constructed for one period of time must be reconstructed for another period of time. Different methodologies of soft computing are used for prediction with temporal data, e.g., neural networks, rough sets. In our approach, we are interested in rough sets. We assume that modeled systems are described by temporal information systems (objects of such systems are ordered in time). We observe the behavior of modeled systems in consecutive time windows which temporal information systems are split into. Observation of changes enables us to determine the so-called prediction rules which can be used to predict future changes of models. In some of our earlier papers we considered models in the form of colored Petri nets built on the basis of decomposed information systems (see [3], [8]). Decomposition of an information system S is a division of S into smaller, relatively independent subsystems. Components obtained in this way represent some modules of S linked inside by means of

J.T. Yao et al. (Eds.): RSKT 2007, LNAI 4481, pp. 468–475, 2007.
© Springer-Verlag Berlin Heidelberg 2007

functional dependencies (cf. [6]). Components of information systems are defined on the basis of functional relative reducts. Therefore, an important thing is to determine how functional relative reducts change in the consecutive time windows. On the basis of the obtained knowledge we can predict future changes of functional relative reducts. At the beginning, we assume that all the time windows (which a given temporal information system S is split into) include the same number of objects and the shift between any two consecutive time windows is constant. In order to determine prediction rules of changes of components we carry out the steps briefly described as follows. We split a given temporal information system $S = (U, A)$ into time windows of the same length preserving a constant shift between two consecutive time windows. We obtain a set $\mathcal{S}$ of all the time windows. Next, for each time window from the set $\mathcal{S}$ and each attribute $a \in A$, we compute a set of all functional relative reducts. We obtain a data table (called a temporal table of functional reducts) whose columns are labeled with attributes from A whereas rows - with consecutive time windows from $\mathcal{S}$. The cells of such a table contain sets of functional relative reducts. For each attribute $a \in A$, we build a temporal decision system. Attributes of this system are labeled with the consecutive time windows (the last attribute is treated as a decision). The number of consecutive time windows taken into consideration is set by us. Each row represents a sequence of sets of functional relative reducts which appeared in consecutive time windows. For each attribute $a \in A$, we mine prediction rules from the temporal decision system. In order to do it we can use prediction matrices described in [7]. In this paper, we propose flow graphs as a tool for mining such rules.

2 Preliminaries

In this section, we present some notions used in the paper.

Temporal Information Systems. A temporal information system is a kind of an information system $S = (U, A)$, with a set U of objects ordered in time, i.e., $U = \{u_t : t = 1, 2, \ldots, n\}$, where u_t is the object observed in time t. By a time window on S of the length λ in a point τ we understand an information system $S' = (U', A')$, where $U' = \{u_\tau, u_{\tau+1}, \ldots, u_{\tau+\lambda-1}\}$, $1 \leq \tau, \tau + \lambda - 1 \leq n$, and A' is a set of all attributes from A defined on the domain restricted to U'. The length λ of S' is defined as $\lambda = card(U')$. In the sequel, the set A' of all attributes in any time window $S' = (U', A')$ on $S = (U, A)$ will be marked, for simplicity, with the same letter A like in S. By $start(S')$ we denote the start point of S', i.e., $start(S') = \tau$, and by $end(S')$ we denote the end point of S', i.e., $end(S') = \tau + \lambda - 1$. If we have two time windows S_i, S_j on the temporal information system, then a shift between S_j and S_i is defined as $\rho = start(S_j) - start(S_i)$.

Decomposition of Information Systems. Each information system can be decomposed into subtables called components of this system. The components define, in a sense, the strongest functional modules of the system linked inside by means of functional dependencies. Decomposition of information systems was discussed in [3], [6], [8]. Here, we recall only the most relevant notions. Let $S = (U, A)$ be an information system. In S there can exist a functional dependency between attributes from $A - \{a\}$ and the attribute a, where $a \in A$, i.e., the values of attributes from $A - \{a\}$ uniquely determine all values of the attribute a. We denote this fact by $(A - \{a\}) \Rightarrow \{a\}$. Sometimes, it may occur that $\{a\}$ depends functionally not on the whole set $A - \{a\}$ but on its subset $B \subset (A - \{a\})$. If B is minimal with respect to inclusion, then B is called a functional $\{a\}$-reduct (relative reduct) of the set of attributes $A - \{a\}$. The set of all functional $\{a\}$-reducts of $A - \{a\}$ in the information system $S = (U, A)$ will be denoted as $FunRed^a(S)$. By subsystem $S' = (U', A')$ of the information system $S = (U, A)$ we understand a system satisfying the following requirements: $U' \subseteq U$, $A' = \{a' : a \in B \subseteq A\}$, $a'(u) = a(u)$ for $u \in U'$ and $V_{a'} = V_a$ for $a \in A$. Let $S = (U, A)$ be an information system. An information system $S^* = (U^*, A^*)$ is called a normal component of S if and only if S^* is a subsystem of S and $A^* = X \cup Y$, where $Y = \{a \in A : X \in FunRed^a(S)\}$. An information system $S^* = (U^*, A^*)$ is called a degenerated component of S if and only if S^* is a subsystem of S and $A^* = \{a\}$ and there does not exist any normal component $S' = (U', X' \cup Y')$ of S such that $a \in X' \cup Y'$. By components we understand both normal components and degenerated components.

Flow Graphs. Flow graphs have been defined by Z. Pawlak [4], [5] as a tool for reasoning from data. A flow graph is a directed, acyclic, finite graph $G = (N, B, \sigma)$, where N is a set of nodes, $B \subseteq N \times N$ is a set of directed branches and $\sigma : B \to [0, 1]$ is a flow function. An input of a node $x \in N$ is the set $I(x) = \{y \in N : (y, x) \in B\}$, whereas an output of a node $x \in N$ is the set $O(x) = \{y \in N : (x, y) \in B\}$. $\sigma(x, y)$ is called a strength of a branch $(x, y) \in B$ and it is also denoted as $str(x, y)$. We define the input and the output of the graph G as $I(G) = \{x \in N : I(x) = \emptyset\}$ and $O(G) = \{x \in N : O(x) = \emptyset\}$, respectively. The input and the output of G consist of external nodes of G. The remaining nodes of G are its internal nodes. For each internal node $x \in N$ the throughflow of x is determined as $\delta(x) = \delta_+(x) = \delta_-(x)$, where $\delta_+(x) = \sum\limits_{y \in I(x)} \sigma(y, x)$ and $\delta_-(x) = \sum\limits_{y \in O(x)} \sigma(x, y)$. For each branch $(x, y) \in B$ we define also $cer(x, y) = \frac{\sigma(x,y)}{\delta(x)}$ (certainty) and $cov(x, y) = \frac{\sigma(x,y)}{\delta(y)}$ (coverage), where $\delta(x) \neq 0$ and $\delta(y) \neq 0$. A directed path $[x \ldots y]$ between nodes x and y in G, where $x \neq y$, is a sequence of nodes $x_1, x_2, \ldots, x_n$ such that $x_1 = x$, $x_n = y$ and $(x_i, x_{i+1}) \in B$, where $1 \leq i \leq n - 1$. For each path $[x_1 \ldots x_n]$ we define $cer[x_1 \ldots x_n] = \prod\limits_{i=1}^{n-1} cer(x_i, x_{i+1})$ (certainty), $cov[x_1 \ldots x_n] = \prod\limits_{i=1}^{n-1} cov(x_i, x_{i+1})$ (coverage) and $str[x_1 \ldots x_n] = \delta(x_1)cer[x_1 \ldots x_n] = \delta(x_n)cov[x_1 \ldots x_n]$ (strength). A connection $< x, y >$ from the node x to the node y is a set of all paths from x to y in G. For each connection $< x, y >$ we define $cer < x, y >= \sum\limits_{[x \ldots y] \in <x,y>} cer[x \ldots y]$

$(certainty)$, $cov < x, y >= \sum\limits_{[x...y] \in <x,y>} cov[x...y]$ $(coverage)$ and $str < x, y >= \sum\limits_{[x...y] \in <x,y>} str[x...y]$ $(strength)$. We can associate a rule r in the form of $x \Rightarrow y$ with each branch $(x, y) \in B$ in G, where x is a predecessor of r, whereas y is a successor of r. The rule r is characterized by three factors: $str(x \Rightarrow y)$ $(strength)$, $cer(x \Rightarrow y)$ $(certainty)$ and $cov(x \Rightarrow y)$ $(coverage)$. Each path $[x_1 ... x_n]$ in G determines a sequence of rules $x_1 \Rightarrow x_2$, $x_2 \Rightarrow x_3$, ..., $x_{n-1} \Rightarrow x_n$. This sequence can be interpreted by a single rule in the form of $x_1 \wedge x_2 \wedge x_3 \wedge ... \wedge x_{n-1} \Rightarrow x_n$, characterized by three factors: $cer(x_1 \wedge x_2 \wedge x_3 \wedge ... \wedge x_{n-1} \Rightarrow x_n) = cer[x_1 ... x_n]$, $cov(x_1 \wedge x_2 \wedge x_3 \wedge ... \wedge x_{n-1} \Rightarrow x_n) = cov[x_1 ... x_n]$ and $str(x_1 \wedge x_2 \wedge x_3 \wedge ... \wedge x_{n-1} \Rightarrow x_n) = \delta(x_1)cer[x_1 ... x_n] = \delta(x_n)cov[x_1 ... x_n]$. Analogously, each connection $< x, y >$ in G determines a single rule in the form of $x \Rightarrow y$, characterized by three factors: $cer(x \Rightarrow y) = cer < x, y >$, $cov(x \Rightarrow y) = cov < x, y >$ and $str(x \Rightarrow y) = \delta(x)cer < x, y >= \delta(y)cov < x, y >$.

3   Mining Prediction Rules of Component Changes

In this section, we present a new approach to mining prediction rules of component changes in the consecutive time windows. Let $S = (U, A)$ be a temporal information system, where $A = \{a_1, a_2, ..., a_m\}$, $m = card(A)$. We assume that time windows (which S is split into) satisfy the following conditions: (1) the first object of the first time window S_1 is simultaneously the first object of S, i.e., $start(S_1) = start(S)$, (2) all time windows have the same length λ, (3) the shift between any two consecutive time windows is constant and equal to ρ. Assuming these conditions we denote by $Wind_\lambda^\rho(S)$ the set of all time windows which S is split into. Therefore, we have $Wind_\lambda^\rho(S) = \{S_1, S_2, ..., S_\omega\}$, where $S_1 = (U_1, A)$, $S_2 = (U_2, A)$, ..., $S_\omega = (U_\omega, A)$. $Wind_\lambda^\rho(S)$ is also ordered in time, i.e., $S_1, S_2, ..., S_\omega$ are consecutive time windows.

In order to mine prediction rules of component changes in the consecutive time windows in S we can execute stages given below. At the beginning, we set the following input parameters: λ - the length of time windows, ρ - the shift between two consecutive time windows, κ - the period of prediction (the length of the sequence of time windows taken into consideration).

Stage 1. We split a given temporal information system S into time windows with length λ and shift ρ. As a result of that, we obtain a set of time windows $Wind_\lambda^\rho(S) = \{S_1, S_2, ..., S_\omega\}$. Obviously, the time windows can overlap.

Stage 2. For each time window $S_i \in Wind_\lambda^\rho(S)$, where $1 \leq i \leq \omega$, for each attribute $a \in A$, we compute the set $FunRed^a(S_i)$ of all functional $\{a\}$-reducts of $A - \{a\}$ in S_i. As a result of that, we obtain a data table called a temporal relative reduct table of S, denoted by $TRRT(S)$. A scheme of such a table is shown in Table 1. The rows of $TRRT(S)$ are labeled with names of time windows from $Wind_\lambda^\rho(S)$, whereas the columns - with attributes from A. Entries of $TRRT(S)$ include the families of adequate relative reducts.

Table 1. A scheme of temporal relative reduct table $TRRT(S)$

$Wind_\lambda^\rho(S)/A$	a_1	$\ldots$	a_m
S_1	$FunRed^{a_1}(S_1)$	$\ldots$	$FunRed^{a_m}(S_1)$
S_2	$FunRed^{a_1}(S_2)$	$\ldots$	$FunRed^{a_m}(S_2)$
$\ldots$	$\ldots$	$\ldots$	$\ldots$
S_ω	$FunRed^{a_1}(S_\omega)$	$\ldots$	$FunRed^{a_m}(S_\omega)$

Stage 3. For each attribute $a \in A$, we determine the set $\bigcup FunRed^a(S)$ of all functional $\{a\}$-reducts appearing in time windows from the set $Wind_\lambda^\rho(S)$, i.e., $\bigcup FunRed^a(S) = FunRed^a(S_1) \cup FunRed^a(S_2) \cup \ldots \cup FunRed^a(S_\omega)$.

Stage 4. For each attribute $a \in A$, we build the so-called temporal decision system $TDS_a = (U_s, A_t)$, shown in Table 2, as follows. A_t includes $\kappa-1$ condition attributes marked with $t_1, t_2, \ldots, t_{\kappa-1}$ and one decision attribute marked with t_κ. For each sequence $\langle S_\delta, S_{\delta+1}, \ldots, S_{\delta+\kappa-1} \rangle$ of consecutive time windows in $Wind_\lambda^\rho(S)$, where $1 \leq \delta \leq \omega - \kappa + 1$ and $\omega = card(Wind_\lambda^\rho(S))$, we create one object s_δ in U_s such that $t_1(s_\delta) = FunRed^a(S_\delta)$, $t_2(s_\delta) = FunRed^a(S_{\delta+1})$, $\ldots$, $t_\kappa(s_\delta) = FunRed^a(S_{\delta+\kappa-1})$. The values of attributes in TDS_a are the sets of functional relative reducts. Obviously, $\omega = \kappa + \eta - 1$.

Table 2. A scheme of temporal decision system TDS_a

U_s/A_t	t_1	t_2	$\ldots$	t_κ
s_1	$FunRed^a(S_1)$	$FunRed^a(S_2)$	$\ldots$	$FunRed^a(S_\kappa)$
s_2	$FunRed^a(S_2)$	$FunRed^a(S_3)$	$\ldots$	$FunRed^a(S_{\kappa+1})$
$\ldots$	$\ldots$	$\ldots$	$\ldots$	$\ldots$
s_η	$FunRed^a(S_\eta)$	$FunRed^a(S_{\eta+1})$	$\ldots$	$FunRed^a(S_{\kappa+\eta-1})$

Stage 5. For each attribute $a \in A$, we build the so-called temporal flow graph TFG_a as follows. Let us denote by $\mathbf{FunRed}_a^{t_1}$ the family of sets of functional relative reducts which is a value set of an attribute t_1 in TDS_a. Analogously, let us assume notation $\mathbf{FunRed}_a^{t_2}, \ldots, \mathbf{FunRed}_a^{t_\kappa}$ for $t_2, \ldots, t_\kappa$, respectively. We can write a set of nodes of TFG_a as $N = N_1 \cup N_2 \cup \ldots \cup N_\kappa$, where N_1 contains nodes representing sets of functional relative reducts from $\mathbf{FunRed}_a^{t_1}$, N_2 contains nodes representing sets of functional relative reducts from $\mathbf{FunRed}_a^{t_2}$, and so on, N_κ contains nodes representing sets of functional relative reducts from $\mathbf{FunRed}_a^{t_\kappa}$. For each branch $(FunRed_i^{t_\delta}, FunRed_j^{t_{\delta+1}})$ of TFG_a, where $FunRed_i^{t_\delta} \in \mathbf{FunRed}_a^{t_\delta}$ and $FunRed_j^{t_{\delta+1}} \in \mathbf{FunRed}_a^{t_{\delta+1}}$, we can compute a strength factor $str(FunRed_i^{t_\delta}, FunRed_j^{t_{\delta+1}}) = \frac{card(X \cap Y)}{card(U_s)}$, a certainty factor $cer(FunRed_i^{t_\delta}, FunRed_j^{t_{\delta+1}}) = \frac{card(X \cap Y)}{card(X)}$, and a coverage factor $cov(FunRed_i^{t_\delta}, FunRed_j^{t_{\delta+1}}) = \frac{card(X \cap Y)}{card(Y)}$, where $X = \{s \in U_s : t_\delta(s) = FunRed_i^{t_\delta}\}$ and $Y = \{s \in U_s : t_{\delta+1}(s) = FunRed_j^{t_{\delta+1}}\}$ for $\delta = 1, 2, \ldots, \kappa - 1$. Each branch $(FunRed_i^\delta, FunRed_j^{\delta+1})$ of TFG_a represents a prediction rule of the form: IF

Table 3. An information system S describing weather processes

U/A	$temp$	dew	hum	$wind$	$press$	U/A	$temp$	dew	hum	$wind$	$press$
d_1	50	31	40	8	2982	d_8	50	47	77	6	2906
d_2	58	31	61	10	2906	d_9	50	40	70	6	2906
d_3	50	47	77	0	2906	d_{10}	50	40	77	8	2906
d_4	58	47	70	0	2906	d_{11}	50	40	77	6	2906
d_5	58	47	61	10	2906	d_{12}	50	47	70	6	2982
d_6	58	47	70	10	2906	d_{13}	50	40	77	4	2982
d_7	58	47	77	10	2906	d_{14}	38	31	61	8	3004

$FunRed_i^\delta$, THEN $FunRed_j^{\delta+1}$. The meaning of such a rule is the following. If a set of functional relative reducts $FunRed_i^\delta$ for an attribute a appears in the time window S_δ, then a set of functional relative reducts $FunRed_j^{\delta+1}$ for an attribute a will appear in the time window $S_{\delta+1}$ with a certainty $cer(FunRed_i^{t_\delta}, FunRed_j^{t_{\delta+1}})$. This rule is also characterized by a strength $str(FunRed_i^{t_\delta}, FunRed_j^{t_{\delta+1}})$ and a coverage $cov(FunRed_i^{t_\delta}, FunRed_j^{t_{\delta+1}})$. Analogously, one can build prediction rules corresponding to the paths or the connections in TFG_a taking the definitions recalled in Section 2 into consideration.

The presented approach has been implemented in the ROSECON system [9] - a computer tool for automatized discovering Petri net models from data tables as well as predicting their changes in time. Computing functional relative reducts has an exponential time complexity with respect to the number of attributes in information systems. Constructing flow graphs has a polynomial time complexity with respect to the number of functional relative reducts.

4 An Example

This section includes an example illustrating the proposed approach. Let us consider an information system describing weather processes: temperature (marked with $temp$), dew point (marked with dew), humidity (marked with hum), wind speed (marked with $wind$), and pressure (marked with $press$). Global states observed in our system are collected in Table 3 representing an information system $S = (U, A)$, for which: a set of objects (global states) $U = \{d_1, d_2, \ldots, d_{14}\}$, a set of attributes (processes) $A = \{temp, dew, hum, wind, press\}$, sets of attribute values (local states of processes): $V_{temp} = \{38, 50, 58\}$ [F], $V_{dew} = \{31, 40, 47\}$ [F], $V_{hum} = \{40, 61, 70, 77\}$ [%], $V_{wind} = \{0, 4, 6, 8, 10\}$ [mph], $V_{press} = \{2906, 2982, 3004\}$ [$100 \times in$]. We assume the following values of parameters: $\lambda = 7$, $\rho = 1$ and $\kappa = 4$. A fragment of a temporal table of functional relative reducts $TRRT(S)$ is shown in Table 4. A temporal decision system TDS_{temp} for the attribute $temp$ is shown in Table 5. Each cell with $\emptyset$ means that the functional relative reduct is an empty set (i.e., values of a suitable attribute does not depend on the values of the remaining attributes). A fragment of a flow graph for the attribute $temp$ is shown in Figure 1. This graph enables us to mine different rules concerning

Table 4. A temporal table of functional relative reducts (fragment)

$Wind_\lambda^\rho(S)/A$	temp	...	press
S_1	$\{hum, wind\}$	...	$\{temp, dew\}, \{hum\}, \{wind\}$
S_2	$\{hum, wind\}$	...	$\emptyset$
S_3	$\{hum, wind\}$	...	$\emptyset$
S_4	$\{wind\}$	...	$\emptyset$
S_5	$\{wind\}$	...	$\emptyset$
S_6	$\{wind\}$	...	$\{temp, dew, hum\}, \{dew, hum, wind\}$
S_7	$\{wind\}$	...	$\{dew, hum, wind\}$
S_8	$\{dew\}, \{hum\}, \{press\}$	...	$\{dew, hum, wind\}$

Table 5. A temporal decision system for the attribute *temp*

U_s/A_t	t_1	t_2	t_3	t_4
s_1	$\{hum, wind\}$	$\{hum, wind\}$	$\{hum, wind\}$	$\{wind\}$
s_2	$\{hum, wind\}$	$\{hum, wind\}$	$\{wind\}$	$\{wind\}$
s_3	$\{hum, wind\}$	$\{wind\}$	$\{wind\}$	$\{wind\}$
s_4	$\{wind\}$	$\{wind\}$	$\{wind\}$	$\{wind\}$
s_5	$\{wind\}$	$\{wind\}$	$\{wind\}$	$\{dew\}, \{hum\}, \{press\}$

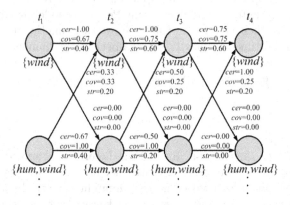

Fig. 1. A flow graph expressing prediction rules for the attribute *temp* (fragment)

component changes in the consecutive time windows. In order to do it we can take branches, paths or connections in the obtained graph (see Section 2). For example, if we take the first branch, then we get a rule as follows. If a functional relative reduct of the attribute *temp* is $\{wind\}$ in the given time window, then a functional relative reduct of *temp* is also $\{wind\}$ in the next time window with certainty equal to 1, coverage equal to 0.67 and strength equal to 0.4. On the basis of the obtained rule, we can predict that a fragment (corresponding to the functional relative reduct $\{wind\}$) of the structure of a current net model will be preserved in a net model constructed for the next time window.

5   Conclusions

In this paper we have presented the use of flow graphs to mine prediction rules of component changes in the consecutive time windows of a temporal information system. Such rules can be helpful to predict future behavior of a modeled system. In general, flow graphs are a suitable tool for mining rules describing changes of different kinds of elements (such as cores, reducts, components, etc.) in time windows of temporal information systems. The future goals include, among others, reduction of obtained flow graphs and building flow graphs for complex temporal information systems.

Acknowledgments

This paper has been partially supported by the grant No. 3 T11C 005 28 from Ministry of Scientific Research and Information Technology of the Republic of Poland.

References

1. Butz, C.J., Yan, W., Yang, B.: The Computational Complexity of Inference Using Rough Set Flow Graphs. In: D. Ślęzak et al. (Eds.), *Proc. of the RSFDGrC'2005*, Regina, Canada, LNAI 3641, Springer-Verlag, Berlin Heidelberg (2005) 335-344.
2. Greco, S., Pawlak, Z., Słowiński, R.: Generalized Decision Algorithms, Rough Inference Rules, and Flow Graphs. In: J.J. Alpigini et al. (Eds.), *Proc. of the RSCTC'2002*, Malvern, USA, LNAI 2475, Springer-Verlag, Berlin Heidelberg, (2002) 93-104.
3. Pancerz, K., Suraj, Z.: An Application of Rough Set Methods to Concurrent Systems Design. In: P. Grzegorzewski, M. Krawczak, S. Zadrożny (Eds.), *Soft Computing. Tools, Techniques and Applications*, EXIT, Warsaw (2004) 229-244.
4. Pawlak, Z.: In Pursuit of Patterns in Data Reasoning from Data - The Rough Set Way. In: J.J. Alpigini et al. (Eds.), *Proc. of the RSCTC'2002*, Malvern, USA, LNAI 2475, Springer-Verlag, Berlin Heidelberg (2002) 1-9.
5. Pawlak, Z.: Flow Graphs and Data Mining. In: J.F. Peters, A. Skowron (Eds.), *Transactions on Rough Sets III*, LNCS 3400, Springer-Verlag, Berlin Heidelberg (2005) 1-36.
6. Suraj, Z.: Rough Set Methods for the Synthesis and Analysis of Concurrent Processes. In: L. Polkowski, S. Tsumoto, T.Y. Lin (Eds.), *Rough Set Methods and Applications*, Physica-Verlag, Berlin (2000) 379-488.
7. Suraj, Z., Pancerz, K., Świniarski R.W.: Prediction of Model Changes of Concurrent Systems Described by Temporal Information Systems. In: H.R. Arabnia et al. (Eds.), *Proc. of the DMIN'2005*, Las Vegas, USA, CSREA Press (2005) 51-57.
8. Suraj, Z., Pancerz, K.: Reconstruction of Concurrent System Models Described by Decomposed Data Tables. *Fundamenta Informaticae*, **71**(1), IOS Press, Amsterdam (2006) 121-137.
9. Suraj, Z., Pancerz, K.: The ROSECON System - a Computer Tool for Modelling and Analysing of Processes. In: M. Mohammadian (Ed.), *Proc. of the CIMCA'2005*, Vol. II, Vienna, Austria, IEEE Computer Society (2006) 829-834.

Approximation Space-Based Socio-Technical Conflict Model

Sheela Ramanna[1], Andrzej Skowron[2], and James F. Peters[3]

[1] Department of Applied Computer Science,
University of Winnipeg,
Winnipeg, Manitoba R3B 2E9 Canada
s.ramanna@uwinnipeg.ca
[2] Institute of Mathematics,
Warsaw University
Banacha 2, 02-097 Warsaw, Poland
skowron@mimuw.edu.pl
[3] Department of Electrical and Computer Engineering,
University of Manitoba
Winnipeg, Manitoba R3T 5V6 Canada
jfpeters@ee.umanitoba.ca

Abstract. Rough set theory and approximation spaces introduced by Zdzisław Pawlak provide a framework for modelling social as well as technical conflicts. This is especially relevant in domains such as requirements engineering, an essential phase of software development. The *socio − technical* conflict model makes it possible to represent social conflicts (as conflict degrees) and technical conflicts (interaction between issues) in a unified framework. Reasoning about conflict dynamics is made possible by approximation spaces and conflict patterns. An illustrative example of such a framework is presented. The contribution of this paper is a formal *socio − technical* model and two approaches to reasoning: vectors of conflict degrees and approximation spaces.

Keywords: Approximation spaces, conflicts, rough sets, requirements engineering.

1 Introduction

Rough set theory and approximation spaces introduced by Zdzisław Pawlak provide frameworks for modelling conflicts. It is customary to represent conflicts arising between agents having differing opinions on various issues (ex: in government, industry) in terms of a voting framework [7,10]. However, there is a need to consider whether the issues themselves are in conflict (i.e., contradictory issues). This is particularly acute in requirements engineering, which provides an appropriate mechanism for understanding what a customer wants, analyzing need, negotiating a reasonable solution, specifying a solution unambiguously, validating a specification, and managing requirements that are transformed into an

J.T. Yao et al. (Eds.): RSKT 2007, LNAI 4481, pp. 476–483, 2007.

operational system[1]. In other words, we need to model a combination of complex conflict situations where there are social conflicts (due to differing stakeholder views) and technical conflicts (due to inconsistent requirements). A requirements interaction framework for social and technical conflicts and conflict dynamics assessment with risk patterns was introduced in [15]. In this paper, we present a formal socio-technical conflict model $STCM$ that facilitates i) representation of the two types of conflicts, ii) introduction of vectors of conflict degrees, iii) reasoning about conflicts with approximation spaces and conflict patterns. Conflict dynamics assessment provides a means of determining the scope of system functionality relative to social and technical conflicts.

This paper is organized as follows. In Sect. 2, we introduce a rough set based socio-technical conflict model and define a vector of conflict degrees. The basic architecture describing the methodology for constructing a decision table for requirements conflicts is presented in Sect. 3. We illustrate the socio-technical model in the context of a home lighting automation system in Sect. 3.1, followed by a discussion of approaches to conflict dynamics with approximation spaces in Sect. 3.2. In Sect. 3.3, we outline a new approach based on vectors of conflict degrees in the context of requirements engineering.

2   Socio-Technical Conflict Model

In this section, we introduce a model for social and technical conflicts. Such a model would be useful in applications such as requirements engineering, there are conflicts arising due to a) differing opinions about what requirements are necessary to be developed and b) requirements that are in conflict since they are contradictory. This is quite common since there could be many subsystems and specification of a requirement in one subsystem may contradict another requirement in a different subsystem. A conflict framework that includes i) conflicts arising due to differing opinions by agents regarding issues also called social conflicts and ii) conflicts arising due to inconsistent issues also called technical conflicts was first presented [15]. Formally, a socio-technical conflict model is denoted by $STCM = (U, B, Ag, V)$, where U is a nonempty, finite set called the *universe*, where elements of U are objects, B is a nonempty, finite set of *conflict features of objects*, $Ag = \{ag_1, \ldots, ag_n\}$ represents agents that vote, V denotes a finite set of *voting function v* on conflict features, where $v : Ag \rightarrow \{-1, 0, 1\}$. Elements of U and conflict features B are domain dependent. For example, in requirements engineering, the objects can be software requirements and conflict features can be scope negotiation parameters such as risk, or priority. Basic conflict theory concepts and an in-depth discussion of the social conflict model can be found in [10,13,14].

In a rough set approach to conflicts, $STCM$ can be represented as a decision system where (U, B, sc, tc) with two decision features sc and tc representing social conflict degree and technical conflict degree. The two decision features are

---

[1] Thayer, R.H and Dorfman, M: Software Requirements Engineering, IEEE Computer Society Press (1997).

complex decisions. The social conflict is derived as a result of voting [15] where $sc = Con(CS)$ of the conflict situation $CS = (Ag, v)$ defined by

$$Con(CS) = \frac{\sum_{\{(ag, ag'):\ \phi_v(ag, ag') = -1\}} |\phi_v(ag, ag')|}{2\lceil \frac{n}{2} \rceil \times (n - \lceil \frac{n}{2} \rceil)} \tag{1}$$

where $n = Card(Ag)$. For a more general conflict situation $CS = (Ag, V)$ where $V = \{v_1, \ldots, v_k\}$ is a finite set of voting functions each for a different issues the *conflict degree* in CS (*tension generated by V*) can be defined by (2).

$$Con(CS) = \frac{\sum_{i=1}^{k} Con(CS_i)}{k}, \tag{2}$$

Equations 1 and 2 offers deeper insight into structure of conflicts, enables analysis of relationship between stakeholders and requirements being debated. In particular, coalitions (like-minded team members) and conflict degrees amongst coalitions are important to find ways for reducing tension and negotiation. The technical conflict tc represents the degree of inconsistency and is determined by domain experts or by an automated tool to detect inconsistency in specification of issues. An illustration of determination of tc and sc is given in Section 3.1.

2.1   Vectors of Conflict Degrees

Let us consider a decision table $DT = (U, B, sc, tc)$ introduced in the previous section. We can consider this table as a table with a vector of two decisions sc, tc. The value vectors of these decisions are partially ordered (using the component wise ordering $\leq$). For a given vector v of conflict degrees, we consider sets $\{x \in U : Inf_B(x) \leq v\}$ and $\{x \in U : Inf_B(x) \geq v\}$, called the $\leq$-class relative to v and the $\geq$-class relative to v, respectively (see, e.g., [4]). If the measurements associated with the members of the sets of features from B are also , e.g., linearly ordered, then one can search, e.g., for patterns defined by the left hand sides of the rules with the following form:

$$\bigwedge_{j=1}^{l} b_i \geq v_i \longrightarrow (sc, tc) \geq (v, v').$$

Such patterns can be used to define approximations of the above defined classes [4]. Another possibility is to define a metric on a vector of conflict degrees and use the values of such a metric as a cumulative conflict degree. We now give an exemplary $STCM$ and conflict dynamics assessment in the context of requirements engineering.

3   Illustration: Requirements Engineering

In a typical requirements engineering process, one of the main goals is to formulate functional and non-functional requirements for the subsequent development

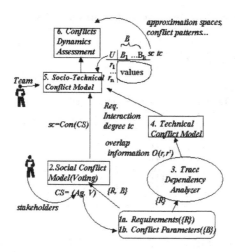

Fig. 1. Requirements Framework

of the system. The basic architecture for requirements framework will be briefly explained in this section. The methodology shown in Fig. 1 is composed of the following steps:

- *Step* 1: Derive Requirements R which constitute U and conflict features B such as effort, importance
- *Step* 2: Compute conflict degree sc for each requirement from R as a result of voting by stakeholders on conflict features B
- *Step* 3: Determine degree of overlap between requirements based on trace dependency In this paper, we assume that this information is obtained by an requirements management tools such as those listed in [2].
- *Step* 4: Determine tc based on the degree of overlap, type of requirements and the extent to which requirements conflict or cooperate
- *Step* 5: Construct decision table $STCM$
- *Step* 6: Assess conflict dynamics with approximation spaces, conflict patterns

Note that in theory, steps 2 and 3 can be performed in parallel. However, in practice, it is better to get agreements between stakeholders about the specific requirements before exploring inconsistencies between them. A complete example of the problem of achieving agreement (minimal social conflict) on high-level system requirements for a home lighting automation system[3] was described in [10]. Due to space restrictions, we refer the reader to [15] for a detailed discussion on the process of construction of $STCM$ decision table for a single high-level requirement (R1- Custom Lighting Scenes).

---

[2] See http://www.incose.org

[3] D. Leffingwell, D. Widrig: Managing Software Requirements, Addision-Wesley, 2003.

3.1   System Requirements: Socio Technical Model

Table 1 represents the $STCM$ with two complex decision features sc and tc derived from two sources of conflicts.

Table 1. Socio-Technical Conflict Model

$R1$	E	I	S	R	T	sc	tc
$r1.1$	M	H	N	L	Y	L	WC
$r1.2$	M	H	N	L	Y	M	VWC
$r1.3$	H	M	N	M	Y	L	NC
$r1.4$	L	H	Y	L	Y	L	SC
$r1.5$	M	H	P	H	Y	M	WC
$r1.6$	M	L	P	H	N	H	VWC

Elements of U are detailed requirements for $R1$, the conflicts features B are scope negotiation parameters: Effort, Importance, Stability, Risk and Testability. The conflict degree sc is classified as follows: L (*low conflict degree* $\leq < 0.3$), M ($0.3 \leq$ *medium conflict degree* ≤ 0.7) and H (*conflict degree* > 0.7). The assessment of $tc = \{SC, WC, VWC, NC\}$ represents the degree of requirements interaction: strong conflict, weak conflict, very weak conflict and no conflict respectively. Briefly, the approach is based on a generic model of potential conflict and cooperation which highlights the nature of added requirements on other features of the system [3]. For example, if a requirement adds *new functionality* to the system, it may have i) no effect(0) on the overall *functionality* ii) negative effect (-) on *efficiency* iii) positive effect(+) on *usability* iv) negative effect(-) on *reliability* v) negative effect(-) on *security* vi) no effect(0) on *recoverability* vii) no effect(0) on *accuracy* and viii) no effect(0) on *maintainability*. This model is very general and is a worst best-case scenario. In practice, one must take into account the degree of overlap between requirements and the type of requirement since it has a direct bearing on the degree of conflict or cooperation. Trace dependencies based on scenarios and observations are used to arrive at the degree of overlap [2]. Technical conflict model is designed to capture information from requirements traceability[4]. Requirements traceability involves defining and maintaining relationships with artifacts created as a part of systems development such as architectural designs, requirements, and source code to name a few. In this paper, we restrict the *artifact information* to other requirements as an aid to identifying conflicting (or cooperating) requirements [2]. The exemplary domain for degree of overlap and conflict degrees are due to [3]. In addition, we assume that an automated requirements traceability tool makes it possible to automatically extract i) conflicts and cooperation information amongst requirements and ii) trace dependencies. The degree of overlap between requirements and the conflict degrees are to a large extent manually assessed.

---

[4] IEEE Std. 830-1998.

3.2   Conflict Dynamics Assessment with Approximation Spaces

This section introduces social and technical conflict analysis and conflict resolution with approximation spaces. Let $DS = (U_{req}, B, sc, tc)$, where U_{req} is a non-empty set of requirements, B is a non-empty set of scope negotiation parameters (conflict parameters), and decision sc denotes social conflict, decision tc denotes technical conflict (see Table 1). By SC_i we denote the decision class corresponding to the decision i, i.e., $SC_i = \{u \in U_{req} : sc(u) = i\}$. Hence, SC_i consists of the requirements from U_{req} with the social conflict level i. By TC_j we denote the decision class corresponding to the decision j, i.e., $TC_j = \{u \in U_{req} : tc(u) = j\}$. Hence, TC_j consists of the requirements from U_{req} with the technical conflict level j. For any boolean combination of descriptors over DS and α, the semantics of α in DS is denoted by $\|\alpha\|_{DS}$, i.e., the set of all objects from U satisfying α [8]. In what follows, $i = L$ and SC_L denote a decision class representing a low degree of social conflict between stakeholders. A generalized approximation space $GAS_{SC} = (U_{req}, N_B, \nu_B^{SC})$ with the neighborhood function $N_B(r)$ and the coverage function ν_B^{SC} for social conflict can be defined as in (3) [13]. Assuming that the lower approximation $B_* SC_L$ represents an acceptable (standard) level of social conflict during negotiation, we are interested in the values ν_B^{SC} in (3).

$$\nu_B^{SC}(N_B(r), B_* SC_L) = \frac{|N_B(r) \cap B_* SC_L|}{|B_* SC_L|}, \tag{3}$$

considered in the context of a decision system DS for neighborhoods $N_B(r)$ and standard $B_* SC_L$ for conflict negotiation. Analogously, we define GAS_{TC} with coverage function ν_B^{TC} for technical conflicts. What follows is a simple example of how to set up a lower approximation space relative to two decisions shown in Table 1:

$B = \{Effort, Risk, Testability\}$, $SC_L = \{r \in U : sc(r) = L\} = \{r1.1, r1.3, r1.4\}$,
$B_* SC_L = \{r1.3, r1.4\}$, $N_B(r1.1) = \{r1.1, r1.2\}$, $N_B(r1.3) = \{r1.3\}$,
$N_B(r1.4) = \{r1.4\}$, $N_B(r1.5) = \{r1.5\}$, $N_B(r1.6) = \{r1.6\}$,
$\nu_B^{SC}(N_B(r1.1), B_* SC_L) = 0$, $\nu_B^{SC}(N_B(r1.4), B_* SC_L) = 0.5$,
$\nu_B^{SC}(N_B(r1.5), B_* SC_L) = 0$, $\nu_B^{SC}(N_B(r1.3), B_* SC_L) = 0.5$,
$\nu_B^{SC}(N_B(r1.6), B_* SC_L) = 0$.

$B = \{Effort, Risk, Testability\}$, $TC_{VWC} = \{r \in U : tc(r) = VWC \vee tc(r) = NC\} = \{r1.2, r.1.3, r1.6\}$,
$B_* TC_{VWC} = \{r1.3, r1.6\}$, $\nu_B^{TC}(N_B(r1.1), B_* TC_{VWC}) = 0$,
$\nu_B^{TC}(N_B(r1.3), B_* TC_{VWC}) = 0.5$, $\nu_B^{TC}(N_B(r1.4), B_* TC_{VWC}) = 0$,
$\nu_B^{TC}(N_B(r1.5), B_* TC_{VWC}) = 0$, $\nu_B^{TC}(N_B(r1.6), B_* TC_{VWC}) = 0.5$.

Based on the experimental rough coverage values, we can set a threshold th for acceptance of r such that acceptable neighborhoods have the following property:

$$\forall N_B \in GAS_{sc}, GAS_{tc} \; . \; (\nu_B^{sc} > th) \wedge (\nu_B^{tc} > th).$$

In our example we assume $th = 0.3$. Consequently, only one requirement $r1.3$ satisfies the criteria with almost no disagreement between stakeholders and requiring no change to the specification of requirement $r1.3$. On the other hand, there is also minimal disagreement on requirements $r1.4$. However, its specification is inconsistent with other requirements in the system (or other subsystems) and needs to be changed. Requirement $r1.6$ has to be renegotiated due to high social conflict.

3.3 Conflict Dynamics Assessment with Vectors of Conflict Degrees

In Sect. 2.1 we have outlined of the problem of approximation of conflict degree classes by some patterns defined by features from the set B. The approach requires using the ordering on feature value sets and methods developed in, e.g., [4]. In particular, the method for the so called tr-reducts discussed in [14] can be extended on this case. This approach makes it possible to analyze the dynamical changes of conflict degrees defined by patterns when some conditions from them are dropped. In this way, one can analyze the changes of vectors of conflict degrees defined in the socio-technical model. In particular, one can select important features which can not be eliminated if one would like to keep the vector of conflict degrees below a given threshold.

4   Related Works

Basic ideas of conflict theory in the context of rough sets are due to [7]. The relationships between the approach to conflicts and information systems as well as rough sets are illustrated in [10,13,14]. Inconsistent requirements (technical conflicts) using classifiers based on rough sets can be found in [6]. Recent research with approximate reasoning about vague concepts in conflict resolution and negotiations between agents (information sources) [5,9], requirements negotiation decision model [1], trace-dependency for identifying conflicts and cooperation among requirements [3], requirements interaction management [17] provide a basis for comparison of the proposed approach and also points to the usefulness of a unified framework for software requirement conflict analysis and negotiation.

5   Conclusion

This paper presents an approximation-space based socio-technical conflict model $STCM$ for two complementary types of conflicts. We also introduce vectors of conflict degrees which take into account dual complex decisions. We suggest reasoning about conflicts with approximation spaces and conflict patterns. The conflict dynamics assessment approach aids in i) selecting important features and ii) provides a mechanism to determine the scope of system functionality that takes into account social and technical conflict.

Acknowledgments. The research of Sheela Ramanna and James F. Peters is supported by NSERC Canada grants 194376 and 185986 respectively. The research of Andrzej Skowron is supported by grant 3 T11C 002 26 from the Ministry of Scientific Research and Higher Education of the Republic of Poland.

References

1. Boehm, B., Grünbacher, P., Kepler, J.: Developing Groupware for Requirements Negotiation: Lessons Learned, IEEE Software, May/June (2001) 46-55.
2. Egyed, A., Grünbacher, P.: Automating Requirements Traceability: Beyond the Record and Replay Paradigm, Proc. of the 17th International Conference on Automated Software Engineering, EEE CS Press,(2002) 163-171.
3. Egyed, A., Grünbacher, P., : Identifying requirements Conflicts and Cooperation: How Quality features and Automated Traceability Can Help, IEEE Software, November/December (2004) 50-58.
4. Greco, S., Matarazzo, B., Słowiński, R.: Rough set theory for multicriteria decision analysis. European Journal of Operational Research **129**(1) (2001) 1–47.
5. Lai, G. Li,C. Sycara, K.and Giampapa, J.: Literature review on multi-attribute negotiations, Technical Report CMU-RI-TR-04-66 (2004) 1-35.
6. Li, Z. and Ruhe. G.: Uncertainty Handling in Tabular-Based Requirements Using Rough Sets, LNAI **3642**, Springer, Berlin, 2005, 678-687.
7. Pawlak, Z.: On Conflicts. *Int. J. of Man-Machine Studies*,Vol. 21 (1984) 127-134.
8. Pawlak, Z.: Rough Sets – Theoretical Aspects of Reasoning about Data. Kluwer Academic Publishers (1991).
9. Kraus, S.: Strategic Negotiations in Multiagent Environments, The MIT Press, 2001.
10. Skowron, A., Ramanna, S., Peters, J.F.: Conflict Analysis and Information Systems: A Rough Set Approach. LNCS **4062**, Springer, Heidelberg, 2006, 233-240.
11. Skowron, A. Rauszer, C.: The Discernibility Matrices and Functions in Information Systems. In: Słowiński, R (ed.): Intelligent Decision Support - Handbook of Applications and Advances of the Rough Sets Theory, System Theory, Knowledge Engineering and Problem Solving **11**, Kluwer, Dordrecht (1992) 331-362.
12. Skowron, A.: Extracting Laws from Decision Tables. *Computational Intelligence: An International Journal* **11**(2) (1995) 371-388.
13. Ramanna, S., Peters, J.F., Skowron, A.: Generalized Conflict and Resolution Model with Approximation Spaces. LNAI **4259**, Springer, Heidelberg, 2006, 274–283.
14. Ramanna, S., Peters, J.F., Skowron, A.: Approaches to Conflict Dynamics based on Rough Sets. *Fundamenta Informaticae* **75**(2006) 1-16.
15. Ramanna, S., Skowron, A.: Requirements Interaction and Conflicts: A Rough Set Approach. Proceedings of the First IEEE Symposium on Foundations of Computational Intelligence, Hawaii, 2007 [to appear].
16. Ślęzak, D.: Approximate Entropy Reducts. *Fundamenta Informaticae* **53** (2002) 365–387.
17. Robinson, W, N., Pawlowski, D, S., Volkov, V.: Requirements Interaction Management, ACM Computing Surveys, 35(2) (2003) 132-190.
18. Skowron, A., Stepaniuk, J,: Tolerance approximation spaces. Fundamenta Informaticae 27(2-3) (1996) 245–253.

Improved Quantum-Inspired Genetic Algorithm Based Time-Frequency Analysis of Radar Emitter Signals*

Gexiang Zhang and Haina Rong

School of Electrical Engineering, Southwest Jiaotong University,
Chengdu 610031 Sichuan, China
{zhgxdylan,ronghaina}@126.com

Abstract. This paper uses an improved quantum-inspired genetic algorithm (IQGA) based time-frequency atom decomposition to analyze the construction of radar emitter signals. With time-frequency atoms containing the detailed characteristics of a signal, this method is able to extract specific information from radar emitter signals. As IQGA has good global search capability and rapid convergence, this method can obtain time-frequency atoms of radar emitter signals in a short span of time. Binary phase shift-key radar emitter signal and linear-frequency modulated radar emitter signal are taken for examples to analyze the structure of decomposed time-frequency atoms and to discuss the difference between the two signals. Experimental results show the huge potential of extracting fingerprint features of radar emitter signals.

Keywords: Quantum-inspired genetic algorithm, time-frequency atom decomposition, radar emitter signal, feature analysis.

1 Introduction

Feature extraction takes a very important part in pulse train deinterleaving, type recognition and individual identification of radar emitters. In [1,2], feature extraction for deinterleaving radar emitter pulse trains was discussed. In [3,4], intra-pulse modulation features were extracted from several advanced radar emitter signals. With the rapid development of modern electronic warfare technology, specific radar emitter identification arouses great interest. In [5-7], several feature extraction methods for radar emitter signals were presented. These features are characteristic of the overall view of radar emitter signals.

Time-frequency atom decomposition (TFAD), also known as matching pursuit or adaptive Gabor representation, is able to decompose any signal into a linear expansion of waveforms selected from a redundant dictionary of time-frequency atoms that well localized both in time and frequency [8,9]. Unlike Wigner and

---

* This work was supported by the Scientific and Technological Development Foundation of Southwest Jiaotong University (2006A09) and by the National Natural Science Foundation of China (60572143).

J.T. Yao et al. (Eds.): RSKT 2007, LNAI 4481, pp. 484–491, 2007.
© Springer-Verlag Berlin Heidelberg 2007

Cohen class distributions, the energy distribution obtained by TFAD does not include interference terms. Different from Fourier and Wavelet orthogonal transforms, the information in TFAD is not diluted across the whole basis [8,10,11]. Therefore, TFAD can be used to capture the natural features (local and detailed characteristics) of radar emitter signals, thus, TFAD can be applied to extract the key features from advanced radar emitter signals to identify some specific radar emitters. However, the necessary dictionary of time-frequency atoms being very large, the computational load is extremely high and consequently becomes the main problem of TFAD.

In this paper, an improved quantum-inspired genetic algorithm (IQGA) is introduced into TFAD to extract some detailed features from advanced radar emitter signals. Based on the concepts of quantum computing, quantum-inspired genetic algorithm (QGA) falls into the latest category of unconventional computation. Due to some outstanding advantages such as good global search capability, rapid convergence and short computing time [10,11], QGA is able to accelerate greatly the process of searching the most satisfactory time-frequency atom in each iteration of TFAD. So this method makes TFAD easy to extract time-frequency atom features from radar emitter signals. In the next section, IQGA is presented briefly. Next, time-frequency analysis of radar emitter signals is discussed. In Section 4, some experiments are conducted on radar emitter signals. Finally, some conclusions are listed.

2   IQGA

In [10], an improved quantum-inspired genetic algorithm was presented. IQGA uses quantum bit (Q-bit) phase to update the rotation angels of quantum gates (Q-gates) to generate the individuals at the next generation. The evolutionary strategy is simple and has only one parameter to adjust. Also, migration and catastrophe operators are employed to strengthen search capability and to avoid premature phenomena. The structure of IQGA is shown in Algorithm 1 and the brief description is as follows.

(i) Some parameters, including population size n, the number v of variables, the number m of binary bit of each variable and evolutionary generation g, are set. The initial value of g is set to 0.

(ii) Population $\boldsymbol{P}(g)=\{\boldsymbol{p}_1^g, \boldsymbol{p}_2^g, \cdots, \boldsymbol{p}_n^g\}$, where $\boldsymbol{p}_i^g (i = 1, 2, \cdots, n)$ is an arbitrary individual in population and $\boldsymbol{p}_i^g$ is represented as

$$\boldsymbol{p}_i^g = \begin{bmatrix} \alpha_{i1}^g | \alpha_{i2}^g | \cdots | \alpha_{i(mv)}^g \\ \beta_{i1}^g | \beta_{i2}^g | \cdots | \beta_{i(mv)}^g \end{bmatrix} . \tag{1}$$

where $\alpha_{ij}^g = \beta_{ij}^g = 1/\sqrt{2}$ $(j = 1, 2, \cdots, mv)$, which means that all states are superposed with the same probability.

(iii) According to probability amplitudes of all individuals in $\boldsymbol{P}(g)$, observation states $\boldsymbol{R}(g)$ is constructed by observing $\boldsymbol{P}(g)$.

Algorithm 1. *Algorithm of IQGA*

```
        Begin
(i)     Initial values of parameters; % Evolutionary generation g=0;
(ii)    Initializing P(g); %
(iii)   Generate R(g) by observing P(g); %
(iv)    Fitness evaluation; %
(v)     Store the best solution among P(g) into B(g);
        While (not termination condition) do
            g=g+1;
(vi)        Generate R(g) by observing P(g-1); %
(vii)       Fitness evaluation; %
(viii)      Update P(g) using quantum rotation gate; %
(ix)        Store the best solution among P(g) and B(g-1) into B(g);
            If (migration condition)
(x)         Migration operation;
            End if
            If (catastrope condition)
(xi)          Catastrope operation;
            End if
        End
    End
```

(iv) Each individual is evaluated.

(v) The best solution in $P(g)$ at generation g is stored into $B(g)$.

(vi) According to probability amplitudes of all individuals in $P(g-1)$, observation states $R(g)$ is constructed by observing $P(g-1)$. This step is similar to step (iii).

(vii) This step is the same as step (iv).

(viii) In this step, the probability amplitudes of all Q-bits in population are updated by using Q-gates given in (2).

$$G = \begin{bmatrix} \cos\theta & -\sin\theta \\ \sin\theta & \cos\theta \end{bmatrix}. \tag{2}$$

where θ is the rotation angle of Q-gate and θ is defined as $\theta = k \cdot f(\alpha, \beta)$, where k is a coefficient whose value has a direct effect on convergent speed of IQGA. $f(\alpha, \beta)$ is a function for determining the search direction of IQGA to a global optimum. The look-up table of $f(\alpha, \beta)$ is shown in Table 1, in which $\xi_1 = \arctan(\beta_1/\alpha_1)$ and $\xi_2 = \arctan(\beta_2/\alpha_2)$, where α_1, β_1 are the probability amplitudes of the best solution stored in $B(g)$ and α_2, β_2 are the probability amplitudes of the current solution.

(ix) The best solution among $P(g)$ and $B(g-1)$ is stored into $B(g)$.

(x) Migration operation is used to introduce new and better individuals to quicken the convergence of IQGA. Migration operation is performed only on the stored best individual instead of all individuals.

(xi) If the best solution stored in $B(g)$ is not changed in some generations, catastrophe operation should be performed.

Table 1. Look-up table of function $f(\alpha, \beta)$ (Sign is a symbolic function)

$\xi_1 > 0$	$\xi_2 > 0$	$f(\alpha, \beta)$	
		$\xi_1 \geq \xi_2$	$\xi_1 < \xi_2$
True	True	+1	-1
True	False	$\text{sign}(\alpha_1 \cdot \alpha_2)$	
False	True	$-\text{sign}(\alpha_1 \cdot \alpha_2)$	
False	False	$\text{sign}(\alpha_1 \cdot \alpha_2)$	$-\text{sign}(\alpha_1 \cdot \alpha_2)$
$\xi_1, \xi_2 = 0$ or $\pi/2$		± 1	

3   Time-Frequency Analysis of Radar Emitter Signals

In our prior work [10,11], the effectiveness of IQGA was validated by using radar emitter signals, and some useful guidelines for setting the parameters were drawn up from extensive experiments, and the detailed steps of IQGA based TFAD was described. In this section, we aim at time-frequency atom analysis of radar emitter signals. Figure 1 shows a binary-phase shift-key (BPSK) radar emitter signal with noise.

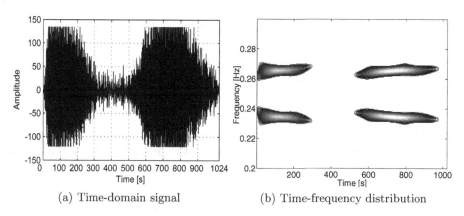

(a) Time-domain signal   (b) Time-frequency distribution

Fig. 1. A BPSK radar emitter signal with noise

IQGA based TFAD is applied to decompose the BPSK radar emitter signal and resemblance coefficient method [3] is used to the correlation ratio C_r to compute the correlation between the original signal f and the restored signal f_r with parts of decomposed time-frequency atoms.

$$C_r = \frac{\langle f, f_r \rangle}{\|f\| \cdot \|f_r\|} . \tag{3}$$

Figure 2(a) illustrates the correlation ratio curve of the first 100 iterations. As can be seen from the correlation ratio curve, there is a steep increase at the first 10 iterations, after which the correlation ratio curve rises slowly. At iteration 10, the correlation ratio amounts to 0.97. We uses the first 10 decomposed

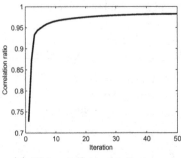

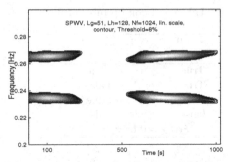

(a) The correlation ratio curve

(b) Time-frequency distribution of the first 10 atoms

Fig. 2. Experimental results of BPSK radar emitter signal

time-frequency atoms to reconstruct the radar emitter signal and give the time-frequency distribution of the reconstructed signal in Fig.2(b). In this paper, time-frequency atom uses Gabor function:

$$g_\gamma(t) = \frac{1}{\sqrt{s}} g(\frac{t-u}{s}) \cos(\nu t + \omega) . \tag{4}$$

where the index $\gamma=(s, u, \nu, \omega)$ is a set of parameters and s, u, ν, ω are scale, translation, frequency and phase, respectively. $g(\cdot)$ is a Gauss-modulated window function as

$$g(t) = e^{-\pi t^2} . \tag{5}$$

The setting of parameters of Gabor time-frequency atom can be referred to [11]. To demonstrate the detailed structure of the first 10 atoms, we list their parameters in Table 2, in which the 10 atoms are labelled as 1 to 10.

Table 2. Parameters of 10 Gabor atoms of BPSK radar emitter signal

	1	2	3	4	5	6	7	8	9	10
s	57.67	57.67	57.67	57.67	57.67	57.67	57.67	57.67	57.67	57.67
u	0	0	0	0	0	0	0	0	0	0
ν	106.89	106.29	107.49	105.47	106.84	107.92	106.02	104.82	105.36	109.12
ω	2.09	4.19	5.76	5.76	4.71	1.05	4.71	5.76	4.71	4.71

To analyze further time-frequency atoms of radar emitter signals, a linear-frequency modulated (LFM) radar emitter signal is employed to conduct the next experiment. The noised LFM radar emitter signal is shown in Fig.3. We use IQGA based TFAD to decompose the LFM radar emitter signal and resemblance coefficient method to evaluate the correlation between the original signal and the restored signal with parts of decomposed time-frequency atoms. The change of correlation ratio of the first 100 iterations is illustrated in Fig.4(a), in which the

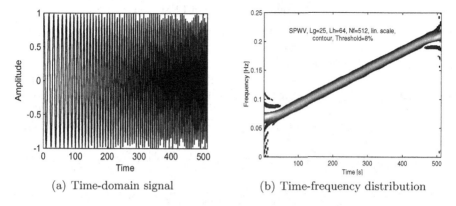

(a) Time-domain signal   (b) Time-frequency distribution

Fig. 3. A LFM radar emitter signal with noise

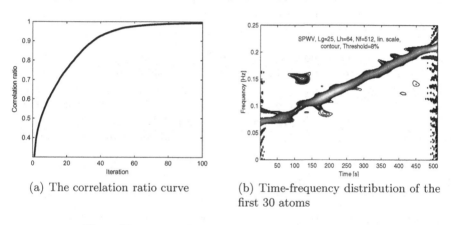

(a) The correlation ratio curve   (b) Time-frequency distribution of the
                                         first 30 atoms

Fig. 4. Experimental results of LFM radar emitter signal

correlation ratio climbs to 0.86 at iteration 30. The first 30 decomposed time-frequency atoms are applied to reconstruct the LFM radar emitter signal and Fig.4(b) shows the time-frequency distribution of the reconstructed signal. In Table 3, the parameters of Gabor atoms are given to demonstrate the detailed structure of the first 30 atoms whose labels are from 1 to 30.

As can be seen from Fig.1 to Fig.4, Tables 2 and 3, several conclusions can be drawn. First of all, IQGA based TFAD is able to decompose radar emitter signals into linear expansion of waveforms selected from redundant dictionary of time-frequency atoms. Next, BPSK radar emitter signal is approximated by only 10 time-frequency atoms, while LFM radar emitter signal need at least 30 time-frequency atoms. BPSK radar emitter signal is easier to represent by time-frequency atoms than LFM radar emitter signal. Also, each atom characterizes a part of radar emitter signal. Therefore, these time-frequency atoms are a certain features of radar emitter signals. Finally, Table 2 and Table 3 show the much difference between BPSK radar emitter signal and LFM radar emitter signal. In

Table 3. Parameters of 30 Gabor atoms of LFM radar emitter signal

	1	2	3	4	5	6	7	8	9	10
s	17.09	17.09	25.63	7.59	7.59	11.39	7.59	11.39	38.44	38.44
u	34.17	17.09	25.63	0	0	0	0	0	38.44	38.44
ν	41.19	38.80	37.26	26.06	24.82	27.86	24.00	30.34	36.20	35.47
ω	2.62	4.71	6.28	2.62	1.05	3.67	5.76	2.09	4.71	1.05
	11	12	13	14	15	16	17	18	19	20
s	11.39	7.59	11.39	11.39	38.44	38.44	38.44	11.39	25.63	25.63
u	0	0	0	0	38.44	38.44	38.44	0	0	0
ν	28.96	23.17	32.27	31.44	34.57	33.59	32.93	27.03	21.33	22.43
ω	3.14	3.67	4.71	0.52	4.71	1.57	3.14	3.14	5.24	2.09
	21	22	23	24	25	26	27	28	29	30
s	38.44	38.44	38.44	38.44	38.44	38.44	38.44	38.44	38.44	17.09
u	0	0	0	0	0	0	0	0	0	0
ν	14.71	15.04	15.85	19.94	14.38	15.45	20.92	17.24	17.57	21.88
ω	3.14	5.76	4.71	6.28	0.52	2.09	3.67	5.76	1.57	0.52

Table 2, there is nearly no change of s and u. The values of s and ν are more than 50 and 100, respectively. In contrast, Table 3 shows great difference from Table 2. The values of s and u vary in a certain range in Table 3. All values of s and u are below 40. There is no value above 45 in parameter ν. It is the difference that indicates the difference between BPSK and LFM radar emitter signals. So we can use the difference to recognize different radar emitter signals.

4   Concluding Remarks

By introducing IQGA into TFAD, time-frequency atoms of radar emitter signals are easy to obtain. Analyzing the structure of time-frequency atoms of BPSK radar emitter signal and LFM radar emitter signal, this paper discusses the difference between BPSK radar emitter signal and LFM radar emitter signal. Containing plenty of detailed characteristics, time-frequency atoms of radar emitter signals may be used as the features to recognize different radar emitters. In the future work, we will use this method to extract fingerprint features from specific radar emitters and to analyze other signals such as fault signals and harmonic signals in power system.

References

1. Milojevic, D.J., Popovic, B.M.: Improved Algorithm for the Deinterleaving of Radar Pulses. IEE Proceedings, Part F, Radar and Signal Processing **139** (1992) 98-104
2. Rong, H.N., Jin, W.D., Zhang, C.F.: Application of Support Vector Machines to Pulse Repetition Interval Modulation Recognition. In: Proceedings of the 6th International Conference on ITS Telecommunications (2006) 1187-1190

3. Zhang, G.X., Rong, H.N., Jin, W.D., Hu, L.Z.: Radar Emitter Signal Recognition Based on Resemblance Coefficient Features. In: Tsumoto,S., et al., (eds.): Lecture Notes in Artificial Intelligence, Vol.3066. Springer-Verlag, Berlin Heidelberg New York (2004) 665-670

4. Zhang, G.X.: Intra-pulse Modulation Recognition of Advanced Radar Emitter Signals Using Intelligent Recognition Method. In: Wang, G., et al., (eds.): Lecture Notes in Artificial Intelligence, Vol.4062. Springer-Verlag, Berlin Heidelberg New York (2006) 707-712

5. Langley, L.E.: Specific Emitter Identification (SEI) and Classical Parameter Fusion Technology. In: Proceedings of the WESCON (1993) 377-381

6. Dudczyk, J., Matuszewski, J., Wnuk, M.: Applying the Radiated Emission to the Specific Emitter Identification. In: Proceedings of 15th International Conference on Microwaves, Radar and Wireless Communications (2004) 431-434

7. Zhang, G.Z., Huang, K.S., Jiang, W.L., Zhou, Y.Y.: Emitter Feature Extract Method Based on Signal Envelope. Systems Engineering and Electronics **28** (2006) 795-797

8. Mallat, S.G., Zhang, Z.F.: Matching Pursuits with Time-Frequency Dictionaries. IEEE Transactions on Signal Processing **41** (1993) 3397-3415

9. Liu, Q.S., Wang, Q., Wu, L.N.: Size of the Dictionary in Matching Pursuits Algorithm. IEEE Transactions on Signal Processing **52** (2004) 3403-3408

10. Zhang, G.X., Rong, H.N., Jin, W.D.: An Improved Quantum-Inspired Genetic Algorithm and Its Application to Time-Frequency Atom Decomposition. Dynamics of Continuous, Discrete and Impulsive Systems (2007) (to appear)

11. Zhang, G.X., Rong, H.N.: Quantum-Inspired Genetic Algorithm Based Time-Frequency Atom Decomposition. In: Shi, Y., et al., (eds.): Lecture Notes in Computer Science, Springer-Verlag, Berlin Heidelberg New York (2007) (to appear)

Parameter Setting of Quantum-Inspired Genetic Algorithm Based on Real Observation*

Gexiang Zhang and Haina Rong

School of Electrical Engineering, Southwest Jiaotong University,
Chengdu 610031 Sichuan, China
{zhgxdylan,ronghaina}@126.com

Abstract. Parameter setting, especially the angle of Q-gate, has much effect on the performance of quantum-inspired evolutionary algorithm. This paper investigates how the angle of Q-gate affects the optimization performance of real-observation quantum-inspired genetic algorithm. Four methods, including static adjustment methods, random adjustment methods, dynamic adjustment methods and adaptive adjustment methods, are presented to bring into comparisons to draw some guidelines for setting the angle of Q-gate. Comparative experiments are carried out on some typical numerical optimization problems. Experimental results show that real-observation quantum-inspired genetic algorithm has good performance when the angle of Q-gate is set to lower value.

Keywords: Quantum computing, genetic algorithm, quantum-inspired genetic algorithm, parameter setting.

1 Introduction

Quantum-inspired genetic algorithm (QGA) is a novel unconventional computation method. QGA is a combination of genetic algorithm and quantum computing. It uses the concepts and principles of quantum computing, such as quantum-inspired bit (Q-bit), quantum-inspired gate (Q-gate) and quantum operators including superposition, entanglement, interference and measurement. In the literature, there are two main types of QGA. One is the binary-observation QGA (BQGA) presented in [1]. The other is the real-observation quantum-inspired genetic algorithm (RQGA) introduced in [2]. Extensive experiments verifies that QGA has the advantages of much better global search capability and much faster convergence only with a small population size over conventional genetic algorithm [1-4]. However, parameter setting has much effect on QGA. In [5,6], some experiments were conducted to discuss the choice of parameters in BQGA. But there is relatively little or no research on the effects of different settings for the parameters of RQGA.

This paper presents four methods to investigate the effects of parameter setting of Q-gate on the performance of RQGA. The four methods include static

---

* This work was supported by the Scientific and Technological Development Foundation of Southwest Jiaotong University (2006A09).

J.T. Yao et al. (Eds.): RSKT 2007, LNAI 4481, pp. 492–499, 2007.

adjustment, random adjustment, dynamic adjustment and adaptive adjustment. We use some typically numerical optimization problems to do comparative experiments to obtain some useful guidelines for setting the angle of Q-gate in RQGA.

2 Parameter Adjustment Methods

Extending two states '1' and '0' to an arbitrary pair of states between '1' and '0' in quantum system, a real-observation quantum-inspired genetic algorithm was proposed to solve globally numerical optimization problems with continuous variables [2]. RQGA uses a Q-gate to drive the individuals toward better solutions and eventually toward a single state corresponding to a real number varying between 0 and 1. The value k of Q-gate angle has a direct effect on convergent speed of RQGA. If k is too large, the search grid of RQGA would be very large and the solution may diverge or converge prematurely to a local optimum. On the contrary, if it is too small, the search grid of RQGA would be very small and RQGA may fall in stagnant state. Hence, the choice of k is very important for RQGA. In the following description, we will present four parameter adjustment methods, including static adjustment, random adjustment, dynamic adjustment and adaptive adjustment, to investigate the effect.

2.1 Static Adjustment Method

According to Def.2 in [2], Q-gate angle k varies between 0 and $\pi/2$. In static adjustment method, Q-gate angle k is a constant in the whole process of searching a globally optimal solution. But how to preset the constant value is the subsequent problem. Intuitively, we always want to choose the best value in the range $[0, \pi/2]$ as the value of k. To determine the best Q-gate angle k, this paper uses some test functions to carry out experiments, in which k changes from 0 to $\pi/2$ and the other condition keep unchangeable. In each test, Q-gate angle k is preset to a constant value. The terminal condition of RQGA is the maximal evolutionary generation. In the experiments, the mean best values and standard deviations for each value of k are recorded and finally two curves of the mean best values and standard deviations for all values of k can be obtained. According to the experimental results, the best value of k is the one that achieves the optimal solution and the smallest standard deviation, which is chosen as the value of k.

2.2 Random Adjustment Method

RQGA is a kind of probabilistic optimization methods, in which there must be randomness. This randomness is not correspondent with fixed k in static adjustment method. Hence, random adjustment method is used to determine Q-gate angle k. In this method, instead of presetting a fixed value, Q-gate angle k can be chosen randomly between 0 and $\pi/2$ in the process of searching the optimal solution. In each experiment, the value of k is generated by using a random function.

2.3 Dynamic Adjustment Method

In random adjustment method, there is a sort of blindness for setting Q-gate angle k. To overcome the blindness of random adjustment method, another method called dynamic adjustment method is employed to determine Q-gate angle k. In this method, Q-gate angle k is defined as a variable. Of course, there are many definitions for k. Here, an example for defining Q-gate angle k is as follows.

$$k = \frac{\pi}{2} e^{-\frac{mod(g,b)}{a}} . \tag{1}$$

where $mod(g,b)$ is a function for calculate the remainder after g is divided by the constant b and g is the evolutionary generation. a is also a constant. In this method, the initial value of k is equal to $\pi/2$ and then k varies from $\pi/2$ to 0, which indicates RQGA begins with large grid to search the optimal solution and then the search grid declines gradually to 0 as the evolutionary generation g increases to b. When the evolutionary generation g amounts to b, the value of k come back to $\pi/2$. In this paper, b and a are set to 100 and 10, respectively.

2.4 Adaptive Adjustment Method

Dynamic adjustment method gives RQGA a changing trend of searching the optimal solution from coarse to fine. In dynamic adjustment method, Q-gate angle k varies as evolutionary generation. As a matter of fact, Q-gate angle k should be adjusted by the diversity of population in RQGA, that is to say, Q-gate angle k should be adjusted by the convergent state of RQGA. When there is much difference between the individuals, Q-gate angle k may change slowly, in contrast, Q-gate angle k may change quickly when little difference exists between the individuals. Therefore, adaptive adjustment method is applied to determine the value of k in the process of searching the optimal solution.

In RQGA, the diversity of population is related to probability amplitudes of Q-bits. In [2], the probability amplitude of a Q-bit is defined by a pair of numbers (α, β) as

$$[\alpha \ \beta]^T . \tag{2}$$

where α and β satisfy normalization equality $|\alpha|^2 + |\beta|^2 = 1$. $|\alpha|^2$ and $|\beta|^2$ denote the probabilities that the qubit will be found in A state and in B state in the act of observing the quantum state, respectively. The probability amplitudes of n Q-bits are represented as

$$\begin{bmatrix} \alpha_1 | \alpha_2 | \cdots | \alpha_n \\ \beta_1 | \beta_2 | \cdots | \beta_n \end{bmatrix} . \tag{3}$$

If there are n individuals in RQGA, we just need n Q-bits to represent the n individuals. In the n Q-bits, the standard deviation of probability amplitudes can embody well the diversity of population.

According to the definition of Q-bit, α and β range between 0 and 1. So the standard deviation of α or β of n individuals also varies between 0 and 1. Thus, a function of the standard deviation of probability amplitudes of n individuals can be used to adjust the Q-gate angle k. The function is written as

$$k = \frac{\pi}{2}(1 - S).$$

(4)

where S is the standard deviation of α of n individuals.

3   Experiments

Some comparative experiments are carried out to investigate the effects of the above four parameter adjustment methods on the performance of RQGA. In this paper, 10 functions $f_1 - f_{10}$, whose global minima are 0, are applied to conduct the comparative experiments. These functions are
  (I) Sphere function

$$f_1(\mathbf{x}) = \sum_{i=1}^{N} x_i^2, \quad -100.0 \le x_i \le 100.0, \ N = 30.$$

(5)

(II) Ackley function

$$f_2(\mathbf{x}) = 20 + e - 20\exp\left(-0.2\sqrt{\tfrac{1}{N}\sum_{i=1}^{N} x_i^2}\right) - \exp\left(\tfrac{1}{N}\sum_{i=1}^{N}\cos\left(2\pi x_i\right)\right).$$

(6)

$$-32.0 \le x_i \le 32.0, \quad N = 30$$

(III) Griewank function

$$f_3(\mathbf{x}) = \frac{1}{4000}\sum_{i=1}^{N} x_i^2 - \prod_{i=1}^{N}\left(\frac{x_i}{\sqrt{i}}\right) + 1, \quad -600.0 \le x_i \le 600.0, \ N = 30.$$

(7)

(IV) Rastrigin function

$$f_4(\mathbf{x}) = 10N + \sum_{i=1}^{N}\left(x_i^2 - 10\cos\left(2\pi x_i\right)\right), \quad -5.12 \le x_i \le 5.12, \ N = 30.$$

(8)

(V) Schwefel function

$$f_5(\mathbf{x}) = 418.9829N - \sum_{i=1}^{N}\left(x_i\sin\left(\sqrt{|x_i|}\right)\right), \quad -500.0 \le x_i \le 500.0, \ N = 30.$$

(9)

(VI) Schwefel's problem 2.22

$$f_6(\mathbf{x}) = \sum_{i=1}^{N}|x_i| + \prod_{i=1}^{N}|x_i|, \quad -10 \le x_i \le 10, \ N = 30.$$

(10)

(VII) Schwefel's problem 1.2

$$f_7(\mathbf{x}) = \sum_{i=1}^{N} \left(\sum_{j=1}^{i} x_j \right)^2, \quad -100 \le x_j \le 100, \ N = 30 \ . \tag{11}$$

(VIII) Schwefel's problem 2.21

$$f_8(\mathbf{x}) = \max_{i=1} \{ |x_i|, 1 \le i \le 30 \}, \quad -100 \le x_i \le 100 \ . \tag{12}$$

(IX) Step function

$$f_9(\mathbf{x}) = \sum_{i=1}^{N} (\lfloor x_i + 0.5 \rfloor)^2, \quad -100 \le x_i \le 100, \ N = 30 \ . \tag{13}$$

(X) Quartic function, i.e. noise

$$f_{10}(\mathbf{x}) = \sum_{i=1}^{N} i x_i^4 + random[0, 1), \quad -1.28 \le x_i \le 1.28, \ N = 30 \ . \tag{14}$$

In RQGA, both the population size and the parameter C_g are set to 20. Static adjustment method is first employed to determine the best value of Q-bit angle k. In the range $[0, \pi/2]$, we conduct the experiment 50 times repeatedly every 0.02π interval. The maximal generation is set to 500. The mean best solutions and the standard deviations are recorded for each experiment. Experimental results are given in Fig.1, in which solid-line and dash line represent the mean best solutions (MBS) and standard deviations (STD), respectively. The given values in the abscissa in Fig.1 should be multiplied by π.

In the same environment as static adjustment method, RQEA are performed 50 independent runs for random adjustment method (RAM), dynamic adjustment method (DAM) and adaptive adjustment method (AAM), for each test function, respectively. The mean best values and the standard deviations are recorded for each test function. Experimental results are listed in Table 1, in which m, σ, g and p represent the mean best, the standard deviation, the maximal number of generations and the population size, respectively. The results are averaged over 50 runs. The best results shown in Fig.1 of static adjustment method (SAM) for each test function are also listed in Table 1 so as to bring into comparison with the other three methods.

Figure 1 shows that different values of Q-gate angles make RQGA obtain different mean best solutions and standard deviations for different functions. Generally speaking, RQGA has good performance when Q-gate angle is less than 0.1π. According to this conclusion, we decrease the value $\pi/2$ to 0.1π in random adjustment method, dynamic adjustment method and adaptive adjustment method, and redo the above experiments. Table 2 lists the experimental results. From Tables 1 and 2, it can be seen that dynamic adjustment method is superior to random adjustment method and is inferior to adaptive adjustment method. The experimental results in Table 2 are better than those in Table 1.

Table 1. Comparisons of four adjustment methods

	g	p		SAM	RAM	DAM	AAM
f_1	500	20	m	0.04	12.33	20.35	8.24
			σ	0.05	27.12	49.48	19.23
f_2	500	20	m	0.03	1.29	1.06	0.85
			σ	0.03	2.43	1.01	0.99
f_3	500	20	m	0.14	0.81	0.92	0.83
			σ	0.18	0.40	0.31	0.37
f_4	500	20	m	0.02	3.65	2.95	2.08
			σ	0.05	6.86	6.19	3.19
f_5	500	20	m	0.19	43.97	7.58	3.68
			σ	0.53	254.16	21.38	4.92
f_6	500	20	m	0.14	1.09	1.02	1.75
			σ	0.17	1.03	1.05	6.60
f_7	500	20	m	0.16	21.17	11.28	12.68
			σ	0.33	38.05	20.98	25.90
f_8	500	20	m	0.03	0.35	0.32	0.32
			σ	0.03	0.40	0.28	0.29
f_9	500	20	m	0	10.80	5.40	3.60
			σ	0	25.54	11.64	9.85
f_{10}	500	20	m	2.0×10^{-3}	0.03	0.01	0.01
			σ	1.8×10^{-3}	0.13	9.6×10^{-3}	0.01

Table 2. Experimental results after decreasing k

	g	p		SAM	RAM	DAM	AAM
f_1	500	20	m	0.04	7.37	10.04	2.60
			σ	0.05	14.70	21.55	4.33
f_2	500	20	m	0.03	0.63	0.54	0.74
			σ	0.03	1.01	0.48	0.97
f_3	500	20	m	0.14	0.62	0.75	0.68
			σ	0.18	0.49	0.37	0.55
f_4	500	20	m	0.02	2.38	2.72	1.90
			σ	0.05	2.54	4.62	2.17
f_5	500	20	m	0.19	4.08	2.72	2.81
			σ	0.53	6.92	4.34	5.85
f_6	500	20	m	0.14	0.91	0.73	1.04
			σ	0.17	0.73	0.62	1.11
f_7	500	20	m	0.16	17.30	1.88	10.98
			σ	0.33	40.52	3.07	20.00
f_8	500	20	m	0.03	0.28	0.17	0.30
			σ	0.03	0.30	0.14	0.28
f_9	500	20	m	0	9.60	0	2.73
			σ	0	20.50	0	9.05
f_{10}	500	20	m	2.0×10^{-3}	0.01	0.01	0.01
			σ	1.8×10^{-3}	8.1×10^{-3}	5.3×10^{-3}	8.9×10^{-3}

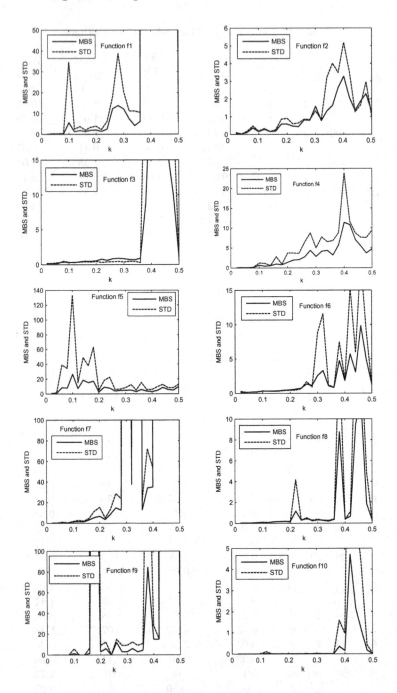

Fig. 1. Experimental results of functions $f_1 - f_{10}$

4   Concluding Remarks

As an unconventional evolutionary computation algorithm, real-observation quantum inspired genetic algorithm has better performance than conventional genetic algorithm, but the performance of RQGA depends a considerable extent on its parameter setting. Q-gate angle is the most important parameter in RQGA. So this paper presents four methods for setting the Q-gate angle. Apart from the four methods, including static adjustment method, random adjustment method, dynamic adjustment method and adaptive adjustment method, are discussed in detail, some comparative experiments are carried out to investigate the advantages and disadvantages of them. This work is helpful to improve further the performance of RQGA.

References

1. Han, K.H., Kim, J.H.: Genetic Quantum Algorithm and Its Application to Combinatorial Optimization Problem. In: Proc. of IEEE Congress on Evolutionary Computation, **2** (2000) 1354-1360
2. Zhang, G.X., Rong, H.N.: Real-Observation Quantum-Inspired Evolutionary Algorithm for a Class of Numerical Optimization Problems. Lecture Notes in Computer Science, Berlin, Springer (2007) (to appear)
3. Zhang, G.X., Jin, W. D. and Li, N.: An Improved Quantum Genetic Algorithm and Its Application. Lecture Notes in Artificial Intelligence, Berlin, Springer, **2639** (2003) 449-452
4. Zhang, G.X., Hu, L.Z. and Jin, W. D.: Quantum Computing Based Machine Learning Method and Its Application in Radar Emitter Signal Recognition. Lecture Notes in Artificial Intelligence, Berlin, Springer, **3131** (2004) 92-103
5. Han, K.H., Kim, J.H.: On Setting the Parameters of Quantum-Inspired Evolutionary Algorithms for Practical Applications. In: Proc. of the 2003 Congress on Evolutionary Computation (2003) 427-428
6. Khorsand, A.R. and Akbarzadeh-T, M.R.: Quantum Gate Optimization in a Meta-Level Genetic Quantum Algorithm. In: Proc. of IEEE Int. Conf. on Systems, Man and Cybernetics, **4** (2005) 3055-3062

A Rough Set Penalty Function for Marriage Selection in Multiple-Evaluation Genetic Algorithms

Chih-Hao Lin and Char-Chin Chuang

Department of Management Information Systems
Chung Yuan Christian University
No. 200, Jhongbei Rd., Jhongli City 320, Taiwan
{linch,g9494003}@cycu.edu.tw

Abstract. Penalty functions are often used to handle constrained optimization problems in evolutionary algorithms. However, most of the penalty adjustment methods are based on mathematical approaches not on evolutionary ones. To mimic the biological phenomenon of the values judgment, we introduce the rough set theory as a novel penalty adjustment method. Furthermore, a new marriage selection is proposed in this paper to modify the multiple-evaluation genetic algorithm. By applying rough-penalty and marriage-selection methods, the proposed algorithm generally is both effective and efficient in solving several constrained optimization problems. The experimental results also show that the proposed mechanisms further improve and stabilize the solution ability.

Keywords: Rough set theory, penalty function, marriage selection, genetic algorithm, MEGA.

1 Introduction

Recently, genetic algorithms (GAs) have become well-known stochastic methods of global optimization based on the evolution theory of Darwin [1]. It would seem that further investigations are needed in order to deal with the real-world constraint problems. One of the most common approaches is the penalty function approach. By introducing penalty terms into the objective function, a constrained optimization problem can be transformed into an unconstrained one [2]. Many recent studies focus on combining GAs and penalty functions to solve the constrained optimization problems, such as the static penalty (SP) [3], the dynamic penalty (DP) [4] and the adaptive penalty (AP) methods [5]. However, most of the modified GAs are based on mathematical strategies but neglect the biological evolutionary approaches.

By mimicking the biological processes of the values judgment and the spouse selection, this paper proposes two novel mechanisms to enhance the multiple-evaluation genetic algorithm (MEGA) [6]. The first mechanism is a rough penalty (RP) method which applies the rough set theory (RST) as a new penalizing strategy. Blending the RPs with the traditional global penalty can imitate the libertarianism and enlarge the genetic diversity. The other modification is a marriage selection operation which combines with the conventional GA selection to mimics the volition mating process.

J.T. Yao et al. (Eds.): RSKT 2007, LNAI 4481, pp. 500–507, 2007.

Compared with the experimental results in the literatures, the proposed algorithm can enhance the exploration and exploitation abilities of the traditional GAs.

The rest of this paper is organized as follows. Section 2 briefly describes the RST and penalizing strategies. The main processes of the modified MEGA are described in Section 3. In Section 4, the experiment results and performance assessments are presented. Finally, the paper concludes with a summary of the results in Section 5.

2 Rough Set Theory and Penalizing Strategy

2.1 Rough Set Theory

Zdzislaw Pawlak proposed the RST in the early 1980's [7]. The rough set analysis approximates the main components of concepts to assist decision making [8].

Definition 1: *Information system.* An information system (IS) is denoted as a triplet $T = (U, A, f)$, where U is a non-empty finite set of objects and A is a non-empty finite set of attributes. Information function f maps an object to its attribute, i.e. $f_a : U \rightarrow V_a$ for $\forall a \in A$, where V_a is called domain of an attribute a. A posteriori knowledge is expressed by one distinguished attributed and denoted by d. A decision system (DS) is a IS of the form $DT = (U, A \cup \{d\}, f)$, where $d \notin A$ is used as supervised learning. The elements of A are called conditional attributes.

Definition 2: *Indiscernibility.* For any attribute set $B \subseteq A$, an equivalence relation induced by B *is* determined as $IND_T(B) = \{(x, y) \in U^2 | \forall a \in B, f_a(x) = f_a(y)\}$. The relation $IND_T(B)$ is called the *B-indiscernibility relation.* The equivalence classes of the B-indiscernibility relation are denoted as $I_B(x)$.

Definition 3: *Set Approximation.* Let $X \subseteq U$ and $B \subseteq A$ in an IS T, the *B-lower* and *B-upper approximations* of X are defined as $\underline{B}X = \{x \in U | I_B(x) \subseteq X\}$ and $\overline{B}X = \{x \in U | I_B(x) \cap X \neq \phi\}$ respectively. $\underline{B}X$ is the set of objects that belongs to X with certainty, whereas $\overline{B}X$ is the set of objects that possibly belongs to X.

Definition 4: *Reducts.* If $X_{DT}^1, X_{DT}^2, \cdots, X_{DT}^r$ are the decision class of DT, the set $\underline{B}X_1 \cup \cdots \cup \underline{B}X_r$ is called the *B-positive region of DT* and is denoted by $POS_B(d)$. A subset $B \subseteq A$ is a set of relative reducts of DT if and only if $POS_B(d) = POS_C(d)$ and $POS_{B-\{b\}}(d) \neq POS_C(d), \forall b \in B$.

2.2 Conventional Penalty Methods

This paper adopts a two-stage penalizing mechanism: one is called global penalty (GP) and the other is called rough penalty (RP). The GP can be one of the conventional penalty methods used to penalize constraint violations of each infeasible solution. The proposed RP applies the RST to enhance the exploration ability of the MEGA.

For illustration, the following minimization problem is used in this section:

$$Minimize \quad F(\vec{X}_j),$$ (1)

$$s.t. \quad g_k(\vec{X}_j) \leq 0, \quad k = 1,2,...,m.$$ (2)

The jth chromosome $\vec{X}_j = (x_{j1}, x_{j2}, \cdots, x_{jn}) \in R^n$ encodes a vector of n variables for numerical problems. Function $F(\vec{X}_j)$ is the fitness of $\vec{X}_j$. The kth constraint violation of $\vec{X}_j$ denoted as $g_k(\vec{X}_j)$.

Several penalizing strategies have been proposed in literatures. According to the degree of constraint violation, penalty coefficient should be determined carefully. The SP applies each static penalty coefficient to each constraint and adjusts penalty coefficients empirically [3]. To adjust penalty coefficient automatically, the DP modifies penalty coefficient only depending on the number of generation [4]. Furthermore, the AP decreases the penalty coefficient if all the best individuals of the past H iterations are feasible and increases otherwise. That is, by reducing the penalty coefficient, the algorithm can increase search ability and find better solutions in infeasible space [5].

In this paper, the GP mixes the DP with AP and formulates the fitness function as:

$$GP(\vec{X}_j) = F(\vec{X}_j) + \lambda_t \times \sum_{k=1}^{m} \left(\max[0, g_k(\vec{X}_j)] \right)^2,$$ (3)

where notation $\lambda_t > 0$ is an adaptive penalty coefficient for the tth generation.

2.3 Penalty Method Based on the Rough Set Theory

The proposed RP method enhances the genetic diversity by introducing a rough penalty concept to each individual. It is clear that penalty adjustment is difficult to adapt suitably. Therefore, this paper finds the better penalizing strategy by using the attribute reduction concept in the RST.

To construct a decision table of rough set, the population in the tth generation can be denoted as $X(t) = \{\vec{X}_1, \vec{X}_2, \cdots, \vec{X}_p\}$, where p is the number of individuals in a population. Each individual is treated as an object in DT, and each constraint violation, $g_k(\vec{X}_j)$, $k = 1, \cdots, m$, is a condition attribute. Thus, the $DT = (U, G \cup \{d\}, f)$ consists of a condition set $G = \{g_1, g_2, \cdots, g_m\}$ and a decision attribute $V_d = \{0,1\}$. For minimization problem, the decision value of each object is assigned by:

$$f_d(\vec{X}_j) = \begin{cases} 1, & \text{if } GP(\vec{X}_j) < GP_{av}(X(t)) \\ 0, & \text{if } GP(\vec{X}_j) \geq GP_{av}(X(t)) \end{cases}$$ (4)

where $GP_{av}(X(t)) = \frac{1}{p} \sum_{j=1}^{p} GP(\vec{X}_j)$.

According to the definition of a reduct of the RST, it is a minimal subset of penalized constraints that enables us to classify objects with better fitness (decision values are 1) and those with worse fitness (decision values are 0). Thus, significant constraints can be decided by a constraint reduction algorithm which is a drop-and-add heuristic and is similar to the attribute reduction algorithm proposed by Fan Li *et al.* [8]. The fitness function with RP, denoted as $RP(\vec{X}_j)$, is composed of zero-one relationship variables μ_{tk} and penalty coefficients α_j and is formulated as:

$$RP(\vec{X}_j) = F(\vec{X}_j) + \lambda_t \alpha_j \times \sum_{k=1}^{m} \mu_{tk} \left(\max[0, g_k(\vec{X}_j)] \right)^2 , \tag{5}$$

$$\alpha_j = \exp\left(-\frac{1}{r} \sum_{k=1}^{r} g_k(\vec{X}_j) \right). \tag{6}$$

The adjustment criterion of penalty coefficient α_j is that a higher degree of feasibility in individual j implies a larger value of penalty coefficient α_j.

3   Modified Multiple-Evaluation Genetic Algorithm

To solve constrained optimization problems, we proposed a modified MEGA. The MEGA is an effective and efficient GA for solving numerical problems [6]. By synthesizing the rough penalty method and marriage selection, the modified MEGA can effectively solve constrained optimization problems. The main components of the modify MEGA are described in this section.

3.1   Coding Mechanism

Because of the numerical property of test functions, the coding mechanism represents a problem solution as a chromosome $\vec{X}_j = (x_{j1}, x_{j2}, \cdots, x_{jn}) \in \Re^n$, where x_{ji} is the value of the ith decision variable in the jth solution [6].

3.2   Two-Stage Selection Operation

The two-stage selection operation consists of a traditional global selection and a proposed marriage selection. The global selection mimics the biologic selective pressure

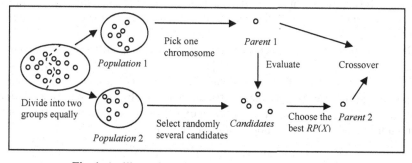

Fig. 1. An illustration of the marriage selection operation

proposed by Darwin and uses fitness $GP(\vec{X}_j)$ as a discriminator to guide the evolution process. Higher quality chromosomes are therefore selected for recombination. The stochastic universal selection is used in the modified MEGA.

To imitate the libertarianism of spouse selection, this paper proposes a marriage selection to assist with the global selection. In the nature world, several male animals always pay court to a female animal. The female evaluates all of the candidates according to its own judgment criterion $RP(\vec{X}_j)$ and then selects the strongest male to become its spouse. The judgment criteria of the female animals may not be the same as the evolution criteria of Nature; however, the judgment criteria provide the genetic diversity. An illustration of the marriage selection is depicted in Fig. 1.

3.3 Recombination Operation

The MEGA proposes an improved *evaluation-based crossover* that combines two selected parents to produce an offspring by randomly introducing isolation, manipulation and reintroduction of gene splicing techniques [6]. Furthermore, the MEGA sequentially applies the uniform and Gaussian mutations become the *two-stage mutation* to enhance both exploration and exploitation abilities [6].

3.4 Replacement Operation and Termination of Evolution

In this paper, the successive generation is generated by three processes. Firstly, the modified MEGA uses the replacement-with-elitism scheme that reduces genetic drift by ensuring that the best two chromosomes are allowed to pass their traits to the next generation. Secondly, the 50% of child chromosomes are produced by the crossover operation. Finally, the mutation operation constructs other child chromosomes.

4 Numerical Experiments

4.1 Implementation and Comparison

Eleven well-known numerical constrained problems are adopted to evaluate the effectiveness and efficiency of the modified MEGA [9]. In this paper, the GP blending the DP ($C=500$, $\alpha=2$) and the AP ($c_1=2$, $c_2=5$, $H=0.04\times$expected iterations, and initial penalty $= 100$) adjusts penalty coefficient iteratively [4] [5]. Environment settings are that the population size is 20, the maximum iteration is 2000, the crossover rate is 0.5 and the elitism size is 2.

Table 1 depicts the experiment results obtained by the modified MEGA. Compared with three existing algorithms [10] [11] [12], the modified MEGA can find the near-optimal solutions for 8 test functions out of 11 constrained problems. Especially, the proposed algorithm can solve these difficult problems by using purely evolutionary mechanisms. That is, this study succeeds in exploring the evolutionary effect on GAs without using any enforcement optimization technique.

Table 1. Comparison of experiment results among the four algorithms

Function	Optimum value	Runarsson and Yao [10]	Farmani and Wright [11]	Venkatraman and Yen[12]	Our Algorithm
Min G1	-15	-15.0000	-15.0000	-14.9999	-15.0000
Max G2	0.803533	0.803015	0.802970	0.803190	0.782294
Max G3	1.0	1.0000	1.0000	1.0000	0.987980
Min G4	-30665.5	-30665.539	-30665.500	-30665.5312	-30665.5381
Min G5	5126.4981	5126.4970	5126.9890	5126.5096	5126.5999
Min G6	-6961.8	-6961.814	-6961.800	-6961.1785	-6961.8139
Min G7	24.306	24.307	24.480	24.41098	24.816882
Max G8	0.095825	0.095825	0.095825	0.095825	0.095825
Min G9	680.63	680.63	680.64	680.7622	680.648016
Min G10	7049.33	7054.32	7061.34	7060.5529	7116.2168
Min G11	0.75	0.75	0.75	0.7490	0.7499

Table 2. Comparison of computational effort among the four algorithms

Function Evaluation	Runarsson and Yao [10]	Farmani and Wright [11]	Venkatraman and Yen [12]	Our Algorithm
Complexity	350000	1400000	50000	40000

Table 3. Comparison of experiment results between the MEGA with different selections

Function	Without Marriage Selection Mean	(Standard deviation)	With Marriage Selection Mean	(Standard deviation)
Min G1	-13.8094	(1.832577)	-14.4082	(0.959480)
Max G2	0.563624	(0.176093)	0.590444	(0.163281)
Max G3	0.790955	(0.253056)	0.925572	(0.003991)
Min G4	-30658.5348	(11.641451)	-30662.6311	(8.871769)
Min G5	7394.9556	(4167.4932)	5236.6209	(168.5924)
Min G6	-6959.9666	(1.915056)	-6960.3141	(1.443794)
Min G7	28.1753	(2.579041)	27.8582	(1.923737)
Max G8	0.095825	(1.34e-17)	0.095825	(1.96e-17)
Min G9	680.9923	(0.120896)	680.9796	(0.183393)
Min G10	10353.997	(1230.0397)	9624.551	(1405.4830)
Min G11	0.833126	(0.083372)	0.795698	(0.068678)

For further comparison, the four functions are also analyzed by the standard complexity evaluation defined as (population size)*(number of generations). We depict the result of complexity evaluations in Table 2. The time complexity of the modified MEGA is significantly lower than that of others. To summarize, the comparisons emphasize that the modified MEGA is both effective and efficient in solving constrained numerical problems by introducing a rough penalty function and a marriage selection.

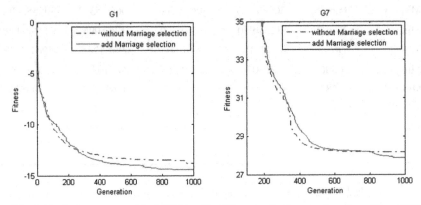

Fig. 2. Evolution curves of the MEGA with different selections on functions G1 and G7

4.2 Performance Assessment of the Modified MEGA

To evaluate the effect of the proposed marriage selection, we manipulate these experiments by using the MEGA without/with the marriage selection mechanism and reveal the results in Table 3. The evolution curves of the MEGA with/without marriage selection on two sampling problems are also depicted in Fig. 2.

The results in Table 3 show that introducing the marriage selection mechanism can decrease the standard deviation and achieve better solution. That is, using the RP marriage selection can improve the solution ability of the MEGA and stabilize the solution quality. In Fig. 2, we can observe that the MEGA without marriage selection always converges quicker than that with marriage selection in the earlier generations. However, the former algorithm easily jumps into a local optimal solution and stops to explore in the later generations. To summarize, the RP marriage selection mechanism can enlarge the genetic diversity to improve the exploration ability of the MEGA.

5   Conclusion

To the best of our knowledge, the proposed algorithm is the first research that synthesizes the rough set theory and the penalty method in a genetic algorithm to solve constrained problems. In this paper, we also imitate the biologic spouse selection phenomenon and propose a novel marriage selection operation to enhance genetic diversity. The proposed algorithm is tested on eleven benchmark functions and compared with three existing algorithms. The experiment results indicate that the modified

MEGA is able to find most of the near-global solution and outperform other three algorithms dramatically. Furthermore, the performance assessments also conclude that introducing the proposed marriage selection in the MEGA can stabilize and enhance the searching ability for the constrained optimization problems.

References

1. Holland, J.H.: Adaptation in Natural and Artificial Systems. MIT Press (1975)
2. Gen M., Cheng R.: A Survey of Penalty Techniques in Genetic Algorithms. Evol. Comput. (1996) 804-809
3. Homaifar, A., Qi, C.X., Lai, S.H.: Constrained Optimization via Genetic Algorithms. Simulation. (1994) 242–254
4. Joines, J.A., Houck, C.R.: On the Use of Nonstationary Penalty Functions to Solve Nonlinear Constrained Optimization Problems with GA's. Proc. 1st IEEE Conf. Evolutionary Computation. (1994) 579–584
5. Hadj-Alouane, A.B., Bean J.C.: A Genetic Algorithm for the Multiple-choice Integer Program. Operations Research. (1997) 92–101
6. Lin, C.H., He, J.D.: A Multiple-Evaluation Genetic Algorithm for Numerical Optimization Problems. Proc. Computability in Europe: Computation and Logic in the Real World. (2007)
7. Pawlak, Z.: Rough Sets. International Journal of Computer and Information Sciences 11, (1982) 341-356.
8. Li, F., Liu, Q-H., Min, F., Yang, G-W.: A New Crossover Operator Based on the Rough Set Theory for Genetic Algorithms. Proc. of 4th International Conference on Machine Learning and Cybernetics, Guangzhou. (2005) 18-21
9. Michalewicz, Z., Schoenauer, M.: Evolutionary Algorithms for Constrained Parameter Optimization Problems. Evol. Comput, vol. 4. (1996) 1–32
10. Runarsson, T., Yao, X.: Stochastic Ranking for Constrained Evolutionary Optimization. IEEE Trans. Evol. Comput. Vol. 4. (2000) 344–354
11. Farmani, R., Wright, J.: Self-adaptive Fitness Formulation for Constrained Optimization. IEEE Trans. Evol. Comput. Vol. 7, No. 5. (2003) 445–455
12. Venkatraman, S., Yen, G.G.: A Generic Framework for Constrained Optimization Using Genetic Algorithms. IEEE Trans. on evolutionary computation, Vol. 9, No. 4. (2005) 424-435

Multiple Solutions by Means of Genetic Programming: A Collision Avoidance Example

Daniel Howard

Company Fellow, QinetiQ, UK
Visiting Scholar, Bio-computing and Developmental Systems Group,
University of Limerick, Ireland
dr.daniel.howard@gmail.com

Abstract. Seldom is it practical to completely automate the discovery of the Pareto Frontier by genetic programming (GP). It is not only difficult to identify all of the optimization parameters *a-priori* but it is hard to construct functions that properly evaluate parameters. For instance, the "ease of manufacture" of a particular antenna can be determined but coming up with a function to judge this on all manner of GP-discovered antenna designs is impractical. This suggests using GP to discover many diverse solutions at a particular point in the space of requirements that are quantifiable, only *a-posteriori* (after the run) to manually test how each solution fares over the less tangible requirements e.g. "ease of manufacture". Multiple solutions can also suggest requirements that are missing. A new toy problem involving collision avoidance is introduced to research how GP may discover a diverse set of multiple solutions to a single problem. It illustrates how emergent concepts (linguistic labels) rather than distance measures can cluster the GP generated multiple solutions for their meaningful separation and evaluation.

Keywords: Genetic Programming, Multiple Solutions.

1 Introduction

The need to discover multiple solutions to a problem is ubiquitous. Mathematics can serve to illustrate that not all problems can be recast as the search for the global minimum and unique solution. Many problems have multiple solutions all of which equivalently satisfy the question(s). For example:

- Finite in number: "how can two lines co-exist?" with three answers: parallel and not touching; crossing and touching at one point; or perfectly aligned and touching at every single point.
- Countably infinite as when the problem is to find whole numbers (integers) that satisfy $x^2 + y^2 = z^2$.
- Uncountably infinite as with linear system of equations $Ax = b$ for m equations and n unknowns when $n > m$ if it is recognized to be under-determined (if over-determined, $m > n$, it may be regularized [1]).

J.T. Yao et al. (Eds.): RSKT 2007, LNAI 4481, pp. 508–517, 2007.

Questions of this world are hardly specific and admit a great number of possible answers. For instance, a university president wants to get more students to use the library. He is not particularly bothered how this is done provided certain norms are kept to (e.g. library must not introduce unacceptable literature). The likelihood is always for the existence of many acceptable alternatives that attract the same result.

Moreover, much problem solving activity is directed at defining the problem, e.g. helping a child with autism can feel this way. Also a problem can become solved without its clear definition. In the personal investment known as classical Freudian psychoanalysis [2] a therapist listens and ventures only very rarely to reflect connections that induce insights over years of daily hourly sessions, while the patient on the couch talks (with no visibility of the analyst sat behind him). Eventually the patient experiences the fantasies that sabotage his behaviour and in doing so the brain's neuronal network rearranges to allow him to take advantage of opportunities, restoring optimism and happiness. Psychology is a next frontier for application of genetic programming (GP) mixed with natural language processing, virtual reality, pattern matching, and wearable devices. At singularity [3] this setup may advance a portable Freudian analyst.

Multiple solutions are present in design optimization when Quality Function Deployment (QFD) links up to the needs of the customer. If one is to use GP or a similar tool to search for the optimal design then one has to recognize that:

1. The literature [4] is vast on multiple solutions and the Pareto Front. However, even when all of the requirements are agreed *a-priori* to form a multi-dimensional space for solution evaluation, even at a single point in this space there will exist multiple and diverse solutions that are equivalent because dimensions (considerations) which matter always escape the analysis.
2. In practice, some of the most important requirements can never be quantified for use by the fitness function of GP.

The first point is illustrated with soccer. Match requirements may be: entertainment value at 8, fair play at 9, and winning, yet important differences in play strategy can and will achieve this same targets mix. Their differences get studied *a-posteriori* after their creation. One needs to obtain the solutions first to see that a new consideration makes them less equivalent, or more plausible in the case of scientific theories.

To explain the second point consider that almost anything can be turned into an antenna, although only a few antenna designs (such as the Yagi-Uda and helical) are listed in the Johnson handbook [5]. The requirements space may include: range, gain at frequency; weight; impedance matching, most of which are handled with a simulator (e.g. with the Livermore code [6]). However, a very important parameter is its "ease of manufacture" and while one can determine the ease of manufacture of a particular antenna design: should it be made in China? does it require a soldering process in France? it is difficult to produce a function for GP to evaluate this for an arbitrary antenna. A poor function is not desirable as it can restrict the creativity of GP. This suggests that GP produce a great number of solutions to review the intangible requirements *a-posteriori* and manually.

Environmental Modelling also needs discovery of multiple solutions. Flooding models are aggregations of many approximate sub-models (soil, run-off, etc.). When an event is not predicted by the overall model, it is often necessary to back track and to calibrate the component sub-models. Immediately, one is faced with many plausible ways in which the flood might have occurred, and the required information to determine this conclusively needs to be identified (so that it may be gathered next time around). Bioinformatics also needs discovery of multiple explanations and models as living systems have evolved so much redundancy.

The rest of this paper: (a) introduces a new "toy problem" that has a finite number of multiple solutions; (b) arranges GP to discover many of these solutions; (c) examines how are the multiple solutions to be clustered when there are so many of them? This paper is a first attempt at research that looks to bring the repository of information that is the population in GP and the concept of genetic diversity in GP for the discovery of multiple solutions.

2   An Illustrative Toy Problem

Table 1 completely describes the toy problem which has a railway track with numbered *positions* and locomotives or *engines* as pieces that are moved on a board game as in Fig. 1. A number of fast and slow engines co-exist on the track. The passage of time is uniform and discrete, and its unit is the *epoch*. Fast engines need one epoch whereas slow engines need two epochs to advance one position. The track contains a number of railroad *points* (switches) that can be set to allow trains to travel along *shortcuts* to another part of the track. A shortcut is activated by setting its entry point enabling engines to escape collision with other engines. The program evolved by GP: (a) takes information about the state of the points at shortcuts and about the location of the engines; (b) it decides whether to set the points - it may set many points in different parts of the track as it is being executed; (c) the same program is evaluated at each epoch but the results can differ because engine positions and the state of points will change from epoch to epoch.

The solution to the toy problem is to avoid all engine collisions for a target number of epochs. This is achieved by evolving a program that sets points at each epoch. If there is no collision, then the corresponding GP generated strategy is a solution. The toy problem is only devised as an illustration.

Points indicated by a black number on a white background (see Fig. 1) are *active* whereas points indicated by a white number on a black background are *passive*. A passive point is automatically set not to upset the engine running on the track and to admit trains coming onto the track. However, an active point is set by the control program that is evolved by GP to divert the engine to the short cut or not. For example, only an engine travelling against the numbering system may use 222 to exit at 79 (see Fig. 1). However, only when travelling with the direction of the track numbering system may an engine use 20 to travel to 112.

Initial conditions are crucial, and care must be taken not to set up the problem to have no solution as engines may be headed for an inevitable crash. This

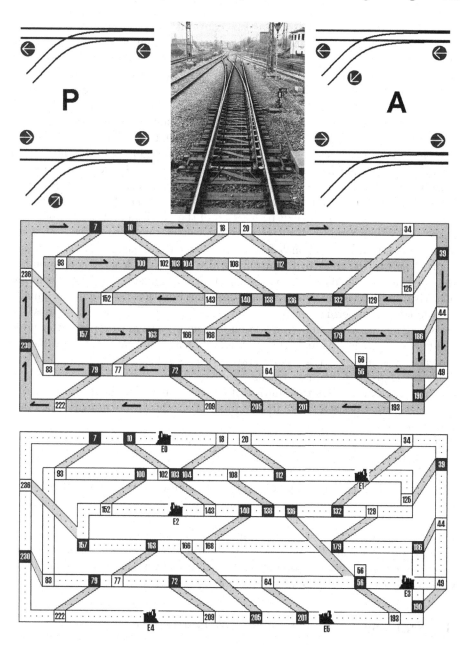

Fig. 1. (Top) Behaviour of the point types with direction of travel: passive (left) and active (right). For an explanation see text. (Middle) The railway track. (Bottom): engine starting positions and directions of motion.

work considered engines that ran at two constant speeds. These start from set locations and directions of travel (with or against the numbering system of the track). Engines occupy one position only as illustrated in Figure 1 and prescribed in Table 1. In the table, if engine i is positioned to move with increasing track numbering then $gtrain[i][0] = 1$ else $gtrain[i][0] = -1$. The array element $gtrain[i][2]$ indicates whether the engine is fast ('0') or slow ('1').

Each point enabled shortcut to another part of the track has a length in track segments that is not shown in Fig. 1. An engine using the shortcut will cover a fixed number of track segments and epochs before emerging from the shortcut to rejoin the track. Information about shortcuts is held in $glink$ [shortcut][property] with initial values in Table 1. Properties are: an ID number; a track direction of allowed access (1 is with increasing position numbering and -1 is with decreasing numbering); entry position; exit or position of the passive point; length between entry and exit in number of discrete track segments: note from Fig. 1 that for example shortcut at position 5 has to be longer than the others as it crosses another part of the track by means of a bridge (generally, lengths correspond to the diagram but not particularly: e.g. shortcut ID 12); 'swap' indicates whether on arrival the engine travels in a reverse direction from how it was going originally (in the track numbering sense), e.g. to enter shortcut ID 2 the engine must be travelling against the numbering system but emerges from the shortcut at node 132 travelling with the numbering system, and this is indicated by the value '1'; indicates whether or not the point starts with its value set, indicated by '1', to guide the engine off the main track.

The fitness measure is the number of epochs achieved without crashing n_{EA} minus the target desired 'crash free' number of epochs n_{ET} (e.g. $n_{ET} = 30,000$): $f = n_{EA} - n_{ET}$.

The fitness is computed as in the pseudo-code in table 2 which considers two options. In the first, the program is halted when the solution is achieved. The second is the simplest method for generating new solutions: when the problem is solved, all previous (phenotype or expressed) solutions are checked. If the solution is different and therefore a new solution, it is written to file. The individual is given the worst possible fitness to discourage it from genetically participating in further evolution. Note that a new solution may differ by the number and programme of execution of points but may not be considered a different solution, as we are only interested in operational differences (differences in the motion of engines) checking for differences in phenotypes not genotypes.

3   Numerical Experiments and Discussion

A typical solution takes around 600 epochs to organize the engines such that they are all moving in the same direction, and the path of the slow engine is isolated from the path of the other four as in Fig. 3. However, in some observed cases, after some 2600 epochs the control program succumbs to a flaw. The particular position of some of the trains causes the control program to open the point at 236 (Fig. 1) allowing two fast engines to move into the inside track in a direct

Table 1. Summary of the toy problem. Array $gtrain$[engine][property] holds information on an engine's direction of motion, position; speed; in-junction (position counter inside the shortcut); and the 56 GP terminals $TrainInfo(0:55)$ can pick this up. Array $glink$ [shortcut][property] holds information about points: start node; end node; point setting; and the 221 GP terminals $SignalInfo(0:220)$ pick this up. Which engine or signal this information is for depends on the value of a that is randomly selected for each terminal created to make up the GP population at run start. Function $JunctionSet$(A,B) uses the difference between its first argument and the location of points (array elements $glink$[0..21][1]) to select a point, and the sign of its second argument tells whether to flip the point or leave it unaltered.

Parameters for engines: array $gtrain$[engine][property].							
$gtrain$ [i][j]		i=0	i=1	i=2	i=3	i=4	i=5
direction of motion	j=0	-1	1	1	1	-1	-1
cell position	j=1	13	119	146	52	214	199
speed: fast 0; slow 1)	j=2	1	0	0	0	0	0
in-junction (working counter)	j=3	0	0	0	0	0	0

Description of each junction: array $glink$ [junction][property] (see text).							GP Parameters	
$glink$ [i][j]	j=0 access	j=1 start	j=2 end	j=3 length	j=5 swap?	j=4 set?	parameter	description
i=0	-1	18	104	3	0	0	pop size	10000
i=1	1	20	112	3	0	0	mate radius	10000
i=2	-1	34	132	6	1	0	max. nodes	1000
i=3	1	44	186	2	0	0	max. init. depth	6
i=4	1	49	190	2	0	0	steady-st. tourn.	kill=2; breed=4
i=5	1	56	136	6	0	0	fitness measure	$f = n_{EA} - n_{ET}$
i=6	-1	64	201	3	0	0	terminals	$TrainInfo$
i=7	-1	77	163	3	1	0		$SignalInfo$
i=8	1	83	230	2	0	0		reals in .01 steps:
i=9	1	93	7	3	0	0		+ve range (.01,1.0)
i=10	-1	102	10	3	0	0		-ve range (-1.0,-.01)
i=11	1	108	138	3	1	0	functions	$JunctionSet(a,b)$
i=12	-1	125	39	6	0	0		protected % + - *
i=13	1	129	179	3	1	1		$avg(a,b) = (a+b)/2$
i=14	1	143	103	3	1	0	no elitism	70% cross-over
i=15	-1	152	100	3	1	0		20% mutation
i=16	1	166	205	6	1	1	$JunctionSet$(A,B)	
i=17	1	168	140	3	1	0	C is jct closest to fabs(A)	
i=18	1	193	56	3	0	0	IF B > 0 THEN	
i=19	1	209	72	3	0	0	IF $glink$[C][4] = 0 THEN	
i=20	-1	222	79	3	0	0	$glink$[C][4] = 1	
i=21	-1	236	157	5	1	0	ELSE	
							glink[C][4] = 0	
							END IF	
							END IF	

$TrainInfo$		$SignalInfo$	
atom value	corresponding array value	atom value	corresponding array value
$0 \le a < 10$	$gtrain$ [a/2][0]	$0 \le a < 66$	$glink$ [a/3][1]
$10 \le a < 30$	$gtrain$ [(a-10)/4][1]	$66 \le a < 110$	$glink$ [(a-66)/2][2]
$30 \le a < 40$	$gtrain$ [(a-30)/2][2]	$110 \le a < 220$	$glink$ [(a-110)/5][4]
$40 \le a < 55$	$gtrain$ [(a-40)/3][3]		

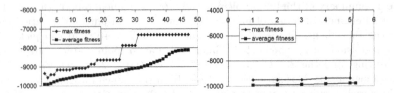

Fig. 2. Fitness versus generation (10,000 generates in steady state GP) plots for two parallel independent runs (pid). Typical progression of maximum and average fitness in a pid shows that fitness improves in jumps. The right pid illustrates an "eureka" moment, when GP suddenly discovers an outstanding solution (fitness = 0.0) at an early generation of the pid.

collision with the slow moving engine. Thus, for a solution to hold for all time it must consider all possible and relevant conditions (as in an exhaustive test, or in a formal software proof).

The solution described in the left side of Fig. 3 (top) is by no means the only way to avoid collision for thousands of epochs. Multiple solutions exist. Searching for a single solution to this toy problem gives an idea as to the nature of the search. This was undertaken using the fitness evaluation measure on the left hand side of Table 2. Observation of the maximum convergence and average convergence of different parallel independent runs over the course of 50 generations reveals that the fitness for the best of generation individual is characterized by flat periods and sudden jumps while the progression of the average fitness is soother as illustrated for two parallel independent runs in Fig. 2. Sometimes the evolution experiences an "Eureka moment" when it discovers a

Table 2. Pseudo-code describing two options for the fitness evaluation of a new individual in steady-state GP: (left) aiming for one solution; (right) aiming for many equivalent solutions. `Visualisation` stores the position of every engine at every epoch in a file. `NotFoundAlready` effects a complete numerical comparison of files to determine if the same solution has appeared. $MAXFLOAT$ is a very large number.

```
InitialiseTrains                        InitialiseTrains
InitialiseRailway                       InitialiseRailway
BEGIN LOOP (i_E = 1 to n_ET)            BEGIN LOOP (i_E = 1 to n_ET)
  MoveTrains                              MoveTrains
  IF CheckforCrash THEN                   IF CheckforCrash THEN
    f = i_E - n_ET                          f = i_E - n_ET
    RETURN                                  RETURN
  END IF                                  END IF
  EvaluateGPchrome                        EvaluateGPchrome
END LOOP                                END LOOP
Visualisation                           IF NotFoundAlready THEN Visualisation
f = 0.0   solved!                       f = -MAXFLOAT   solved and punished!
RETURN                                  RETURN
```

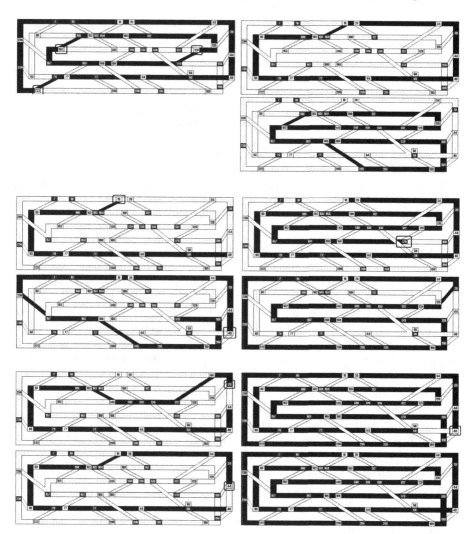

Fig. 3. (Top) Left: Desirable and robust this solution first gets engines to follow one another and next isolates the slow one from segregating both groups in independent tours (tours are indicated in black). Right: In an alternative solution which was observed to avoid a crash for 2121 epochs, slow and fast sets of engines move back and forth between the tours on the left and those on the right side of the figure. (Middle) Solution discovered is good for at least 3000 epochs. Follow in order from top left, bottom left to top right is the complex tour of the fast engine (the circled node starts that portion of the tour). The figure on the bottom right is the tour of the slower engine. This solution has excellent timing with point switching that allows all engines to cover a great part of the track. (Bottom) This more involved discovered solution holds for at least 3000 epochs. From top left, bottom left, to top right is the tour by the fast engine. The tours may be repeated or switched to according to the position of the slow engine. It covers the entire track (figure on the bottom right) 'unaware' of the existence of the other engines.

solution that holds probably for ever (e.g. for at least thirty thousand epochs). The classic solution sees the fast engines travelling the inner tour and the slow engine travelling the outer tour. Very rarely such a solution will appear in the initial population in one of the 10,000 individuals generated by the ramp half and half method at maximum depth of 6 (table 1). On average, at most two to three new (operationally different) solutions are obtained inside a parallel independent run over fifty generations. In Fig. 2 such solutions correspond to the jumps in the maximum fitness.

Searching for multiple solutions is accomplished in a crude way by the fitness evaluation scheme on the right hand side of Table 2. The toy problem has n_{ET}, an objective or free parameter: desired number of 'collision free' epochs. When this was set to 1500 epochs, an intricate solution emerged which was verified for at least 3000 epochs (Fig. 3 (middle)). Another solution also emerged which sees the slow engine covering the entire track (no use of junctions) while the other four faster engines stay out of its way (Fig. 3 (bottom)).

When the free parameter was set to 10,000 epochs this produced five unique solutions, all of them valid for at least 30,000 epochs (probably valid for all time). Upon visualization they contain all of the tactics illustrated in Fig. 3.

It was found (cannot present details for reasons of space) that even by multiple independent runs the standard GP search for one solution cannot be relied on to produce a number of unique solutions (especially at a value of the free parameter $\geq 10,000$). Unique solutions can occur by chance in the initial population, and may or may not arrive in parallel independent runs. However, using the fitness evaluation measure on the right hand side of Table 2 is much more successful at producing a large number of unique solutions in a single run of GP. Even if the free parameter n_{ET} is increased to 10,000 epochs (essentially asking for all time) it is possible to obtain five unique solutions with a single run of GP (many more unique solutions probably exist). As n_{ET} is increased not only are other phenotypic variants produced, but also the phenotypes use completely different strategies to avoid engine collision.

4   Conclusions

This paper explained why multiple solutions are important and introduced a toy problem that is particularly useful for understanding how to efficiently generate many solutions: GP must set points on a railway track to prevent an engine collision for a number of time steps.

It demonstrated: (a) generating multiple solutions using one run of GP; (b) using the phenotype to check for the uniqueness of the solution, and (c) how emergent behaviour of the solutions can help to cluster these solutions. The concept of distance between solutions that equally solve the problem is subtle. It is submitted that the concept of a norm or simple distance measure is not very meaningful. Utility is more meaningful. Namely, should a problem be solved by a large number of solutions, then as users of these solutions we tend to cluster them into groups to discuss their relative merits. Any particular clustering may be motivated by a need, or a mental model that alerts us to certain differences or combined difference and not to others. In the toy problem example, solution visualization triggers linguistic descriptions in the mind of an observer for clusters of solutions such as "solutions where the slow engine covers the whole track, pretends as if the fast engines did not exist, and these stay out of its way". Linguistic descriptions [10] help to identify apparent tactics or strategies that can be attributed to recognizable and emergent features of the solution.

GP formulations capable of maximizing discovery of solutions probably should consider techniques that aim to maintain diversity in a run, e.g. the method in [7] is motivated by the objective of sustaining innovation throughout a run; the method in [8] uses 'operational equivalence' and multi-run to harvest promising solution components and [9] extended this idea to subroutines. These and other methods should be adapted to satisfy the goal of diverse multiple solution generation.

References

1. Penrose, Roger (1955). A generalized inverse for matrices. Proceedings of the Cambridge Philosophical Society 51: 406-413.
2. Mark Solms (2004). "Freud Returns", Scientific American, April 26, 2004.
3. Ray Kurzweil (2005), The Singularity is Near: when humans transcend biology, Viking Press.
4. Deb, K. and Gupta, H. (2005) Searching for Robust Pareto-Optimal Solutions in Multi-objective Optimization. EMO 2005: 150-164.
5. Johnson, R. C. (editor) (1993). Antenna Engineering Handbook, McGraw-Hill (3rd Edition).
6. Burke, G.J. Miller, E.K. Poggio, A.J. (2004). The Numerical Electromagnetics Code (NEC) - a brief history, IEEE Antennas and Propagation Society International Symposium, Vol.3, 2871- 2874.
7. Hu, J., Goodman, E.D. and Seo, K. (2003) Continuous Hierarchical Fair Competition Model for Sustainable Innovation in Genetic Programming. Chapter 6 in Genetic Programming Theory and Practice, Kluwer Academic Publishers.
8. Howard, D. (2003) Modularization by Multi-Run Frequency Driven Subtree Encapsulation. Chapter 10 in Genetic Programming Theory and Practice, Kluwer Academic Publishers.
9. Ryan, C., Keijzer, M. and Cattolico, M. (2004) Favorable Biasing of Function Sets Using Run Transferable Libraries. Chapter 7 in Genetic Programming Theory and Practice II, Kluwer Academic Publishers.
10. Zadeh, L. (1999) From Computing with Numbers to Computing with Words – From Manipulation of Measurements to Manipulation of Perceptions, IEEE Transactions on Circuits and Systems, 45, 105-119, 1999.

An Approach for Selective Ensemble Feature Selection Based on Rough Set Theory[*]

Yong Yang[1,2], Guoyin Wang[2], and Kun He[2]

[1] School of Information Science and Technology,
Southwest Jiaotong University,
Chengdou, 610031, P.R.China
[2] Institute of Computer Science & Technology,
Chongqing University of Posts and Telecommunications,
Chongqing, 400065, P.R.China
{yangyong,wanggy}@cqupt.edu.cn, pandahe_916@yahoo.com.cn

Abstract. Rough set based knowledge reduction is an important method for feature selection. Ensemble methods are learning algorithms that construct a set of base classifiers and then classify new objects by integrating the prediction of the base classifiers. In this paper, an approach for selective ensemble feature selection based on rough set theory is proposed, which meets the tradeoff between the accuracy and diversity of base classifiers. In our simulation experiments on the UCI datasets, high recognition rates are resulted.

Keywords: Rough set, ensemble learning, selective ensemble, feature selection.

1 Introduction

Rough set is a valid theory for data mining. The most advantage of rough set is its great ability for attribute reduction (feature selection). It has been successfully used in many domains such as machine learning, pattern recognition, intelligent data analyzing and control algorithm acquiring, etc [1][2][3][4][5]. Based on the feature selection, some classifiers can be built. There are always over one reduction for an information system. Thus, it is a problem to choose a suitable reduction or integrate several reductions into a system to get better performance.

Ensemble learning has been a hot research topic in machine learning since 1990s'[6]. Ensemble methods are learning algorithms that construct a set of base classifiers and then classify new objects by integrating the prediction of the base classifiers. An ensemble system is often much more accurate than each base classifier. Ditterrich proved the effectiveness of ensemble methods from the viewpoint

---

[*] This paper is partially supported by National Natural Science Foundation of China under Grant No.60373111 and No.60573068, Program for New Century Excellent Talents in University (NCET), Natural Science Foundation of Chongqing under Grant No.2005BA2003.

J.T. Yao et al. (Eds.): RSKT 2007, LNAI 4481, pp. 518–525, 2007.

of statistic, computation and representation in [7]. As a popular machine learning method, ensemble methods are often used in pattern recognition, network security, medical diagnosis, etc [7][8][9][10].

A necessary and sufficient condition for an ensemble system to be more accurate than any of its base classifiers is that all base classifiers are accurate and diverse. An accurate classifier is one that has an error rate less than random guessing on new instances. Two classifiers are diverse if they make different prediction errors on unseen objects [7]. Besides accuracy and diversity, another important issue for building an efficient ensemble system is the choice of the function for combining the predictions of the base classifiers. There are many techniques for the integration of an ensemble system, such as majority voting, weighted voting, reliability-based weighted voting, etc [8].

There are many methods proposed for ensemble. The most popular way for ensemble is to get different subset of the original dataset by resampling the training data set many times. Bagging [9], boosting [10] and cross-validation are all such ensemble methods. These methods work well especially for unstable learning algorithms, such as decision trees, neural network. Some other methods are also studied, such as manipulating the output targets [11], injecting randomness into classifiers [12]. Besides these methods, there is another effective approach for ensemble, which uses different feature subsets, and is usually called ensemble feature selection [13]. Ensemble feature selection (EFS) is also a classical ensemble method. It takes different feature subset as the input features for a base classifier construction.

There are two methods for generating base classifiers and integrating the predictions of base classifiers. One is called direct strategy, the other is called over producing and choosing strategy. The direct strategy aims to generate an ensemble of base classifiers directly in the training period. The over producing and choosing strategy is also called selective ensemble, which creates a lot of base classifiers at first, and then chooses a subset of the most suitable base classifiers and generates the final prediction.

In this paper, based on rough set theory and ensemble learning theory, a selective ensemble feature selection method is proposed. The rest of this paper is organized as follows. In Section 2, based on the basic concepts and methods of rough set theory and the diversity of ensemble learning, an algorithm for selective ensemble feature selection is proposed. Simulation experiment results are illustrated in Section 3. Finally, conclusion and future works are discussed in Section 4.

2 Ensemble Feature Selection Based on Rough Set Theory

2.1 Basic Concept of Rough Set Theory

Rough set (RS) is a valid mathematical theory for dealing with imprecise, uncertain, and vague information. It has been applied successfully in such fields as machine learning, data mining, pattern recognition, intelligent data analyzing and control algorithm acquiring, etc, since it was developed by Professor Pawlak in 1980s [1][2]. Some basic concepts of rough set are introduced here for the convenience of following discussion.

Def.1 A decision information system is a formal representation of a data set to be analyzed. It is defined as a pair $s = (U, R, V, f)$, where U is a finite set of objects and $R = C \cup D$ is a finite set of attributes, C is the condition attribute set and $D = \{d\}$ is the decision attribute set. With every attribute $a \in R$, set of its values V_a is associated. Each attribute a determines a function $f_a : U \to V_a$.

Def.2 For a subset of attributes $B \subseteq A$, an indiscernibility relation is defined by $Ind(B) = \{(x, y) \in U \times U : a(x) = a(y), \forall a \in B\}$.

Def.3 The lower approximation $B_-(X)$ and upper approximation $B^-(X)$ of a set of objects $X \subseteq U$ with reference to a set of attributes $B \subseteq A$ may be defined in terms of equivalence classes as follows:
$B_-(X) = \bigcup \{E \in U/Ind(B)|E \subseteq X\}$, $B^-(X) = \bigcup \{E \in U/Ind(B)|E \cap X \neq \Phi\}$.
     They are also called as the B_- lower and B^- upper approximation respectively. They can also be defined as follows:
$B_-(X) = \{x \in U|[x]_B \subseteq X$, $B^-(X) = \{x \in U|[x]_B \cap X \neq \Phi\}$.
     Where, $[x]_B \in U/Ind(B)$ is the equivalence class of object induced by the set of attributes $B \subseteq A$.

Def.4 $POS_P(Q) = \bigcup_{x \in U/Ind(B)} P_-(X)$ is the P positive region of Q, where P and Q are both attribute sets of an information system.

Def.5 A reduction of P in an information system is a set of attributes $S \subseteq P$ such that all attributes $a \in P - S$ are dispensable, all attributes $a \in S$ are indispensable and $POS_S(Q) = POS_P(Q)$. We use the term $RED_Q(P)$ to denote the family of reductions of P. $CORE_Q(P) = \bigcap RED_Q(P)$ is called as the Q-core of attribute set P.

Def.6 The discernibility matrix $M_DC = \{c_{ij}\}_{n*n}$ of an information system S is defined as:

$$c_{ij} = \begin{cases} \{a \in C : x(_i) \neq x(_j)\}, D(x_i) \neq D(x_j) \\ \qquad\qquad 0 \qquad\qquad , D(x_i) = D(x_j) \end{cases} \quad i = 1, 2, ..., n. \qquad (1)$$

Where, $D(x_i)$ is the value of the decision attribute. Based on the dicernibility matrix, all possible reducts can be generated. An attribute reduction algorithm based on dicernibility and logical operation is proposed in [3]. The detailed algorithm is introduced in Algorithm 1.
     Any attributes combination of C_0 as well as a conjunctive term of P' can be an attribute reduction of the original information system. All possible reducts of the original information system can be generated with the Algorithm 1. Using these reducts, classifiers could be built which have the same classification ability

Algorithm 1. Attribute reduction algorithm based on dicernibility matrix and logical operation

Input   : An information system S with its discernibility matrix $M_D C$.
Output: Reduction of S.
1. Find all core attributes (C_0) in the discernibility matrix $M_D C$. $Redu = C_0$.
2. Find the set (T) of elements $(C_{ij}$'s) of $M_D C$ that is nonempty and does not contain any core attribute. $T = \{C_{ij} : C_{ij} \cap C_0 = \Phi \wedge C_{ij} \neq \Phi\}$.
3. Each attribute is taken as a Boolean variable. A logic function (P) is generated as the Boolean conjunction of disjunctions of each components belonging to element (C_{ij}) of T. That is, $P = \wedge_{ij} \{\vee_k \{a_k\} : a_k \in C_{ij} \wedge C_{ij} \in T\}$.
4. Express the logic function P in a simplified form (P') of a disjunction of minimal conjunctive expressions by applying the distributivity and absorption laws of Boolean algebra.
5. Choose suitable reduction for the problem.

as the original whole decision table. Therefore, these classifiers can be taken as candidate base classifiers of an ensemble system.

2.2   Diversity in Ensemble Method

2.2.1 Measurement of Diversity in Ensemble

Theoretically speaking, if base classifiers are more diverse between each other, an ensemble system will be more accurate than its base classifiers. There are a number of ways to measure the diversity of ensemble methods. Some of them are called pairwise diversity measures, which are able to measure the diversity in predictions of a pair of classifiers, and the total ensemble diversity is the average of all the classifier pairs of the ensemble. For example, plain disagreement, fail/non-fail disagreement, Q statistic, correlation coefficient, kappa statistic and double fault measures [14][15][16]. Some others are called non-pairwise diversity measures, which measure the diversity in predictions of the whole ensemble only. For example, the entropy measure, measure of difficult, coincident failure diversity, and generalized diversity [14][15][16].

The double fault measure (DF) can characterize the diversity between base classifiers. It is the ratio between the number of observations on which two classifiers are both incorrect. It was proposed by Giacinto in [17]. It is defined as follows.

$$Div_{i,j} = \frac{N^{00}}{N^{11} + N^{10} + N^{01} + N^{00}}. \tag{2}$$

Where, N^{ab} is the number of instances in the data set, classified correctly $(a=1)$ or incorrectly $(a=0)$ by classifier i, and correctly $(b=1)$ or incorrectly $(b=0)$ by classifier j. The denominator in (2) is equal to the total number of instances N.

2.2.2 Relationship Between Diversity Measure and Integration Method

As discussed above, there are many diversity measure and integration methods. What is the relationship between a diversity measure and an integration method? Is it effective when we choose a diversity measure and an integration method randomly for an ensemble system? The relationship between diversity measure and integration method is a hot research topic in ensemble learning [14][15][16][18]. In [18], the relationship between 10 diversity measures and 5 integration methods were discussed. It was found that there was little correlation between the integration methods and diversity measures. In fact, most of them showed independent relationship. Only the double fault measure and the measure of difficult showed some correlations greater than 0.3. The measure of difficult showed stronger correlation with the integration methods than the double fault measure. Unfortunately, it was more computationally expensive.

In this paper, the double fault measure and integration method of majority are used in this proposed ensemble method since they showed higher correlation and they both have lower computation complexity.

2.3 Selective Ensemble Feature Selection Base on Rough Set Theory (SEFSBRS)

Based on rough set theory and the diversity measure of the double fault measure, a selective ensemble feature selection method based on rough set theory is proposed here.

Firstly, all possible reducts are generated based on the discernibility matrix of a training set. All candidate base classifiers are generated with the reducts. Secondly, based on the diversity measure defined in Equation (2), all base classifiers are clustered on validation set, and then, a pair of base classifiers, which are the most diverse among two clusters, are chosen from each two clusters. Therefore, the classifiers which are more accurate and more diverse among all the classifiers are chosen for the ensemble system. At last, the majority voting is taken as the integration method for ensemble, and the final prediction can be taken on the testing set. The detailed algorithm is introduced in Algorithm 2.

Algorithm 2. Selective ensemble feature selection base on rough set theory

---

Input   : Decision tables of the training set, validation set and testing set.
Output: Final ensemble prediction.
Apply Algorithm 1 on the training set to generate its all reducts.
Construct all the classifiers using the reducts.
for *each classifier* **do**
  | Calculate $div_{(i,j)}$ of each two classifiers on the validation set according
  | to Equation (2).
end
Based on all $div(i, j)$, all classifiers are clustered.
for *each two clusters* **do**
  | Select a pair of classifiers which are the most diversity among all
  | pairwise classifiers of the two clusters.
end
Generate the final prediction of the ensemble system based on the majority voting of the selected classifiers on the testing set.

---

3 Experiment Results

Several experiments have been done on UCI data sets to test the validity of the proposed method. The data sets used are shown in Table 1.

Table 1. Datasets used in the experiments

Dataset	Data size	Concept	Condition attribute
Breast Cancer Wisconsin	699	2	9
Vote	435	2	16
Iris	150	3	4
Credit screening	690	2	15

Several comparative experiments are done.

The first experiment is done using the proposed method (SEFSBRS).

The second experiment uses an ensemble strategy based on all the classifiers. It is named ensemble all in this paper. It includes the following steps. All the possible classifiers are created based on all reductions generated from the dicernibility matrix at first. Then, final prediction is obtained based on all the classifiers.

The third experiment is based on the feature selection algorithm of MIBARK [19]. A reduct is generated using the MIBARK algorithm. Then, a classifier is constructed according to the reduct. The final prediction is gotten using the single classifier only.

The forth experiment is based on the feature selection algorithm proposed in [20], the detailed experiment process is similar to the third one.

The fifth experiment is based on SVM, and the detailed experiment process is similar to the third one too.

Each dataset is divided into a training set, a validation set and a testing set for all the experiments. Each set contains 60%, 20%, and 20% of the total data set respectively. The 5- fold validation method is carried out for each dataset. The correct recognition rates of each method for these datasets are shown in Table 2.

From the experiment results, we can find that the proposed method (SEFSBRS) is valid. It can get high recognition rate. By comparing SEFSBRS and Ensemble all, we can find that the selective ensemble is almost as accurate as

Table 2. Experiment results

Dataset	SEFSBRS	Ensemble all	MIBARK	Feature selection	SVM
Breast Cancer Wisconsin	96.37%	96.23%	70.97%	87.23%	86.40%
Vote	94.10%	94.00%	91.11%	78.98%	87.14%
Iris	84.13%	86.44%	72.37%	50.83%	94%
Credit screening	68.60%	68.87%	41.63%	19.69%	56%
Average	85.80%	86.39%	69.02%	59.18%	80.89%

the ensemble using all possible candidate classifiers. Sometimes, it can get higher accuracy. However, SEFSBRS often use less classifiers than Ensemble all. So, it will be more efficient in real application. By comparing SEFSBRS, MIBARK, and Feature selection, we can find that the ensemble is more accurate. It proves that the ensemble of base classifiers is more effective than a single classifier. We can also find that the ensemble method even over performs SVM in most cases. Thus, the method can be taken as a useful method in machine leaning, pattern classification, etc.

4   Conclusion and Future Works

In this paper, a selective ensemble feature selection method based on rough set theory is proposed. All candidate classifiers for ensemble are produced based on the disernibiltiy matrix. The classification ability of each base classifiers is guaranteed. For the purpose of selecting uncorrelated base classifiers for ensemble, cluster method is used. It can ensure more diversity on the selective classifiers. Experiment results show its validity.

In the future, the proposed method will be used in real pattern recognition problems, such as emotion recognition. At the same time, improvement of the proposed method should be discussed too.

References

1. Pawlak, Z.: Rough sets. *International J. Comp. Inform. Science.* 11 (1982) 341-356.
2. Pawlak, Z.: Rough Classification. *International Journal of Man-Machine Studies.* 5 (1984) 469-483.
3. Wang, G.Y., Wu, Y., Fisher, P.S.: Rule Generation Based on Rough Set Theory. In: Proceedings of SPIE, Washington (2000) 181-189.
4. Skowron, A., Polkowski, L.: Decision Algorithms: A Survey of Rough Set - Theoretic Methods. *Fundamenta Informaticae.* 3-4 (1997) 345-358.
5. Zhong, N., Dong, J.Z., Ohsuga, S.: Using Rough sets with Heuristics for Feature Selection. *Journal of Intelligent Information Systems.* 3 (2001) 199-214.
6. Ditterrich, T.G.: Machine learning research: four current direction. *Artificial Intelligence Magzine.* 4 (1997) 97-136.
7. Ditterrich, T.G.: Ensemble methods in machine learning. In: Kittler, J., Roli, F., (Eds.), *Multiple Classifier Systems.* LNCS 1857, Springer, Berlin (2001) 1-15.
8. Tsymbal, A., Pechenizkiy, M., Cunningham, P.: Diversity in search strategies for ensemble feature selection. *Information Fusion.* 1 (2005) 83-98.
9. Breiman, L.: Bagging predictors. *Machine Learning.* 2 (1996) 123-140.
10. Freund, Y.: Boosting a weak algorithm by majority. *Information and Computation.* 2 (1995) 256-285.
11. Ditterrich, T.G., Bakiri, G.: Solving multi-class learning problem via error-correcting output codes. *Journal of Artificial Intelligence Research.* 2 (1995) 263-286.
12. Ditterrich, T.G.: An Experimental Comparison of Three Methods for Constructing Ensembles of Decision Trees: Bagging, Boosting, and Randomization. *Machine Learning.* (1998).

13. Opitz, D.: Feature selection for ensembles. In: Proceedings of 16th National Conference on Artificial Intelligence, AAAI Press, Florida (1999) 379-384.
14. Kuncheva, L.I., Whitaker, C.J.: Measures of diversity in classifier ensembles and their relationship with the ensemble accuracy. *Machine Learning*. 2(2003) 181-207.
15. Kuncheva, L.I.: That elusive diversity in classifier ensembles. In: Proceedings of First Iberian Conference on Pattern Recognition and Image Analysis (IbPRIA), Mallorca (2003) 1126-1138.
16. Brow, G., Wyatt, J., Harris, R., Yao, X.: Diversity Creation Methods: A Survey and Categorisation. *Journal of Information Fusion*. 1 (2005) 1-28.
17. Giacinto, G., Roli, F.: Design of effective neural network ensembles for image classification processes. *Journal of Image Vision and Computing*. 9 (2001) 699-707.
18. Shipp, C.A., Kuncheva, L.I.: Relationships between combination methods and measures of diversity in combining classifiers. *Journal of Information Fusion*. 2 (2002) 135-148.
19. Miao, D.Q., Hu, G.R.: A Heuristic Algorithm for Reduction of Knowledge. *Journal of Computer Research and Development*. 6 (1999) 681-684 (in Chinese).
20. Hu, X.H., Cercone, N.: Learning Maximal Generalized Decision Rules via Discretization, Generalization and Rough set Feature Selection. In: Proceedings of 9th International Conference on Tools with Artificial Intelligence (ICTAI '97), California (1997) 548-556.

Using Rough Reducts to Analyze the Independency of Earthquake Precursory Items

Yue Liu[1], Mingzhou Lin[2], Suyu Mei[1], Gengfeng Wu[1], and Wei Wang[2]

[1] School of Computer Engineering & Science, Shanghai University,
Shanghai, 200072, China
{yliu,gfwu}@staff.shu.edu.cn
[2] Seismological Bureau of Shanghai,
Shanghai 200062, China

Abstract. To find earthquake precursory items that are closely relative to earthquake is very important for earthquake prediction. Rough Set Theory is an important tool to process imprecise and ambiguous information. In this paper, the discernibility matrix approach based on Rough Set Theory is optimized to find all possible rough reducts with reduced time and space complexity. Furthermore, this approach is applied to analyze the dependency among earthquake precursory items. After several experiments, some most important precursory items are found, while some items are considered to be redundant. The results maybe provide the researches with the direction on the relationship between earthquake precursory items and earthquake.

Keywords: Rough Reducts, Discernibility Matrix, Earthquake Precursory Items, Earthquake Prediction.

1 Introduction

Earthquake prediction is a difficult problem that has not been solved in the world and is still in experiential stage [1]. So far, the main earthquake prediction method is to draw an analogy between the earthquake cases and the present observed anomalies (exceptional natural phenomena such as water level changes and increases in concentrations of Radon gas in deep wells). However, strong uncertainty exists between anomalies and earthquake. The anomalous situations vary greatly from different earthquakes. One anomaly unnecessarily happened before an earthquake, and an earthquake may not certainly erupt following the anomalies. In addition, the identification of anomalies is also very uncertain. It is difficult to say that the observed anomalies just are the factors caused earthquake because many different factors can cause an earthquake. These uncertainties bring difficulty to earthquake prediction and make the prediction accuracy rather low. Therefore, finding the anomalies correlative to the gestation of earthquake and analyzing the relationship between earthquake and anomalies are very important and necessary. Through researches of various observation means, people have reported dozens or even hundreds of earthquake precursory

J.T. Yao et al. (Eds.): RSKT 2007, LNAI 4481, pp. 526–533, 2007.

items to describe the anomalies. Earthquake Cases in China [2] has reported 41 precursory items of seismometry including seismic belt (band), seismic gap (segment), seismicity pattern, etc. and other 71 precursory items of precursor, for instance, fixed leveling (short leveling), mobile leveling, ranging tilt, apparent resistivity, and so on. These precursory items reflect the time, space and intensity characteristics of earthquake activities from different aspects. Meanwhile, certain relativity may also exist among them, which has been proved by earthquake researches [3,4,5,6,7]. But, to our knowledge, none of them analyze the relationship between earthquakes and all these precursory items and the relationship among the precursory items.

Rough Set is a theory developed in recent years as an important tool to deal with imprecise and ambiguous information. Since Z.Pawlak raised Rough Set Theory (RST) in 1982 [8], more and more researchers began to study the theory and have achieved a great deal of successes. Not only did they build the mathematical model of RST, but also proposed some valuable algorithms on many areas [9,10]. As a hot topic of RST, attributes reduction aims to find the redundant (unimportant) attributes and then delete them. So, the RST is a suitable tool to analyze the independency of earthquake precursory items. This paper proposes an improved attribute reduction approach based on discernibility matrix and called Optimized Discernibility Matrix based Approach (ODMA). Furthermore, ODMA is applied to analyze the earthquake precursory items, remove the unimportant ones, and then get those independent ones and their relationship. This research maybe provides the guidance on the direction of seismic study and the relationships among the physical parameters. As a result, the accuracy of earthquake prediction are bound to improve.

The rest of this paper is organized as follows. In Section 2, details of ODMA are described. In Section 3, experiments are performed on earthquake dataset. Finally, conclusions are drawn in Section 4.

2   Optimized Discernibility Matrix Based Approach

Popular attribute reduction algorithms are mainly summarized to four categories: attribute reduction algorithm based on discernible matrix and logical calculation, attribute reduction algorithm based on mutual information, heuristic algorithm of attribute reduction, and attribute reduction algorithm based on feature selection [11]. The main aim of them is to find minimal or optimal reducts for the sake of satisfying application requirements or avoiding so-called combination explosion of reducts search. However, in order to analyze the independency of earthquake precursory items, all the attribute reduction sets need to be found out. Therefore, the so-called Optimized Discernibility Matrix based Approach (ODMA) is proposed based on the existed discernibility matrix approach [11,12,13]. Here, the definition of discernibility matrix is given firstly.

Definition 1. *An information table DT is the tuple:*

$$DT = (U, AT, V)$$

where $AT = C \cup D$ is a finite nonempty set of attributes, $C = \{a_k \mid k=1,2,...,m \}$ is the set of the condition attributes, $D = \{d\}$ is the set of the decision attributes, $d \notin C$, $U = (x_1, x_2, ..., x_n)$ is a finite nonempty set of objects. V is a nonempty set of values for the attributes in AT. Suppose $a_k(x_i) \in V$ is the value on attribute a_k of the i-th object x_i, then the discernibility matrix, which stores attributes that differentiate any two objects of the universe, can be defined by:

$$d_{ij} = \begin{cases} \{a_k \in C \mid a_k(x_i) \neq a_k(x_j)\}, & if \ d(x_i) \neq d(x_j), \\ 0, & otherwise. \end{cases} \quad (1)$$

where d_{ij} denotes matrix element (i,j), $i,j \in [1..n]$.

The ODMA can be described as the following steps.

Step 1: Rearrange dataset/decision table so that the objects with identical class label are neighborly distributed.

It is time-consuming to calculate discernible matrix. In ODMA, the objects with identical class label are neighborly distributed so that the objects in the same clustered area need not be compared and the corresponding element is zero-valued in the matrix. As a result, the time cost is reduced to some extent.

Step 2: Calculate the discernible matrix defined as formula (1), if empty element is detected, report inconsistency.

Step 3: Express each non-empty element of discernible matrix in the form of disjunctive item defined as follows:

$$c_{ij} = \bigvee_{a \in d_{ij} \neq 0} a_i. \quad (2)$$

where a_i denotes the i-th attribute in d_{ij} .

Step 4: Convert discernible matrix into conjunctive normal forms expressed by discernibility function defined as follows:

$$f(a_1, a_2, ..., a_m) = \wedge \{\forall \ c_{ij} \mid 1 \leqslant j \leqslant i \leqslant |U|, c_{ij} \neq \emptyset\}. \quad (3)$$

Step 5: Calculate core attribute sets and remove all core-containing disjunctive items in conjunctive normal form (Core attribute refers to the disjunctive item in matrix such that $|c_{ij}| = 1$).

Step 6: Remove all supersets in the conjunctive normal form defined in formula (3).

Step 7: Convert the conjunctive normal form into disjunctive normal form by using heuristic searching techniques as follows in order to eliminate combinational search.

1. Heuristic knowledge 1: Once the attribute subset on the current search path is a superset of existing ones, stop the depth search and redirect to next attribute on the same layer. After having completed the searching on the current layer, go back to the upper layer and continue searching.

2. Heuristic knowledge 2: Suppose the completed search path from the root node to current layer is 'path'. Check the current attribute. If it's already exist in 'path', stop searching at the following attribute of this layer, which has not appeared in 'path'. Generate a conjunction item once searching to the bottom of the tree.

3. Heuristic knowledge 3: Sort the disjunctive items in ascending order by their number of attributes and the attributes of each item in descending order by their appeared frequency.

Step 8: Extract all conjunctive items out of the disjunctive normal form generated in Step 7, and then add the core attribute sets into each conjunctive item. All the generated conjunctive items are saved as reducts.

3   Analysis of the Independency of Earthquake Precursory Items by Using ODMA

There are four kinds of relationship between anomalies and earthquake: a. earthquake erupted with anomalies; b. earthquake erupted without anomalies; c. anomalies observed but no earthquake follows; d. no anomalies and no earthquake. Then the four usual situations in practice of prediction can be summarized as follows.

1. Anomalies observed, a and c constitute a complete set.
2. Earthquake erupted, a and b constitute a complete set.
3. No anomalies observed, b and d constitute a complete set.
4. No earthquake erupted, c and d constitute a complete set.

In practice, relation d has no value, so the situations 3 and 4 have no significance to researchers. We just focus on the commonly existing situation 1 and 2. Regrettably, there is so far no counter-cases (i.e. c. anomalies observed but no earthquake erupted) were made and situation 1 contained c is an incomplete set in fact. Therefore, we can only study situation 2.

Universe. After being removed the earthquake cases without anomalies, the total 191 earthquake cases in China from 1966 to 1999 [2] constitute the universe.

Condition Attributes. 41 seismological earthquake precursory items and 71 other precursor ones, with exception of the precursory item number because it is not a physical parameter, are used as condition attributes and their values come from anomalies recorded in each earthquake case. However, some anomalies were really observed before the eruption of earthquakes, some were added after the eruption of earthquake. It is difficult to distinguish. There are only two options: (1) whether the earthquake is predicted or not, if the anomalies are recorded then the value of the responding earthquake precursory item equals 1. (2) the value of the responding earthquake precursory item equals 0 if the earthquake is not predicted. However, there may be some special earthquake cases, which don't belong to the two above situations, that the values of their condition attributes are equal to 0 but prediction type is different. In this paper, we didn't take such special cases into consideration.

Decision Attributes. Take the prediction situation as decision attribute. Two kinds of description are alternative to earthquake prediction: (1) predicted and unpredicted. (2) long-term prediction, medium-term prediction, short-term prediction, imminent earthquake prediction, and no prediction.

3.1   Analysis of Earthquake Precursory Items of Seismometry (1)

Earthquake precursory items of seismometry refer to the precursory information of the strong earthquake in the precursors or earthquake records (earthquake map and earthquake catalog), which are observed by seismology method. In the first analysis method, we divide simply earthquake prediction into two categories, i.e. predicted or not predicted. If predicted, assign the decision attribute to 1, else 0. The condition attributes are the following 41 seismometry earthquake precursory items:

1: seismic belt (band), 2: seismic gap (segment), 3: seismicity pattern (temporal, spacial, quiescence or activation), 4: precursory earthquake (or swarm), 5: swarm activity, 6: index of seismic activity (comprehensive index A, seismic entropy, degree of seismic activity and fuzzy degree of seismic activity), 7: seismic magnitude factor Mf-value, 8: fractal dimension of capacity, 9: earthquake rhythm, 10:strain release (energy release), 11: earthquake frequency, 12: b value, 13: h value, 14: seismic window, 15: earthquake deficiency, 16: induced foreshock, 17: foreshock, 18: exponential (A(b) value) of earthquake situation, 19: seismic concentration (concentration degree, spacial concentration degree, band concentration degree), 20: time interval between earthquakes, 21: composite fault plane solution of small earthquakes, 22: symbolic contradiction ratio of P-wave onsets, 23: stress drop of earthquake, 24: circumstance stress, 25: quality factor (value), 26: wave velocity (wave velocity, wave velocity ratio), 27: S-wave polarization, 28: seismic coda wave (sustained time ratio, attanuation coefficien, attanuation rate), 29: amplitude ratio, 30: microseisms, 31:seismic wave form, 32: total area of fault plane ($\Sigma(t)$), 33: regulatory ratio of small earthquakes, 34: seismic information deficiency (Iq), 35: seismic inhomogenous degree (GL value), 36: Algorithmic Complexity (C(n), Ac), 37: parameter of seismic gap, 38: area of earthquake coverage (A value), 39: E, N and S elements, 40: η value, 41: D value.

If the above-mentioned earthquake precursory item is recorded in the earthquake case, its value equals 1, else 0. For example, in the 16 September 1994 Taiwan Strait earthquake, seismic belt and background seismic and gestation gap with abnormity have been recorded, so we assign 1 to the first and second condition attributes. By this means, for any earthquake case, a sequence with 41 elements whose value is 1 or 0 is produced as an object in the decision table. Then, apply ODMA on the decision table and get the following five reducts:

{1, 2, 3, 4, 6, 10, 11, 12, 13, 14, 15, 17, 26, 28, 29, 31, 33}, {1, 2, 3, 4, 6, 10, 11, 12, 13, 14, 15, 17, 21, 26, 28, 29, 33}, {1, 2, 3, 4, 6, 10, 11, 12, 13, 14, 15, 17,

20, 26, 28, 29, 33}, {1, 2, 3, 4, 6, 10, 11, 12, 13, 14, 15, 17, 18, 26, 28, 29, 33},
{1, 2, 3, 4, 6, 10, 11, 12, 13, 14, 15, 16, 17, 26, 28, 29, 33}.

The attributes of 5, 7, 8, 9, 19, 22, 23, 24, 25, 27, 30, 32, 34, 35, 36, 37, 38, 39, 40, and 41 never appeared are redundant (i.e. absolutely unnecessary attributes). On the contrary, the attributes of 1, 2, 3, 4, 6, 10, 11, 12, 13, 14, 15, 17, 26, 28, 29, and 33 appeared in all the five sets are kernel attributes (i.e. absolutely necessary attributes).

3.2   Analysis of Earthquake Precursory Items of Seismometry (2)

The way to get the condition attributes in this analysis is same as that in Sect. 3.1. But the decision attribute is divided into five types: long-term prediction, medium-term prediction, short-term prediction, imminent earthquake prediction, and no prediction. If none of the first four types is predicted for an earthquake, it means no prediction. So, we can use four binary digits to describe the decision attribute. If one of the prediction type is reported, its corresponding binary digit is assigned to 1, or 0. Take the 16 September 1994 Taiwan Strait earthquake as an example, as only the medium-term prediction is reported, its decision attribute should be 0100. There are only 16 different kinds of combination of these five types: 1111 means all the four former stages are predicted, ..., 0000 means none of the four stages is predicted. A decision table is built based on above rule. This decision table is composed of the objects, which are the 45-length strings of '0' or '1'. Finally, 15 reducts as follows are gained:

{1, 2, 3, 4, 6, 10, 11, 12, 13, 14, 15, 17, 26, 28, 29, 31, 33, 40, 41}, {1, 2, 3, 4, 6, 10, 11, 12, 13, 14, 15, 17, 21, 26, 28, 29, 33, 40, 41}, {1, 2, 3, 4, 6, 10, 11, 12, 13, 14,15, 17, 20, 26, 28, 29, 33, 40, 41}, {1, 2, 3, 4, 6, 10, 11, 12, 13, 14, 15, 17, 18, 26, 28, 29, 33, 40, 41}, {1, 2, 3, 4, 6, 10, 11, 12, 13, 14, 15, 16, 17, 26, 28, 29, 33, 40, 41}, {1, 2, 3, 4, 6, 7, 10, 11, 12, 13, 14, 15, 17, 26, 28, 29, 31, 33, 40}, {1, 2, 3, 4, 6, 7, 10, 11, 12, 13, 14, 15, 17, 21, 26, 28, 29, 33, 40}, {1, 2, 3, 4, 6, 7, 10, 11, 12, 13, 14, 15, 17, 20, 26, 28, 29, 33, 40}, {1, 2, 3, 4, 6, 7, 10, 11, 12, 13, 14, 15, 17, 18, 26, 28, 29, 33, 40}, {1, 2, 3, 4, 6, 7, 10, 11, 12, 13, 14, 15, 16, 17, 26, 28, 29, 33, 40}, {1, 2, 3, 4, 5, 6, 10, 11, 12, 13, 14, 15, 17, 26, 28, 29, 31, 33, 40}, {1, 2, 3, 4, 5, 6, 10, 11, 12, 13, 14, 15, 17, 21, 26, 28, 29, 33, 40}, {1, 2, 3, 4, 5, 6, 10, 11, 12, 13, 14, 15, 17, 20, 26, 28, 29, 33, 40}, {1, 2, 3, 4, 5, 6, 10, 11, 12, 13, 14, 15, 17, 18, 26, 28, 29, 33, 40}, {1, 2, 3, 4, 5, 6, 10, 11, 12, 13, 14, 15, 16, 17, 26, 28, 29, 33, 40}.

The attributes of 8, 9, 19, 22, 23, 24, 25, 27, 30, 32, 34, 35, 36, 37, 38, and 39 are redundant, and the attributes of 1, 2, 3, 4, 5, 6, 7, 10, 11, 12, 13, 14, 15, 17, 26, 28, 29, 33, and 44 are kernel attributes.

3.3   Analysis of Earthquake Precursory Items of Precursors

For the sake of completeness, we do the similar experiments on the earthquake precursory items of precursors as the ones of seismometry. In a similar way to Sect. 3.1, attributes of 1, 8, 9, 11, 12, 22, 30, 31, 33, 55, 59 and 69 are selected as the most necessary attributes. The same result is also obtained in the similar method to Sect. 3.2. The above numbers correspond to the precursors:

1: fixed leveling (shortleveling), 8: ranging tilt, 9: fault creep, 11: roving gravity, 12: apparent resistivity, 22: radon in groundwater, 30: ground water level, 31: ground water level and lake water level, 33: water temperature, 55: electric induction stress, 59: piezo-capacity strain, 69: macroscopic phenomena.

4   Conclusion and Discussion

In this paper, an improved approach called ODMA (Optimized Discernibility Matrix based Approach) is proposed through resorting objects and employing heuristic searching techniques in order to reduce the time cost. Then ODMA is applied to the analysis of earthquake precursory items. Some valuable results are acquired. From the crossover set of the absolutely essential attributes in Sect. 3.1 and Sect. 3.2, only 16 out of the 41 widely used seismometry precursory items are absolutely essential attributes. They are seismic belt (band), seismic gap (segment), seismicity pattern (temporal, spacial, quiescence or activation), precursory earthquake (or swarm), index of seismic activity (comprehensive index A, seismic entropy, degree of seismic activity, and fuzzy degree of seismic activity), strain release (energy release), earthquake frequency, b value, h value, seismic window, earthquake deficiency, foreshock, wave velocity (wave velocity, wave velocity ratio), seismic coda wave (sustained time ratio, attenuation coefficient (a), attenuation rate p), amplitude ratio, and regulatory ratio of small earthquakes. The other 25 items are compatible to the above 16 items or can be included by them. From the results in Sect. 3.3, in the 70 precursory prediction items, 12 items, i.e. fixed leveling (short leveling), ranging tilt, fault creep, roving gravity, apparent resistivity, radon in groundwater, ground water level, ground water level and lake water level, water temperature, electric induction stress, piezo-capacity strain, and macroscopic phenomena are absolutely essential and independently attributes, other 58 precursory items are not essential. Therefore, only 28 of the 111 earthquake precursory items are absolutely essential attributes, and the others are not essential and they are compatible to the above 28 precursory items or can be included. Furthermore, the following conclusions can be drawn:

1. In the gestation stage of earthquake, the above 28 precursory items may show the primary physical characters of earthquake, and they are independent, with the implication that we can know the real reason for the eruption of the earthquake after finding a same physical process that can simultaneously produce all the 28 precursory items.
2. As the above 28 precursory items are the most important factors for the eruption of earthquake, blindness of investigation can be avoided and the cost of earthquake prediction will be reduced sharply by only concentrating on the research of them.

In this paper, we do a primary research on the analysis of the dependency among earthquake precursory items. We know some earthquake precursory items can be ulteriorly divided into several items, such as the index of seismic activity embodies comprehensive index, seismic entropy, degree of seismic activity, and fuzzy degree of seismic activity. These items may be dependent to each other or in hierarchy relationship. But we cannot do an in-depth study based on the materials available from the Earthquake Cases in China. In our future work, more information needs to be collected to solve this problem.

Acknowledgments. This work is supported by the National Natural Science Foundation of China (No. 70502020 and 20503015) and China Earthquake Administration (No. 104090).

References

1. Mei, S., Feng, D., Zhang, G., Zhu, Y., Gao, X., Zhang, Z.: Introduction to Earthquake Prediction in China. 1st edn. Earthquake Press, Beijing (1993)
2. Chen, Q. (ed.): Earthquake Cases in China (1966-1999) (VCD Version). The Center for Analysis and Prediction, China Earthquake Administration, Beijing (2004)
3. Zhang, Z., Zheng, D., Luo, Y.: Researches on Earthquake Cases. Earthquake, **18** (1990) 9–24
4. Wu, F., Xu, J., Zhang, X., Dong, X.: The Statistical Study on Seismological Anomalies in Earthquake Predcition. Earthquakes, **20** (2000) 66–69
5. Fu, W., Zheng, X., Lv, Y., Kong, H.: The Precursory Complexity of Earthquake and Earthquake Prediction. Journal of Heilongjinag August First Land Reclamation University, **13** (2001) 54–57
6. Lu, Y., Yan, L., Guo, R.: Discussion about Correlation of Some Seismological Parameters in Mediun-and-Shor Term Earthquake Prediction. Earthquakes, **19** (1999) 11–18
7. Wang, W., Liu, Z., Song, X., Wang, S.: Seismicity Quantitation and Its Application to Midterm Earthquke Prediction. Earthquake Research in China, **5** (1999) 116–127
8. Pawlak, Z.: Rough sets. International Journal of Computer and Information Science, **11** (1982) 341–356
9. Wang, J., Miao, D., Zhou, Y.: Rough Set Theory and Its Application: A survey. Pattern Recognition and Artificial Intelligence, **9** (1996) 337–344
10. Liu, Q., Huang, Z., Yao, L.: Rough Set Theory: Present State and Prospects. Computer Science, **24** (1997) 1–5
11. Wang, G.: Rough Set Theory and Knowledge Acquisition. Xian Jiaotong University Press, Xian (2001)
12. Skowron, A., Rauszer, C.: The Discernibility Matrices and Functions in Information Systems. In: Slowinski, R. (ed.),: Intelligent Decision Support - Handbook of Applications and Advances of the Rough Sets Theory, System Theory, Knowledge Engineering and Problem Solving, Vol. 11. Kluwer, Dordrecht (1992) 331–362
13. Nguyen, H.S.: Approximate Boolean Reasoning. Foundations and Applications in Data Mining. Transactions on Rough Sets,Lecture Notes in Computer Science, Vol. 4100. Springer-Verlag, Berlin Heidelberg New York (2006) 334–506

Examination of the Parameter Space of a Computational Model of Acute Ischaemic Stroke Using Rough Sets

Kenneth Revett

University of Westminster
Harrow School of Computer Science
Harrow, London
England HA1 3TP
revettk@westminster.ac.uk

Abstract. Complex diseases such as stroke and cancer involve a wide range of biological parameters with respect to the systems involved and disease progression. Computational models of such diseases have led to new insights into the mechanism of action which have resulted in the development of novel therapeutic intervention strategies. Such models are generally quite complex because they incorporate a wide range of relevant biological variables and parameters. In this paper, we examine a biologically realistic computational model of acute ischaemic stroke with respect to the variable and parameter space using rough sets. The aim of this investigation was to extract a set(s) of variables and relevant parameters that predict in a qualitative fashion the final extent of tissue damage caused by a "typical" ischameic stroke.

Keywords: computer simulation, cortical spreading depression, data mining, ischaemic stroke, and rough sets.

1 Introduction

The underlying mechanism for stroke is ischemia characterised by a transient or permanent reduction in cerebral blood flow in an area supplied by a blood vessel to an area of the brain. The World Health Organisation (1998) defined severe cerebral ischaemia (stroke) as: "a syndrome of rapidly developing clinical signs of focal or global disturbance of cerebral function, with symptoms lasting 24 hours or longer or leading to death, with no apparent cause other than of vascular origin". The economic impact of stroke is staggering in both developed and developing countries. Wolfe estimated 4.5 million deaths a year from stroke world-wide with over nine million survivors and an overall incidence rate of 0.2 – 0.25 % [15]. Stroke *per se* is the leading cause of disability in adults, the second most important cause of dementia and third most important cause of death in developed countries [15]. The burden of stroke on healthcare systems around the world are staggering both in terms of the required resources with an estimated costs of 2–30 billion US dollars per year [15]. For these reasons, pathophysiology of stroke has been an area of active research.

The brain is the organ with the highest oxygen demand in the body, consuming 20% of the cardiac output. An ischemic event in the brain results in hypoxia (reduced tissue

J.T. Yao et al. (Eds.): RSKT 2007, LNAI 4481, pp. 534–541, 2007.
© Springer-Verlag Berlin Heidelberg 2007

oxygen levels) and hypoglycaemia (reduced tissue glucose levels) leading to irreversible cell damage (stroke) in a few minutes if perfusion is not restored [1]. Cerebral ischaemia triggers a multitude of events of which three primary pathways have been reported to result in tissue damage: excitotoxicity, inflammation and apoptosis. These pathways form a complex series of pathophysiological events that evolve in time and space (see Figure 1). Together, these damage generating pathways, termed the 'Post Ischemic Cascade,' have been extensively studied using both in vitro and in vivo models to identify possible therapeutic targets in the treatment of stroke [3].

The three major pathways in the Post Ischemic Cascade are brought about by the interactions of numerous cellular components (e.g. transcription factors, caspases) that have been recently identified and grouped by the functional roles these cellular components play in normal physiology [3]. Several laboratories have employed DNA microarray analysis in order to attempt to elucidate a set of genes that may be responsible for the ensuing pathophysiologic pathways [14]. These studies have provided some insight into the genes that are dysregulated – but generally the number of dysregulated genes are too numerous to be useful directly. Other laboratories are employing Magnetic resonance Imaging (MRI) as a tool for investigating the temporal and spatial dynamics of relevant pathophysiological events – both in humans and in animal models of stroke [5,7]. Although some useful information has been gleaned from these studies – they have not as yet lived up to their promise of delivering a comprehensive definition of the events occurring in the PIC. Part of the difficulties encountered in these studies is the inability to control the variables – which are by definition not known with certainty. Which systems are affected – to what degree – the causal relationships between the multitude of systems involved makes this an extremely difficult multi-optimisation task.

The ability to perform computer modelling of complex phenomena has been with us for decades now. There are numerous complex meteorological models that involve massively parallel computational facilities that are in daily use for weather prediction One of the key advantages of a computational model is the ability to control every variable and parameter. Computational models of diseases/pathophysiologies require incorporation of both temporal and spatial evolution of the relevant state variables. A critical task is to identify exactly what the critical state variables should be. This task must be informed by the scientific question at hand – the model should be as simple as necessary – but not too simple.

The first computational model of stroke and related phenomenon was reported in the literature by Tuckwell and Miurma in 1981[11]. This model attempted to capture some of the dynamics of tissue damage expansion – using a simplified diffusion process. Reggia has presented a 2D model which include potassium dynamics suitable for the incorporation of some of the relevant biological events reported in ischemic stroke [4,8]. Revett et al. have proposed a biologically rich model with several key variables that incorporates the relevant spatial and temporal dynamics (see below and [10] for details). Other models have incorporated the same basic strategy employed in [10], incorporating a particular hypothesis into their system [2,11,12,13]. The problem with many of these models is that though they are phenomenologically correct in many of the details – the variable and/or parameter space tends to be large and have remained unexplored. The primary aim of this paper is to explore the rich state variable/parameter space of the stroke model proposed by this author to determine the relationship between particular state variables and the extent of tissue damage resulting from an ischaemic stroke. In the next section of this paper, we describe the stroke model in some detail, followed by a brief discussion of rough sets, followed by a results section. The principal results of this work are discussed in the conclusion section, with the addition of future works.

2   Stroke Simulation

The computational model of stroke employed in this study is a simulation that incorporates both the spatial and temporal evolution of key variables that have been reported in the literature to be critical in the development and evolution of tissue damage [3,5]. Since stroke is confined to brain tissue, we simulated a 2D hexagonally tessellated network of cortex – in essence a 2D cellular automata with drop-off boundary conditions. Embedded within each cell of the CA is a set of state variables that will be updated during the course of the simulation. The dynamics of the system is implemented via a series of equations (ordinary and partial differential) that describe phenomenologically the behaviour of these variables as reported in the literature. Updating of the state equations is performed using a Forward Euler method in order to reduce the computational time (the results have been corroborated using a 4th order Runge-Kutta method). The spatial and temporal steps in the model are such that each cell occupies 0.125 *mm* and each update represents 13 *msec* (see [9,10] for details). The values of each of the state variables are recorded at some particular spatial location(s) and at specific time intervals, which are stored for subsequent graphical display and analysis (see Figure 1). reader is referred to [10,11] for details).

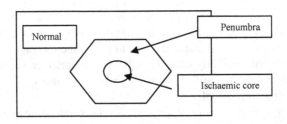

Fig. 1. A depiction of the relationship between the ischaemic core, the surrounding penumbra, and normal tissue. Note these regions are not drawn to scale.

Each state variable is set to their initial value at the start of the simulation and are updated until a stopping criteria has been reached (generally a certain amount of simulated time – typically on the order of 2-4 hours – depending on the particular requirements of the experiment). A time consuming aspect of this type of simulation is the determination of parameter values – which in this case were "discovered" by hand. There are 60 parameters in this model – each modulating a particular facet of the state variable behaviour (due to space constraints, the details are omitted and the interested

Since stroke is caused by a reduction in the blood supply to a particular region of the brain, the primary variable to consider is the cerebral blood flow. Each region of the simulated cortex has a local blood flow level. The literature reports that there are three compartments which differ in their blood flow levels: the ischaemic core, the penumbra, and the surrounding normal tissue. These compartments are depicted in Figure 2. The ischaemic core is the central region in Fig 2 which has essentially no blood supply and hence will die (become infarcted) within 2-3 minutes if the blood supply is not restored. Surrounding the ischaemic core is the penumbra – a region that has reduced blood supply, but sufficient resources to maintain vitality provided there are no further compromising events. The penumbra is surrounded by tissue that has normal blood flow levels which usually survives. In addition to tissue death in the ischaemic core, experimental models of stroke and clinical results indicate that tissue surrounding the ischaemic core (e.g. the penumbral tissue) may also undergo infarction [3,5]. The final amount of tissue damage is largely determined by the extent of the expansion of the infarct into the penumbra. Understanding the mechanisms underlying tissue damage expansion into this region is of paramount clinical importance, as it may lead to new therapeutic measures that reduce post-infarct debilitation. While great progress has been made in this area during recent years [1,3,7], the mechanisms by which focal ischemia evolves into infarction and the factors which determine the ultimate extent of the infarct are still unsettled. The prolonged *reduction* of blood supply in the peri-infarct region may gradually lead to progressive tissue damage via numerous pathological metabolic pathways, yet the relative importance of these factors remains controversial [3,7].

This expansion process is incorporated into the model presented in this paper, yielding results that are consistent with major literature reports on the subject. What one would like to know is which variables are involved – and how are they modulated by relevant parameters. These parameters (and their values) are not free in this model – they represent actual biological processes that have been reported in numerous literature reports. But literature reports provide a snapshot of the variables and parameters – and are valid only under the given experimental conditions in which they were measured.

Clinically relevant information with respect to variables/parameters under a wide operating range that occurs in real populations would be very useful in developing rational therapeutic strategies. This information is potentially available from simulation studies – provided the model it is based on is correct and sufficiently comprehensive. The question is how to extract this information from the simulation.

In this work, we employ rough sets as a methodology for extracting values for variable/parameters (attributes) that influence the behaviour of the simulation in specific ways. Since the variable space is large (16 state variables and 60 parameters), we focused on the relationship between cerebral blood flow and the resulting tissue damage.

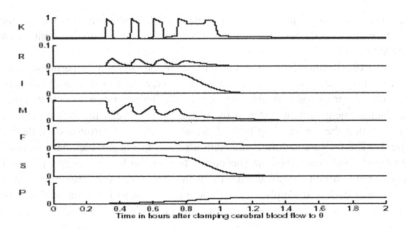

Fig. 2. A sample of the major state variables recorded over time at a particular location (a cell at radius 3 from the edge of the ischaemic core). The x-axis represents time in hours and the y-axis indicates abbreviations for 7 key state variables.

3 Methodology

The first stage of processing the data for use with rough sets is to create a decision table [6]. The attributes of this table were the relevant variables and parameters that relate to blood flow. The binary decision class represented whether or not tissue damage expanded beyond a certain point into the ischaemic core ('1') or not ('0'). To populate the records within the decision table, a large number of simulations were run while varying the values of parameters through automatic programme control. A large amount of processing power was required – since there are a large number of equations that are updated. We simulated a section of brain tissue composed of a square lattice of 256x256 cells (simulating 32x32mm of 2D tissue) – each cell containing the full complement of variables, equations, and parameters. Each time step represents 13 *msec*, and we wanted to allow sufficient time for damage to occur – therefore each simulation iterated for 10^6 iterations, amounting to approximately 3.6 of simulated time. This value was in accordance with our own work with this simulation and consistent with experimental models of stroke [7]. On a fast Pentium IV (3.06 GHz and 1024 MB DDR ram), 10^6 iterations is completed in approximately 90 minutes. We therefore ran the simulations on a cluster of 40 Pentium IVs (equipped as per above) for a total of 96 hours, yielding a total of 6,400 simulations. We then examined the damage resulting from each simulation and classified them as either a ('0') or ('1') based on whether the damage encroached sufficiently into the penumbra. A programme was written that would that determined the extent of the damage and set a threshold that was initially set to a radius of 5 cells (out of 40) around the ischaemic core. The programme then extracted the records and assigned the appropriate decision value. It was desired that the results would yield approximately equal numbers of 0s and 1s. If this was not the case, then the threshold would be changed accordingly (decreased if too many 0s and vice versa for too many 1s). From this particular study, the radius was set to 3 cells. This resulted in 46% (2,944) objects with a '0' decision class and 3,456 objects belonging to decision class '1'. Please note all attributes in the decision table were discretised using the 'Global' method within RSES 2.2. Note the dataset was complete – so imputation was not necessary.

With the ability to create the decision table – the critical issue is to decide which variables to record and which parameters to vary. In order not to bias the decision table, all variables were included in the decision table (yielding 26). In addition, all parameters relevant to blood flow (6 in this case) were varied randomly from an initial control condition. The control condition was one that generated results that were consistent with published reports and were directly acquired from literature – and so forms our gold standard (mean values). Values for parameters were varied plus/minus the mean by the addition of random noise between 10 and 100% of the mean values

4 Results

The principal result from the rough sets analysis was the set of rules that were generated. These rules relate the extent of tissue expansion into the penumbra with respect to parameters controlling the blood flow. In order to estimate the accuracy of the rule set, we validated the results using 10-fold cross validation. The resulting classification accuracy was 92.3% based on the decision rule classification strategy. We also applied several other classification techniques available in RSES 2.2, the results of which are depicted in table 1.

Table 1. Classification accuracy with respect to the classification technique

Decision rules (All)	92.3%
Decision rules (LEM)	90.3%
k-NN	88.9%
LTF-C	89.7%

With an acceptable accuracy level, the critical result from this study is the rule set – which should provide information regarding the specific variables/parameters and their values that predict whether or not tissue damage will expand significantly into the penumbra. The number of rules generated was on the order of 987 – a fairly large number of rules. The rules were filtered based on support and the accuracy was re-tested with the filtered rule set and the values are reported in Table 2.

Table 2. Number of rules as a function of support filtering and the resulting classification accuracy

Support Threshold	Number of Rules	Accuracy
Support: 1-2	1,784	93.1%
Support: 1-3	895	91.8%
Support: 1-4	271	94.6%
Support: 1-5	72	89.3%

As can be seen from Table 2, the accuracy is fairly consistently high, and the number of rules are reasonable - especially when removing rules with low support between 1-5 (89.3%). Although no clear trend exists from the data in Table 2, it is clear that the number of rules can be significantly reduced without a significant drop in classification accuracy. We therefore focus on the rules obtained when removing support between 1-5. The attributes (state variables and/or parameters) yield sets of values that can be used in conjunction with this model to explain under what circumstances stroke progresses into the surrounding penumbra.

5 Conclusions

In this preliminary study, we employed rough sets as a tool to extract the values of attributes that determined whether or not a stroke would progress beyond a certain position within the surrounding penumbra. The decision table that was employed in this was the result of a computer simulation based on a biologically realistic computational model of acute ischaemic stroke. The model incorporates the temporal and spatial dynamics of key state variables that have been reported to occur in animal models of stroke. The parameters in the model were initially selected by a pain-staking manual search. What was sought in this study was to determine which variables/parameters were and their values were important predictors of progressive tissue damage. Rough sets provides – in part through the discretisation process – a range of values that preserves the behaviour of the model.

The results from this study are encouraging – to the author's knowledge this is the first attempt to apply rough sets to a biologically realistic model of a disease. It is a feasible step to apply this same methodology to other biologically realistic model of diseases such as cancer. One key advantage of this approach is the ability to vary the attribute values in a rational way. In most applications of data mining, one works with a snapshot of the data – without the ability to alter the data in any significant way.

References

1. Back, T., Kohno, K., & Hossmann, K.A.. Cortical negative DC deflections following middle cerebral artery occlusion and KCL-induced spreading depression: E.ect on blood flow, Tissue oxygenation, and electroencephalogram. Journal of Cerebral Blood Flow and Metabolism, 14, (1994), pp. 12-19.
2. Chapsuisat, G., grenier, E., Boissel, J.P., Dronne, M.A., Gilquin, H., & Hommel, A Global model for ischemic stroke with stress on spreading depressions,
3. Fisher, M., & Garcia, J.. Evolving stroke and the ischemic penumbra. Neurology, 47(October), (1996), pp. 884-888.
4. Goodall, S., Reggia, J.A., Chen, Y., Ruppin, E., & Whitney, C.. A computational model of acute focal cortical lesions. Stroke 28(1), (1997), 101-109.
5. Iijima, T., Meis, G., & Hossmann, K.A. (1992). Repeated negative de.ections in rat cortex following middle cerebral artery occlusion are abolished by MK-801. E.ect on volume of ischemic injury. Journal of Cerebral Blood Flow and Metabolism, 12, 727-733.

6. Pawlak, Z. Rough Sets, International Journal of Computer and Information Sciences, 11, pp. 341-356, 1982.
7. Pulsinelli , W. Pathophysiology of acute ischaemic stroke. The Lancet, 339, (1992), 533-536.
8. Reggia, J., & Montgomery, D. A computational model of visual hallucinations in migraine.Computers in Biology and Medicine, 26(2), (1996), pp. 133-141.
9. Reggia, J., E. Ruppin, & Berndt, R. (eds),. Neural Modeling of Brain and Cognitive Disorders. United Kingdom: World Scienti.c Publishing. (1996).
10. Revett, K., Ruppin, E., Goodall, S. & Reggia, J. Spreading Depression in Focal Ischemia: A Computational Study, Journal of Cerebral Blood Flow and Metabolism, Vol 18{9), 998-1007,1998.
11. Reshodko, L.V., & Bures, J. Computer simulation of reverberating spreading depression in a network of cell automata. Biol. Cybernetics, 18, (1975), pp. 181-189.
12. Shapiro, B., Osmotic forces and gap junctions in spreading depression: a computational model, J Computational neuroscience, 2001 vol 10(1), 99-120.
13. Tuckwell, H.C., & Miura, R.M.. A mathematical model for spreading cortical depression. Biophysical Journal, 23, (1978), pp. 257-276.
14. Via, N., A Genetic Approach to Ischemic Stroke Therapy, *Cerebrovascular Diseases* 2004;17 (Suppl. 1), pp. 63-69.
15. Wolfe, C " The Burden of Stroke" in Wolfe, C, Rudd, T and Beech, R (eds) Stroke Services and Research (1996) The Stroke Association

Using Rough Set Theory to Induce Pavement Maintenance and Rehabilitation Strategy

Jia-Ruey Chang[1], Ching-Tsung Hung[2], Gwo-Hshiung Tzeng[3], and Shih-Chung Kang[4]

[1] Department of Civil Engineering, MingHsin University of Science & Technology,
No.1, Hsin-Hsing Road, Hsin-Fong, Hsin-Chu, 304, Taiwan
jrchang@must.edu.tw
[2] Institute of Civil Engineering, National Central University,
ChungLi, Taoyuan, 320, Taiwan
92342008@cc.ncu.edu.tw
[3] Distinguished Chair Professor, Kainan University,
No.1 Kainan Road, Luchu, Taoyuan County, 338, Taiwan
ghtzeng@mail.knu.edu.tw
[4] Department of Civil Engineering, National Taiwan University,
No. 1, Sec. 4, Roosevelt Road, Taipei, 10617, Taiwan
sckang@caece.net

Abstract. Rough Set Theory (RST) is an induction based decision-making technique, which can extract useful information from attribute-value (decision) table. This study introduces RST into pavement management system (PMS) for maintenance and rehabilitation (M&R) strategy induction. An empirical study is conducted by using the pavement distress data collected from 7 county roads by experienced pavement engineers of Taiwan Highway Bureau (THB). For each road section, the severity and coverage of existing distresses and required M&R treatment were separately recorded. The analytical database consisting of 2,348 records (2,000 records for rule induction, and 348 records for rule testing) are established to induce M&R strategies. On the basis of the testing results, total accuracy and total coverage for the induced strategies are as high as 88.7% and 84.2% respectively, which illustrates that RST certainly can reduce distress types and remove redundant records to induce the proper M&R strategies.

Keywords: Rough set theory (RST), Pavement management system (PMS), Maintenance and rehabilitation (M&R).

1 Introduction

Various distress types would occur to pavement because of dynamic loading, overweighed trucks, weak foundation, improper mix design, change of climates, etc [1]. Pavement distress survey records the severity and coverage of existing distress types in order to adopt proper maintenance and rehabilitation (M&R) treatments. It is extremely important that if proper M&R treatments can be implemented at right time for specific distress type. Proper M&R treatments can not only save long-term

J.T. Yao et al. (Eds.): RSKT 2007, LNAI 4481, pp. 542–549, 2007.
© Springer-Verlag Berlin Heidelberg 2007

expense but keep the pavement above an acceptable serviceability [2, 3]. However, M&R strategies are usually made by engineers' subjective judgments. The objective of this study is to utilize Rough Set Theory (RST) in dealing with enormous distress data to induce proper M&R strategies for decreasing M&R judgment errors and improving the efficiency of decision-making process in pavement management system (PMS).

2 Rough Set Theory (RST)

In this section, the basic concept of RST is presented. Rough set, originally proposed by Pawlak [4], is a mathematical tool used to deal with vagueness or uncertainty. More detailed discussion about the process of RST can refer to the literatures [5-7]. The original concept of approximation space in rough set can be described as follows. Given an approximation space $apr = (U, A)$, where U is the universe which is a finite and non-empty set, and A is the set of attributes. Then based on the approximation space, we can define the lower and upper approximations of a set.

Let X be a subset of U, and the lower and upper approximation of X in A are conceptualized as Eq. (1) and (2), respectively.

$$\underline{apr}(A) = \{x \mid x \in U, U / Ind(A) \subset X\} . \tag{1}$$

$$\overline{apr}(A) = \{x \mid x \in U, U / Ind(A) \cap X \neq \varnothing\} . \tag{2}$$

where $U / Ind(A) = \{(x_i, x_j) \in U \cdot U, f(x_i, a) = f(x_j, a) \quad \forall a \in A\}$.

Eq. (1) represents the least composed set in A containing X, called the best lower approximation of X in A, and Eq. (2) represents the greatest composed set in A contained in X, called the best upper approximation. After constructing upper and lower approximations, the boundary can be represented as

$$BN(A) = \overline{apr}(A) - \underline{apr}(A) . \tag{3}$$

According to the approximation space, we can calculate reducts and decision rules. Given an information system $I = (U, A)$ then the reduct, $RED(B)$, is a minimal set of attributes $B \subseteq A$ such that $r_B(U) = r_A(U)$ where

$$r_B(U) = \frac{\sum card(\underline{B}X_i)}{card(U)} . \tag{4}$$

denotes the quality of approximation of U by B.

Once the reducts have been derived, overlaying the reducts on the information system can induce the decision rules. A decision rule can be expressed as $\varnothing \Rightarrow \theta$, where $\varnothing$ denotes the conjunction of elementary conditions, $\Rightarrow$ denotes '*indicates*', and θ denotes the disjunction of elementary decisions.

The advantage of the induction based approaches such as RST is that it can provide the intelligible rules for decision-makers (DMs). These intelligible rules can help DMs to realize the contents of data sets. In the following empirical study, enormous records from pavement distress surveys are used to calculate reducts of distress types and to induce M&R strategies.

3 Empirical Study: Pavement M&R Strategy Induction

Although the development and implementation of M&R strategies for pavement is important, literature addressing this issue is limited. Colombrita et al. [8] build a multi-criteria decision model based on Dominance-based Rough Set Approach (DRSA) to provide highway agencies with a decision support system for more efficient M&R budget allocation. In the empirical study, 18 distress types and 4 M&R treatments provide as the attributes and decision variables, respectively. With the data collected from 2,348 asphalt-surfaced road sections, RST is employed to induce M&R strategies. Rough Set Exploration System (RSES) package is utilized to execute analyses [9, 10]. RSES is a software tool that provides the means for analysis of tabular data sets with use of various methods, in particular those based on RST.

3.1 Pavement Distress Survey and M&R Treatments

Pavement distress survey is conducted for the purpose of monitoring the existing pavement condition and making the appropriate M&R decisions. Generally, the severity and coverage should be separately identified and recorded for each distress type on one road section. For accurate, consistent, and repeatable distress survey, one comprehensive distress survey manual is required for clarifying the definition, severity, and coverage of each distress type. In the empirical study, pavement distress survey was carried out following the Distress Identification Manual for the Long-Term Pavement Performance Program issued by Federal Highway Administration (FHWA) in June 2003 [11]. Furthermore, the required M&R treatment for each road section was decided according to the Standardized M&R Guidance which is issued by Taiwan Highway Bureau (THB) and used in various highway authorities in Taiwan.

3.2 Data Description

Pavement distress surveys were conducted on seven county roads with asphalt surface in Chung-Li Engineering Section of THB by 8 experienced pavement engineers. The seven county roads (110, 112, 112A, 113, 113A, 114 and 115) are located in northern Taiwan. Engineers conducted surveys by walking or driving. The distress information and required M&R treatment for each road section were recorded. The totally collected 2,348 records (2,000 records are randomly selected for training dataset; the rest of 348 records are for testing dataset) are utilized in the empirical study, which are integrated as Table 1. The first column in Table 1 shows the record numbers, column 2 to column 19 illustrate the 18 distress types, and the last column refers to the required M&R treatment (decision variable). The details are described as follows:

Table 1. Summary of analytical database

Rec.	D1	D2	D3	D4	D5	D6	D7	D8	D9	D10	D11	D12	D13	D14	D15	D16	D17	D18	M&R
1	2	0	0	0	0	0	0	0	0	0	0	0	0	0	0	5	0	0	1
2	1	0	0	0	0	0	0	0	0	0	0	0	0	0	6	0	0	0	1
3	2	1	1	0	0	0	0	0	1	0	0	0	1	0	0	0	0	0	1
4	2	1	0	0	0	0	0	0	1	0	0	0	1	0	0	0	0	0	1
5	2	0	0	0	1	0	0	0	0	0	0	0	1	0	0	0	0	0	1
6	1	0	1	0	0	0	0	0	0	0	0	0	1	0	0	0	0	0	1
7	1	0	1	0	0	0	0	0	0	0	0	0	1	0	0	0	0	0	1
8	2	0	1	0	0	0	0	0	0	0	0	0	1	0	0	0	0	0	1
9	1	0	0	0	0	0	0	0	0	0	0	0	1	0	0	0	0	0	1
10	1	0	0	0	0	0	0	0	0	0	0	0	1	0	0	0	0	0	1
11	2	0	0	0	0	0	0	0	0	0	0	0	1	0	0	0	0	0	1
12	4	0	0	0	0	0	0	0	0	0	0	0	1	0	0	0	0	0	1
13	5	0	0	0	0	0	0	0	0	0	0	0	2	0	0	0	0	0	1
14	2	0	0	0	0	0	0	0	0	0	2	0	0	0	0	0	0	0	1
15	1	0	0	0	0	0	1	0	0	1	0	0	0	0	0	0	0	0	1
:	:	:	:	:	:	:	:	:	:	:	:	:	:	:	:	:	:	:	:
:	:	:	:	:	:	:	:	:	:	:	:	:	:	:	:	:	:	:	:
2345	0	0	9	0	0	0	0	0	0	0	0	0	0	0	0	0	0	0	4
2346	0	5	0	0	0	0	0	0	0	0	0	0	0	0	0	0	0	0	4
2347	0	5	0	0	0	0	0	0	0	0	0	0	0	0	0	0	0	0	4
2348	0	5	0	0	0	0	0	0	0	0	0	0	0	0	0	0	0	0	4

- The empirical study explores 18 common distress types in Taiwan, which are represented from D1 to D18: D1. Alligator Cracking, D2. Block Cracking, D3. Longitudinal Cracking, D4. Transverse Cracking, D5. Edge Cracking, D6. Reflection Cracking, D7. Pothole, D8. Bleeding, D9. Rutting, D10. Corrugation, D11. Lane/Shoulder Drop-off, D12. Depression, D13. Structure Drop-off, D14. Utility Cut Patching, D15. Shoving, D16. Manhole Drop-off, D17. Patching Deterioration, D18. Raveling. Figure 1 shows examples of distress types.
- The severity levels of distress are classified as L (low), M (moderate), and H (high). The coverage levels of distress are classified as A (local), B (medium), and C (extensive). Therefore, there are nine combinations (LA, LB, LC, MA, MB, MC, HA, HB, HC) of severity and coverage which are represented by number 1 to 9 respectively, and plus 0 represents no distress. For example, "D1 = 2" denotes Alligator Cracking occurs with low severity and medium coverage. Figure 2 shows examples of distress types with different severity and coverage combinations.
- M&R treatments for asphalt pavement used in the empirical study are classified as four types, which are represented as number 1 to 4 referring to no M&R required, localized M&R (such as full-depth patching, crack sealing, etc.), global M&R (such as fog seal, slurry seal, aggregate surface treatment, etc.), and major M&R (such as milling, hot recycling, heater scarifying, AC overlay, reconstruction, etc.) respectively. Figure 3 shows examples of M&R treatments.

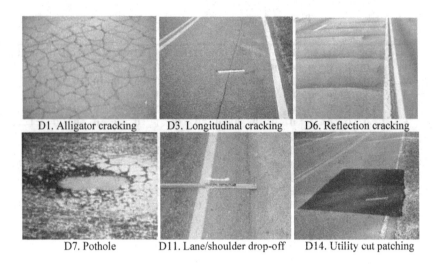

| D1. Alligator cracking | D3. Longitudinal cracking | D6. Reflection cracking |

| D7. Pothole | D11. Lane/shoulder drop-off | D14. Utility cut patching |

Fig. 1. Examples of distress types

(a)   (b)

Fig. 2. (a) Alligator cracking with high severity and extensive coverage ("D1 = 9"); (b) Patching with high severity and medium coverage ("D17 = 8")

(a)   (b)   (c)

Fig. 3. M&R treatments: (a) Full-depth patching; (b) Hot recycling; (c) AC overlay

3.3 Induction of M&R Strategies

First of all, reduct calculation is conducted using exhaustive algorithm [12] in RSES. Two reduct sets with 14 attributes each are obtained and shown below. The 13 attributes (cores) obtained from intersection of the two reduct sets are D1, D2, D3,

D4, D5, D6, D7, D8, D10, D11, D13, D14, D16. It is found that D12. Depression, D17. Patching Deterioration and D18. Raveling are not shown in both reduct sets.

- D1, D2, D3, D4, D5, D6, D7, D8, D9, D10, D11, D13, D14, D16
- D1, D2, D3, D4, D5, D6, D7, D8, D10, D11, D13, D14, D15, D16

Then, M&R strategies are induced based on the calculated reducts by randomly selecting 2,000 out of 2,348 records. The exhaustive algorithm [12] in RSES is chosen again to construct all minimal decision rules. Hence the induced 83 M&R strategies are shown in Table 2. For example, the first and second row in Table 2 represent the first and second M&R strategy, which match with 525 records (most records) and 315 records, respectively:

- IF   (D1=2)
  (That is, Alligator Cracking occurs with low severity and medium coverage.)
  THEN   the required M&R treatment will be no M&R required (450 records) or localized M&R (75 records)
- IF   (D1=1) & (D5=1)
  (That is, both Alligator Cracking and Edge Cracking occur with low severity and local coverage.)
  THEN   the required M&R treatment will be no M&R required (315 records)

The exhaustive algorithm may be time-consuming due to computational complexity. Therefore approximate and heuristic solutions such as genetic or Johnson algorithms [12], which allow setting initial conditions for number of reducts to be calculated, required accuracy, coverage and so on, can be considered in the future.

Table 2. Summary of induced 83 M&R strategies

No. of Rule (1-83)	Match	M&R Strategies
1	525	(D1=2) => {M&R={1[450],2[75]}}
2	315	(D1=1)&(D5=1) => {M&R=1[315]}
3	210	(D1=1)&(D3=1) => {M&R={1[175],2[35]}}
4	192	(D1=1)&(D3=1)&(D4=2) => {M&R={1[132],2[60]}}
5	176	(D1=2)&(D5=1) => {M&R=1[176]}
:	:	:
:	:	:

3.4 Testing of M&R Strategies

The rest of 348 records are used to conduct testing analyses focusing on the induced 83 M&R strategies. The results are shown in Table 3 and discussed as follows. Note that 1, 2, 3, and 4 are represented as no M&R required, localized M&R, global M&R, and major M&R, respectively. Rows in Table 3 correspond to actual (required) 4 M&R treatments while columns represent decision values as returned by induced 83 M&R strategies (classifier).

Table 3. Testing results of induced 83 M&R strategies

| | | Predicted Treatments | | | | | | |
		1	2	3	4	No. of obj.	Accuracy	Coverage
	1	171	2	3	3	201	0.955	0.891
Actual	2	5	69	5	5	111	0.821	0.757
Treat.	3	2	3	15	3	27	0.652	0.852
	4	0	1	1	5	9		0.714	0.778
	True positive rate	0.961	0.920	0.625	0.313			

Total number of tested objects: 348
Total accuracy: 0.887
Total coverage: 0.842

- The values on diagonal represent correctly classified cases. If all non-zero values in Table 3 appear on the diagonal, we conclude that classifier makes no mistakes for the testing data.
- Accuracy: Ratio of correctly classified objects from the class to the number of all objects assigned to the class by the classifier. For instance, 0.955 = 171/(171+2+3+3), 171 represents the number of records whose actual M&R treatment is 1 (no M&R required) as the same with the predicted M&R treatment by 83 M&R strategies. If the predicted treatment is 2 (2 records), 3 (3 records), and 4 (3 records), this must be incorrect.
- Coverage: Ratio of classified (recognized by classifier) objects from the class to the number of all objects in the class. That is, for all M&R treatments, the ratio of strategies which can be recognized to carry out prediction (including incorrect prediction). For instance, 0.891 = (171+2+3+3)/201.
- True positive rate: For each M&R treatment, the ratio of treatment which can be correctly predicted by 83 M&R strategies. For instance, 0.961 = 171/(171+5+2).
- Total accuracy: Ratio of number of correctly classified cases (sum of values on diagonal) by 83 M&R strategies to the number of all tested cases. For instance, 0.887 = (171+69+15+5)/(171+2+3+3+5+69+5+5+2+3+15+3+1+1+5).
- Total coverage: Total coverage equals 1 which means that all objects have been recognized (classified) by 83 M&R strategies. Such total coverage is not always the case, as the induced classifier may not be able to recognize previously unseen object. If some test objects remain unclassified, the total coverage value is less than 1. For instance, 0.842 = (171+2+3+3+5+69+5+5+2+3+15+3+1+1+5)/348.

Note that total accuracy must be defined depending on the correct M&R treatment prediction. From the testing results, the total accuracy is as high as 0.887 and the total coverage is 0.842. Therefore, the 83 M&R strategies can be used to reliably reason M&R treatments in practice.

4 Conclusions

The purpose of pavement distress survey is to assist engineers in making proper M&R decisions. Proper M&R treatments can save long-term expense and keep the pavement above an acceptable serviceability. However, M&R strategies are usually made by engineers' subjective judgments. In the study, we have demonstrated the successful application of RST to the problem of inducing 83 M&R strategies by using 2,348 actual data (2,000 records for rule induction, and 348 records for rule testing). On the basis of the testing results, total accuracy and total coverage for the induced strategies is as high as 88.7% and 84.2% respectively, which illustrates that RST can easily reduce distress types and remove redundant records from enormous pavement distress data to induce proper M&R strategies. The induced M&R strategies can decrease M&R judgment errors and assist engineers to reliably reason M&R treatments. The efficiency of decision-making process in pavement management system (PMS) can be improved as well. The M&R strategies induced in this study provide a good foundation for further refinement when additional data is available.

Acknowledgments. This study is partial results of project NSC 93-2211-E-159-003. The authors would like to express their appreciations to National Science Council of Taiwan for funding support.

References

1. Huang, Y.H.: Pavement Analysis and Design. Prentice Hall Inc. (1993)
2. Eltahan, A.A., Von Quintus, H.L.: LTPP Maintenance and Rehabilitation Data Review–Final Report. Publication No. FHWA-RD-01-019, Federal Highway Administration (FHWA) (2001)
3. Abaza, K.A., Ashur, S.A., Al-Khatib, I.A.: Integrated Pavement Management System with a Markovian Prediction Model. Journal of Transportation Engineering 130(1) (2004) 24–33
4. Pawlak, Z.: Rough Sets. International Journal of Computer and Information Sciences 11(5) (1982) 341–356
5. Pawlak, Z.: Grzymala-Busse, J., Slowinski, R., Ziarko, W.: Rough Sets. Communications of the ACM 38(11) (1995) 89–95
6. Grzymala-Busse, J.: Knowledge Acquisition under Uncertainty - A Rough Set Approach. Journal of intelligent and Robotic Systems 1(1) (1988) 3–16
7. Walczak, B., Massart, D.L.: Rough Sets Theory. Chemometrics and Intelligent Laboratory Systems 47(1) (1999) 1–16
8. Colombrita, R., Augeri, M.G., Lo Certo, A.L., Greco, S., Matarazzo, B., Lagana, G.: Development of a Multi-Criteria Decision Support System for Budget Allocation in Highway Maintenance Activities. Proceedings of Societa Italiana Infrastrutture Viarie (SIIV) 2004, Florence, Italy, Oct. 27-29 (2004)
9. Bazan, J.G., Szczuka, M.S.: The Rough Set Exploration System. In: Peters, J.F., Skowron, A. (eds.): Transactions on Rough Sets III. Lecture Notes in Computer Science, Vol. 3400. Springer-Verlag, Berlin Heidelberg New York (2005) 37–56
10. Rough Set Exploration System (RSES), Institute of Mathematics, Warsaw University (http://logic.mimuw.edu.pl/~rses) (2006)
11. Miller, J.S., Bellinger, W.Y.: Distress Identification Manual for the Long-Term Pavement Performance Program (FHWA-RD-03-031), Federal Highway Administration (2003)
12. Bazan, J.G., Nguyen, H.S., Nguyen, S.H., Synak, P., Wróblewski, J.: Rough Set Algorithms in Classification Problem. In: Polkowski, L., Tsumoto, S., Lin, T. (eds.): Rough Set Methods and Applications. Physica-Verlag, Heidelberg New York (2000) 49–88

Descent Rules for Championships

Annibal Parracho Sant'Anna

Universidade Federal Fluminense
Rua Passo da Patria, 156 bl. D. sl. 309 Niteroi-RJ 24210-240 Brazil
santanna@pesquisador.cnpq.br
http://www.producao.uff.br/pos/site_professores/aps

Abstract. This article proposes alternative classification rules for the clubs in the Brazilian Soccer Championship employing Rough Sets Theory. A feature of the procedure adopted to determine the rough sets is that it allows for attributes associated to intransitive relations. One of such attributes is the result of direct confrontation. The classification procedure follows three stages. First, the condition attributes are ranked according to their importance for the official classification on points. Then the condition attributes are sequentially applied to generate a classification. Finally the result of this sequential classification is compared to the decision partition. The results obtained for different sets of attributes and different years are consistent, with the number of distinct classes obtained increasing with the quality of approximation.

Keywords: index of quality of approximation, importance measure, tournament rules.

1   Introduction

Copying the European model, the Brazilian Soccer Championship has employed during the last years a schedule of full confrontation between all clubs in two turns. At the end of the year, predetermined numbers of clubs are selected, at the upper and lower tails of the Standings Table, to international tournaments and to descent to 2nd division championship. In the year of 2006, a similar structure started to be employed also in the 2nd division. This kind of framework results in a long championship with important decisions in the last rounds affected by possible lack of interest in the matches by clubs already sacrificed by the descent rule or that have already conquered the championship or any other ultimate purpose. The fixed numbers of clubs descending and ascending may also lead to unfair decisions, for instance transferring to the 2nd division a club that always wins his matches against another that is kept in the elite group.

If we are able to manage flexible schedules in terms of number of clubs in the championship, fairer rules may be applied to classify the clubs in such a way that clubs in the same cluster would be kept together in the elite group or sent together to the 2nd division. This article proposes classification rules to be applied in such case. Three classification variables are considered: two attributes more directly linked to the total number of points earned and, as a complementary attribute,

J.T. Yao et al. (Eds.): RSKT 2007, LNAI 4481, pp. 550–557, 2007.

the result of direct confrontation. This last variable does not determine an order relation in the set of clubs, since it lacks the transitive property. In [8] is developed a classification approach that allows for incorporating this kind of attribute.

The new classification procedure follows three stages. First, measures of importance of the condition attributes are determined and the attributes are ranked according to the values of such measures. The evaluation according to each attribute is then successively applied to rank the options, the ties according to the attribute considered the most important being eliminated according to the second most important, the ties remaining after that eliminated by the third attribute and so on. Finally, the classification so generated is compared to that determined by the decision attribute. At this last step, ties according to all condition attributes should be kept but new ties may be generated to eliminate the contradictions between the two classifications.

Different sets of condition variables were applied. Besides the result of direct confrontation and the number of goals scored, were alternatively considered the number of goals taken, the number of losses and the number of wins. As tactical reasons make the number of goals taken increase with the number of goals scored and the championship points computation rule sets a distance of two points between a win and a tie and of only one between a tie and a loss, the number of wins becomes more important for the classification on points than the other attributes considered. The number of losses is the most important attribute when it is employed but the overall quality of the model with it is found to be smaller than that of the model with the number of wins replacing it. And the model with the lowest quality of information is that with the number of goals taken.

In the next section the index of quality of approximation is presented. Section 3 deals with the measurement of importance. Section 4 develops the sequential classification procedure. Finally, in Section 5 the clubs problem is addressed. Concluding comments form Section 6.

2   Related Work

The approach here developed applies the quality of approximation framework of Rough Sets Theory. The main interest of this work is benefiting from the flexibility of the index of quality of approximation applied to employ more general kinds of attributes to expand the initial decision classes.

Recent years have been filled with intense work on the derivation of association and classification rules, especially suited to deal with large databases. Some of these methods employ indicators and aggregation criteria similar to those employed here. The present study shall be followed by analyses comparing its results to results derived from the application of tools like C4.5 [6], CN2 [2], RIPPER [3] and their developments.

On the other hand, the approach here taken may also be applied to larger databases. The comparison to other methods in such context must also be done.

3 The Index of Quality of Approximation

The index of quality of approximation proposed in [8] treats symmetrically the condition and decision partitions. A binary relation is derived from the values of each attribute isolated or set of attributes as in [7]. The index measures the probability of agreement, in the comparison of an arbitrary pair of items, between the two relations, that associated to the condition attributes and that associated to the decision attributes. More precisely, the quality of the approximation of the decision partition **D** by the set of condition attributes **C** is gauged dividing, by the total number of pairs of comparable options, the sum of the number of pairs of indiscernible options according to **C** which belong to the same class according to **D** and the number of pairs discernible according to **C** which belong to distinct classes according to **D**.

This definition may be set in a graph theory framework. Consider two graphs with the same set of nodes. Each node represents an option. In the first graph, two different nodes are linked if and only if the respective options belong to the same class in the decision partition. This graph is represented by the indicator function G_D defined in the domain of pairs of options by $G_D(o_1,o_2) = 1$ if o_1 and o_2 are indiscernible in **D** and $G_D(o_1,o_2) = 0$ otherwise. In the second graph, analogously, two nodes are linked if and only if the respective options are indiscernible according to the set of condition attributes. This second graph is represented by the indicator function G_C defined in the same domain by $G_C(o_1,o_2) = 1$ if o_1 and o_2 are indiscernible according to **C** and $G_C(o_1,o_2) = 0$ otherwise. The quality of information index is then defined by the ratio $I_D(C)$ between the number of pairs of different options o_1, o_2 with $G_C(o_1,o_2) = G_D(o_1,o_2)$ divided by the total number of pairs of different options considered.

This definition may be easily extended to the case of dominance, treated by [5], [7] and others in the context of Rough Sets Theory. The analogous index of quality of approximation for an ordered classification **D** by a set of attributes **C** will be obtained dividing, by the number of pairs of different elements of the universe of options for which there is indiscernibility or there is dominance according to **C** or according to **D**, the sum of the number of such pairs for which there is concomitantly indiscernibility according to **C** and **D** with the number of pairs with dominance in the same direction according to **C** and **D**.

4 The Importance Measures and the Criteria to Determine Joint Dominance

Since the indices of quality of approximation above described, as well as the classical ones, may be calculated for any subset of the set of condition attributes **C**, their values may be combined to determine the effect on the quality of approximation of the inclusion or exclusion of one or more attributes. A measure of this effect is Shapley value [10] brought to measure attributes importance by [4]. It is determined, for attribute **i**, by

$$\phi_i = (\Sigma(p_K * (I_D(K \cup i) - I_D(K)))) / \mathbf{n},$$

where the sum is on **K**, for **K** varying along the subsets of $C \backslash \{i\}$, **n** is the total number of attributes in **C**, I_D is the index of quality of information and the weight p_K is given by the inverse of the number of combinations of **n-1** elements in subsets with the cardinality equal to the cardinality of the subset **K**.

Replacing the weights in the computation of ϕ_i by equal weights, we have Banzhaf values [1]. Shapley and Banzhaf values depend strongly on the value of the index of quality of the approximation by the attribute alone. On the contrary, the weight to be given to the gain of information due to each possible inclusion of the attribute in the model should decrease as increases the number of variables left out of the model when such inclusion occurs. In [9] new importance measures that take this aspect into account were tested. In the application below is employed, besides Shapley and Banzhaf values, a measure with increasing weights derived from those in Shapley value by dividing p_K by the factorial of the cardinality of the complement of K with respect to $\mathbf{C} \backslash \{i\}$. This results in weights inversely proportional to the number of permutations of **n** elements in subsets with the cardinality equal to that of $\mathbf{C} \backslash \mathbf{K}$, instead of the number of combinations.

Another important aspect to be taken into account in the computation of importance values is the form of evaluation of joint dominance. To be able to classify all pairs of options, it will be here assumed joint dominance of one option o_1 over another option o_2 according to a set of attributes if the number of attributes in this set for which o_1 is preferable to o_2 is larger than the number of attributes for which o_2 is preferable to o_1. If these numbers are equal, the options are considered indiscernible. By this way, every pair of different options enters the comparison.

This suits the present application, where the condition attributes are chosen by their ability to contradict each other. If agreement between a larger number of the attributes is required, then the possibility of determining dominance or indiscernibility will be very small. Other definitions of joint dominance may be adopted, labeling as incomparable certain pairs of options. These would not be counted neither in the numerator nor in the denominator of the indices of quality of approximation. This may lead to unsuitably small denominators in some computations.

5 The Sequential Classification Procedure

For the kind of application here envisaged, the modeler looks for condition attributes that together are able to help to explain the decision table but isolated do not provide considerable information about each other. For this reason they are evaluated by their importance value. The importance value of a given condition attribute measures the presence in it of an aspect important for the decision classification and not present in the other condition attributes. Then, taking them together will hardly allow for partitioning the options in a suitable number of classes. Opposing the partition so generated to the decision partition would increase even more the chance of contradictions.

On the other hand, if we consider each condition attribute isolated, it will agree with the decision attribute in the matter concerning that particular aspect that makes it important and the chance of such agreement will be higher as higher its importance value. And, if one such attribute is unable to separate two objects that are discernible in the decision partition, it may be expected that one of the other condition attributes, taken isolatedly, would be able to do it by considering another relevant aspect.

Taking this into account, the classification procedure here adopted is composed of the following three steps: 1) rank the condition attributes according to their importance values; 2) classify according to the most important condition attribute; inside each equivalence class of this first classification, classify according to the condition attribute with the second highest importance value; proceed likewise until all condition attributes with positive importance have been considered. 3) Compare the classification so generated to the decision classification and eliminate the contradictions by considering indiscernible those options that are indiscernible according to each of them or present opposed dominance relations.

This procedure results in larger classes than those in the decision partition. But the successive elimination of the attributes, associated with the entrance in operation of the attributes from the most important to the less important, results in a large chance of useful results. Besides, its sequential character makes it very easy and fast to apply. And the rules it engenders, contemplating the condition attributes in a fixed order, are easy to understand. It applies, also, by the same way in the context of symmetric instead of antisymmetric relations, being enough to substitute discernibility for dominance where it appears.

6   Data Analysis

In this section the techniques described in the previous sections are employed to develop a framework to classify clubs at the end of a championship. It is assumed that the clubs match each other at least once along the championship. The decision variable is the number of points earned. The main goal is to offer a precise classification rule that might substitute for the presently applied rule that select to descend the four last clubs on points. The data analyzed are provided by the results of Brazilian National Championships, from 2003 to 2006. Before 2003, the championship model included a playoffs season. Playoffs result in a very different motivation for scoring goals and winning games along the championship.

Besides classifying on points earned in wins and ties, the League employs the number of wins and the balance of goals as untying criteria. To determine how these variables more essentially affect the performance of the clubs, we may try to identify factors related to offensive and defensive skills. The offensive power is naturally measured by the number of goals scored.

The defensive ability is more difficult to access. To take a symmetric point of view, we may consider first, with a negative sign, the number of goals taken. But a favorable balance of offensive and defensive power may lead to large numbers of goals pro and against the team. The strength of the defense is forged to assure that the team will not be beaten when the offensive power is not enough, so that the ability of avoiding defeats is a more natural way to access it than the number of goals suffered. If there were proportionality between the vectors of wins and ties or if one of these vectors were constant, there would be precise correlation between number of games lost and number of points. However this is not the case.

To explain the final ranking the number of games lost or won leave almost the same space to be filled by the other variables in the model proposed. Because the League assigns 3 points for win and only one for tie, sets of condition attributes with the number of wins replacing the number of losses presents a higher quality of approximation. Nevertheless, this asymmetric punctuation may be exaggerated and, taking into account our goal of contradicting to some extent the decision classification if from such contradiction may result sensible enlarging of the partition classes, models with the number of losses and not the number of wins may be more attractive.

Direct confrontation appears as a complement for the above-considered attributes by bringing into consideration the aspects of the teams structure that affect each pair of clubs particularly. There are patterns of preparation of each team that affect differently their performance against certain others. These performance relations are not transmitted transitively, since the preparation of a given club to face two different clubs does not affect the way these clubs match each other. The importance values here found lead to the conclusion that this factor has an influence in the final points classification distinct from those captured by the other attributes studied.

Table 1 shows the importance measures for the different attributes in the model with losses, goals scored and direct confrontation as condition attributes. Similar results were found for models with other attributes. The values along each column show small variation through time. The application of increasing weights brings, in general, the smaller values and Shapley values are the highest, reflecting the higher or lower importance given by each form of calculation to the index of quality of information of the attribute isolated. It is also noticeable that in all models the increasing weights value for direct confrontation is sometimes higher than Banzhaf value, confirming that, even though less influent isolatedly, this attribute has a specific contribution.

The same ranking of attributes and rough classes determination procedure was applied to the 2nd division championship, that in 2006 started to employ an analogous schedule, with absence of playoffs and ascent and descent decision on points. The results were similar, with smaller importance values suggesting that the attributes considered may be less important for the 2nd division.

As was to be expected, the number of classes in the final classification increases as the attributes with higher importance values enter the model. During the 4 years, the number of classes varies between 1 and 4 for the model with losses replaced by goals taken, between 5 and 9 for the model with losses and between 10 and 15 for the model with wins instead of losses. The number of classes for the unique decision attribute varies between 17 to 20.

Table 1. Attributes Importance

year	value	losses	goals	dir.conf.
2003	increasing	0.29	0,18	0.06
	Banzhaf	0.30	0.19	0.06
	Shapley	0.38	0.27	0.14
2004	increasing	0.24	0.14	0.08
	Banzhaf	0.25	0.15	0.07
	Shapley	0.33	0.23	0.15
2005	increasing	0.28	0.22	0.03
	Banzhaf	0.29	0.23	0.03
	Shapley	0.37	0.31	0.11
2006	increasing	0.28	0.17	0.08
	Banzhaf	0.29	0.18	0.08
	Shapley	0.37	0.26	0.16

Table 2. Size of Descent Classes

year	goals taken	losses	wins
2003	0	1	6
2004	0	2	1
2005	1	3 or 5	4
2006	1	2	4

Finally, the results of the classification process are exemplified in Table 2. This table presents for each model the number of clubs in the lower class with size closer to 4, the predetermined number of clubs presently selected to descent. For the years of 2003, 2004 and 2006, the decision rule characterizing the descent for the model with losses, goals scored and direct confrontation as condition attributes, is: proportion of losses above .5.

For 2005 this rule identifies as the lowest a unitary class. The next class, with 3 clubs, would be formed by adding two clubs satisfying the rule: proportion of losses equal to .5 and less than 1.25 goals scored per game. The following class, with 5 clubs, would be formed adding those determined by the rule: .45 to .5 of losses and less than 1.25 goals scored per game. These are simple rules and the constancy of the bound of .5 of losses seems to make this model able to offer a perfectly feasible system. It may be interesting to notice that the 2005 season, which produced the different rule, was affected by a scandal involving referees manipulation of results that provoked invalidation and repetition of eleven matches.

7 Final Comments

A new index of quality of approximation was here employed. It has been used to rank the attributes in terms of its importance in the presence of competing attributes, difference importance evaluation formulae being applied. The ranks were then explored in a sequential strategy to combine preference criteria into global dominance relations.

This approach was successfully applied to review the classification of the clubs in the Brazilian Soccer Championship. A feature of the procedure employed is its ability to deal with intransitive attributes, in the case studied represented by the result of direct confrontation.

The models applied included three condition attributes. This number can be raised without any computational problem as the ranking of the attributes to enter the final classification procedure in a sequential manner makes this procedure very fast. Measurement of importance is also very simple, what stimulates the computation of different forms to access importance. It requires that the index of quality of approximation be computed for all subsets of the set of condition attributes, but this computation, based on the direct confrontation of binary tables, is straightforward.

Acknowledgments. This work was partially funded by a CNpq productivity grant. I am grateful to two anonymous referees for useful suggestions for improvement and future development.

References

1. Banzhaf, J. F.: Weighted Voting doesn't work: a Mathematical Analysis, Rutgers Law Review **19** (1965) 317–343
2. Clark, P. and Nibblett, T.: The CN2 Induction Algorithm, Machine Learning Journal **3** (1989) 261-283
3. Cohen, W.: Fast Effective Rule Induction. Proceedings of ICML95 (1995) 115-123
4. Greco, S., Matarazzo, B., and Slowinski, R.: Fuzzy measures as a technique for rough set analysis. Proceedings of the EUFIT'08 **1** (1998) 99–103
5. Greco, S., Matarazzo, B. and Slowinski, R.: Rough Approximation of a Preference Relation by Dominance Relations, EJOR (1999) **119** 63–83
6. Quinlan, J. R.: C4.5: Programs for Machine Learning, Morgan Kaufmann, San Francisco (1993)
7. Sai, Y, Yao, Y. Y. and Zhong, N.: Data analysis and mining in ordered information tables. Proceedings of the IEEE ICDM 2001 (2001) 497–504
8. Sant'Anna, A. P.: Probabilistic Indices of Quality of Approximation. In Rough Computing: Theories, Technologies and Applications, Idea Group Inc. (2007) in press.
9. Sant'Anna, A. P. and Sant'Anna, L. A. F. P.: A Probabilistic Approach to Evaluate the Exploitation of the Geographic Situation of Hydroelectric Plants. Proceedings of ORMMES'06 (2006)
10. Shapley, L.: A value for n-person games. In Kuhn, H., Tucker, A., Contributions to the Theory of Games II, Princeton University Press, Princeton (1953) 307–317

Rough Neuro Voting System for Data Mining: Application to Stock Price Prediction

Hiromitsu Watanabe[1], Basabi Chakraborty[2], and Goutam Chakraborty[2]

[1] Graduate School of Software and Information Science
Iwate Prefectural University, Japan
g236d012@edu.soft.iwate-pu.ac.jp
[2] Faculty of Software and Information Science
Iwate Prefectural University, Japan
basabi@soft.iwate-pu.ac.jp, goutam@soft.iwate-pu.ac.jp

Abstract. This work proposes a rough neuro voting system with modified definitions of rough set approximations for knowledge discovery from complex high dimensional data. Proposed modification of rough set concepts has been used for attribute subset selection. Ensemble of neural networks are used for analysing subspaces of data in parallel and a voting system is used for final decision. The rough neuro voting system is used for stock price prediction with considering other influencing factors in addition to day-to-day stock data. The proposed approach shows effective in predicting increment or decrement of the nextday's stock price from simulation experiment.

Keywords: Data Mining, Stock Price Prediction, Rough Neuro Voting System, Neural Network Ensemble.

1 Introduction

Classical data anylysis techniques based mainly on statistics and mathematics are no longer adequate for analyzing increasingly huge collection of data in variety of domains. New intelligent data analysis methodologies are evolving for discovery of knowledge from complex data bases. Though many successful applications of the above tools are reported in the literature [1], advanced hybrid techniques are necessary for better understanding of the inherent knowledge in the form of simple rules for high dimensional complex data. Soft computing methodologies are widely used for solving real life problems as they provide a flexible information processing capability for handling ambiguity, uncertainty and incompleteness prevalent in real life data. Neural networks and rough set theory are useful tools for classification and rule generation from raw data.

The basic task behind discovery of knowledge from raw data is to divide the data set in the attribute space into different classes, and define the class boundaries by simple yet accurate rules. Most of the real world data with large number of attributes comes with noises some of which are irrelevant to the decision under consideration. In general, one major step is to find which attributes are important. Depending on the method used for rule extraction the real valued data are

J.T. Yao et al. (Eds.): RSKT 2007, LNAI 4481, pp. 558–565, 2007.
© Springer-Verlag Berlin Heidelberg 2007

to be put into a number of discrete levels. So for continuous data, the next step is data discretization. The last step is to find simple rules relating a combination of attribute values to a proper decision. The rules should be simple, small in number and yet accurate enough to express most of the data available.

Rough set theory, proposed by Pawlak in 1981 [2], has shown promise in handling vast amount of noisy data for extracting patterns in them. It can be used in different steps of knowledge discovery, from attribute selection [3], data discretization [4] to decision rule generation [5]. More recently, a comprehensive rough set based knowledge discovery process is proposed in [6]. On the other hand, all phases of knowledge discovery can also be accomplished by evolutionary algorithms [7]. Rough and fuzzy set theory or artificial neural network (MLP), genetic algorithm and their several integration as hybrid approaches have been used for knowledge discovery and rule generation [8] [9] [10]. In this work a rough neuro voting system has been proposed for knowledge extraction from high dimensional complex data. The rough set is used for the selection of subsets of attributes in the first step, an ensemble of neural network is used to extract information from the selected subspaces of multidimensional data and a voting system is used in the final step for decision. The proposed algorithm has been used to analyse and predict stock price fluctuation.

2 Knowledge Extraction from Complex High Dimensional Data

Analysis of real life complex multidimensional data such as stock market (financial) data, clinical or social data for knowledge extraction and prediction of future behaviour is a difficult problem. Artificial neural networks are widely used for such prediction of time series data. For financial time series with high dimension, training a neural network with the whole set of multidimensional data is difficult and time consuming, consequently the trained network cannot produce accurate prediction. It is also true for other ill behaved, highly complex multidimensional time series data. Most of the high dimensional complex data contains important information clustered in a subspace of multidimensional space. Efficient selection of the subspaces is extremely necessary for extracting important information for analysis of the data in a reasonable time and computational complexity.

Now to deal with the problem, a rough neuro voting system has been proposed in this work which contains three steps. In the first step an optimum number of subspaces of multidimensional data for which the data is well behaved and contains information for prediction is selected. To find the boundaries of these subspaces, the concept of rough set theory is used. Rough set approach is one of the efficient approaches for attribute reduction and rule generation from vast noisy data. Unlike statistical correlation-reducing approaches, it relies only on set operations, and requires no intervention. But it does not have the necessary searching technique to find the best rules out of large number of possibilities. This searching is especially important when the number of condition attributes are still many, and each of them have a large number of discrete levels. Attribute subset selection is done here from the dependency and discriminant index defined in rough set theory and a proposed modification of the concept for better searching. In fact this step performs the attribute subset selection.

In the next step each subspace of data is used to train an artificial neural network for knowledge extraction or prediction. An ensemble of neural network is used in this step and in the final step the final decision is calculated from the outputs of the ensemble of neural networks by voting.

3   Proposed Rough Neuro Voting System

3.1   Rough Set Preliminaries

According to *rough set* theory an *information system* is a four-tuple $S = (U, Q, V, f)$ where U, Q and V, non-empty finite sets, represents the universe of objects, the set of attributes and the set of possible attribute values respectively. f is the information function which, given an object and an attribute, maps it to a value, i,e., $f : U \times Q \to V$. The set of attributes Q can be partitioned by two disjoint subsets, the *condition* attributes C and the *decision* attributes D.

An *indiscernibility relation* is an equivalence relation with respect to a set of attributes which partitions the universe of objects into a number of classes in such a manner that the member of same classes are indiscernible while the member of different classes are distinguishable (discernible) with respect to the particular set of attributes. P being a subset of Q, *P-indiscernibility relation*, denoted by $IND(P)$, is defined as,

$IND(P) = \{(x, y) \in U \times U : f(x, a) = f(y, a)$, for every feature $a \in P\}$

$U/IND(P)$ denotes the set of equivalence classes of U induced by $IND(P)$, also denoted as P^*. The various terms defined above is explained below with the example of an information system described in Table 1.

If $P = \{ht\}$, $P^* = \{\{B, D, G\}, \{A, F, I\}, \{C, H\}, \{E\}\}$. Similarly, when $P = \{ht, hs\}$, $P^* = \{\{B\}, \{D, G\}, \{F, I\}, \{A\}, \{C\}, \{H\}, \{E\}\}$. Here the objects are divided into two classes - good swimmer and bad swimmer. We will denote the partitions, induced by condition attributes, as $A^* = \{X_1, X_2, \ldots\}$, and the partitions induced by decision or concept attributes as $B^* = \{Y_1, Y_2, \ldots\}$.

For any decision partition Y induced by a subset of decision attributes B, and for any subset of condition attributes A, the *A-lower* ($\underline{A}$) and the *A-upper* ($\overline{A}$) approximation of Y are defined as $\underline{A}(Y) = \bigcup\{X \in A^* : X \subseteq Y\}$ and $\overline{A}(Y) = \bigcup\{X \in A^* : X \cap Y \neq \phi\}$ The boundary region for the decision Y with respect to the subset of attributes A is defined as $BND_A(Y) = \overline{A}(Y) - \underline{A}(Y)$. $POS_A(Y) = \underline{A}(Y)$ and $NEG_A(Y) = U - \overline{A}(Y)$ are known as *A-positive* region of Y and *A-negative* region of Y. Using the above example, when the objects are partitioned by $A = \{ht\}$, and the concept is good swimmer, then $Y = \{B, C, D, F, G, I\}$, $\underline{A}(Y) = \{\{B, D, G\}\}$, $\overline{A}(Y) = \{\{B, D, G\}, \{C, H\}, \{A, F, I\}\}$, $POS_A(Y) = \{\{B, D, G\}\}$, $BND_A(Y) = \{\{C, H\}, \{A, F, I\}\}$, and $NEG_A(Y) = \{\{E\}\}$.

Table 1. An Example of an Information System

Universe of	Attributes (Q)			
Objects (U)	Condition Attributes (C)			Decision Attribute (D)
Name	Height (ht)	Hand Span (hs)	Pulse Rate (pr)	Swimming Ability (sa)
Andy (A)	180 (*tall*)	160 (*medium*)	84 (*high*)	not good
Bill (B)	192 (*very tall*)	190 (*very long*)	64 (*low*)	good
Conrad (C)	170 (*medium*)	189 (*very long*)	66 (*low*)	good
David (D)	193 (*very tall*)	180 (*long*)	83 (*high*)	good
Elliot (E)	162 (*short*)	159 (*medium*)	65 (*low*)	not good
Fahad (F)	181 (*tall*)	191 (*very long*)	82 (*high*)	good
Glenn (G)	194 (*very tall*)	179 (*long*)	86 (*high*)	good
Harvey (H)	172 (*medium*)	161 (*medium*)	68 (*low*)	not good
Inoue (I)	179 (*tall*)	192 (*very long*)	81 (*high*)	good

Inexactness of a rough set is due to the existence of boundary region. The greater the boundary region, the lower is the accuracy of any condition attribute to infer a decision. In real life data, we usually end up only with boundary sets. To deal with inexactness, two more terms, *dependency* ($\gamma_A(B)$) and *discriminant index* ($\beta_A(B)$) have been defined in rough set literature. $\gamma_A(B)$ means dependency of decision B on condition A. $\beta_A(B)$ means to what extent the partition induced by attribute A matches the concept partition induced by B, both positive and negative portions.

$\gamma_A(B)$ is mathematically defined as

$$\gamma_A(B) = \frac{|POS_A(B^*)|}{|U|} \tag{1}$$

$\beta_A(B)$ is defined as

$$\beta_A(B) = \frac{|POS_A(B^*) \cup NEG_A(B^*)|}{|U|} = \frac{|U - BND_A(B^*)|}{|U|} \tag{2}$$

For the simple problem described in Table. 1, we have only one decision attribute, $D = \{sa\}$. When $A = \{ht\}$, $POS_{ht}(sa) = \{\{B, D, G\}\}$, $BND_{ht}(sa) = \{\{C, H\}, \{A, F, I\}\}$, and $NEG_{ht}(sa) = \{\{E\}\}$. Then $\gamma_{ht}(sa) = 0.333$ and $\beta_{ht}(sa) = 0.444$. As the boundary region is there, the rule between height and swimming ability can only be partially defined for the positive and negative regions. We can partition the objects more precisely by adding one more attribute, say *hand span*. Then $A = \{\{ht\}, \{hs\}\}$, and the boundary region vanishes improving $\beta_{\{ht,hs\}}(sa)$ to its maximum value 1.0. It can be easily seen that the discriminant index does not improve if we use $A=\{ht, pr\}$. It is also interesting to note that $\beta_{\{hs\}}(sa)$ is 1.0, or in other words the hand span attribute alone is sufficient to make the decision of good or bad swimmer. Corresponding to the three partitions of the objects by *hand span* attribute, i.e., $\{B, C, F, I\}$, $\{D, G\}$, and $\{A, E, H\}$, there are three rules which cover the whole data set. With real world data, a β value of 1.0 is, in general, not possible.

3.2   Proposed Measure for Attribute Selection

With real life data for a particular dicision we do not get a complete positive or negative region. To extract some meaningful information even out of this situation, we propose terms $n_D(A \backslash B)$ and $n_N(A \backslash B)$, for a partition induced by the set of condition attributes A and a decision attribute B where $B^* = \{Y_1, Y_2\}$, as follows:

$$n_D(A \backslash B) = \sum_{X_i \in A^*} |X_i| : \left\{ \left(\frac{|X_i \cap Y_1|}{|X_i \cap Y_2|} \vee \frac{|X_i \cap Y_2|}{|X_i \cap Y_1|} \right) < \epsilon \right\} \tag{3}$$

$$n_N(A \backslash B) = N - n_D(A \backslash B) \tag{4}$$

ϵ is a fraction whose value is set to nearly 0 (say, $\epsilon = 0.2$). $|X_i|$ denotes the cardinality of set $|X_i|$ and N denotes the total number ofsample points. Though the samples do not form a clear decision, they do describe some rules with high degree of accuracy, n_D means the number of samples that helps in decision making and n_N represent the samples that are not consistent with decision making. If $n_D(A \backslash B)$ is a good proportion of N, we can say that the partitions induced by the set of condition attributes in A are good features to take the decision B. When it is not, we can conclude that the condition attributes in A has nothing to do with the decision B. Those attribute combination are irrelevant. Following conventional meaning of the term, we define *confidence* for the partition induced by condition attribute subset A for decision B as,

$$confidence = \beta_A(B) = \frac{n_D(A \backslash B)}{n_D(A \backslash B) + n_N(A \backslash B)}$$

3.3   Attribute Subset Selection

For attribule subset selection, we calculate the value of β for different combinations of condition attributes and pick up those subsets of attributes with high values of β. For high dimensional data, we need to consider a certain number of attributes grouped together to restrict the search problem, we call this grouping of attributes, as merging. If many attributes are merged, the partition becomes smaller and the *confidence* is low. On the other hand, if we partition the samples using only single attributes the partition do not have any discriminating power. So for a particular problem we need to consider the minimum number of attributes to be merged to have a meaningful partition. This number is dependent on the characteristic of the data and is heuristically chosen by some experimentation.

3.4   Rule Extraction from Artificial Neural Network

Once these subsets of attributes are identified, the relation between the input attributes and the output target is represented by feedforward multilayer artificial neural networks, one network for each such subset representing a subregion of the data. Neural network is chosen because it has the ability to learn complex boundaries. The neural networks are trained using error back propagation with data from their respective partitions. A particular neural network is like an expert for knowledge extraction for prediction/ classification for data falling in that partition. First we check if the data falls within such discriminating partitions or not. If not, we conclude that discriminatory knowledge extraction is not possible from that partition.

3.5   Voting System from Ensemble of Neural Network

Finally the aggregate of the outputs from the ensemble of neural networks are taken as the final decision. The aggregation has to be done by averaging or voting.

4   Simulation Experiments and Results

The proposed method has been used to predict fluctuation in share price.

4.1   Data Preparation

We chose consecutive 1000 days' stock data of the automobile company TOYOTA, Japan. As day-to-day stock data cannot be predicted only from the stock-price itself we consider several factors which give rise to a multidimensional data set. The factors considered are:

1. The recent change of value of the company
2. The Nikkei stock average
3. The New York Dow Jones average
4. The exchange rate between Japanese yen and US $
5. The stock value of companies with same or similar product line
6. The Nikkei average in terms of US$
7. The deviation of long (or medium) term average from the present stock value

Altogether the data consists of 11 attributes, which are as follows:

- The variation of the stock value, today's increase/decrease with respect to previous day, i.e., our target. $= C_0$
- The variation of the stock value, previous day's increase/decrease of value of the our target with respect to day before yesterday $= C_1$
- The variation of stock value of Nissan on the previous day $= C_2$
- The variation of US$ and Japanese Yen exchange rate on the previous day $= C_3$
- The variation of Dow Jones stock average on the previous day $= C_4$
- The variation of the Nikkei stock average on the previous day $= C_5$
- The variation of the nikkei average in terms of US$ value $= C_6$
- The deviation of long term average in positive direction $= C_7$

- The deviation of long term average in negative direction $= C_8$
- The deviation of short term average in positive direction $= C_9$
- The deviation of short term average in negative direction $= C_{10}$

The task is to predict whether the stock price will be up or down from today to tomorrow by using the past values of the multidimensional time series.

4.2 Simulation Experiment

We compared the performance of prediction using only neural network trained by error-backpropagation and our proposed rough neuro voting system.

ANN parameters were as follows:

- Number of input nodes: 11
- Number of output node: 1
- Number of hidden nodes: 15
- Training rate: 0.01
- Number of train data: 998
- Number of test data: 200
- Number of training epochs: 10000

We used all the available data to train the ANN. Out of them 200 randomly selected data were used for testing. Three sets of test data were used, and the experiment has been repeated for 50 times. We used the same data set for prediction by our proposed system. For attributes C_1 to C_6, we divide the data into 4 discrete levels, large positive, positive, negative, and large negative. C_7 to C_{10} are already discretized to values 1 and 0. We merged 3 attributes and considered $_{11}C_3 = 165$ subsets of attributes for subset selection. The parameters are as follows:

- Number of data available: 998
- Number of test data: 200
- Number of condition attributes: 11
- Number of decision attribute: 1
- Number of attributes marged to create partition: 3
- Number of data partitions: 5
- Number of input nodes of ANN: 3
- Number of output node of ANN; 1
- Number of hidden nodes: 5
- Training rate :0.01
- Number of training epochs: 100000

From the results of this experiment it is found that the percentage of correct classification improved for the rough neuro voting (RNV) system (78% for all data partitions) than ANN (Average 67% with a maximum of 64% and minimum 52.5%) and the result is much more stabilized. For any financial system, as we need to predict only for the next day, where as all the previous data are available, we can rely with a very high degree on RNV system output.

5 Conclusion

In this work we proposed a rough neuro voting system for extraction of knowledge for prediction or discrimination of high dimensional complex data. We applied our proposed scheme for stock market prediction. Forecasting stock price is a complex problem and does not depends on the past stock values only. There are other factors which demand analysis of multidimensional data with different chracteristics of the attributes. Though artificial neural networks are powerful for extracting information from raw data, large complex high dimensional data poses serius problems for computational time and complexity.

In our work we used concepts from rough set theory and proposed some modifications to handle real world data for finding proper subset of attributes. This facilitates to divide the data into smaller subspaces for analysing individually and in parallel with lesser computation time and complexity. Finally from the ensemble of neural networks the final decision can be achieved by aggragation or voting of the outputs from the individual network. It is shown that the prediction result with our proposed scheme is better than the result from neural network alone. Though there are various points to be explored further for more correct prediction, the simulation results show that the proposed rough neuro voting system is a practical approach to tackle problems with complex high dimensional data like stock price analysis.

References

1. Mitra, S.,Pal, S.K.and Mitra, P.: Data Mining in soft Computing Framework: A survey,*IEEE Trans. on NN* **13** (1) (2002) 3-14.
2. Pawlak, Z. (1991): Rough sets, Theoretical Aspects of Reasoning About Data, Kluwer Academic Publishers (1991).
3. Modrejewski M.: Feature Selection using rough sets theory in it Proc. of the European Conference on Machine Learning (1993) 213–226.
4. Grzymala-Busse, J. W., Stefanowski, J. (2001): Three Discretization Methods for Rule Induction, *International Journal of Intelligent Systems* **16** (1) (2001) 29–38.
5. Komorowski, J., Pawlak, Z., Polkowski, L.and Skowron, A.: Rough Sets: A Tutorial. In: S. K. Pal and A. Skowron(Eds.) Rough Fuzzy Hybridization: A New Trend in Decision Making Springer-Verlag (1999) 3-98.
6. Zhong, N. and Skowron, A.(2001): A rough set-based knowledge discovery process,*Int. J. Appl. Math. Comput. Sci.* **11** (3) (2001) 603–619.
7. Freitas, A. A. : Data Mining and Knowledge Discovery with Evolutionary Algorithms, Springer (2002).
8. Banerjee, M.et al.: Rough fuzzy MLP: Knowledge encoding and Classification, i*IEEE Trans. on NN* **9** (6) (1998) 1203–1215.
9. Chakraborty, B.: Feature Subset selection by Neuro-Rough Hybridization, *Rough Sets and Current Trends in Computing (Lecture Notes in Computer Science)* **2005** (2001) 519–526.
10. Chakraborty, G., and Chakraborty, B.: Hybrid Rough-Genetic Algorithm for Knowledge Discovery from Large data, *Proc. of IEEE International Workshop on Soft Computing as Transdisciplinary Science and Technology (WSTST05)* (2005) 904–913.

Counting All Common Subsequences to Order Alternatives*

Hui Wang[1], Zhiwei Lin[1], and Günther Gediga[2]

[1] School of Computing and Mathematics
University of Ulster at Jordanstown,
BT37 0QB, Northern Ireland, UK
{h.wang,z.lin}@ulster.ac.uk
[2] Institut für Evaluation und Marktanalysen
Brinkstr. 19, 49143 Jeggen, Germany
gediga@eval-institut.de

Abstract. Many real world tasks involve the need to order alternatives based on specifications of preference over subsets of the alternatives, the problem of *preference based alternative ordering*. Examples include dancing championship adjudication, Eurovision Song Contest decision-making, collaborative filtering and meta-search engines. One usual solution to this problem consists in allocating *scores* to the alternatives and aggregating the scores to generate ranks for all alternatives. Examples of this solution include competition adjudication. Another solution involves generating *ranks* from different sources for all the alternatives and then adding the rank values of each alternative to give the Borda scores for this alternative. The Borda scores are then used to order the alternatives. Examples include elections and meta-search engines. The problem with these two approaches is, the scores or ranks are sometimes hard to determine (e.g., collaborative filtering).

In this paper we take the view that *relative preferences over alternatives* (e.g., one alternative is preferred to another) are easier to obtain than absolute scores or ranks. We consider an alternative approach to this problem where, instead of using absolute scores or ranks, we use relative preferences over subsets of the alternatives to generate a total ordering that is maximally agreeable with the given preferences.

We consider a set of preference specifications over all or part of the alternatives. For every pair of alternatives we calculate the probability that the two alternatives should be placed in an order. Then the ORDER-BY-PREFERENCE algorithm is used to construct a total ordering for all the alternatives, which is guaranteed to be approximately optimal.

Keywords: Alternative ordering, weak order, preference, neighbourhood counting.

1 Introduction

In many real world tasks we need to rank order things, *alternatives*, by a group of people, *voters*. The rank ordering by an individual voter is a *preference* over all

---

* The idea presented in this paper was discussed at EU COST Action 274 (TARSKI) Workshop in Belfast, 2005.

J.T. Yao et al. (Eds.): RSKT 2007, LNAI 4481, pp. 566–573, 2007.

or part of the alternatives and the aim is to find a *total ordering*[1] over all alternatives that maximally agrees with all individual preferences, where every pair of alternatives is ordered with respect to their ranks. We call this the problem of *preference based alternative ordering*. Some examples include:

- adjudicating competitions: where a group of adjudicators each produces a rank ordering of the competitors, the preference of one adjudicator, and the overall rank ordering, a total ordering, is decided based on the preferences of all adjudicators.
- Eurovision Song Contest: the judging committee of each participating country rank order only 10 other countries, leaving the remaining countries equally at the bottom. Here the preferences are over a subset of the alternatives, and the aim is to generate a total ordering of all participating countries that maximally agrees with all the preferences.
- Cooperative filtering or recommendation systems: where preferences over a selection of products in a category are given by the recommenders, and the aim is once again to rank order the products based on preferences.
- Meta-search engine (whose goal is to combine the rankings of several WWW search engines): where, given a search query, N search engines each produces a ranking of documents resulting in N rankings altogether. The goal is to generate a single ranking that agrees best with all the N rankings. This application is similar to the Eurovision application.

One solution to this problem allocates absolute scores to the alternatives. One example is the adjudication of competitions. Another solution involves generating *rank* values from different sources for all the alternatives and then adding the rank values of each alternative to give the Borda scores [2,3] for this alternative. The Borda scores are then used to order the alternatives. Examples include elections and meta-search engines. The problem with these two solutions, where absolute numbers (scores or ranks) are used, is the numbers are usually hard to determine and are sometimes given without absolute certainty.

In this paper we take the view that relative preferences over alternatives are easier to obtain than absolute numbers. So we consider an alternative approach to this problem where, instead of using absolute numbers, we use relative and qualitative preferences over subsets of the alternatives to generate a total ordering that is approximately maximally agreeable with the given preferences. In our approach, we consider a set of preference specifications over all or part of the alternatives. For a pair of alternatives we calculate the probability that the two alternatives should be placed in this order. Then the ORDER-BY-PREFERENCE algorithm [4] is used to construct a total ordering for all the alternatives, which is guaranteed to be approximately optimal.

---

[1] We use the term "total order" here only to emphasise that the alternatives are pairwise comparable by a voting system, although in reality some alternatives may not be comparable. In this sense it may be more appropriate to understand the term as "weak order" [1].

2   Learning to Order Things

One very related work is the paper [4]. This paper presents an online (or incremental) algorithm to learn a preference function from voters, and a greedy ordering algorithm to generate an approximately optimal total order (i.e., approximately maximally agrees with a preference function).

The preference function used in [4] is a linear combination of a set of N primitive preference functions:

$$PREF(u,v) = \sum_{i=1}^{N} w_i R_i(u,v) \tag{1}$$

The primitive preference functions R_i are provided by *ranking voters*. Online learning framework was adopted in which the weight w_i assigned to each ranking voter R_i is updated incrementally.

If $PREF(u,v) > PREF(v,u)$, then u is preferred to v. In this paper we will consider another preference function – the probability that u is preferred to v. The rest of the paper will discuss how to formulate and calculate this probability.

A special type of preference function is *rank ordering*. Let S be a totally ordered set with '>' as the comparison operator. An ordering function into S is a function $f : X \rightarrow S$. The function f induces the preference function R_f, defined as

$$R_f(u,v) \stackrel{\text{def}}{=} \begin{cases} 1 & \text{if } f(u) > f(v) \\ 0 & \text{if } f(u) < f(v) \\ \frac{1}{2} & \text{otherwise} \end{cases} \tag{2}$$

R_f is called a *rank ordering* for X into S. If $R_f(u,v) = 1$, we say that u is preferred to v, or u is ranked higher than v.

The greedy ordering algorithm used in [4] is cited in Algorithm 2.1.

Algorithm 2.1. ORDER-BY-PREFERENCE: generate an approximately optimal total order from pair-wise preferences.

Input. an instance set X; a preference function PREF
Output. an approximately optimal total ordering function ρ
1 Let $V = X$
2 **for** *each* $v \in V$ **do** $\phi(v) = \sum_{u \in V} PREF(v,u) - \sum_{u \in V} PREF(u,v)$
3 **while** V *is not empty* **do**
4 | Let $t = \arg \max_{u \in V} \phi(u)$
5 | Let $\rho(t) = |V|$
6 | Let $V = V - \{t\}$
7 | **for** *each* $v \in V$ **do** $\phi(v) = \phi(v) + PREF(t,v) - PREF(v,t)$
8 **end**

---

3   A Probability Function

A central element in our approach is calculating the probability that one alternative is preferred to another. In this section we present a probability framework for this purpose.

Let V be a set or *data space*, and $\mathcal{F}$ be a σ-field [2] on V or *extended data space*.

Consider a probability function P over V, which is a mapping from $\mathcal{F}$ to $[0, 1]$ satisfying the three axioms of probability [5]. For $X \in \mathcal{F}$ let $f(X)$ be a non-negative measure of X satisfying $f(X_1 \cup X_2) = f(X_1) + f(X_2)$ if $X_1 \cap X_2 = \emptyset$. As an example, we can take $f(X)$ for the cardinality of X.

A probability function is defined in [6] as follows: $G : \mathcal{F} \to [0, 1]$ such that, for $X \in \mathcal{F}$,

$$G(X) = \sum_{E \in \mathcal{F}} P(E)f(X \cap E)/K \qquad (3)$$

where $K = \sum_{E \in \mathcal{F}} P(E)f(E)$.

Each $E \in \mathcal{F}$ is called a *neighbourhood* and, if E overlaps with X (i.e. $f(X \cap E) \neq 0$), then E is called a *neighbourhood of X*.

Let D be a sample of data drawn from V according to probability distribution P. It was shown in [6] that, if the *principle of indifference* is assumed and if $f(X) = |X|$ for $X \in \mathcal{F}$, G can be estimated from D as follows:

$$\hat{G}(t) = \frac{1}{nK} \sum_{x \in D} ncm(t, x) \qquad (4)$$

where $ncm(t, x)$ is the number of such $E \in \mathcal{F}$ that covers both t and x, or, the number of common neighbourhoods of t and x. Therefore the task of estimating G probability is transformed into one of *counting common neighbourhoods*.

To compute $ncm(t, x)$, we need to specify $\mathcal{F}$ in a way so that necessary information (e.g., preference ordering information) can be taken into consideration in the definition of neighbourhoods. For multivariate data, where the data space V is defined by multiple attributes, a formula has been discovered to compute $ncm(t, x)$ [7,6] where both t and x are data tuples. For sequence data a dynamic program has been developed to compute $ncm(t, x)$ where t and x are sequences [8].

Our work in this paper is based on the premise that a preference specification is treated as a sequence. Therefore we review the body of work on neighbourhood counting for sequence – *all common subsequences*. This review is based mainly on [8].

---

[2] A σ-field $\mathcal{F}$ on V is a class of subsets of V, such that:

1. $V \in \mathcal{F}$;
2. If $A \in \mathcal{F}$ then $A' \in \mathcal{F}$, where A' is the complement of A;
3. If $A, B, \cdots \in \mathcal{F}$, then $A \cup B \cup \cdots \in \mathcal{F}$.

One σ-field on V is the power set of V if V is finite.

3.1   All Common Subsequences – A Brief Review

We let $\mathcal{A}$ be a finite set of symbols – an alphabet. A sequence α is an ordered set $\{a_i\}_1^n = \{a_1, a_2, \cdots, a_n\}$, where $a_i \in \mathcal{A}$. For simplicity we also write the above sequence as $a_1 a_2 \cdots a_n$. A special sequence, *empty sequence*, is identified and denoted by ϵ, which is an empty set. Therefore a data space V can be defined as the set of all possible sequences generated by alphabet $\mathcal{A}$.

Consider two sequences α and β. If α can be obtained by deleting zero or more symbols from β, we say α is a *subsequence* of β. A *neighbourhood of a sequence* is one of its subsequences. Readers are invited to consult [8] for the justification of such a definition of neighbourhood.

If we want to use the G probability we need to compute the number of all common subsequences of two sequences. A brute force approach is exponential in the length of the sequence. A dynamic program is proposed in [8] through which the number of all common subsequences can be computed in polynomial time. This dynamic program is based on the following theorem.

Theorem 1. *Consider two sequences,* $\alpha = \{a_1, \cdots, a_m\}$ *and* $\beta = \{b_1, \cdots, b_n\}$. *Let* $N(i, j)$ *be the number of common subsequences of* $\{a_1, \cdots, a_i\}$ *and* $\{b_1, \cdots, b_j\}$, *i.e., the prefixes of sequences* α *and* β *of lengths* i *and* j. *Then*

$$N(i, j) = N(i - 1, j - 1) \times 2, \qquad\qquad \text{if } a_i = b_j$$
$$N(i, j) = N(i - 1, j) + N(i, j - 1) - N(i - 1, j - 1), \qquad \text{if } a_i \neq b_j$$

Consequently $ncm(\alpha, \beta) = N(m, n)$.

4   The Probability That One Alternative Is Before Another

Let $\mathcal{A}$ be a finite set of alternatives, i.e., an alphabet; V be the set of all possible preference orderings over all subsets of the alternatives, i.e., sequences without repetition that are constructed of $\mathcal{A}$; D be a subset of such preferences. Our aim in this section is to estimate the probability that one alternative is before another based on D. Such probability can be used as a preference function, as discussed in Section 2. With a preference function available we can generate a total ordering on the alternatives using the ORDER-BY-PREFERENCE algorithm.

Let $a, b \in \mathcal{A}$ be two alternatives. We construct a two-word sequence $\alpha = ab$. Our task can then be formulated as calculating the G probability of the two-word sequence α on the basis of a sample D of sequences.

According to Eq.(4) we have

$$\hat{G}(\alpha) = \frac{1}{nK} \sum_{x \in D} ncm(\alpha, x) \tag{5}$$

where K is a normalisation factor independent of α, n is the number of sequences in D, and $ncm(\alpha, x)$ is the number of common subsequences of α and x.

5   Examples

In this section we present two examples. A toy example illustrates our approach and a realistic example illustrates the capability and limitation of our approach.

5.1   A Toy Example

Suppose there are three voters: v_1, v_2, v_3, and there are five alternatives a, b, c, d, e. The preferences by individual votes are $v_1 : abcde$, $v_2 : ab, aced$, $v_3 : bc, aed$. Assuming equal weighting [3] of the voters we have a preference set $D = \{abcde, ab, aced, bc, aed\}$.

Consider a query sequence $q = ac$. To calculate $\hat{G}(q)$ we need to calculate $ncm(q, x)$ for every $x \in D$, which is the number of all common subsequences of q and x.

Consider $x = abcde$. The set of all subsequences of x is $\{\emptyset, a, b, c, d, e, ab, ac, ad, ae, bc, bd, be, cd, ce, de, abc, abd, abe, acd, ace, ade, bcd, bce, bde, cde, abcd, abce, abde,$ $acde, bcde, abcde\}$. The set of all subsequences of ac is $\{\emptyset, a, c, ac\}$. Therefore $ncm(ac, abcde) = 4$. Similarly we have $ncm(ac, ab) = 2$, $ncm(ac, aced) = 4$, $ncm(ac, bc) = 2$, and $ncm(ac, aed) = 2$. Therefore $\hat{G}(ac) = \frac{1}{NK}(4+2+4+2+2) = \frac{14}{NK}$.

Similarly $\hat{G}(ab) = \frac{14}{NK}$, $\hat{G}(ad) = \frac{15}{NK}$, $\hat{G}(ae) = \frac{15}{NK}$, $\hat{G}(bc) = \frac{13}{NK}$, $\hat{G}(bd) = \frac{12}{NK}$, $\hat{G}(be) = \frac{12}{NK}$, $\hat{G}(cd) = \frac{13}{NK}$, $\hat{G}(ce) = \frac{13}{NK}$, and $\hat{G}(de) = \frac{12}{NK}$. Applying the ORDERBYPREFERENCE algorithm we get the approximately optimal total order [4]: a, b, c, e, d, which agrees maximally with the individual preferences by all three voters. A decision can then be made on the basis of this total order.

Note that the preference set $D = \{abcde, ab, aced, bc, aed\}$ is the basis of coming up with the preferred ordering, by using the ORDER-BY-PREFERENCE algorithm. To use this algorithm we need pair-wise preferences, which are calculated from the preference set by our formalism presented in the paper.

5.2   Eurovision Song Contest

Table 1 is the score board of Eurovision Song Contest Final in 2005. 39 countries took part and 24 went into the final as contestants. During the final the 24 contestants performed, and then the 39 countries cast their votes. A vote by a country is a distribution of scores $\{1, 2, \cdots, 8, 10, 12\}$ to 10 contestants. The final ranking of contestants is based on the aggregation of scores awarded by the voting countries to the contestants. This is an application of a positional voting method similar to the Borda score.

The "total" row is the aggregate scores gained by the contestants. The contestants are ordered from left to right in a descending order of the score. It is clear the first three places were won by Greece, Malta and Romania.

---

[3] In the case of un-equal weighting, frequency can be introduced so that higher weighted voters are given higher frequency values.

[4] This is a property of the ORDERBYPREFERENCE algorithm.

Table 1: Scoreboard of Eurovision Song Contest 2005 – final

	GR	MA	RO	IS	LA	MD	SM	SZ	DE	NO	CR	HU	TU	BH	RU	AL	FY	CY	SW	UA	SP	UK	FR	GE
TOTAL	230	192	158	154	153	148	137	128	125	125	115	97	92	79	57	53	52	46	30	30	28	18	11	4
AL	12	4	5	3	0	0	6	0	0	0	2	0	8	0	0		10	7	0	0	0	0	1	0
AN	4	0	7	8	10	0	0	1	3	2	0	6	0	0	0	0	0	0	0	0	12	0	5	0
AU	4	5	6	1	0	2	12	0	0	8	0	7	10	0	3	0	0	0	0	0	0	0	0	0
BE	0	5	1	8	6	7	3	10	0	4	0	2	0	0	12	0	0	0	0	0	0	0	0	0
BG	12	8	7	6	5	1	0	0	4	3	0	2	10	0	0	0	0	0	0	0	0	0	0	0
BH	6	0	2	0	0	4	10	0	0	3	12	1	8		0	5	7	0	0	0	0	0	0	0
BU	12	0	8	0	0	6	4	0	0	1	2	5	3	0	0	0	7	10	0	0	0	0	0	0
CR	5	4	0	0	7	1	12	3	0	0		6	0	10	0	2	8	0	0	0	0	0	0	0
CY	12	6	8	0	1	2	10	4	0	3	0	7	0	0	0	0	0		0	0	0	5	0	0
DE	2	10	3	5	6	0	0	1		12	0	0	8	4	0	0	0	7	0	0	0	0	0	0
ES	0	4	0	1	10	6	0	12	5	8	2	3	0	0	7	0	0	0	0	0	0	0	0	0
FY	7	0	2	0	0	5	10	0	0	8	1	4	3	0	12		0	6	0	0	0	0	0	0
FI	3	8	0	5	4	0	0	10	2	12	1	0	0	0	7	0	0	0	6	0	0	0	0	0
FR	8	7	5	10	0	2	6	0	0	0	0	3	12	0	0	1	0	0	0	0	0	4		0
GE	12	8	0	5	7	1	3	4	6	0	2	0	10	0	0	0	0	0	0	0	0	0	0	
GR		8	5	0	1	7	6	3	0	4	0	2	0	0	0	10	0	12	0	0	0	0	0	0
HU	5	10	8	1	4	2	3	6	0	7	0		0	0	0	0	0	0	0	0	0	0	0	0
IC	2	4	5	0	3	8	0	7	10	12	1	6	0	0	0	0	0	0	0	0	0	0	0	0
IR	2	10	5	6	12	0	0	3	7	4	0	0	0	1	0	0	0	0	0	0	0	8	0	0
IS	7	10	12		6	4	0	2	3	1	0	8	0	0	0	0	0	0	0	0	0	0	5	0
LA	0	5	0	2		8	1	12	4	6	7	3	0	0	10	0	0	0	0	0	0	0	0	0
LI	1	2	0	3	12	10	0	8	4	5	6	0	0	0	7	0	0	0	0	0	0	0	0	0
MA	6		7	8	10	2	0	1	3	5	0	0	0	0	0	0	12	0	0	0	4	0	0	0
MD	4	0	7	6	12		1	0	0	3	0	0	0	0	10	0	0	0	5	8	0	0	0	2
MO	0	5	4	12	0	0	6	8	10	0	7	0	0	0	0	0	1	0	3	0	0	0	0	2
NE	10	5	3	7	0	0	4	0	8	1	2	0	12	6	0	0	0	0	0	0	0	0	0	0
NO	4	10	6	5	8	0	0	3	12		2	0	7	0	0	0	0	1	0	0	0	0	0	0
PL	1	0	7	0	4	3	0	6	5	8	2	10	0	0	0	0	0	0	12	0	0	0	0	0
PO	0	0	12	5	6	10	0	4	1	0	0	2	0	0	0	0	0	0	0	7	8	0	0	0
RO	10	2		7	0	12	6	0	4	0	5	8	3	0	0	0	0	1	0	0	0	0	0	0
RU	4	12	0	8	5	10	6	7	0	0	1	3	0	0		0	0	0	2	0	0	0	0	0
SM	12	0	3	0	0	5		0	0	2	10	6	0	4	0	8	7	1	0	0	0	0	0	0
SL	2	1	0	0	7	3	10	6	0	4	12	0	0	8	0	0	5	0	0	0	0	0	0	0
SP	8	7	12	6	0	4	0	0	10	3	0	5	0	0	0	0	0	0	2	1		0	0	0
SW	12	6	2	1	3	0	4	5	10	8	0	0	0	7	0	0	0	0		0	0	0	0	0
SZ	7	3	0	1	0	0	12		0	0	8	0	6	5	0	10	2	0	0	0	4	0	0	0
TU	12	8	4	3	0	7	0	0	0	0	0	6		10	0	2	5	0	0	0	0	1	0	0
UA	0	10	0	7	1	12	3	5	0	6	8	2	0	0	4	0	0	0	0		0	0	0	0
UK	12	10	0	7	6	2	0	0	8	5	0	0	1	4	0	0	0	3	0	0	0		0	0

We applied our approach presented in the previous section and obtained a ranking very similar to Table 1: $GR, MA, TU, RU, SM, RO, LA, DE, IS, BH,$ $AL, SP, FY, MD, CY, UA, SZ, CR, HU, FR, NO, SW, UK, GE$.

There is some difference between the two rankings. One possible explanation is that the ORDER-BY-PREFERENCE algorithm [4] is only able to produce an approximately optimal ordering. However the benefit of our approach is that there is no need to allocate scores, which can be hard to do.

6   Summary and Future Work

In this paper we consider the problem of *preference based alternative ordering*. We argue that relative preferences over alternatives (e.g., one alternative is preferred to another) are easier to obtain than absolute scores or ranks. We consider an approach to this problem where, instead of using absolute scores or ranks, we use relative preferences over subsets of the alternatives to generate a total ordering that is maximally agreeable with the given preferences.

We consider a set of preference specifications over all or part of the alternatives. For every pair of alternatives we calculate the probability that the two alternatives should be placed in an order. Then the ORDER-BY-PREFERENCE algorithm [4] is used to construct a total ordering for all the alternatives, which is guaranteed to be approximately optimal. This approach is demonstrated through two examples: one toy example that illustrates how the method works and another example that illustrates how effective the method is.

In future work we will compare this approach with various voting systems, including other preference based alternative ranking methods (see, e.g., [9,1]). We will also investigate how may Arrow axioms [10] our ranking method satisfies.

Acknowledgement. The authors would like to thank the anonymous reviewers for their comments, which help shape the present form of this paper.

References

1. Janicki, R., Koczkodaj, W.: A weak order approach to group ranking. Computers and Mathematics with Applications **32**(2) (1996) 51–59
2. Saari, D.G.: Chaotic Elections! A Mathematician Looks at Voting. American Mathematical Society (2001)
3. Saari, D.G.: Decisions and Elections: Explaining the Unexpected. Cambridge University Press (2001)
4. Cohen, W.W., Schapire, R.E., Singer, Y.: Learning to order things. Journal of Artificial Intelligence Research **10** (1999) 243–270
5. Ash, R.: Real Analysis and Probability. Academic Press, New York (1972)
6. Wang, H., Dubitzky, W.: A flexible and robust similarity measure based on contextual probability. In: Proceedings of IJCAI'05. (2005) 27–32
7. Wang, H.: Nearest neighbors by neighborhood counting. IEEE Transactions on Pattern Analysis and Machine Intelligence **28**(6) (2006) 942–953
8. Wang, H.: All common subsequences. In: Proc. IJCAI-07. (2007) 635–640 Oral presentation.
9. Bouysso, D., Vincke, P.: Ranking alternatives on the basis of preference relations: a progress report with special emphasis on outranking relations. Journal of Multi-Criteria Decision Analysis **6** (1997) 7785
10. Arrow, K.: Social Choice and Individual Values. John Wiley, Philadelphia, PA (1951)

Lecture Notes in Artificial Intelligence (LNAI)

Author Index